Renault 25 Owners Workshop Manual

Peter G Strasman

Models covered
Renault 25 TS & GTS; 1995 cc
Renault 25 GTX; 2165 cc
Renault 25 V6 Turbo; 2458 cc
Renault 25 V6 Injection & Limousine; 2664 cc

Does not cover Diesel engine

Haynes Publishing
Sparkford Nr Yeovil
Somerset BA22 7JJ England

Haynes Publications, Inc
861 Lawrence Drive
Newbury Park
California 91320 USA

Acknowledgements

Thanks are due to the Champion Sparking Plug Company Limited who supplied the illustrations showing the spark plug conditions. Certain other illustrations are the copyright of Renault UK Limited, and are used with their permission. Thanks are also due to Sykes-Pickavant, who supplied some of the workshop tools and to all those people at Sparkford who helped in the production of this manual.

© **Haynes Publishing Group 1987**

A book in the **Haynes Owners Workshop Manual Series**

Printed by J. H. Haynes & Co. Ltd, Sparkford, Nr Yeovil, Somerset BA22 7JJ, England

All rights reserved. No part of this book may be reproduced or transmitted in any form or by any means, electronic or mechanical, including photocopying, recording or by any information storage or retrieval system, without permission in writing from the copyright holder.

ISBN 1 85010 228 7

British Library Cataloguing in Publication Data
Strasman, Peter G.
 Renault 25 owners workshop manual
 1. Renault automobile
 I. Title
 629.28'722   TL215.R4
 ISBN 1-85010-228-7

Whilst every care is taken to ensure that the information in this manual is correct, no liability can be accepted by the authors or publishers for loss, damage or injury caused by any errors in, or omissions from, the information given.

Restoring and Preserving our Motoring Heritage

Few people can have had the luck to realise their dreams to quite the same extent and in such a remarkable fashion as John Haynes, Founder and Chairman of the Haynes Publishing Group.

Since 1965 his unique approach to workshop manual publishing has proved so successful that millions of Haynes Manuals are now sold every year throughout the world, covering literally thousands of different makes and models of cars, vans and motorcycles.

A continuing passion for cars and motoring led to the founding in 1985 of a Charitable Trust dedicated to the restoration and preservation of our motoring heritage. To inaugurate the new Museum, John Haynes donated virtually his entire private collection of 52 cars.

Now with an unrivalled international collection of over 210 veteran, vintage and classic cars and motorcycles, the Haynes Motor Museum in Somerset is well on the way to becoming one of the most interesting Motor Museums in the world.

A 70 seat video cinema, a cafe and an extensive motoring bookshop, together with a specially constructed one kilometre motor circuit, make a visit to the Haynes Motor Museum a truly unforgettable experience.

Every vehicle in the museum is preserved in as near as possible mint condition and each car is run every six months on the motor circuit.

Enjoy the picnic area set amongst the rolling Somerset hills. Peer through the William Morris workshop windows at cars being restored, and browse through the extensive displays of fascinating motoring memorabilia.

From the 1903 Oldsmobile through such classics as an MG Midget to the mighty 'E' type Jaguar, Lamborghini, Ferrari Berlinetta Boxer, and Graham Hill's Lola Cosworth, there is something for everyone, young and old alike, at this Somerset Museum.

Haynes Motor Museum

Situated mid-way between London and Penzance, the Haynes Motor Museum is located just off the A303 at Sparkford, Somerset (home of the Haynes Manual) and is open to the public 7 days a week all year round, except Christmas Day and Boxing Day.

Telephone 01963 440804.

Contents

Introductory pages

About this manual	5
Introduction to the Renault 25	5
General dimensions, weights and capacities	6
Jacking, towing and wheel changing	7
Buying spare parts and vehicle identification numbers	8
General repair procedures	9
Tools and working facilities	10
Safety first!	12
Routine maintenance	13
Recommended lubricants and fluids	19
Conversion factors	20
Fault diagnosis	21

Chapter 1
Engine — 25

Chapter 2
Cooling, heating and air conditioning — 103

Chapter 3
Fuel system — 118

Chapter 4
Ignition system — 163

Chapter 5
Clutch — 174

Chapter 6
Manual transmission — 183

Chapter 7
Automatic transmission — 210

Chapter 8
Driveshafts, hubs, roadwheels and tyres — 224

Chapter 9
Braking system — 236

Chapter 10
Suspension and steering — 255

Chapter 11
Bodywork and fittings — 272

Chapter 12
Electrical system — 290

Index — 356

Spark plug condition and bodywork repair colour section between pages 32 and 33

Renault 25 V6i

About this manual

Its aim

The aim of this manual is to help you get the best value from your vehicle. It can do so in several ways. It can help you decide what work must be done (even should you choose to get it done by a garage), provide information on routine maintenance and servicing, and give a logical course of action and diagnosis when random faults occur. However, it is hoped that you will use the manual by tackling the work yourself. On simpler jobs it may even be quicker than booking the car into a garage and going there twice, to leave and collect it. Perhaps most important, a lot of money can be saved by avoiding the costs a garage must charge to cover its labour and overheads.

The manual has drawings and descriptions to show the function of the various components so that their layout can be understood. Then the tasks are described and photographed in a step-by-step sequence so that even a novice can do the work.

Its arrangement

The manual is divided into twelve Chapters, each covering a logical sub-division of the vehicle. The Chapters are each divided into Sections, numbered with single figures, eg 5; and the Sections into paragraphs (or sub-sections), with decimal numbers following on from the Section they are in, eg 5.1, 5.2, 5.3 etc.

It is freely illustrated, especially in those parts where there is a detailed sequence of operations to be carried out. There are two forms of illustration: figures and photographs. The figures are numbered in sequence with decimal numbers, according to their position in the Chapter – eg Fig. 6.4 is the fourth drawing/illustration in Chapter 6. Photographs carry the same number (either individually or in related groups) as the Section or sub-section to which they relate.

There is an alphabetical index at the back of the manual as well as a contents list at the front. Each Chapter is also preceded by its own individual contents list.

References to the 'left' or 'right' of the vehicle are in the sense of a person in the driver's seat facing forwards.

Unless otherwise stated, nuts and bolts are removed by turning anti-clockwise, and tightened by turning clockwise.

Vehicle manufacturers continually make changes to specifications and recommendations, and these, when notified, are incorporated into our manuals at the earliest opportunity.

Whilst every care is taken to ensure that the information in this manual is correct, no liability can be accepted by the authors or publishers for loss, damage or injury caused by any errors in, or omissions from, the information given.

Introduction to the Renault 25

The car is undoubtedly well-built and fully equipped to a luxurious standard.

Overhaul and repair operations generally present few problems for the home mechanic, with the exception of certain work on the ignition and fuel systems which, even here, can be accomplished with patience and perseverence.

The wide use of electronic components and associated circuitry will mean that, without sophisticated test equipment, any work in this area will have to be entrusted to the professional technician.

General dimensions, weights and capacities

Dimensions
Overall length:
    Four-cylinder models .... 4623 mm (182.0 in)
    V6 models (except Turbo and Limousine) .... 4649 mm (183.0 in)
    V6 Turbo .... 4699 mm (185.0 in)
    V6 Limousine .... 4923 mm (193.8 in)
Overall width .... 1772 mm (69.8 in)
Overall height:
    All models except Limousine .... 1405 mm (55.3 in)
    Limousine .... 1415 mm (55.7 in)
Ground clearance (kerb wt) .... 119 mm (4.7 in)
Wheelbase:
    All models except Limousine .... 2723 mm (107.2 in)
    Limousine .... 2949 mm (116.1 in)
Front track:
    Four-cylinder models .... 1491 mm (58.7 in)
    Six-cylinder models .... 1496 mm (58.9 in)
Rear track .... 1461 mm (57.5 in)

Weights
Kerb weight (with oil, coolant and full fuel tank):
TS .... 1120 kg (2469 lb)
GTS .... 1155 kg (2546 lb)
GTS automatic .... 1190 kg (2623 lb)
GTX .... 1200 kg (2646 lb)
GTX automatic .... 1185 kg (2612 lb)
V6i .... 1280 kg (2822 lb)
V6i automatic .... 1290 kg (2844 lb)
V6 Turbo .... 1366 kg (3011 lb)
V6 Limousine .... 1351 kg (2978 lb)
V6 Limousine automatic .... 1361 kg (3000 lb)
Maximum towing weight:
    Braked trailer:
        TS and GTS .... 1311 kg (2890 lb)
        GTX .... 1351 kg (2978 lb)
        V6i .... 1591 kg (3507 lb)
        V6 Turbo .... 1501 kg (3309 lb)
        V6 Limousine .... 1471 kg (3243 lb)
    Unbraked trailer:
        TS and GTS .... 615 kg (1356 lb)
        GTX .... 591 kg (1303 lb)
        V6i .... 641 kg (1413 lb)
        V6 Turbo .... 661 kg (1457 lb)
        V6 Limousine .... 656 kg (1446 lb)
Maximum roof rack load .... 60 kg (132 lb)

Capacities
Engine oil with filter change:
    Four-cylinder engine .... 5.5 litre (9.7 pints)
    Six-cylinder engine (except Turbo) .... 6.0 litre (10.6 pints)
    V6 Turbo .... 7.9 litre (13.9 pints)
Manual transmission:
    Type NG3 .... 2.0 litre (3.5 pints)
    Type UN1 .... 3.4 litre (6.0 pints)
Automatic transmission:
    From dry .... 6.0 litre (10.6 pints)
    At fluid renewal .... 2.5 litre (4.4 pints)
Final drive (Type 4141 automatic transmission only) .... 1.6 litre (2.8 pints)
Cooling system*:
    1995 cc engine .... 7.5 litre (13.2 pints)
    2165 cc engine .... 8.0 litre (14.1 pints)
    2458 cc engine .... 9.8 litre (17.2 pints)
    2664 cc engine .... 9.5 litre (16.7 pints)
* Add 0.28 litre (0.5 pint) to models with automatic transmission or air conditioning
Fuel tank:
    TS and GTS models .... 66.0 litre (14.5 gal)
    All other models .... 70.0 litre (15.4 gal)

Jacking, towing and wheel changing

1 To avoid repetition, the procedure for raising the vehicle, in order to carry out work under it, is not included before each relevant operation described in this Manual.
2 It is to be preferred, and it is certainly recommended, that the vehicle is positioned over an inspection pit or raised on a lift. Where these facilities are not available, use ramps or jack up the vehicle strictly in accordance with the following guide. Once the vehicle is raised, supplement the jack with axle stands.

Wheel changing
3 To change a roadwheel, remove the jack and wheelbrace from the compartment on the left-hand side within the boot (photo).
4 Pull the spare wheel cradle release ring and the spare wheel will drop to the floor.
5 Using the wheelbrace, lever off the roadwheel trim plate, if fitted.
6 Make sure that the handbrake is fully applied and if possible chock the roadwheel opposite to the one being removed.
7 Release but do not remove the roadwheel bolts.
8 Locate the jack under the sill jacking point nearest to the wheel being changed. Be careful on fuel injection models that the head of the jack does not foul the fuel filter which is located just behind the rear end of the right-hand sill.
9 Using the wheelbrace, turn the jack screw and raise the wheel off the floor.
10 Remove the bolts and wheel, and bolt on the spare wheel.
11 Lower the jack and fully tighten the wheel bolts. Fit the trim plate.

Jacking
12 When raising the car for repair or maintenance work, preferably use a trolley jack, or a hydraulic bottle or heavy screw type jack.
13 Place the jack under the sill jacking points using a shaped wooden block with a groove into which the weld flange of the sill will locate.
14 The front end may be raised if a substantial wooden baulk of timber is placed under both side-members just to the rear of the crossmember. Place the jack under the centre of the timber.
15 **Never** attempt to jack the car under the front crossmember or front or rear suspension arms.

Towing
16 The towing hooks may be used for towing or being towed on normal road surfaces (photos).
17 When being towed, remember to unlock the steering column by turning the ignition key.
18 Vehicles equipped with automatic transmission should preferably be towed with front wheels off the ground. However, towing with front wheels on the ground is permissible if an extra 2.0 litre (3.5 pints) of specified transmission fluid is poured into the transmission, the speed is restricted to 30 km/h (18 mph) and the distance towed kept below 50 km (30 miles).
19 Keep the selector lever in 'N' during towing and remove the surplus oil after the vehicle has been made serviceable again; a syphon being the best way to do this.

Wheelbrace and jack

Front towing hook

Rear towing hook

Buying spare parts and vehicle identification numbers

Buying spare parts

Spare parts are available from many sources, for example Renault garages, other garages and accessory shops, and motor factors. Our advice regarding spare part sources is as follows.

Officially appointed Renault garages – This is best source of parts which are peculiar to your car and are not generally available (eg complete cylinder heads, internal gearbox components, badges, interior trim etc). It is also the only place at which you should buy parts if your vehicle is still under warranty – non-Renault components may invalidate the warranty. To be sure of obtaining the correct parts it will always be necessary to give the storeperson your car's vehicle identification number, and if possible, to take the 'old' part along for positive identification. Many parts are available under a factory exchange scheme – any parts returned should always be clean. It obviously makes good sense to go straight to the specialists on your car for this type of part for they are best equipped to supply you.

Other dealers and accessory shops – These are often very good places to buy materials and components needed for the maintenance of your car (eg oil filters, spark plugs, bulbs, drivebelts, oils and grease, touch-up paint, filler paste etc). They also sell general accessories, usually have convenient opening hours, charge lower prices and can often be found not far from home.

Motor factors – Good factors will stock all of the more important components which wear out relatively quickly (eg clutch components, pistons, valves, exhaust systems, brake pipes/seals and pads etc). Motor factors will often provide new or reconditioned components on a part exhange basis – this can save a considerable amount of money.

Vehicle identification numbers

Modifications are a continuing and unpublicised process in vehicle manufacture, quite apart from major model changes. Spare parts manuals and lists are compiled upon a numerical basis, the individual vehicle numbers being essential to correct identification of the component required.

When ordering spare parts, always give as much information as possible. Quote the car model, year of manufacture, body and engine numbers as appropriate.

Two plates are used on the Renault 25, and are located on the top rail above the radiator; an oval plate and a rectangular plate (photo).

Engine number

This is located on a plate fixed adjacent to the exhaust manifold on four-cylinder engines, or close to the engine oil dipstick on six-cylinder engines (photo).

Vehicle identification plates

Engine number (2165 cc)

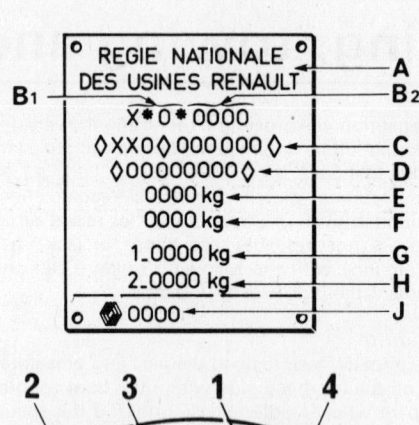

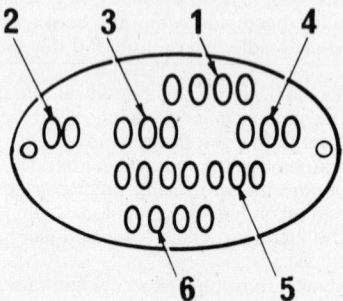

Deciphering vehicle identification plates

A   Manufacturer's name
B1   EEC number
B2   Vehicle type number
C   Manufacturer's identification code
D   Chassis number
E   Gross vehicle weight
F   Gross train weight
G   Maximum front axle weight
H   Maximum rear axle weight
J   Model year (some countries)
1   Vehicle factory symbol
2   First figure indicates transmission type, second figure indicates special equipment
3   Basic specification level
4   Extra factory-fitted equipment
5   Fabrication number
6   Model year (some countries)

General repair procedures

Whenever servicing, repair or overhaul work is carried out on the car or its components, it is necessary to observe the following procedures and instructions. This will assist in carrying out the operation efficiently and to a professional standard of workmanship.

Joint mating faces and gaskets

Where a gasket is used between the mating faces of two components, ensure that it is renewed on reassembly, and fit it dry unless otherwise stated in the repair procedure. Make sure that the mating faces are clean and dry with all traces of old gasket removed. When cleaning a joint face, use a tool which is not likely to score or damage the face, and remove any burrs or nicks with an oilstone or fine file.

Make sure that tapped holes are cleaned with a pipe cleaner, and keep them free of jointing compound if this is being used unless specifically instructed otherwise.

Ensure that all orifices, channels or pipes are clear and blow through them, preferably using compressed air.

Oil seals

Whenever an oil seal is removed from its working location, either individually or as part of an assembly, it should be renewed.

The very fine sealing lip of the seal is easily damaged and will not seal if the surface it contacts is not completely clean and free from scratches, nicks or grooves. If the original sealing surface of the component cannot be restored, the component should be renewed.

Protect the lips of the seal from any surface which may damage them in the course of fitting. Use tape or a conical sleeve where possible. Lubricate the seal lips with oil before fitting and, on dual lipped seals, fill the space between the lips with grease.

Unless otherwise stated, oil seals must be fitted with their sealing lips toward the lubricant to be sealed.

Use a tubular drift or block of wood of the appropriate size to install the seal and, if the seal housing is shouldered, drive the seal down to the shoulder. If the seal housing is unshouldered, the seal should be fitted with its face flush with the housing top face.

Screw threads and fastenings

Always ensure that a blind tapped hole is completely free from oil, grease, water or other fluid before installing the bolt or stud. Failure to do this could cause the housing to crack due to the hydraulic action of the bolt or stud as it is screwed in.

When tightening a castellated nut to accept a split pin, tighten the nut to the specified torque, where applicable, and then tighten further to the next split pin hole. Never slacken the nut to align a split pin hole unless stated in the repair procedure.

When checking or retightening a nut or bolt to a specified torque setting, slacken the nut or bolt by a quarter of a turn, and then retighten to the specified setting.

Locknuts, locktabs and washers

Any fastening which will rotate against a component or housing in the course of tightening should always have a washer between it and the relevant component or housing.

Spring or split washers should always be renewed when they are used to lock a critical component such as a big-end bearing retaining nut or bolt.

Locktabs which are folded over to retain a nut or bolt should always be renewed.

Self-locking nuts can be reused in non-critical areas, providing resistance can be felt when the locking portion passes over the bolt or stud thread.

Split pins must always be replaced with new ones of the correct size for the hole.

Special tools

Some repair procedures in this manual entail the use of special tools such as a press, two or three-legged pullers, spring compressors etc. Wherever possible, suitable readily available alternatives to the manufacturer's special tools are described, and are shown in use. In some instances, where no alternative is possible, it has been necessary to resort to the use of a manufacturer's tool and this has been done for reasons of safety as well as the efficient completion of the repair operation. Unless you are highly skilled and have a thorough understanding of the procedure described, never attempt to bypass the use of any special tool when the procedure described specifies its use. Not only is there a very great risk of personal injury, but expensive damage could be caused to the components involved.

Tools and working facilities

Introduction

A selection of good tools is a fundamental requirement for anyone contemplating the maintenance and repair of a motor vehicle. For the owner who does not possess any, their purchase will prove a considerable expense, offsetting some of the savings made by doing-it-yourself. However, provided that the tools purchased meet the relevant national safety standards and are of good quality, they will last for many years and prove an extremely worthwhile investment.

To help the average owner to decide which tools are needed to carry out the various tasks detailed in this manual, we have compiled three lists of tools under the following headings: *Maintenance and minor repair*, *Repair and overhaul*, and *Special*. The newcomer to practical mechanics should start off with the *Maintenance and minor repair* tool kit and confine himself to the simpler jobs around the vehicle. Then, as his confidence and experience grow, he can undertake more difficult tasks, buying extra tools as, and when, they are needed. In this way, a *Maintenance and minor repair* tool kit can be built-up into a *Repair and overhaul* tool kit over a considerable period of time without any major cash outlays. The experienced do-it-yourselfer will have a tool kit good enough for most repair and overhaul procedures and will add tools from the *Special* category when he feels the expense is justified by the amount of use to which these tools will be put.

It is obviously not possible to cover the subject of tools fully here. For those who wish to learn more about tools and their use there is a book entitled *How to Choose and Use Car Tools* available from the publishers of this manual.

Maintenance and minor repair tool kit

The tools given in this list should be considered as a minimum requirement if routine maintenance, servicing and minor repair operations are to be undertaken. We recommend the purchase of combination spanners (ring one end, open-ended the other); although more expensive than open-ended ones, they do give the advantages of both types of spanner.

Combination spanners - 10, 11, 12, 13, 14 & 17 mm
Torx bits
Adjustable spanner - 9 inch
Engine sump/gearbox drain plug key
Spark plug spanner (with rubber insert)
Spark plug gap adjustment tool
Set of feeler gauges
Brake bleed nipple spanner
Screwdriver - 4 in long x $1/4$ in dia (flat blade)
Screwdriver - 4 in long x $1/4$ in dia (cross blade)
Combination pliers - 6 inch
Hacksaw (junior)
Tyre pump
Tyre pressure gauge
Oil can
Fine emery cloth (1 sheet)
Wire brush (small)
Funnel (medium size)

Repair and overhaul tool kit

These tools are virtually essential for anyone undertaking any major repairs to a motor vehicle, and are additional to those given in the *Maintenance and minor repair* list. Included in this list is a comprehensive set of sockets. Although these are expensive they will be found invaluable as they are so versatile - particularly if various drives are included in the set. We recommend the ½ in square-drive type, as this can be used with most proprietary torque wrenches. If you cannot afford a socket set, even bought piecemeal, then inexpensive tubular box spanners are a useful alternative.

The tools in this list will occasionally need to be supplemented by tools from the *Special* list.

Sockets (or box spanners) to cover range in previous list
Reversible ratchet drive (for use with sockets)
Extension piece, 10 inch (for use with sockets)
Universal joint (for use with sockets)
Torque wrench (for use with sockets)
'Mole' wrench - 8 inch
Ball pein hammer
Soft-faced hammer, plastic or rubber
Screwdriver - 6 in long x $5/16$ in dia (flat blade)
Screwdriver - 2 in long x $5/16$ in square (flat blade)
Screwdriver - $1^1/2$ in long x $1/4$ in dia (cross blade)
Screwdriver - 3 in long x $1/8$ in dia (electricians)
Pliers - electricians side cutters
Pliers - needle nosed
Pliers - circlip (internal and external)
Cold chisel - $1/2$ inch
Scriber
Scraper
Centre punch
Pin punch
Hacksaw
Valve grinding tool
Steel rule/straight-edge
Allen keys
Selection of files
Wire brush (large)
Axle-stands
Jack (strong trolley or hydraulic type)

Special tools

The tools in this list are those which are not used regularly, are expensive to buy, or which need to be used in accordance with their manufacturers' instructions. Unless relatively difficult mechanical jobs are undertaken frequently, it will not be economic to buy many of these tools. Where this is the case, you could consider clubbing together with friends (or joining a motorists' club) to make a joint purchase, or borrowing the tools against a deposit from a local garage or tool hire specialist.

Tools and working facilities

The following list contains only those tools and instruments freely available to the public, and not those special tools produced by the vehicle manufacturer specifically for its dealer network. You will find occasional references to these manufacturers' special tools in the text of this manual. Generally, an alternative method of doing the job without the vehicle manufacturers' special tool is given. However, sometimes, there is no alternative to using them. Where this is the case and the relevant tool cannot be bought or borrowed, you will have to entrust the work to a franchised garage.

- Valve spring compressor
- Piston ring compressor
- Balljoint separator
- Universal hub/bearing puller
- Impact screwdriver
- Micrometer and/or vernier gauge
- Dial gauge
- Tachometer
- Universal electrical multi-meter
- Cylinder compression gauge
- Lifting tackle
- Trolley jack
- Light with extension lead

Buying tools

For practically all tools, a tool factor is the best source since he will have a very comprehensive range compared with the average garage or accessory shop. Having said that, accessory shops often offer excellent quality tools at discount prices, so it pays to shop around.

There are plenty of good tools around at reasonable prices, but always aim to purchase items which meet the relevant national safety standards. If in doubt, ask the proprietor or manager of the shop for advice before making a purchase.

Care and maintenance of tools

Having purchased a reasonable tool kit, it is necessary to keep the tools in a clean serviceable condition. After use, always wipe off any dirt, grease and metal particles using a clean, dry cloth, before putting the tools away. Never leave them lying around after they have been used. A simple tool rack on the garage or workshop wall, for items such as screwdrivers and pliers is a good idea. Store all normal wrenches and sockets in a metal box. Any measuring instruments, gauges, meters, etc, must be carefully stored where they cannot be damaged or become rusty.

Take a little care when tools are used. Hammer heads inevitably become marked and screwdrivers lose the keen edge on their blades from time to time. A little timely attention with emery cloth or a file will soon restore items like this to a good serviceable finish.

Working facilities

Not to be forgotten when discussing tools, is the workshop itself. If anything more than routine maintenance is to be carried out, some form of suitable working area becomes essential.

It is appreciated that many an owner mechanic is forced by circumstances to remove an engine or similar item, without the benefit of a garage or workshop. Having done this, any repairs should always be done under the cover of a roof.

Wherever possible, any dismantling should be done on a clean, flat workbench or table at a suitable working height.

Any workbench needs a vice: one with a jaw opening of 4 in (100 mm) is suitable for most jobs. As mentioned previously, some clean dry storage space is also required for tools, as well as for lubricants, cleaning fluids, touch-up paints and so on, which become necessary.

Another item which may be required, and which has a much more general usage, is an electric drill with a chuck capacity of at least 5/16 in (8 mm). This, together with a good range of twist drills, is virtually essential for fitting accessories such as mirrors and reversing lights.

Last, but not least, always keep a supply of old newspapers and clean, lint-free rags available, and try to keep any working area as clean as possible.

Spanner jaw gap comparison table

Jaw gap (in)	Spanner size
0.250	1/4 in AF
0.276	7 mm
0.313	5/16 in AF
0.315	8 mm
0.344	11/32 in AF; 1/8 in Whitworth
0.354	9 mm
0.375	3/8 in AF
0.394	10 mm
0.433	11 mm
0.438	7/16 in AF
0.445	3/16 in Whitworth; 1/4 in BSF
0.472	12 mm
0.500	1/2 in AF
0.512	13 mm
0.525	1/4 in Whitworth; 5/16 in BSF
0.551	14 mm
0.563	9/16 in AF
0.591	15 mm
0.600	5/16 in Whitworth; 3/8 in BSF
0.625	5/8 in AF
0.630	16 mm
0.669	17 mm
0.686	11/16 in AF
0.709	18 mm
0.710	3/8 in Whitworth; 7/16 in BSF
0.748	19 mm
0.750	3/4 in AF
0.813	13/16 in AF
0.820	7/16 in Whitworth; 1/2 in BSF
0.866	22 mm
0.875	7/8 in AF
0.920	1/2 in Whitworth; 9/16 in BSF
0.938	15/16 in AF
0.945	24 mm
1.000	1 in AF
1.010	9/16 in Whitworth; 5/8 in BSF
1.024	26 mm
1.063	1 1/16 in AF; 27 mm
1.100	5/8 in Whitworth; 11/16 in BSF
1.125	1 1/8 in AF
1.181	30 mm
1.200	11/16 in Whitworth; 3/4 in BSF
1.250	1 1/4 in AF
1.260	32 mm
1.300	3/4 in Whitworth; 7/8 in BSF
1.313	1 5/16 in AF
1.390	13/16 in Whitworth; 15/16 in BSF
1.417	36 mm
1.438	1 7/16 in AF
1.480	7/8 in Whitworth; 1 in BSF
1.500	1 1/2 in AF
1.575	40 mm; 15/16 in Whitworth
1.614	41 mm
1.625	1 5/8 in AF
1.670	1 in Whitworth; 1 1/8 in BSF
1.688	1 11/16 in AF
1.811	46 mm
1.813	1 13/16 in AF
1.860	1 1/8 in Whitworth; 1 1/4 in BSF
1.875	1 7/8 in AF
1.969	50 mm
2.000	2 in AF
2.050	1 1/4 in Whitworth; 1 3/8 in BSF
2.165	55 mm
2.362	60 mm

Safety first!

Professional motor mechanics are trained in safe working procedures. However enthusiastic you may be about getting on with the job in hand, do take the time to ensure that your safety is not put at risk. A moment's lack of attention can result in an accident, as can failure to observe certain elementary precautions.

There will always be new ways of having accidents, and the following points do not pretend to be a comprehensive list of all dangers; they are intended rather to make you aware of the risks and to encourage a safety-conscious approach to all work you carry out on your vehicle.

Essential DOs and DON'Ts

DON'T rely on a single jack when working underneath the vehicle. Always use reliable additional means of support, such as axle stands, securely placed under a part of the vehicle that you know will not give way.

DON'T attempt to loosen or tighten high-torque nuts (e.g. wheel hub nuts) while the vehicle is on a jack; it may be pulled off.

DON'T start the engine without first ascertaining that the transmission is in neutral (or 'Park' where applicable) and the parking brake applied.

DON'T suddenly remove the filler cap from a hot cooling system – cover it with a cloth and release the pressure gradually first, or you may get scalded by escaping coolant.

DON'T attempt to drain oil until you are sure it has cooled sufficiently to avoid scalding you.

DON'T grasp any part of the engine, exhaust or catalytic converter without first ascertaining that it is sufficiently cool to avoid burning you.

DON'T allow brake fluid or antifreeze to contact vehicle paintwork.

DON'T syphon toxic liquids such as fuel, brake fluid or antifreeze by mouth, or allow them to remain on your skin.

DON'T inhale dust – it may be injurious to health (see *Asbestos* below).

DON'T allow any spilt oil or grease to remain on the floor – wipe it up straight away, before someone slips on it.

DON'T use ill-fitting spanners or other tools which may slip and cause injury.

DON'T attempt to lift a heavy component which may be beyond your capability – get assistance.

DON'T rush to finish a job, or take unverified short cuts.

DON'T allow children or animals in or around an unattended vehicle.

DO wear eye protection when using power tools such as drill, sander, bench grinder etc, and when working under the vehicle.

DO use a barrier cream on your hands prior to undertaking dirty jobs – it will protect your skin from infection as well as making the dirt easier to remove afterwards; but make sure your hands aren't left slippery. Note that long-term contact with used engine oil can be a health hazard.

DO keep loose clothing (cuffs, tie etc) and long hair well out of the way of moving mechanical parts.

DO remove rings, wristwatch etc, before working on the vehicle – especially the electrical system.

DO ensure that any lifting tackle used has a safe working load rating adequate for the job.

DO keep your work area tidy – it is only too easy to fall over articles left lying around.

DO get someone to check periodically that all is well, when working alone on the vehicle.

DO carry out work in a logical sequence and check that everything is correctly assembled and tightened afterwards.

DO remember that your vehicle's safety affects that of yourself and others. If in doubt on any point, get specialist advice.

IF, in spite of following these precautions, you are unfortunate enough to injure yourself, seek medical attention as soon as possible.

Asbestos

Certain friction, insulating, sealing, and other products – such as brake linings, brake bands, clutch linings, torque converters, gaskets, etc – contain asbestos. *Extreme care must be taken to avoid inhalation of dust from such products since it is hazardous to health.* If in doubt, assume that they *do* contain asbestos.

Fire

Remember at all times that petrol (gasoline) is highly flammable. Never smoke, or have any kind of naked flame around, when working on the vehicle. But the risk does not end there – a spark caused by an electrical short-circuit, by two metal surfaces contacting each other, by careless use of tools, or even by static electricity built up in your body under certain conditions, can ignite petrol vapour, which in a confined space is highly explosive.

Always disconnect the battery earth (ground) terminal before working on any part of the fuel or electrical system, and never risk spilling fuel on to a hot engine or exhaust.

It is recommended that a fire extinguisher of a type suitable for fuel and electrical fires is kept handy in the garage or workplace at all times. Never try to extinguish a fuel or electrical fire with water.

Note: *Any reference to a 'torch' appearing in this manual should always be taken to mean a hand-held battery-operated electric lamp or flashlight. It does NOT mean a welding/gas torch or blowlamp.*

Fumes

Certain fumes are highly toxic and can quickly cause unconsciousness and even death if inhaled to any extent. Petrol (gasoline) vapour comes into this category, as do the vapours from certain solvents such as trichloroethylene. Any draining or pouring of such volatile fluids should be done in a well ventilated area.

When using cleaning fluids and solvents, read the instructions carefully. Never use materials from unmarked containers – they may give off poisonous vapours.

Never run the engine of a motor vehicle in an enclosed space such as a garage. Exhaust fumes contain carbon monoxide which is extremely poisonous; if you need to run the engine, always do so in the open air or at least have the rear of the vehicle outside the workplace.

If you are fortunate enough to have the use of an inspection pit, never drain or pour petrol, and never run the engine, while the vehicle is standing over it; the fumes, being heavier than air, will concentrate in the pit with possibly lethal results.

The battery

Never cause a spark, or allow a naked light, near the vehicle's battery. It will normally be giving off a certain amount of hydrogen gas, which is highly explosive.

Always disconnect the battery earth (ground) terminal before working on the fuel or electrical systems.

If possible, loosen the filler plugs or cover when charging the battery from an external source. Do not charge at an excessive rate or the battery may burst.

Take care when topping up and when carrying the battery. The acid electrolyte, even when diluted, is very corrosive and should not be allowed to contact the eyes or skin.

If you ever need to prepare electrolyte yourself, always add the acid slowly to the water, and never the other way round. Protect against splashes by wearing rubber gloves and goggles.

When jump starting a car using a booster battery, for negative earth (ground) vehicles, connect the jump leads in the following sequence: First connect one jump lead between the positive (+) terminals of the two batteries. Then connect the other jump lead first to the negative (–) terminal of the booster battery, and then to a good earthing (ground) point on the vehicle to be started, at least 18 in (45 cm) from the battery if possible. Ensure that hands and jump leads are clear of any moving parts, and that the two vehicles do not touch. Disconnect the leads in the reverse order.

Mains electricity and electrical equipment

When using an electric power tool, inspection light etc, always ensure that the appliance is correctly connected to its plug and that, where necessary, it is properly earthed (grounded). Do not use such appliances in damp conditions and, again, beware of creating a spark or applying excessive heat in the vicinity of fuel or fuel vapour. Also ensure that the appliances meet the relevant national safety standards.

Ignition HT voltage

A severe electric shock can result from touching certain parts of the ignition system, such as the HT leads, when the engine is running or being cranked, particularly if components are damp or the insulation is defective. Where an electronic ignition system is fitted, the HT voltage is much higher and could prove fatal.

Routine maintenance

Maintenance is essential for ensuring safety, and desirable for the purpose of getting the best in terms of performance and economy from your car. Over the years the need for periodic lubrication – oiling, greasing and so on – has been drastically reduced, if not totally eliminated. This has unfortunately tended to lead some owners to think that because no such action is required, components either no longer exist, or will last forever. This is a serious delusion. It follows therefore that the largest initial element of maintenance is visual examination and a general sense of awareness. This may lead to repairs or renewals, but should help to avoid roadside breakdowns.

Although fluid levels and brakepad wear are monitored by sensors, there is still no real substitute for visual inspection. This is a precautionary operation to be able to anticipate the illumination of a warning lamp which may well otherwise occur at a time when topping-up, rectification of a leak, or brake overhaul may at least be inconvenient if not expensive or dangerous.

Every 400 km (250 miles) or weekly (whichever comes first)

Check engine oil level (Chapter 1)
Check coolant level (Chapter 2)
Check tyre pressures (Chapter 8)
Check operation of all lights and horn (Chapter 12)
Check washer bottle level (Chapter 12)

After first 1600 km (1000 miles) – new cars or after major engine overhaul

Renew engine oil and filter (Chapter 1)
Adjust valve clearances (after overhaul only) (Chapter 1)
Check torque of cylinder head bolts (after overhaul only) (Chapter 1)
Renew manual transmission oil (Chapter 6)

Every 8000 km (5000 miles) or six months, (whichever comes first)

Renew engine oil (Chapter 1)
Clean the crankcase ventilation system hoses and jets (Chapter 1)
Check tension of drivebelt(s) (Chapter 2)
Clean and regap spark plugs (Chapter 4)
Check manual transmission oil level (Chapter 6)
Check automatic transmission fluid level (Chapter 7)
Check final drive oil level (Type 4141 automatic transmission) (Chapter 7)
Check brake master cylinder fluid level (Chapter 9)
Check power steering fluid level (Chapter 10)

Every 16 000 km (10 000 miles) or annually (whichever comes first)

In addition to, or in place of, the 8000 km (5000 mile) service tasks

Renew engine oil filter (Chapter 1)
Renew air cleaner element (Chapter 3)
Check idle speed and exhaust CO content (Chapter 3)
Check exhaust system for corrosion (Chapter 3)
Renew spark plugs (Chapter 4)
Check clutch cable adjustment (if applicable) (Chapter 5)
Check driveshaft bellows for condition (Chpater 8)
Check condition of brake pipes and hoses (Chapter 9)
Check disc pad wear (Chapter 9)
Check handbrake cable adjustment (Chapter 9)
Check rear brake shoe wear (if applicable) (Chapter 9)
Check steering and suspension balljoints, bushes and bellows for condition (Chapter 10)
Check front wheel alignment (Chapter 10)
Check underbody for corrosion (Chapter 11)
Check headlamp beam alignment (Chapter 12)

Every 32 000 km (20 000 miles) or two years (whichever comes first)

In addition to, or in place of, the 16 000 km (10 000 mile) service tasks

Renew coolant (Chapter 2)
Renew fuel filter (carburettor models) (Chapter 3)
Clean fuel pump (carburettor models) (Chapter 3)
Clean EGR (emission control system) nozzles (Chapter 3)
Clean inside distributor cap, and check cap contacts and rotor for erosion (Chapter 4)
Renew brake hydraulic fluid (Chapter 9)

Every 65 000 km (40 000 miles) or four years (whichever comes first)

In addition to, or in place of, the 32 000 km (20 000 mile) service tasks

Renew timing belt (Chapter 1)
Renew rocker shaft oil filters (Chapter 1)
Renew fuel filter (fuel injection models) (Chapter 3)
Renew EGR (emission control system) valve (Chapter 3)
Renew manual transmission oil (Chapter 6)
Renew automatic transmission oil (Chapter 7)
Renew final drive oil (Type 4141 automatic transmission) (Chapter 7)

Engine compartment (2165 cc)

1. Heater control vacuum unit
2. Front strut top mounting
3. Brake fluid reservoir
4. Brake servo unit
5. Clutch master cylinder
6. Coolant expansion tank
7. Battery
8. Bonnet lock striker
9. Exhaust manifold
10. Engine oil filler cap
11. Power steering fluid reservoir
12. Ignition coil
13. Inlet manifold
14. Fuel injector manifold
15. Air regulating (idle speed) valve
16. Air cleaner
17. Computer casing
18. Radiator
19. Electric fan motor

Engine compartment (2664 cc)

1. Heater control vacuum unit
2. Front strut top mounting
3. Brake fluid reservoir
4. Brake servo unit
5. Brake master cylinder
6. Fuel filter
7. Ignition coil
8. Diagnostic socket
9. Coolant expansion tank
10. Battery
11. Supplementary air device
12. Engine oil filler/breather cap
13. Crankcase vent oil separator
14. Cruise control diaphragm unit
15. Bonnet lock striker
16. Radiator
17. Throttle cable reel
18. Air distributor front manifold
19. Air cleaner
20. Computer casing
21. Alternator
22. Clutch master cylinder
23. Power steering fluid reservoir

View from under front end of 2165 cc model

1 Engine oil drain plug
2 Brake caliper
3 Suspension track control arm (wishbone)
4 Tie-rod end balljoint
5 Tie-rod
6 Driveshaft inboard joint
7 Transmission drain plug
8 Earth strap
9 Exhaust pipe
10 Gearchange rod
11 Gearchange rod
12 Exhaust suspension ring

View from under front end of 2664 cc model

1. Horn compressor
2. Engine sump pan
3. Engine oil drain plug
4. Crossmember
5. Suspension track control arm
6. Brake caliper
7. Exhaust pipes
8. Anti-roll bar
9. Driveshaft inboard joint
10. Oil filter cartridge
11. Headlamp
12. Transmisson oil drain plug
13. Steering arm
14. Tie-rod
15. Gearchange rod
16. Gearchange rod

View from under rear end

1. Radius rod
2. Track control arm
3. Reaction member
4. Exhaust pipe
5. Handbrake cable equaliser
6. Fuel tank
7. Shock absorber
8. Silencer
9. Spare wheel carrier
10. Brake pipe

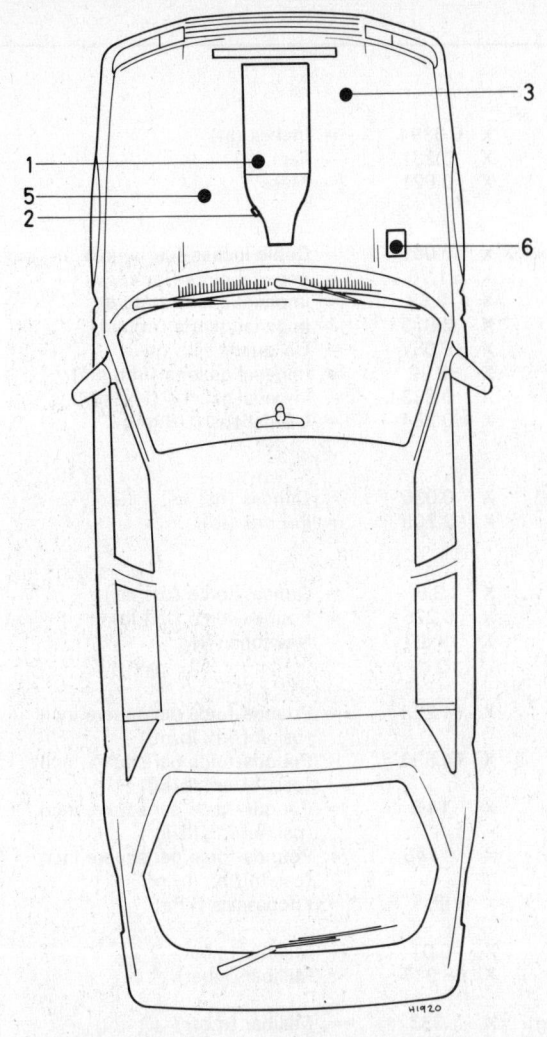

4-cylinder models

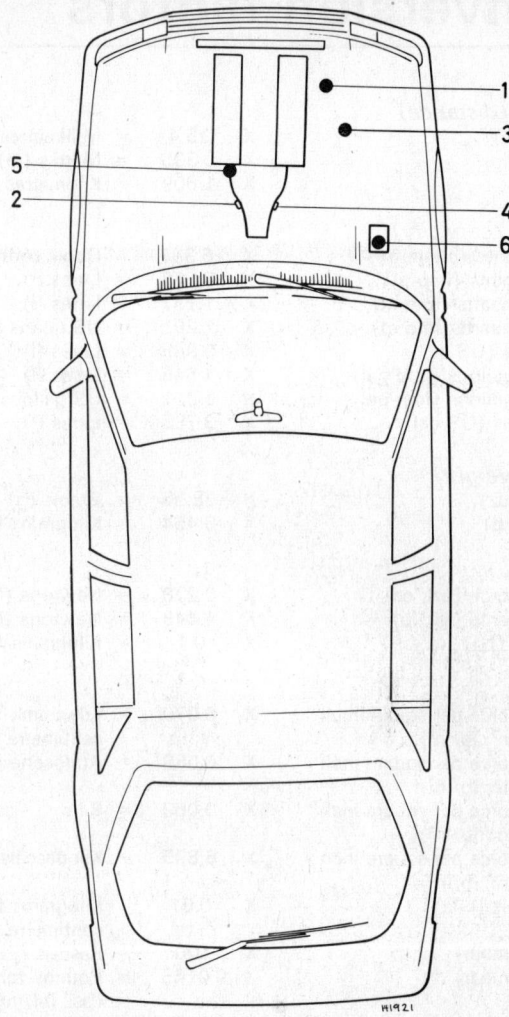

6-cylinder models

Recommended lubricants and fluids

Component or system	Lubricant type or specification
Engine (1)	SAE 15W/40, 20W/40 or 20W/50 multigrade engine oil (API SE)
Manual transmission (2)	SAE 80W oil (API GL5)
Automatic transmission (3)	Dexron® ATF
Final drive (Type 4141 automatic transmission only) (4)	SAE 80W oil (API GL5)
Power-assisted steering (5)	Dexron® ATF
Steering gear (general)	Semi-fluid or molybdenum disulphide grease
Brake/clutch system (6)	Hydraulic fluid (Dot 3 or 4, or SAE J1703)
General greasing	Lithium-based or molybdenum disulphide grease

Conversion factors

Length (distance)
Inches (in)	X	25.4	= Millimetres (mm)	X 0.0394	= Inches (in)
Feet (ft)	X	0.305	= Metres (m)	X 3.281	= Feet (ft)
Miles	X	1.609	= Kilometres (km)	X 0.621	= Miles

Volume (capacity)
Cubic inches (cu in; in^3)	X	16.387	= Cubic centimetres (cc; cm^3)	X 0.061	= Cubic inches (cu in; in^3)
Imperial pints (Imp pt)	X	0.568	= Litres (l)	X 1.76	= Imperial pints (Imp pt)
Imperial quarts (Imp qt)	X	1.137	= Litres (l)	X 0.88	= Imperial quarts (Imp qt)
Imperial quarts (Imp qt)	X	1.201	= US quarts (US qt)	X 0.833	= Imperial quarts (Imp qt)
US quarts (US qt)	X	0.946	= Litres (l)	X 1.057	= US quarts (US qt)
Imperial gallons (Imp gal)	X	4.546	= Litres (l)	X 0.22	= Imperial gallons (Imp gal)
Imperial gallons (Imp gal)	X	1.201	= US gallons (US gal)	X 0.833	= Imperial gallons (Imp gal)
US gallons (US gal)	X	3.785	= Litres (l)	X 0.264	= US gallons (US gal)

Mass (weight)
Ounces (oz)	X	28.35	= Grams (g)	X 0.035	= Ounces (oz)
Pounds (lb)	X	0.454	= Kilograms (kg)	X 2.205	= Pounds (lb)

Force
Ounces-force (ozf; oz)	X	0.278	= Newtons (N)	X 3.6	= Ounces-force (ozf; oz)
Pounds-force (lbf; lb)	X	4.448	= Newtons (N)	X 0.225	= Pounds-force (lbf; lb)
Newtons (N)	X	0.1	= Kilograms-force (kgf; kg)	X 9.81	= Newtons (N)

Pressure
Pounds-force per square inch (psi; lbf/in^2; lb/in^2)	X	0.070	= Kilograms-force per square centimetre (kgf/cm^2; kg/cm^2)	X 14.223	= Pounds-force per square inch (psi; lbf/in^2; lb/in^2)
Pounds-force per square inch (psi; lbf/in^2; lb/in^2)	X	0.068	= Atmospheres (atm)	X 14.696	= Pounds-force per square inch (psi; lbf/in^2; lb/in^2)
Pounds-force per square inch (psi; lbf/in^2; lb/in^2)	X	0.069	= Bars	X 14.5	= Pounds-force per square inch (psi; lbf/in^2; lb/in^2)
Pounds-force per square inch (psi; lbf/in^2; lb/in^2)	X	6.895	= Kilopascals (kPa)	X 0.145	= Pounds-force per square inch (psi; lbf/in^2; lb/in^2)
Kilopascals (kPa)	X	0.01	= Kilograms-force per square centimetre (kgf/cm^2; kg/cm^2)	X 98.1	= Kilopascals (kPa)
Millibar (mbar)	X	100	= Pascals (Pa)	X 0.01	= Millibar (mbar)
Millibar (mbar)	X	0.0145	= Pounds-force per square inch (psi; lbf/in^2; lb/in^2)	X 68.947	= Millibar (mbar)
Millibar (mbar)	X	0.75	= Millimetres of mercury (mmHg)	X 1.333	= Millibar (mbar)
Millibar (mbar)	X	0.401	= Inches of water (inH$_2$O)	X 2.491	= Millibar (mbar)
Millimetres of mercury (mmHg)	X	0.535	= Inches of water (inH$_2$O)	X 1.868	= Millimetres of mercury (mmHg)
Inches of water (inH$_2$O)	X	0.036	= Pounds-force per square inch (psi; lbf/in^2; lb/in^2)	X 27.68	= Inches of water (inH$_2$O)

Torque (moment of force)
Pounds-force inches (lbf in; lb in)	X	1.152	= Kilograms-force centimetre (kgf cm; kg cm)	X 0.868	= Pounds-force inches (lbf in; lb in)
Pounds-force inches (lbf in; lb in)	X	0.113	= Newton metres (Nm)	X 8.85	= Pounds-force inches (lbf in; lb in)
Pounds-force inches (lbf in; lb in)	X	0.083	= Pounds-force feet (lbf ft; lb ft)	X 12	= Pounds-force inches (lbf in; lb in)
Pounds-force feet (lbf ft; lb ft)	X	0.138	= Kilograms-force metres (kgf m; kg m)	X 7.233	= Pounds-force feet (lbf ft; lb ft)
Pounds-force feet (lbf ft; lb ft)	X	1.356	= Newton metres (Nm)	X 0.738	= Pounds-force feet (lbf ft; lb ft)
Newton metres (Nm)	X	0.102	= Kilograms-force metres (kgf m; kg m)	X 9.804	= Newton metres (Nm)

Power
Horsepower (hp)	X	745.7	= Watts (W)	X 0.0013	= Horsepower (hp)

Velocity (speed)
Miles per hour (miles/hr; mph)	X	1.609	= Kilometres per hour (km/hr; kph)	X 0.621	= Miles per hour (miles/hr; mph)

Fuel consumption*
Miles per gallon, Imperial (mpg)	X	0.354	= Kilometres per litre (km/l)	X 2.825	= Miles per gallon, Imperial (mpg)
Miles per gallon, US (mpg)	X	0.425	= Kilometres per litre (km/l)	X 2.352	= Miles per gallon, US (mpg)

Temperature

Degrees Fahrenheit = (°C x 1.8) + 32   Degrees Celsius (Degrees Centigrade; °C) = (°F - 32) x 0.56

*It is common practice to convert from miles per gallon (mpg) to litres/100 kilometres (l/100km), where mpg (Imperial) x l/100 km = 282 and mpg (US) x l/100 km = 235

Fault diagnosis

Introduction

The vehicle owner who does his or her own maintenance according to the recommended schedules should not have to use this section of the manual very often. Modern component reliability is such that, provided those items subject to wear or deterioration are inspected or renewed at the specified intervals, sudden failure is comparatively rare. Faults do not usually just happen as a result of sudden failure, but develop over a period of time. Major mechanical failures in particular are usually preceded by characteristic symptoms over hundreds or even thousands of miles. Those components which do occasionally fail without warning are often small and easily carried in the vehicle.

With any fault finding, the first step is to decide where to begin investigations. Sometimes this is obvious, but on other occasions a little detective work will be necessary. The owner who makes half a dozen haphazard adjustments or replacements may be successful in curing a fault (or its symptoms), but he will be none the wiser if the fault recurs and he may well have spent more time and money than was necessary. A calm and logical approach will be found to be more satisfactory in the long run. Always take into account any warning signs or abnormalities that may have been noticed in the period preceding the fault – power loss, high or low gauge readings, unusual noises or smells, etc – and remember that failure of components such as fuses or spark plugs may only be pointers to some underlying fault.

The pages which follow here are intended to help in cases of failure to start or breakdown on the road. There is also a Fault Diagnosis Section at the end of each Chapter which should be consulted if the preliminary checks prove unfruitful. Whatever the fault, certain basic principles apply. These are as follows:

Verify the fault. This is simply a matter of being sure that you know what the symptoms are before starting work. This is particularly important if you are investigating a fault for someone else who may not have described it very accurately.

Don't overlook the obvious. For example, if the vehicle won't start, is there petrol in the tank? (Don't take anyone else's word on this particular point, and don't trust the fuel gauge either!) If an electrical fault is indicated, look for loose or broken wires before digging out the test gear.

Cure the disease, not the symptom. Substituting a flat battery with a fully charged one will get you off the hard shoulder, but if the underlying cause is not attended to, the new battery will go the same way. Similarly, changing oil-fouled spark plugs for a new set will get you moving again, but remember that the reason for the fouling (if it wasn't simply an incorrect grade of plug) will have to be established and corrected.

Don't take anything for granted. Particularly, don't forget that a 'new' component may itself be defective (especially if it's been rattling round in the boot for months), and don't leave components out of a fault diagnosis sequence just because they are new or recently fitted. When you do finally diagnose a difficult fault, you'll probably realise that all the evidence was there from the start.

Electrical faults

Electrical faults can be more puzzling than straightforward mechanical failures, but they are no less susceptible to logical analysis if the basic principles of operation are understood. Vehicle electrical wiring exists in extremely unfavourable conditions – heat, vibration and chemical attack – and the first things to look for are loose or corroded connections and broken or chafed wires, especially where the wires pass through holes in the bodywork or are subject to vibration.

All metal-bodied vehicles in current production have one pole of the battery 'earthed', ie connected to the vehicle bodywork, and in nearly all modern vehicles it is the negative (–) terminal. The various electrical components – motors, bulb holders etc – are also connected to earth, either by means of a lead or directly by their mountings. Electric current flows through the component and then back to the battery via the bodywork. If the component mounting is loose or corroded, or if a good path back to the battery is not available, the circuit will be incomplete and malfunction will result. The engine and/or gearbox are also earthed by means of flexible metal straps to the body or subframe; if these straps are loose or missing, starter motor, generator and ignition trouble may result.

Assuming the earth return to be satisfactory, electrical faults will be due either to component malfunction or to defects in the current supply. Individual components are dealt with in Chapter 12. If supply wires are broken or cracked internally this results in an open-circuit, and the easiest way to check for this is to bypass the suspect wire temporarily with a length of wire having a crocodile clip or suitable connector at each end. Alternatively, a 12V test lamp can be used to verify the presence of supply voltage at various points along the wire and the break can be thus isolated.

If a bare portion of a live wire touches the bodywork or other earthed metal part, the electricity will take the low-resistance path thus formed back to the battery: this is known as a short-circuit. Hopefully a short-circuit will blow a fuse, but otherwise it may cause burning of the insulation (and possibly further short-circuits) or even a fire. This is why it is inadvisable to bypass persistently blowing fuses with silver foil or wire.

Spares and tool kit

Most vehicles are supplied only with sufficient tools for wheel changing; the *Maintenance and minor repair* tool kit detailed in *Tools and working facilities*, with the addition of a hammer, is probably sufficient for those repairs that most motorists would consider attempting at the roadside. In addition a few items which can be fitted without too much trouble in the event of a breakdown should be carried. Experience and available space will modify the list below, but the following may save having to call on professional assistance:

Spark plugs, clean and correctly gapped
HT lead and plug cap – long enough to reach the plug furthest from the distributor
Distributor rotor
Drivebelt(s) – emergency type may suffice
Spare fuses
Set of principal light bulbs
Tin of radiator sealer and hose bandage
Exhaust bandage
Roll of insulating tape
Length of soft iron wire
Length of electrical flex
Torch or inspection lamp (can double as test lamp)
Battery jump leads
Tow-rope
Ignition waterproofing aerosol
Litre of engine oil
Sealed can of hydraulic fluid
'Jubilee' clips
Tube of filler paste

Fault diagnosis

If spare fuel is carried, a can designed for the purpose should be used to minimise risks of leakage and collision damage. A first aid kit and a warning triangle, whilst not at present compulsory in the UK, are obviously sensible items to carry in addition to the above.

When touring abroad it may be advisable to carry additional spares which, even if you cannot fit them yourself, could save having to wait while parts are obtained. The items below may be worth considering:

Clutch and throttle cables
Cylinder head gasket
Alternator brushes
Tyre valve core

One of the motoring organisations will be able to advise on availability of fuel etc in foreign countries.

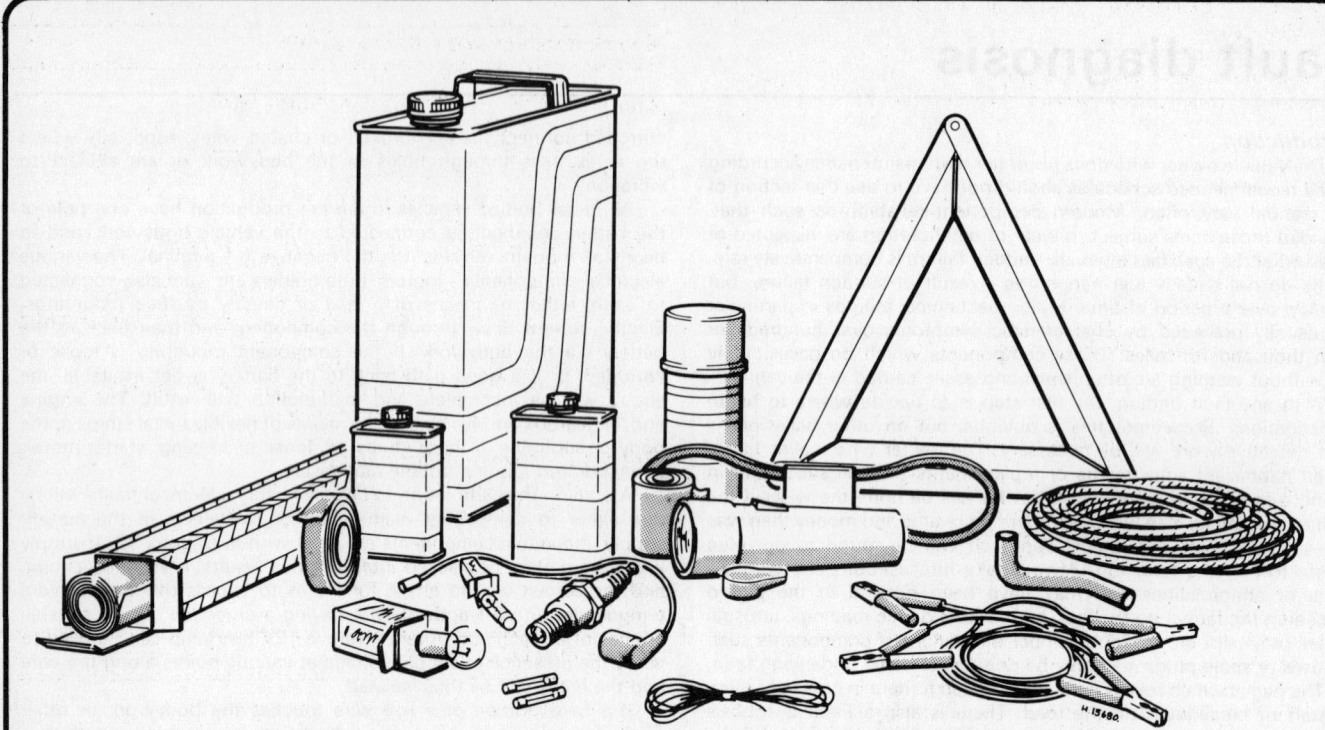

Carrying a few spares can save you a long walk!

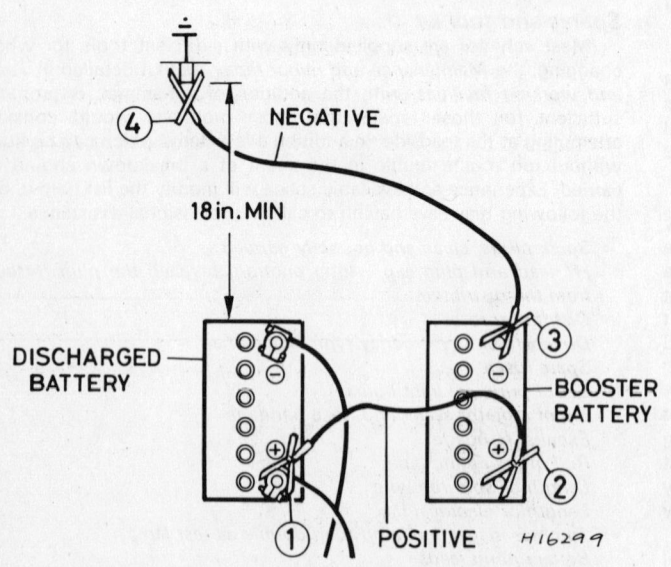

Jump start lead connections for negative earth vehicles – connect leads in order shown

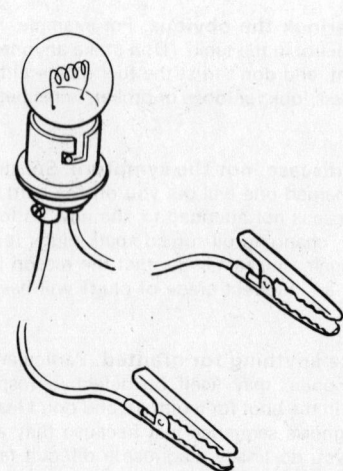

Simple test lamp is useful for tracing electrical faults

Fault diagnosis

Engine will not start

Engine fails to turn when starter operated
  Flat battery (recharge, use jump leads, or push start)
  Battery terminals loose or corroded
  Battery earth to body defective
  Engine earth strap loose or broken
  Starter motor (or solenoid) wiring loose or broken
  Automatic transmission selector in wrong position, or inhibitor switch faulty
  Ignition/starter switch faulty
  Major mechanical failure (seizure)
  Starter or solenoid internal fault (see Chapter 12)

Starter motor turns engine slowly
  Partially discharged battery (recharge, use jump leads, or push start)
  Battery terminals loose or corroded
  Battery earth to body defective
  Engine earth strap loose
  Starter motor (or solenoid) wiring loose
  Starter motor internal fault (see Chapter 12)

Starter motor spins without turning engine
  Flat battery
  Starter motor pinion sticking on sleeve
  Flywheel gear teeth damaged or worn
  Starter motor mounting bolts loose

Engine turns normally but fails to start
  Damp or dirty HT leads and distributor cap (crank engine and check for spark)
  No fuel in tank (check for delivery at carburettor)
  Excessive choke (hot engine) or insufficient choke (cold engine)
  Fouled or incorrectly gapped spark plugs (remove, clean and regap)
  Other ignition system fault (see Chapter 4)
  Other fuel system fault (see Chapter 3)
  Poor compression (see Chapter 1)
  Major mechanical failure (eg camshaft drive)

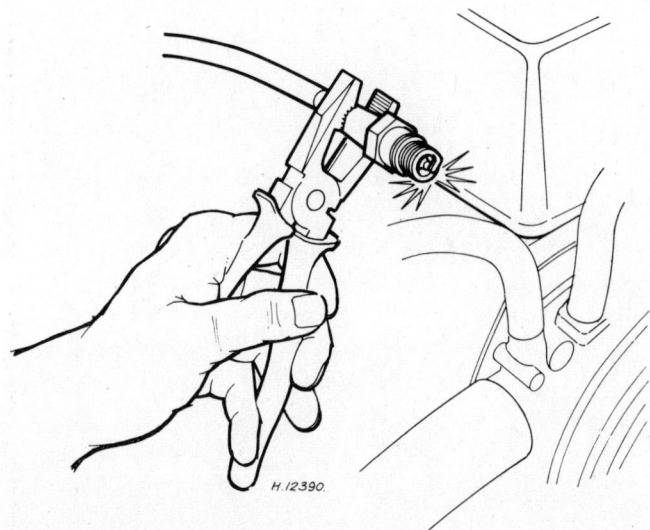

Crank engine and check for a spark. Note use of insulated tool

Engine fires but will not run
  Insufficient choke (cold engine)
  Air leaks at carburettor or inlet manifold
  Fuel starvation (see Chapter 3)
  Ignition fault (see Chapter 4)

Engine cuts out and will not restart

Engine cuts out suddenly – ignition fault
  Loose or disconnected LT wires
  Wet HT leads or distributor cap (after traversing water splash)
  Coil failure (check for spark)
  Other ignition fault (see Chapter 4)

Engine misfires before cutting out – fuel fault
  Fuel tank empty
  Fuel pump defective or filter blocked (check for delivery)
  Fuel tank filler vent blocked (suction will be evident on releasing cap)
  Carburettor needle valve sticking (1995 cc)
  Carburettor jets blocked (fuel contaminated) (1995 cc)
  Other fuel system fault (see Chapter 4)

Engine cuts out – other causes
  Serious overheating
  Major mechanical failure (eg camshaft drive)

Engine overheats

Ignition (no-charge) warning light illuminated
  Slack or broken drivebelt – retension or renew (Chapter 4)

Ignition warning light not illuminated
  Coolant loss due to internal or external leakage (see Chapter 2)
  Thermostat defective
  Low oil level
  Brakes binding
  Radiator clogged externally or internally
  Electric cooling fan not operating correctly
  Engine waterways clogged
  Ignition timing incorrect or automatic advance malfunctioning
  Mixture too weak

Note: *Do not add cold water to an overheated engine or damage may result*

Low engine oil pressure

Gauge reads low or warning light illuminated with engine running
  Oil level low or incorrect grade
  Defective gauge or sender unit
  Wire to sender unit earthed
  Engine overheating
  Oil filter clogged or bypass valve defective
  Oil pressure relief valve defective
  Oil pick-up strainer clogged
  Oil pump worn or mountings loose
  Worn main or big-end bearings

Note: *Low oil pressure in a high-mileage engine at tickover is not necessarily a cause for concern. Sudden pressure loss at speed is far more significant. In any event, check the gauge or warning light sender before condemning the engine.*

Engine noises

Pre-ignition (pinking) on acceleration
Incorrect grade of fuel
Ignition timing incorrect
Distributor faulty or worn
Worn or maladjusted carburettor
Excessive carbon build-up in engine

Whistling or wheezing noises
Leaking vacuum hose
Leaking carburettor or manifold gasket
Blowing head gasket

Tapping or rattling
Incorrect valve clearances
Worn valve gear
Worn timing chain or belt
Broken piston ring (ticking noise)

Knocking or thumping
Unintentional mechanical contact (eg fan blades)
Worn drivebelt
Peripheral component fault (generator, water pump etc)
Worn big-end bearings (regular heavy knocking, perhaps less under load)
Worn main bearings (rumbling and knocking, perhaps worsening under load)
Piston slap (most noticeable when cold)

Chapter 1 Engine

Contents

Part A: Four-cylinder engines

Big-end bearings – renewal	14
Camshaft – removal and refitting	7
Crankcase ventilation system	4
Cylinder head – dismantling and decarbonising	26
Cylinder head – removal and refitting	8
General description	1
Engine – complete dismantling	25
Engine – complete reassembly	29
Engine – dismantling (general)	23
Engine – method of removal	18
Engine – reassembly (general)	28
Engine – refitting (automatic transmission in car)	33
Engine – refitting (manual transmission in car)	31
Engine – removal (leaving automatic transmission in car)	21
Engine – removal (leaving manual transmission in car)	19
Engine ancillary components – refitting	30
Engine ancillary components – removal	24
Engine/automatic transmission – reconnection and refitting	34
Engine/automatic transmission – removal and separation	22
Engine/manual transmission – reconnection and refitting	32
Engine/manual transmission – removal and separation	20
Engine oil and filter	2
Engine/transmission mountings – renewal	17
Examination and renovation	27
Fault diagnosis – four-cylinder engines	36
Induction and injection manifolds (2165 cc engine) – removal and refitting	13
Initial start up after major overhaul	35
Inlet and exhaust manifolds (1995 cc) – removal and refitting	12
Intermediate shaft – removal and refitting	9
Major operations possible without removing the engine	5
Oil pump – removal and refitting	11
Piston/connecting rod/cylinder liner assembly – removal and refitting	16
Piston rings – renewal	15
Sump pan – removal and refitting	10
Timing belt – removal and refitting	6
Valve clearances – adjustment	3

Part B: V6 engines

Big-end bearings – renewal	49
Camshaft front and rear oil seals – renewal	48
Crankcase ventilation system	40
Crankshaft front oil seal – renewal	47
Cylinder head – dismantling and decarbonising	61
Cylinder heads – removal and refitting	44
Cylinder head bolts – retightening	71
Engine – complete dismantling	60
Engine – complete reassembly	64
Engine – dismantling (general)	58
Engine – method of removal	53
Engine – reassembly (general)	63
Engine – refitting (automatic transmission in car)	68
Engine – refitting (manual transmission in car)	66
Engine – removal (leaving automatic transmission in car)	56
Engine – removal (leaving manual transmission in car)	54
Engine ancillary components – refitting	65
Engine ancillary components – removal	59
Engine/automatic transmission – reconnection and refitting	69
Engine/automatic transmission – removal and separation	57
Engine/manual transmission – reconnection and refitting	67
Engine/manual transmission – removal and separation	55
Engine oil and filter	38
Engine/transmission mountings – renewal	52
Examination and renovation	62
Fault diagnosis – V6 engines	72
General description	37
Initial start-up after major overhaul	70
Major operations possible without removing the engine	41
Oil pump – removal and refitting	42
Piston/connecting rod/cylinder liner assembly – removal and refitting	51
Piston rings – renewal	50
Rocker shafts – removal and refitting	45
Sump pan and anti-emulsion plate – removal and refitting	46
Timing cover, chains and sprockets – removal and refitting	43
Valve clearances – adjustment	39

Chapter 1 Engine

Specifications

Part A: Four-cylinder engines

General

Engine type .... Four-cylinder, in-line mounted longitudinally at front of car. Light alloy construction with wet cylinder liners. Belt-driven overhead camshaft

Identification and application:
TS and GTS models J6R
GTX models J7T
Capacity:
   J6R .... 1995 cc (121.7 cu in)
   J7T .... 2165 cc (132.1 cu in)
Bore .... 88.0 mm (3.467 in)

	J6R	J7T
Stroke	82.0 mm (3.231 in)	89.0 mm (3.507 in)
Compression ratio	9.2:1	9.9:1
Maximum power (DIN)	74 kW (103 bhp) at 5500 rpm	89 kW (123 bhp) at 5250 rpm
Maximum torque	162 Nm (119 lbf ft)	182 Nm (134 lbf ft)

Compression pressure:
   Minimum .... 10.3 bar (150 lbf/in²)
   Maximum difference between cylinders .... 1.4 bar (20 lbf/in²)
Firing order .... 1-3-4-2 (No 1 at flywheel end)

Cylinder block

Material .... Light alloy
Overall depth .... 149.25 to 149.75 mm (5.880 to 5.900 in)

Cylinder liners

Height:
   J6R .... 143.50 mm (5.654 in)
   J7T .... 148.50 mm (5.851 in)
Bore .... 88.0 mm (3.467 in)
Base locating diameter .... 93.60 mm (3.688 in)
Protrusion without O-ring .... 0.008 to 0.150 mm (0.003 to 0.006 in)

Crankshaft

Number of main bearings 5
Main bearing journal diameter .... 62.892 mm (2.478 in)
Main bearing journal minimum regrind diameter (0.050 mm undersize) ... 62.642 mm (2.468 in)
Crankpin diameter:
   J6R .... 52.296 mm (2.060 in)
   J7T .... 56.296 mm (2.218 in)
Crankpin minimum regrind diameter (0.025 mm undersize):
   J6R .... 52.271 mm (2.059 in)
   J7T .... 56.271 mm (2.217 in)
Crankshaft endfloat:
   J6R .... 0.07 to 0.25 mm (0.003 to 0.01 in)
   J7T .... 0.13 to 0.30 mm (0.005 to 0.012 in)

Connecting rods and gudgeon pins

Small-end play .... 0.31 to 0.57 mm (0.012 to 0.022 in)
Gudgeon pins (press fit in small-end):
   Length .... 75.0 mm (2.955 in)
   Diameter .... 23.0 mm (0.906 in)

Piston rings

Piston ring thickness:
   Top compression .... 1.75 mm (0.069 in)
   Second compression .... 2.0 mm (0.079 in)
   Oil control .... 4.0 mm (0.158 in)

Camshaft

Number of bearings 5
Endfloat:
   J6R .... 0.07 to 0.13 mm (0.003 to 0.005 in)
   J7T .... 0.07 to 0.15 mm (0.003 to 0.006 in)

Cylinder head

Material .... Light alloy
Overall height .... 111.6 mm (4.397 in)
Maximum permitted surface distortion .... 0.05 mm (0.002 in)

Valves

Stem diameter	8.0 mm (0.315 in)
Head diameter:	
Inlet	44.0 mm (1.734 in)
Exhaust	38.5 mm (1.517 in)
Valve seat angle:	
Inlet	120°
Exhaust	90°
Valve seat width:	
Inlet	1.8 mm (0.071 in)
Exhaust	1.6 mm (0.063 in)
Valve guides:	
Bore	8.0 mm (0.315 in)
Outside diameter (nominal)	13.0 mm (0.512 in)
Repair oversizes:	
With one groove	13.10 mm (0.516 in)
With two grooves	13.25 mm (0.522 in)
Valve springs:	
Free length:	
Earlier models	47.2 mm (1.859 in)
Later models	46.0 mm (1.812 in)
Valve clearances (cold):	
Inlet	0.10 mm (0.004 in)
Exhaust	0.25 mm (0.009 in)
Valve timing (with rocker arm clearance 0.35 mm/0.014 in):	
J6R:	
Inlet valve opens	20° BTDC
Inlet valve closes	60° ABDC
Exhaust valve opens	60° BBDC
Exhaust valve closes	20° ATDC
J7T:	
Inlet valve opens	17° BTDC
Inlet valve closes	63° ABDC
Exhaust valve opens	63° BBDC
Exhaust valve closes	17° ATDC

Lubrication system

Oil pump:	
Type	Gear, driven from intermediate shaft
Clearances:	
Gear teeth to body	0.05 to 0.12 mm (0.002 to 0.005 in)
Gear endfloat	0.02 to 0.10 mm (0.0008 to 0.004 in)
Oil capacity:	
With filter renewal	5.5 litres (9.7 pints)
Oil pressure (at 80°C 176°F):	
At idle	0.8 bar (11.6 lbf/in²)
At 3000 rpm	3.0 bar (43.5 lbf/in²)

Torque wrench setting

	Nm	lbf ft
Cylinder head bolts:		
Stage 1	49	36
Stage 2	79	58
Stage 3	95	70
Main bearing cap bolts	95	70
Big-end cap nuts:		
J6R	49	36
J7T	64	47
*Flywheel bolts	58	43
*Driveplate-to-crankshaft bolts	68	50
Driveplate-to-torque converter bolts	30	22
Crankshaft pulley bolt:		
J6R	79	58
J7T	131	97
Intermediate shaft sprocket bolt	49	36
Camshaft sprocket bolt	49	36
Oil pump fixing bolts	44	32
Sump pan bolts	10	7
Flywheel or torque converter housing-to-engine bolts	54	40
Rocker shaft filter bolt	20	15

*Renew bolts at reassembly

Chapter 1 Engine

Part B: V6 engines

General

Engine type	Six-cylinder, 90° vee configuration mounted longitudinally at front of car. Light alloy construction with wet cylinder liners. Chain-driven overhead camshafts
Identification and application:	
V6 Turbo	Z7U
V6i	Z7V
Capacity:	
Z7U	2458 cc (150.0 cu in)
Z7V	2664 cc (162.5 cu in)

	Z7U	**Z7V**
Bore	91.0 mm (3.585 in)	88.0 mm (3.467 in)
Stroke	63.0 mm (2.482 in)	73.0 mm (2.876 in)
Compression ratio	8.6:1	9.2:1
Maximum power (DIN)	133 kW (182 bhp) at 5500 rpm	104 kW (144 bhp) at 5500 rpm
Maximum torque	280 Nm (207 lbf ft)	220 Nm (162 lbf ft)

Compression pressure:	
Minimum	10.3 bar (150 lbf/in^2)
Maximum difference between cylinders	1.4 bar (2.0 lbf/in^2)
Firing order	1-6-3-5-2-4 (No 1 at flywheel end of left-hand cylinder bank)

Cylinder liners

Bore	88.0 mm (3.467 in)
Base locating diameter	93.48 mm (3.683 in)
Liner protrusion	0.16 to 0.23 mm (0.006 to 0.009 in)
Maximum difference in protrusion between two adjacent liners	0.04 mm (0.0016 in)
Liner base seal thickness:	
Blue	0.087 mm (0.0034 in)
White	0.102 mm (0.0040 in)
Red	0.122 mm (0.0048 in)
Yellow	0.147 mm (0.0058 in)

Crankshaft

Number of main bearings	4
Main bearing journal diameter	70.062 mm (2.7604 in)
Main bearing journal minimum regrind diameter (Z7V only – no regrinding permitted for Z7U)	69.762 mm (2.7486 in)
Crankpin diameter:	
Z7U	60.010 to 60.029 mm (2.3644 to 2.3651 in)
Z7V	52.296 mm (2.0605 in)
Crankpin minimum regrind diameter (Z7V only – no regrinding permitted for Z7U)	51.996 mm (2.0486 in)
Crankshaft endfloat	0.07 to 0.27 mm (0.0028 to 0.0106 in)
Thrust washer thicknesses available	2.30 mm (0.091 in), 2.40 mm (0.095 in), 2.45 mm (0.097 in), 2.50 mm (0.099 in)

Connecting rods and gudgeon pins

Small-end play	0.20 to 0.38 mm (0.008 to 0.015 in)
Gudgeon pins (press fit in small-end)	
Length	72.0 mm (2.837 in)
Outside diameter	23.5 mm (0.926 in)
Inside diameter	14.0 mm (0.552 in)

Piston rings

Piston ring thickness:	
Top compression	1.5 mm (0.059 in)
Second compression	2.0 mm (0.079 in)
Oil control	4.0 mm (0.158 in)

Camshaft

Number of bearings	4
Endfloat	0.07 to 0.14 mm (0.0028 to 0.0055 in)

Cylinder head
Material .... Light alloy
Overall height .... 110.92 to 111.22 mm (4.370 to 4.382 in)
Maximum permitted surface distortion .... 0.05 mm (0.0019 in)

Valves
Stem diameter .... 8.0 mm (0.315 in)
Head diameter:
   Inlet .... 44.0 mm (1.734 in)
   Exhaust .... 37.0 mm (1.458 in)
Valve seat angle:
   Inlet .... 120°
   Exhaust .... 90°
Valve seat width:
   Inlet .... 1.7 to 2.1 mm (0.067 to 0.083 in)
   Exhaust .... 2.0 to 2.4 mm (0.079 to 0.095 in)
Valve guides:
   Bore .... 8.0 mm (0.315 in)
   Outside diameter (nominal) .... 13.0 mm (0.512 in)
   Repair oversizes:
      With one groove .... 13.10 mm (0.516 in)
      With two grooves .... 13.25 mm (0.522 in)
Valve springs:
   Free length .... 47.2 mm (1.859 in)
Valve clearances (cold):
   Inlet .... 0.10 mm (0.004 in)
   Exhaust .... 0.25 mm (0.009 in)
Valve timing (with rocker arm clearance 0.35/0.014 in):
   Z7U:
      Inlet valve opens .... 8°BTDC
      Inlet valve closes .... 40°ABDC
      Exhaust valve opens .... 40°BBDC
      Exhaust valve closes .... 8°ATDC
   Z7V: **LH bank**   **RH bank**
      Inlet valve opens .... 21°BTDC   19°BTDC
      Inlet valve closes .... 45°ABDC   55°ABDC
      Exhaust valve opens .... 57°BBDC   55°BBDC
      Exhaust valve closes .... 21°ATDC   19°ATDC

Lubrication system
Oil pump:
   Type .... Gear-driven by chain from crankshaft pulley
Oil capacity with filter renewal:
   Z7U .... 7.9 litres (13.9 pints)
   Z7V .... 6.0 litres (10.6 pints)
Oil pressure (at 80°C/176°F):
   At idle .... 2.2 bar (32 lbf/in²)
   At 4000 rpm .... 4.4 bar (64 lbf/in²)

Torque wrench settings

	Nm	lbf ft
Cylinder head bolts:		
Stage 1	20	15
Stage 2	61	45
Stage 3: slacken in sequence, then tighten to	20	15
Stage 4	angular tighten through 115°	angular tighten through 115°
Main bearing cap nuts:		
Stage 1	31	23
Stage 2	angular tighten through 75°	angular tighten through 75°
Big-end bearing cap nuts	46	34
Camshaft sprocket bolt	79	58
*Flywheel bolts	46	34
Crankshaft pulley nut	184	135
Sump pan bolts	20	15
*Driveplate (automatic transmission) to crankshaft bolts	71	52
Driveplate-to-torque converter bolts	30	22
Oil pump mounting bolts	15	11
Oil pump sprocket bolts	7	5
Timing cover bolts	15	11
Camshaft pulley bolt	98	72
Bellhousing bolts	54	40

Renew bolts at reassembly

PART A: FOUR-CYLINDER ENGINES

1 General description

The ohc engine is a four-cylinder in-line unit mounted at the front of the vehicle. It incorporates a crossflow design head, having the inlet valves and manifold on the left-hand side of the engine and the exhaust on the right. The inclined valves are operated by a single rocker shaft assembly which is mounted directly above the camshaft. The rocker arms have a stud and locknut type of adjuster for the valve clearances, providing easy adjustment. No special tools are requuired to set the clearances. The camshaft is driven via its sprocket from the timing belt, which in turn is driven by the crankshaft sprocket.

This belt also drives an intermediate shaft which actuates the fuel pump on the 1995 cc engine which is a carburettor version. The 2165 cc engine is fuel injected.

The intermediate shaft drives the oil pump by means of a short driveshaft geared to it.

The distributor is driven from the rear end of the camshaft by means of an offset dog.

A spring-loaded jockey wheel assembly provides the timing belt tension adjustment. A single, twin or triple pulley is mounted on the front of the crankshaft and this drives the alternator/water pump drivebelt, the power steering pump drivebelt and the air conditioning compressor drivebelt, as applicable. The crankshaft runs in the main bearings which are shell type aluminium/tin material. The crankshaft endfloat is taken up by side thrust washers. The connecting rods also have aluminium/tin shell bearing type big-ends.

Aluminium pistons are employed, the gudgeon pins being a press fit in the connecting rod small-ends and a sliding fit in the pistons. The No 1 piston is located at the flywheel end of the engine (at the rear). As is typical with Renault engines, removable wet cylinder liners are employed, each being sealed in the crankcase by a flange and O-ring. The liner protrusion above the top surface of the crankcase is crucial; when the cylinder head and gasket are tightened down they compress the liners to provide the upper and lower seal of the engine coolant circuit within the engine. The cylinder head and crankcase are manufactured in light alloy.

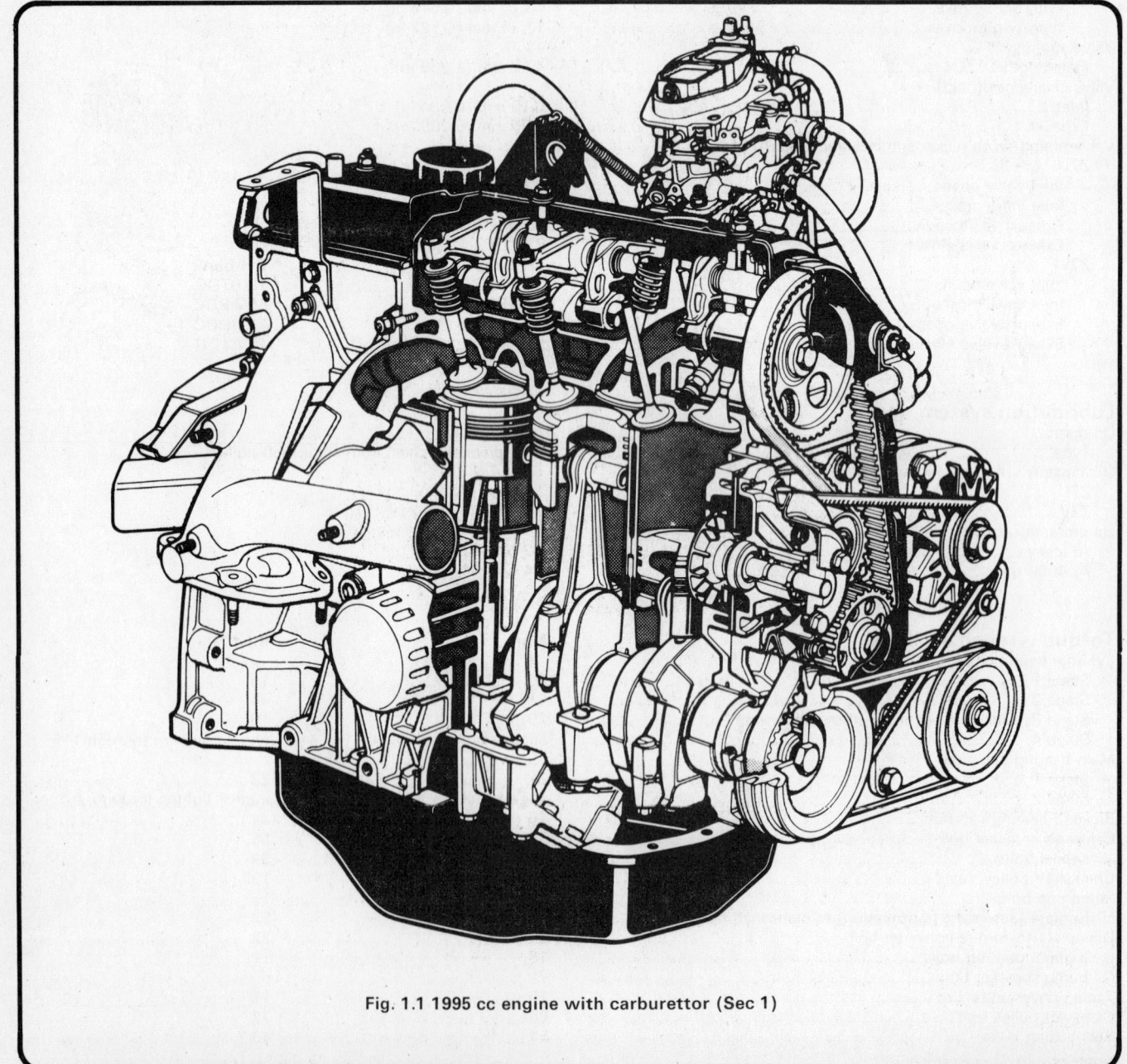

Fig. 1.1 1995 cc engine with carburettor (Sec 1)

Chapter 1 Engine

Fig. 1.2 2165 cc engine with fuel injection (Sec 1)

2 Engine oil and filter

Note: *The use of an oil filter removal tool is recommended when renewing the filter.*

1 At weekly intervals, check the engine oil level, preferably when the engine is cold.
2 Withdraw the dipstick, wipe it cleann and re-insert it. Withdraw it for the second time and read off the oil level which should be between the low and high marks.
3 Top up if necessary through the cap on the rocker cover. To raise the oil level from the 'MIN' to 'MAX' levels requires 1.0 litre (1.76 pints) of oil.
4 At the intervals specified in *'Routine Maintenance'* change the oil and filter. Drain the oil hot into a suitable container after unscrewing the socket type drain plug which accepts a 3/8 in square drive wrench (photo).
5 When the oil has drained, unscrew and discard the oil filter. A filter removal tool will be required for this. If one is not available, drive a large screwdriver through the filter canister and use it as a lever to unscrew the filter (photo).
6 Smear the rubber sealing ring of the new filter with engine oil and tighten it fully by hand only (photo).
7 Refill the engine with the specified type and quantity of oil (photo).
8 Start the engine, and run it for a few seconds to allow time for the new filter to fill with oil.
9 Switch off, wait a few minutes and check the oil level on the dipstick. Top up as necessary to the full mark.

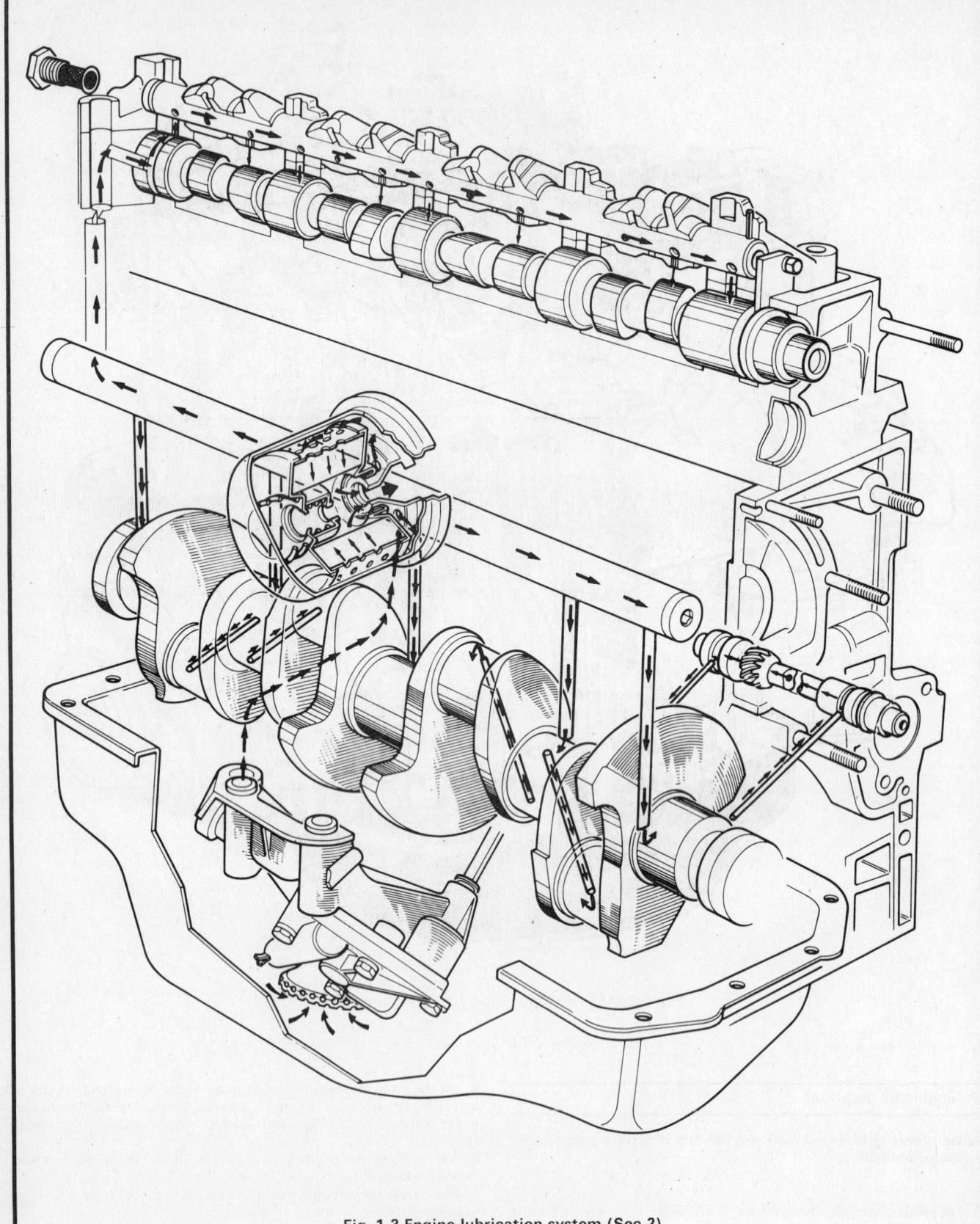

Fig. 1.3 Engine lubrication system (Sec 2)

Are your plugs trying to tell you something?

Normal.
Grey-brown deposits, lightly coated core nose. Plugs ideally suited to engine, and engine in good condition.

Heavy Deposits.
A build up of crusty deposits, light-grey sandy colour in appearance.
Fault: Often caused by worn valve guides, excessive use of upper cylinder lubricant, or idling for long periods.

Lead Glazing.
Plug insulator firing tip appears yellow or green/yellow and shiny in appearance.
Fault: Often caused by incorrect carburation, excessive idling followed by sharp acceleration. Also check ignition timing.

Carbon fouling.
Dry, black, sooty deposits.
Fault: over-rich fuel mixture.
Check: carburettor mixture settings, float level, choke operation, air filter.

Oil fouling.
Wet, oily deposits. Fault: worn bores/piston rings or valve guides; sometimes occurs (temporarily) during running-in period.

Overheating.
Electrodes have glazed appearance, core nose very white – few deposits. Fault: plug overheating. Check: plug value, ignition timing, fuel octane rating (too low) and fuel mixture (too weak).

Electrode damage.
Electrodes burned away; core nose has burned, glazed appearance. Fault: pre-ignition.
Check: for correct heat range and as for 'overheating'.

Split core nose.
(May appear initially as a crack). Fault: detonation or wrong gap-setting technique.
Check: ignition timing, cooling system, fuel mixture (too weak).

WHY DOUBLE COPPER IS BETTER FOR YOUR ENGINE.

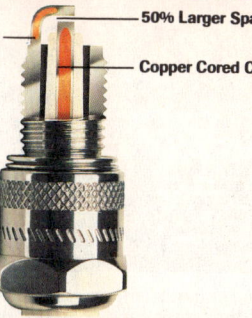

Unique Trapezoidal Copper Cored Earth Electrode — 50% Larger Spark Area
Copper Cored Centre Electrode

Champion Double Copper plugs are the first in the world to have copper core in both centre <u>and</u> earth electrode. This innovative design means that they run cooler by up to 100°C – giving greater efficiency and longer life. These double copper cores transfer heat away from the tip of the plug faster and more efficiently. Therefore, Double Copper runs at cooler temperatures than conventional plugs giving improved acceleration response and high speed performance with no fear of pre-ignition.

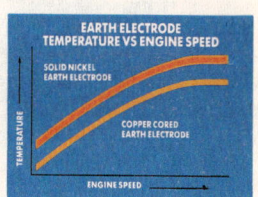

Champion Double Copper plugs also feature a unique trapezoidal earth electrode giving a 50% increase in spark area. This, together with the double copper cores, offers greatly reduced electrode wear, so the spark stays stronger for longer.

 FASTER COLD STARTING

 FOR UNLEADED OR LEADED FUEL

 ELECTRODES UP TO 100°C COOLER

 BETTER ACCELERATION RESPONSE

 LOWER EMISSIONS

 50% BIGGER SPARK AREA

THE LONGER LIFE PLUG

Plug Tips/Hot and Cold.
Spark plugs must operate within well-defined temperature limits to avoid cold fouling at one extreme and overheating at the other.
Champion and the car manufacturers work out the best plugs for an engine to give optimum performance under all conditions, from freezing cold starts to sustained high speed motorway cruising.
Plugs are often referred to as hot or cold. With Champion, the higher the number on its body, the hotter the plug, and the lower the number the cooler the plug.

Plug Cleaning
Modern plug design and materials mean that Champion no longer recommends periodic plug cleaning. Certainly don't clean your plugs with a wire brush as this can cause metal conductive paths across the nose of the insulator so impairing its performance and resulting in loss of acceleration and reduced m.p.g.
However, if plugs are removed, always carefully clean the area where the plug seats in the cylinder head as grit and dirt can sometimes cause gas leakage.
Also wipe any traces of oil or grease from plug leads as this may lead to arcing.

This photographic sequence shows the steps taken to repair the dent and paintwork damage shown above. In general, the procedure for repairing a hole will be similar; where there are substantial differences, the procedure is clearly described and shown in a separate photograph.

First remove any trim around the dent, then hammer out the dent where access is possible. This will minimise filling. Here, after the large dent has been hammered out, the damaged area is being made slightly concave.

Next, remove all paint from the damaged area by rubbing with coarse abrasive paper or using a power drill fitted with a wire brush or abrasive pad. 'Feather' the edge of the boundary with good paintwork using a finer grade of abrasive paper.

Where there are holes or other damage, the sheet metal should be cut away before proceeding further. The damaged area and any signs of rust should be treated with Turtle Wax Hi-Tech Rust Eater, which will also inhibit further rust formation.

For a large dent or hole mix Holts Body Plus Resin and Hardener according to the manufacturer's instructions and apply around the edge of the repair. Press Glass Fibre Matting over the repair area and leave for 20-30 minutes to harden. Then ...

... brush more Holts Body Plus Resin and Hardener onto the matting and leave to harden. Repeat the sequence with two or three layers of matting, checking that the final layer is lower than the surrounding area. Apply Holts Body Plus Filler Paste as shown in Step 5B.

For a medium dent, mix Holts Body Plus Filler Paste and Hardener according to the manufacturer's instructions and apply it with a flexible applicator. Apply thin layers of filler at 20-minute intervals, until the filler surface is slightly proud of the surrounding bodywork.

For small dents and scratches use Holts No Mix Filler Paste straight from the tube. Apply it according to the instructions in thin layers, using the spatula provided. It will harden in minutes if applied outdoors and may then be used as its own knifing putty.

Use a plane or file for initial shaping. Then, using progressively finer grades of wet-and-dry paper, wrapped round a sanding block, and copious amounts of clean water, rub down the filler until glass smooth. 'Feather' the edges of adjoining paintwork.

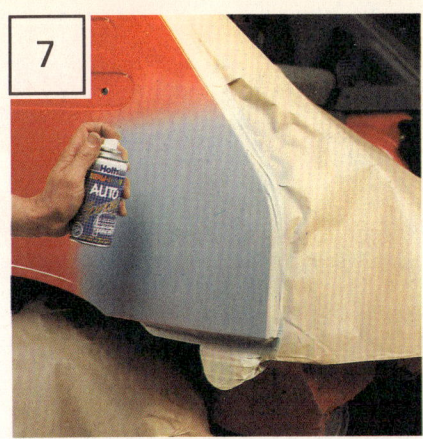

Protect adjoining areas before spraying the whole repair area and at least one inch of the surrounding sound paintwork with Holts Dupli-Color primer.

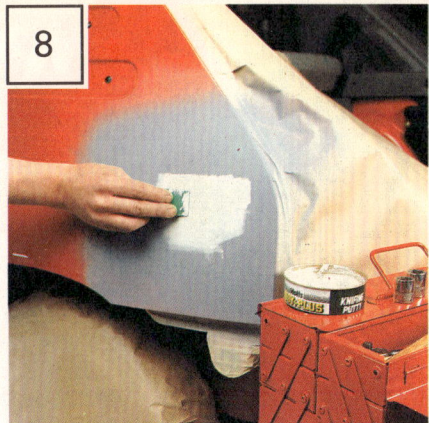

Fill any imperfections in the filler surface with a small amount of Holts Body Plus Knifing Putty. Using plenty of clean water, rub down the surface with a fine grade wet-and-dry paper – 400 grade is recommended – until it is really smooth.

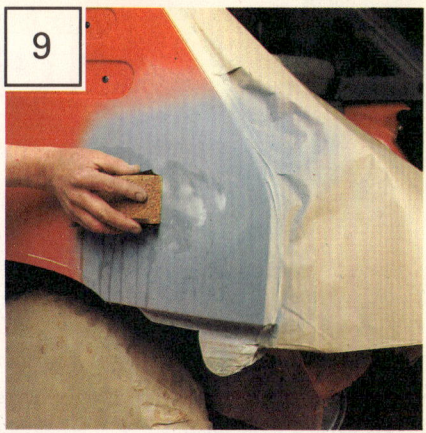

Carefully fill any remaining imperfections with knifing putty before applying the last coat of primer. Then rub down the surface with Holts Body Plus Rubbing Compound to ensure a really smooth surface.

Protect surrounding areas from overspray before applying the topcoat in several thin layers. Agitate Holts Dupli-Color aerosol thoroughly. Start at the repair centre, spraying outwards with a side-to-side motion.

If the exact colour is not available off the shelf, local Holts Professional Spraymatch Centres will custom fill an aerosol to match perfectly.

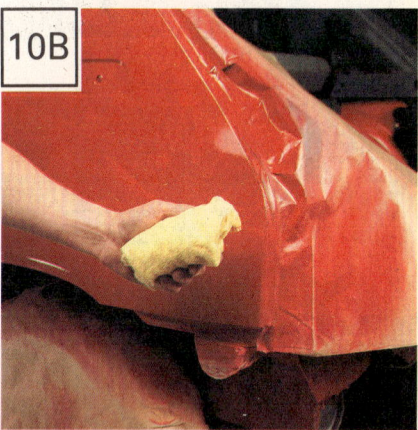

To identify whether a lacquer finish is required, rub a painted unrepaired part of the body with wax and a clean cloth.

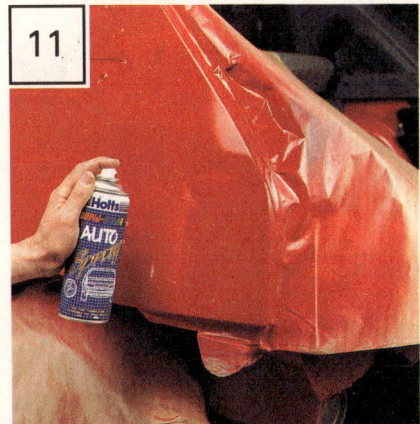

If *no* traces of paint appear on the cloth, spray Holts Dupli-Color clear lacquer over the repaired area to achieve the correct gloss level.

The paint will take about two weeks to harden fully. After this time it can be 'cut' with a mild cutting compound such as Turtle Wax Minute Cut prior to polishing with a final coating of Turtle Wax Extra.

When carrying out bodywork repairs, remember that the quality of the finished job is proportional to the time and effort expended.

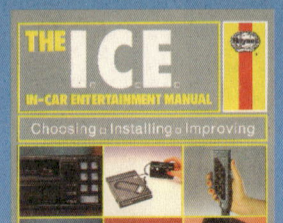

HAYNES No1 for DIY

Haynes publish a wide variety of books besides the world famous range of *Haynes Owners Workshop Manuals*. They cover all sorts of DIY jobs. Specialist books such as the *Improve and Modify* series and the *Purchase and DIY Restoration Guides* give you all the information you require to carry out everything from minor modifications to complete restoration on a number of popular cars. In addition there are the publications dealing with specific tasks, such as the *Car Bodywork Repair Manual* and the *In-Car Entertainment Manual*. The *Household DIY* series gives clear step-by-step instructions on how to repair everyday household objects ranging from toasters to washing machines.

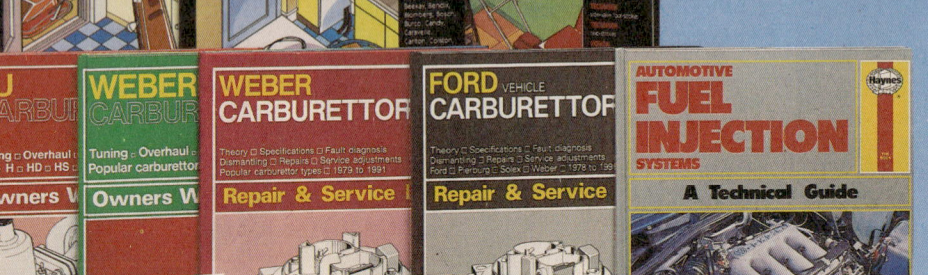

Whether it is under the bonnet or around the home there is a Haynes Manual that can help you save money. Available from motor accessory stores and bookshops or direct from the publisher.

Chapter 1 Engine

2.4 Engine oil drain plug

2.5 Using a strap wrench to unscrew the oil filter

2.6 Fitting a new oil filter

2.7 Filling the engine with oil

3 Valve clearances – adjustment

1 The precise adjustment of the valve/rocker clearances is of utmost importance for two main reasons. The first, to enable the valves to be opened and fully closed at the precise moment required by the cycle of the engine. The second, to ensure quiet operation annd minimum wear of the valve gear components.

2 Settings made when the engine is on the bench will require rotation of the engine and this may be done by turning the exposed crankshaft pulley bolt. If the engine is in the car and a manual gearbox fitted, remove the spark plugs, select top gear, then jack-up the front so the one front wheel is clear of the ground and can be turned. With automatic transmission, this method is not possible and 'inching' the engine using the starter motor will have to be resorted to. Alternatively the crankshaft pulley bolt can be turned as described previously.

3 Turn the engine using one of the methods described until the exhaust valve of No 1 cylinder is fully open. The clearances for No 3 cylinder inlet valve and No 4 cylinder exhaust valve can now be checked and adjusted using a feeler blade.

4 Remember No 1 cylinder is at the flywheel end of the engine and of course the inlet and exhaust valves are on the side adjacent to their respective manifolds.

5 If the clearance requires adjustment, loosen the locknut, and with the feeler in position turn the adjuster screw until the feeler blade is

3.5 Adjusting a valve clearance

Chapter 1 Engine

nipped and will not move. Now unscrew the adjuster until the feeler blade is a stiff sliding fit. Tighten the locknut and recheck the clearance. Refer to the Specifications for the correct clearances (photo).
6 Repeat the adjustment procedure using the following sequence:

Exhaust valve fully fully open on cylinder:	Inlet valve to adjust on cylinder:	Exhaust valve to adjust on cylinder:
1	3	4
3	4	2
4	2	1
2	1	3

4 Crankcase ventilation system

1 The system is designed to extract oil fumes and blow-by gas from within the crankcase and burn them in the combustion chambers.
2 Periodically clean out the connecting hoses and jets and make sure that the hoses are a tight fit on their connectors.

5 Major operations possible without removing the engine

The following operations are possible without the need to remove the engine from the car:

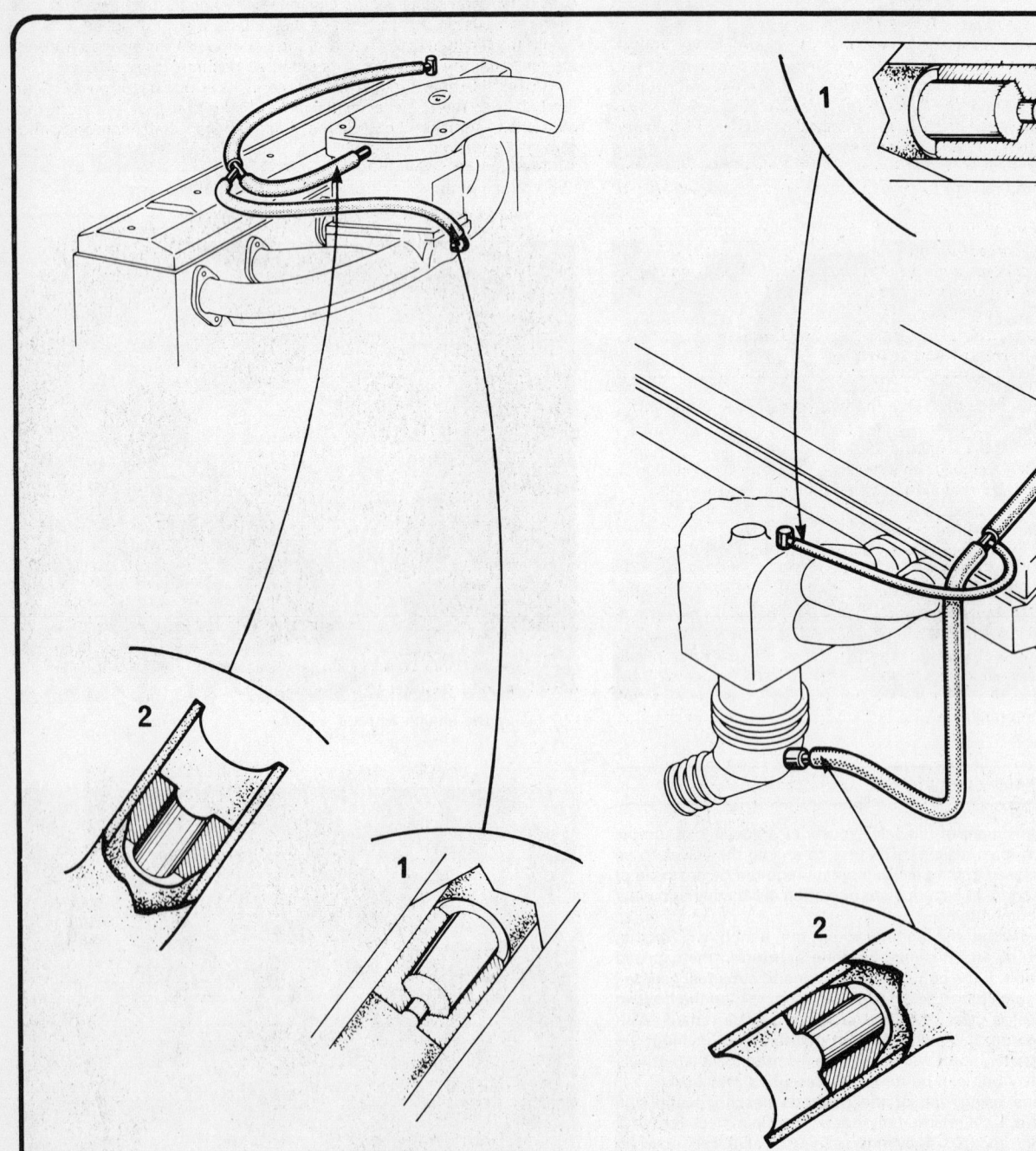

Fig. 1.4 Crankcase ventilation system 1995 cc engine (Sec 4)

1 Jet – 2.0 mm diameter   2 Jet – 5.5 mm diameter

Fig. 1.5 Crankcase ventilation system – 2165 cc engine (Sec 4)

1 Jet – 2.0 mm diameter   2 Jet – 5.5 mm diameter

Chapter 1 Engine

Timing belt – renewal
Camshaft – removal and refitting
Cylinder head – removal and refitting
Intermediate shaft – removal and refitting
Sump pan – removal and refitting
Oil pump – removal and refitting
Inlet and exhaust manifolds (1995 cc engine) – removal and refitting
Injection and induction manifolds (2165 cc engine) – removal and refitting
Big-end bearings – renewal
Piston rings – renewal
Piston/connecting rod/cylinder liner – removal and refitting
Engine/transmission mountings – renewal

Warning: *Vehicles equipped with air conditioning:*

Whenever overhaul of a major nature is being undertaken to the engine, and components of the air conditioning system obstruct the work, and such items of the system cannot be unbolted and moved aside sufficiently far within the limits of their flexible connecting pipes, to avoid such obstruction the system should be discharged by your dealer or a competent refrigeration engineer.

As the system must be completely evacuated before recharging, the necessary vacuum equipment to do this is only likely to be available at a specialist.

The refrigerant fluid is Freon 12 and although harmless under normal conditions, contact with the eyes or skin must be avoided. If Freon comes into contact with a naked flame, then a poisonous gas will be created which is injurious to health.

6 Timing belt – removal and refitting

1 Raise the front of the car so that the wheels clear the ground and support it with suitable stands.
2 Unbolt and remove the engine undertray, if fitted.
3 On 216 cc models, remove the power steering pump drivebelt.
4 Loosen the alternator retaining bolts and remove its drivebelt.
5 Remove the rocker cover.
6 Detach the battery earth terminal.
7 Remove the spark plugs and keep them in order for refitting.
8 Remove the timing belt cover.
9 Turn the crankshaft (using a suitable spanner on the crankshaft pulley bolt) until No 1 piston (flywheel end) is at TDC. The correct setting will be established when the square cut-out in the camshaft sprocket is at its lowest point and the dimple is in line with the rocker cover stud. The dimple and keyway on the crankshaft sprocket will be at their highest point. An alternative way to set TDC is to remove the brass plug from the crankcase and insert a rod to engage in the slot which is located in the crankshaft counterbalance weight. Obviously, to find this slot, the engine No 1 piston must be approaching TDC. This can be checked by removing No 1 spark plug and feeling the compression being generated. Yet a further means of checking TDC is to observe the timing mark on the flywheel is in line with the 'O' mark on the graduated scale at the bellhousing aperture (photos).
10 Lock the flywheel or driveplate (auto. trans.) starter ring gear teeth with a suitable tool, unscrew the crankshaft pulley bolt and take off the pulley. Never join the TDC sensor teeth on the flywheel as a means of preventing it rotating. If the timing belt is to be used again, mark its running direction.
11 Loosen the tensioner retaining bolts and remove the belt.
12 Take care when removing the tensioner as it is operated by a spring and piston in the crankcase side and the tensioner is under considerable pressure. As the tensioner is unbolted, take care that the spring and piston don't fly out of their aperture at high speed.
13 Fit the tensioner spring and plunger, compress the spring and bolt the tensioner into position in its retracted position (photos).
14 Check that the sprocket timing marks have not been moved. Slip the belt over the sprocket teeth, making sure that it is taut from the crankshaft sprocket around the intermediate shaft sprocket and camshaft sprocket.
15 Refit the crankshaft pulley.

6.9A Camshaft sprocket (No 1 piston at TDC)

6.9B Camshaft sprocket dimple aligned with rocker cover stud

6.9C Crankshaft sprocket (No 1 piston at TDC)

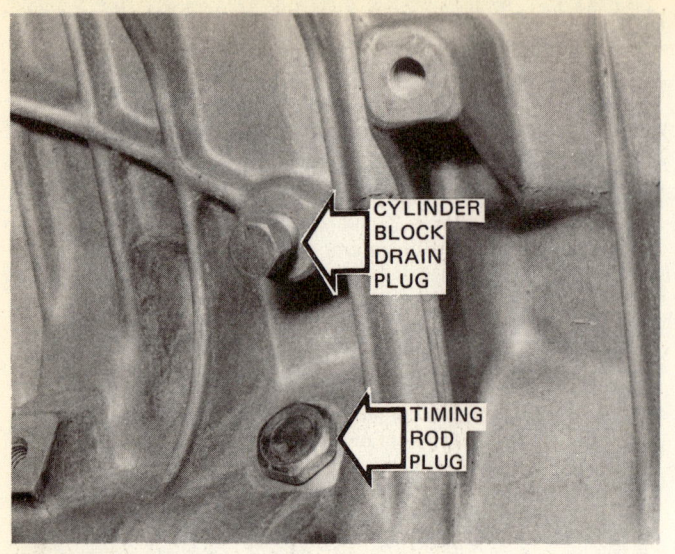

6.9D Timing rod plug

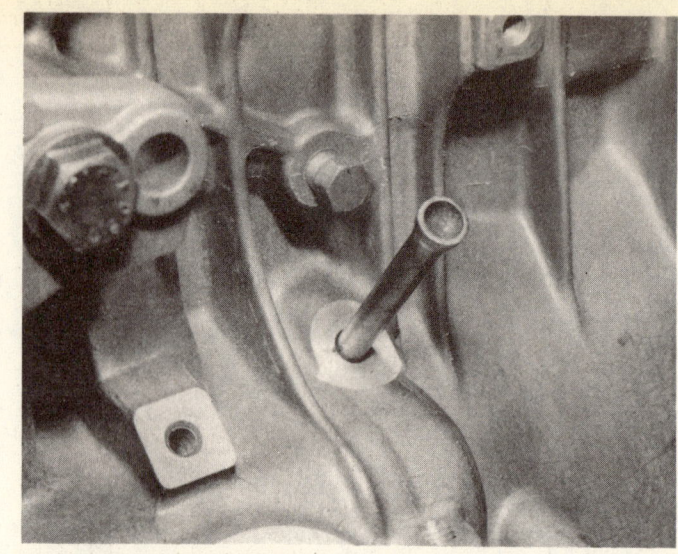

6.9E Timing rod inserted into plug hole

6.9F Timing rod slot in crankshaft counterweight (sump removed)

6.9G Flywheel timing mark at TDC position

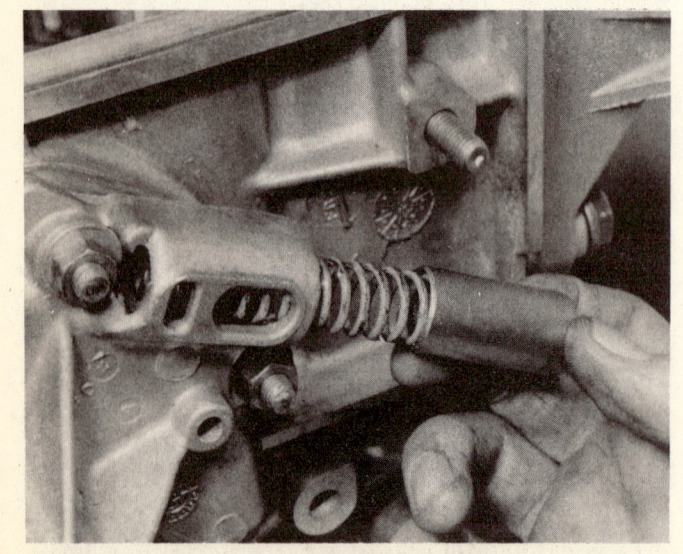

6.13A Timing belt tensioner spring and plunger

6.13B Fitting tensioner pulley

Chapter 1 Engine

6.13C Tensioner pulley fully retracted

16 Release the tensioner nuts so that the tensioner pulley is spring-loaded against the timing belt, retighten the tensioner nuts (photo).
17 Now turn the crankshaft two complete turns in a clockwise direction, never turn it anti-clockwise.
18 Release the tensioner nuts again, then retighten them.
19 Refit all the removed components and lower the car to the ground.

7 Camshaft – removal and refitting

1 Disconnect the battery earth lead.
2 Drain the engine coolant (see Chapter 2).
3 Detach the throttle cable from the carburettor (where applicable) and bracket on the rocker cover.
4 Remove the radiator and the electric cooling fan (see Chapter 2).
5 Unscrew and withdraw the front grille panel.
6 Remove the engine undertray, if one is fitted.
7 Detach the warm air trunking.
8 Remove the drivebelts to the alternator and, where fitted, the power steering pump and air conditioning pump.
9 Unbolt and remove the timing cover.
10 Disconnect the respective HT leads and remove the sparking plugs. Next, rotate the crankshaft to locate the No 1 piston (flywheel end) at TDC or firing stroke. This can be checked by ensuring the timing mark on the flywheel is in line with the 'O' graduation marked on the aperture in the clutch housing. Rotate the crankshaft by means of a spanner applied to the crankshaft pulley retaining bolt.
11 Remove the timing belt as described in Section 6.
12 To remove the camshaft sprocket, pass a suitable rod or screwdriver through a hole in the sprocket and jam against the top surface of the cylinder head to prevent the camshaft from turning. Unscrew the retaining bolt and remove the sprocket. Take great care as the sprocket is manufactured in sintered metal and is therefore relatively fragile. Also, take care not to damage or distort the cylinder head. Remove the Woodruff key.
13 Remove the rocker cover and unscrew the cylinder head retaining bolts in a progressive manner in the sequence shown in Fig. 1.6. With the bolts extracted, remove the rocker assembly and relocate some of the bolts with spacers to the same depth as the rocker pedestals fitted under the bolt heads, to ensure that the cylinder head is not disturbed during subsequent operations. If for any reason the head is disturbed, then the head gasket and respective liner seals will have to be renewed.
14 Unclip the distributor cap and place it to one side.
15 Remove the rotor arm.
16 Unbolt and remove the distributor body.
17 Remove the camshaft retaining thrustplate (two bolts).
18 Unbolt and remove the timing belt rear cover plate.
19 The camshaft front oil seal can now be extracted from its housing. Carefully prise out from the front or preferably drift out from the rear.

6.16 Tightening belt tensioner pulley nut

Take great care not to damage the seal location bore in the head.
20 The camshaft can now be withdrawn carefully through the seal aperture in the front of the cylinder head. Take care during its removal not to snag any of the lobe corners on the bearings as they are passed through the cylinder head.
21 Check that the respective camshaft location bearings in the cylinder head are perfectly clean and lubricate with some engine oil. Similarly lubricate the camshaft journals and lobes.
22 Insert the camshaft carefully into the cylinder head, guiding the cam sections through the bearing apertures so as not to score the bearing surfaces (photo).

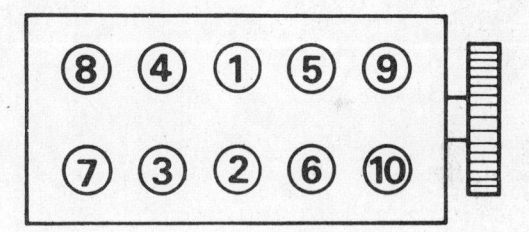

Fig. 1.6 Cylinder head bolt loosening and tightening sequence (Sec 7)

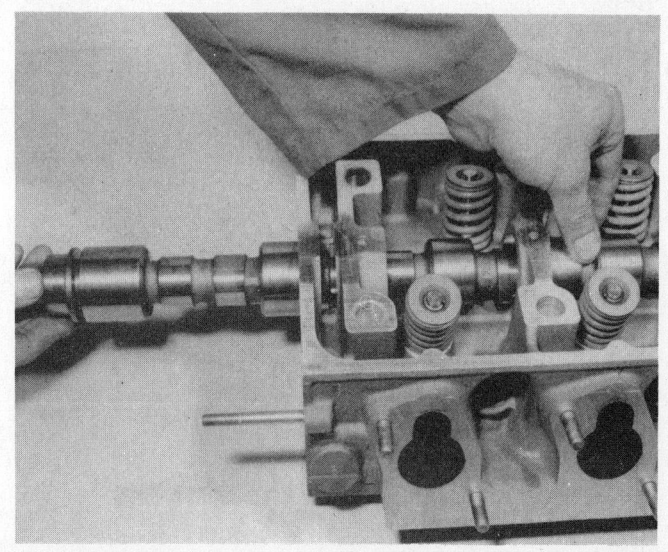

7.22 Inserting camshaft into cylinder head

23 With the camshaft in position, the front oil seal can be carefully drifted into position. Lubricate the seal lips with oil and drive into its location using a suitable tube drift (photo).
24 Refit the rocker assembly and tighten the cylinder head bolts to the specified torque in the sequence shown in Fig. 1.6. Bolt on the timing belt rear cover plate (photo).
25 Refit the camshaft sprocket with Woodruff key and tighten the retaining bolt to the specified torque. Note that the offset (d) in Fig. 1.7 faces towards the cylinder head. The square cut-out in the sprocket will be at its lowest point when No 4 cam lobes are pointing downwards. Bolt on the camshaft thrust plate (photos).
26 Refit the timing belt as described in Section 6.
27 Refit the distributor body, rotor and cap.
28 Refit and connect the spark plugs.
29 Refit the timing cover.
30 Refit and tension the drivebelts.
31 Fit the radiator and grille.
32 Reconnect the throttle cable.
33 Refill the cooling system.
34 Refit the warm air trunking.
35 Check and adjust the valve clearances as described in Section 3.
36 Refit the engine undertray.
37 Reconnect the battery.

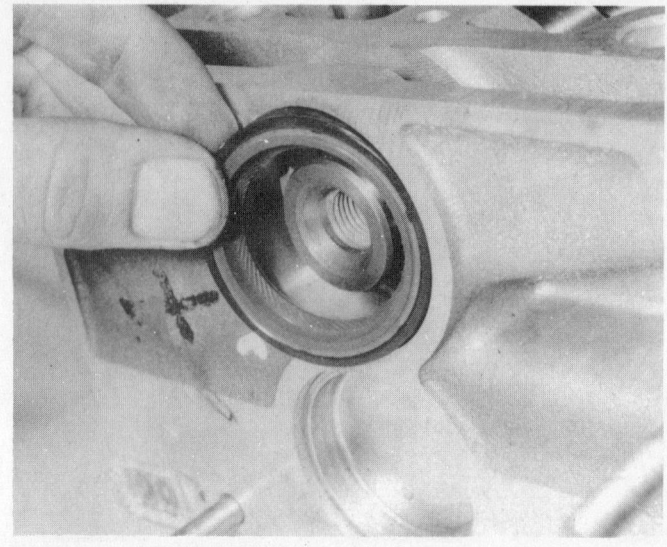

7.23 Camshaft front oil seal

7.24 Timing belt rear cover plate

7.25A Camshaft Woodruff key

7.25B Tightening the camshaft sprocket bolt

7.25C Tightening camshaft thrust plate bolt

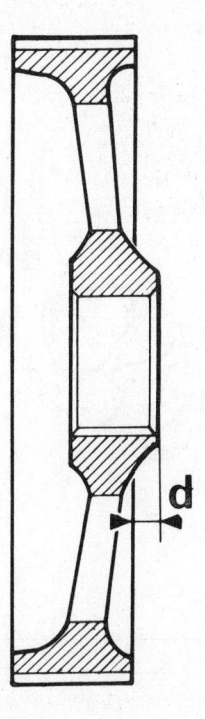

Chapter 1 Engine

11 The cylinder head retaining bolts can now be progessively loosened, and with the exception of the front right-hand bolt, removed. Loosen off the bolts in the sequence shown in Fig. 1.6.

12 The cylinder head is now ready for removal from the block, and it es essential that it is detached in the correct manner. **Do not** lift the head directly upwards from the block. The head must be swivelled from the rear in a horizontal manner, pivoting on the remaining head dowel and bolt at the front. You will probably have to tap the head round on its rear corner, using a soft-headed mallet or wood block to break the seal between the head, gasket and block. Once the head is pivoted round and the seal broken, it can be lifted clear. This action is necessary because if the head were to be lifted directly from the block, the seal between the head gasket and block would be unbroken and the linerrs would thus be disturbed. Each liner has a seal around its lower flange; where the liners are not being removed, it is essential that this seal is not broken or disturbed. If the liners are accidentally disturbed they must be removed and new lower seals fitted.

13 The rocker shaft assembly and final bolt can be removed from the cylinder head once it is withdrawn from the engine.

14 To prevent the possibility of movement by the respective cylinder liners whilst the head is removed, it is advisable to place a clamp-plate over the top edges of the liners. A suitable plate (or bar) can be fastened tempoerarily in position using the existing head bolt holes, using shorter bolts of the desired thread and diameter with large, flat washers under the heads. Alternatively, use a block of wood with two bolts and spacers, clamping it in position in diagonal fashion.

15 Refer to Section 26 for details of dismantling and decarbonising the cylinder head.

16 Before refitting the cylinder head, check that all the mating surfaces are perfectly clean. Loosen the rocker arm adjuster screws fully back. Also ensure that the cylinder head bolt holes in the crankcase are clean and free of oil. Syringe or soak up any oil left in thhe bolt holes, and in the oil feed hole on the rear left-hand corner of the block. This is most important in order that the correct bolt tightening torque can be applied.

17 Prior to removing the liner clamps, rotate the crankshaft to locate the psitons halfway down the bores. Check that the location dowel is in postition at the front right-hand corner.

18 Remove the liner clamps.

19 Fit the cylinder head gasket onto the cylinder block upper face and ensure that it is exactly located. If possible, screw a couple of guide studs into position. They must be long enough to pass through the cylinder head so that they can be removed when it is in position (photo).

20 Lower the cylinder head into position, engaging with the dowel, and then locate the rocker assembly. Make sure that the coolant pump hose is connected (photos).

21 Lubricate the cylinder head bolt threads and washers with engine oil, then screw them into position. Tighten them progressively in the sequence given in Fig. 1.6. Tighten all the bolts to the initial torque specified then further tighten the second torque specified. Now tighten to the final torque as given in the Specifications (photo).

22 Refit the timing belt (Section 6).

23 Reconnect the hoses and electrical wiring plugs.

24 Refit the exhaust manifold.

Fig. 1.7 Camshaft sprocket (Sec 7)

(d) Offset towards cylinder head

8 Cylinder head – removal and refitting

1 Disconnect the battery.
2 Drain the cooling system.
3 Remove the air cleaner.
4 On 1995 cc models, disconnect cables and hoses from the carburettor and remove it complete with inlet manifold.
5 On 2165 cc models, remove the fuel injection and induction manifolds as described in Chapter 3.
6 Remove the exhaust manifold.
7 Disconnect all coolant and vacuum hoses and electrical wiring plugs from the cylinder head.
8 Remove the timing belt as described in Section 6.
9 Detach the diagnostic socket and tie it to one side.
10 Unclip and tie the distributor cap aside complete with HT leads. Remove the rotor.

8.19 Locating cylinder head gasket

8.20A Lowering cylinder head onto block

8.20B Locating rocker assembly

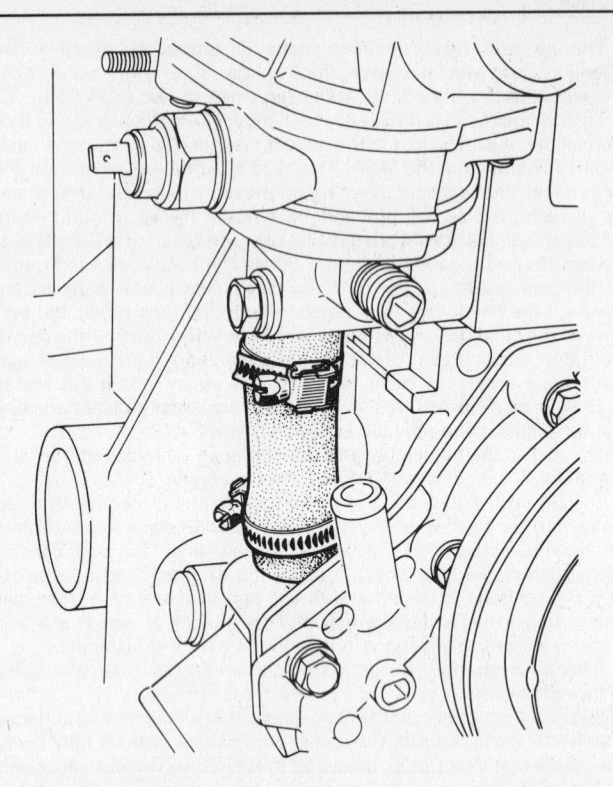

Fig. 1.8 Coolant hose between pump and cylinder head (Sec 8)

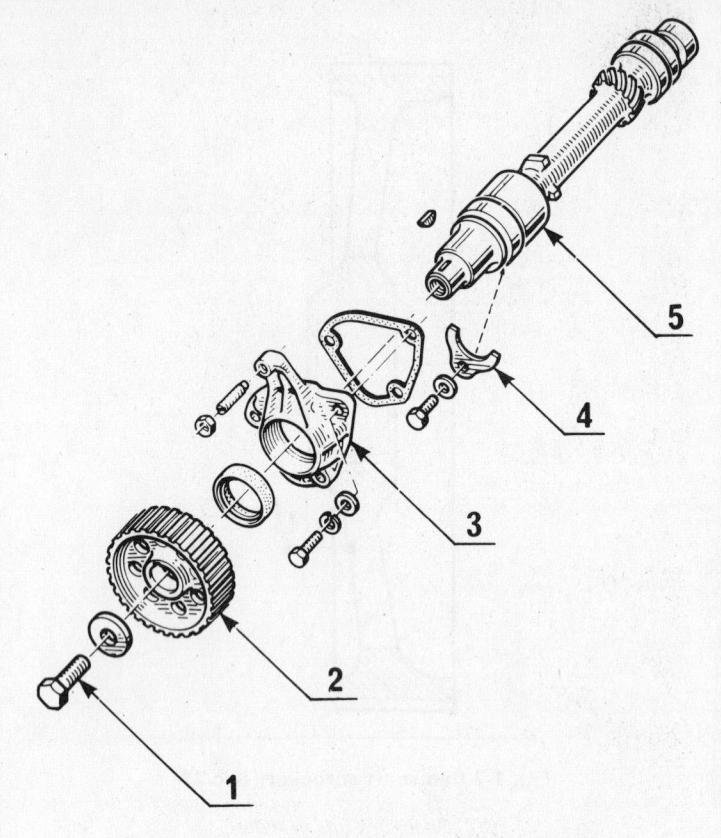

Fig. 1.9 Intermediate shaft components (Sec 9)

1 Bolt
2 Sprocket
3 Front housing
4 Forked retaining plate
5 Shaft

8.21 Tightening cylinder head bolt

25 On 2165 cc models, refit the induction and fuel injection manifolds.
26 On 1995 cc models, refit the inlet manifold and carburettor and reconnect all hoses and cables.
27 Adjust the valve clearances as described in Section 3.
28 Refit the air cleaner.
29 Refill the cooling system and reconnect the battery.

9 Intermediate shaft – removal and refitting

1 On 2165 cc engined models, remove the power steering pump drivebelt (Chapter 2).
2 Remove the alternator drivebelt (Chapter 2).
3 Disconnect the battery.
4 Clamp the inlet manifold heat hose to prevent circulation. Clamp it as close as possible to the inlet manifold and then detach the outlet hose on the manifold.
5 Loosen the timing belt tensioner and remove the timing belt, taking care not to get oil or grease on it.
6 Unscrew the intermediate shaft sprocket bolt, passing a screwdriver shaft or rod through one of the holes in the sprocket to lock against the front of the block and retain the sprocket in position whilst it is being unscrewed.
7 Withdraw the sprocket and remove the Woodruff key from the shaft.
8 Unbolt and remove the intermediate shaft housing cover and the forked lockplate.
9 The oil seal can now be extracted. Prise it free with a screwdriver, but take care not to damage the housing.
10 Before the intermediate shaft can be withdrawn, the oil pump drive pinion must first be extracted. To do this, unscrew the two side cover retaining bolts and remove the cover. The oil pump drive pinion is now accessible and can be extracted.
11 On 1995 cc engines, unbolt and remove the fuel pump.
12 The intermediate shaft can now be removed from the front of the crankcase.
13 Lubricate the shaft and insert it through the front of the crankcase.
14 Slide the lockplate fork into the protruding shaft location groove and secure the plate with the bolt and washer. Check that the shaft is free to rotate on completion (photos).
15 Fit the new oil seal into the intermediate shaft front cover and lubricate its lips (photos).
16 Locate the sprocket and bolt it into position (photos).
17 Refit the oil pump driveshaft and gear and on 1995 cc engines, the fuel pump.
18 Fit and tension the timing belt (Section 6).
19 Reconnect all hoses and drivebelts and reconnect the battery.

9.14A Intermediate shaft lockplate fork

9.14B Lockplate fork bolted in position

9.15A Intermediate shaft cover and gasket

9.15B Fitting intermediate shaft cover bolt

9.16A Fitting intermediate shaft sprocket

9.16B Fitting sprocket bolt and washer

10 Sump pan – removal and refitting

1 Drain the engine oil as described in Section 2.
2 Unbolt and remove the flywheel or torque converter housing front lower cover plate. Remove the engine oil sensor.
3 Unbolt and remove the wiring harness duct which runs under and across the front of the sump pan.
4 Raise the front of the car, support it securely so that the front roadwheels hang free. This will have the effect of moving the anti-roll bar towards the front of the car to make extraction of the sump rear screws easier.
5 Unscrew the securing screws and remove the sump pan. If it is stuck tight, tap it gently with a wooden or plastic-faced hammer.
6 Remove the original gasket and clean the mating surfaces.
7 Check that the mating surfaces of the sump and crankcase are perfectly clean, with no sections of old gasket remaining.
8 Smear an even layer of sealant round the two mating flange surfaces and locate the gasket.
9 Fit the sump carefully into position and locate the retaining bolts.
10 Tighten the bolts progressively by hand in a diagonally opposed sequence.
11 Fit and tighten the sump drain plug, and where applicable refit the oil level indicator sensor.
12 Refill with engine oil.

11 Oil pump – removal and refitting

1 Remove the sump as desribed in Section 10.
2 Unbolt and remove the distributor drive cover plate from the left-hand side of the crankcase.
3 Withdraw the drive pinion and then the oil pump driveshaft, noting that the circlip is at its lower end.
4 Unbolt and remove the oil pump. The pump flange is located on dowels so pull it straight off.
5 Refitting is a reversal of removal. Apply instant (RTV) gasket to the distributor drive cover plate before bolting it into position.
6 Refit the sump pan as described in Section 10.

12 Inlet and exhaust manifolds (1995 cc engine) – removal and refitting

The operations are described in Chapter 3.

13 Induction and injection manifolds (2165 cc engine) – removal and refitting

The operations are described in Chapter 3.

14 Big-end bearings – renewal

1 Remove the sump pan (Section 10) and the oil pump (Section 11).
2 The big-end bearing caps and their connecting rods are not very clearly numbered, so it is best to centre punch them at adjacent points and on the side which is towards the oil filter so that there is no doubt as to their position and orientation in the block. Number the cap and the adjacent machined surface on the connecting rod 1 to 4, numbering from the flywheel end.
3 Unbolt the big-end bearing caps and remove them and discard the shells.
4 Push each connecting rod upwards by not more than 1 mm (0.039 in) or just enough to be able to push the shell bearing in the connecting rod round the crankpin in and remove it.
5 Refit the new shells, making sure that the recess in the cap is perfectly clean, also the backs of the shells themselves.
6 Bolt on the caps (adjacent numbers matching) and tighten the bolts to the specified torque.
7 Refit the oil pump and the sump pan as previously described. Fill the engine with oil.

15 Piston rings – renewal

Note: *A piston ring compressor will be required for this operation.*

1 Although the pistons cannot be withdrawn through the top of the block independently of the cylinder liners (see next Section), new piston rings can be fitted in the following way. The normal process of piston bore glaze removal cannot be carried out to enable the new rings to bed in properly. Despite this, the procedure is useful if a broken ring has to be renewed.
2 Remove the cylinder head (Section 8).
3 Remove the sump and oil pump (Sections 10 and 11).
4 Disconnect the big-end bearing caps (Section 14).
5 Push each connecting rod upwards until the piston rings are out of the bore. If a wear ridge is evident, then this must be removed by careful scraping or using a ridge reamer.
6 Remove the piston rings by sliding two or three feeler blades behind the top ring and removing it upwards using a twisting motion. Refit using the same method.
7 If the top compression ring is being renewed, a stepped replacement should be obtained to prevent it making contact with the wear ridge.
8 A piston ring compressor should be fitted to the rings (well lubricated with engine oil) and the wooden handle of a hammer placed on the piston crown while the head of the hammer is given a sharp blow with the hand. This will drive the piston/rod assembly down the cylinder bore out of the compressor which will le left standing on the liner rim.
9 Reconnect the big-ends, refit the cylinder head as described earlier.

16 Piston/connecting rod/cylinder liner assembly – removal and refitting

1 These assemblies can be removed while the engine is in the car, once the cylinder head and sump pan have been removed as described in earlier Sections.
2 The operations for removal are as described in Section 25 and for refitting in Section 29.

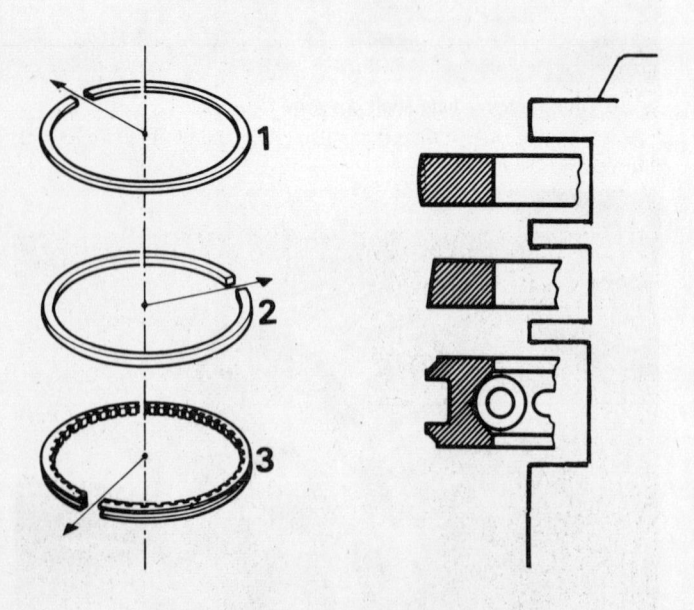

Fig. 1.10 Piston ring fitting diagram (Sec 15)

1 Top compression ring 3 Oil control ring
2 Second compression ring

17 Engine/transmission mountings – renewal

1 The engine and transmission flexible mountings may be renewed with the units in the car.
2 Support the engine or transmission securely on a jack using a block of wood as an insulator.
3 Renew only one mounting at a time by unscrewing the through-bolt nuts and unbolting the mounting bracket from the crankcase, transmission casing or body member (photo).

18 Engine – method of removal

1 The engine may be removed on its own or withdrawn complete with transmission for separation later.
2 If the transmission is not in need of overhaul, it will be easier to remove the engine and leave the transmission in the car.

19 Engine – removal (leaving manual transmission in car)

Note: *Suitable lifting gear will be required for this operation.*

1 Open the bonnet, mark the position of the hinges on its underside using a pencil or masking tape and with the help of an assistant, unbolt and remove the bonnet.
2 Disconnect the battery.
3 Drain the engine oil.
4 Drain the cooling system (Chapter 2).
5 Disconnect the leads from the radiator fan and thermostatic switch.
6 Remove the radiator and side deflectors as described in Chapter 2.
7 Unbolt and remove the radiator top rail.
8 Disconnect the throttle cable (Chapter 3).
9 Unbolt the exhaust downpipes from the manifold and at the first coupling.
10 Remove the air cleaner and intake pipe.
11 Disconnect all vacuum hoses including the one for the brake servo. Identify each hose with masking tape using a different number or letter if there is likely to be any doubt as to its reconnection.
12 Disconnect the coolant hoses, including those for the heater at the engine compartment rear bulkhead.
13 On 2165 cc engines, unclip the computer box from the left-hand wing valance (near the rear of the headlamp) and rest it on top of the engine. The wiring plugs will have to be disconnected before this can be done. Disconnect all engine electrical and earth leads (photo), identifying them with numbered masking tape if necessary.
14 Disconnect the HT lead from the ignition coil and the TDC sensor from the clutch housing.

17.3 Engine flexible mounting

15 Unbolt the loom protector from under the front end of the sump pan and place it to one side (photo).
16 Disconnect the fuel lines.
17 Disconnect the starter motor leads.
18 On 2165 cc engined cars, release the steering pump fluid hoses from their clips, but do not disconnect them. Unbolt the pump and tie it to one side of the engine compartment under the left-hand body member.
19 Disconnect the earth strap which runs between the clutch bellhousing and the body member. Unbolt and remove the starter motor (Chapter 12) (photos).
20 Attach a suitable hoist to the engine lifting lugs and just take its weight.
21 Unscrew the engine-to-transmission connecting bolts.
22 Unbolt and remove the reinforcement brackets at the base of the bellhousing, then unbolt and remove the bellhousing lower cover plate (photo).
23 Place a jack under the transmission.
24 Unscrew the engine mounting through-bolt nuts.
25 Make a final check to ensure that any components, hoses etc. affecting engine removal have been removed or disconnected as applicable.
26 Raise the engine to clear the mounting brackets, then withdraw it forward to disconnect it from the transmission and lift it up and out of the engine compartment.

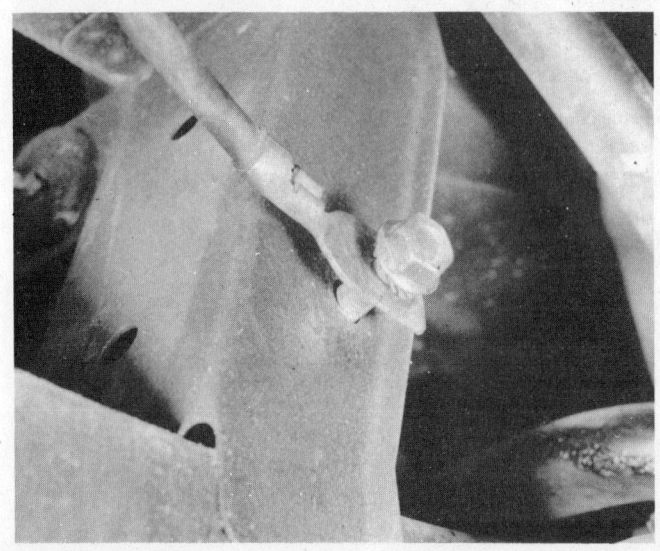

19.13 Engine earth lead

19.15 Wiring loom protector

Chapter 1 Engine

20 Engine/manual transmission – removal and separation

Note: *Suitable lifting gear and a balljoint splitter will be required for this operation.*

1 The operations include those described in paragraphs 1 to 19 of the preceding Section plus the following additional work.
2 Drain the transmission oil.
3 Disconnect the gearchange link rods and the reverse interlock operating cable as described in Chapter 6.
4 Disconnect the reversing lamp switch leads.
5 Disconnect the speedometer sensor.
6 On 1995 cc models, disconnect the clutch cable from the release arm.
7 On 2165 cc models, remove the shield and unbolt the clutch slave cylinder. Tie it up out of the way without disconnecting the fluid hose.
8 Raise the front of the car so that the roadwheels are clear of the ground, support securely on stands and remove the roadwheels.
9 Using a suitable drift, drive out the dual rollpins which secure the driveshafts to the transmission.
10 Unscrew the nuts from the suspension upper balljoint and tie rod end balljoint taper pins and using a balljoint splitter, disconnect the balljoints from the hub carriers and the steering arms.
11 Grip each hub carrier in turn and swivel it downwards while an assistant disconnects the driveshafts from the transmission. Do not strain the flexible brake hoses.
12 With the hoist attached to the engine lifting lugs, and taking the weight of the engine and transmission, unscrew the engine mounting through-bolts and those for the transmission. To provide clearance during removal, unbolt and remove the transmission mounting brackets (photos).
13 Withdraw the engine/transmission forward and upwards from the engine compartment (photo).
14 Clean away external dirt using paraffin and a stiff brush or a water soluble solvent.
15 To separate the engine from the transmission, proceed as follows.
16 Unscrew the bolts which hold the reinforcement brackets and cover plate to the lower face of the clutch bellhousing.
17 Unscrew and remove the engine-to-transmission connecting bolts. Note the location of the various brackets, wiring harness clips and other attachments held by some of the bolts (photo).
18 Support the engine with wooden blocks under the sump pan so that the transmission has a clear gap underneath, and then withdraw the transmission from the engine. **Do not** allow the weight of the transmission to hang upon the clutch shaft while the latter is still engaged with the clutch driven plate.

21 Engine – removal (leaving automatic transmission in the car)

Note: *Suitable lifting gear will be required for this operation.*

1 Carry out the operations described in paras 1 to 18 of Section 19.
2 Unbolt and remove the starter motor.
3 Working through the starter motor aperture, lock the teeth of the ring gear using a suitable tool and unscrew the driveplate-to-torque converter fixing bolts.
4 In order to bring each bolt into view, the crankshaft will have to be turned by means of a socket on its pulley bolt while the ring gear locking tool is temporarily removed.
5 Unscrew and remove the bolts which hold the torque converter lower cover plate. The cover plate will drop off its dowels only after the engine and transmission are separated.
6 Unscrew and remove the engine-to-transmission bolts.
7 Attach a hoist to the engine lifting lugs and just take the weight of the engine.
8 Place a jack under the transmission.
9 Unscrew the engine mounting through-bolt nuts.
10 Raise the engine to clear the mounting brackets, then withdraw it forward to disconnect it from the transmission and lift it up and out of the engine compartment.
11 Bolt a retaining bar to one of the torque converter housing flange bolts in order to retain the converter in full engagement with the transmission and to prevent damage to the oil seal.

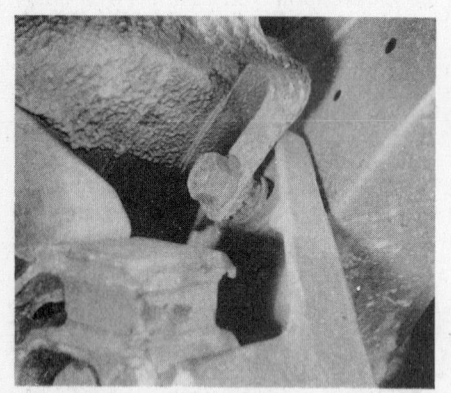

19.19 Transmission earth strap

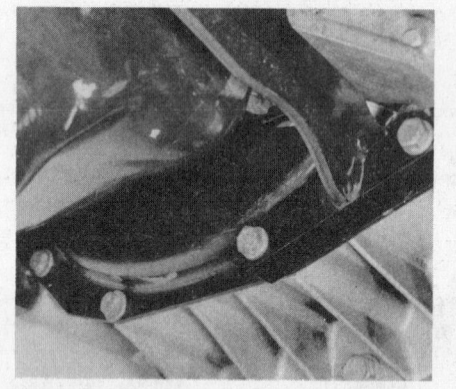

19.22 Bellhousing reinforcement bracket and lower cover plate

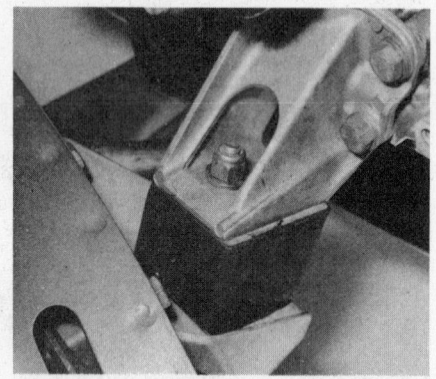

20.12A Right-hand engine mounting

20.12B Transmission mounting

20.13 Removing engine/transmission

20.17 Unscrewing engine/transmission bolt

Chapter 1 Engine

22 Engine/automatic transmission – removal and separation

Note: *Suitable lifting gear and a balljoint splitter will be required for this operation.*

1 Carry out the operations described in paras 1 to 18 of Section 19.
2 Disconnect the earth straps.
3 Unbolt and remove the starter motor.
4 Remove the transmission dipstick tube.
5 Disconnect the fluid hoses from the transmission fluid cooler and plug them.
6 Disconnect the TDC sensor and the speedometer sensor, also the reversing lamp switch leads.
7 Disconnect the driveshafts from the transmission as described in Section 20, paragraphs 8 to 11.
8 Disconnect the speed selector control linkage or cable and kick-down cable after reference to Chapter 7, also the vacuum pipe and computer wiring plugs.
9 With the hoist attached to the lifting lugs and taking the weight of the engine and transmission, remove the transmission mountings and unscrew the nuts from the engine mounting through-bolts. Prepare to guide the computer out with the transmission.
10 Withdraw the engine/transmission forward and upwards out of the engine compartment.
11 Clean away external dirt using paraffin and a stiff brush or a water soluble solvent.
12 Separate the transmission from the engine as described in paragraphs 6 to 8 of Section 21. Pull the transmission from the engine and then fit a torque converter retaining bar as described in Section 21 paragraph 11.

25.10 Main bearing cap number

23 Engine – dismantling (general)

1 Stand the engine on a strong bench so as to be at a comfortable working height. Failing this it can be stripped down on the floor, but at least stand it on a sheet of hardboard.
2 During the dismantling process, the greatest care should be taken to keep the exposed parts free from dirt. As the engine is stripped, clean each part in a bath of paraffin.
3 Never immerse parts with oilways in paraffin, e.g. the crankshaft, but to clean, wipe down carefully with a paraffin dampened rag. Oilways can be cleaned out with a piece of wire. If an air line is available, allparts can be blown dry and the oilways blown through as as added precaution.
4 Re-use of old gaskets is false economy and can give rise to oil and water leaks, if nothing worse. To avoid the possibility of trouble after the engine has been reassembled always use new gaskets throughout.
5 Do not throw the old gaskets away as it sometimes happens that an immediate replacement cannot be found and the old gasket is then very useful as a template. Hang up the gaskets on a suitable nail or hook as they are removed.
6 To strip the engine, it is best to work from the top downwards. The engine oil sump proves a firm base on which the engine can be supported in an upright position. When the stage is reached where the pistons are to be removed, turn the engine on its side. Turn the block upside down to remove the crankshaft.
7 Wherever possible, replace nuts, bolts and washers finger-tight from whenever they were removed. This helps avoid later loss and muddle. If they cannot be replaced then lay them out in such a fashion that it is clear from where they came.

24 Engine ancillary components – removal

1 Before complete engine dismantling begins, remove the following items:

Exhaust manifold (Chapter 3)
Alternator (Chapter 12)
Distributor cap and plug leads (Chapter 4)
Engine mounting and reinforcement brackets
Timing belt cover
Coolant pump (Chapter 2)
Clutch (Chapter 5)
Inlet manifold and carburettor – 1995 cc (Chapter 3)
Injection and induction manifold – 2165 cc (Chapter 3)
Fuel pump – 1995 cc (Chapter 3)

25 Engine – complete dismantling

1 Remove the timing belt as described in Section 6.
2 On 1995 cc engines, unbolt and remove the fuel pump from the left-hand side of the crankcase. On 2165 cc engines, unbolt the inlet manifold support strut.
3 Unbolt and remove the belt tensioner.
4 Unscrew and remove the spark plugs.
5 Remove the cylinder head as described in Section 8.
6 Jam the flywheel teeth and unscrew the flywheel bolts (or driveplate – automatic transmission). The flywheel bolt holes are offset so it can only be fitted one way. Renew the flywheel bolts at refitting.
7 Remove the intermediate shaft as described in Section 9.
8 Turn the engine on its side and remove the sump pan as described in Section 10.
9 Identify each big-end cap and rod to enable refitting in the correct position.
10 The main and big-end bearing caps are numbered from 1 to 4, from the flywheel end of the block. The main bearing cap numbers are read from the oil filter side of the crankcase (photo).
11 The big-end bearing caps and their connecting rods are not very clearly marked, so it is best to centre punch them at adjacent points and on the oil filter side so that there is no doubt as to their position and orientation in the block.
12 The big-end caps will not pass out through the cylinder liners so each liner/piston/connecting rod must be removed as an assembly out of the top of the cylinder block. Before doing this however, centre punch each liner rim and adjacent surface on the top of the cylinder block so that they will be refitted in their original position and orientation position and orientation if of course they are not being renewed completely.
13 Unbolt the big-end bearing caps keeping the bearing shells taped to their original cap or connecting rod if they are to be used again, although this is not to be recommended (see Section 14 for bearing renewal).
14 Withdraw the cylinder liner/piston/rod assemblies.
15 Unbolt and remove the main bearing caps, again keeping their bearing shells with their respective caps if they are to be refitted, again not to be recommended.
16 Note that the front and rear main bearing caps are located on dowels so if they are tight, tap them straight off, **not** in a sideways direction.
17 Withdraw the crankshaft, noting the two semi-circular thrust washers located in the crankcase. Remove the bearing shells and

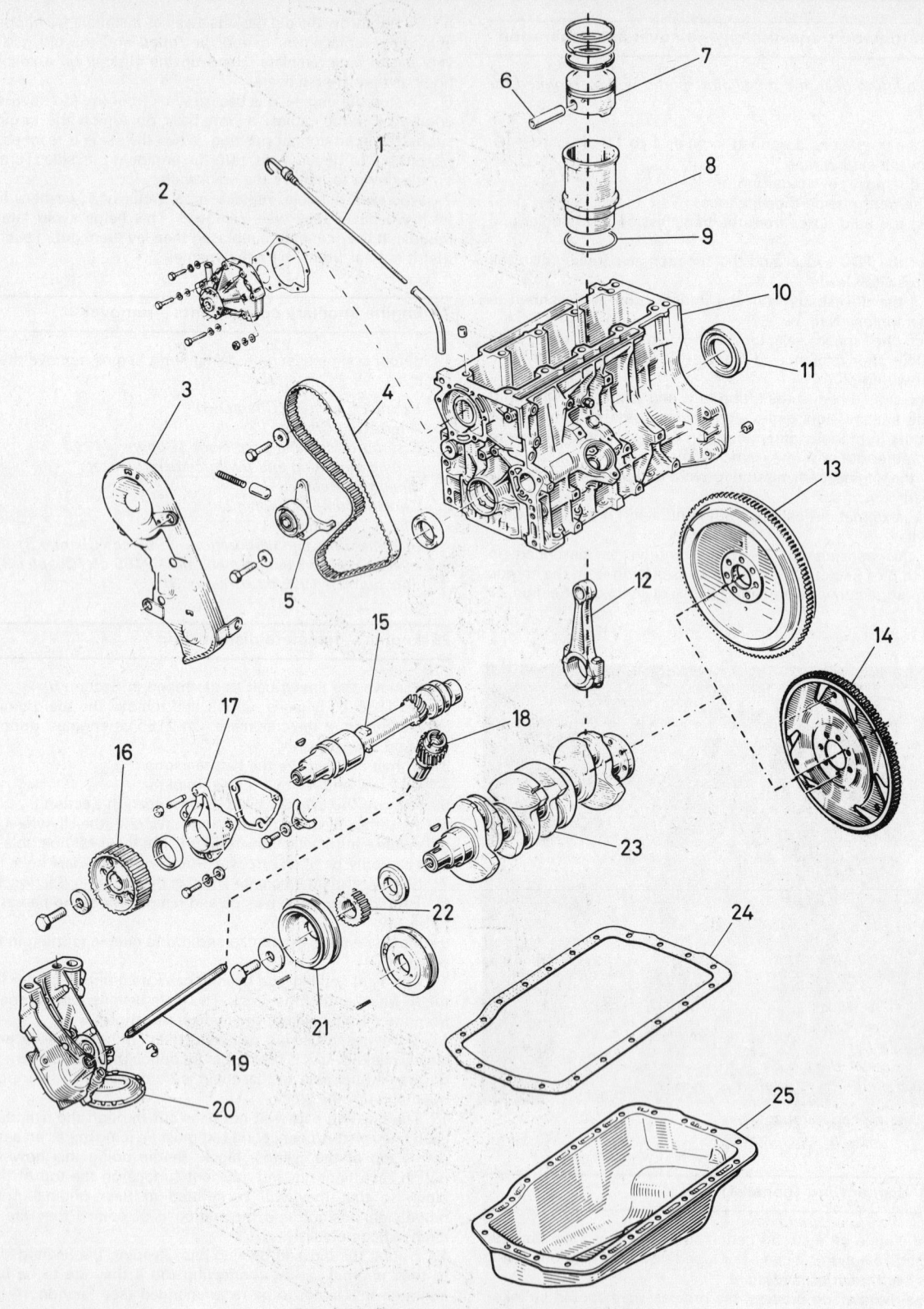

Fig. 1.11 Crankcase components (Sec 25)

1	Dipstick	8	Cylinder liner	14	Driveplate (automatic transmission)	20	Oil pump
2	Coolant pump	9	O-ring seal			21	Crankshaft pulley
3	Timing belt cover	10	Crankcase	15	Intermediate shaft	22	Crankshaft timing belt sprocket
4	Timing belt	11	Oil seal	16	Sprocket		
5	Timing belt tensioner	12	Connecting rod	17	Housing	23	Crankshaft
6	Gudgeon pin	13	Flywheel (manual transmission)	18	Oil pump drive gear	24	Gasket
7	Piston			19	Oil pump driveshaft	25	Sump pan

47

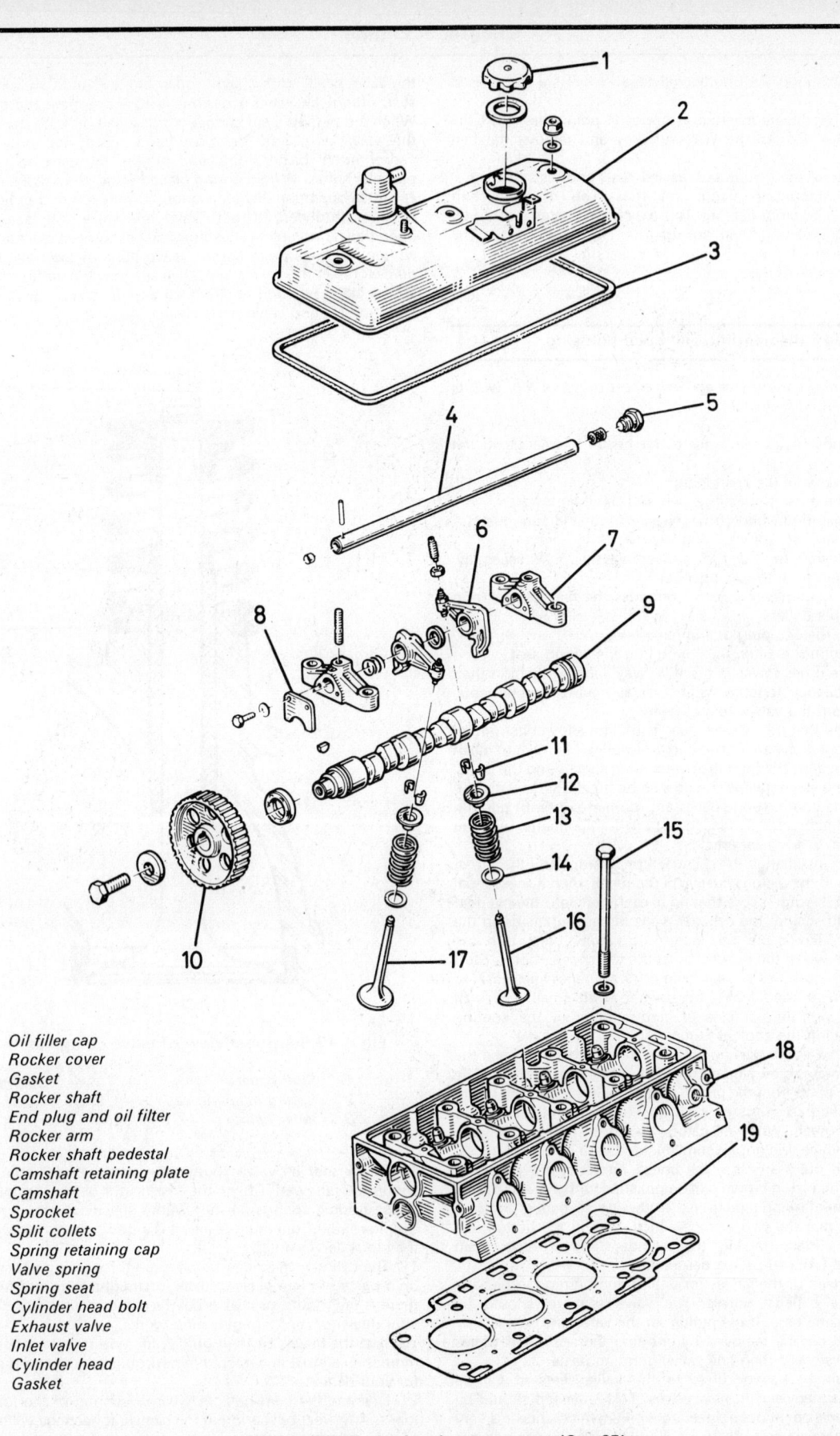

1 Oil filler cap
2 Rocker cover
3 Gasket
4 Rocker shaft
5 End plug and oil filter
6 Rocker arm
7 Rocker shaft pedestal
8 Camshaft retaining plate
9 Camshaft
10 Sprocket
11 Split collets
12 Spring retaining cap
13 Valve spring
14 Spring seat
15 Cylinder head bolt
16 Exhaust valve
17 Inlet valve
18 Cylinder head
19 Gasket

Fig. 1.12 Cylinder head components (Sec 25)

identify them by marking with numbered tape if they are to be used again.

18 Using two screwdrivers inserted at opposite points, lever off the crankshaft sprocket. Extract the Woodruff key and remove the belt guide.

19 Remove the crankshaft front and rear oil seals and discard them.

20 The flywheel mounting flange on the 2165 cc crankshaft incorporates a sealed spigot bearing. This may be extracted if worn by filling it full of grease and then driving in a close fitting rod. The hydraulic pressure created will force out the bearing.

26 Cylinder head – dismantling and decarbonising

Note: A valve spring compressor and valve seat grinding tool will be required to complete this operation.

1 With the cylinder head removed to the bench, dismantle in the following way.
2 Unscrew and remove the spark plugs.
3 Remove the thermostat housing cover and the thermostat.
4 Remove the distributor body from the rear of the cylinder head.
5 Remove the camshaft sprocket and key.
6 Extract the camshaft front oil seal, and remove the rocker assembly.
7 Lift the camshaft from the cylinder head.
8 Using a valve spring compressor, compress the first valve spring and remove the split collets.
9 GEntly release the compressor and remove it.
10 Take off the spring retainer, the spring and the spring seat.
11 Remove all the other valves in a similar way keeping them in their original fitted sequence together with their associated components. Remove and discard the valve stem oil seals.
12 Bearing in mind that the cylinder head is of light alloy construction and is easily damaged use a blunt scraper or rotary wire brush to clean all traces of carbon deposits from the combustion spaces and the ports. The head stems and valve guides should also be freed from any carbon deposits. Wash the combustion spaces and ports down with paraffin and scrape the cylinder ehad surface free of any foreign matter with the side of a steel rule, or a similar article.
13 If the engine is installed in the car, clean the pistons and the top of the cylinder bores. If the pistons are still in the block, then it is essential that great care is taken to ensure that no carbon gets into the cylinder bores as this could scratch the cylinder walls or cause damage to the piston and rings. To ensure this does not happen, first turn the crankshaft so that two of the pistons are at the top of their bores. Stuff rag into the other two bores or seal them off with paper and masking tape. The waterways should also be covered with small pieces of masking tape to prevent particles of carbon entering the cooling system and damaging the coolant pump.
14 Press a little grease into the gap between the cylinder walls and the two pistons which are to be worked on. With a blunt scraper carefully scrape away the carbon from the piston crown, taking great care not to scratch the aluminium. Also scrape away the carbon from the surround lip of the cylinder wall. When all carbon has been removed, scrape away the grease which will not be contaminated with carbon particles, taking care not to press any into the bores. To assist prevention of carbon build-up the piston crown can be polished with a metal polish. Remove the rags or masking tape from the other two cylinders and turn the crankshaft so that the two pistons which were at the bottom are now at the top. Place rag in the cylinders which have been decerbonised, and proceed as just described.
15 Examine the head of the valves for pitting and burning, especially the heads of the exhaust valves. The valve seatings should be examined at the same time. If the pitting on the valve and seat is very slight, the marked can be removed by grinding the seats and valves together with coarse, and then fine, valve grinding paste.
16 Where bad pitting has occurred to the valve seats it will be necessary to recut them and fit new valves. This latter job should be entrusted to the local agent or engineering worked. In practice it is very seldom that the seats are so badly worn. Normally it is the valve that is too badly worn for refitting, and the owner can easily purchase a new set of valves and match them to the seats by valve grinding.
17 Valve grinding is carried out as follows. Smear a trace of coarse carborundum paste on the seat face and apply a suction grinder tool to the valve head. With a semi-rotary motion, grind the valve head to its seat, lifting the valve occasionally to redistribute the grinding paste. When a dull matt even surface is produced on both the valve seat and the valve, wipe off the paste and repeat the process with fine carborundum paste, lifting and turning the valve to redistribute the paste as before. A light spring placed under the valve head will greatly ease this operation. When a smooth unbroken ring of light greay matt finish is produced, on both valve and valve seat faces, the grinding operation is complete. Carefully clean away every trace of grinding compound, take great care to leave none in the pots or in the valve guides. Clean the valves and valve seats with paraffin soaked rag, then with a clean rag, and finally, if an air line is available, blow the valve, valve guides and valve ports clean.

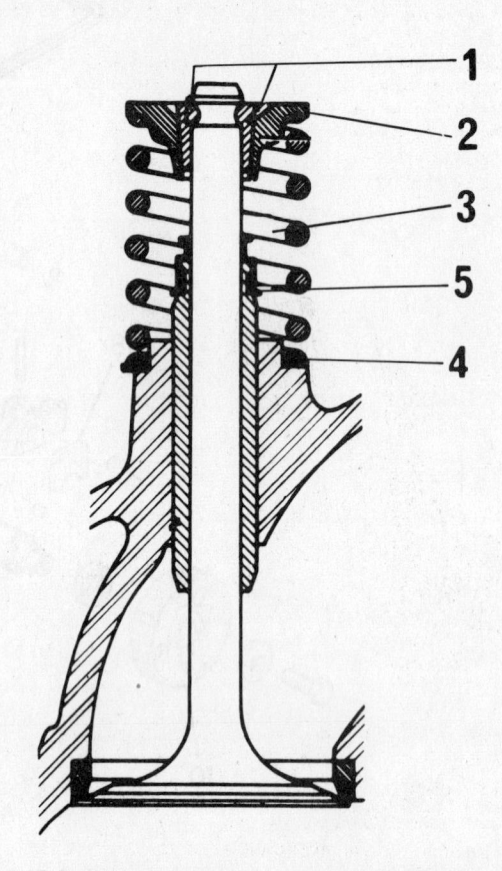

Fig. 1.13 Sectional view of valve components (Sec 26)

1	Split collets	4	Spring seat
2	Spring retaining cap	5	Oil seal
3	Valve spring		

18 Check that all valve springs are intact. If any one is broken, all should be renewed. Check the free height of the springs against new ones. If some springs are not within specifications, replace them all. Springs suffer from fatugue and it is a good idea to renew them even if they look serviceable.
19 The cylinder head can be checked for warping either by placing it on a piece of plate glass or using a straight-edge and feeler blades. If there is any doubt or if its block face is corroded, have it re-faced by your dealer or motor engineering works.
20 Test the valves in their guides for side to side rock. If this is any more than almost imperceptible new guides must be fitted, again a job for your dealer.
21 Examine the camshaft for wear or scoring or the journals or cam lobes. Any wear or scoring in the camshaft bearings will mean renewal of the cylinder head as the bearings are line-bored and cannot be replaced independently. It may be possible to have worn cam lobes reprofiled by a specialist firm.
22 Check the rocker shafts and arms and renew any components which are worn, see Section 27.

Chapter 1 Engine

23 Commence reassembly by oiling the stem of the first valve and pushing it into its guide which should have been fitted with a new oil seal (photos).
24 Fit the spring seat, the valve spring so that the close coils are towards the cylinder head and then the spring retaining cap (photos).
25 Compress the valve spring and using a little grease locate the split cotters in the valve stem cut-out (photo).
26 Gently release the compressor, checking to see that the collets are not displaced.
27 Fit the remaining valves in the same way.
28 Tap the end of each valve stem with a plastic or copper-faced hammer to settle the components.
29 Oil the camshaft bearings in the cylinder head.
30 Insert the camshaft carefully into the cylinder head, guiding the cam sections through the bearing apertures so as not to score the bearing surfaces.
31 Fit a new camshaft oil seal, the Woodruff key and camshaft sprocket and then tighten its bolt to the specified torque.
32 Refit the distributor body, tighten the bolts.
33 Refit the rocker assembly.
34 Refit the thermostat housing cover and thermostat.
35 Refit the spark plugs.

26.23 Fitting a valve into its guide

26.24A Valve stem oil seal and spring seat

26.24B Fitting a valve spring

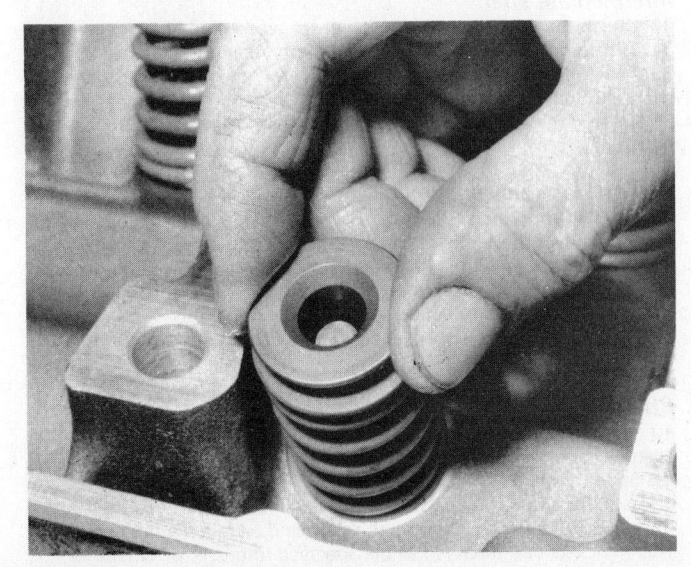

26.24C Valve spring cap

26.25 Compressing a valve spring

27 Examination and renovation

1 With the engine stripped down and all parts thoroughly clean, it is now time to examine everything for wear. The following items should be checked and where necessary renewed or renovated as described in the following Sections.

Cylinder block and crankcase

2 Clean away all old gasket material and then examine the casting for cracks particularly about bolt holes. If they are found, specialist welding or cold repair will be required.
3 Clean out the oilways and galleries with compressed air or wire.
4 If the cylinder bores are worn, this will be evident by the emission of exhaust smoke and general deterioration in engine performance together with increased oil consumption. A good way to test the condition of the engine when it is still in the car is to have it at normal operating temperature with the spark plugs removed. Screw a compression tester (available from mosy modern accessory stores) into the first plug hole. Hold the accelerator pedal fully depressed and crank the engine on the starter motor for several revolutions. Record the reading. Zero the tester and check the remaining cylinders in the same way. All four compression figures should be approximately equal and within the tolerance given in the Specifications. If they are all low, suspect piston ring or cylinder bore wear. If only one reading is down, suspect a valve not seating.
5 The cylinder bores must be checked for taper, ovality, scoring and scratching. Start by examining the top of the cylinder bores. If they are at all worn, a ridge will be felt on the thrust side. This ridge marks the limit of piston ring travel.
6 An internal micrometer or dial gauge can be used to check bore wear and taper against Specifications, but this is a pointless operation if the engine is obviously worn as indicated by excessive oil consumption.
7 The engine is fitted with renewable 'wet' cylinder liners and these are supplied complete with piston, rings and gudgeon pin.

Pistons and connecting rods

8 The gudgeon pin is an interference fit in the connecting rod small end and removal or refitting and changing a piston is a job best left to your dealer or engine reconditioner. This is owing to the need for a press and jig and careful heating of the connecting rod.
9 Removal and refitting piston rings is described in the following paragraphs.
10 Remove the piston rings from the top of the piston. To avoid breaking a ring either during removal or refitting, slide two or three old feeler blades at equidistant points behind the top ring and slide it up them. Remove the other rings in a similar way.
11 Clean carbon from the ring grooves, a segment of old piston ring is useful for this purpose.
12 Clean out the oil return holes in the piston ring grooves and fit the new piston rings.
13 If proprietary rings are being fitted to old pistons, the top ring will be supplied stepped so that it does not impinge on the wear ridge.
14 Insert each piston ring in turn squarely into its bore and check the ring end gap. If it is not within the specified tolerance, carefully grind the end-face of the ring. This does not apply to new rings supplied as part of Renault piston/liner sets which are pre-gapped and must not be altered.
15 Now check each compression ring in its groove and measure the clearances with a feeler gauge. If it is tight the ring may be dubbed flat on a sheet of wet and dry paper laid flat on a piece of plate glass.
16 Fit the rings to the piston using the feeler blade method as described for removal. Work from the top of the piston, fitting the oil control ring first.
17 Lubricate the piston rings and locate their end gaps at 120° apart.

Crankshaft

18 Examine the crankpin and main journal surfaces for signs of scoring or scratches, and check the ovality and taper of the crankpins and main journals. If the bearing surface dimensions do not fall within the tolerance ranges given in the Specifications at the beginning of this Chapter, the crankpins and/or main journals will have to be reground.
19 Big-end and crankpin wear is accompanied by distinct metallic knocking, particularly noticeable when the engine is pulling from low revs, and some loss of oil pressure.
20 Main bearing and main journal wear is accompanied by severe engine vibration rumble – getting progressively worse as engine revs increase – and again by loss of oil pressure.
21 If the crankshaft requires regrinding take it to an engine reconditioning specialist, who will machine it for you and supply the correct undersize bearing shells.

Big-end and main bearing shells

22 Inspect the big-end and main bearing shells for signs of general wear, scoring, pitting and scratches. The bearings should be matt grey in colour. With lead-indium bearings, should a trace of copper colour be noticed, the bearings are badly worn as the lead bearing material has worn away to expose the indium underlay. Renew the bearings if they are in this condition or if there are any signs of scoring or pitting. **You are strongly advised to renew the bearings – regardless of their condition at time of major overhaul. Refitting used bearings is a false economy.**
23 The undersizes available are designed to correspond with crankshaft regind sizes. The bearings are in fact, slightly more then the stated undersize as running clearances have been allowed for during their manufacture.
24 Main and big-end bearing shells can be identified as to size by the marking on the back of the shell. Standard size shell bearings are marked STD or .00, undersize shells are marked with the undersize such as 0.020 u/s. This marking method applies only to replacement bearing shells and not to those used during production.

Flywheel/driveplate and starter ring gear

25 If the starter ring gear teeth on the flywheel (manual transmission) or torque converter driveplate (automatic transmission) are excessively worn, it will be necessary to obtain complete new assemblies. It is not possible to obtain separate ring gears.
26 On manual transmission models, examine the clutch mating surface of the flywheel and renew the flywheel if scoring or cracks are evident.
27 On automatic transmission models, the driveplate face should be checked for run-out, using a dial gauge. The maximum permissible run-out is 0.3 mm (0.012 in). Renew the plate if this figure is exceeded.
28 The flywheel/driveplate retaining bolts must be renewed on assembly.

Camshaft

29 Check the bearings in the cylinder head. If worn, a new head will be required.
30 Inspect the camshaft journals and lobes, scoring or general wear will indicate the need for new parts.
31 The camshaft sprocket teeth should be unchipped and free from wear.
32 The camshaft thrust plate should be free from scoring otherwise camshaft endfloat will be excessive.

Intermediate shaft

33 Check the journals and gear teeth for wear, chipping or scoring.
34 The sprocket teeth should be free from wear and damage.
35 The forked thrust plate should be unworn, otherwise excessive shaft endfloat will occur.
36 Wear in the shaft bearings can only be rectified by the purchase of a new crankcase.

Timing belt and tensioner

37 The belt should be without any sign of fraying, cuts or splits and no deformation of the teeth.
38 Even if the timing belt appears to be in good condition, it is recommended that it is renewed at the intervals specified in the Routine Maintenance Section at the beginning of this Manual.
39 Check the belt tensioner pulley. It should spin freely without noise. If it does not, renew it.

Rocker gear

40 Slight wear in the heels of the rocker arms can be removed using an oilstone, but ensure that the contour is maintained.
41 Scoring or wear in the components can only be rectified by dismantling and renewing the defective components.
42 Unscrew the end plug from the shaft and extract the plug and filter. This filter must be renewed at the specified intervals, or whenever it is removed.

Chapter 1 Engine

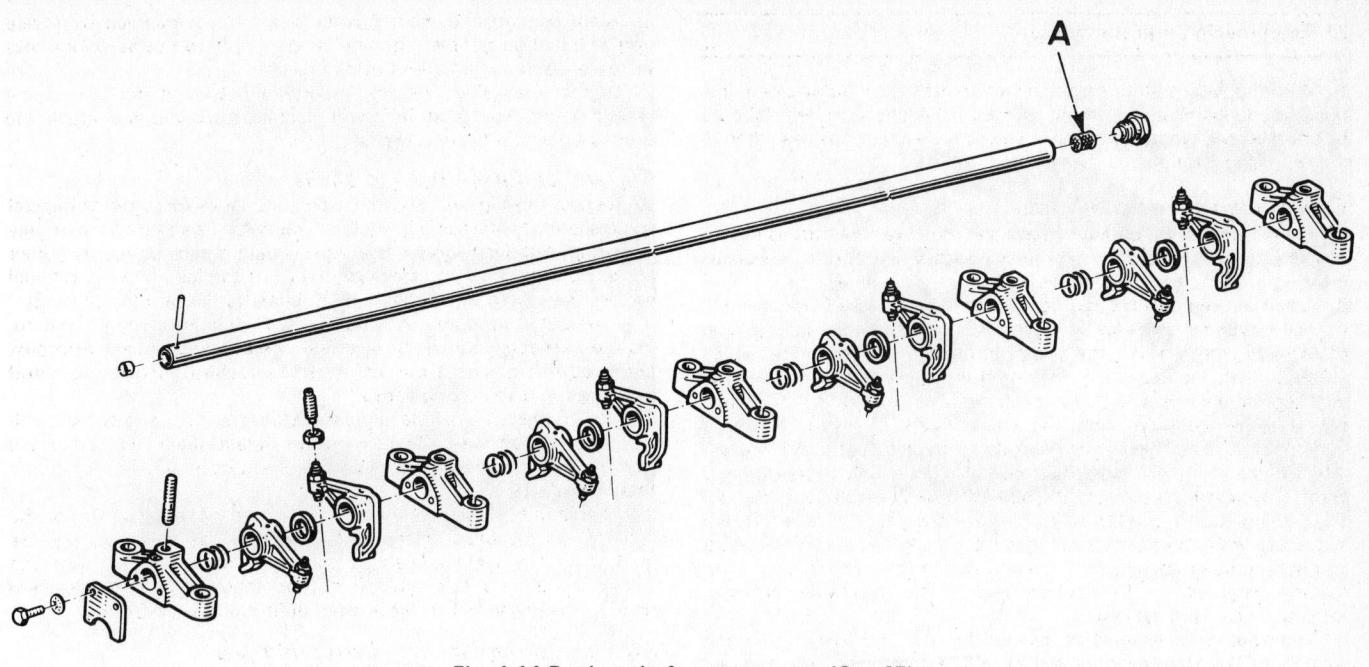

Fig. 1.14 Rocker shaft components (Sec 27)

A   Oil filter

43 Number the rocker arms and pedestals/bearings. Note that bearing No 5 has two threaded holes to retain the thrust plate which controls the camshaft endfloat, and a hole for the roll pin which locates the shaft and pedestal. Renew the roll pin if it is not the solid type.
44 Keep the respective parts in order as they are removed from the shaft and note their respective locations. Note also that the machined flat section on top of pedestals 1 to 4 all face towards the camshaft sprocket.
45 Lubricate each components as it is assembled with engine oil. Lay the pedestals, spacers, springs and rockers out in order.
46 Support the rocker shaft in a soft-jawed vice and insert the new filter into the end of it, fit the retaining bolt and tighten it to the specified torque.
47 Assemble the respective pedestals, rocker arms, springs and spacers onto the shaft. When the shaft assembly is complete, compress the last pedestal to align the retaining pin hole in the shaft and pedestal. Drive a new pin into position to secure it. Early models fitted with a hollow type roll pin should have the later solid type pin fitted on reassembly.

Oil pump

48 It is essential that all parts of the pump are in good condition for the pump to work effectively.
49 To dismantle the pump, remove the cover retaining bolts and detach the cover.
50 Extract the gears and clean the respective components. (photos).
51 Inspect for any signs of damage or excessive wear. Use a feeler gauge and check the clearance between the rotor (gear) tips and the inner housing.
52 Also, check the gear endfloat using a straight-edge rule laid across the body of the pump and feeler gauge inserted between the rule and gears.
53 Compare the clearances with the allowable tolerances given in the Specifications at the start of this Chapter and, if necessary, renew any defective parts, or possibly the pump unit.
54 Do not overlook the relief valve assembly. To extract it, remove the split pin and withdraw the cup, spring, guide and piston. Again, look for signs of excessive wear or damage and renew as applicable.
55 Check the pump driveshaft for signs of wear or distortion and renew if necessary.

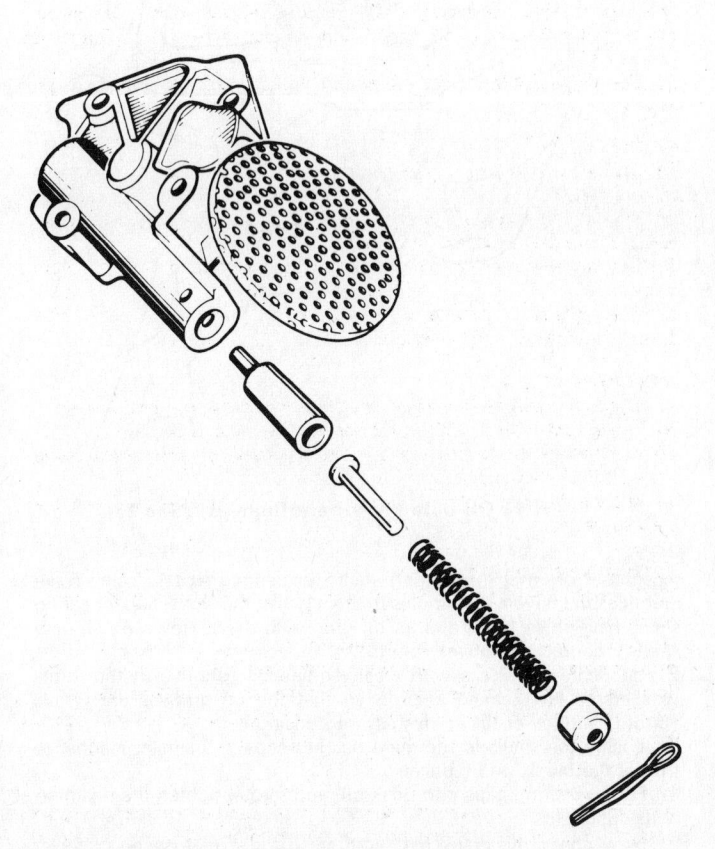

Fig. 1.15 Oil pump pressure relief valve (Sec 27)

27.50A Oil pump gears

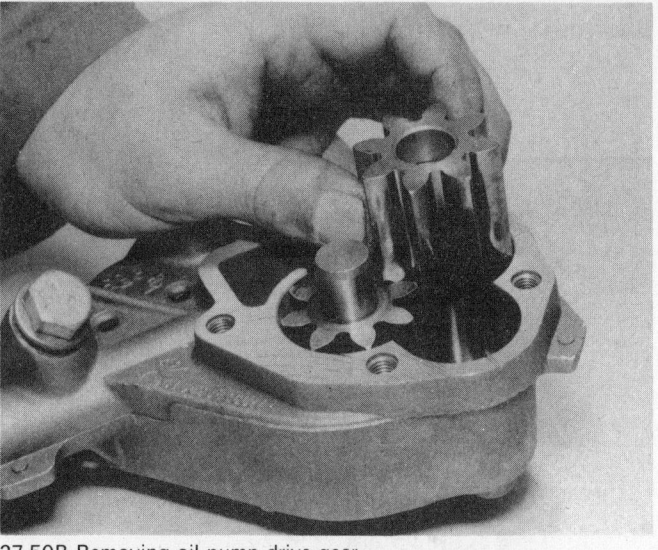

27.50B Removing oil pump drive gear

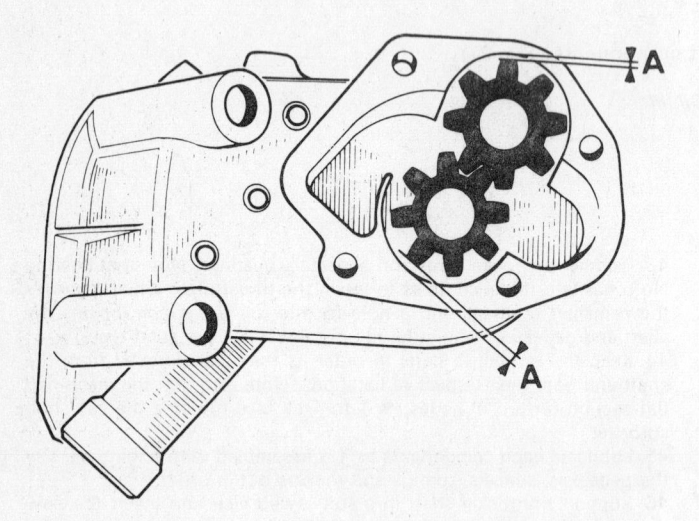

Fig. 1.16 Oil pump rotor-to-body clearance (A) (Sec 27)

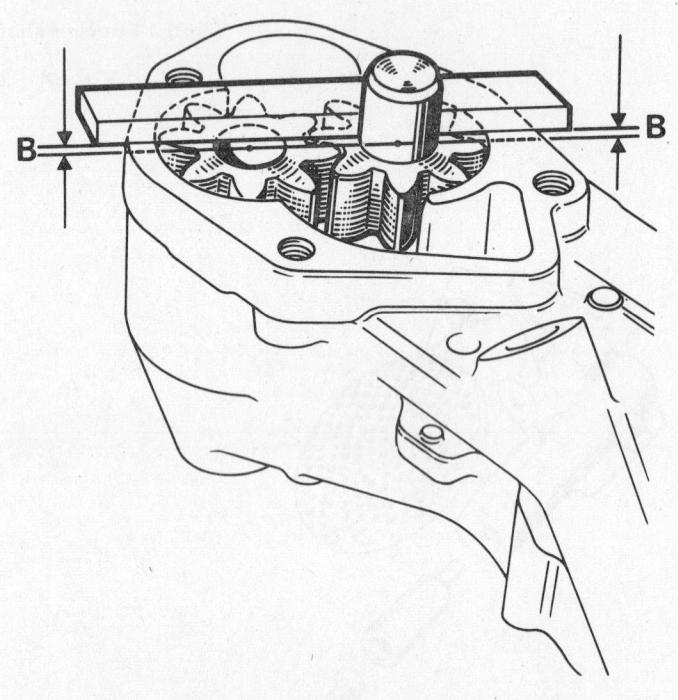

Fig. 1.17 Oil pump rotor endfloat (B) (Sec 27)

28 Engine – reassembly (general)

1 To ensure maximum life with minimum trouble from a rebuilt engine, not only must everything be correctly assembled, but everything must be spotlessly clean, all the oilways must be clear, locking washers and spring washers must always be fitted where indicated and all bearing and other working surfaces must be thoroughly lubricated during assembly.
2 Before assembly begins renew any bolts or studs, the threads of which are in any way damaged, and whenever possible use new spring washers.
3 Apart from your normal tools, a supply of clean rag, an oil can fitted with engine oil (an empty plastic detergent bottle thoroughly cleaned and washed out, will do just as well), a new supply of assorted spring washers, a set of new gaskets, and a torque wrench, should be collected together.

29 Engine – complete reassembly

Crankshaft and main bearings

1 Invert the block and locate the main bearing upper shells into position, engaging the lock tabs into the cut-outs in the bearing recesses. Note that the bearing shells for bearing Nos 1, 3 and 5 are identical and have two oil holes in them whilst the Nos 2 and 4 bearing shells have three holes and an oil groove in them. However, all new shells incorporate grooves (photo).
2 Lubricate the shells with clean engine oil (photo), fit the thrust washers to No 2 main bearing so that the oil grooves are visible (photo) and lower the crankshaft into position.
3 Locate the shells in the main bearing caps in a similar manner to that of the block, and lubricate.
4 Fit the bearing caps into position, and torque tighten the retaining bolts (photo).
5 Now check the crankshaft endfloat using a dial gauge or feeler gauge (photo). Select a thrust washer of suitable thickness to provide the correct endfloat (see Specifications).
6 Remove the main bearing caps and crankshaft, and insert the selected thrust washers, with their grooved side towards the crankshaft.

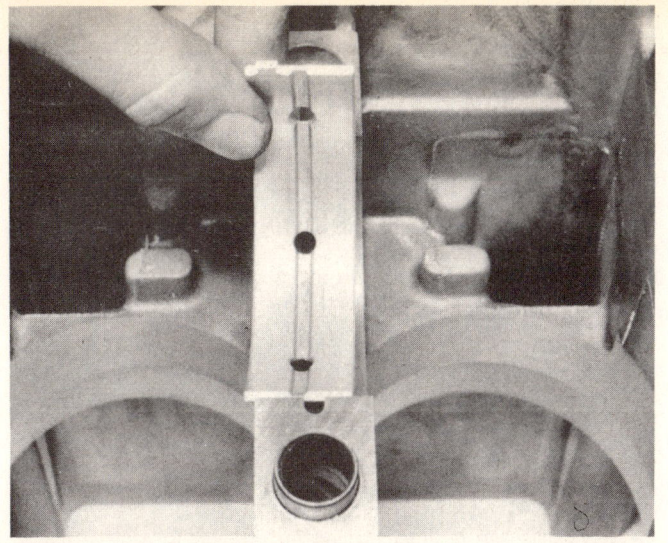

29.1 Main bearing shell in crankcase

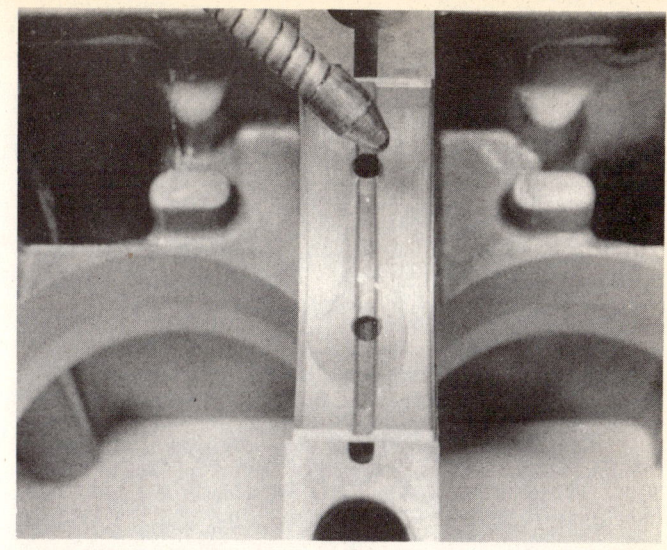

29.2A Oiling main bearing shell

29.2B Crankshaft thrust washer

29.2C Lowering crankshaft into crankcase

29.4 Fitting a main bearing cap. Note the locating dowel (arrowed)

29.5 Checking crankshaft endfloat

Chapter 1 Engine

7 Refit the main bearing caps again and tighten their bolts to the specified torque (photo).
8 The grooves at the sides of the front and rear main bearing caps must now be sealed with RTV type sealant (instant gasket). Make sure that the sealant is forced in under sustained pressure to prevent any air pockets. Cease injecting when the sealant begins to seep from the joints (photo).
9 Lubricate the new front and rear crankshaft oil seals and carefully locate them into their apertures, tapping them fully into position using a tubular drift of a suitable diameter. Ensure that the seals face the correct way round, with the cavity/spring side towards the engine. Should the seal lip accidentally become damaged during fitting, remove and discard it and fit another new seal (photos).
10 If a new clutch spigot bearing is being fitted now is the time to do it. Drive it home using a suitable diameter tubular drift. **Note**: *When fitting this bearing to the crankshaft, smear the outer bearing surface with a suitable thread locking compound.* To the crankshaft front end fit the belt guide, the Woodruff key and sprocket (rollpins visible) (photos).

Flywheel or driveplate (automatic transmission)
11 Locate the flywheel or driveplate on the crankshaft mounting flange. The bolt holes are offset so it will only go on one way (photo).
12 Use new fixing bolts, and having cleaned their threads, apply thread locking fluid to them. Refit the clutch (Chapter 5) (photos).

Cylinder liners/pistons/connecting rods
Note: *A piston ring compressor will be required for this operation.*

13 Before fitting the piston and connecting rod assemblies into the liners, the liners must be checked in the crankcase for depth of fitting. This is carried out as follows.
14 Although the cylinder liners fit directly onto the crankcase inner flange, O-ring seals are fitted between the chamfered flange and the lower cylinder section, as shown in Fig. 1.18. New O-rings must always be used once the cylinders have been disturbed from the crankcase.
15 First, insert a liner into the crankcase *without its O-ring* and measure how far it protrudes from the top face of the crankcase. Lay a straight-edge rule across its top face and measure the gap to the top face of the cylinder block with feeler gauges. It should be as given in the Specifications.
16 Now check the height on the other cylinders in the same way and note each reading. Check that the variation in protrusion on adjoining liners does not exceed 0.0016 in (0.04 mm).
17 New liners can be interchanged for position to achieve this if necessary, and when in position should be marked accordingly 1 to 4 from the flywheel end.
18 Remove each liner in turn and position an O-ring seal onto its lower section so that it butts into the corner, taking care not to twist or distort it.

29.7 Tightening a main bearing cap bolt

29.8 Sealing main bearing cap groove

29.9A Crankshaft front oil seal

29.9B Crankshaft rear oil seal

29.10A Clutch spigot bearing in crankshaft rear flange

29.10B Timing belt spacer/guide

29.10C Crankshaft sprocket Woodruff key

29.10D Fitting crankshaft sprocket

29.11 Fitting flywheel

29.12A Applying thread locking fluid to a flywheel bolt

Chapter 1 Engine

29.12B Tightening a flywheel bolt

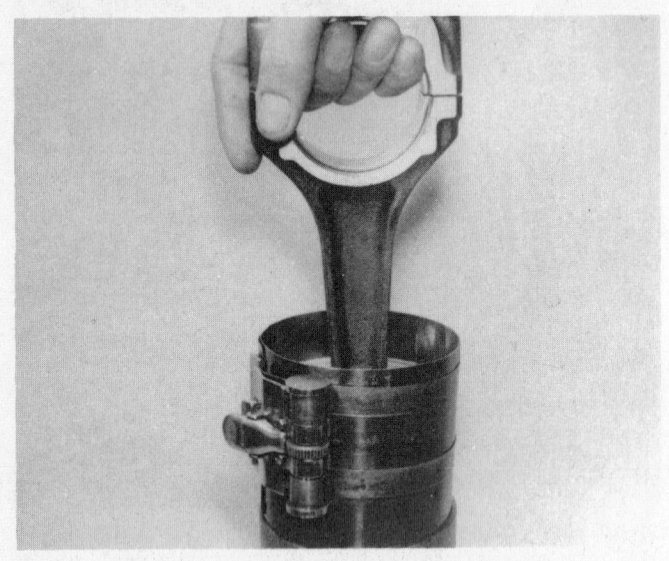

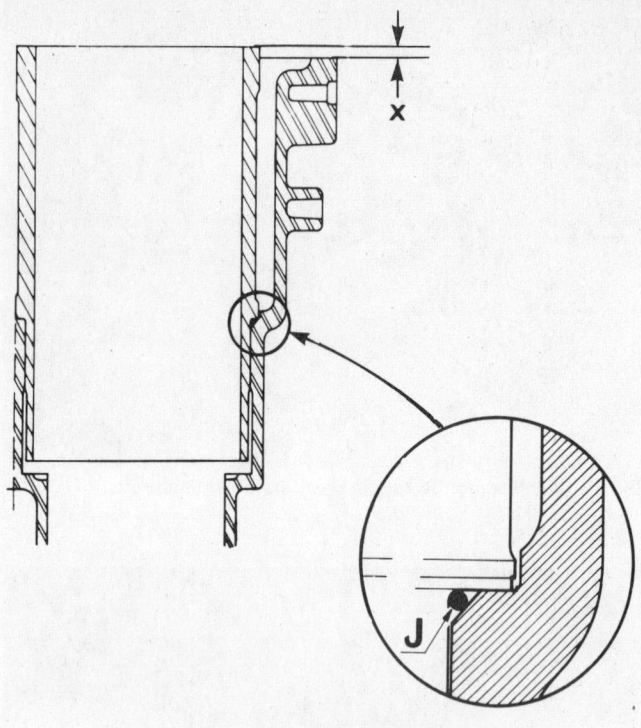

Fig. 1.18 Cylinder liner projection (Sec 29)

J   O-ring   X   Protrusion above block

19 Wipe the liners and pistons clean and smear with clean engine oil, prior to their respective fitting.

20 Refit the pistons to the liners, using a piston ring compressor tightened around well oiled rings to install the piston into the lower end of the cylinder liners. Fit the liners/pistons into the cylinder block. Fit the big-end caps with shells (photos).

21 Observe the following important points:

   (a) The arrows on top of the pistons must point towards the flywheel (photo)

   (b) The connecting rod and cap bolts must be tightened to the specified torque with the numbered markings in alignment (photo)

   (c) When assembled, reclamp the liners and rotate the crankshaft to ensure it rotates smoothly (photo).

29.20A Fitting a piston/connecting rod into cylinder liner

29.20B Fitting a cylinder liner/piston/rod assembly into the block

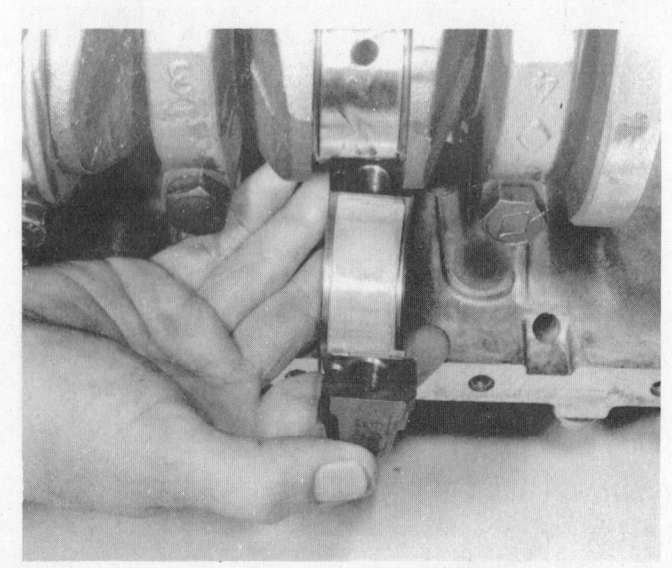

29.20C Fitting a big-end cap

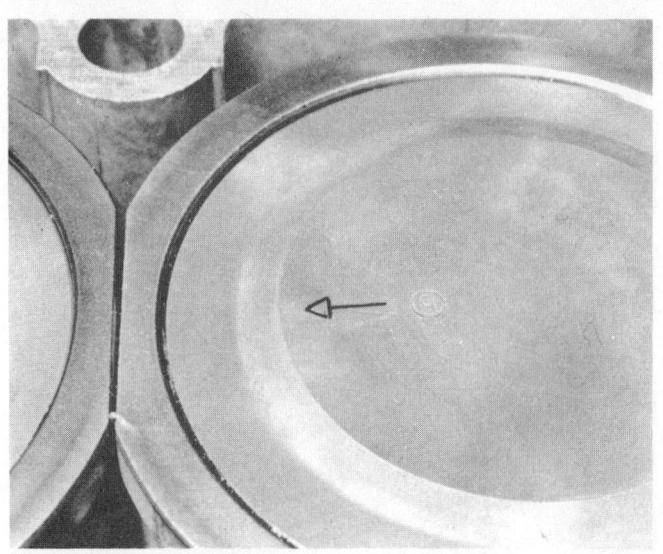

Chapter 1 Engine

Oil pump
22 Lubricate the respective parts of the oil pump and reassemble.
23 Insert the rotors and refit the cover. No gasket is fitted on this face.
24 Tighten the retaining bolts to secure the cover.
25 Insert the oil pressure relief valve assembly, fitting the piston into the spring and the cup over the spring at the opposing end. Compress into the cylinder and insert a new split pin to retain the valve assembly in place.
26 Fit the assembled pump into position. Tighten the retainig bolts (photos).

Intermediate shaft and timing belt tensioner
27 Lubricate the shaft and insert it through the front of the crankcase (photo).
28 Slide the lockplate fork into the protruding shaft location groove and secure the plate with the bolt and washer. Check that the shaft is free to rotate on completion.
29 Fit the new oil seal into the intermediate shaft front cover and lubricate its lips.
30 Fit the coolant pump (Chapter 2).
31 Fit the Woodruff key into its groove in the shaft and carefully locate the intermediate shaft drive sprocket into position with its large offset inner face towards the crankcase. Use a suitable diameter drift to tap the sprocket into position over the key.

29.21A Piston directional arrow

29.21B Tightening a big-end cap bolt

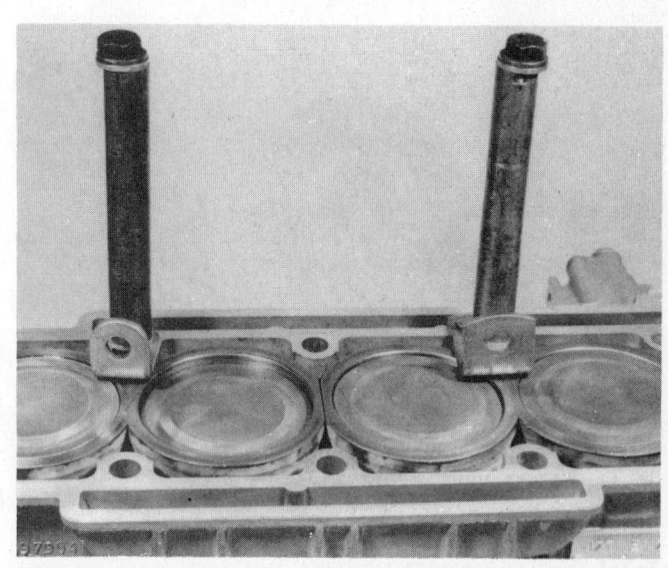

29.21C Typical cylinder liner clamps in position

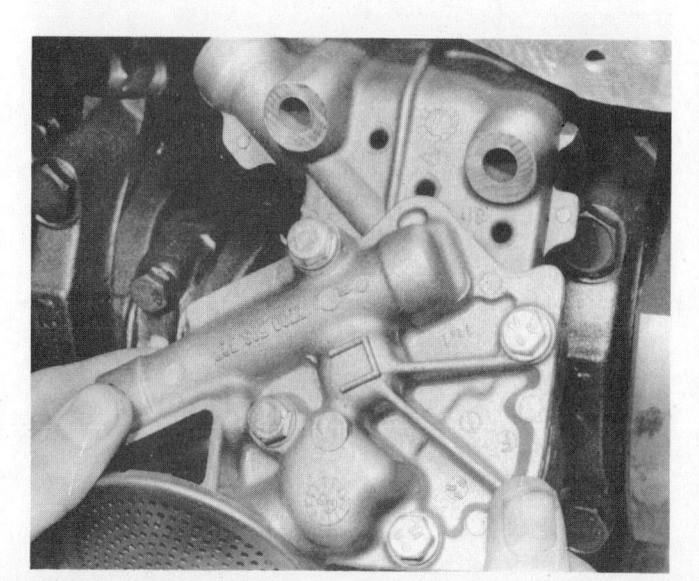

29.26A Fitting the oil pump

29.26B Tightening an oil pump mounting bolt

Chapter 1 Engine

32 Prevent the sprocket from rotating by inserting a screwdriver blade or similar through a sprocket hole, and tighten the retaining nut (complete with flat washer) to the specified torque.

33 The timing belt tensioner can also be fitted at this stage. Insert the spring into its housing in the side of the crankcase and locate the plunger over it. Compress the spring and locate the tensioner jockey wheel arm, retaining it with bolts. The spring tension is quite strong and an assistant will probably be required here. Fully retract the tensioner.

34 If not already located, the oil pump drive pinion and shaft can now be inserted through the side cover hole in the crankcase. Make sure that the limiting circlip is in position on the oil pump end of the shaft (photos).

35 Once in position lubricate with engine oil to prevent pinion 'pick-up' on restarting the engine.

36 Ensure that the intermediate shaft and oil pump drive rotate freely, then refit the side cover. This should be sealed by applying a thick bead of RTV sealant (instant gasket) to its underside flange edge. On 1995 cc engines, fit the fuel pump. On 2165 cc engines use a new gasket and bolt the manifold support strut over the pump hole (photos).

Sump pan

37 Use a new gasket and smear both sump and crankcase mating faces with gasket cement (photo).

29.27 Fitting the intermediate shaft

29.34A Oil pump driveshaft and circlip

29.34B Oil pump drive pinion

29.36A Crankcase side cover being fitted

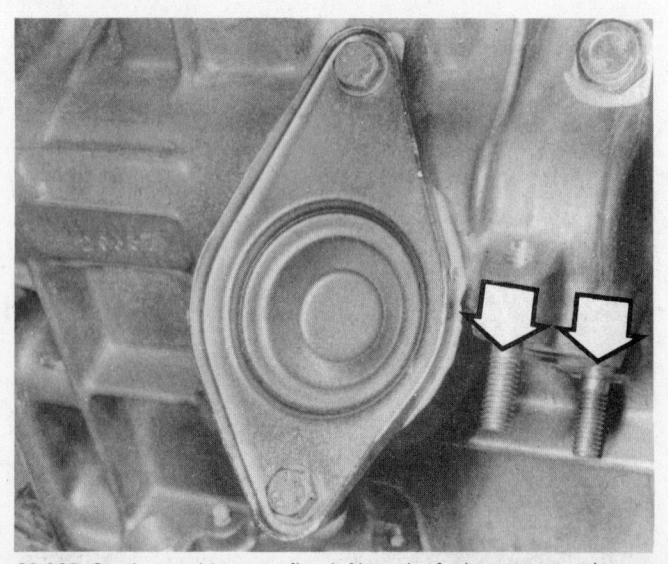

29.36B Crankcase side cover fitted. Note the fuel pump mounting studs – 1995 cc engine (arrowed)

Chapter 1 Engine

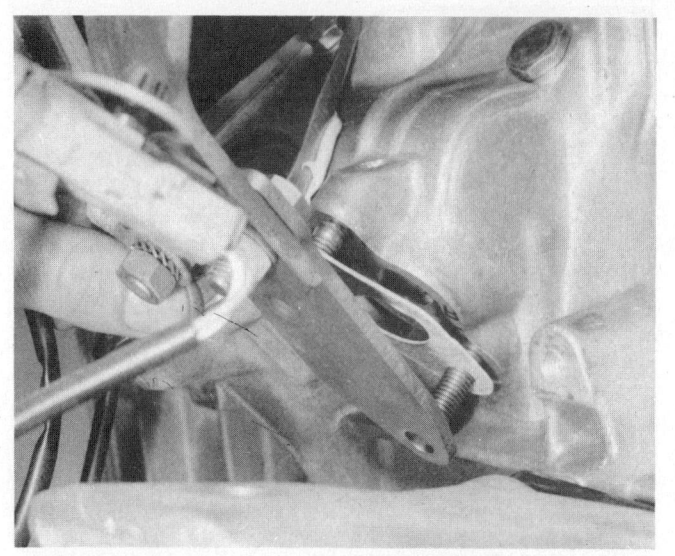

29.36C Manifold support strut and gasket (2165 cc engine)

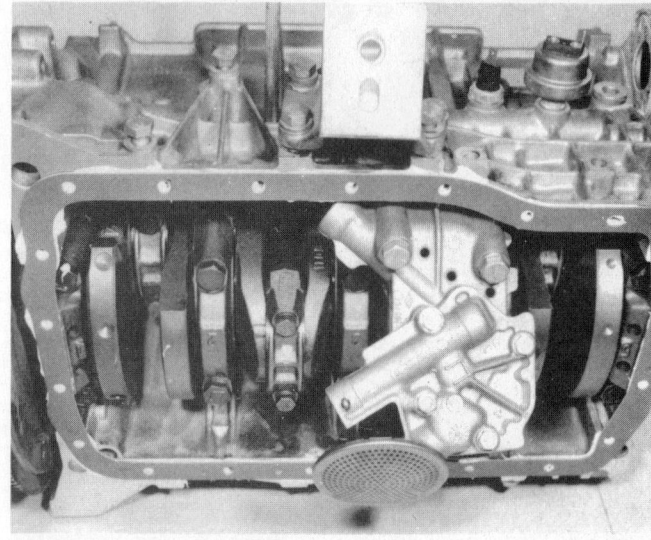

29.37 Sump gasket in position

38 Locate the sump pan, screw and tighten the fixing bolts.
39 Fit the oil level sensor and tighten the oil drain plug (photo).

Camshaft
40 Refit the camshaft to the cylinder head as described in Section 7.

Cylinder head and rocker gear
41 Refit the cylinder head by referring to Section 8.

Timing belt
42 Refit and tension the timing belt by referring to Section 6.

Ancilliaries
43 Check and adjust the valve clearances as described in Section 3.
44 Refit the rocker cover using a new gasket (photos).
45 Fit the timing belt cover with distance sleeves (photos).
46 Fit the crankshaft pulley and bolt. Tighten the bolt to specified torque locking the flywheel starter ring gear teeth to prevent rotations (photos).
47 Fit the coolant distribution tube to the rear of the coolant pump (photo).
48 Fit the exhaust manifold using a new gasket.
49 Use a new gasket and fit the thermostat housing which

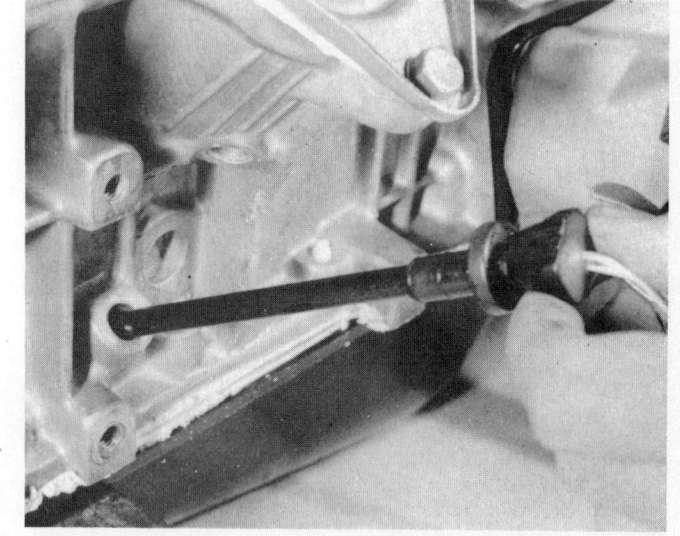

29.39 Oil level sensor

29.44A Fitting rocker cover

29.44B Rocker cover stud seal

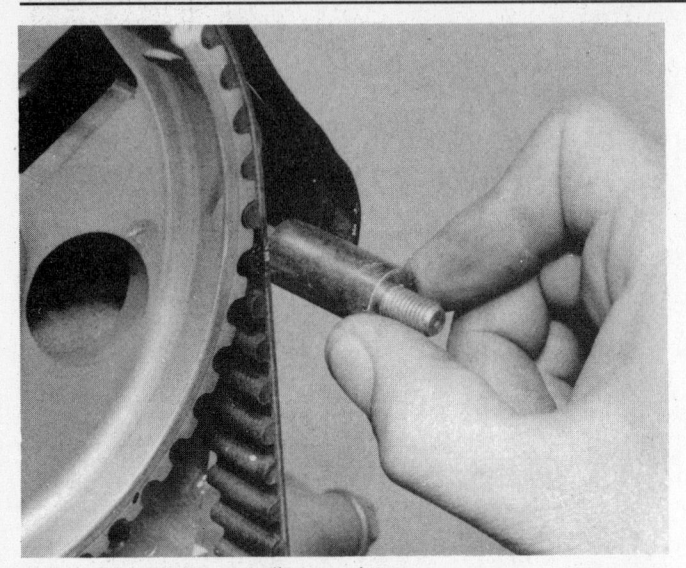

Chapter 1 Engine

29.45A Timing belt cover distance sleeve

29.45B Fitting timing belt cover

29.46A Fitting crankshaft pulley

29.46B Tightening crankshaft pulley bolt

29.47 Coolant distribution tube

incorporates the sensors. Connect the short coolant hose (photos).
50 Fit a new thermostat, 'O' ring and the thermostat housing cover.
51 Oil the sealing ring of a new engine oil filter and screw it into position using hand pressure only.
52 Fit the right-hand reinforcement plate and mounting bracket (photo).
53 Fit the left-hand reinforcement plate. This is retained with socket-headed screws (photo).
54 Fit the inlet manifold and carburettor (1995 cc engine models) or induction and injection manifolds (2165 engine models cc) after reference to Chapter 3.
55 Fit the engine oil pressure and temperature sensors (photo).
56 Attach the starter motor to the crankcase by means of its front mounting bracket.
57 Fit the alternator to its mounting bracket.
58 Correctly route the wiring harness, then connect it and clip it in position (photo).
59 Clip in position and connect as many as possible of the coolant, vacuum and breather hoses.
60 Fit and tension the alternator/coolant pump drivebelt on 1995 cc models. On 2165 cc models equipped with power-assisted steering, the drivebelt should be fitted after the engine is in the car with the steering pump refitted.
61 Fit the distributor to the rear end of the camshaft. The drive dog is offset and the distributor is not adjustable.
62 Fit the spark plugs and connect the HT leads.

29.49A Thermostat housing gasket

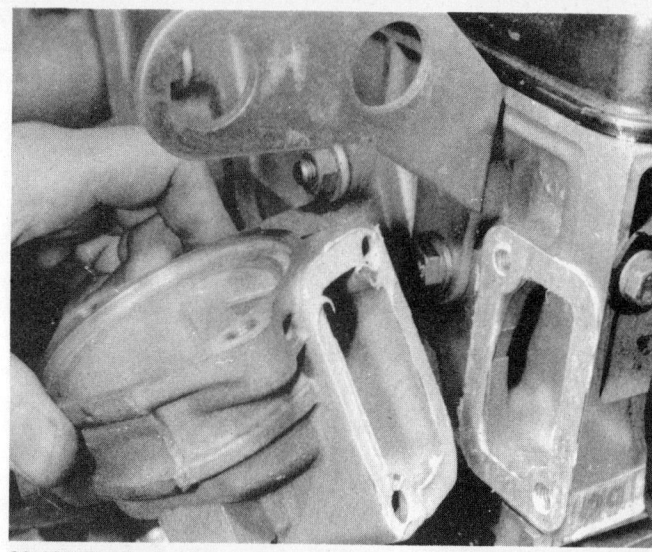

29.49B Fitting the thermostat housing

29.52 Right-hand reinforcement plate and mounting bracket

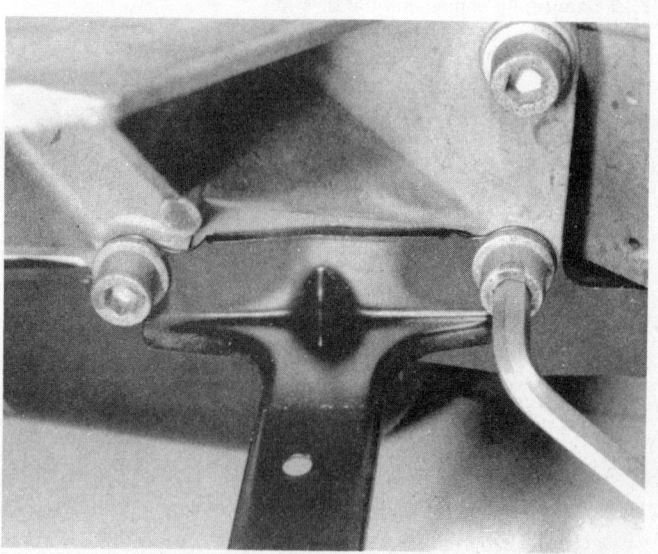

29.53 Tightening left-hand reinforcement plate socket-headed screws

29.55 Engine oil temperature and pressure sensors

29.58 Wiring harness protective sleeve and clips

32.2 Connecting engine and transmission

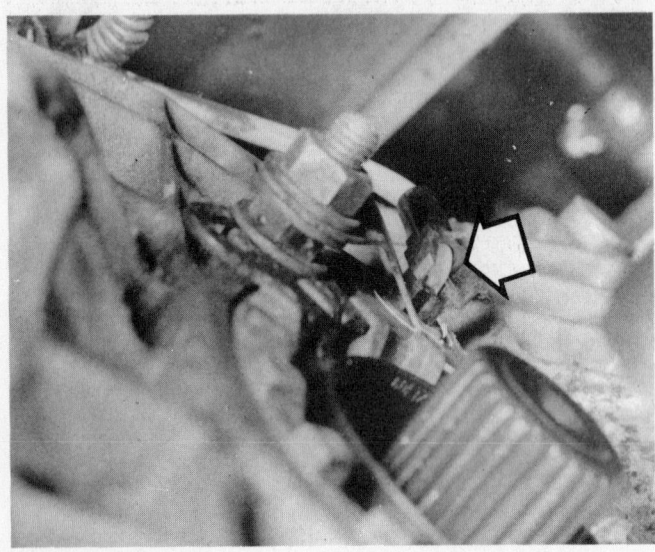

32.14A Speedometer sensor (arrowed)

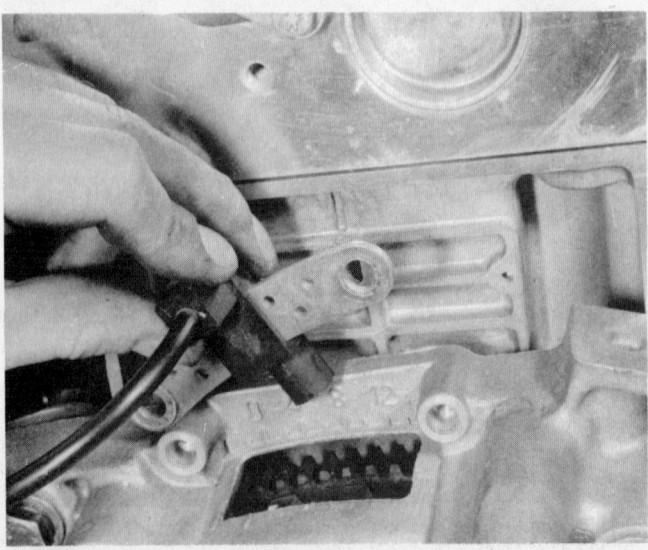

32.14B TDC sensor

32.19A Throttle cable support bracket

30 Engine ancillary components – refitting

The operations are briefly covered in the preceding Section but for more detailed instructions, use the Chapter references in Section 24.

31 Engine – refitting (manual transmission in car)

1 The operations are reversals of those described in Section 19 but observe the following points.
2 Make sure that the clutch driven plate has been centralised, and smear the clutch shaft splines with a little molybdenum disulphide grease.
3 Refill the engine with oil and coolant.
4 Ensure that all wires and hoses have been reconnected in their correct positions.

32 Engine/manual transmission – reconnection and refitting

1 Check that the clutch driven plate is centralised, and then apply a smear of grease to the clutch input shaft.
2 Offer the transmission to the engine and locate it on the positioning dowels (photo).
3 Screw in the connecting bolts, noting the locations of the lifting lugs and clips.
4 Refit the cover plate to the front face of the clutch bellhousing and screw in the reinforcement bracket bolts.
5 The remaining operations are a reversal of removal but observe the following points.
6 It will be found easier to install the engine if the right-hand mounting bracket bolts are loosened.
7 Arrange the lifting gear so that the transmission is slanting downwards at a steep angle and the engine/transmission is canted over to the left.
8 Once in position, connect the engine and transmission mountings.
9 Connect the gearchange rods, reverse interlock cable and the earth straps.
10 Reconnect the driveshafts using new double roll pins. Seal their ends with sealant.
11 Reconnect the suspension balljoints.
12 On 2165 cc models relocate the power steering pump, fit and tension the drivebelt.
13 Reposition the computer and reconnect all wiring plugs.
14 Connect the speedometer and TDC sensor to the transmission housing (photos).
15 Fit the exhaust downpipe using a new gasket.

Chapter 1 Engine

16 Fit the radiator, and connect the coolant hoses and the grille.
17 On 1995 cc models, reconnect the clutch cable and adjust it.
18 On 2165 cc models, bolt on the clutch slave cylinder and heat shield.
19 Reconnect the throttle cable (photos).
20 Fit the air cleaner.
21 With the help of an assistant, refit the bonnet.
22 Connect the battery.
23 Fill the engine and transmission with oil. Refill the cooling system.

33 Engine – refitting (automatic transmission in car)

1 Apply a smear of molybdenum disulphide grease to the locating boss on the torque converter.
2 Remove the temporary holding bar from the torque converter.
3 Lower the engine into the car, making sure that the torque converter does not move forward as the engine is connected and bolted to the transmission. The upper bellhousing bolts must be in position before the engine is connected to the transmission as they cannot be fitted afterwards.
4 The bellhousing lower cover must be located on the dowels before the engine and transmission are brought together, and the alignment mark on the driveplate must be between the two marks on the torque converter.
5 The remaining reconnection and refitting operations are reversals of removal and separation (Section 21) but observe the following points.
6 Reconnect the vacuum pipe and the control linkage before connecting the transmission mountings. The control cable must have the notches towards the balljoint.
7 Refill the engine with oil and coolant and top up the transmission fluid if necessary.

34 Engine/automatic transmission – reconnection and refitting

1 The reconnection operations are as described in the preceding Section.
2 Once the engine/automatic transmission has been installed, refill with oil and coolant and adjust controls and cables as described in Chapter 7.
3 Top up the transmission fluid if necessary.

32.19B Throttle cable bracket clip

35 Initial start-up after major overhaul

1 Before starting the engine, check that all hoses, controls and electrical leads have been connected.
2 Make sure that tools and rags have been removed from the engine compartment.
3 Starting may take a little longer than usual as the fuel pump and carburettor must first fill with fuel (1995 cc engine).
4 Have the throttle speed screw turned in an extra turn to increase the engine idle speed. This will help to offset the stiffness of the new engine components.
5 If the majority of internal components have been renewed, treat the engine as a new one and restrict speed for the first 1000 km (600 miles).
6 After the first 800 km (500 first miles), check the tightness of the cylinder head bolts. To do this, unscrew bolt number 1 in Fig. 1-6 through a quarter turn and then tighten to the specified Stage 3 torque. Repeat this procedure for each of the remaining bolts in numerical sequence.
7 It is recommended that the engine oil is renewed at the end of the first 1000 km (600 miles). Also the tension of the drivebelts should be checked and the idle speed and mixture adjusted if necessary.

36 Fault diagnosis – four-cylinder engines

Symptom	Reason(s)
Engine will not crank or cranks very slowly	Discharged battery Poor battery connections Starter motor fault
Engine cranks but will not start	No fuel Ignition circuit fault Fuel system fault Leak in crankcase vent system hoses Leak in intake manifold
Engine stalls or rough idle	Leak in crankcase vent system hoses Leak in intake manifold Very weak mixture Incorrect valve clearances
Hesitation or poor acceleration	Incorrectly adjusted mixture Clogged air cleaner Incorrect valve clearances

Chapter 1 Engine

Symptom	Reason(s)
Excessive oil consumption	Worn piston rings or cylinder bores Worn oil seals or leaking gaskets
Excessive mechanical noise from engine	Incorrect valve clearances General internal wear
Pinking on acceleration	Fuel octane too low Overheating Carbon build up in engine Excessive oil vapour being drawn into crankcase breather system Upper cylinder lubricant being used Weak mixture

Refer also to Fault diagnosis, Chapters 3 and 4.

PART B: V6 engines

37 General description

The V6 ohc engine fitted to the Renault 25 range of models is the product of collaboration between manufacturers Renault, Peugeot and Volvo. Its basic design is as follows.

The cylinder heads are of the crossflow type: that is to say, the inlet ports are on one side of the head and the exhaust ports on the other. There are separate alloy-cast iron valve seats fitted and the cylinder head bolts also retain the rocker arm mechanism.

The camshafts, journalled in four bearings of different sizes, are chain-driven from the front of the crankshaft; each chain being tensioned separately. The oil pump too is chain-driven off the crankshaft at the front end, and is located in the front of the block. The whole driving system is covered by a single section light alloy cover plate.

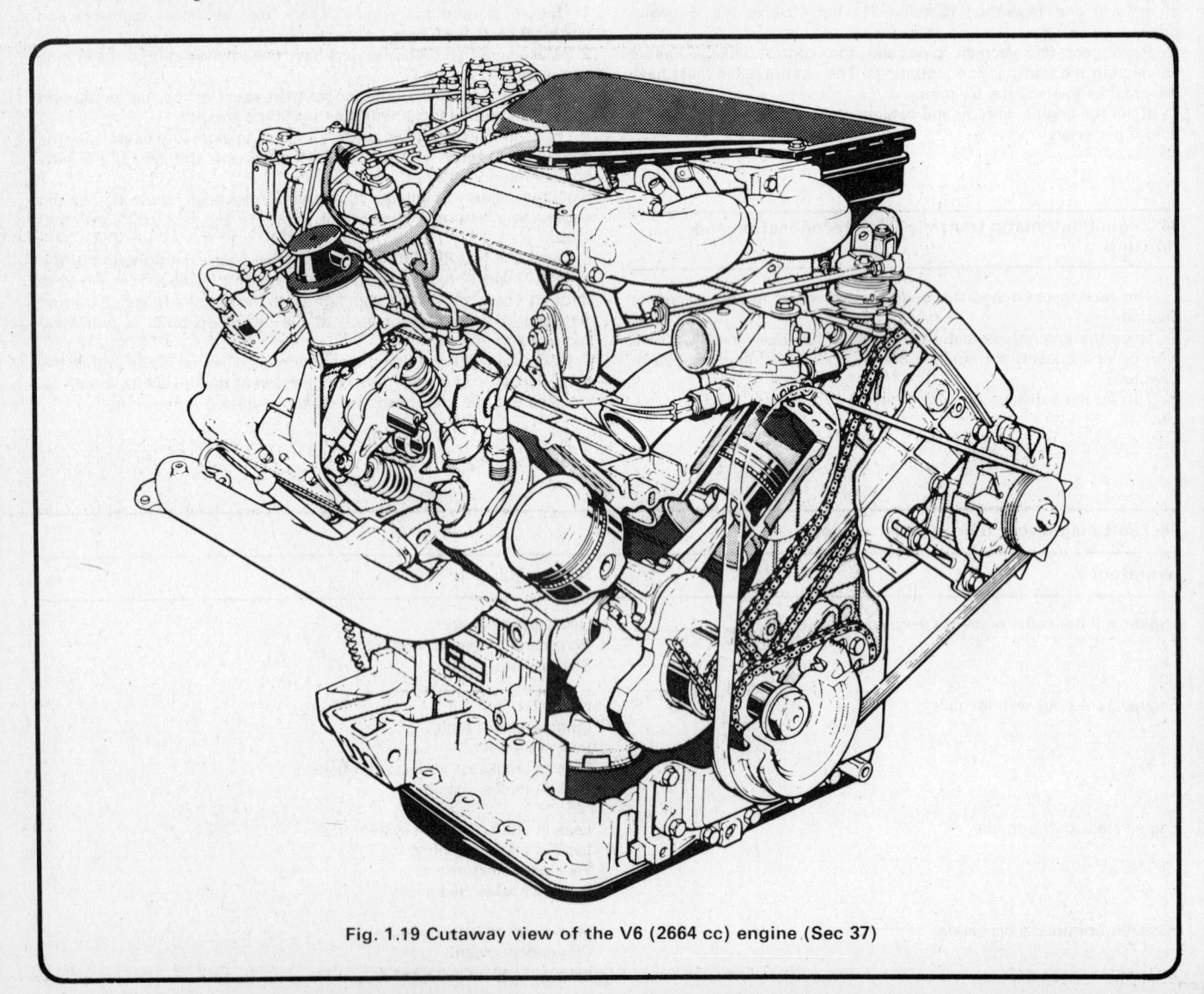

Fig. 1.19 Cutaway view of the V6 (2664 cc) engine (Sec 37)

Chapter 1 Engine

At the end of each camshaft there is an extra driving arrangement: on 2664 models, the distributor is worm driven from the rear right-hand side, while on 2458 cc Turbo models, the distributor is driven from the front left-hand side. The power steering pump unit is also driven from the rear end of the left-hand camshaft, by means of an external pulley.

The cylinder block has wet replaceable cast iron liners, which are class matched to the very light aluminium alloy pistons. These have two compression rings and one oil scraper ring.

The crankshaft is short, strong and has ground, surface-hardened bearing journals; it runs in four well-proportioned main bearings in the cylinder block. The main bearing caps and lower half of the crankcase are separate units, but the nuts for the main bearing caps are underneath (outside) the lower crankcase. This helps give added torsional strength to the light alloy engine construction. There are only three big-end crankpins, two connecting rods being mounted on each. The big-end bearings and main bearings run in renewable shells and all the crankpins can be ground to a limited undersize. The endfloat on the crankshaft is goverened by variable size thrust washers.

The connecting rods are made of drop-forged steel; the gudgeon pins are pressed into them and journalled in the pistons.

On Turbo versions, although the specified engine overhaul and servicing details generally apply, special reference should be made to the Specifications at the beginning of this Supplement and to relevant paragraphs detailing Turbo differences in some Sections of this Chapter.

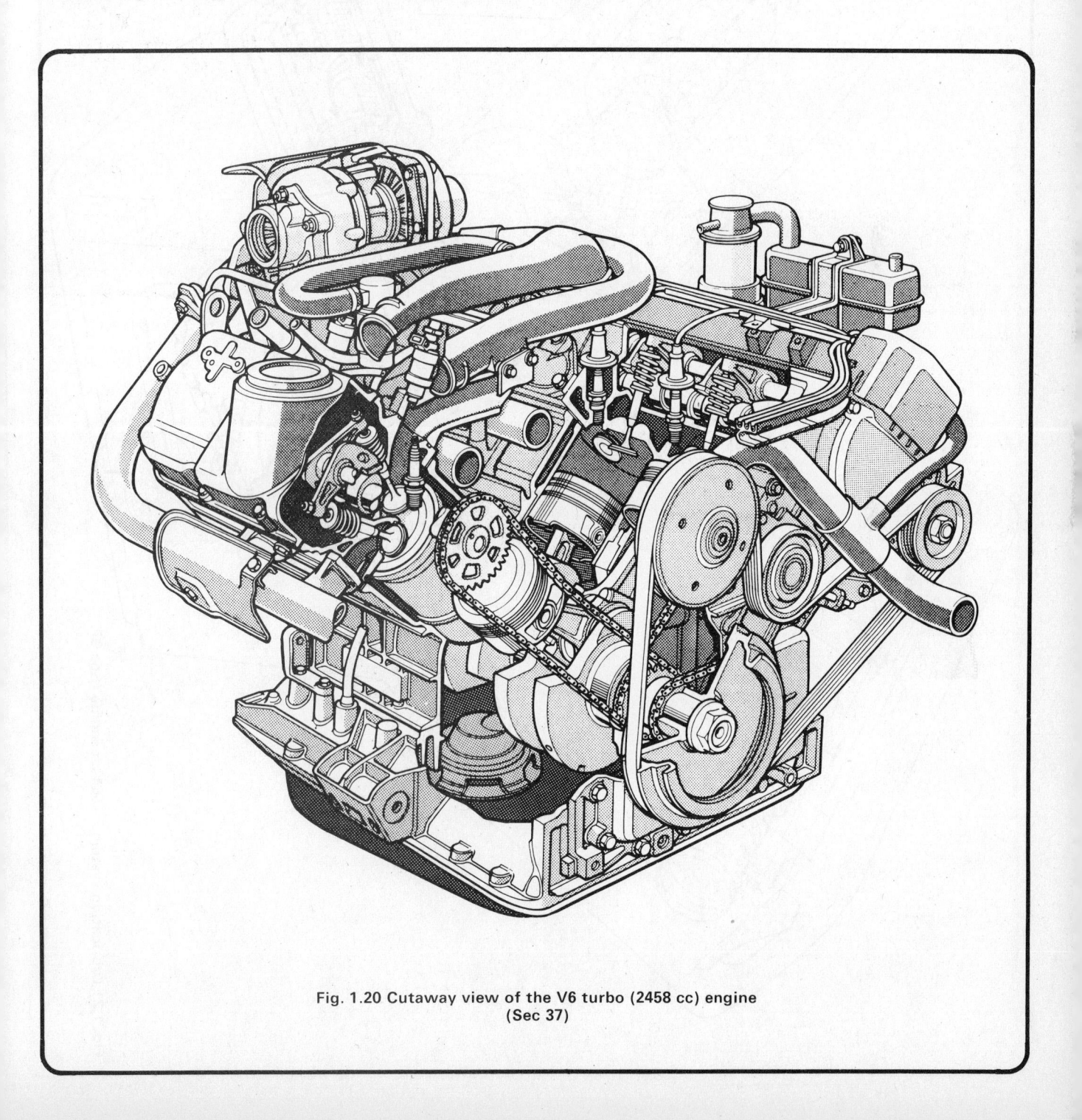

The engine lubrication system relies on a crankshaft driven oil pump and incorporates a full-flow filter.

On Turbo models, the system incorporates an oil cooler and supplies oil to the turbocharger.

Fig. 1.20 Cutaway view of the V6 turbo (2458 cc) engine
(Sec 37)

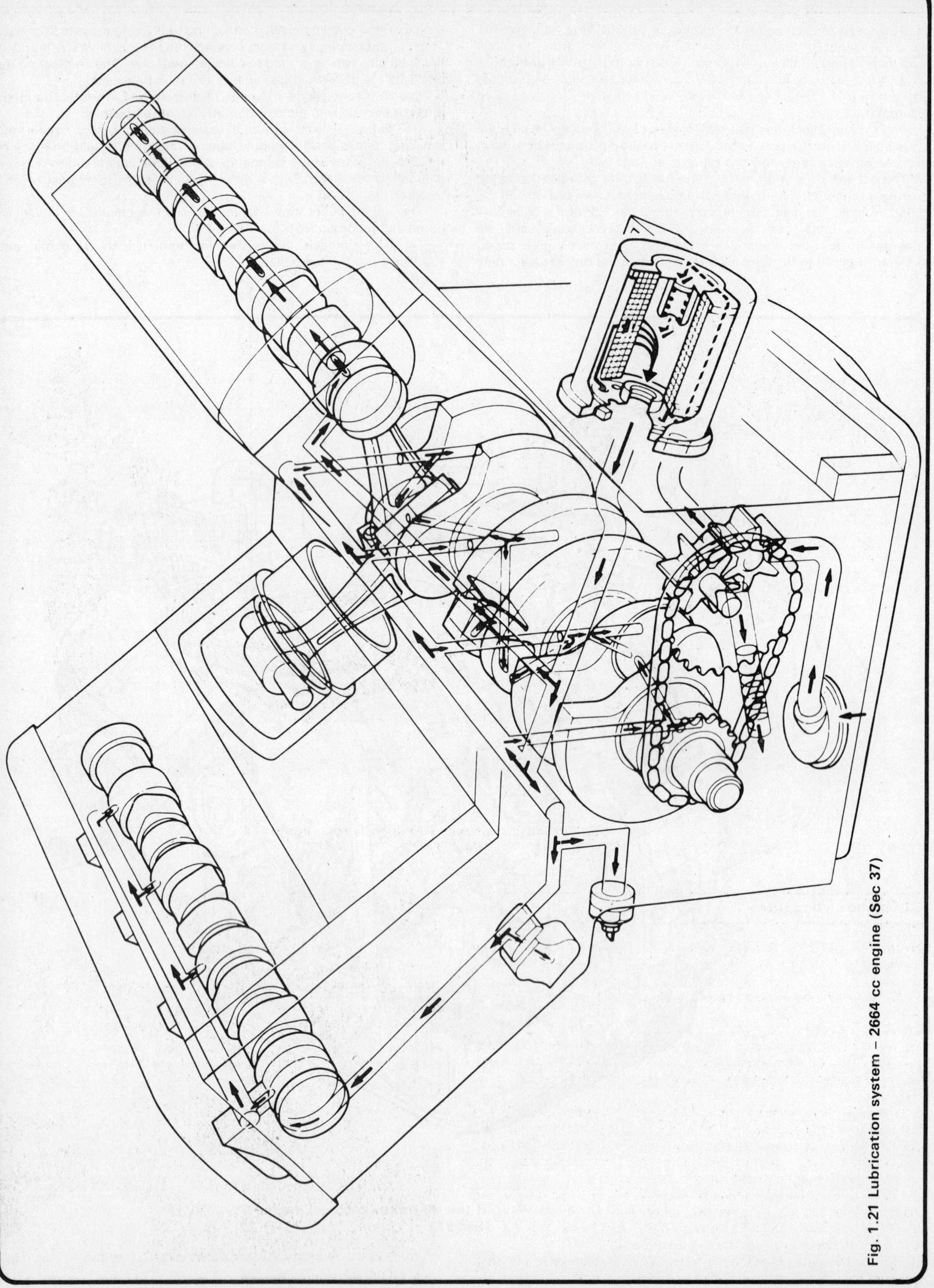

Fig. 1.21 Lubrication system – 2664 cc engine (Sec 37)

Chapter 1 Engine 67

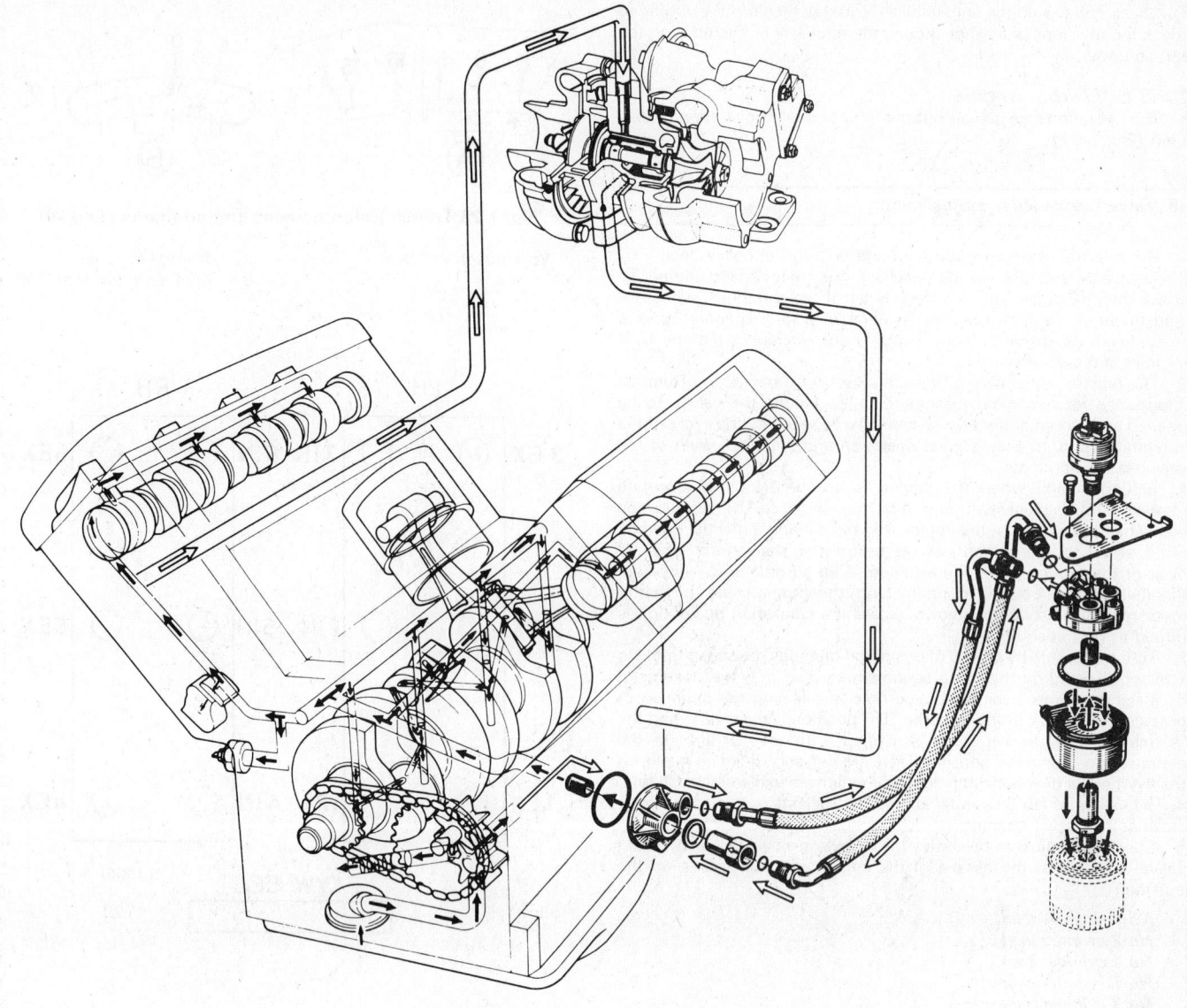

Fig. 1.22 Lubrication system – 2458 cc Turbo engine
(Sec 37)

38 Engine oil and filter

Note: *The use of an oil filter removal tool is recommended when changing the filter.*

1 Refer to Section 2, paragraphs 1 to 4 (photos).

2664 cc engine

2 The oil filter is located on the left-hand side of the cylinder block towards the front. Accessibility is poor and necessitates removal of the engine undertray. Alternatively, it may be possible to reach the filter from above.
3 If the old filter proves too tight to be removed by hand pressure, use a strap wrench to unscrew it. Alternatively, a large worm drive hose clip (or two joined together to make up the required diameter) strapped around the canister will provide a better grip point. If necessary, the canister can be tapped round using a small hammer.
4 Once the old filter is removed, wipe clean the joint area in the cylinder block and lubricate the new seal ring.
5 Screw the filter carefully into position by hand, ensuring that the seal does not twist or distort as it is tightened.
6 Do not overtighten the filter; it need only be tightened by firm hand pressure for half a turn after it seats against the block.

38.1 Oil filler/breather cap

7 Check and top up the engine oil as required, then run the engine to check for any signs of leakage around the filter seal before refitting the engine undertray.

2458 cc (Turbo) engine
8 The operations are similar but the filter and oil cooler are remotely sited (Fig. 1–22).

39 Valve clearances – adjustment

1 Remove the retaining screws and lift each rocker cover clear.
2 The clearances given in the Specifications are for a cold engine. To check the clearances you will need a set of feeler gauges, whilst for adjustment a 13 mm spanner (preferably a ring spanner) and a screwdriver are required. Note that different clearances are specified for inlet and exhaust valves.
3 The precise adjustment of the valve/rocker clearances is of utmost importance for two main reasons. First, to enable the valves to be opened and closed at the precise moments required by the cycle of the engine. Second, to ensure quiet operation and minimum wear of the valvegear components.
4 Settings made when the engine is on the bench will require rotation of the crankshaft and this may be done by turning the crankshaft pulley bolt. If the engine is in the car and a manual gearbox fitted, select top gear, then jack up the front so that a front wheel is clear of the ground and can be turned. With automatic transmission, this method is not possible and 'inching' the engine using the starter motor will have to be resorted to, unless the crankshaft pulley bolt is turned as described above.
5 Turn the engine by means of one of the methods described until No 1 piston is rising on the compression stroke. This may be ascertained by placing a finger over No 1 plug hole and feeling the build-up of pressure. Carry on turning until the TDC position is reached, when the 'A' mark shown in Fig. 1.23 is aligned with the 'O' line on the clutch/torque converter housing. There are two sets of timing marks on the flywheel or driveplate, the details of which are explained in Chapter 4. The valves of No 5 cylinder should be 'rocking' – ie exhaust valve closing, inlet valve opening.
6 Check and adjust as necessary the following valve clearances. Inlet valves are nearest the middle of the engine, exhaust valves on the outside:

No 1 cylinder inlet
No 2 cylinder inlet
No 4 cylinder inlet
No 1 cylinder exhaust
No 3 cylinder exhaust
No 6 cylinder exhaust

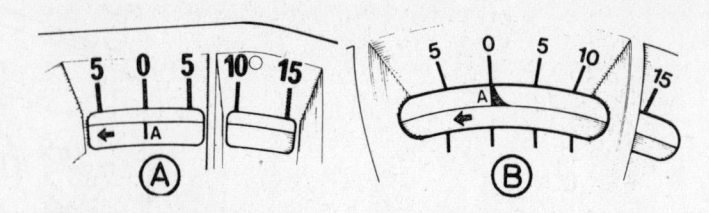

Fig. 1.23 Transmission housing timing marks (Sec 39)

No 1 piston at TDC A *Manual transmission*
 B *Automatic transmission*

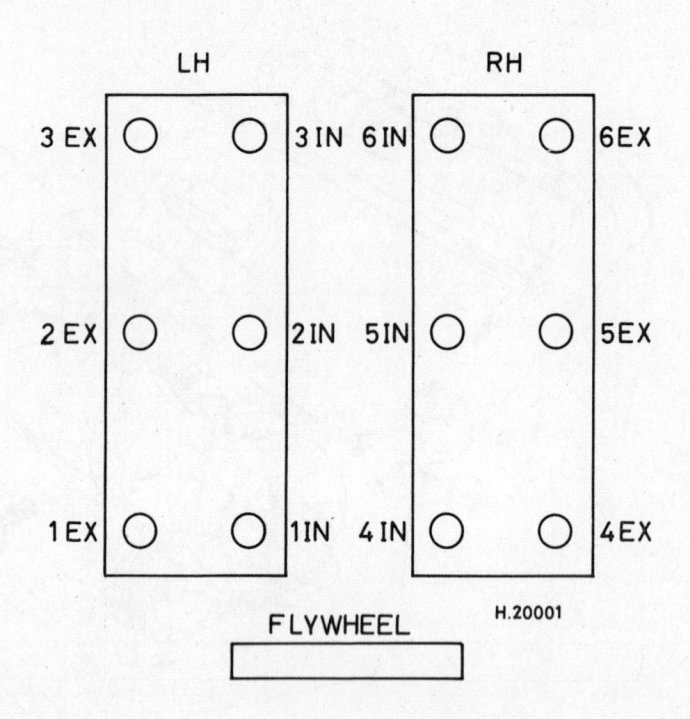

Fig. 1.24 Valve location diagram (Sec 39)

7 If the clearance requires adjustment, loosen the locknut, and with the feeler in position turn the adjuster screw until the feeler blade is nipped and will not move. Now unscrew the adjuster until the feeler blade is a firm sliding fit. Tighten the locknut without moving the screw and recheck the clearance (photo).
8 Now rotate the crankshaft by one complete turn so that the flywheel 'A' mark is again aligned with the 'O' mark on the housing. Check that No 1 cylinder rocker arms are 'rocking' and then check/adjust the following valves:

No 3 cylinder inlet
No 5 cylinder inlet
No 6 cylinder inlet
No 2 cylinder exhaust
No 4 cylinder exhaust
No 5 cylinder exhaust

9 When all the valve clearances have been checked and where necessary adjusted, refit the rocker covers, renewing the gaskets if the old ones are damaged or are brittle. Ensure that each cover retaining screw has the special flat washer fitted (photos).
10 When the engine is restarted check around the rocker cover edges for any signs of oil leaks.

39.7 Adjusting a valve clearance

Chapter 1 Engine 69

39.9A Fitting a rocker cover

39.9B Tightening a rocker cover screw

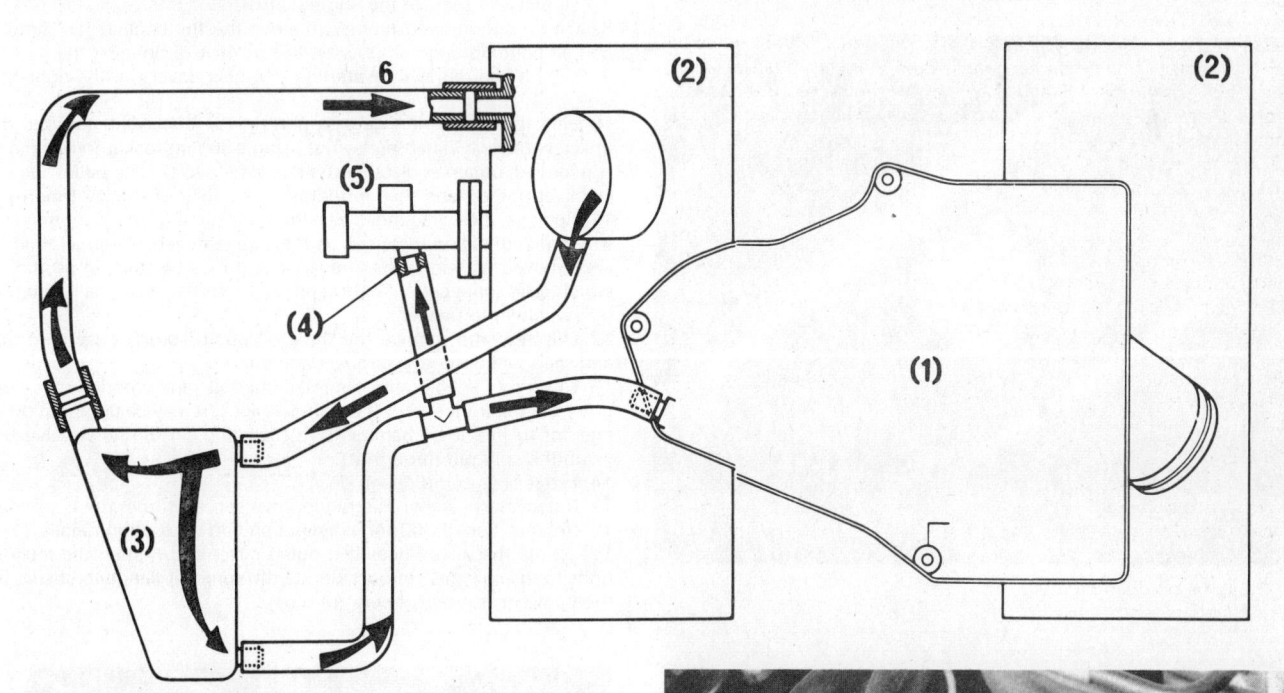

Fig. 1.25 Crankcase ventilation system (2664 cc engine) (Sec 40)

1 Air cleaner
2 Rocker cover
3 Oil separator
4 Jet – 2.75 mm dia
5 Cold start injector
6 Oil return to sump

40 Crankcase ventilation system

1 The purpose of the system is as described in Section 4, but note the different system layouts (photo).
2 On 1986 and later models, a further modified system is fitted in conjunction with the remotely sited air cleaner (see Chapter 3). This new crankcase ventilation system incorporates some modified and re-routed hoses.

40.1 Oil separator in crankcase vent system

Chapter 1 Engine

41 Major operations possible without removing the engine

Where major overhaul or servicing is required it will be wise to remove the complete engine and obtain the ease of access provided by mounting the unit on a bench. However, the following components may be removed and refitted with the engine still in place in the vehicle:

Oil pump
Timing cover, chains and sprockets
Cylinder heads
Rocker shafts
Sump pan and anti-emulsion plate
Crankshaft front oil seal
Camshaft front and rear oil seals
Big-end bearings
Piston rings
Piston/connecting rod/cylinder liner
Engine/transmission mountings

Warning: *Vehicles equipped with air conditioning. Whenever overhaul of a major nature is being undertaken to the engine, and components of the air conditioning system obstruct the work, and such items of the system cannot be unbolted and moved aside sufficiently far within the limits of their flexible connecting pipes, then to avoid such obstruction the system should be discharged by your dealer or a competent refrigeration engineer.*

As the system must be completely evacuated before recharging, the necessary vacuum equipment is only likely to be available at a specialist.

The refrigerant fluid is Freon 12 and although harmless under normal conditions, contact with the eyes and skin must be avoided. If Freon comes into contact with a naked flame, then a poisonous gas will be created which is injurious to health.

42 Oil pump – removal and refitting

1 The oil pump is located in the front face of the cylinder block. Access to it is gained by removing the timing cover. The pump is driven by a chain from the crankshaft.
2 Remove the sump drain plug and drain the engine oil (photo).
3 Drain the cooling system and remove the radiator (see Chapter 2).
4 Remove the air filter unit (see Chapter 3).
5 Remove the respective front end drivebelts and the alternator adjusting strap.
6 Remove the rocker covers and the three coolant pump bolts.
7 Unbolt and remove the engine undertray.
8 On air conditioned models, remove the the cooling fan motor(s) and air conditioning compressor, but **do not** disconnect the pipes.
9 Unbolt and remove the starting ring gear cover (on the right-hand side).
10 Jam the starter ring gear to prevent the crankshaft turning, then unscrew the crankshaft pulley nut using a 36 mm socket. When the nut is removed, unjam and rotate the crankshaft so that its pulley locating key is at the top and then withdraw the crankshaft pulley (photo).
11 Remove the retaining bolts from the timing cover, noting their varying lengths and locations as they are removed. When all the bolts are removed withdraw the timing case. It may be stuck in position, in which case some careful light taps with a soft-headed mallet from the rear will help to unseal it.
12 Unscrew and remove the three oil pump sprocket securing bolts and then withdraw the sprocket and chain.
13 Unscrew the oil pump unit retaining bolts and withdraw the pump unit from the front end of the cylinder block. It may be tight, but do not attempt to free it by hammering or levering on the relief valve boss (with the split pin through it).
14 Extract the pump driven gear.
15 If necessary remove the pump drive sprocket from the crankshaft.
16 Refer to Section 62 for examination and renovation details.
17 Locate the driven gear and pump cover and tighten the retaining bolts to the specified torque. Rotate the sprocket flange to ensure that the pump gear rotates freely (photos).

42.2 Engine oil drain plug

42.10 Flywheel starter ring gear jamming tool

42.17A Oil pump driven gear

Chapter 1 Engine

42.17B Oil pump drive gear and cover

42.18 Priming the oil pump

42.20A Crankshaft distance piece

40.20B Crankshaft sprocket Woodruff key

42.20C Crankshaft sprocket

42.20D Oil pump drive chain and sprockets

18 Prime the pump on completion by squirting some engine oil through the aperture shown with the oil filter removed (photo).
19 The pump driven sprocket bolt threads should be smeared with thread locking compound when fitting the sprocket.
20 Ensure that the drive sprocket distance piece is in position on the crankshaft before refitting the sprocket and chain (photos).
21 Refit the timing cover and the crankshaft pulley as described in earlier paragraphs.

43 Timing cover, chains and sprockets – removal and refitting

1 Remove the rocker covers.
2 Unscrew and remove the three coolant pump bolts.
3 Remove the timing cover as described in the preceding Section.
4 Rotate the crankshaft until its Woodruff key is uppermost. This will set No 1 piston 15.0 mm down its bore BTDC to prevent piston-to-valve contact.
5 Remove the timing sprockets with the oil pump drive chain. Retain the key and spacer.
6 Unscrew and remove the camshaft sprocket bolts.
7 Using quick drying paint, mark all timing components LH or RH if they are to be used again, so that they can be refitted in their original positions.
8 Do not disturb the chain tensioner retaining lock (1) in Fig. 1.26, but retract both chain tensioners by turning the ratchet mechanism anti-clockwise using a screwdriver – see Fig. 1.27.
9 Remove the right-hand and then the left-hand timing chains.
10 Remove the chain tensioners, the chain slippers and filters.
11 Refer to Section 62 for details of examination and renovation.
12 Commence refitting by turning the left-hand camshaft until the slot

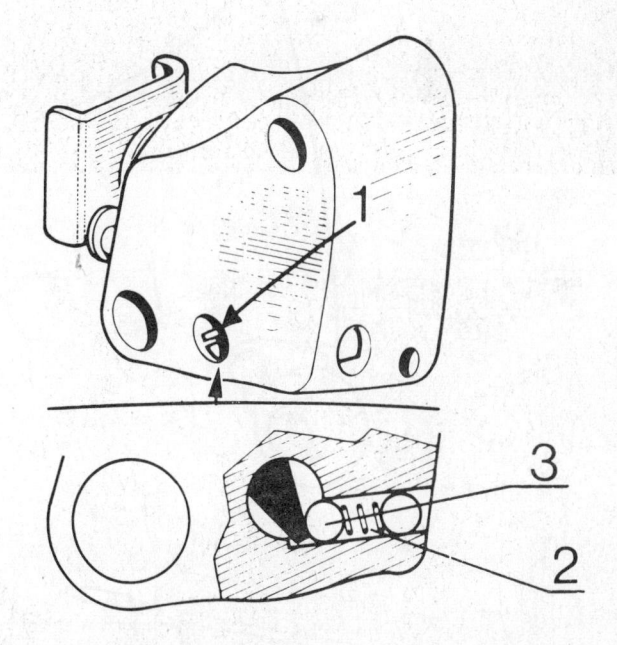

Fig. 1.26 Chain tensioner details (Sec 43)

1 Retaining lock
2 Spring
3 Ball

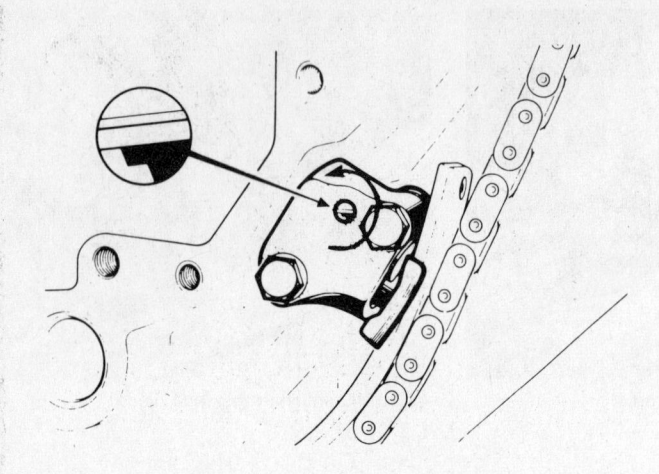

is aligned with the retaining plate as shown in Fig. 1.28. In this position the valves of No. 1 cylinder will be 'rocking' (one valve seating as the other one starts to rise).

13 Now turn the right-hand camshaft until the slot is aligned with the retaining plate as shown in Fig. 1.29. In this position, the valves of No 6 cylinder will be 'rocking'.

14 Clean the oil filters (B) in Fig. 1.30 and fit them.

15 Refit the chain tensioners, the chain slippers and guides.

16 Refit the crankshaft sprocket so that the timing mark is visible (photos).

17 The left-hand side timing chain assembly is fitted first. Rotate the crankshaft so that its keyway is facing upwards to the centre-line of the left-hand cylinder bank.

18 Fit the left-hand camshaft sprocket into the chain so that the timing mark on the sprocket front face is between the two marked links of the chain. Supporting the chain and sprocket in this position, fit the chain lower end over the crankshaft sprocket and align the single marking on the chain link with the timing mark on the front face of the sprocket. The chain is fitted over the *rear* teeth of the double timing sprocket (photos).

19 As the cylinder head and camshaft are in position, then the camshaft sprocket can be fitted to the camshaft so that the driving pin engages its slot in the shaft flange. Locate the camshaft retaining bolt and tighten it to the specified torque whilst preventing the crankshaft from turning (photo).

20 To fit the right-hand timing chain assembly, first turn the crankshaft through 150° so that the timing mark on the crankshaft sprocket aligns with the oil pump cover lower retaining bolt (photo).

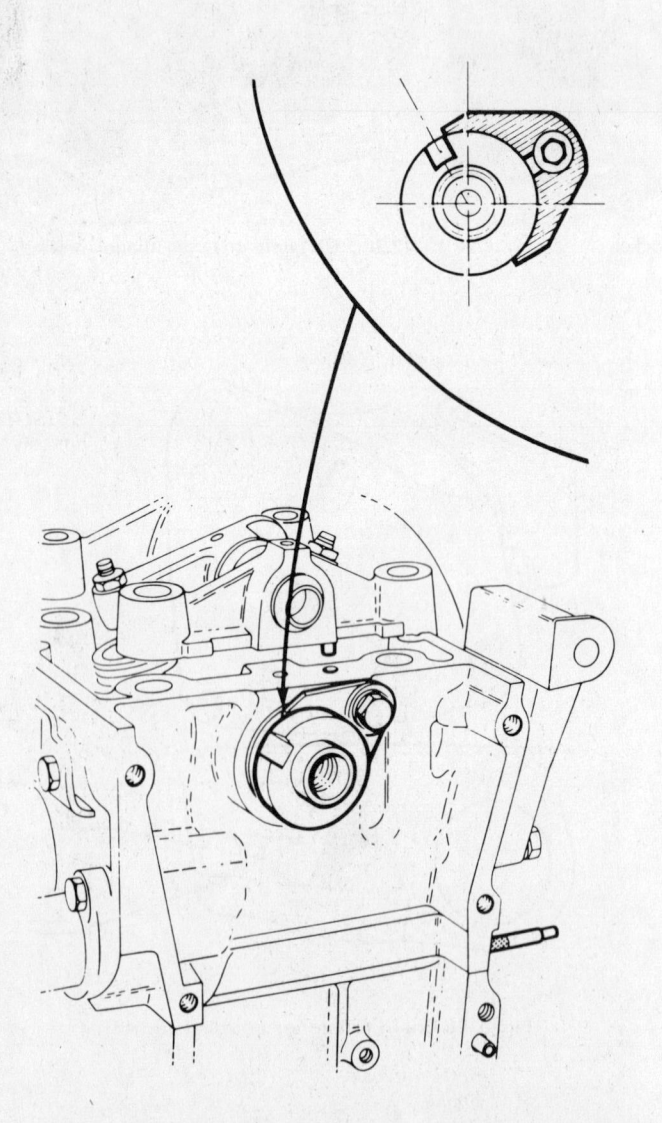

Fig. 1.27 Turn the ratchet mechanism anti-clockwise to retract chain tensioner (Sec 43)

Fig. 1.28 Left-hand camshaft slot aligned with retaining plate (Sec 43)

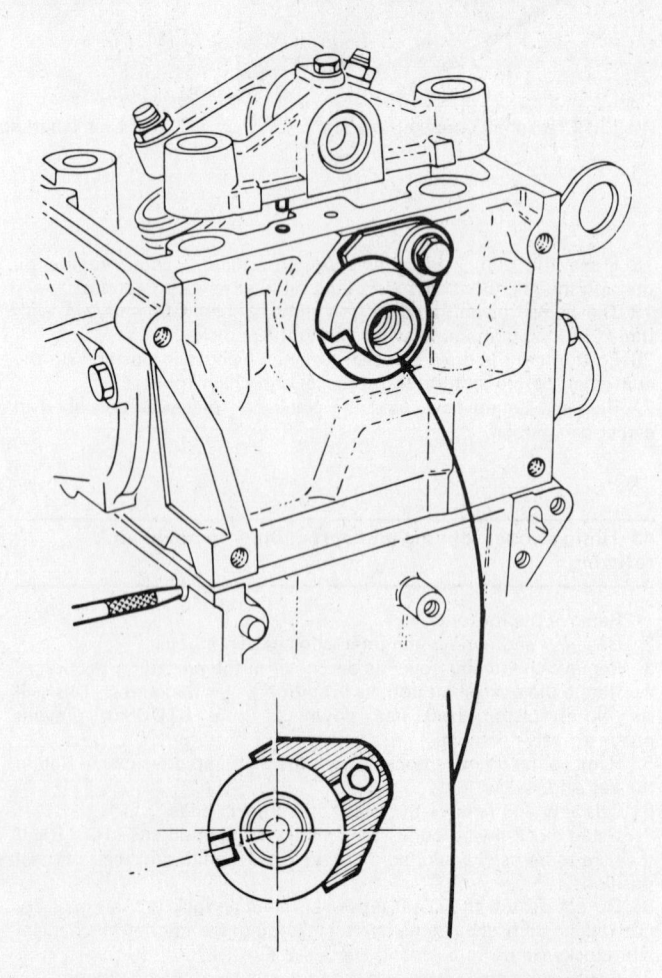

Fig. 1.29 Right-hand camshaft slot alignment diagram (Sec 43)

Fig. 1.30 Chain tensioner oil filters (B) (Sec 43)

Fig. 1.31 Crankshaft sprocket timing mark (Sec 43)

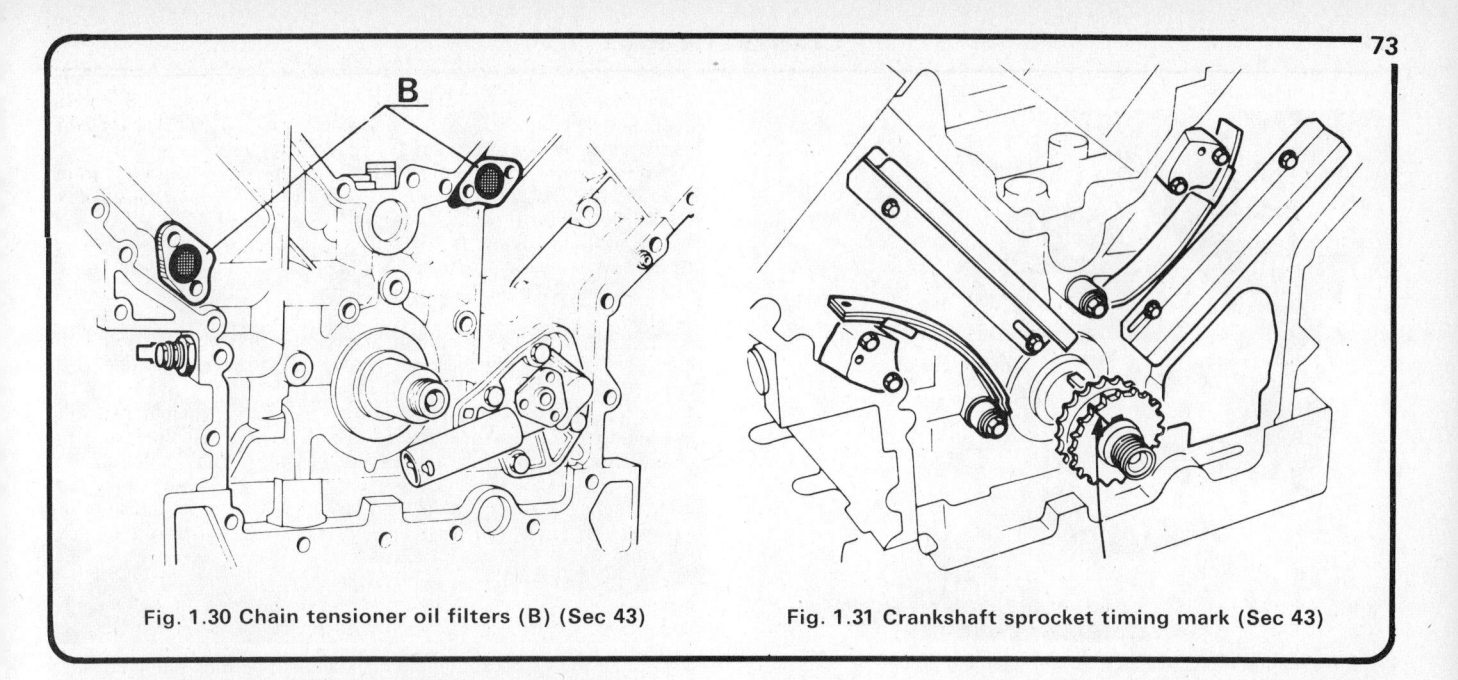

43.16 Crankshaft timing chain sprocket

43.18A Chain-to-camshaft sprocket alignment mark (left-hand bank)

43.18B Crankshaft sprocket-to-left-hand timing chain alignment

43.18C Left-hand cylinder bank timing chain and sprockets assembled

Chapter 1 Engine

43.19 Tightening the camshaft sprocket retaining bolt

43.20 Crankshaft sprocket timing mark aligned with oil pump cover bolt hole

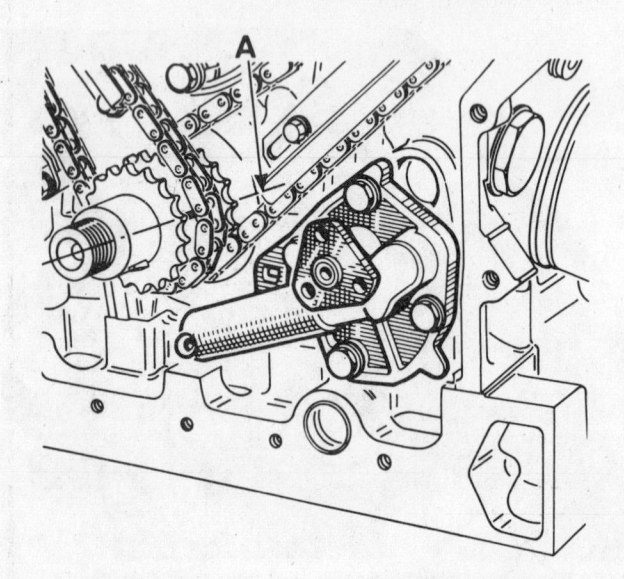

Fig. 1.32 Crankshaft sprocket timing mark aligned with edge of left-hand chain slipper (Sec 43)

43.22 Right-hand timing chain alignment marks

21 On Turbo models, turn the crankshaft through 120° so that the timing mark on the crankshaft sprocket aligns with the edge of the left-hand fixed chain slipper (Mark A, Fig. 1.32).
22 Now fit the right-hand camshaft sprocket into its chain so that its timing mark is between the twin markings on the chain links. Supporting the chain and sprocket in this position, fit the chain over the crankshaft sprocket so that the single marking on the chain link aligns with the crankshaft sprocket timing mark. This chain is fitted over the front teeth of the crankshaft sprocket (photo).
23 The camshaft sprocket can be fitted to the camshaft, engaging its drive pin into the flange slot. The camshaft sprocket retaining bolts should then be refitted. Prevent the crankshaft from turning and tighten the retaining bolt to the specified torque setting.
24 The timing chain tensioners can be reset when the respective camshaft sprockets are bolted in position on their camshafts and the chains fitted. Press the tensioner shoe in so that it touches the tensioner body, then release the shoe but do not assist the spring action.
25 If the timing chains and sprockets have been correctly assembled, then the timing marks on the crankshaft and camshaft sprockets will be aligned as shown in Fig. 1.34. Turn the crankshaft only clockwise when checking.
26 Refit the oil pump after reference to Section 42.
27 Locate the timing gear cover gaskets (dry) and then refit the timing cover, engaging it over the location dowels. When fitting the cover retaining bolts, the lower section bolts should be smeared with a thread sealing compound. Trim off any protruding gasket from the cylinder head faces (photos).
28 Using a suitable diameter drift, carefully drive the crankshaft pulley oil seal into position. Lubricate the seal prior to fitting.
29 Refit the crankshaft pulley and then fit the retaining nut. Jam the flywheel to prevent the crankshaft from turning and tighten the pulley nut to the specified torque (photos).
30 Refit and tension the drivebelts (Chapter 2).

44 Cylinder heads – removal and refitting

Note: Renault special tool Mot. 589 and a set of cylinder liner clamps will be required for this operation.

If one or both cylinder heads are to be removed with the engine in the vehicle or where the rest of the engine is not being dismantled, you will need to keep the timing chain taut throughout the operation. If the chain tension is slackened for any reason you will have to remove the timing cover in order to reset the chain tensioner(s). A separate chain is used to drive each camshaft.

RH (WITH DISTANCE PIECE)　　　　　　　LH

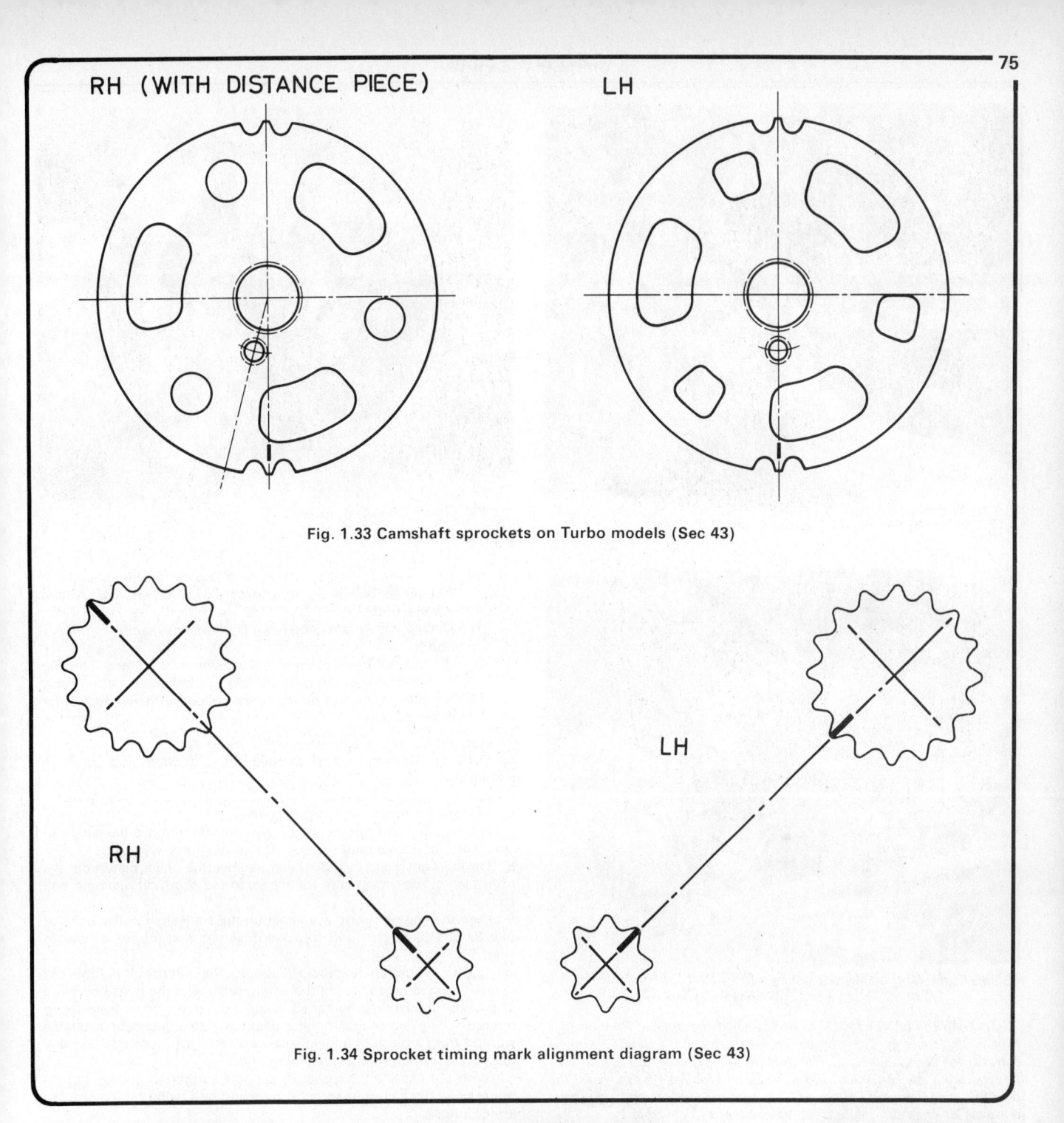

Fig. 1.33 Camshaft sprockets on Turbo models (Sec 43)

RH　　　　　　　LH

Fig. 1.34 Sprocket timing mark alignment diagram (Sec 43)

43.27A Locating timing cover gaskets 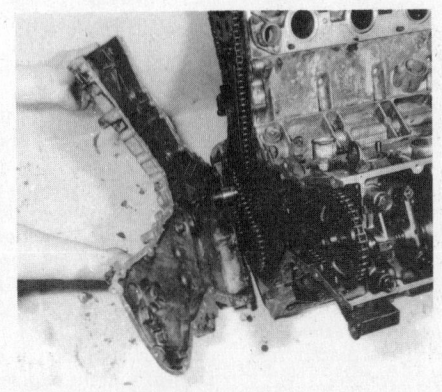　　43.27B Refitting timing chain covers 　　43.27C Tightening timing cover bolts

Chapter 1 Engine

43.27D Trimming protruding gasket

43.29A Fitting crankshaft pulley

43.29B Tightening crankshaft pulley nut

To keep the chains taut you will need to use special Renault tool Mot. 589 (see Fig. 1.36 or be able to fabricate a similar support (photo) for the sprocket(s). The dummy bearing bracket (F) will only be required if the crankshaft has to be turned for any reason whilst the head(s) are removed, such as for pistons/connecting rods/liners removal, in order to maintain the valve timing.

1 Proceed as follows for each cylinder head in turn. The only difference between the two are the front end of the right-hand cylinder head camshaft bolt access aperture is covered by a plate secured by two bolts, and there is no drive pulley at the rear end (as with the left-hand camshaft) so this is also covered by a similar plate.
2 Where the engine is still in the vehicle, carry out the following preliminary operations:

 (a) Drain the cooling system and disconnect the appropriate cooling and heating hoses as necessary, referring to Chapter 2 if required for further details. Note their respective connections
 (b) Disconnect the battery earth lead
 (c) Disconnect the ignition leads and other associated electrical wires from the cylinder head(s), noting their respective positions. Also detach the diagnostic socket and bracket for removal of the right-hand cylinder head
 (d) Remove the alternator (left-hand cylinder head)
 (e) On Turbo engines, refer to Chapter 3 for details of disconnection of turbo charger and its cooling system components
 (f) Detach the exhaust downpipe(s) from the manifold connection(s)
 (g) On the left-hand cylinder head, remove the power steering drivebelt and pump unit, but do not disconnect the hoses. Rest the pump unit on the bulkhead out of the way so that it does not leak

3 Refer to Chapter 4 and remove the distributor.
4 Refer to Chapter 3 and remove the induction and injection manifolds.
5 Remove the dipstick tube clip on the right-hand cylinder head.
6 Unbolt and remove the rocker cover(s).
7 On the left-hand cylinder head, unscrew and remove the camshaft sprocket bolt access plug using a 10 mm Allen key (photo).
8 On the right-hand cylinder head, remove the cover plate from the front face of the timing case for access to the camshaft retaining bolt (photo).
9 Turn the engine (using a spanner on the crankshaft pulley bolt) so that the left-hand camshaft sprocket drive pin is at the top as shown (Fig. 1.37).
10 Locate the camshaft sprocket support (Renault tool Mot. 589(S)) in position on the top face of the timing case, with the bolts holding it in position moderately tightened in the two rocker cover front fixing holes. In the event of this tool not being available, fabricate a suitable support from a piece of angled steel as shown. Cut it to size and drill the necessary holes for the retaining bolts.
11 With the camshaft sprocket(s) suitably supported, loosen but do not remove the camshaft sprocket retaining bolt using a 10 mm Allen key (photos).
12 Progressively loosen and remove the cylinder head retaining bolts in the sequence shown in Fig. 1.38, then lift clear the rocker assembly. Mark them accordingly to avoid confusion (right and left).
13 Unscrew and remove the camshaft flange bolt and remove the flange from its recess.
14 Slowly loosen off the camshaft sprocket bolt until the camshaft is disengaged from its sprocket.
15 Using a suitable rod (an old pushrod – from another engine! – is ideal) press down one cylinder head location dowel – the positions of which are shown in Fig. 1.39.
16 Unscrew and remove the timing cover-to-cylinder head retaining bolts. There are four bolts to each cylinder head.
17 The cylinder head(s) are now ready for removal. To prevent the cylinder liners being disturbed and unseated when removing the cylinder heads, lightly tap the head with a soft-faced mallet so that it pivots sideways on its remaining dowel.
Do not turn the engine to lift the cylinder head(s), or the liners will be unseated.

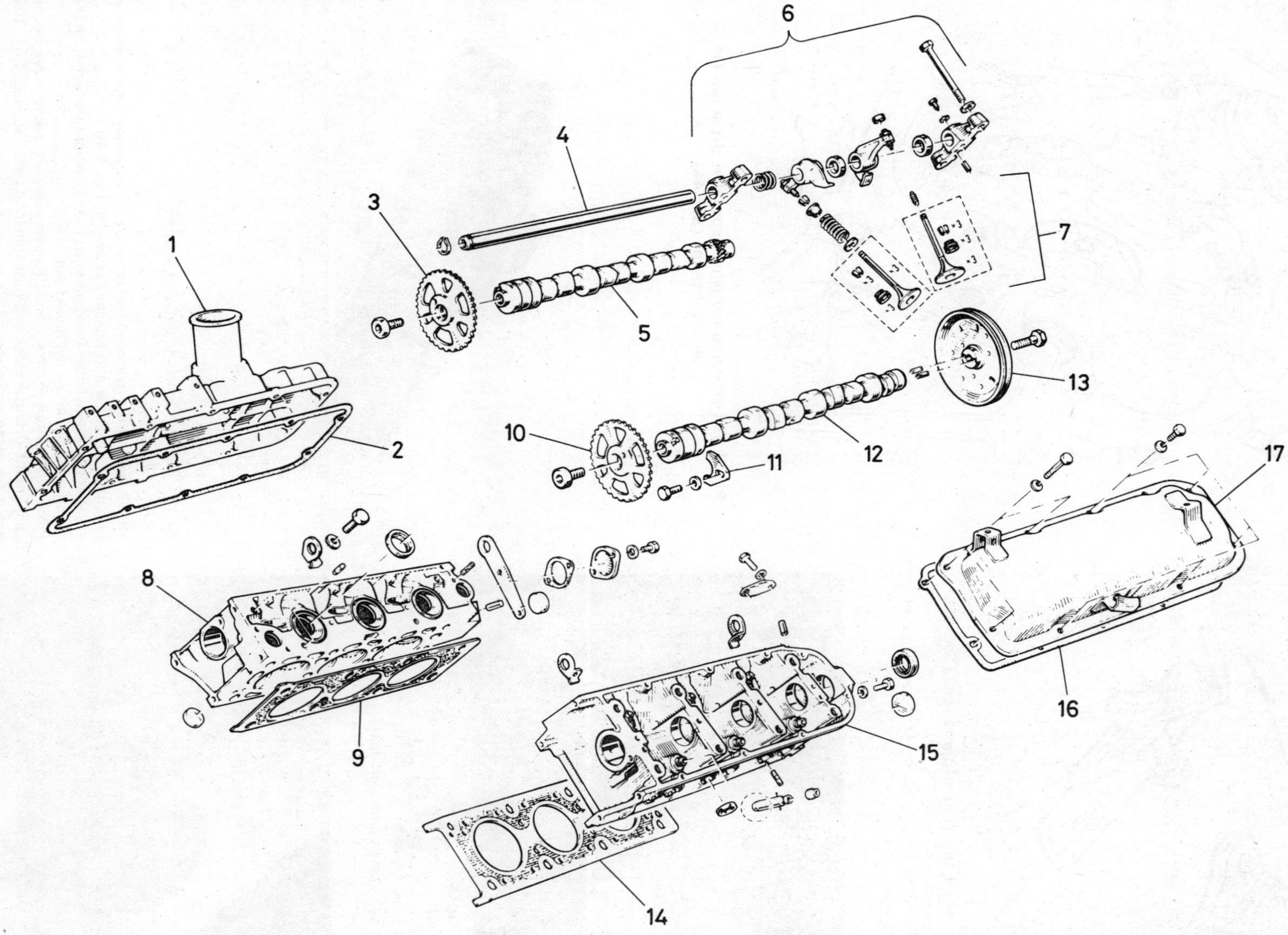

Fig. 1.35 Cylinder head components (Sec 44)

1 Rocker cover
2 Gasket
3 Camshaft sprocket
4 Rocker shaft
5 Right-hand camshaft
6 Rocker components
7 Valve components
8 Right-hand cylinder head
9 Gasket
10 Camshaft sprocket
11 Camshaft retaining plate
12 Left-hand camshaft
13 Power steering pump drive pulley
14 Gasket
15 Left-hand cylinder head
16 Gasket
17 Rocker cover

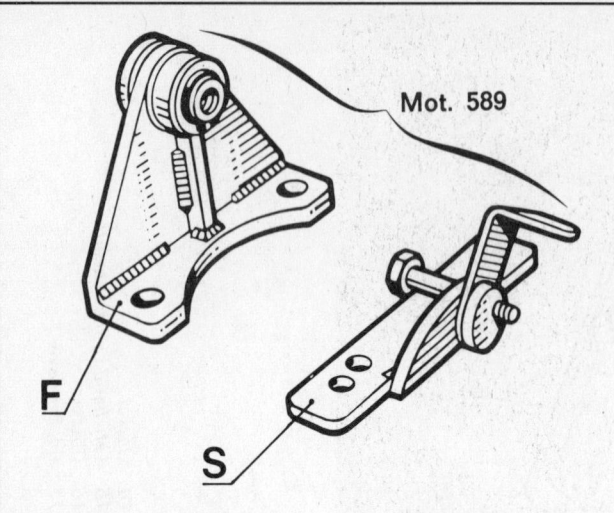

Chapter 1 Engine

Fig. 1.36 Renault special sprocket support tool (Sec 44)

F Dummy support bearing
   (crankshaft turning)
S Sprocket support bracket

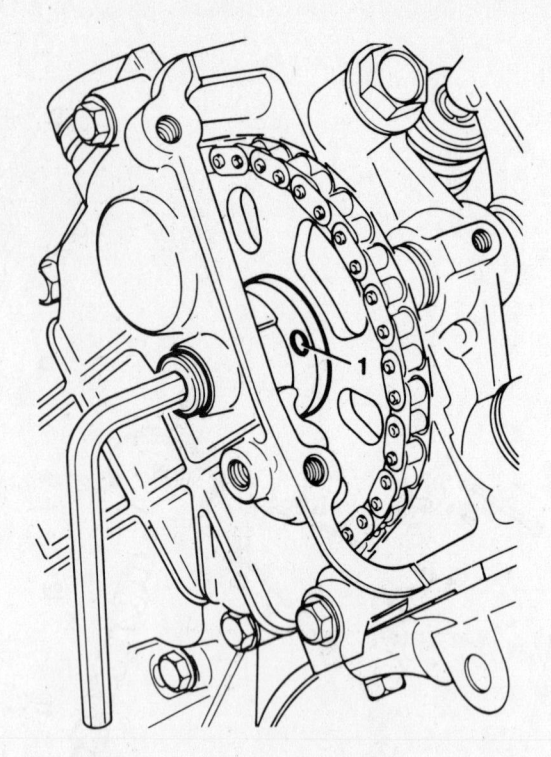

Fig. 1.37 Left-hand camshaft sprocket drive pin (1) at highest point (Sec 44)

44.7 Camshaft sprocket bolt access plug (left-hand)

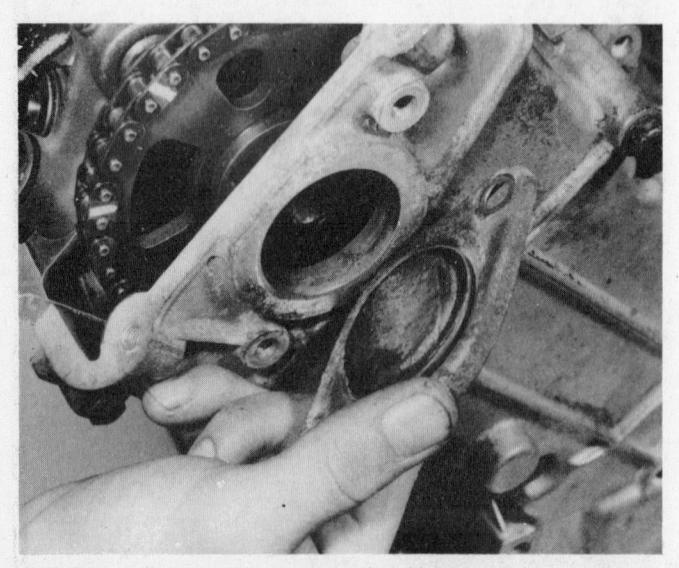

44.11A Typical camshaft sprocket support

44.8 Camshaft sprocket bolt access plug (right-hand)

18 As soon as the cylinder head is removed, fit cylinder liner clamps as described in Part A and typically shown in Fig. 1.40.
19 Remove the cylinder head gasket(s). Any coolant remaining in the cylinder block can be syphoned out.
Prise out the two cylinder head location dowels, taking care not to damage the cylinder block and gasket mating face – see Fig. 1.41. This will make cleaning easier.
20 The respective mating faces of both the cylinder block and cylinder head(s) must be cleaned prior to reassembly. Particular care must be taken when cleaning the cylinder block face to protect the two oil holes shown in Fig. 1.39. Plug these holes temporarily whilst cleaning, but don't forget to remove the plugs before refitting the new gasket!
21 For cylinder head dismantling and overhaul details, refer to Section 61.

44.11B Camshaft sprocket support in position

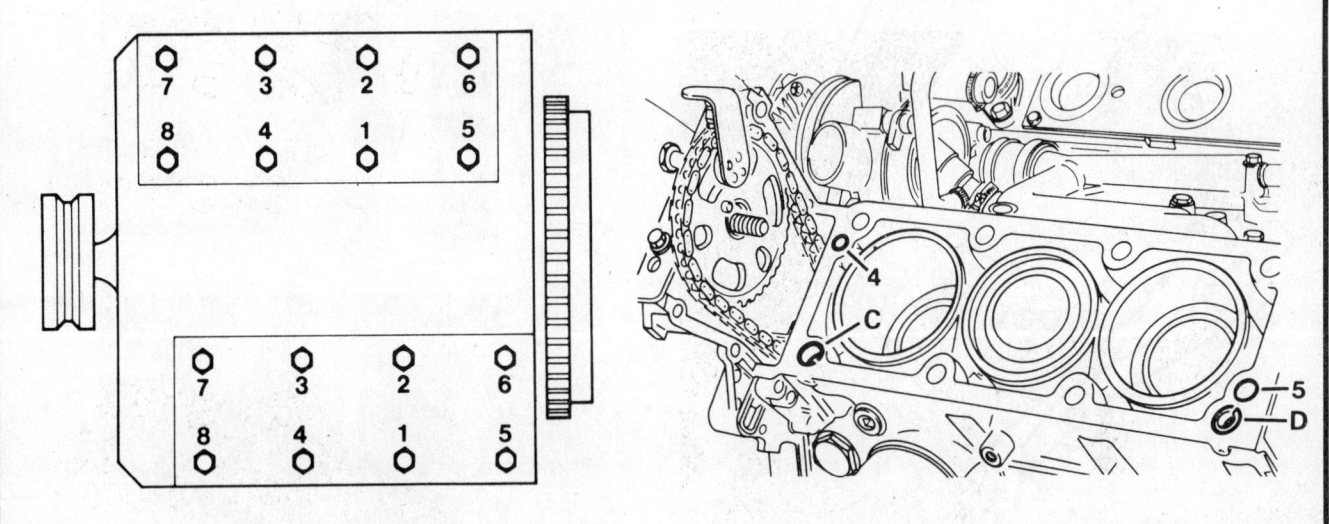

Fig. 1.38 Cylinder head bolt loosening and tightening sequence (Sec 44)

Fig. 1.39 Dowel locations (C and D) and oil feed holes (4 and 5) (Sec 44)

Fig. 1.40 Typical cylinder liner retaining clamps (Sec 44)

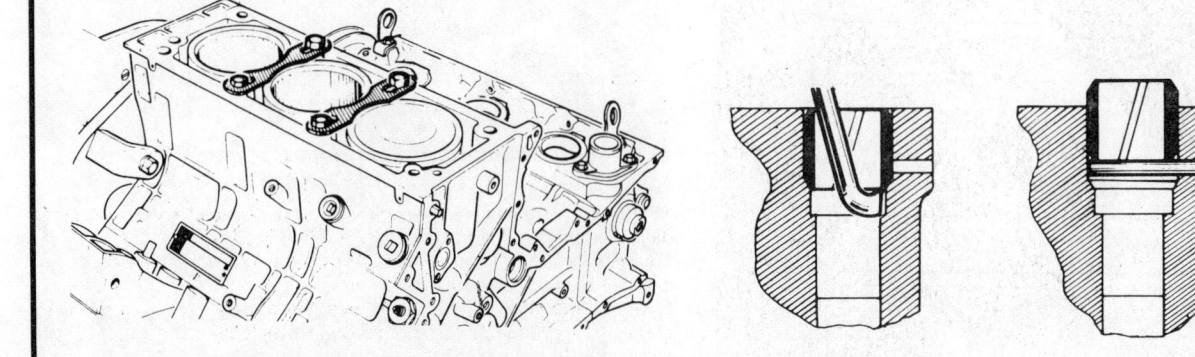

Fig. 1.41 Tools for removing cylinder head dowels (Sec 44)

Chapter 1 Engine

22 When the cylinder head(s) only are removed from the engine it is important not to turn the engine unless:

(a) Liner clamp plates are fitted to prevent the cylinder liners from being disturbed
(b) Camshaft sprocket bearing bracket(s) are fitted – see the introduction to this Section

23 Remove the liner retainers.

24 Where the cylinder liners have not been removed, their respective protrusions above the top face of the cylinder block must be checked before refitting the cylinder head(s).

25 To do this, place a rule across the top of the liners and check the clearance between the rule lower edge and the cylinder block top face using feeler gauges. The liner protrusions must be within the limits given in the Specifications. The nearer the protrusion is to 0.009 in (0.23 mm) the better.

26 If the liner protrusions are outside this tolerance, the liner base seals require renewal so that the correct protrusion can be attained. Refer to Section 51.

27 Refit one cylinder head at a time in the following manner, taking care not to allow the timing chain to become slack at any time during assembly. If it does, you will have to remove the timing cover after the head has been fitted in order to reset the chain tension and check the valve timing – see Section 42. A separate tensioner is fitted to each timing chain.

28 New timing cover-to-cylinder head gasket sections must be used. Clean off any of the old gasket still remaining and then cut to size and locate new gasket sections as shown in Fig. 1.42. Stick them in position by smearing with a thin layer of RTV sealant.

29 Fit the cylinder head locating dowels so that they stand well proud of the head (photo).

30 Remove the liner retaining plates from the side of the engine on which you are working, and wipe over the cylinder block mating face.

31 Check that you have the correct gasket for the side on which you are working – they are different. Place it in position and make sure all the bolt holes and oil holes line up. The two tongues should follow the profile of the block at the front end. Do not use any jointing compound (photo).

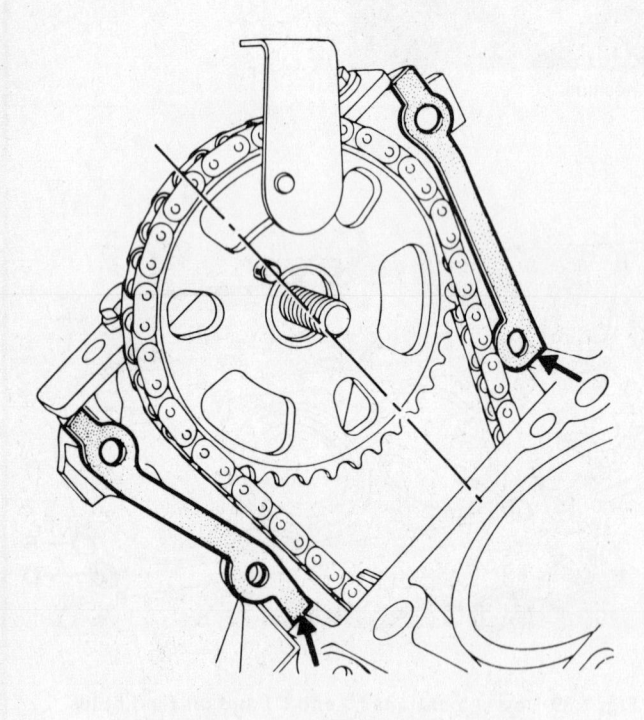

Fig. 1.42 Timing cover gaskets (Sec 44)

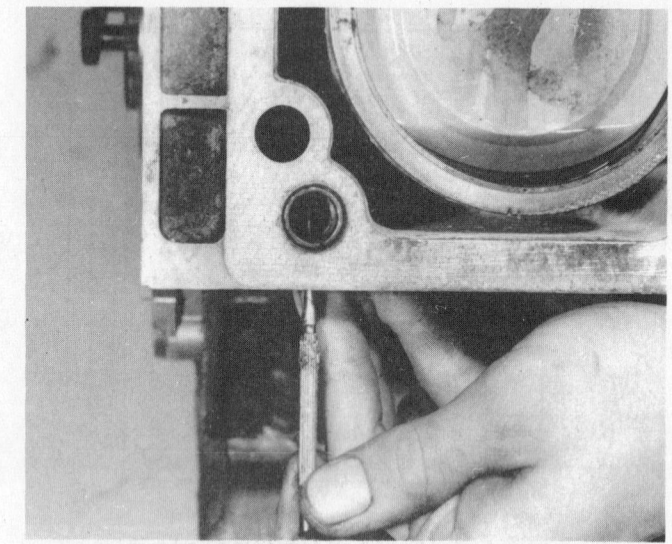

44.29 Raising a cylinder head dowel

Fig. 1.43 Cylinder head gaskets (Sec 44)

44.31 Cylinder head gasket correctly located

Chapter 1 Engine

32 Place the cylinder head in position, having first checked that its face is completely clean. Locate it on the dowels and make sure it seats correctly (photo).
33 Locate the timing cover-to-cylinder head retaining bolts, but hand tighten them only at this stage.
34 Align the camshaft drive peg hole and mate the camshaft with the sprocket. Relocate the camshaft retaining plate and fit and tighten the securing bolt.
35 Fit and hand tighten the camshaft sprocket retaining bolt.
36 Now pick up the correct rocker assembly for the side of the engine you are working on. They were marked appropriately when they were removed. Remember the assemblies are similar, but that when refitting them, the circlip on the end of the rocker shaft is at the front of the engine for the left-hand side assembly, and at the rear of the engine for the right-hand side assembly. Place the assembly in position on the head.
37 Oil the cylinder head bolts and screw them loosely into position (photo).
38 Check that the camshaft and rocker arms are correctly lined up. With the left-hand side cylinder head, the camshaft should be lined up so that the valves for No 1 cylinder (at the rear) rock (exhaust closing, inlet opening). If rebuilding the right-hand side cylinder head, the camshaft should be rotated to a position where the valves for No 6 cylinder (front) rock.
39 The cylinder head bolts can now be tightened in the sequence shown in Fig. 1.38. Initially tighten them in the sequence given to the first stage specified torque, then to the second and third stages, finally tightening using the angular method. Do this by fixing a graduated disc to the bolt head and a pointer to the socket extension.
40 Check and adjust the valve clearances (Section 39).
41 Refit the rocker covers using new gaskets.
42 Fit the camshaft sprocket bolt access plate or plug.
43 Reconnect the coolant hoses and vacuum hoses and and electrical leads.

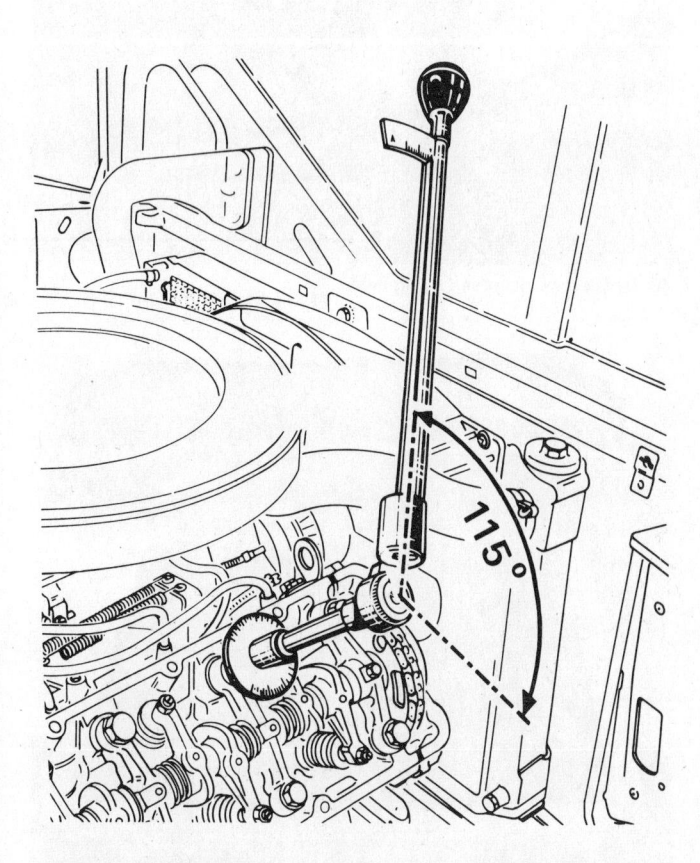

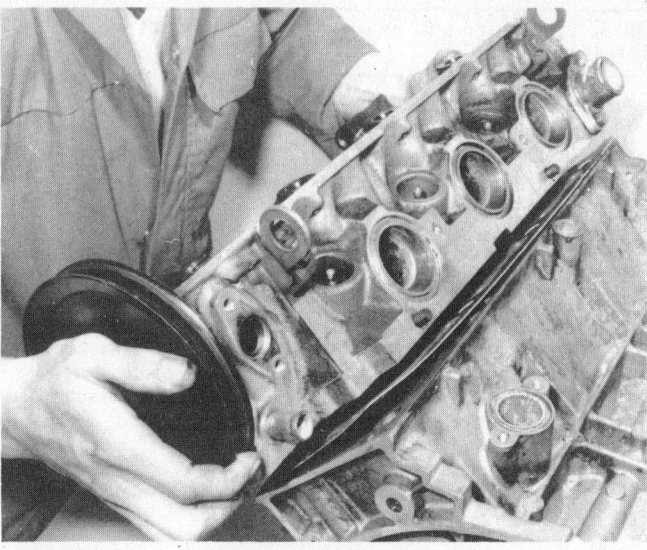

44.32 Lowering a cylinder head onto the block

44.37 Rocker assembly refitted

44 Locate the alternator and power steering pump and tension the drivebelts.
45 Fit the distributor.
46 Fit the induction and injection manifolds (Chapter 3).
47 Reconnect the exhaust pipe.
48 Refit and connect the turbocharger and its cooling system components (if applicable).
49 Refill and bleed the cooling system (Chapter 2).
50 Refit the dipstick tube clip.
51 Reconnect the battery earth lead.

45 Rocker shafts – removal and refitting

1 The rocker assemblies are removable once the cylinder head bolts are withdrawn and have no separate fixings.
2 Refer to Section 61 for details of dismantling and reassembly.

Fig. 1.44 Tightening a cylinder head bolt through 115° (Sec 44)

46 Sump pan and anti-emulsion plate – removal and refitting

1 Drain the engine oil (Section 38).
2 Raise the front of the car so that the front roadwheels hang free. This will cause the anti-roll bar to move forward sufficiently far to enable the rear sump pan screws to be extracted (photo).
3 Unscrew and remove the screws and remove the sump pan. If it is stuck tight, cut around the gasket using a sharp knife.
4 Unscrew the three bolts and remove the oil suction pipe and gauge.
5 Unscrew the six bolts and remove the anti-emulsion plate.
6 Relocate the anti-emulsion plate and secure with bolts and washers (photo).
7 Locate a new O-ring onto the bottom end of the oil suction pipe, then refit the suction pipe assembly into position (photos).
8 Fit the new sump gasket into position and then refit the sump pan. Locate and tighten its retaining bolts to the specified torque. Make sure that the sump drain plug is tight (photos).
9 Lower the car to the ground and fill the engine with oil.

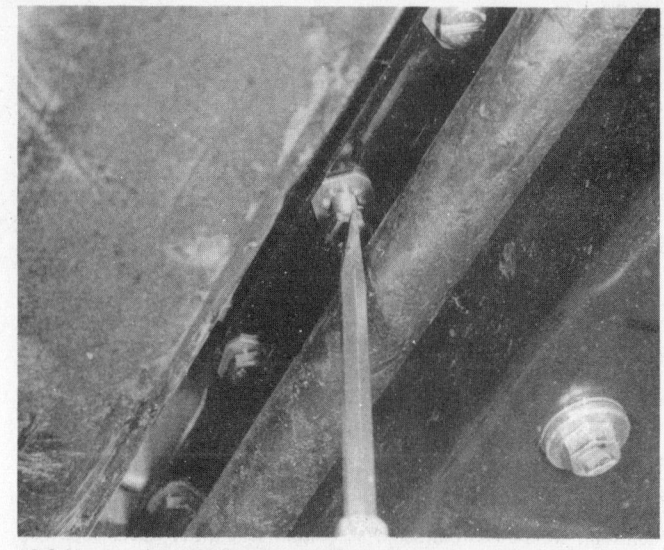

46.2 Unscrewing a sump pan rear bolt

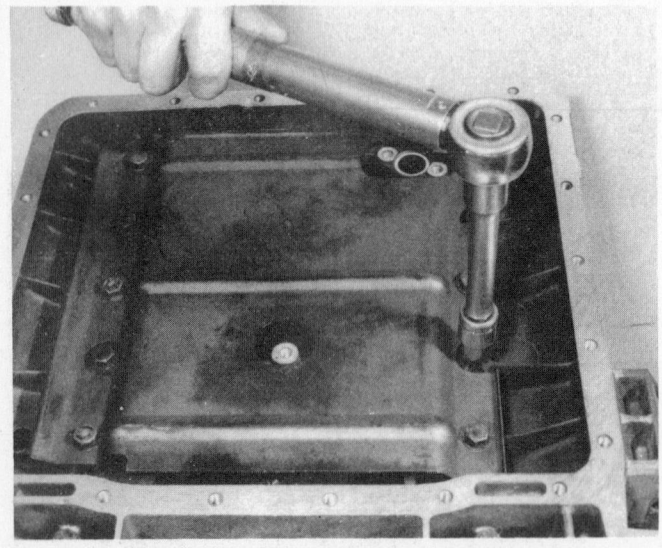

46.6 Tightening an anti-emulsion plate fixing bolt

46.7A Oil pick-up pipe O-ring seal

46.7B Oil pick-up pipe and strainer refitted

46.8A Refitting sump pan and gasket

Chapter 1 Engine

46.8B Tightening the sump pan bolts

47.6 Timing case oil seal

48.2 Power steering pulley removed to expose camshaft rear oil seal

47 Crankshaft front oil seal – renewal

1 Raise the vehicle at the front end and support it on safety stands or ramps.
2 Unbolt and remove the engine undertray.
3 Loosen the air compressor unit and remove the compressor drivebelt (air conditioned models only).
4 Loosen the alternator and remove its drivebelt. On Turbo models remove the steering pump belt.
5 Remove the starter ring gear cover plate and then locate a locking plate to prevent the flywheel from turning. Unscrew the crankshaft pulley retaining nut, and then remove the flywheel (or driveplate) locking plate. Turn the crankshaft so that its keyway points to the top and then withdraw the pulley.
6 The oil seal is now accessible and can be prised free from the timing case using a screwdriver. Take care not to score or damage the housing (photo).
7 Lubricate the new seal prior to fitting. Carefully tap or press it into position, ensuring that it is inserted squarely and is not distorted once in position. A pipe drift of suitable diameter is ideal for seal fitment.
8 Reassembly of the crankshaft pulley and the respective drivebelt is a reversal of the removal process. Be sure to tighten the pulley retaining nut to the specified torque. The drivebelt tension must be adjusted (see Chapter 2).

48 Camshaft front and rear oil seals – renewal

Left-hand rear
1 Loosen the power steering pump and remove its drivebelt, except on Turbo models.
2 Remove the camshaft pulley retaining bolt and then withdraw the pulley (photo).
3 The oil seal can now be carefully eased out of its housing using a screwdriver.
4 Lubricate the new seal before fitting and then insert it into its housing, using a suitable diameter tube drift to ensure that it is fitted correctly and without distortion. When in position the new seal should be flush with the cylinder head rear face.
5 Relocate the pulley and tighten its retaining bolt to the specified torque. Adjust the power steering pump drivebelt tension to complete.

Left-hand front (Turbo models)
6 Remove the distributor cap, rotor and body.
7 Pivot the oil seal by driving one side of it inwards with a drift, then remove the seal.
8 Grease the new seal and tap it squarely into position.
9 Refit the distributor components.

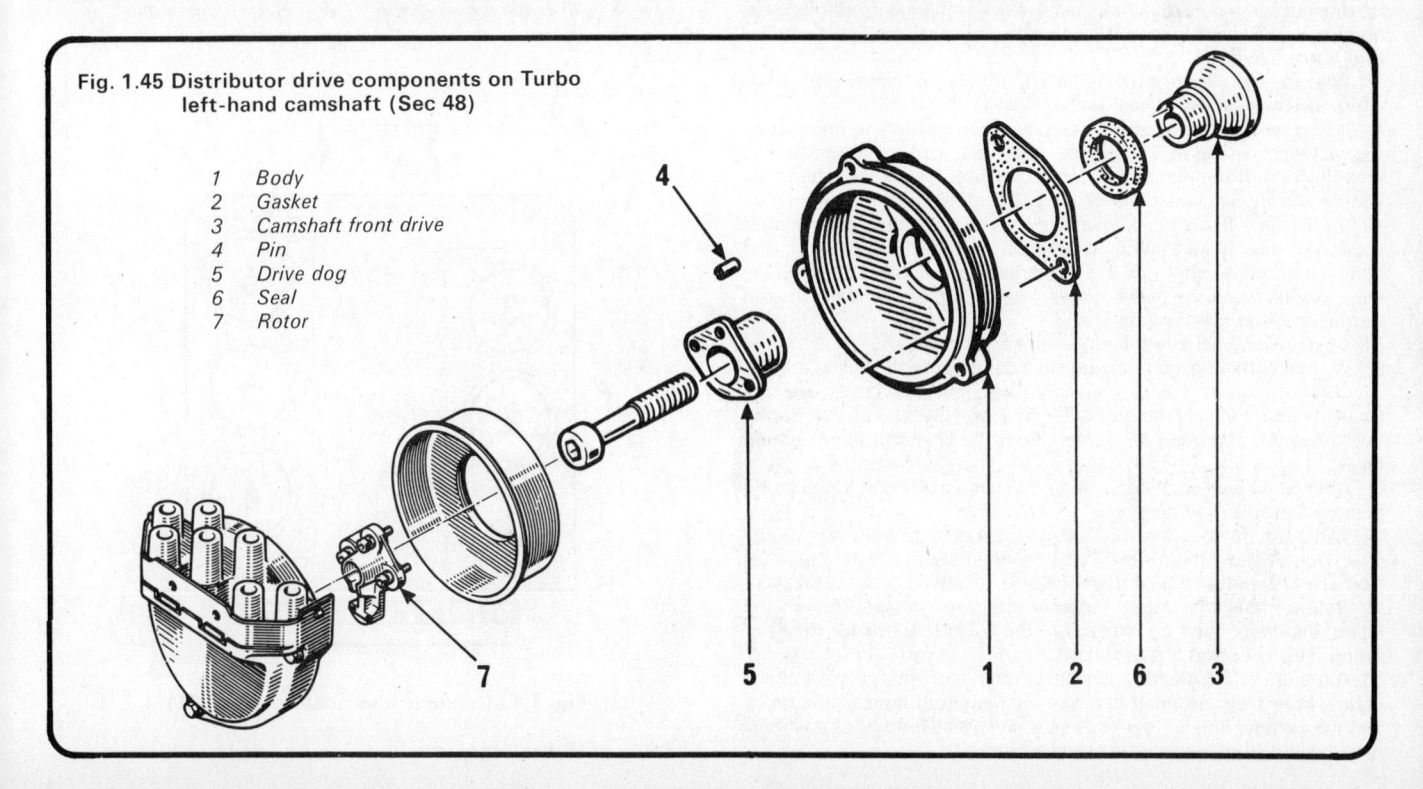

Fig. 1.45 Distributor drive components on Turbo left-hand camshaft (Sec 48)

1 Body
2 Gasket
3 Camshaft front drive
4 Pin
5 Drive dog
6 Seal
7 Rotor

Chapter 1 Engine

49 Big-end bearings – renewal

1 Drain the engine oil, remove the sump pan, oil pick-up, anti-emulsion plate and bottom casing as described in Section 46.
2 The operations are very similar to those described in Part A, Section 14 of this Chapter, but observe the following points.
3 Mark the big-end caps and rods numerically so that the bearing cap will be refitted the correct way round when the mating marks are adjacent.
4 Although each connecting rod and cap is die-stamped with a letter and number, they are numbered 1 to 6 from the rear of the engine. *These numbers relate only to the rod positions on the crankshaft and **not** to the cylinder numbers.*
5 The crankshaft will have to be turned during removal of the rods to obtain access to the cap nuts.

50 Piston rings – renewal

Refer to Part A Section 15 of this Chapter, but note the location of the rings and gaps – see Fig. 1.46.

51 Piston/connecting rod/cylinder liner assembly – removal and refitting

Note: *Renault special tool Mot. 589 may be required for this operation. A set of cylinder liner clamps will be required.*

1 As the pistons, connecting rods and cylinder liners are to be removed with the engine in position in the car remove the cylinder heads as described in Section 44. As mentioned in that Section, you will need to support the camshaft sprocket in order to keep the timing chain taut, and you will require special tool Mot. 589, dummy bearing 'F' for this purpose. This supports the camshaft sprocket whilst allowing the crankshaft to be turned for access to the respective big-end bolts. Note that when fitting the dummy bearing, it may be necessary to fit shims between the bearing support and the cylinder block in order to keep the timing chain taut, so have some shims at hand.
2 If all six piston and liner assemblies are to be removed or the engine is removed for general overhaul, remove the timing gear assemblies as described in Section 43. (In this case the camshaft sprocket support tool is not needed).
3 With the cylinder head(s) removed, unbolt and remove the sump and anti-emulsion plate (see Section 46).
4 It is essential that where the existing liner, piston and connecting rod assemblies are to be refitted, they are marked for positional identification. If the engine has previously been dismantled then they should already be marked, in which case note the marking orientation relative to the cylinder block to ensure correct reassembly. If marking is necessary, use quick-drying paint and mark the piston, liner and connecting rod number (No 1 is in the left-hand bank at the flywheel end) in turn. Mark the piston crown and the top face of each liner in turn at corresponding points.
5 Unscrew and remove the big-end bearing cap nuts.
6 Withdraw the big-end bearing caps complete with shell bearings.
7 Withdraw each cylinder liner/piston/connecting rod assembly upwards from the cylinder block. The liner seals will be broken during this operation and must be cleaned from the liner and block mating faces.
8 After removal of each connecting rod, temporarily refit its matching bearing cap and shell bearing.
9 Withdraw each connecting rod/piston assembly from its cylinder liner. Do not allow the piston rings to spring outwards during removal from the liners, but restrain them with the fingers to avoid breakage.
10 Before fitting the piston and connecting rod assemblies into the liners, the liners must be checked in the cylinder block for depth of fitting. This is carried out as follows:
11 Although the cylinder liners fit directly onto the cylinder block flange, base seals are fitted between the flanges in the cylinder block and the cylinder liners. New seals must always be used once the liners have been disturbed from the cylinder block.

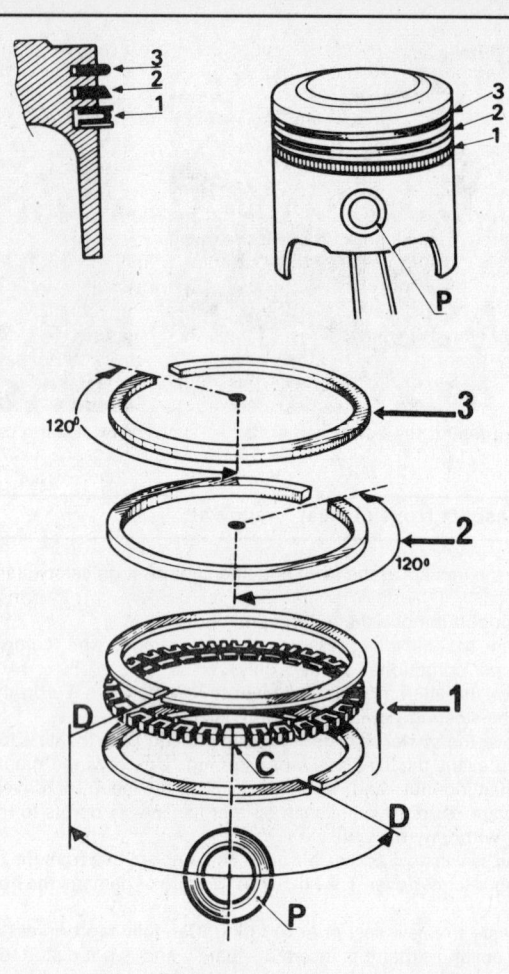

Fig. 1.46 Piston ring arrangement (Sec 50)

1 Oil control ring
2 Second compression ring
3 Top compression ring
C End gap
D End gap
P Gudgeon pin

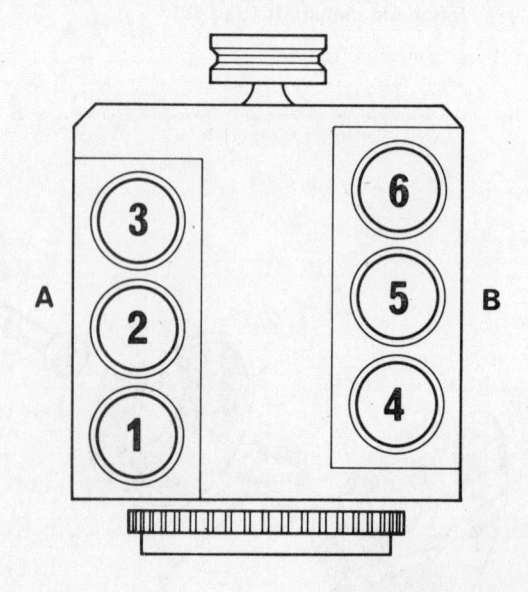

Fig. 1.47 Cylinder liner location (Sec 51)

A Left-hand bank
B Right-hand bank

Chapter 1 Engine

12 Varying thicknesses of liner base seals are available in order to obtain the correct liner protrusion above the top face of the cylinder block faces on each bank. The liner base seals are available in four thicknesses (see Specifications) and are colour coded for identification (photo).

13 Selection of the base seal thicknesses required is made by inserting the respective liners into position in the cylinder block without the base seals fitted. The respective liner protrusions in each cylinder block bank face in turn are then checked. Lay a straight-edge across the top face of a liner and measure the gap to the top face of the cylinder block with feeler gauges. Repeat this with the two other liners in the bank concerned and make a note of the respective protrusions (photos).

14 Now subtract the largest measurement reading taken from 0.009 in (0.23 mm) to determine the theoretical seal thickness requirement.

15 Seal thickness equal to, or just less than, the calculated thickness should then be selected. As an example, if the maximum reading taken from a liner protrusion is 0.004 in (0.10 mm), subtract this from 0.009 in (0.23 mm) to give 0.005 in (0.13 mm). In this instance a red coded base seal is required, which has a thickness of 0.0048 in (0.122 mm).

16 One seal of the same thickness is fitted to each liner. When in position the seal castellations must be engaged in the recess as shown (Fig. 1.48).

17 When the liners are refitted into the cylinder block, locate them so that the base seal tags are visible, then recheck their protrusions above the cylinder block face.

51.12 Cylinder liner base seal

51.13A Fitting cylinder liner into block

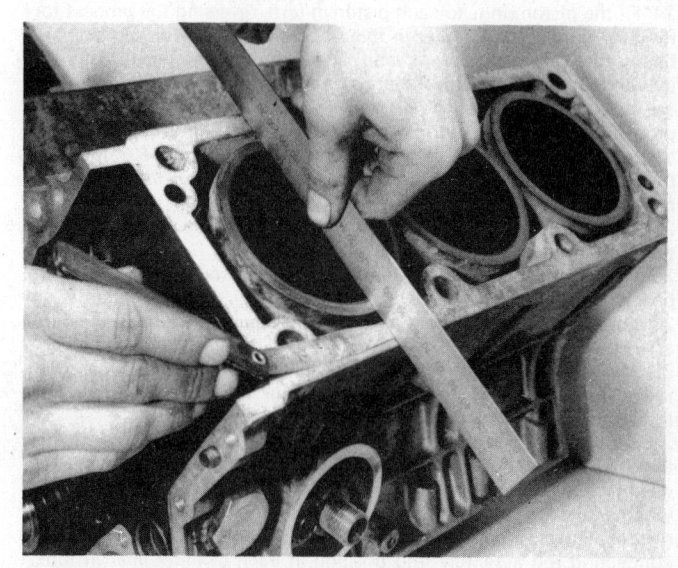

51.13B Checking cylinder liner protrusion

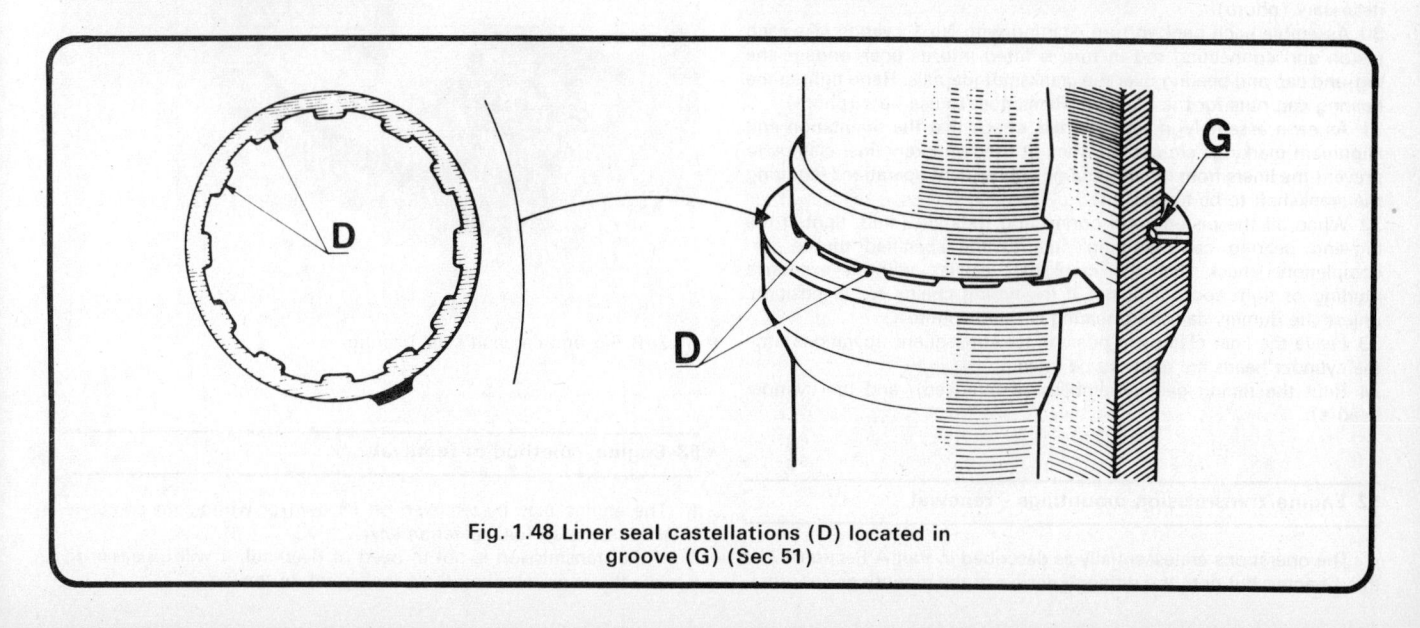

Fig. 1.48 Liner seal castellations (D) located in groove (G) (Sec 51)

18 The protrusional difference between adjacent liners when fitted must not exceed 0.0016 in (0.04 mm). In addition, the protrusional differences (where applicable) must be stepped in one direction as shown (Fig. 1.49). Where necessary, change the thickness of one or more liner seals accordingly whilst keeping the protrusions within the permitted tolerance.

19 When the respective liner positions and base seals are selected, mark their positions accordingly. When re-using the original liners this should not be necessary, since they should have been marked for position when dismantling. It is essential that they be refitted with their respective pistons and connecting rods. The assemblies are numbered 1 to 6 as shown in Fig. 1.47.

20 Each cylinder liner is now ready for fitment of the connecting rod and piston assemblies.

21 As previously explained, the fitting of the pistons to the connecting rods will have been carried out by the Renault agent due to the difficulty of removing and inserting the gudgeon pin. Check that the pistons have been fitted correctly, however, by noting that the arrows on the piston crowns will face towards the timing gear end, and that the connecting rod/cap number markings are on the side noted during dismantling.

22 If new piston and liner assemblies have been supplied, they must be cleaned of the anti-rust coating. Soak them in a suitable solvent to remove the protective coating – do not scrape it off.

23 Keep the respective pistons and rings with their mating cylinder. They are carefully matched sets.

24 Fit the piston rings to each piston in turn, reversing the process for removal as described earlier in this Chapter. Where new ring sets are being fitted to old pistons and liners, follow the manufacturer's instructions supplied with the rings. *It is most important that the rings are assembled in the correct manner.* Once fitted into their ring grooves, check that the rings are free to rotate around the piston and that they do not bind in the groove when fully compressed. If they do, then it is likely that a piece of carbon still remains on the ring land and it needs further cleaning.

25 Liberally oil the rings and ring grooves and turn the rings so that the gaps are at three different points of circle.

26 Locate the shell bearings in the connecting rod and cap big-end bearings (photos).

27 Lubricate each liner bore and piston assembly in turn as they are assembled.

28 Make sure the piston ring gaps are still correctly spread out around the piston. Fit the piston ring compressor around the piston and tighten it up. Take care not to scratch the piston (photo).

29 Offer the piston (and connecting rod) to its respective bore, noting that it can only fit one way round. The arrow on the top of the piston must face towards the front of the engine. It is easy therefore to check whether the pistons have been refitted correctly; the arrows on the pistons must point uniformly. Progressively press the piston/connecting rod assembly into position in each cylinder liner in turn. Excessive force should not be required, firm hand pressure only being necessary (photo).

30 Assemble each bank in turn, starting with No 1 cylinder. As each piston and connecting rod in turn is fitted into its liner, engage the big-end cap and bearing over the crankshaft journals. Hand tighten the bearing cap nuts for the moment. Remember to use nuts (photo).

31 As each assembly in turn is fitted, check that the orientation and alignment markings are correct, then fit the temporary liner clamps to prevent the liners from moving during subsequent operations requiring the crankshaft to be turned.

32 When all the pistons and connecting rods are fitted, tighten the big-end bearing cap retaining nuts to the specified torque. On completion, check that the crankshaft rotates without excessive binding or tight spots (but not if the timing chains are in position, unless the dummy camshaft bearing is fitted (photo).

33 Leave the liner clamps in position for subsequent operations until the cylinder heads are ready to be fitted.

34 Refit the timing gear assemblies (if removed), and the cylinder head(s).

52 Engine/transmission mountings – renewal

The operations are essentially as described in Part A Section 17 of this Chapter, but note the different design of the mountings (photos).

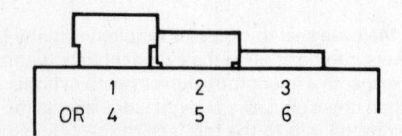

Fig. 1.49 Step cylinder liners as shown to compensate for protrusional differences (Sec 51)

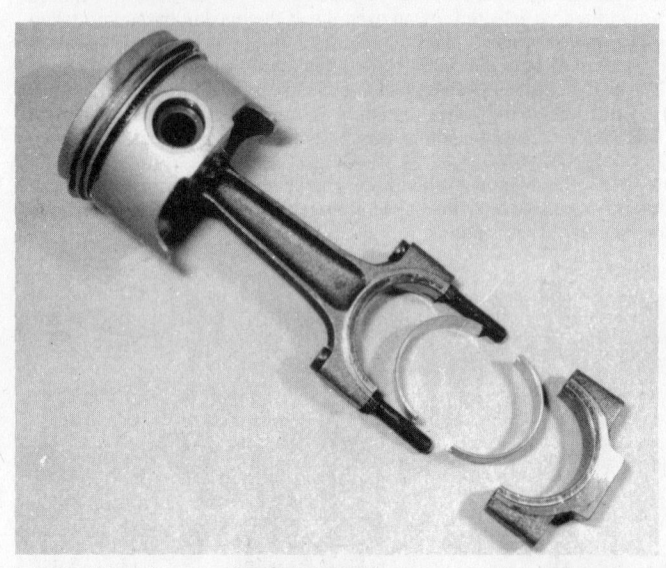

51.26A Piston/connecting rod components

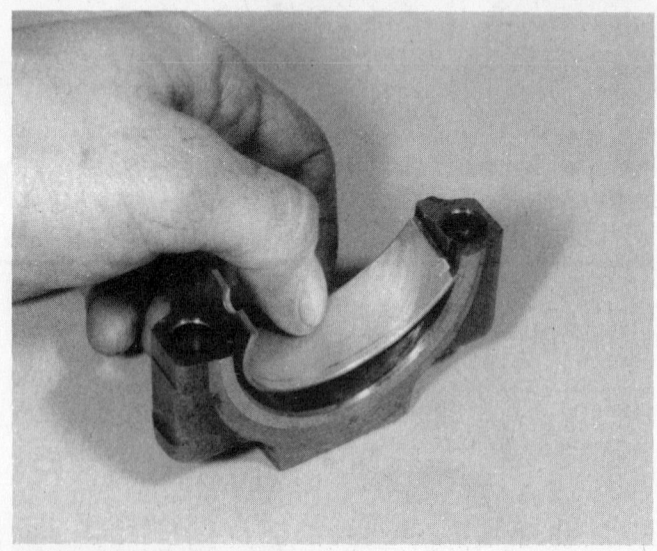

51.26B Big-end cap and shell bearing

53 Engine – method of removal

1 The engine may be removed on its own or withdrawn complete with transmission for separation later.

2 If the transmission is not in need of overhaul, it will be easier to remove the engine leaving the transmission in the car.

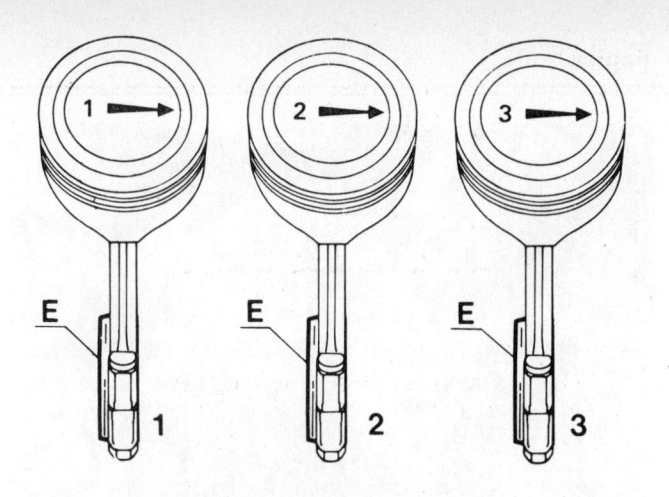

Fig. 1.50 Left-hand bank of pistons with rod flange (E) offset to rear (Sec 51)

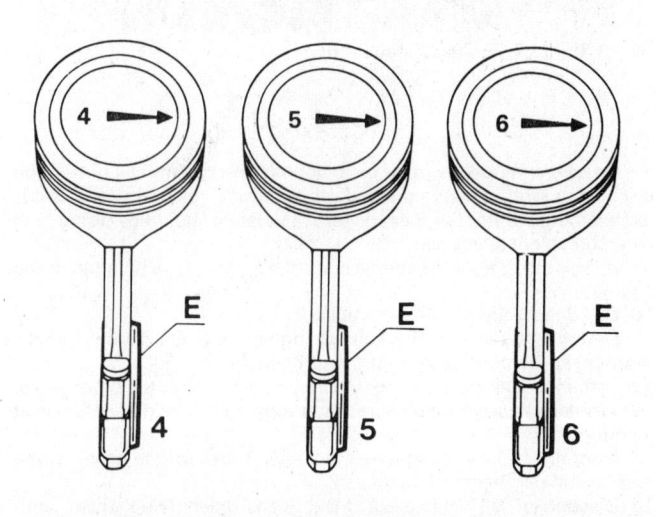

Fig. 1.51 Right-hand bank of pistons with rod flange (E) offset to front (Sec 51)

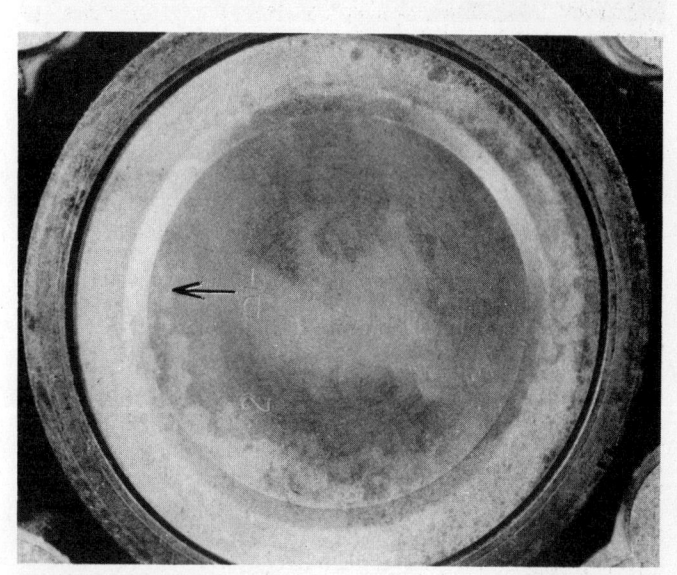

51.28 Using a piston ring compressor

51.29 Piston directional mark

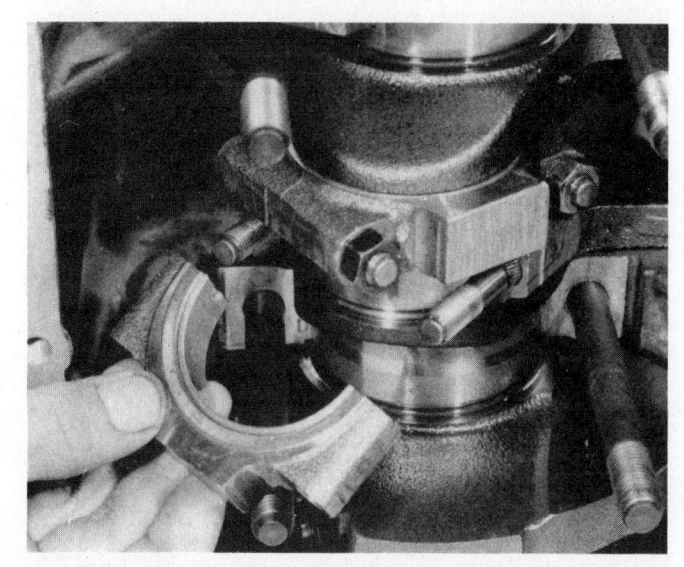

51.30 Fitting big-end cap and bearing shell

51.32 Tightening connecting rod big-end nuts

88　　　　　　　　　　　　　　　　　　　Chapter 1 Engine

52.1A Engine flexible mounting

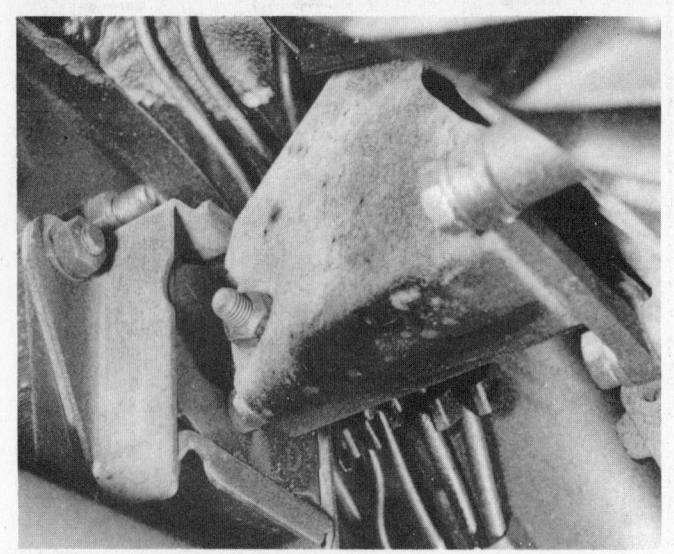

52.1B Transmission flexible mounting

54 Engine – removal (leaving manual transmission in car)

Note: *Suitable lifting gear will be required for this operation.*

1　Open the bonnet, mark the position of its hinges on its underside and with the help of an assistant, unbolt and remove the bonnet.
2　Disconnect the battery and remove the air cleaner.
3　Drain the engine oil.
4　Drain the cooling system.
5　Remove the radiator grille and headlamp wipers (if fitted).
6　Unbolt the radiator top rail and remove it with the side deflectors.
7　Disconnect the electrical leads and remove the radiator/fan assembly.

8　On cars equipped with air conditioning, the system will have to be discharged and the compressor moved aside, also the condenser removed from in front of the radiator. The system should be discharged by a competent specialist.
9　Unbolt the ignition electronic control box and place it on top of the engine.
10　Disconnect the accelerator cable.
11　Unbolt and remove the exhaust downpipes. On Turbo models, disconnect the downpipe from the turbocharger.
12　Disconnect all vacuum hoses including the one for the brake servo. Identify each hose with numbered tape so there will be no confusion at reconnection.
13　Disconnect the coolant hoses including those for the heater at the engine compartment rear bulkhead.
14　Disconnect all engine electrical leads, identifying them with numbered tape if necessary.
15　Disconnect the HT lead from the ignition coil, and the alternator leads.
16　Unbolt the loom protector from the front end of the sump pan and place it to one side.
17　Disconnect the fuel lines.
18　Disconnect the starter motor leads, and unbolt and remove the starter motor.
19　On Turbo models, remove the alternator, the turbocharger heat shield and the intake scoop.
20　Release the power steering pump fluid hoses from their clips, but do not disconnect them. Unbolt the pump and tie it to one side of the engine compartment.
21　Unbolt the engine to body earth strap.
22　On Turbo models, disconnect the oil pipe bracket.
23　Remove the clutch slave cylinder heat shield, the TDC sensor and cover plate.
24　On Turbo models, disconnect the cylinder block oil hoses – see Fig. 1.55, the absolute pressure sensor pipes and the pressure indicator connection at the front strut turret (Fig. 1.56).
25　Remove the left-hand engine damper and its bracket (photo).
26　Disconnect the right-hand engine damper bottom mounting.
27　Release the cruise control diaphragm unit.
28　Unbolt the auxiliary air valve.
29　Remove the cover plate from the lower part of the flywheel housing.
30　Attach a hoist to the engine lifting lugs and just take its weight.
31　Unscrew the engine-to-transmission connecting bolts.
32　Place a jack under the transmission.
33　Unscrew the engine mounting through-bolt nuts and disconnect the mountings.
34　Pull the engine forwards until it is clear of the clutch shaft and then hoist it up out of the engine compartment.

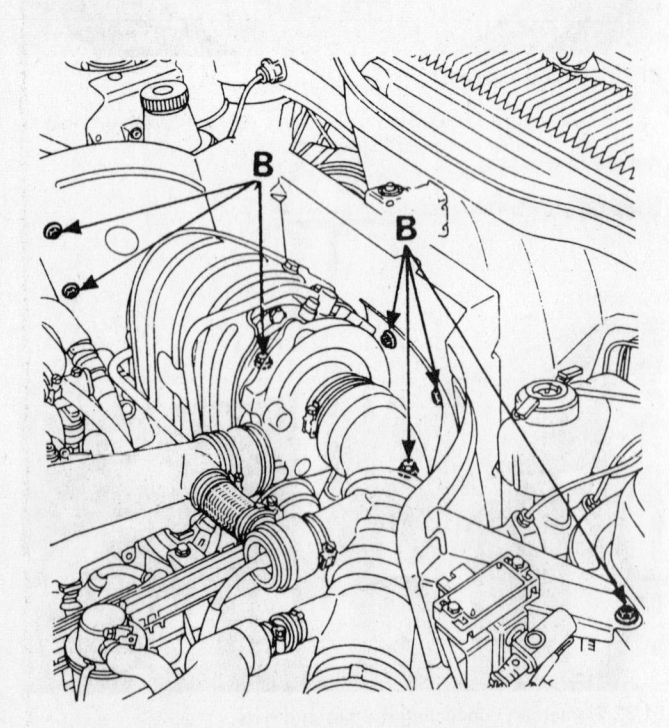

Fig. 1.52 Turbocharger heat shield and intake scoop screws (B) (Sec 54)

89

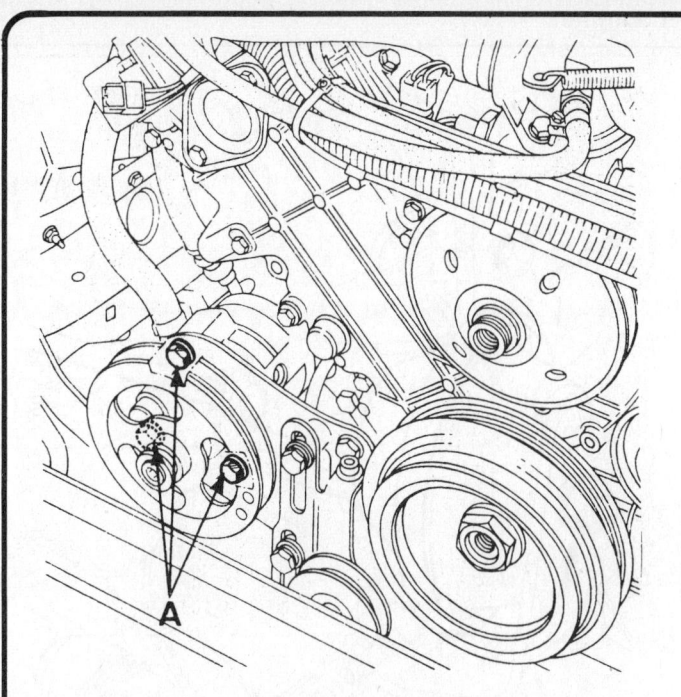

Fig. 1.53 Power steering pump mounting bolts (A) on Turbo models (Sec 54)

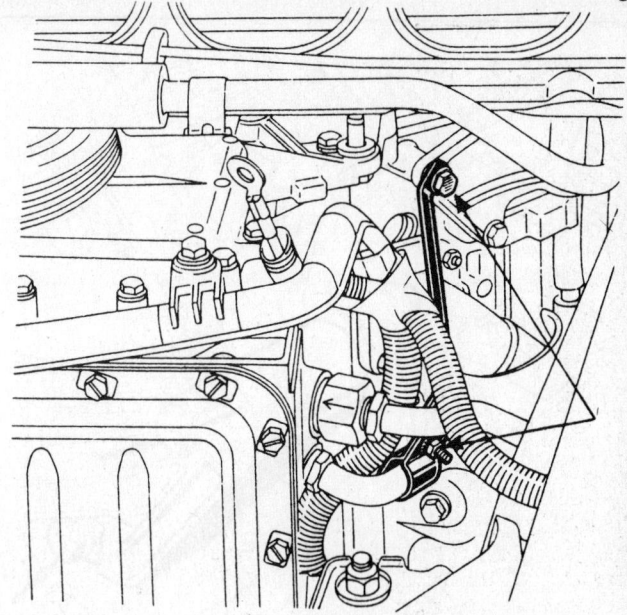

Fig. 1.54 Oil pipe bracket fixings on Turbo models (Sec 54)

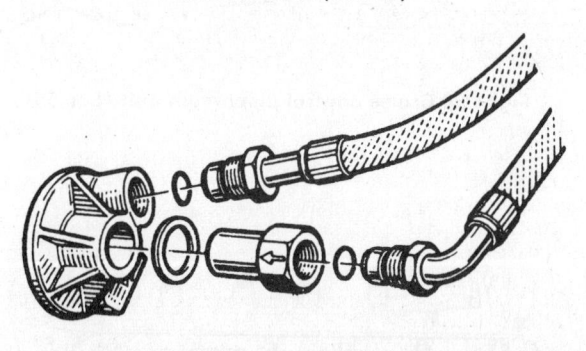

Fig. 1.55 Cylinder block oil hoses on Turbo models (Sec 54)

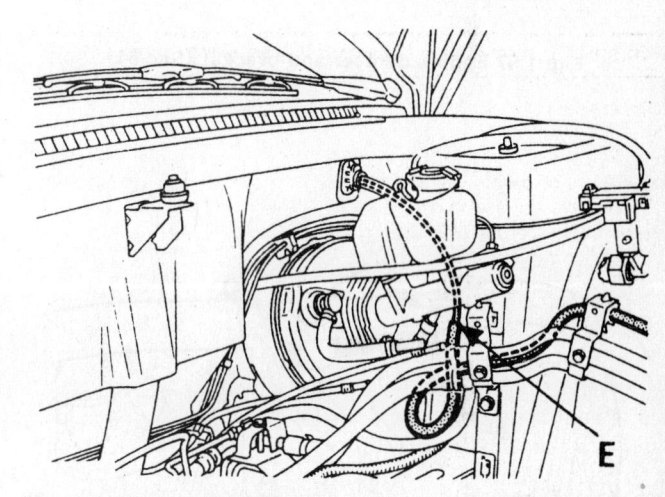

Fig. 1.56 Pressure indicator connection (E) on Turbo models (Sec 54)

54.25 Left-hand engine damper

Chapter 1 Engine

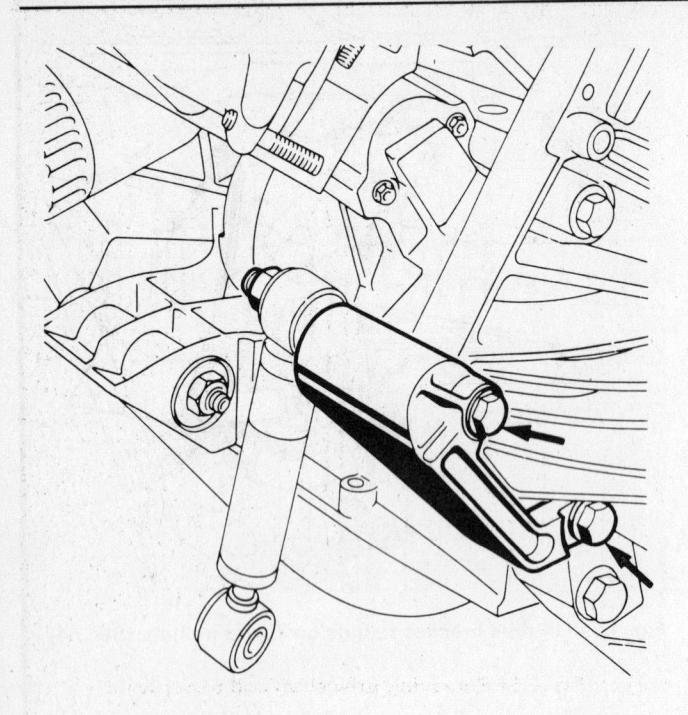

Fig. 1.57 Engine damper and bracket (Sec 54)

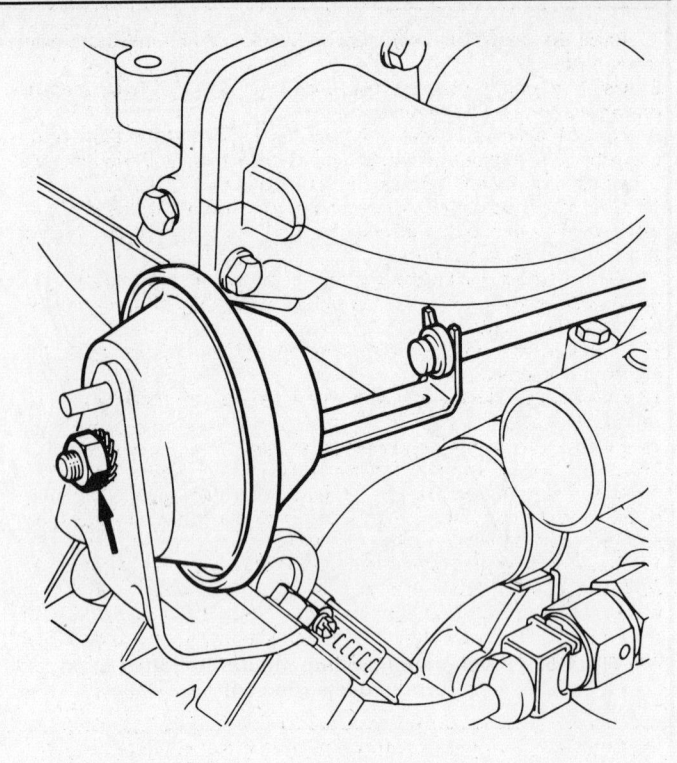

Fig. 1.58 Cruise control diaphragm unit (Sec 54)

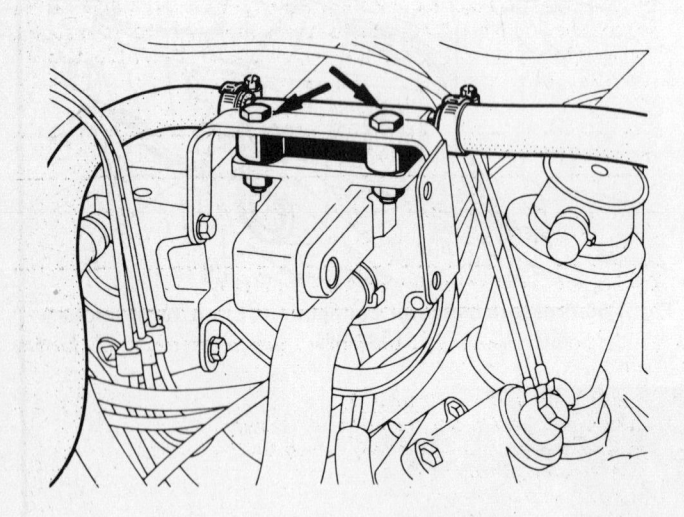

Fig. 1.59 Auxiliary air valve (Sec 54)

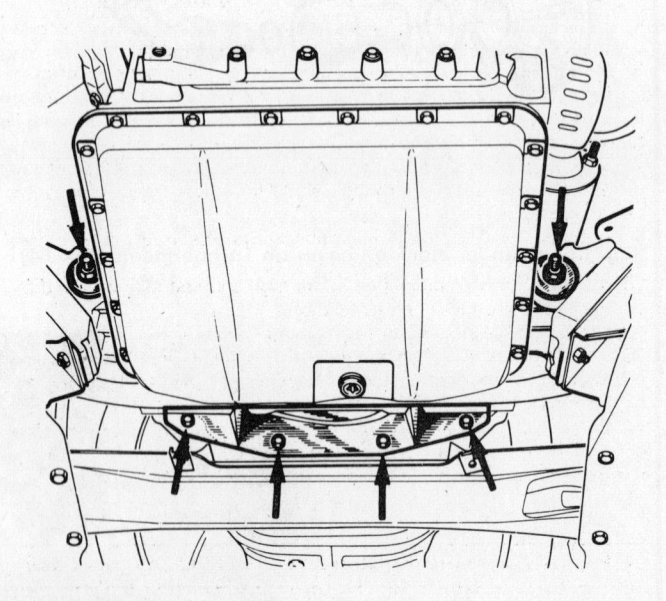

Fig. 1.60 Engine mounting through-bolts and flywheel cover plate (Sec 54)

55 Engine/manual transmission – removal and separation

Note: *Suitable lifting gear and a balljoint splitter will be required for this operation.*

1 The operations include those described in the preceding Section plus the following additional work. Do not remove the engine-to-transmission connecting bolts nor the bellhousing lower cover plate prior to removal of the engine/transmission.
2 Drain the transmission oil.
3 Disconnect the gearchange link rods and the reverse interlock operating cable as described in Chapter 6.
4 Disconnect the reversing lamp switch leads.
5 Disconnect the speedometer sensor.
6 Remove the heat shield, unbolt the clutch slave cylinder and tie it up out of the way without disconnecting the fluid hose.

Chapter 1 Engine

7 Raise the front of the car, support securely and remove the front roadwheels.
8 Using a drift, drive out the dual roll pins which secure the driveshafts to the transmission.
9 Using a balljoint splitter, unscrew the nuts from the suspension upper balljoint and tie-rod end balljoint taper pins, and disconnect the balljoints from the hub carriers and steering arms.
10 Grip each hub carrier in turn and swivel it downwards while an assistant disconnects the driveshafts from the transmission. Do not strain the flexible brake hoses.
11 With the hoist taking the weight of the engine and transmission, unscrew the engine mounting through-bolts and unbolt the transmission mounting brackets.
12 Withdraw the engine/transmission forwards and upwards from the engine compartment.
13 Clean away external dirt using paraffin and a stiff brush or a water soluble solvent.
14 Remove the bellhousing lower cover plate.
15 Unbolt and remove the starter motor.
16 Unscrew and remove the engine-to-transmission connecting bolts. Note the location of the various brackets, wiring harness clips and other attachments held by some of the bolts.
17 Support the engine on wooden blocks placed under the sump pan so that the transmission has a clear gap underneath it, and then withdraw the transmission from the engine. Do not allow the weight of the transmission to hang upon the clutch shaft, while the latter is still engaged with the clutch driven plate.

56 Engine – removal (leaving automatic transmission in car)

Note: *Suitable lifting gear will be required for this operation.*

1 Carry out the operations described in Section 54 excluding those applicable to Turbo models.
2 The following additional tasks must also be carried out.
3 Disconnect the earth straps between transmission and body.
4 Unbolt and remove the starter motor.
5 Disconnect the vacuum pipe from the inlet manifold.
6 Disconnect the kick-down switch cable at the throttle housing.
7 Remove the lower cover plate from the torque converter housing.
8 Lock the teeth of the driveplate starter ring gear using a suitable tool and, then unscrew the driveplate-to-torque converter fixing bolts. In order to bring each bolt into view, the crankshaft will have to be turned by means of its pulley bolt while the ring gear locking tool is temporarily removed.
9 Unscrew and remove the transmission bellhousing-to-engine connecting bolts.
10 Attach a hoist to the engine lifting lugs and just take its weight.
11 Place a jack under the transmission.
12 Remove the engine mounting through-bolts.
13 Raise the engine to clear the mounting brackets and then withdraw it forward to disconnect it from the transmission, and lift it up and out of the engine compartment.
14 Bolt a retaining bar to one of the torque converter housing flange bolts in order to retain the converter in full engagement with the transmission, and to prevent damage to the oil seal and loss of fluid.

57 Engine/automatic transmission – removal and separation

Note: *Suitable lifting gear and a balljoint splitter will be required for this operation.*

1 Carry out the operations described in paragraphs 1 to 28 of Section 54 excluding those applicable to Turbo models.
2 Drain the final drive lubricant.
3 Refer to Chapter 7 and disconnect all the items listed for removal of the Type 4141 automatic transmission.
4 Disconnect the driveshafts from the transmission as described in Section 55, paragraphs 7 to 10.
5 Remove the left-hand engine damper and bracket.
6 Take the weight of the engine and transmission and disconnect the engine and transmission mountings.

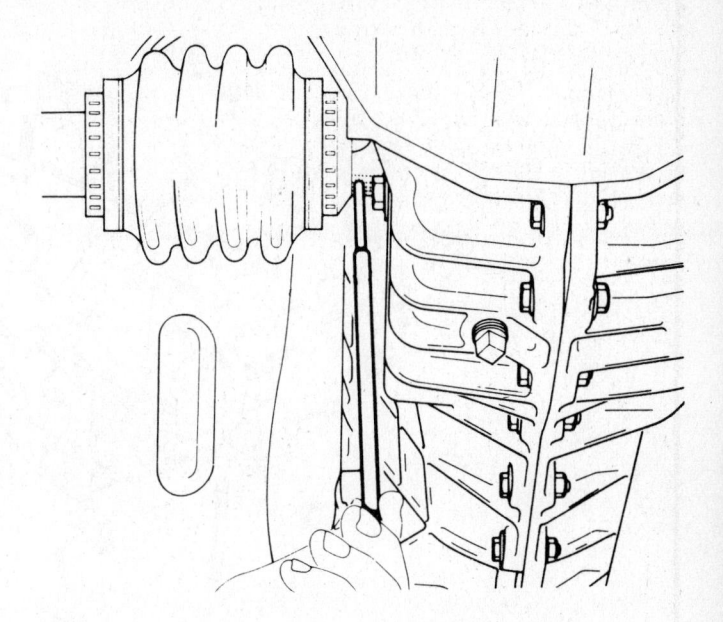

Fig. 1.61 Removing driveshaft roll pins (Sec 55)

7 Lift the engine/transmission forward and then up and out of the engine compartment.
8 Clean away external dirt using paraffin and a stiff brush or a water soluble solvent.
9 Separate the transmission from the engine as described in paragraphs 7 to 9 of Section 56. Pull the transmission from the engine and then fit a torque converter retaining bar as described in Section 56, paragraph 14.

58 Engine – dismantling (general)

Refer to Part A Section 23, of this Chapter.

59 Engine ancillary components – removal

Before complete engine dismantling begins, remove the following items:

Exhaust manifolds (Chapter 3)
Induction and injection manifolds (Chapter 3)
Turbocharger (Turbo models) – Chapter 3
Alternator (Chapter 12)
Engine mounting brackets
Coolant pump (Chapter 2)
Distributor (Chapter 4)
Clutch (Chapter 5)
Oil filter

60 Engine – complete dismantling

1 With the engine standing securely on wooden blocks placed under its sump pan, remove the rocker covers and cylinder heads as described in Section 44.
2 Remove the oil pump (Section 42) and the timing chains and sprockets (Section 43).
3 Turn the engine on its side and remove the sump pan and anti-emulsion plate.
4 Remove the piston/connecting rod/liner assemblies (Section 51).
5 Remove the flywheel or driveplate (auto. trans).
6 On manual transmission models remove the clutch unit, referring to Chapter 5.

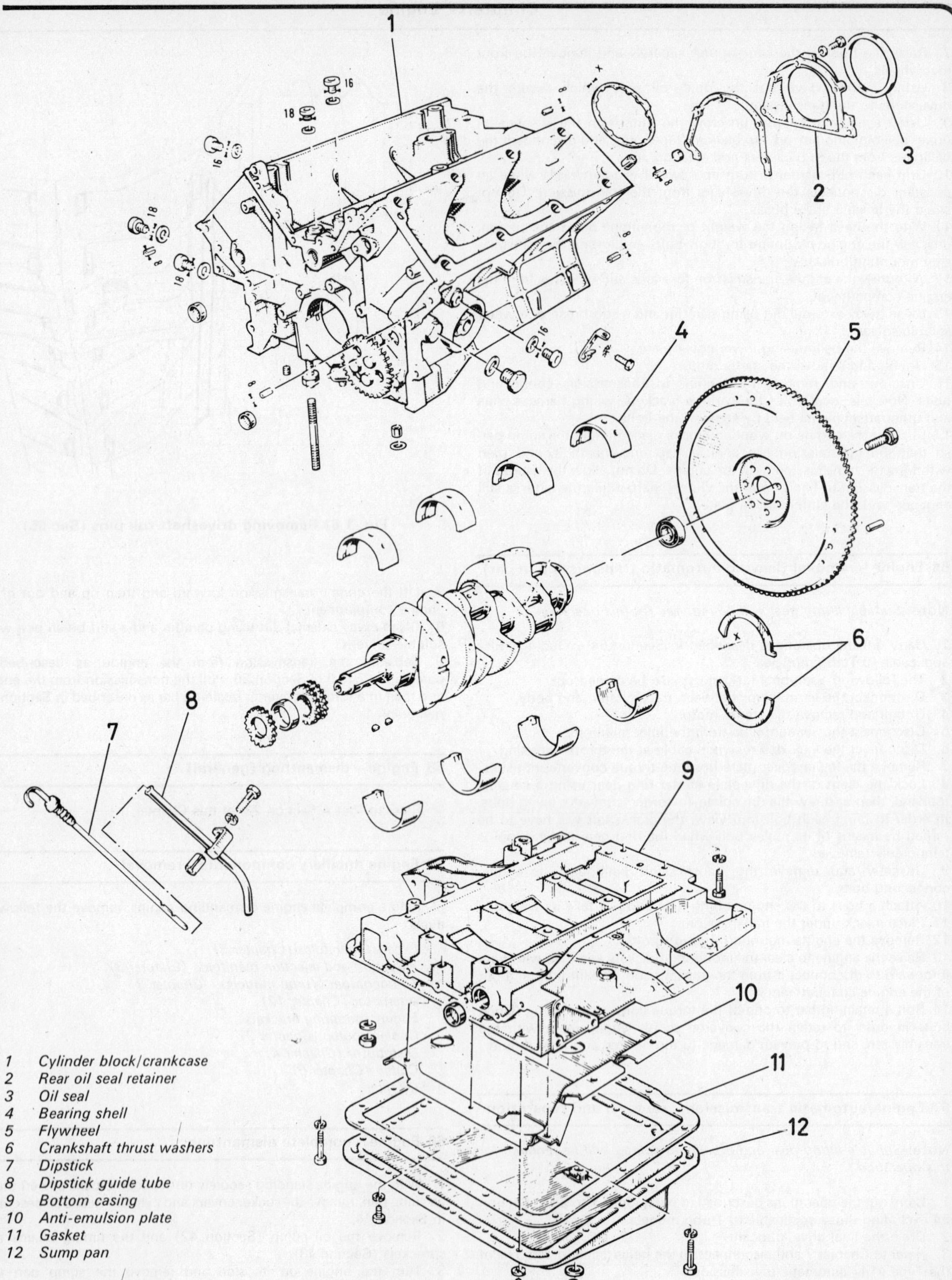

1 Cylinder block/crankcase
2 Rear oil seal retainer
3 Oil seal
4 Bearing shell
5 Flywheel
6 Crankshaft thrust washers
7 Dipstick
8 Dipstick guide tube
9 Bottom casing
10 Anti-emulsion plate
11 Gasket
12 Sump pan

Fig. 1.62 Cylinder block and crankcase components (Sec 60)

Fig. 1.63 Timing components (Sec 60)

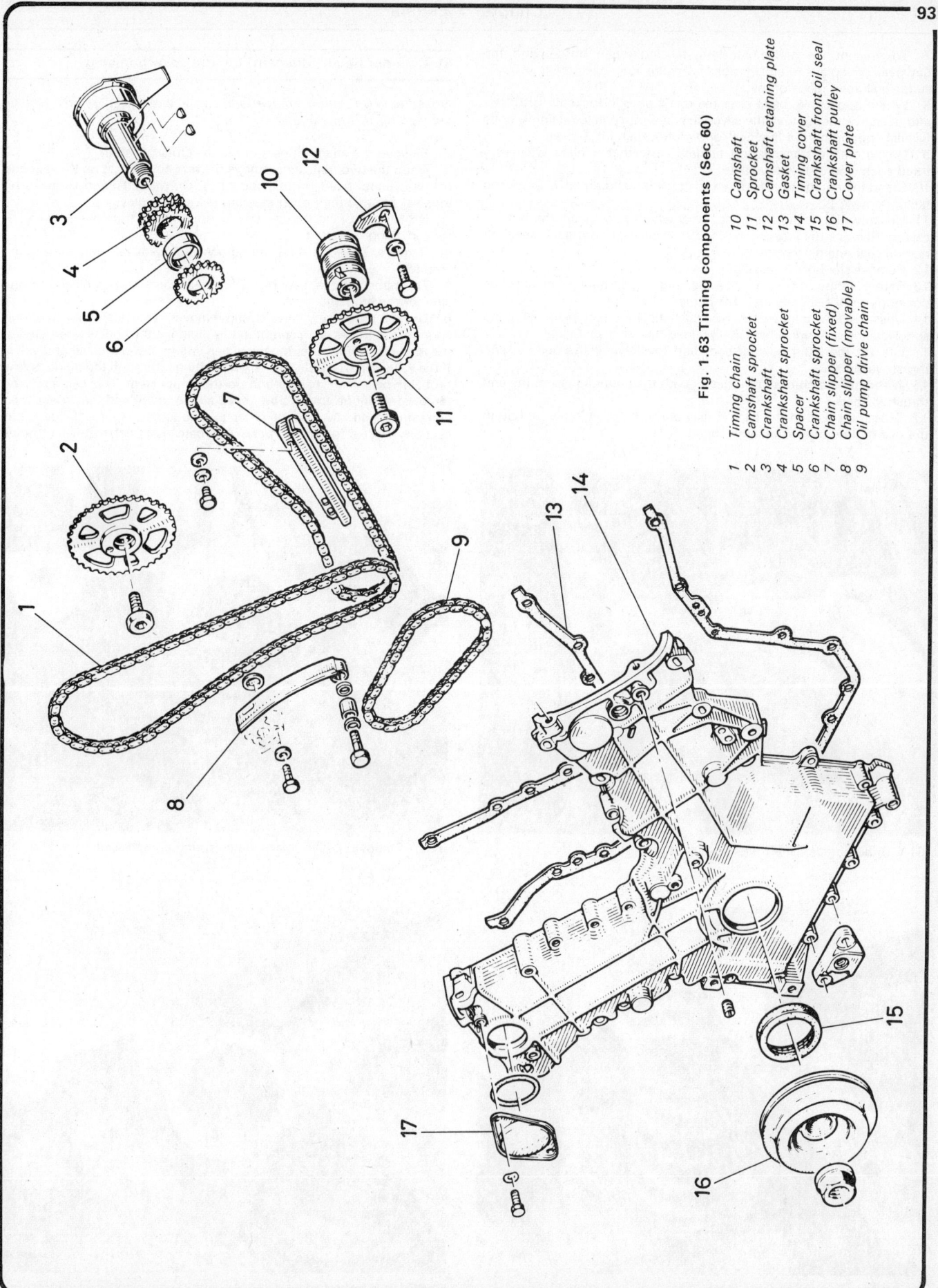

1 Timing chain
2 Camshaft sprocket
3 Crankshaft
4 Crankshaft sprocket
5 Spacer
6 Crankshaft sprocket
7 Chain slipper (fixed)
8 Chain slipper (movable)
9 Oil pump drive chain
10 Camshaft
11 Sprocket
12 Camshaft retaining plate
13 Gasket
14 Timing cover
15 Crankshaft front oil seal
16 Crankshaft pulley
17 Cover plate

Chapter 1 Engine

7 To prevent the crankshaft from turning when unscrewing the flywheel/driveplate retaining bolts, jam the ring gear teeth using a suitable shaped piece of metal.
8 Where applicable, bend over the retaining bolt locktabs (if fitted) and unscrew and remove the seven flywheel/driveplate retaining bolts (whilst supporting the flywheel/driveplate), then lift it clear.
9 On automatic transmission models, note that a plate washer is fitted each side of the driveplate.
10 Stand the engine upside down so that it is standing on its block top face.
11 Unscrew the main bearing cap nuts which also secure the bottom casing. Remove the washers and the smaller bolts from the crankshaft rear oil seal retaining plate.
12 Remove the bottom casing.
13 Before lifting out the main bearing caps, mark their relative positions (if not already marked) numerically.
14 Unscrew and remove the rear main bearing seal plate retaining screws using a 6 mm Allen key. Remove the main seal plate.
15 Lift out the main bearing caps and the lower halves of the end thrust washers.
16 Withdraw the crankshaft together with the upper halves of the end thrust washers.
17 Tape the shells to the caps if they are to be used again, although the use of new ones is recommended.

61 Cylinder head – dismantling and decarbonising

Note: *A valve spring compressor and a valve seat grinder will be required for this operation.*

1 Remove the spark plugs and the injectors (if fitted).
2 Undo the two bolts and remove the rear cover plate on the rear end of the cylinder head (right-hand side). On the left-hand (non-Turbo models) remove the power steering drivebelt pulley.
3 On Turbo models, remove the distributor drive from the front end of the left-hand camshaft.
4 Remove the bolt and withdraw the camshaft retaining plate (photo).
5 The camshaft can now be fed out rearwards and be placed to one side for the moment.
6 Using a proprietary valve spring remover, remove all the valves and springs. Compress each spring in turn, remove the collets, then release the tool and remove the upper washer, spring, lower washer and valve. If the valve seems a little stiff when pushing it through the guide, this is probably because it has carbon on the upper stem. This can easily be removed with fine emery cloth. Keep all the valves and their associated components in the correct order and together. A box divided into sections is ideal for this purpose. Alternatively punch a series of holes

61.4 Unbolting camshaft retaining plate

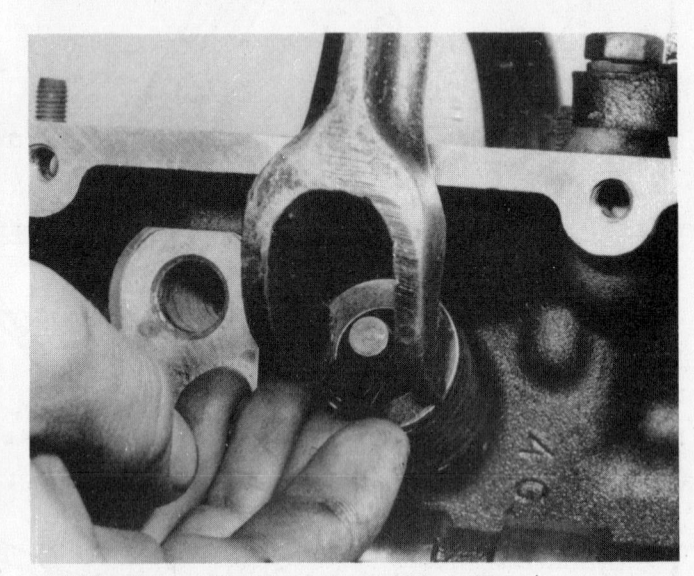

61.6A Removing split collets, valve spring compressed

61.6B Valve spring and retaining cap

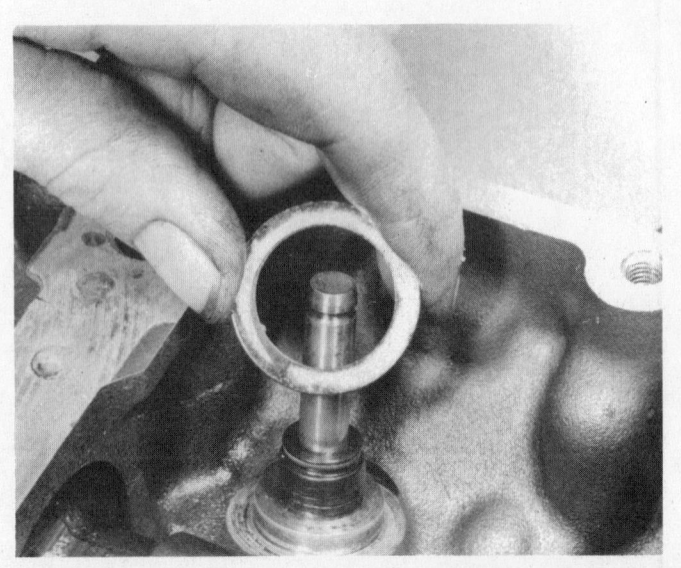

61.6C Valve spring seat

Chapter 1 Engine

61.6D Valve stem oil seal

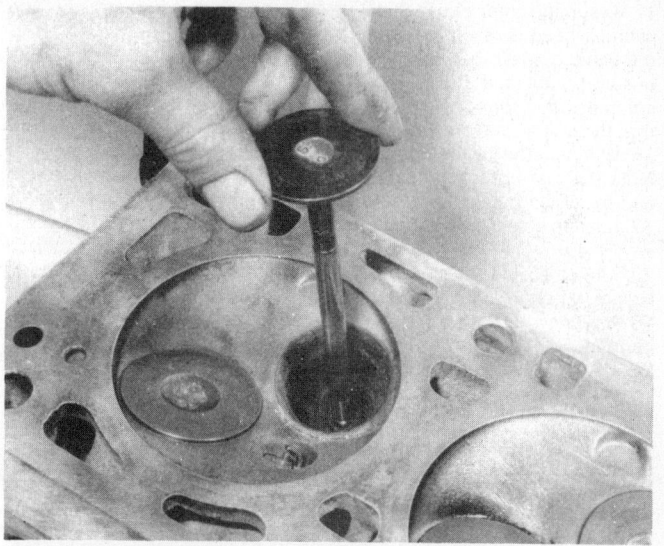

61.6E Removing (smaller) exhaust valve

in a piece of cardboard and push each valve in order through the cardboard, stem upwards; finally, place the spring and washers over it (photos).
7 Remove the valve stem seals from the valve guides and place them with their respective valves.
8 With the valves removed, clean out any carbon from the ports, the cylinder head face and the combustion chambers. Examine the valve seats; if they are only slightly pitted, it is possible to make them smooth again by grinding the valves against the seats, as described in Section 26. Where pitting is very deep, they will have to be recut; a job for the specialist with the necessary equipment.
9 Check the valve guides with the valve stems to ensure that they are a good fit: the stems should move easily in the guides without side play. Worn guides can be extracted and new ones refitted but this should be left to your Renault dealer.
10 Examine the valves, checking them for straightness and the condition of the face. Slight pitting can be removed by grinding-in, but if the pitting is deep the valve will have to be machine ground by a specialist. Grinding-in is covered in Part A, Section 26 of this Chapter.
11 The camshaft and its bearings in the cylinder head are part of the Renault engine which seems virtually everlasting. If there should be noticeable play between the camshaft and its bearings, or if the bearings are damaged, the cylinder head will have to be renewed as the bearings are machined in the head. If very light scratches are present on the camshaft, these can be removed by gently rubbing down with a very fine grade emery cloth or oilstone. The greatest care must be taken to keep the cam profiles smooth.
12 Refit the camshaft in its journals and rotate it by hand. Check that it runs smoothly and that there is no trace of warping. Check the endfloat (see Specifications) with the camshaft retaining plate in position. If it is excessive, fit a new retainer of different thickness. These are available from 0.1556 to 0.1565 in (3.952 to 3.976 mm) (photos).
13 If there is any doubt about the condition of the valve springs, renew them.
14 Check the cylinder heads and valve seats for cracks, or damage. Then check the surface condition of the cylinder head face. Use a straight-edge. The measurements must be made down the length of the block and diagonally across it. The maximum permitted deviation or warp is given in the Specifications. If the head is out of true to this sort of degree, it must be renewed.

61.12 Checking camshaft endfloat

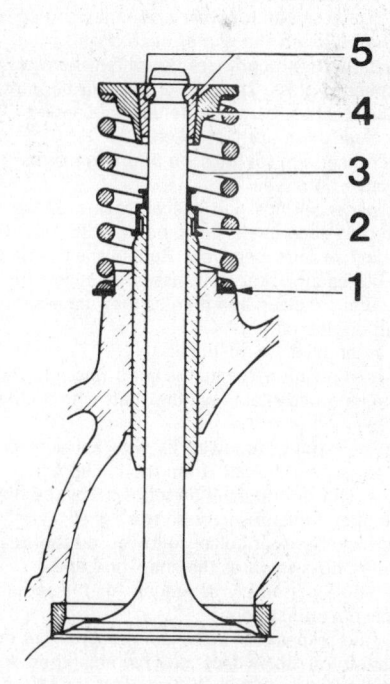

Fig. 1.64 Valve components (Sec 61)

1 Spring seat
2 Valve stem oil seal
3 Spring
4 Spring retaining cap
5 Split collets

96   Chapter 1 Engine

15 Assuming the valves have been ground in as necessary, the cylinder head is ready for reassembly. Start by fitting new oil seals to the valve guides, then refit the valves. Place each valve in turn in its position, refit the lower washer, spring and upper washer, then compress the spring. Refit the collets and release the tool, ensuring that the collets stay in position correctly (photo).
16 When all the valves are refitted, slide the camshaft in from the rear. Note that the right-hand side camshaft has a worm drive on the rear end for driving the distributor on non-Turbo models.
17 Lock the camshaft in position with the retaining plate and secure it with the bolt and washer.
18 The oil seal at the rear of the left-hand camshaft must be renewed before refitting the power steering pump drive pulley (see Section 48).
19 Use new O-ring seals when refitting the right-hand camshaft endplate.
20 Decarbonising of the piston crowns is covered in Part A Section 26 of this Chapter.

62 Examination and renovation

1 With the engine stripped down and all components thoroughly clean, examine everything for wear. The following items should be checked and where necessary renewed or renovated as described in the following Sections.

Rocker assemblies
2 Although the right-hand and left-hand bank rocker assemblies are identical, they should be dismantled separately and their respective components laid out in order of fitting and kept separate from each other.
3 Unscrew the bolt from the end bearing pedestal and withdraw the pedestals, rocker arms, spacers and springs.
4 The oilway plugs at the end of each rocker shaft are press fitted and should not be removed. Note the direction of fitting of each rocker shaft to ensure correct reassembly. If refitted incorrectly, the oil feed holes will be blocked. Note that when fitted the oil holes face downwards.
5 Check the rocker arm wear faces (where they contact the camshaft lobes). Where they are only slightly worn they can be refaced. More excessive wear will necessitate their renewal.
6 Inspect the rocker shafts for wear and renew if necessary. Do not try to remove the plugs on the end of each shaft.
7 Check the shafts for straightness by rolling them on a flat surface such as a flat sheet of glass. They are unlikely to be bent, but if this is the case they must either be straightened or renewed. Check the oil feed holes and clear them out if blocked.
8 Check each rocker arm for wear on an unworn part of the shaft to check for excessive bore wear.
9 Check the spacer springs and spacer collars. There are collars of two widths fitted to each shaft, these being 5.35 mm (0.210 in) and 8.2 mm (0.322 in) wide respectively. Renew them if badly worn.
10 To reassemble each rocker shaft assembly ready for refitting, first ensure that all parts are clean and then as they are assembled lubricate them with clean engine oil.
11 Both shaft assemblies are identical.
12 If not already in position, locate the snap-ring into the end groove, then slide the bearing pedestal down the shaft with the flat on the boss to the snap-ring.
13 Locate a spacer spring, followed by a rocker arm located with its tappet adjuster screw on the left (from the circlip end).
14 Fit the thinner spacer into position followed by the next rocker arm, fitted with its tappet adjuster screw on the right.
15 Locate the thicker spacer collar, followed by the bearing pedestal with the flat on its boss towards the snap-ring end.
16 Repeat the above assembly sequence for the remaining pairs of rocker arms, spacers and springs.
17 Ensure that the respective rocker components are correctly fitted and that the shaft oil holes face downwards, then insert the end pedestal bolt with washer to secure.
18 Repeat this procedure with the second rocker shaft assembly.

Cylinder liners
19 The cylinder bores must be examined for taper, ovality, scoring and scratches. Start by carefully examining the top of the cylinder bores. If they are at all worn a very slight ridge will be found on the thrust side.

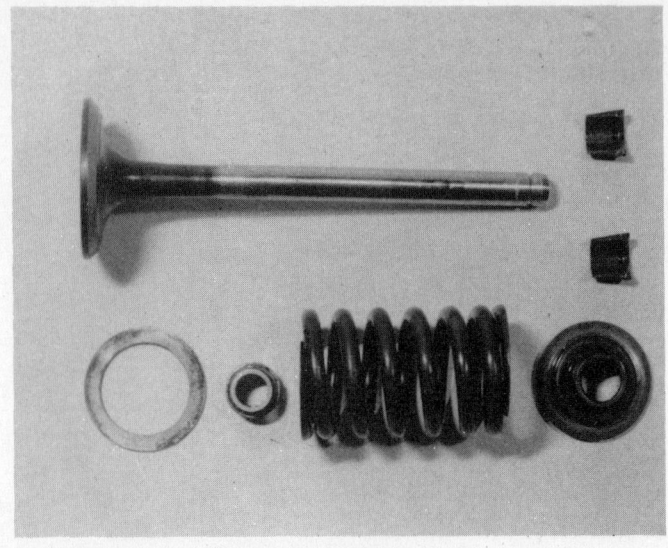

61.15 Components of one valve assembly

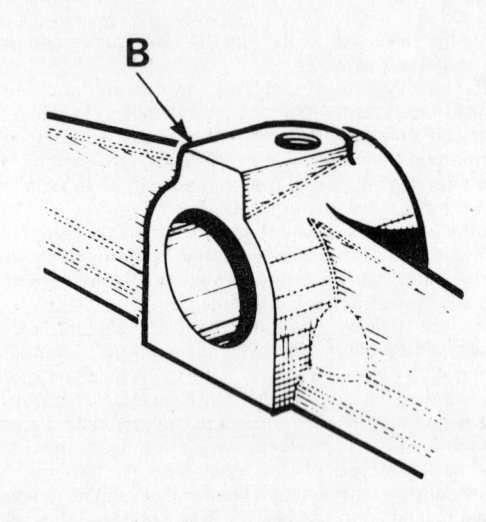

Fig. 1.65 Flat face of boss (B) towards circlip on rocker shaft (Sec 62)

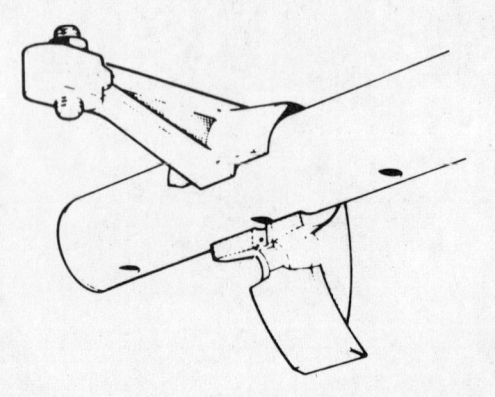

Fig. 1.66 Rocker shaft oil lubrication holes facing downwards (Sec 62)

This marks the top of the piston ring travel. The owner will have a good indication of the bore wear prior to dismantling the engine, or removing the cylinder head. Excessive oil consumption accompanied by blue smoke from the exhaust is a sure sign of worn cylinder bores and piston rings.

20 Measure the bore diameter just under the ridge with a micrometer and compare it with the diameter at the bottom of the bore, which is not subject to wear. If the difference between the two measurements is more than 0.15 mm (0.006 in) then it will be necessary to fit new pistons and liner assemblies. If no micrometer is available remove the rings from a piston and place the piston in each bore in turn about 3/4 in (18 mm) below the top of the bore. If a 0.25 mm (0.010 in) feeler gauge can be slid between the piston and the cylinder wall on the thrust side of the bore then remedial action must be taken.

21 If the liners are re-usable, deglaze the bores using coarse emery cloth. The upper ridge is best removed using a de-ridging tool. Clean all traces of abrasive from the bores on completion.

22 Should the liners have been disturbed they must be completely removed from the cylinder block and new seals fitted, otherwise once the seals have been disturbed the chances are that water will leak into the sump. Refer to Section 51 for cylinder liner installation details.

Pistons and piston rings

23 If the old pistons are serviceable and are to be refitted, carefully remove the piston rings as given in Part A Section 15 of this Chapter.

24 With the rings removed from the pistons, clean out the grooves but take care not to scratch the aluminium in any way. Pieces of old piston ring are useful for this task, but protect your fingers – piston rings are sharp!

25 If new rings are to be fitted to the old pistons then the top ring should be stepped so as to clear the ridge left in the bore above the previous top ring. If a normal but unstepped new ring is fitted it will hit the ridge and break because the new ring will not have worn in the same way as the old, which will have worn in unison with the ridge. The piston rings are supplied pre-gapped for the particular liner diameter and it is not necessary to check them in the liners. It is not necessary to measure the side clearance in the piston ring grooves with the rings fitted as the groove dimensions are accurately machined during manufacture. When fitting new oil control rings to old pistons it may be necessary in this instance to have the groove widened by machining to accept the new wider rings.

26 New pistons and liners are only supplied as complete sets. The respective components are carefully matched to each other during production. Accordingly, do not interchange components from one set to another.

Connecting rods and gudgeon pins

27 A visual check only can be carried out to observe whether any movement or play can be seen when the piston is held still and the connecting rod pushed.

28 If there has been evidence of small-end knock with the engine at normal working temperature, then the connecting rod/piston assembly should be taken to a Renault dealer as special tools are required to dismantle and refit these components.

29 Have the connecting rods checked for alignment whenever the gudgeon pins are renewed.

Timing chains, tensioners and sprockets

30 After a considerable mileage the timing chains and sprockets will have worn and possibly be in need of renewal.

31 Clean and examine the camshaft sprockets and also the crankshaft and oil pump drive sprockets. If they are worn, indicated by hooked teeth, then they must be renewed. It is possible to renew the chains at time of major overhaul; if new sprockets are being fitted, new chains **must** be used.

32 Check the respective tensioner blades and pivots for wear. The tensioner block itself should not be dismantled; if suspect, renew it. If the retaining lock in the tensioner is accidentally removed then it must be renewed since there is no way of checking the lock finger position relative to the thrust ball during assembly, and it is possible to jam the tip of the finger with the spring.

Cylinder block

33 Check for cracks. The cost of welding the alloy material must be weighed against a new casting. Threaded holes which have stripped, may have proprietary thread installed to rectify, the only exception to this being the oilway plug in the cylinder block next to the oil filter.

Oil pump

34 Check the pump rotor and its housing for any signs of wear or damage. The complete pump assembly must be renewed if found to be defective.

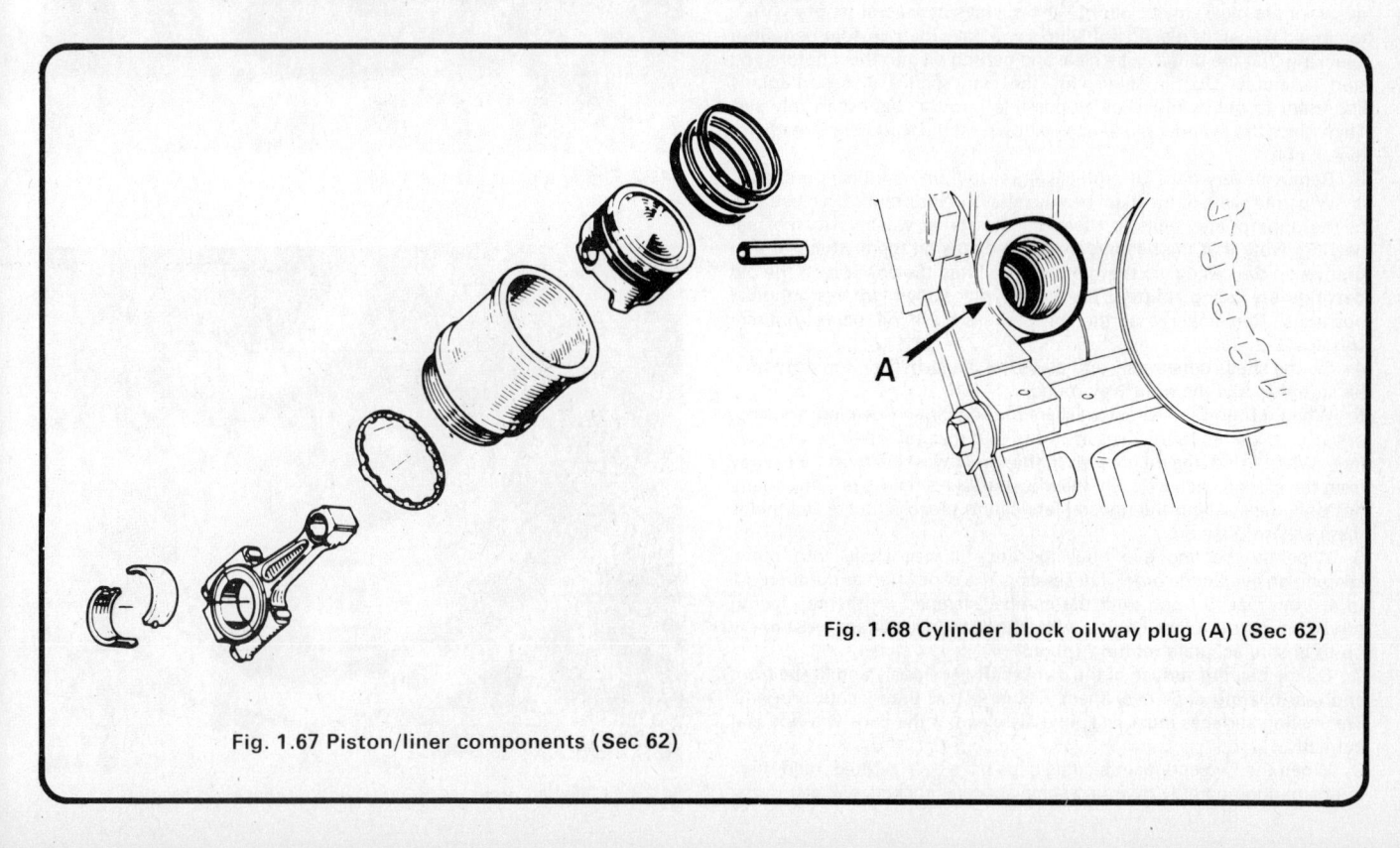

Fig. 1.67 Piston/liner components (Sec 62)

Fig. 1.68 Cylinder block oilway plug (A) (Sec 62)

Chapter 1 Engine

35 To inspect the oil pressure relief valve, extract the split pin and withdraw the spring seat, spring and release valve plunger. Check the end face of the plunger and its contact face within the body for signs of scoring or damage, and renew if necessary.
36 The drive chain and sprockets should also be checked for wear and renewed if worn.
37 Reassembly of the pump unit is a reversal of the dismantling procedure.
38 When assembling the relief valve plunger, ensure that it is fitted the correct way round.

Engine/transmission flexible mountings
39 The engine mounting rubbers are often ignored simply because they do not normally present any problems. However their work rate is probably equal to any other engine component and therefore if the engine is removed at any time, it is worthwhile checking the condition of the mounting rubbers. If they show signs of deterioration due to oil impregnation, heat or simply age, they should be renewed. Mountings that have lost their resistance to shocks will cause engine/transmission vibrations and increase the fatigue rate of other associated engine/transmission connections, as well as that of the driver and occupants of the car.

Crankshaft main and big-end bearings and flywheel/driveplate
40 For details of examination and renovation of these components, refer to Part A Section 27 of this Chapter.

63 Engine – reassembly (general)

Refer to Part A, Section 28 of this Chapter.

64 Engine – complete reassembly

Note: *New flywheel/driveplate bolts and new big-end cap nuts will be required when reassembling the engine.*

Crankshaft and main bearings
1 Make sure that the crankcase is thoroughly clean and that all the oil-ways are clear. Inject oil into the oilways at several points with a forcefeed oil can or plastic bottle: this will have the two-fold benefit of checking that the oilways are clear and getting oil into them before you start assembly. Do the same with the crankshaft – it is particularly important to get as much oil as possible into the crankshaft oilways. Then place the cylinder block upside down on the workbench ready for reassembly.
2 Remove every trace of protective grease from new bearing shells.
3 Wipe the seats of the main bearing shells in the crankcase clean and fit the appropriate shells in them. Fit the thrust washers to the rear bearing. Note that the bearing shells have tabs on them which fit into grooves in the casing, so they can only be fitted the one way. If the old bearings are being refitted, be sure to place them in their original positions. Remember that the inner shells have oil holes in them (photos).
4 Oil the shells generously and place the crankshaft on top of them – be sure that it is the right way round.
5 When refitting the thrust washers to the rear end bearing housing, note that there are four halves in all and that the upper two halves have tags. When fitted, the oil grooves in the thrust washers must face away from the cap on each side. Oil the thrust washers before refitting them; not only does it help the general lubrication process, but it also helps them stay in position.
6 Wipe the bearing cap housings and fit their shells into them, keeping an eye on the order if necessary. The caps fitted as numbered 1 to 4, from rear to front, with the number stamped on the cap facing towards the front of the engine. Also, the bearing shell tags must marry up to ensure accurate refitting (photo).
7 Oil the bearing surface of the crankshaft generously and fit the front and rear bearing caps over them, ensuring that they locate properly. The mating surfaces must be spotlessly clean or the caps will not seat correctly.
8 When the two end main bearing caps have been refitted, hold them in position temporarily by fitting some spacers (sockets will do) under

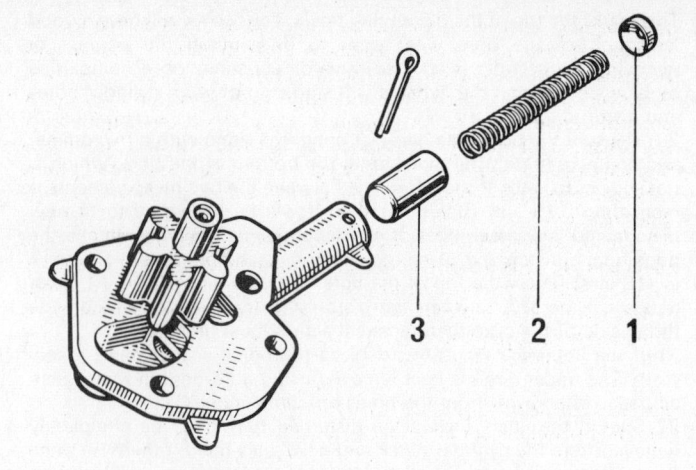

Fig. 1.69 Oil pump (Sec 62)

1 Cap
2 Spring
3 Plunger

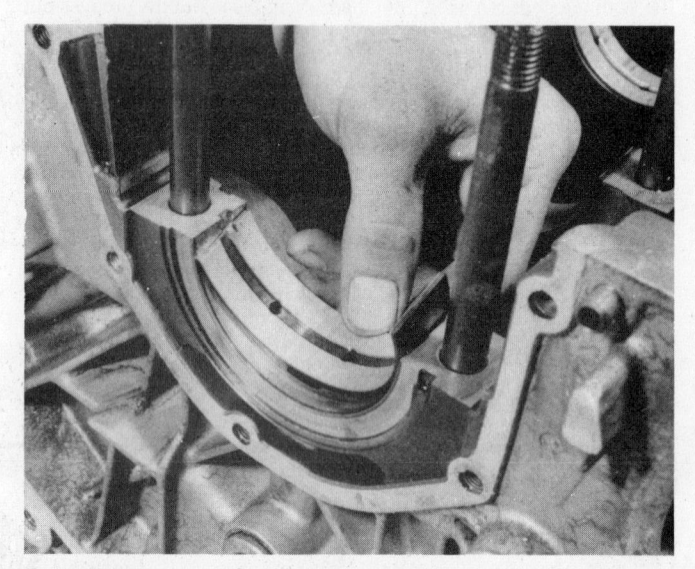

64.3A Fitting a main bearing shell

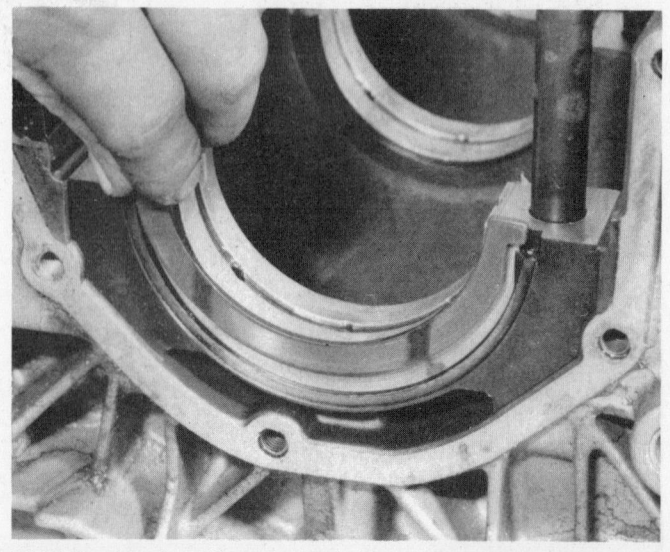

64.3B Crankshaft endfloat thrust washer at rear main bearing

Chapter 1 Engine

the retaining nuts which should be tightened to the specified torque. Check that the crankshaft rotates freely. Some stiffness is normal with new shells, but there should be no binding or tight spots.

9 Lever the crankshaft forwards and backwards, checking the amount of endfloat with the Specifications given at the beginning of this Chapter (photo). The clearance can be reduced or increased by altering the thickness of the thrust washers which are available in three different oversizes, apart from the standard size, as quoted in the Specifications.

10 When satisfied that the endfloat is acceptable, leave the two end main bearing caps bolted in position temporarily until you are ready to refit the bottom casing. Locate the connecting rod big-end bearings before refitting the bottom casing, to avoid having to work through its confined access apertures.

Pistons/connecting rods/cylinder liners
11 The operations are described in Section 51.

Bottom casing and main bearing cap nuts
12 Remove the nuts and temporary spacers from the front and rear main bearing cap studs.
13 Oil Nos 2 and 3 main bearing cap shells and fit the caps (photo).
14 Locate the rear seal cover gasket, smearing it with grease to hold it in position, then fit the seal housing so that its bottom face is flush with the cylinder block bottom face. Fit and tighten the Allen screws (photos).
15 Trim off the gasket ends protruding from the lower facings, and check that they are flush by placing a straight-edge across them.
16 Locate a new O-ring seal in the channel between the cylinder block and bottom casing.
17 Check that the mating surfaces of both the cylinder block and lower casing are clean and oil free.
18 Apply a thin layer of RTV sealant to the cylinder block mating surface, then lower the bottom casing into position (photos).
19 Locate the main bearing cap washers and nuts, but do not tighten them yet (photo).

64.9 Checking crankshaft endfloat with main bearing cap temporary retainers in position

64.13 Locating a main bearing cap

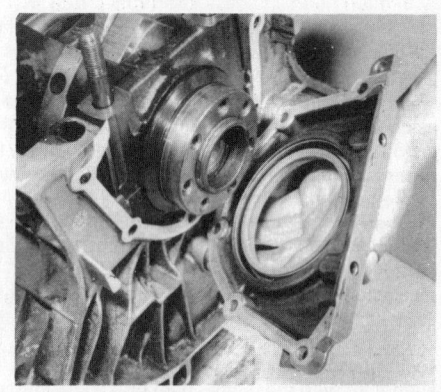

64.14A Crankshaft rear oil seal and housing

64.14B Checking crankshaft rear oil seal retainer is flush with crankcase

64.14C Tightening crankshaft rear oil seal retainer fixing screws

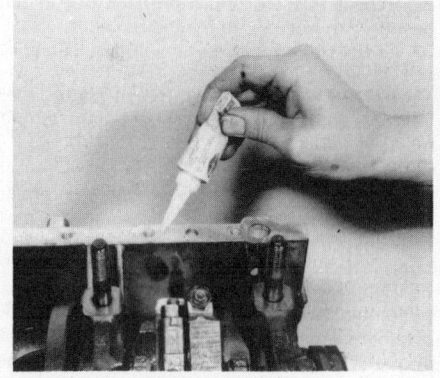

64.18A Applying RTV sealant to crankcase mating surface

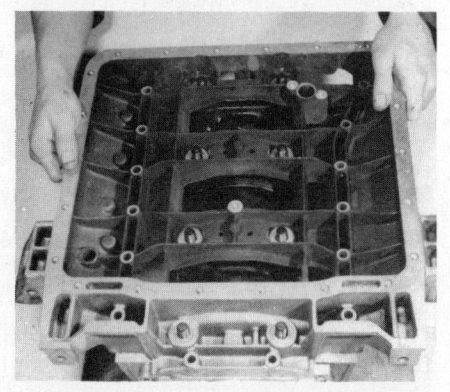

64.18B Locating lower casing on crankcase

64.19 Main bearing washers and nuts

64.20 Checking lower casing and adjacent surfaces for flush fitting

64.21 Tightening main bearing nuts

64.28 Chain tensioner oil filter gauge

20 Check the alignment of the cylinder block with the bottom casing rear face, and the alignment of the bellhousing retaining brackets, using a straight-edge (photo). It is essential that they be aligned, or distortion could result when refitting the transmission. Align if necessary and then tighten the two Allen screws which secure the rear of the bottom casing to the rear seal housing.
21 Tighten the main bearing cap nuts to the specified torque in the sequence shown in Fig. 1.70 (photo). Each nut must then be further tightened (in the sequence given) a further 75°. To do this, first make a protractor or angle template using a piece of stiff cardboard. Mark one edge as the 0° line and then using a geometrical protractor, mark a line or cut the cardboard at an angle of 75° from the 0° line. Locate the socket onto the nut to be tightened and mark a line on the socket in alignment with the 0° line, then tighten the nut until the socket line is in alignment with the 75° line. An assistant will be required to hold the cardboard template in position whilst the nut is being tightened, as the template must not move at all.
22 Repeat this procedure on each nut in turn until the eight main bearing cap nuts are tightened correctly.
23 Fit the bottom casing retaining bolts and tighten them evenly.

Sump pan and anti-emulsion plate
24 Refit after reference to Section 46.

Cylinder heads
25 With the bottom half of the engine rebuilt, it is time to turn the engine over and rest it on the sump. To make working on it more pleasant, place a block under the rear end of the lower crankcase to bring it level.
26 Refitting both the cylinder heads is basically the same operation performed twice. Start by rotating the crankshaft so that the No 1 cylinder piston is 0.59 in (15 mm) before the peaks of its compression stroke (TDC). Keep the crankshaft in this position whilst fitting the cyinder heads and rocker assemblies to avoid the possibility of the pistons touching the valves during the subsequent operations.
27 The fitting procedure is described in Section 44, but ignore the reference to the timing chain and sprocket.

Timing chains, sprockets and tensioners
28 Relocate the timing chain tensioner oil filter gauzes (photo).
29 Refit the chain guides and tensioners into position on the cylinder block front face.
30 Locate and bolt the tensioner blades into position.
31 Engage the timing sprocket key into its keyway in the crankshaft and then refit the timing sprocket so that its timing mark faces outwards.
32 Refer to Section 42 paragraphs 17 to 25 for the remaining refitting procedure.

Oil pump
33 Refit after reference to Section 42, paragraphs 17 to 20.

Timing cover and crankshaft pulley
34 Refer to Section 43, paragraph 27 to 29.

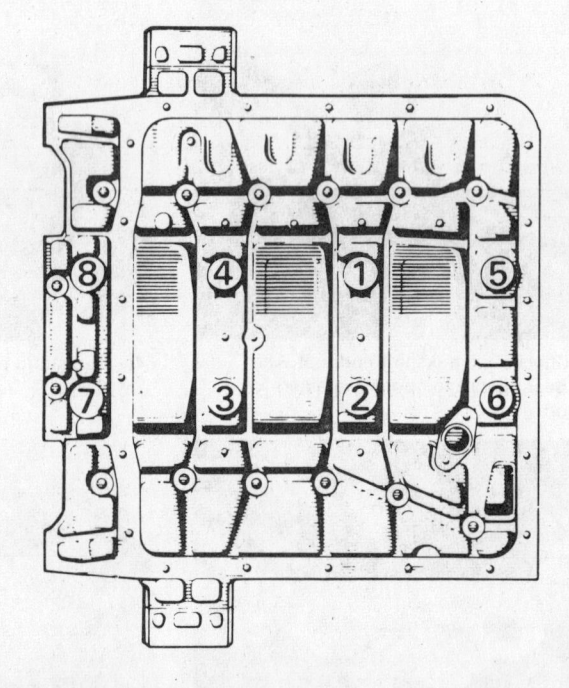

Fig. 1.70 Main bearing cap nut tightening sequence (Sec 64)

Flywheel/driveplate
35 On manual gearbox models if the spigot bearing was removed or if a new crankshaft has been fitted, the spigot bearing must be fitted into position before the flywheel is bolted to the crankshaft mounting flange.
36 Grease the bearing with heat-resistant grease, but do so sparingly.
37 Clean out the bearing housing in the end of the crankshaft, though this should already be clean if the engine has been fully dismantled and thoroughly cleaned.
38 Place the bearing squarely in the flange and tap if firmly home into position.
39 New flywheel or driveplate retaining bolts must always be used when refitting to the crankshaft.
40 Locate the flywheel or driveplate into position on the crankshaft rear flange so that the respective bolt holes are aligned. In the case of the driveplate a spacer washer must be fitted on each side of the plate as shown (Fig. 1.71).
41 Smear the retaining bolt threads with thread locking compound before fitting. Tighten the bolts progressively to the specified torque.
42 Check the valve clearances (Section 39), then fit the rocker covers using new gaskets.

Chapter 1 Engine 101

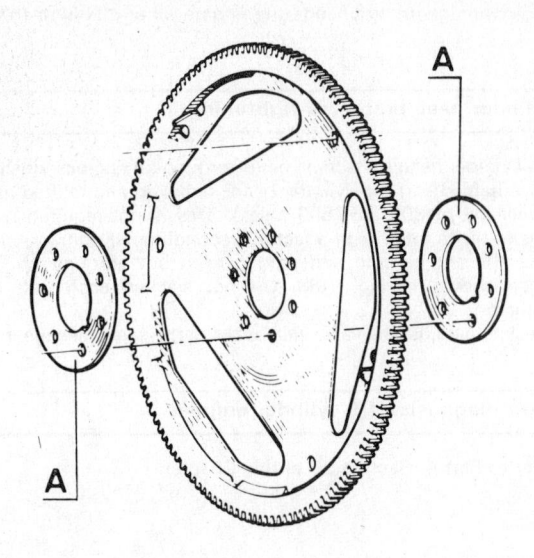

Fig. 1.71 Driveplate with spacer (A) on each side (Sec 64)

plate splines and finally with the flywheel bush, it may be necessary to turn the gearbox slightly or to raise or lower it fractionally, but on no account allow the weight of the gearbox to hang upon the clutch assembly while the primary shaft is passing through it, and support it at all times.
4 Refit the bellhousing attachment bolts. Refit the starter motor and its heat shield at this stage, rather than when the engine is in the car.

Refitting

5 Reverse the operations described in Section 55, tighten all fixings to specified torque and apply sealant to the threads of the reverse interlock cable union before screwing it into the transmission casing.
6 Fill the cooling system, transmission and engine.

68 Engine – refitting (automatic transmission in car)

1 The operations are a reversal of those described in Section 56, but observe the following points.
2 Grease the torque converter location in the crankshaft rear flange.
3 Make sure that the top bellhousing bolts are in position before the engine is offered to the automatic transmission.
4 Tighten all bolts and nuts to specified torque.
5 Fill the engine with coolant and oil.

65 Engine ancillary components – refitting

1 The refitting of all the ancillary equipment to the engine, before it is itself refitted, is mainly a matter of following in reverse order the procedure laid down in Section 59.
2 However, there are one or two points that should be mentioned before the refitting procedure begins. Firstly, when refitting the exhaust manifolds, always use new gaskets and make sure they are the correct way round. When refitting the right-hand manifold also refit the engine dipstick guide tube. Where the engine only has been removed, the exhaust manifold on the left-hand cylinder bank is best fitted after the engine is installed and the starter motor located in its transmission housing aperture, as access is restricted.
3 Do not refit the oil filter until after the starter motor is fitted.
4 Refer to Chapter 4 for details of fitting the distributor.
5 Locate new O-rings for the cooling pipe flange connections to the cylinder block on each side.
6 Refit the Y-pipe and its associated coolant hose connections.
7 The coolant pump and its associated pipe fittings can now be fitted to the front end of the cylinder block. Ensure that all the hose connections are secure, also the pump mounting bolts.
8 Refit the cooling hoses at the rear of the cylinder block.
9 Make sure that the clutch driven plate has been centralised (Chapter 5).
10 Tension the drivebelts for the alternator and coolant pump.

66 Engine – refitting (manual transmission in car)

1 The operations are a reversal of the removal operations described in Section 53.
2 Refill and bleed the cooling system (Chapter 2).
3 Fill the engine with oil.
4 Check the adjustment of the accelerator cable.

67 Engine/manual transmission – reconnection and refitting

Reconnection
1 Check that the release bearing and lever are correctly located and lubricate the clutch shaft splines sparingly with grease.
2 Adequately support the engine and gearbox at the same relative height so that the gearbox primary shaft will slide horizontally into the splined hub of the friction disc.
3 The help of an assistant to keep the engine still will be useful. Keep the gearbox quite level and push it into engagement with the engine. In order to engage the primary shaft of the gearbox with the driven

69 Engine/automatic transmission – reconnection and refitting

1 Reconnection and refitting operations are a reversal of those described in Section 57.
2 Fill the engine and final drive with their correct grades of oil.
3 Fill and bleed the cooling system.
4 Adjust the governor cable as described in Chapter 7.

70 Initial start-up after major overhaul

1 Make sure the battery is fully charged – it may have to work hard before the engine starts. Check oil and coolant levels.
2 On Turbo models, as the turbocharger oil pipes will have been disconnected, it is essential to prime the oil system.
3 To do this, disconnect the junction block (A) in Fig. 1.72 of the ignition power module.

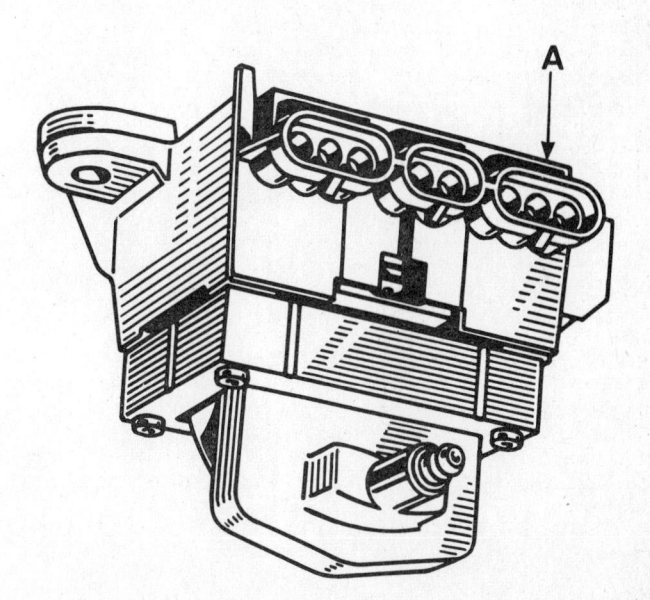

Fig. 1.72 Ignition power module on Turbo models (Sec 70)

A   *Junction block*

4 Disconnect the turbocharger oil inlet tube and fill with engine oil.
5 Operate the starter motor until oil is seen to flow through the inlet pipe. Reconnect the pipe and the junction block.
6 Allow the engine to idle for 30 seconds and switch off (to re-establish oil flow to the turbocharger).
7 Start the engine on all models.
8 As soon as the engine fires and runs, keep it going at a fast tickover only (no faster) and bring it up to normal working temperature.
9 As the engine warms up, there will be odd smells and some smoke from parts getting hot and burning off oil deposits. Look for water or oil leaks which will be obvious if serious, particularly as the engine is clean.
10 When the engine running temperature has been reached, adjust the idling speed as described in Chapter 3.
11 Do not forget that although the old pistons and crankshaft may still have been used, new bearing shells will have been fitted. Treat is as a new engine and run-in at reduced revolutions for at least 500 miles (800 km).
12 After the first 500 miles or so, retighten the cylinder head bolts as described in the next Section. If new bearings etc have been fitted, it will be advantageous to change the engine oil and filter at the same time.

71 Cylinder head bolts – re-tightening

1 The cylinder head bolts must be re-torqued on engines which have been overhauled and the cylinder heads removed and refitted after an initial mileage of 800 km (500 miles). This is not required on new vehicles or those fitted with a factory-reconditioned unit.
2 Have the engine cold and tighten each bolt through 45° in the sequence shown in Fig. 1.38. Do not slacken each bolt before re-tightening.
3 Check and adjust the valve clearances at the same mileage interval.

72 Fault diagnosis – V6 cylinder engines

Refer to Part A, Section 36 of this Chapter.

Chapter 2 Cooling, heating and air conditioning

Contents

Air conditioning system – description and precautions	14
Air conditioning system – maintenance	15
Air conditioning system (1986 on) – description, removal and refitting of components	17
Air conditioning system (up to 1986) – removal and refitting of main components	16
Coolant mixtures	4
Coolant pump – removal and refitting	10
Coolant temperature switch – testing	9
Cooling system – draining, flushing and refilling	3
Description	1
Drivebelts – tensioning, removal and refitting	8
Fault diagnosis – cooling, heating and air conditioning	18
Heater – removal and refitting	12
Heating and ventilation system – description	11
Maintenance	2
Radiator – removal, renovation and refitting	6
Radiator fan and switch – removal and refitting	7
Rear passenger compartment heater	13
Thermostat – removal, testing and refitting	5

Specifications

System type .... No-loss, semi-sealed with expansion tank, thermostat and belt-driven coolant pump. Electric radiator fan. Heater with ducts to rear passenger compartment. Air conditioner as factory-fitted option

Thermostat
Opens:
- All models but Turbo .... 88°C (190°F)
- Turbo .... 86°C (187°F)

Full open .... 100°F (212°F)
Travel .... 7.5 mm (0.30 in)

Cooling fan thermostatic switch
Cuts in at .... 87 to 97°C (189 to 207°F)
Cuts out at .... 77 to 87°C (171 to 189°F)

Coolant capacity
1995 cc (J6R) engine .... 7.5 litre (13.2 pints)
2165 cc (J7T) engine .... 8.0 litre (14.1 pints)
2458 cc (Z7U) engine .... 9.8 litre (17.2 pints)
2664 cc (Z7V) engine .... 9.5 litre (16.7 pints)

For automatic and air conditioned models add 0.28 litre (0.5 pint) to capacities shown

Antifreeze concentration
For protection down to:
- −23°C (−10°F) 35%
- −40°F (−40°F) 50%

Torque wrench settings
	Nm	lbf ft
Cylinder block drain plug	17	13
Thermostat housing cover bolts	13	10
Coolant pump mounting bolts	15	11
Alternator mounting bolt	40	30
Alternator adjuster link bolt	20	15

Chapter 2 Cooling, heating and air conditioning

1 Description

The system is of pressurised semi-sealed type with the inclusion of an expansion bottle to accept coolant displaced from the system when hot and to return it when the system cools.

Coolant is circulated by thermosyphon action and is assisted by means of the impeller in the belt-driven coolant pump.

A thermostat is fitted in the outlet of the coolant pump. When the engine is cold, thermostat valve remains closed so that the coolant flow which occurs at normal operating temperatures through the radiator matrix is interrupted.

As the coolant warms up, the thermostat valve starts to open and allows the coolant flow through the radiator to resume.

The engine temperature will always be maintained at a constant level (according to the thermostat rating) whatever the ambient air temperature.

The coolant circulates around the engine block and cylinder head and absorbs heat as it flows, then travels out into the radiator to pass across the matrix. As the coolant flows across the radiator matrix, air flow created by the forward motion of the car cools it and it returns via the bottom tank of the radiator to the cylinder block. This is a continuous process, assisted by the coolant pump impeller.

All models are fitted with an electric cooling fan which is actuated by the thermostat switch according to coolant temperature.

The car interior heater operates by means of coolant from the engine cooling system.

The carburettor is fitted with coolant connections to permit coolant from the cooling system to circulate to the carburettor base. On four-cylinder models coolant also circulates to the automatic choke operating mechanism.

On models with automatic transmission, a fluid cooler is built into the cooling system circuit.

An auxiliary air valve is fitted in connection with the fuel injection cold start system (Chapter 3).

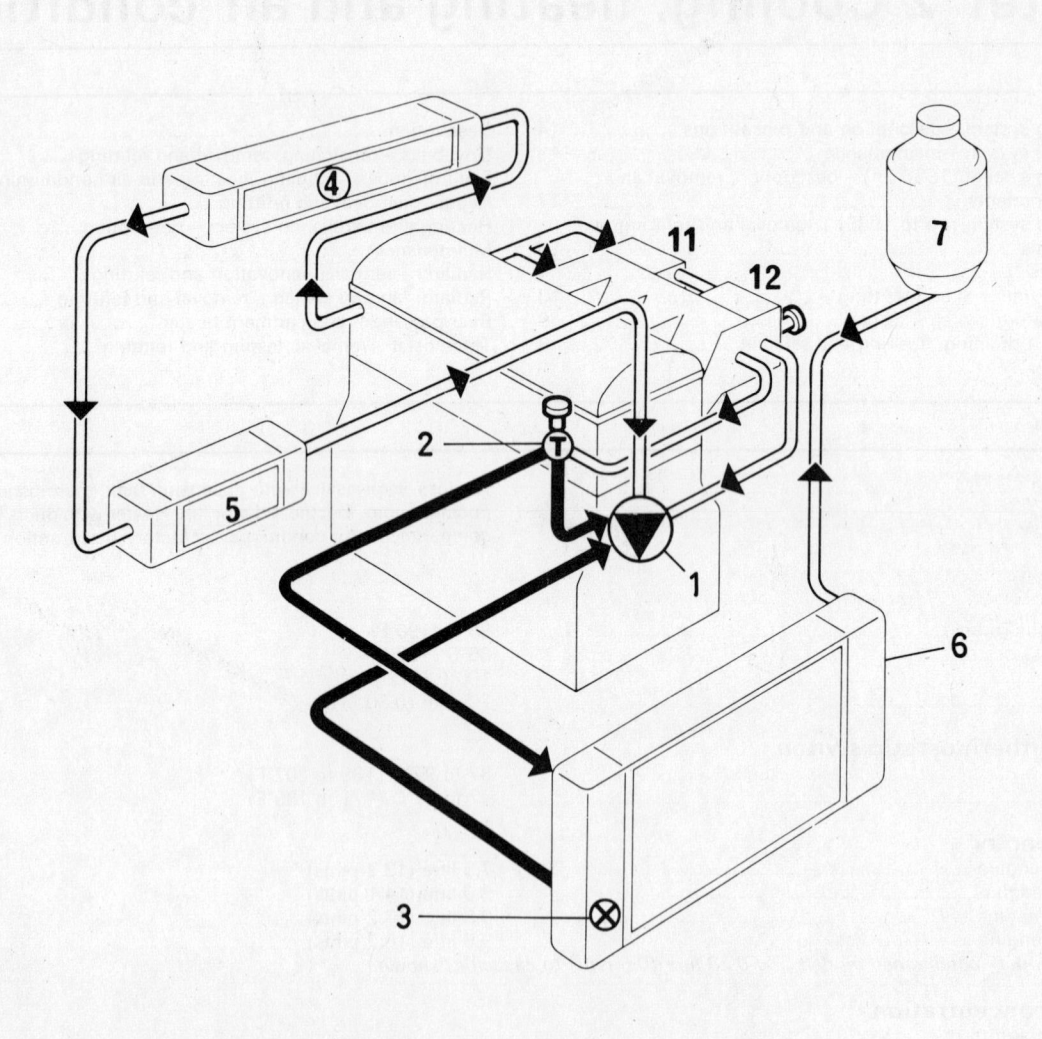

Fig. 2.1 Cooling system on four-cylinder models (Sec 1)

1 Coolant pump
2 Thermostat
3 Fan thermal switch
4 Heater matrix
5 Automatic transmission fluid cooler
6 Radiator
7 Expansion bottle
11 Carburettor base flange (1995 cc)
12 Automatic choke (1995 cc)

1 Coolant pump
2 Thermostat
3 Expansion tank
4 Heater matrix
5 Automatic transmission fluid cooler
6 Fan thermal switch
7 Radiator
8 Auxiliary air valve
9 Engine oil cooler
10 Calibrated orifice
A Non-turbo models
B Turbo models

Fig. 2.2 Cooling system on six-cylinder models (Sec 1)

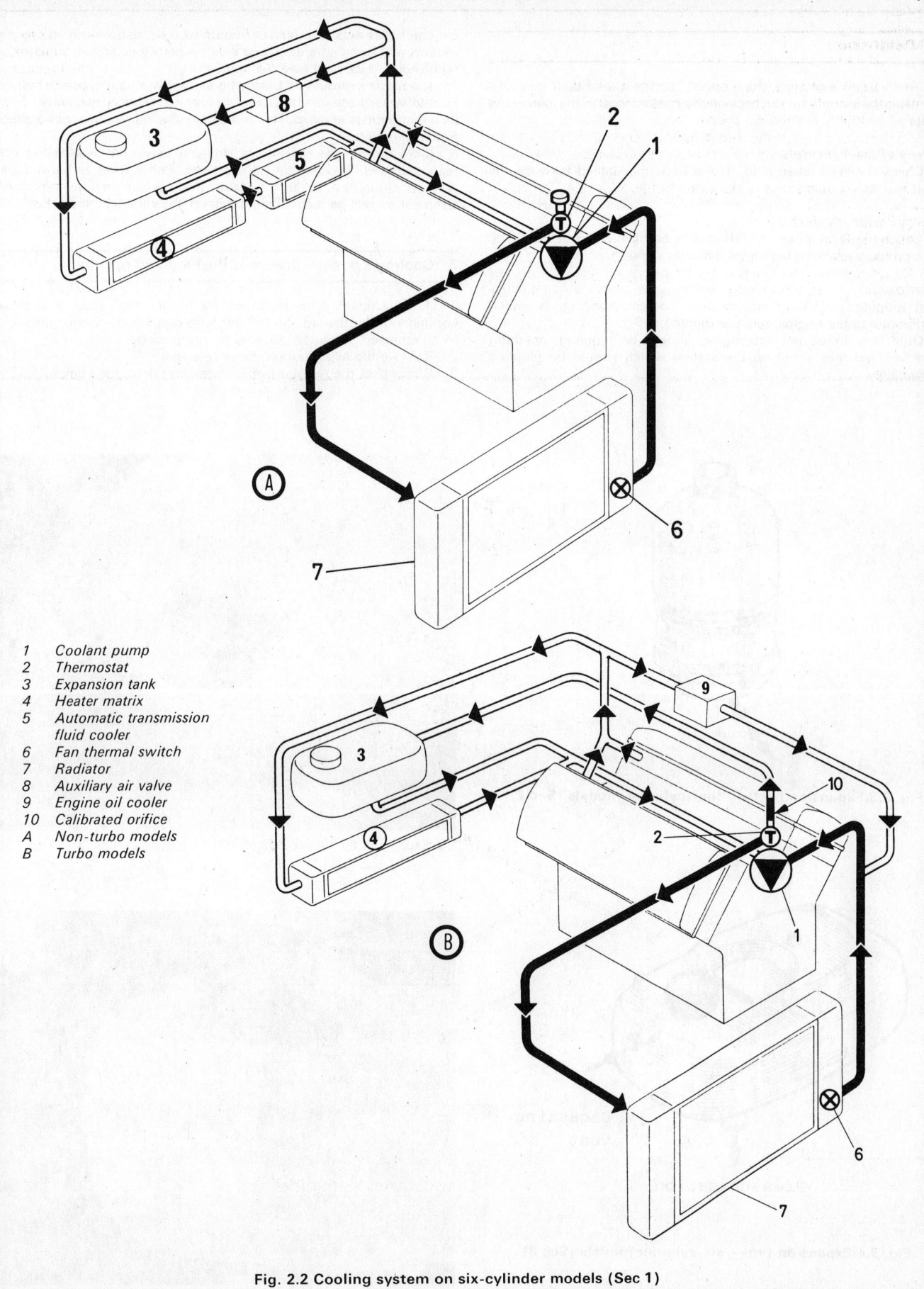

Chapter 2 Cooling, heating and air conditioning

2 Maintenance

1 Apart from renewing the coolant at the prescribed intervals, maintenance is confined to checking the coolant level in the expansion bottle.

Four-cylinder models
2 Check the level when cold. It should be maintained between the MINI and MAXI marks on the expansion bottle.

Six-cylinder models
3 Check the level when cold. It should be maintained at the FULL mark on the expansion tank level indicator (photo).

All models
4 If topping-up is required, use only coolant made up in similar proportions to the original mixture (photo).
5 Only very infrequent topping-up should be required, anything more will indicate a leak in the system which should be rectified immediately.
6 The hoses and their clamps should also be inspected regularly for security and good condition, and the drivebelt checked and adjusted or renewed as necessary (see Section 8).
7 The radiator should be brushed periodically, or a compressed air jet or cold water hose directed onto it to remove insects and leaves from its cooling tubes and fins. This will ensure the maximum cooling effect from the radiator.
8 Blanking off the radiator or grille in very cold weather is not recommended. Provided the correct thermostat is used with a suitable antifreeze mixture in the system, the correct engine operating temperature will be maintained whatever the climatic conditions.

3 Cooling system – draining, flushing and refilling

1 If the coolant is to be saved for further use, place a suitable container under the radiator bottom hose connection. If the coolant is to be renewed, it may be allowed to run to waste.
2 Remove the expansion bottle or tank cap.
3 Disconnect the radiator bottom hose and drain the coolant.

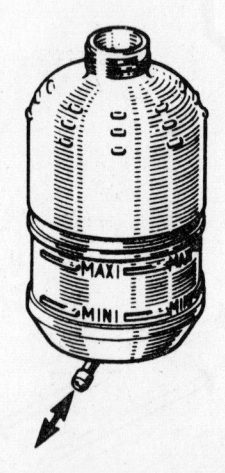

Fig. 2.3 Expansion bottle – four-cylinder models (Sec 2)

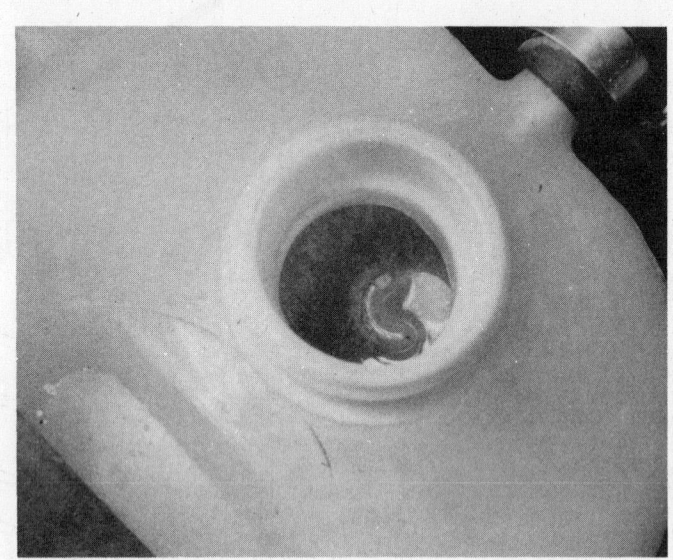

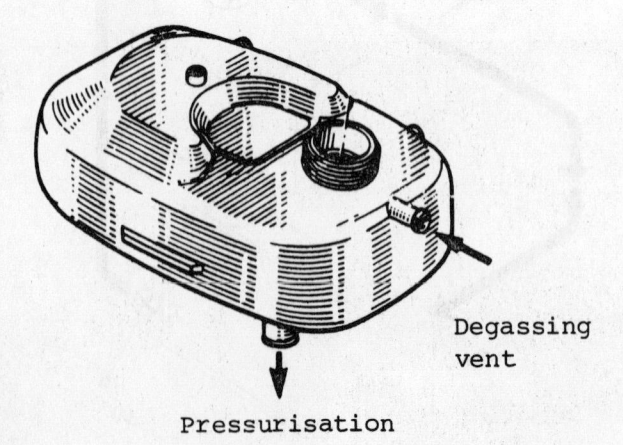

Fig. 2.4 Expansion tank – six-cylinder models (Sec 2)

2.3 Expansion tank level indicator

2.4 Filling expansion tank

Chapter 2 Cooling, heating and air conditioning

4 If the coolant has been renewed regularly, then the system may be refilled immediately. If the cooling system has been neglected, remove the cylinder block drain plug (photo) and place a cold water hose in the radiator top hose. Flush the system through until the water comes out clear from the drain plug hole and the radiator hose. If contamination is severe, the radiator may have to be removed and cleaned out; see Section 6.
5 Reconnect the radiator hose and tighten its clip. Tighten the cylinder block drain plug if removed.
6 Open the system bleed screws.
7 On four-cylinder models, release the expansion bottle and have an assistant hold it as high as possible above the engine.
8 Fill the system slowly through the expansion bottle or tank opening. Use only coolant made up as an antifreeze mixture as described in the next Section.
9 As soon as coolant is ejected from the bleed screws, close them. Relocate the expansion bottle.
10 Fill the expansion bottle to MAXI or the tank to FULL mark.
11 Fit the expansion bottle or tank cap.
12 Start and run the engine for fifteen minutes at a fast idle until the cooling fan cuts in.
13 Switch off the engine and allow it to cool completely, preferably overnight.
14 Top up the expansion bottle or tank to the MAXI or FULL marks again using specified antifreeze mixture.

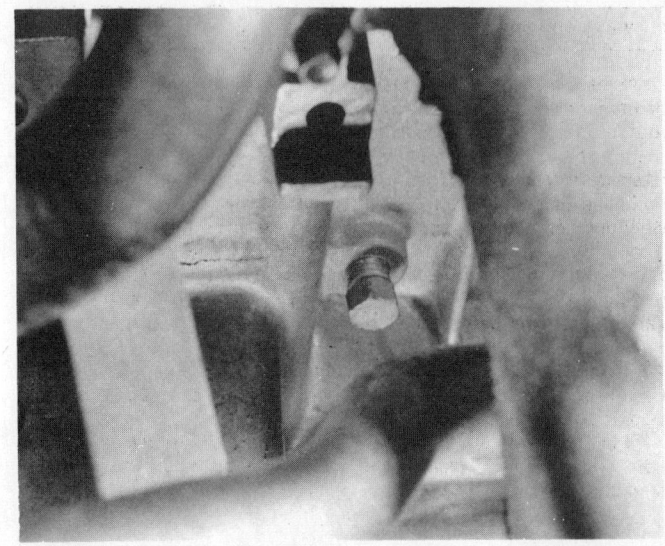

3.4 Cylinder block drain plug

4 Coolant mixtures

1 It is essential that an approved type of antifreeze is employed, in order that the necessary antifreeze and anticorrosion proportions are maintained.
2 The radiator matrix of some models is made of aluminium and it is essential to use an antifreeze product suitable for this material.
3 Whilst the life of the coolant originally used in the vehicle is stated to be 3 years or 40 000 miles (60 000 km), owners are recommended to consider removing the coolant yearly to ensure that all the essential properties of the solution are fully maintained.
4 Make up the solution, ideally using distilled water or rain water, in the proportions necessary to give protection in the prevailing climate. Percentages of antifreeze necessary are usually found on the container. Do not use too low a percentage of antifreeze, or the anticorrosion properties will not be sufficiently effective.
5 If it is suspected that the coolant strength is unsatisfactory, a check may be made using a hydrometer. Employ the instrument as instructed by the manufacturer, and using the correction tables normally supplied. If the protection is found to be insufficient, drain off some of the coolant and replace it with pure antifreeze. Recheck the coolant with the hydrometer.
6 Even in territories where the climate does not require the use of antifreeze, never use plain water in the cooling system; always add a suitable corrosion inhibitor.

5.1 Thermostat housing (four-cylinder models)

5 Thermostat – removal, testing and refitting

Four-cylinder models
1 The thermostat housing is located at the timing belt end of the cylinder head with a coolant hose connecting to the radiator. The unit is separate from the coolant pump (photo).

Six-cylinder models
2 The thermostat is located between the cylinder banks at the timing belt end of the engine with a coolant hose connecting to the radiator (photo).
3 The thermostat housing is an integral part of the coolant pump on six-cylinder engines.

All models
4 Drain enough coolant to reduce the level in the system below the thermostat housing. About one third of the total capacity of the system should be sufficient.
5 Remove the thermostat housing cover bolts and tie the cover up out of the way with hose still attached.

5.2 Thermostat housing (six-cylinder models)

Chapter 2 Cooling, heating and air conditioning

6 Withdraw the thermostat. If it is stuck tight, do not lever it by means of its bridge piece but cut around its periphery using a sharp knife (photo).
7 Clean away all the old gasket from the mating faces of the thermostat housing and cover (photo).
8 Test the thermostat by suspending it in a pan of water, together with a thermometer. Commence warming the water, watch when the thermostat begins to open, and check the temperature. Compare the opening temperature with the temperature stamped on the thermostat.
9 Transfer the thermostat to cold water, and check that it closes promptly.
10 If the thermostat does not operate as outlined, it should be renewed.
11 Fit the thermostat into its housing, locate a new gasket and bolt on the cover.
12 Top up the cooling system and bleed as described in Section 3.

6 Radiator – removal, renovation and refitting

1 Open the bonnet and remove the radiator grille (Chapter 11) and the headlamp wipers (Chapter 12). On four-cylinder models remove the radiator blanking plate. This plate should be permanently removed if the car is used regularly for trailer towing (photo).
2 Disconnect the battery and the leads from the cooling fan and the thermal switch.
3 Drain the cooling system as described in Section 3.
4 Disconnect all the radiator hoses (photo).
5 On models with air conditioning, release the condenser mountings but on no account uncouple the refrigerant pipes.
6 Remove the radiator upper retaining brackets remove the radiator/fan assembly from the engine compartment complete with side deflectors (photos).

5.6 Removing thermostat

5.7 Thermostat housing with cover and thermostat removed (four-cylinder models)

6.1 Radiator blanking plate and clip (four-cylinder models)

6.4 Releasing coolant hose clip

6.6A Unscrew radiator top bracket bolt

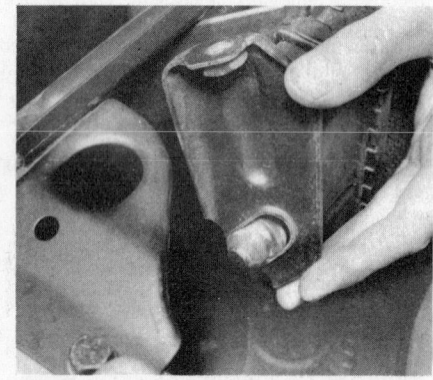

6.6B Radiator top bracket

6.6C Removing radiator (six-cylinder engine)

6.6D Radiator side deflector (arrowed)

6.6E Radiator bottom fixing grommet

Chapter 2 Cooling, heating and air conditioning

7 If the radiator is clogged, reverse flush it with a cold water hose. In severe cases of contamination, use a radiator cleaner but strictly in accordance with the manufacturer's instructions. The product must be suitable for aluminium radiators.
8 If the radiator is leaking, a temporary repair may be made using fibreglass or a similar product. Soldering is not possible as the matrix is aluminium and the tanks plastic, so a permanent solution is to renew the radiator or exchange it for a reconditioned unit.
9 Do not leave the radiator empty for more than 48 hours or the aluminium will start to corrode.
10 Refitting is a reversal of removal.
11 Refill and bleed the cooling system.

7 Radiator fan and switch – removal and refitting

1 The fan assembly is riveted to the radiator and its removal will require withdrawal of the radiator and drilling out the pop rivets (photo).
2 The thermal switch which regulates the operation of the electric cooling fan is fitted into the radiator (photo). Ideally, tests to the switch should be carried out after it has been removed from its location. If this is done, the cooling system will have to be drained and refilled, but the following alternative method may be submitted (photo).
3 Connect a test light to the fan switch. Where the engine is cold (ie switch contacts open), it should not light up.
4 Start the engine and blank off the radiator to ensure quick warming up.
5 Hold a thermometer in contact with the radiator. When the temperature reaches the cut-in point (see Specifications), the lamp should light.
6 Switch off the engine and allow the temperature of the radiator to drop. When the cut-out temperature is reached (see Specifications), the lamp should go out.
7 If the switch does not operate within the test limits, renew it.
8 If the switch does operate correctly, then the fault must lie in the fan assembly or the connecting wiring or relay.
9 Refitting is a reversal of removal. Use a new seal when screwing in the thermal switch and avoid overtightening it. **Note:** On models equipped with air conditioning, twin fans are mounted on the radiator as a means of dissipating the additional heat generated by condenser ahead of the radiator.

8 Drivebelts – tensioning, removal and refitting

1 Vee, multi-toothed or multi-grooved belts are used to drive the coolant pump, alternator, and air conditioner compressor (where fitted) from the crankshaft pulley (photo). The power steering pump on four-cylinder and Turbo models is driven from the crankshaft pulley but on other six-cylinder models it is driven from the rear end of the left-hand camshaft.
2 Check the belts regularly for cuts or fraying, and if evident, renew them.
3 At the intervals specified in Routine Maintenance, check the deflection of the belts at the points indicated in Figs. 2 – 5 to 2 – 10 (photo). If the deflection exceeds 4.5 mm (0.18 in) retension the belt (cold).
4 To tension the crankshaft pulley driven belts (except on Turbo models and the air conditioner belt on other six-cylinder engines) release the driven accessory adjuster link and mounting bolts and move the accessory (photos). Retighten the bolts.
5 To remove a drivebelt, release the driven accessory adjuster and mounting bolts and move the accessory fully in towards the engine. With the drivebelt slack, slip it off the pulleys. If there is any difficulty doing this, turn the crankshaft pulley and at the same time press the belt against the pulley rim when it will ride up and over the rim.
6 To tension the air conditioner compressor, or power steering pump drivebelt on non-Turbo six-cylinder models and the main drivebelt on Turbo models, adjust the position of the tensioner pulley. If the belt is to be removed or refitted, slacken the tensioner pulley right off.
7 It will be obvious that before an inner belt can be removed the outer one will have to be taken off.

7.1 Radiator fan/motor assembly

7.2 Radiator thermostatic switch

8.1 Drivebelts on 2165 cc engine

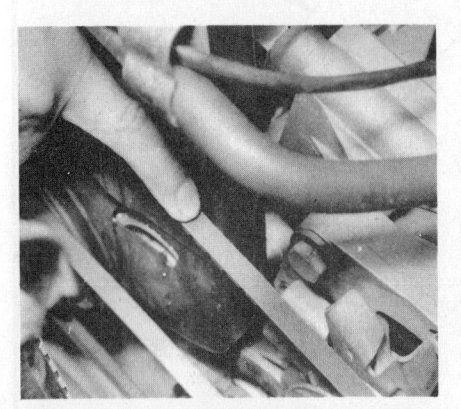

8.3 Checking drivebelt deflection

8.4A Alternator belt tensioner (V6 engine)

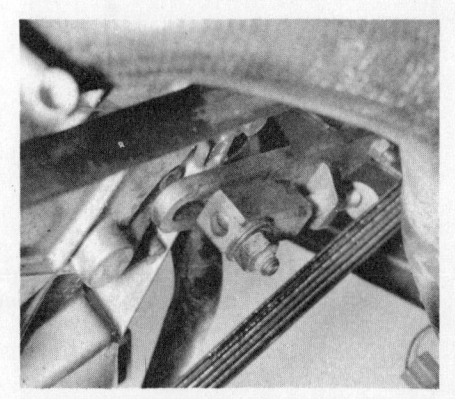

8.4B Alternator adjuster link block (V6 engine)

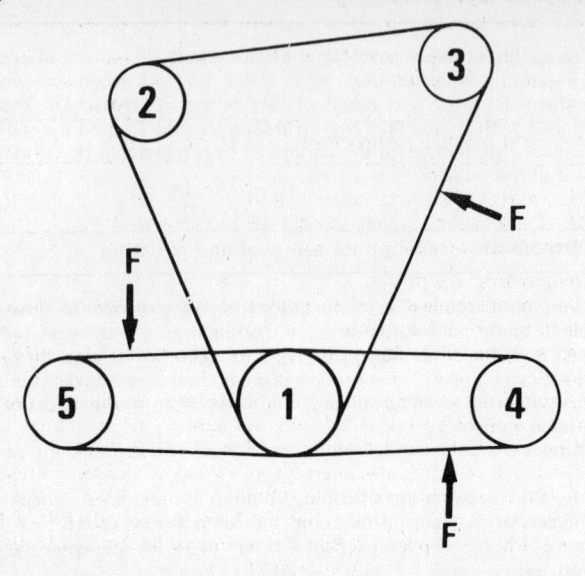

Fig. 2.5 Drivebelt arrangement – four-cylinder models (Sec 8)

1 Crankshaft pulley
2 Coolant pump
3 Alternator
4 Power steering pump
5 Air conditioner compressor
F Belt deflection testing points

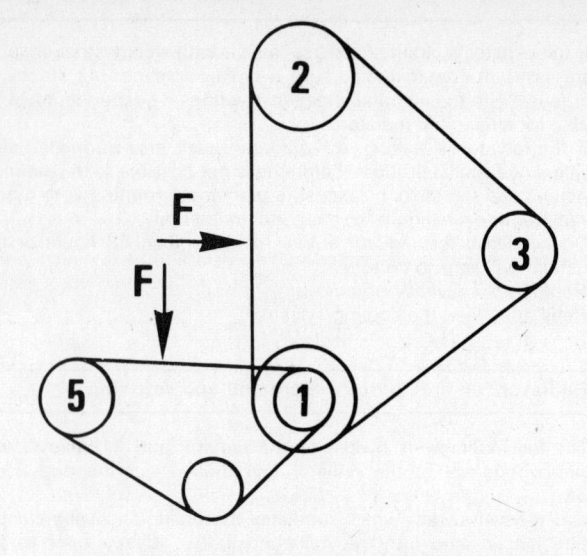

Fig. 2.6 Drivebelt arrangement – six-cylinder models (typical) (Sec 8)

1 Crankshaft pulley
2 Coolant pump
3 Alternator
5 Air conditioner compressor
F Belt deflection testing points

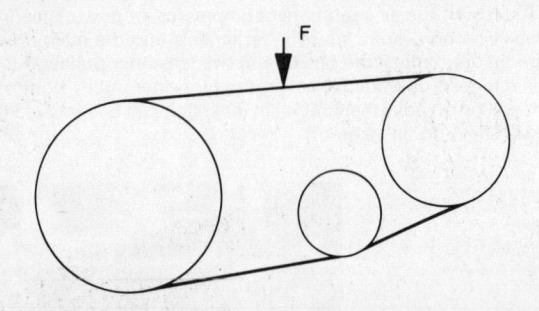

Fig. 2.7 Camshaft driven power steering pump drivebelt (six-cylinder models except Turbo) (Sec 8)

F Belt deflection testing point

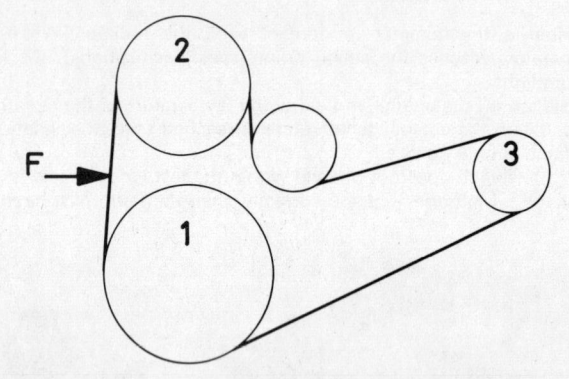

Fig. 2.8 Turbo drivebelt from crankshaft pulley (1) to coolant pump (2) and alternator (3) (Sec 8)

F Belt deflection testing point

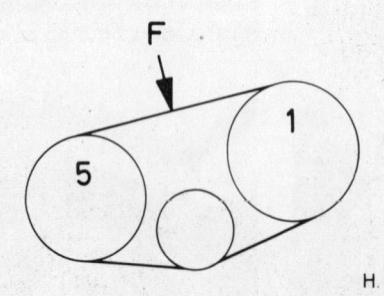

Fig. 2.9 Turbo drivebelt from crankshaft pulley (1) to power steering pump (5) (Sec 8)

F Belt deflection testing point

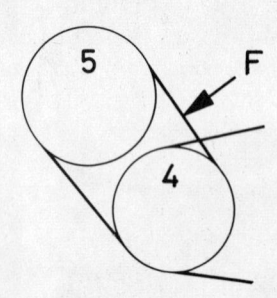

Fig. 2.10 Turbo drivebelt arrangement with power steering pump (5) driven from air conditioner compressor pulley (4) (Sec 8)

F Belt deflection testing point

Chapter 2 Cooling, heating and air conditioning

8 Fit the new belt, again turning the crankshaft pulley if the belt is difficult to prise over the pulley rim.
9 Tension the new belt in accordance with the following deflection table.

	4-cylinder	6-cylinder
Alternator/coolant pump	3.5 mm (0.14 in)	3.0 mm (0.12 in)
Power steering pump	3.5 mm (0.14 in)	3.0 mm (0.12 in)
Air conditioner compressor	4.0 mm (0.16 in)	4.0 mm (0.16 in)

10 The belts should be tensioned preferably when the engine is cold. If it must be done hot, add 1.0 mm (0.39 in) to the foregoing figures.

9 Coolant temperature switch – testing

1 Where an engine coolant transmitter unit is suspected of being faulty, the only simple way to test it is by substitution.
2 The transmitter unit is located in the front of the thermostat housing.
3 Removal necessitates partial draining of the cooling system, whilst the gauge removal necessitates removal of the instrument panel (see Chapter 12).
4 It is not possible to repair the gauge or sender unit and they must therefore be renewed if faulty.
5 A simple test of the gauge may be made by touching the sender unit wire to earth (ignition on) whilst an assistant observes the gauge. The gauge should read 'hot'; if not, either there is a break in the wiring or the gauge itself is at fault.

10 Coolant pump – removal and refitting

1 Disconnect the battery negative lead.
2 Drain the cooling system as described in Section 3.
3 Remove the radiator.

Four-cylinder models
4 Disconnect the lead from the coolant temperature switch, also the hoses from the coolant pump.
5 Slacken the drivebelt and unbolt and remove the coolant pump pulley.
6 Unscrew the coolant pump fixing bolts, tap the pump gently to break its gasket seal and remove the pump, at the same time compressing the timing belt tensioner plunger spring.
7 The pump is of disposable type and if worn or leaking it must be renewed, no repair being possible (photo).
8 Clean away all old gasket material from the cylinder block and locate a new one in position. Bolt the pump into position, holding the timing belt tensioner spring compressed (photos).
9 Refit the pulley, and the drivebelt. Tension the belt (Section 8) (photo).
10 Fill and bleed the cooling system. Reconnect the battery.

10.7 Rear view of coolant pump (four-cylinder engine)

10.8A Fitting coolant pump (four-cylinder engine)

10.8B Tightening coolant pump bolt

10.9 Tightening coolant pump pulley bolt

112   Chapter 2 Cooling, heating and air conditioning

Six-cylinder models
11 Remove the air cleaner. Unbolt and remove the U-shaped manifold from the front of the air distributor housings (see Section 7, Chapter 4).
12 Disconnect the throttle rod balljoints (the lower one is retained by a wire clip).
13 Unbolt the throttle reel and place it to one side with throttle cable attached.
14 Slacken the alternator drivebelt and remove it. Remove the radiator fan and shroud.
15 Disconnect the coolant hoses from the coolant pump, noting particularly those at the rear of the pump, the clips for which are difficult to reach (photo). Disconnect the temperature switch leads.
16 Unscrew the coolant pump fixing bolts which can be removed using a ring spanner inserted behind the pulley (photo). Remove the pump.
17 A special threaded extractor is required to remove the pump pulley but a three-legged extractor or press are alternatives.
18 The pump is of disposable type and if it is faulty or leaking, renew it as it cannot be repaired. The new pump is supplied without the rear cover, thermostat, pulley, thermostat housing cover and temperature sender switch, so remove these items from the old unit.
19 Use new gaskets when reassembling and refitting, also new sealing rings on the front U-manifold of the air distributor housings when they are refitted.
20 Fit and tension the drivebelt, refill the system with coolant and reconnect the battery.

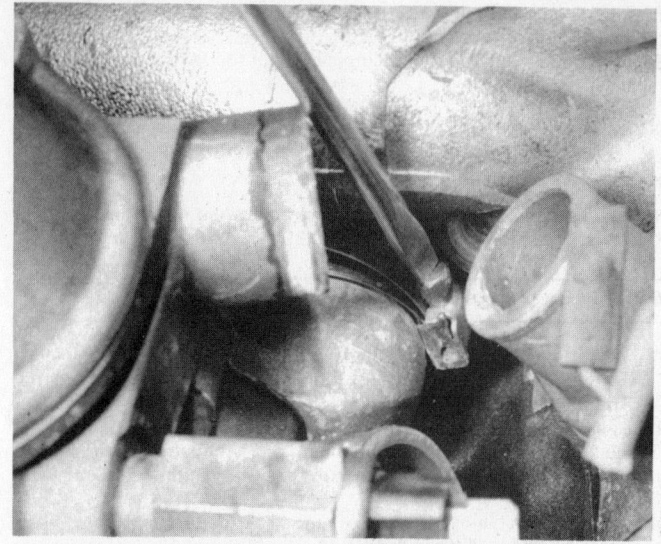

11 Heating and ventilation system – description

1 The heating system utilises the heat from the engine cooling system. Temperature control is not arranged by means of a coolant flow valve but by hot and cold air blending. This means that the heater matrix is maintained at a continuously high temperature and the car interior heating is immediately effective whenever it is switched on.
2 Airflow, which is of fresh air inducted type, is directed to front and rear occupants and to the windscreen and side windows.
3 A four-speed blower fan controls the airflow through the system and this is mounted under the scuttle grille below the windscreen.
4 The thumb wheel type controls on the facia panel operate by vacuum through a vacuum chamber located in the right-hand recess below the windscreen wiper arm.
5 Cables are used for air temperature and demisting control purposes.
6 Stale air from the car interior is exhausted through the rear doors.

12 Heater – removal and refitting

1 Disconnect the battery.
2 Partially drain the cooling system.
3 Disconnect the heater hoses at their engine pipe and stub (photos).
4 Refer to Chapter 11 and remove the facia panel. Disconnect the heater wiring plugs and vacuum hoses.
5 Remove the four heater unit mounting screws and withdraw the heater from the car interior, complete with control panel and cables. Protect the carpets against coolant spillage.
6 With the heater removed, the control cables may be disconnected and the heater casing sections unclipped for access to the matrix.
7 The blower motor/fan may be removed after withdrawing the scuttle grille (Chapter 11) (photo).
8 If the matrix is clogged, try reverse flushing it as described for the radiator. If it is leaking, renew it; a repair seldom proves satisfactory.
9 Refitting is a reversal of removal but adjust the control cables as shown in Figs. 2.12 and 2.13 by altering the position of the cable conduit clips. Check that all seals and gaskets are good.
10 Reconnect the vacuum hose and refit the facia panel as described in Chapter 11.
11 Refill and bleed the cooling system.

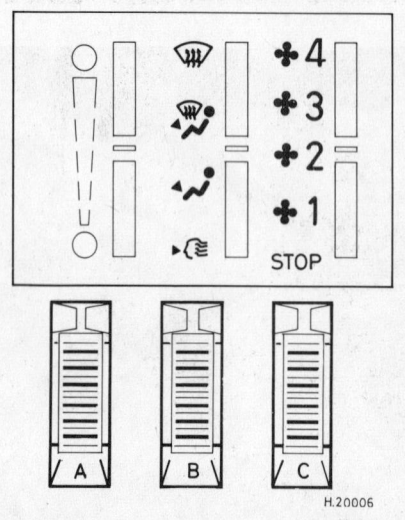

10.15 Releasing hose clip at rear of coolant pump

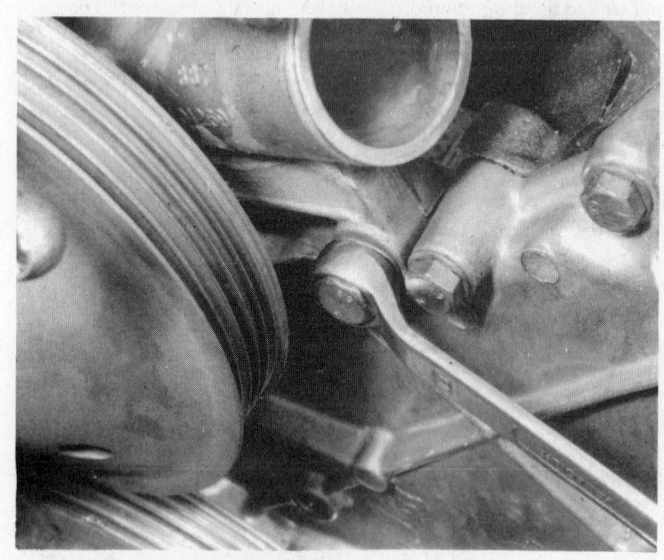

10.16 Unscrewing coolant pump bolt

Fig. 2.11 Heater controls (Sec 11)

A   Air temperature control   C   Air inlet (blower fan)
B   Air distribution control

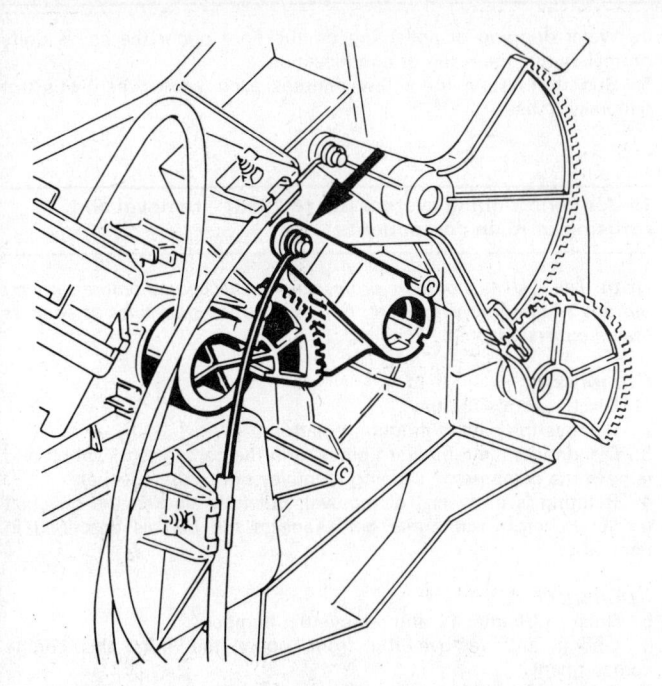

Fig. 2.12 Heater control cable setting – air temperature in cold position (Sec 12)

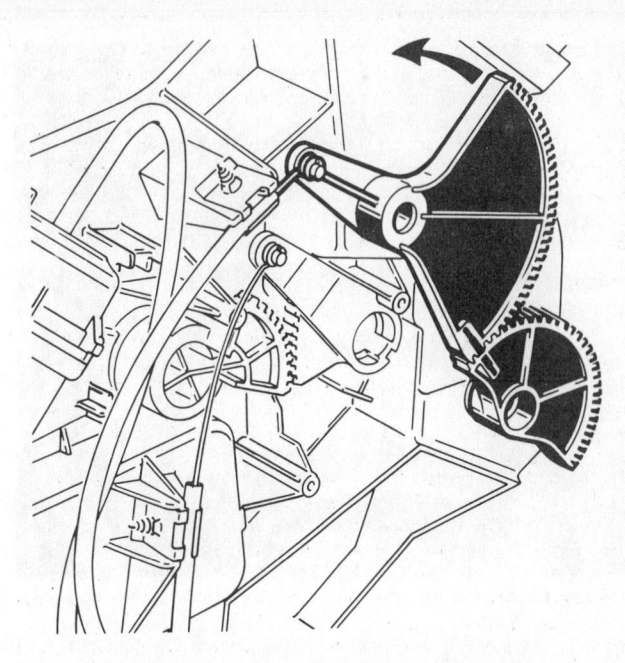

Fig. 2.13 Heater control cable setting – air distribution in windscreen demist position (Sec 12)

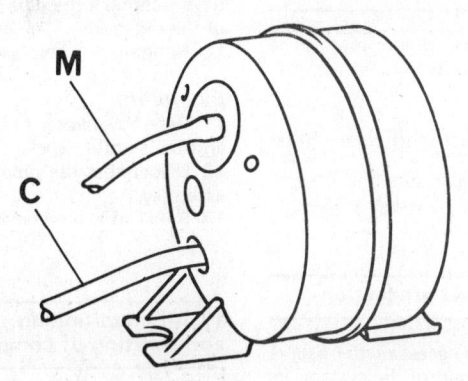

Fig. 2.14 Heater vacuum unit (Sec 12)

M   To brake servo hose   C   To control

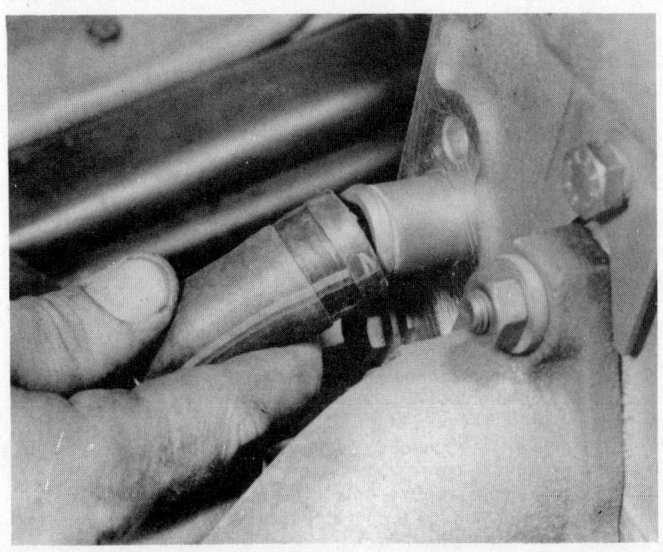

12.3A Disconnecting heater hose from cylinder head

12.3B Disconnecting heater hose

Chapter 2 Cooling, heating and air conditioning

12.7 Heater blower motors

13 Rear passenger compartment heater

1 The rear console incorporates an independent blower motor, the necessary control switches and air distribution ducts.
2 Remove the centre armrest (front seat).
3 Extract the screws which secure the console.
4 Withdraw the console until the blower switch and cigar lighter leads can be disconnected.
5 The air ducts are detachable from the console.

14 Air conditioning system – description and precaution

1 When an air conditioning system is fitted, it is necessary to observe special precautions whenever dealing with any part of the system, its associated components and any items within the engine and heating compartment that necessitate disconnection of the system.
2 The refrigeration circuit contains a liquid gas (Freon) and it is therefore dangerous to disconnect any part of the system without specialised knowledge and equipment. If for any reason the system must be disconnected (engine removal for example), entrust this task to your Renault dealer or a refrigeration engineer.
3 The layout of the air-conditioning system is shown in Fig. 2.15.
4 The refrigerant gas must not be allowed to come in contact with a naked flame or a poisonous gas will be created. Do not allow the fluid to come in contact with the skin or eyes.
5 Whenever the refrigerant lines are disconnected, all pipe lines and openings must be plugged or capped to prevent the entry of moisture.
6 After the system pipe lines have been reconnected, always have the system evacuated and recharged by your dealer or a competent refrigeration engineer.

15 Air conditioning system – maintenance

1 Regularly inspect the condition of hoses and the security of connections.
2 A sight glass is located on top of the receiver/dryer chamber. If bubbles are evident immediately after switching on the system, a fluid leak is indicated.
3 Keep the fins of the condenser free from flies and dirt and keep the compressor drivebelt in good condition and correctly tensioned (Section 8).

4 Water dripping or collecting on the floor under the car is quite normal, and is the result of condensation.
5 Run the system for a few minutes each week when it is not generally in use.

16 Air conditioning system (up to 1986) – removal and refitting of main components

Note: *The system should be professionally discharged before carrying out any of the following work. Cap or plug the pipe lines as soon as they are disconnected.*

Compressor
1 Disconnect the battery.
2 Remove the battery mounting platform.
3 Disconnect the refrigerant pipes from the compressor. Unbolt and remove the compressor, slipping its pulley out of the drivebelt.
4 Refitting is a reversal of removal, but fully slacken the drivebelt tensioner before fitting the belt. Tension the belt as described in Section 8.

Condenser
5 Refer to Chapter 11 and remove the bumper.
6 Unbolt and remove the front upper rail from the engine compartment.
7 Remove the radiator grille (Chapter 11).
8 Remove the cooling fans, supports and cowls.
9 Disconnect the pipe lines and remove the condenser from the front of the radiator.
10 Refitting is a reversal of removal.

Evaporator
11 Refer to Chapter 11 and remove the centre console, the facia panel and the scuttle panel.
12 Disconnect the pipe lines and remove the evaporator/blower fan assembly.
13 Refitting is a reversal of removal.

17 Air conditioning system (1986 on) – description, removal and refitting of components

1 This is a modified system to provide automatic temperature regulation and a recirculating (recycle) mode to maintain a set temperature inside the car during periods of heavy exhaust pollution from other vehicles.
2 Generally, the operations described in the preceding Section apply. Air control flaps are actuated by vacuum motors (Fig. 2.17) and the control panel incorporates a temperature sensor to monitor the passenger compartment temperature level and signal the continuous changes which may be required to maintain the preset temperature level.

Recirculating flap motor
3 Remove the evaporator/blower fan assembly as described in Section 16. Disconnect the battery.
4 Unclip the rear section of the heater casing and remove the motor.
5 When refitting the motor, make sure that when the control panel (facia) indicator is in the recycling mode, the flap is in the correct phase.

Mixing flap micromotor
6 Disconnect the battery.
7 Remove the glove compartment.
8 With the micromotor now accessible, disconnect it and then unbolt and remove it.
9 To refit the motor, reconnect the battery and switch on the ignition. The motor should be connected but not fixed in position.
10 Move the indicator on the facia to the cold (blue) position.
11 Allow the motor to run until it stops and switch off the ignition.
12 Twist the flap toothed sector clockwise and fit the motor in position (Fig. 2.18). Do not overtighten the three fixing screws.

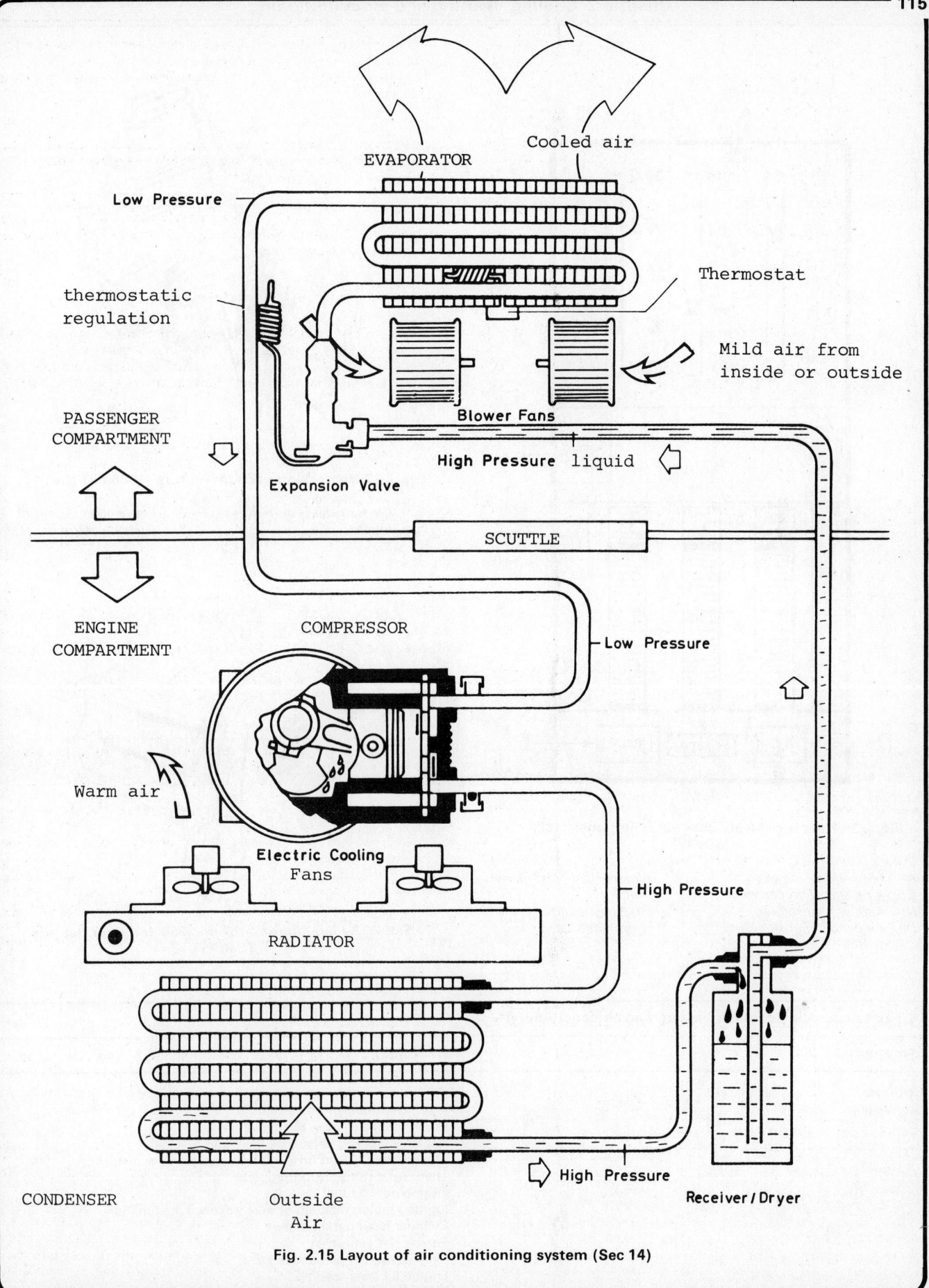

Fig. 2.15 Layout of air conditioning system (Sec 14)

116　　Chapter 2 Cooling, heating and air conditioning

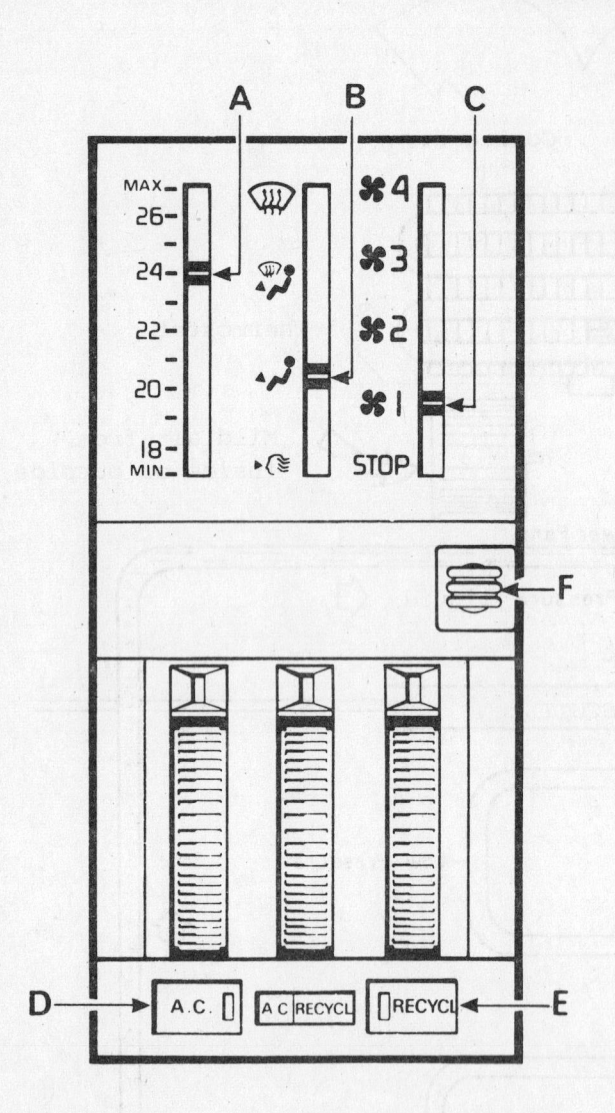

Fig. 2.16 Heater controls with air conditioner (1986 on) (Sec 17)

- A　Temperature control
- B　Air distribution control
- C　Airflow (booster fan)
- D　Air conditioner stop/start switch
- E　Air recirculation stop/start switch
- F　Passenger compartment temperature sensor

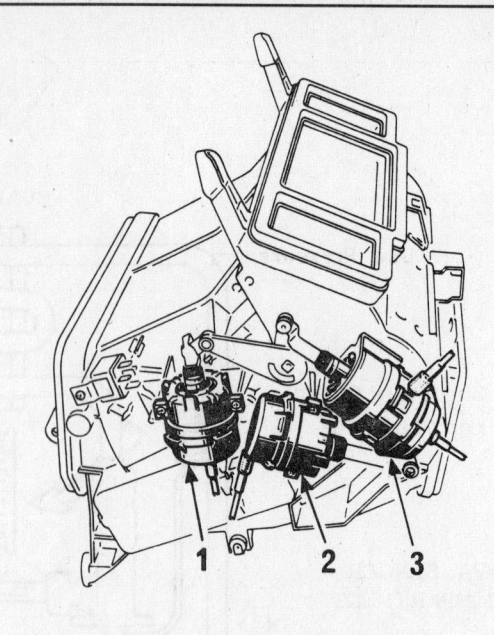

Fig. 2.17 Heater unit vacuum motors (1986 on) (Sec 17)

1　Closes facia vent in demist mode
2　Footwell vent control
3　Demist control

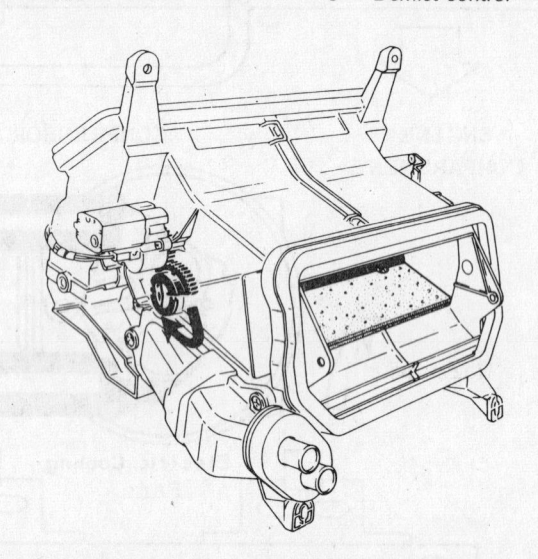

Fig. 2.18 Flap micromotor on 1986 and later heaters (Sec 17)

18 Fault diagnosis – cooling, heating and air conditioning

Symptom	Reason(s)
Cooling Overheating	Insufficient coolant in system Pump ineffective due to slack drivebelt Radiator blocked either internally and externally Kinked or collapsed hose causing coolant flow restriction Thermostat not working properly Engine out of tune Ignition timing retarded or auto advance malfunction Cylinder head gasket blown Engine not yet run-in Exhaust system partially blocked Engine oil level too low Brakes binding

Chapter 2 Cooling, heating and air conditioning

Symptom	Reason(s)
Engine running too cool	Faulty, incorrect or missing thermostat
Loss of coolant	Loose hose clips Hoses perished or leaking Radiator leaking Expansion tank pressure cap defective Blown cylinder head gasket Cracked cylinder block or head

Heater

Symptom	Reason(s)
Heater gives insufficient output	Engine overcooled (see above) Heater matrix blocked Heater controls maladjusted or broken

Air conditioner

Symptom	Reason(s)
Bubbles observed in sight glass of receiver drier	Leak in system Low refrigerant level
No cooling	No refrigerant
Expansion valve frosted over on evaporator	Faulty or clogged expansion valve Thermal bulb leaking
Insufficient cooling	Faulty expansion valve Air in refrigerant circuit Clogged condenser Receiver drier clogged Faulty compressor Compressor overfilled with oil

Chapter 3 Fuel system

Contents

Part A: Fuel systems – general maintenance and overhaul
Air cleaner – removal and refitting	3
Air cleaner element – renewal	2
Description	1
Fuel filter – renewal	4
Fuel level transmitter – removal, testing and refitting	8
Fuel pump (carburettor models) – cleaning, testing, removal and refitting	5
Fuel pump (K Jetronic fuel injection models) – removal and refitting	7
Fuel pump (Type R fuel injection models) – removal and refitting	6
Fuel tank – removal, repair and refitting	9
Throttle cable – removal, refitting and adjustment	10

Part B: Carburettor fuel system
Carburettor – removal and refitting	13
Fault diagnosis – carburettor fuel system	15
Weber DARA carburettor – description	11
Weber DARA carburettor – idle speed and mixture adjustment	12
Weber DARA carburettor – overhaul and adjustment	14

Part C: Type R (Renix) fuel injection system
Component – testing	25
Computer – removal and refitting	21
Description	16
Fault diagnosis – Type R fuel injection system	26
Fuel injection manifold – removal and refitting	19
Fuel injection system components – removal (general)	18
Fuel pressure regulator – removal and refitting	20
Idle mixture potentiometer – removal and refitting	23
Idle speed and mixture – adjustment	17
No load/full load switch – removal, refitting and adjustment	24
Sensors – removal and refitting	22

Part D: Bosch K Jetronic fuel injection system
Airflow sensor – removal and refitting	31
Cold start injector – removal and refitting	37
Control pressure regulator – removal and refitting	35
Description	27
Fault diagnosis – Bosch K Jetronic fuel injection system	39
Fuel injection system components – removal (general)	29
Fuel injection system components – testing	38
Fuel injectors – removal and refitting	34
Idle speed and mixture – adjustment	28
Intake manifold – removal and refitting	33
Lower air casing – removal and refitting	32
Metering/distributor unit – removal and refitting	30
Supplementary air valve – removal and refitting	36

Part E: Turbocharger
Air intercooler – removal and refitting	42
Description	40
Throttle casing – removal and refitting	41
Turbocharger – removal and refitting	43
Turbocharger pressure regulator – adjusting, removal and refitting	44

Part F: Emission control, manifolds and exhaust system
AI system (four-cylinder carburettor engine) – description and testing	47
AI system (V6 engine) – description	49
EGR system (four-cylinder carburettor engine) – description and testing	46
EGR system (V6i engine) – description	48
Emisssion control system – description	45
Exhaust manifold – removal and refitting	51
Exhaust system	52
Inlet manifold – removal and refitting	50

Specifications

Part A: Fuel systems – general

Fuel

Tank capacity:
TS and GTS models	66.0 litre (14.5 gallons)
All other models	70.0 litre (15.4 gallons)
Grade	97 to 99 RON (4-star)

Chapter 3 Fuel system

Part B: Carburettor fuel system

Type .... Weber 28 x 36 DARA or 32 DARA, dual barrel down draught with automatic choke
Application .... TS and GTS models

Calibration and settings

Carburettor 28 x 36 DARA O/OC (manual transmission):

	Primary	Secondary
Venturi	22	29
Main jet	112	155
Idling jet	47	42
Air compensating jet	200	100
Emulsifier	F99	F56
Diffuser	3.5	4
Accelerator pump	50	
Needle valve	225	
Float setting	7.0 mm	
Float stroke	15.0 mm	
Initial throttle opening (medium cold)	0.75 mm	
Vacuum part-open setting:		
Compensator compressed	3.0 mm	
Compensator released	7.5 mm	
Deflooding device	5.5 mm	
Idle speed	650 to 750 rev/min	
CO at idle	1.0 to 2.0%	

Carburettor 28 x 36 DARA 1 (automatic transmission):
Calibration as for 28 x 36 DARA O/OC except for

	Primary	Secondary
Idling jet	47	47
Initial throttle opening (medium cold)	0.95 mm	
Deflooding device	6.5 mm	
Idle speed	750 to 850 rev/min in N	

Carburettor 32 DARA 53 (certain European territories):

	Primary	Secondary
Venturi	25	26
Main jet	130	140
Idling jet	47	42
Air compensating jet	155	150
Emulsifier	F58	F56
Diffuser	3.5	4
Accelerator pump	60	
Needle valve	225	
Float level	7 mm	
Float stroke	15 mm	
Initial throttle opening (medium cold)	0.90 mm	
Vacuum part-open setting:		
Compensator compressed	5.5 mm	
Compensator released	10.0 mm	
Deflooding device	9.0 mm	
Defuming device (throttle flap gap)	0.5 mm	
Choke electrical assistance	30W	
Idling speed	750 to 850 rev/min	
CO at idle	1.0 to 2.0%	

Torque wrench setting | Nm | lbf ft
Carburettor mounting nut .... | 24 | 18

Part C: Type R (Renix) fuel injection system

Type .... Computer-controlled in conjunction with ignition system
Application .... GTX and Turbo models

Pressure regulator

Zero vacuum .... 2.3 to 2.7 bar (33.4 to 39.1 lbf/in^2)
Vacuum (530 mbar) .... 1.8 to 2.2 bar (26.1 to 31.9 lbf/in^2)

Calibration and settings

Idle speed:
  GTX .... 775 to 825 rev/min
  Turbo .... 675 to 725 rev/min
CO at idle .... 1.5 to 2.0%
Fuel injector resistance .... 2.0 to 3.0 ohms

Part D: Bosch K Jetronic fuel injection system

Type .... Electronically-controlled, airflow regulated
Application V6i

Fuel injectors
Opening pressure	3.5 to 4.1 bar (51 to 59 lbf/in²)
Leakproof up to	2.3 bar (33 lbf/in²)
Spray pattern	35°
Input pressure	4.5 to 5.2 bar (65 to 75 lbf/in²)
Minimum residual pressure:	
After ten minutes	1.7 bar (25 lbf/in²)
After twenty minutes	1.5 bar (22 lbf/in²)
Control pressure (hot engine):	
Vacuum disconnected	3.0 to 3.4 bar (44 to 49 lbf/in²)
Vacuum connected (idling)	3.6 to 4.0 bar (52 to 58 lbf/in²)

Settings
Idling speed:	
Manual transmission	850 to 950 rev/min
Automatic transmission	675 to 725 rev/min in D
CO at idle	0.5 to 1.5%

Part E: Turbocharger

Type .... Garrett T3

Calibration
Pressure limiter	950 to 980 mbar for a rod movement of between 0.36 and 0.40 mm (0.014 and 0.016 in)
Full load manifold pressure:	
At 2000 to 3000 rev/min	800 to 900 mbar
At max engine speed	600 to 700 mbar
Cooling fan cut-in temperature	above 107° (225°F)
Engine safety pressure release:	
At pressure of	1100 to 1200 mbar
Bypass valve	Opens at 180 to 220 mbar

Torque wrench setting
	Nm	lbf ft
Pressure	19	14

Part F: Emission control, manifolds and exhaust system

Emission control system
All models	Crankcase ventilation system
Certain territories	Exhaust gas recirculation and air injection systems

Manifolds
Inlet	Light alloy
Exhaust	Cast-iron

Exhaust system
Four-cylinder models	Twin downpipes with expansion box and silencer
Six-cylinder models	Dual front pipes with expansion box and silencer

Torque wrench settings
	Nm	lbf ft
Inlet manifold	24	18
Exhaust manifold nuts	27	20
Exhaust downpipe to manifold	40	30

PART A: FUEL SYSTEMS – GENERAL MAINTENANCE AND OVERHAUL

1 Description

The fuel system varies with the particular vehicle model: see Specifications.

On carburettor models, the unit is a twin choke Weber with automatic choke fed by a mechanically-operated fuel pump actuated by an eccentric cam on the intermediate shaft.

The R (Renix) type fuel injection system works in conjunction with the integral electronic ignition system to provide optimum engine performance.

Other models are equipped with the Bosch K Jetronic fuel injection system.

Warning: *Many of the procedures in this Chapter entail the removal of fuel pipes and connections which may result in some fuel spillage. Before carrying out any operation on the fuel system refer to the precautions given in Safety First! at the beginning of this manual and follow them implicitly. Petrol is a highly dangerous and volatile liquid and the precautions necessary when handling it cannot be over-stressed.*

Remember that fuel lines in the fuel injection system may contain fuel under pressure, even if the engine has not been running for some time. **Do not** *smoke or allow any naked flame nearby when there is any risk of fuel spillage. Take care to avoid spilling fuel onto a hot engine, and keep a fire extinguisher handy.*

Chapter 3 Fuel system

2 Air cleaner element – renewal

Four-cylinder models with carburettor
1 The air cleaner is located on the left-hand side of the engine compartment and incorporates a wax thermostat to regulate the volume of hot and cold air admitted. A flap control is used to do this.
2 Access to the air cleaner element is obtained by unscrewing the wing nut and removing the end cover. Extract the element and discard it.
3 Wipe out the air cleaner casing and insert the new element, refit the cover.

Four-cylinder models with fuel injection
4 Release the three toggle clips and take off the lid (photo).
5 Extract the filter element and discard it (photo).
6 Wipe out the air cleaner casing and insert the new element, refit the lid.

Six-cylinder models up to 1986 (not Turbo)
7 Release the four toggle clips and extract the three screws which hold the lid to the casing (photos). Remove the lid.
8 Extract the filter element and discard it (photo).
9 Wipe out the air cleaner casing and insert the new element, refit the lid.

2.4 Air cleaner casing toggle clip (2165 cc engine)

2.5 Removing air cleaner element (2165 cc engine)

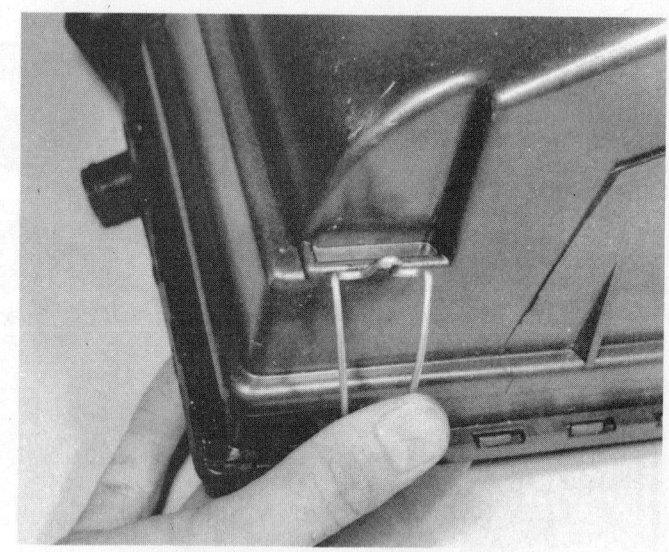

2.7A Air cleaner casing toggle clip (2664 cc engine)

2.7B Air cleaner lid fixing screw (2664 cc engine)

2.8 Air cleaner element (2664 cc engine)

Chapter 3 Fuel system

Six-cylinder models from 1986 and Turbo

10 The filter element is located in a remotely-sited air cleaner casing on these models. A modified crankcase ventilation system is also fitted.
11 Access to the filter element is obtained by releasing the strap (1 – Fig. 3.2) and prising back the toggle clips (4). Remove the cover (3) and extract the filter element.
12 Fit the new element having first wiped the inside of the casing clean.

3 Air cleaner – removal and refitting

1 Remove the filter element as previously described (Section 2).
2 Remove the casing fixing screws and, as the casing is withdrawn, disconnect the crankcase vent hose, the vacuum hose and air ducts, and the sensor connecting plugs (photos).
3 When refitting the air cleaner to six-cylinder engines, make sure that the O-ring at the flowmeter inlet is in good condition.

Fig. 3.1 Air filter and oil separator arrangement on 1986 and later V6 models (Sec 2)

1 Air cleaner
2 Automatic idle adjustment valve
3 Rebreather hose
4 Oil separator
5 Idling microswitch

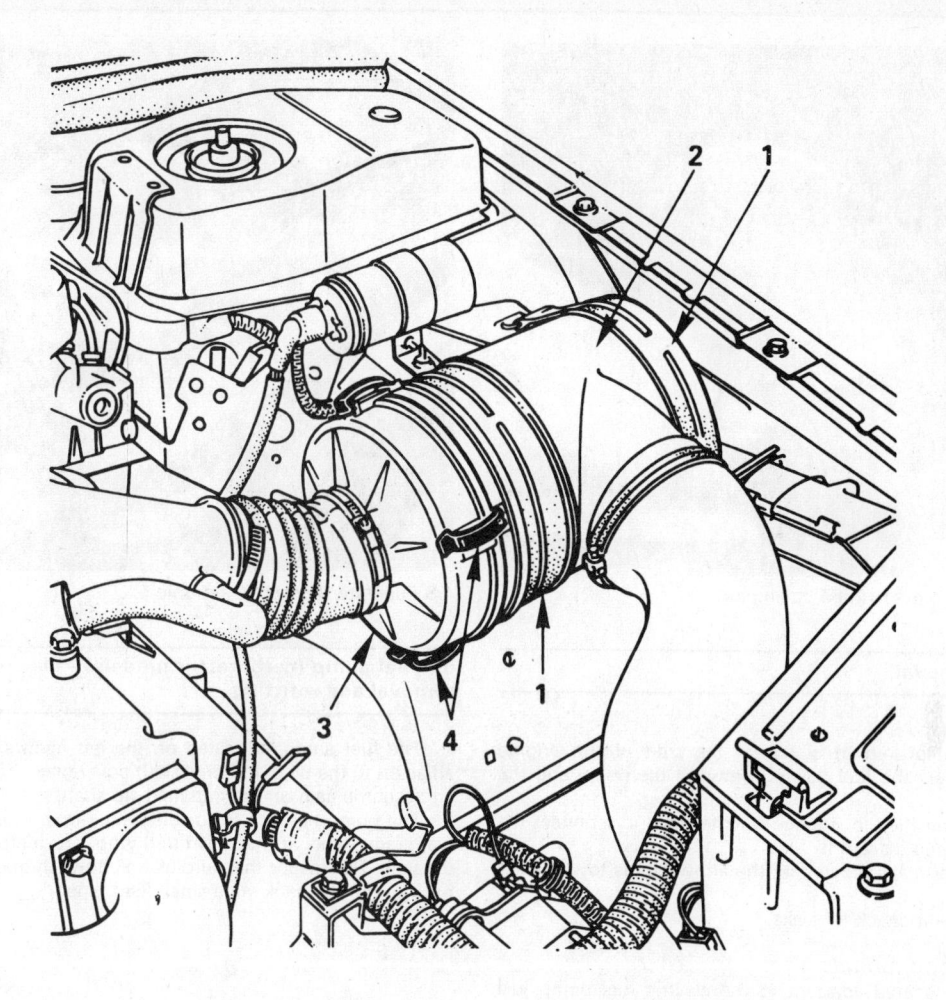

Fig. 3.2 Air cleaner fixings on 1986 V6 models (Sec 2)

1 Fixing strap
2 Cleaner casing
3 Cover
4 Toggle clips

3.2A Air cleaner casing (2165 cc engine)

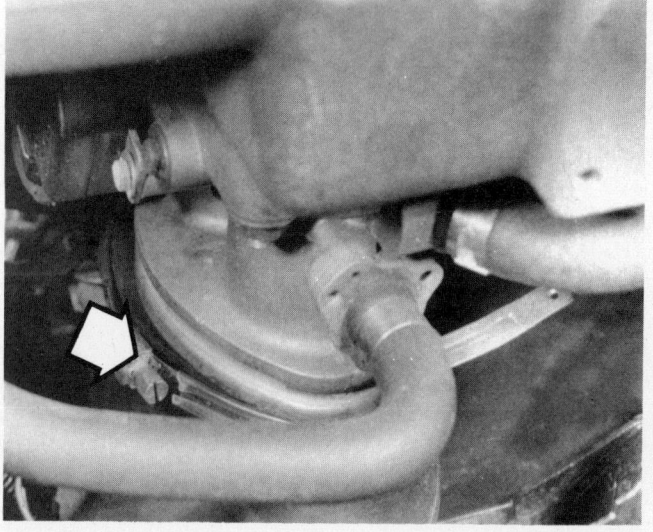

3.2B Air cleaner duct clip (arrowed) at throttle housing (2165 cc engine)

124   Chapter 3 Fuel system

3.2C Air cleaner intake duct (2664 cc engine)

4.8 Fuel filter (Bosch K Jetronic)

4 Fuel filter – removal

Carburettor models

1   The fuel filter is located in a clip at the side of the engine compartment and filters the fuel flowing between the pump and the carburettor.
2   At the intervals specified in Routine Maintenance, disconnect the hoses from the filter and discard it.
3   Refit the new filter making sure that the arrow points towards the carburettor.
4   Start the engine and check for leaks.

Type R fuel injection models

5   The fuel filter is located adjacent to the electric fuel pump just behind the rear of the right-hand body sill.
6   Release the filter from the common pump/filter mounting plate and disconnect the hoses.
7   Fit the new filter making sure that the direction of flow arrow marked on it is towards the engine, and that the plastic sleeve is between the filter and the clamp.

K Jetronic fuel injection models

8   The fuel filter is located in a clip within the engine compartment close to the power steering pump pulley (photo).
9   Renewal of the filter is as described for other types.

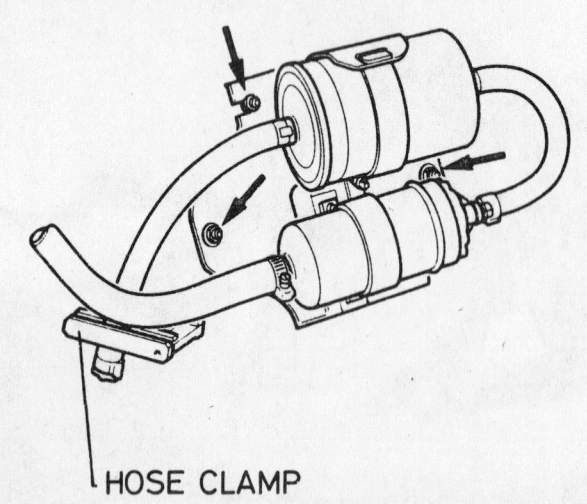

5 Fuel pump (carburettor models) – cleaning, testing, removal and refitting

1   The fuel pump is located on the left-hand side of the crankcase adjacent to the oil pump driveshaft cover plate. The pump is driven by an eccentric cam on the intermediate shaft.
2   The pump has one major difference from most fuel pumps of the diaphragm type, in that it is fitted with a restricted fuel return feed to the tank. This means that surplus fuel flows through the restriction and back to the fuel tank via a small feed pipe.

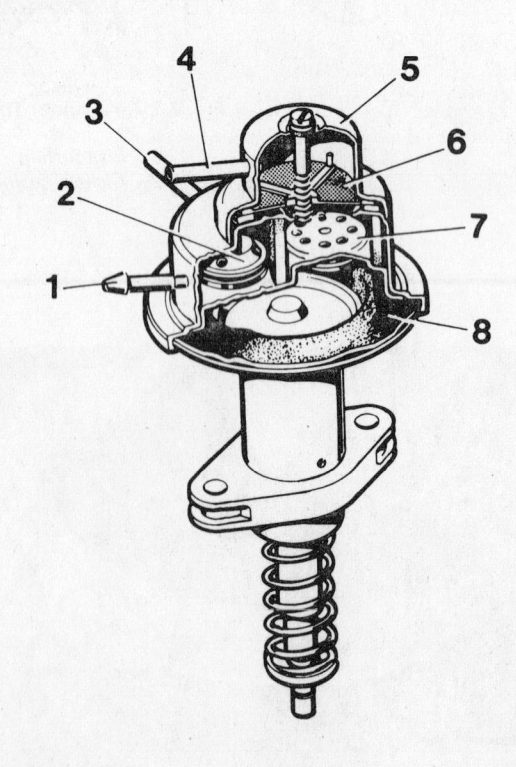

Fig. 3.3 Fuel pump and filter (Type R fuel injection) (Sec 4)

Fig. 3.4 Cutaway view of mechanically-operated fuel pump (Sec 5)

1   Fuel return
2   Outlet valve
3   Outlet
4   Inlet
5   Cover
6   Filter gauge
7   Inlet valve
8   Diaphragm

Chapter 3 Fuel system

3 The fuel pump is of the continuous operation type. This is necessary so that the fuel in the pump is circulating continuously and does not become too warm at any stage. The possibility of air locks occurring in the fuel supply is also thereby avoided.
4 The only maintenance that can be carried out on this type of pump is to clean the filter qauze or strainer which can become blocked with sediment sucked up from the fuel tank. Do this regularly at the intervals specified in Routine Maintenance.
5 To do this, remove the screw in the centre of the top of the pump. Then carefully ease off the top section complete with inlet pipe, and place it carefully to one side (photos).
6 Prise the filter away carefully and wash any sediment off it in a petrol bath. Do not scrub it; damage may result.
7 Refit the filter gauze and cover in the reverse order to that in which they were removed and then check the operation of the pump. If any doubt still exists as to its efficiency, it must be renewed.
8 If the fuel pump is suspected of malfunction, a simple test can be carried out by detaching the fuel feed pipe at the carburettor. Place the disconnected pipe end into a clean container or piece of rag and disable the ignition system. Having ensured that the car is not in gear, turn the engine on the starter motor. Assuming that there is fuel in the fuel tank, there should be a good flow of petrol from the pipe. If there is very little or no flow, remove the pump for inspection.

5.5A Fuel pump cover and seal

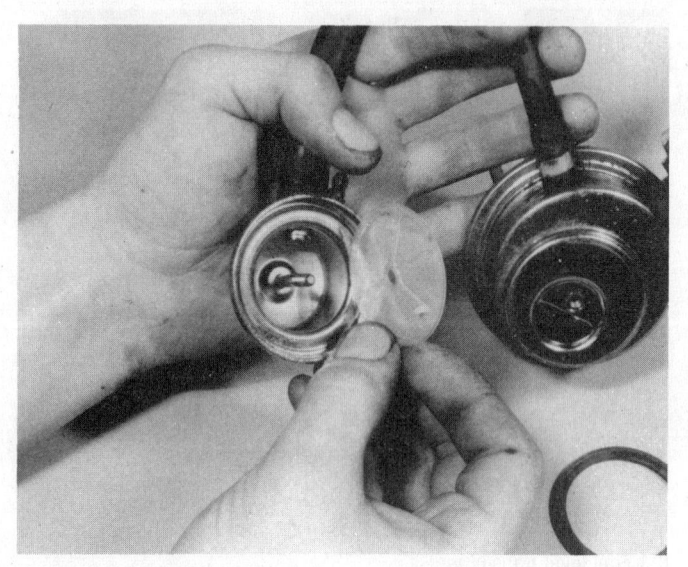

5.5B Fuel pump filter

7.1 Fuel pump and accumulator (Bosch K Jetronic)

9 Remove the inlet and outlet pipes from the fuel pump and the surplus fuel return pipe. Plug the ends of the pipes to stop dirt entering them or fuel escaping.
10 Undo the bolts holding the fuel pump and lift it away.
11 Due to the design of this type of pump there is very little that can be done to it. It is not possible to strip it down in the same way as other diaphragm type pumps: the diaphragm is sealed in. If therefore it goes wrong mechanically, a complete new pump must be fitted.
12 Clean away the old gasket remains and place a new gasket in position.
13 Fit a new fuel pump in the reverse order to that which was used to remove the old one.

3 Disconnect the pump electrical leads.
4 Loosen the mounting plate upper bolt.
5 Slide the plate out from under the upper bolt and remove the pump from the plate.
6 Refitting is a reversal of removal; make sure that the leads are connected correctly to the (+) and (–) terminals as indicated on the pump body.

6 Fuel pump (Type R fuel injection models) – removal and refitting

1 The pump is mounted on the same plate as the fuel filter (see Section 4).
2 Disconnect the hoses from the pump, anticipating some loss of fuel.

7 Fuel pump (K Jetronic fuel injection models) – removal and refitting

1 The fuel pump is mounted, together with the pressure accumulator, on the same mounting plate just behind the body sill on the right-hand side of the fuel tank (photo).
2 Disconnect the electrical leads and the fuel hoses and remove the mounting plate screws. Anticipate some loss of fuel.
3 Refitting is a reversal of removal.

Chapter 3 Fuel system

8 Fuel level transmitter – removal, testing and refitting

Note: *A suitable bending sealant will be needed when refitting the transmitter. An ohmmeter is needed for the test.*

1 Disconnect the battery.
2 Refer to Chapter 11 and remove the rear seat cushion and insulating felt (photo).
3 Prise out the large grommet to expose the fuel tank transmitter (sender) unit (photo).
4 Disconnect the fuel hose and electrical leads.
5 The method of removal and refitting now depends upon whether a cylindrical type unit is fitted or one with a float arm. The latter is identified by a yellow label and the wording 'Attention, hold the top of the unit when loosening the nut'.
6 Loosen the locking ring. On cylindrical type units, a tool similar to the one shown in Fig. 3.6 should be made up to unscrew the ring.
7 When withdrawing the float arm type unit, a hooked piece of wire will be required to lift the float to enable it to pass out through the transmitter aperture.

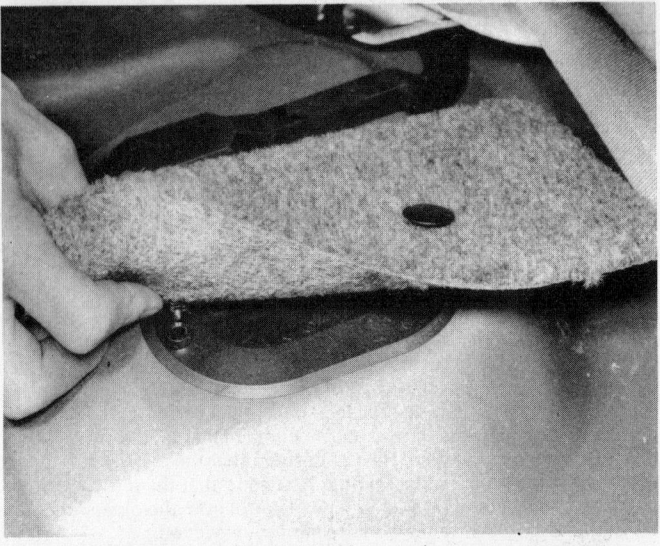

8.2 Lifting carpet to expose fuel level transmitter cover grommet

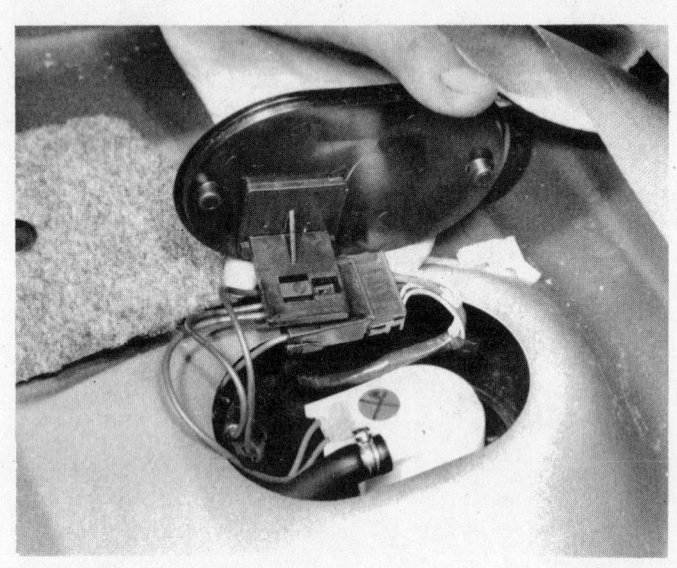

8.3 Fuel level transmitter

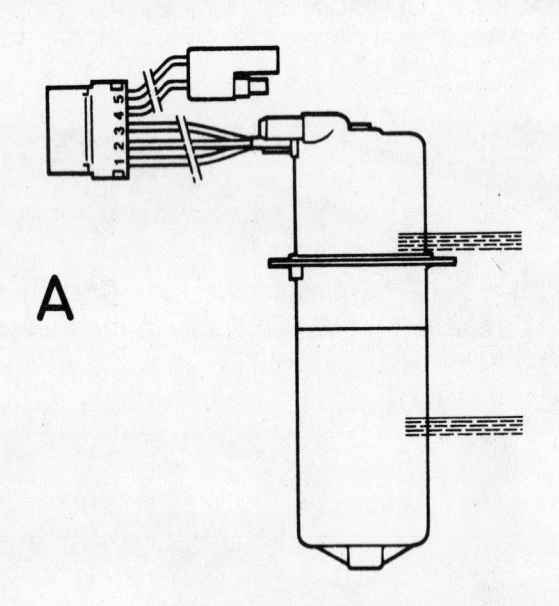

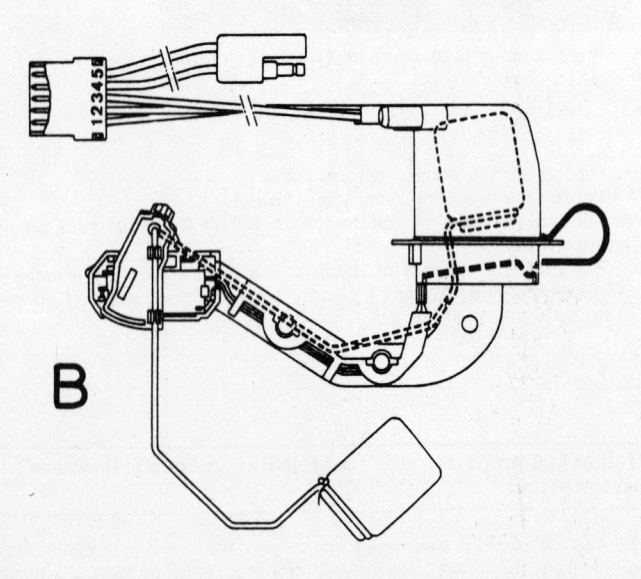

Fig. 3.5 Types of fuel level transmitter (Sec 8)

A Cylindrical type   B Float arm type

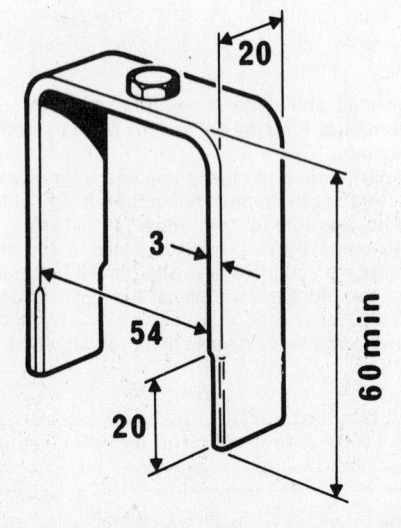

Fig. 3.6 Diagram for making transmitter unit removal tool (Sec 8)

All measurements in mm

Chapter 3 Fuel system

8 The float arm type of transmitter unit may be tested if an ohmmeter is available using the following table:

Float height H	Resistance at terminals 1 and 3
Highest point	350 to 370 ohms
Intermediate	170 to 210 ohms
Lowest point	16 to 36 ohms

9 Refitting is a reversal of removal. The unit is sealed by means of a bead of sealing compound (such as is used for bonding windscreens) applied to the seal.

9 Fuel tank – removal, repair and refitting

1 The fuel tank may be of steel or plastic construction according to model and date of production. Later models have a plastic tank which is also the only kind of tank supplied as a replacement.
2 If a metal tank is being renewed with a plastic one then certain hoses and fittings will also have to be changed.
3 Disconnect the battery.
4 Syphon or pump out the fuel from the tank into a suitable container which can be sealed.
5 Disconnect the connections from the fuel tank transmitter unit as described in the preceding Section.
6 Raise the rear of the car and support it securely.
7 Remove the reaction members which run between the suspension arms and the body under the fuel tank (photo).
8 Disconnect the handbrake cable.
9 Disconnect the fuel filler hose (photo).
10 Support the fuel tank on a jack with a block of wood as an insulator and remove the tank fixing bolts (photo). Carefully lower the tank until the various hoses can be disconnected.
11 If the fuel has been withdrawn to remove water or sediment, remove the transmitter unit and pour in some paraffin, shake the tank vigorously then allow it to drain. Repeat as necessary, giving a final flush out with clean fuel.
12 If a metal tank is being removed to mend a leak, leave any repair to the professionals. Never attempt to solder or weld a fuel tank as it will explode unless all vapour has been purged by steaming out.
13 If a plastic tank is leaking, renew it.
14 Refitting is a reversal of removal but take care not to trap the fuel hoses between the tank and the side-member when bolting it into position.

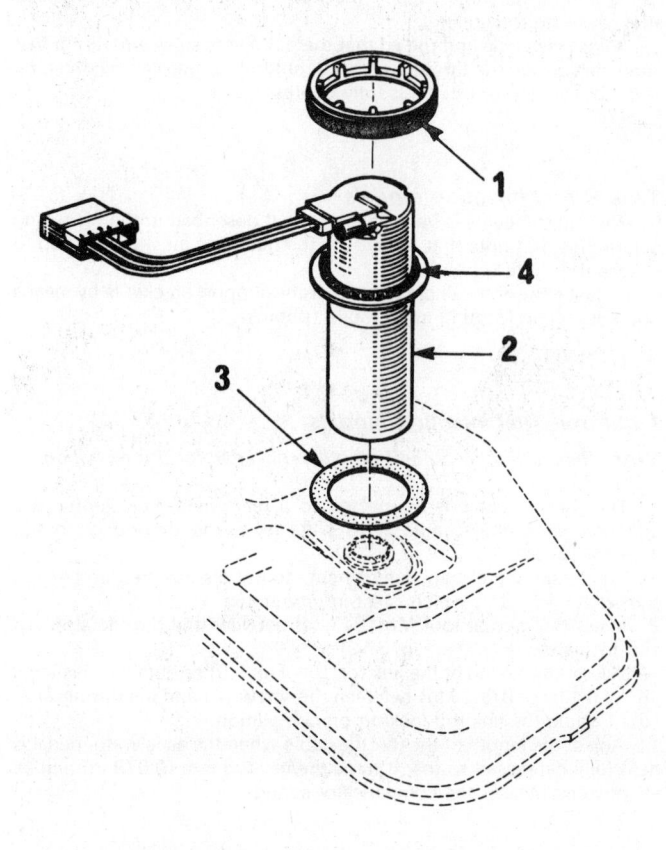

Fig. 3.7 Cylindrical type fuel level transmitter (Sec 8)

1 Ring nut
2 Transmitter
3 Seal
4 Sealing compound

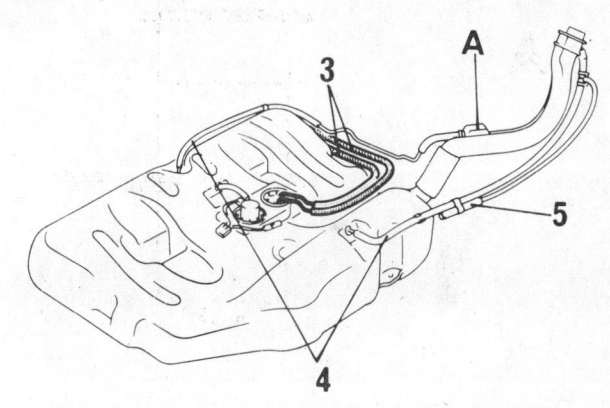

Fig. 3.8 Fuel tank hoses (Sec 9)

A Housing
3 Supply and return hoses
4 Anti blowback and defuming hoses
5 Vent to atmosphere

9.7 Reaction member

9.9 Fuel tank filler and breather pipes under rear wheel arch

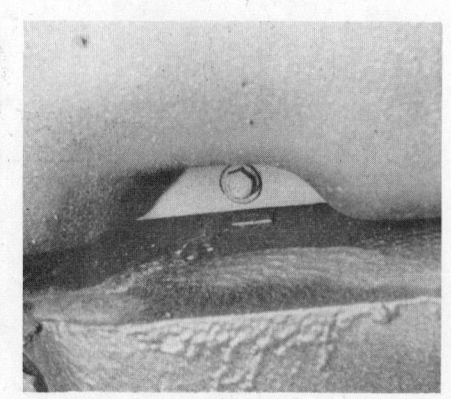

9.10 Fuel tank mounting bolt

Chapter 3 Fuel system

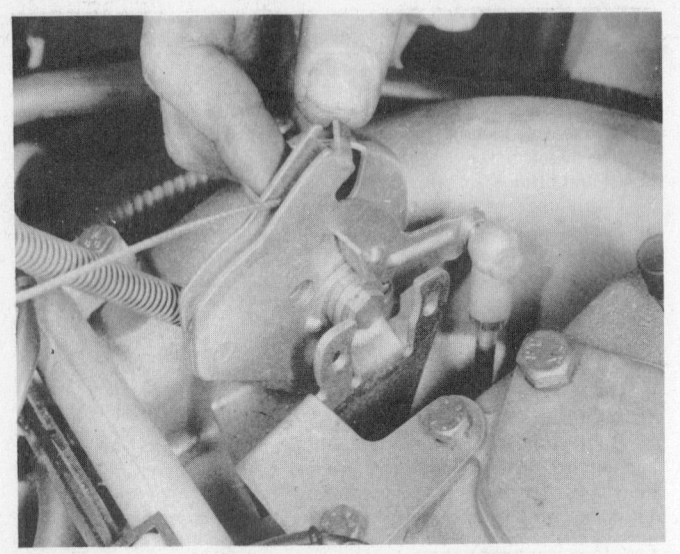

10.5 Disconnecting throttle cable (Type R fuel injection)

10.6 Throttle cable bracket and adjuster (Type R fuel injection)

10.7 Throttle cable, reel and link rod (K Jetronic fuel injection)

10 Throttle cable – removal, refitting and adjustment

Carburettor models

1 The throttle cable is attached to the top of the accelerator pedal arm and retained by a clip.
2 At the carburettor, the throttle cable end fitting engages in the notch of a quadrant, movement then being transmitted to the carburettor throttle lever through a ball-jointed link rod. Once both ends of the cable have been disconnected, it can be withdrawn through the bulkhead grommet and removed.
3 Adjustment to the cable is made by moving the position of the spring clip at the cable support bracket. The link rod can be adjusted after releasing the locknut.
4 Adjust the cable and rod so that there is a very small amount of free movement when the throttle is closed, but full throttle can be obtained when the accelerator pedal is fully depressed.

Type R fuel injection models

5 The throttle cable is very similar to that described in the preceding paragraphs, but note that the tab on the quadrant must be bent up to release the cable (photo).
6 Adjustment of the cable tension at the support bracket is by means of a spring clip locating in a groove (photo).

K Jetronic fuel injection models

Note: *Renault tool Mot. 843-08 will be needed for this operation.*

7 The throttle cable is connected to a reel which has throttle rod attachments to the throttle valve plate lever and the cruise control actuator (photo).
8 When carrying out any adjustment, do not disturb the throttle stop screw (4 – Fig. 3.9) which is set in production.
9 Insert the special tool Mot. 843-08 between the throttle stop (2) and the safety stop (1 – Fig. 3.10).
10 Adjust the length of the link rod (6 – Fig. 3.9) to obtain a clearance (B) of 0.1 mm (0.0039 in) between the screw (4) and the throttle stop (5). Tighten the link rod locknut on completion.
11 Adjust the length of the throttle cable when the accelerator pedal is held fully depressed so that it turns the reel 2.0 mm (0.079 in) and so slightly compresses the compensator spring.

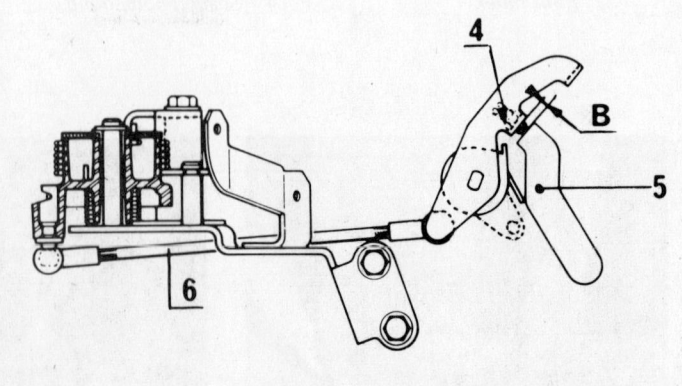

Fig. 3.9 Throttle reel (K Jetronic fuel injection) (Sec 10)

B Clearance – 0.1 mm (0.0039 in)
4 Throttle stop screw
5 Throttle stop
6 Link rod

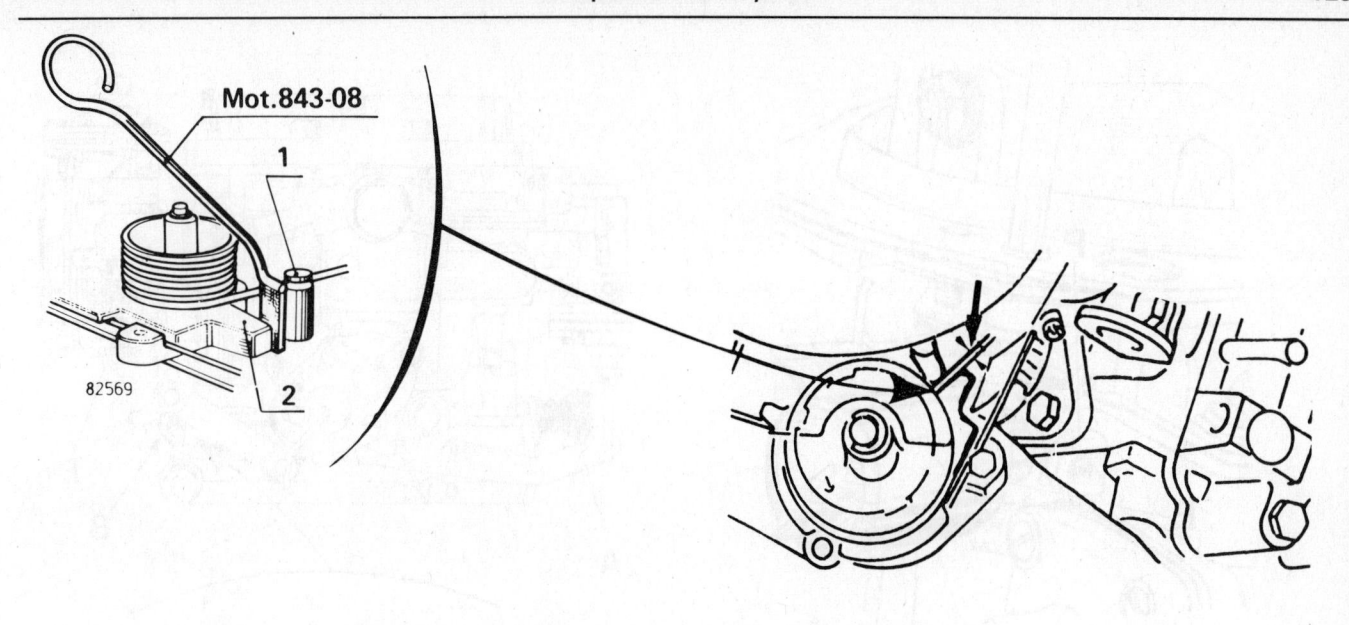

Fig. 3.10 Special tool used to adjust throttle stop clearance (Sec 10)

1 Safety stop *2 Throttle stop*

PART B: CARBURETTOR FUEL SYSTEM

11 Weber DARA carburettor – description

The carburettor is of dual barrel, downdraught type with a coolant-heated automatic choke.

The throttle valve plate block is coolant-heated. On certain versions an additional electric choke heater is fitted.

The carburettor incorporates a constant CO circuit controlled by a mixture screw (B – Fig. 3.14) and a volume (idle speed) screw (A) which operates on a supplementary circuit.

An accelerator pump, power and full throttle enrichment devices are also fitted, together with a secondary barrel lock-out system which uses a diaphragm to prevent the secondary throttle valve from opening when the diaphragm is subject to vacuum.

Two solenoid valves are used, one is an anti-diesel (run-on) valve which shuts off the idle circuit as soon as the ignition is switched off and the other closes the constant CO idling supplementary circuit.

12 Weber DARA carburettor – idle speed and mixture adjustment

Note: *An exhaust gas analyser will be needed for this operaton.*

1 Have the engine at normal operating temperature with the ignition system correctly adjusted, particularly the spark plugs.
2 Remove the tamperproof cap from the mixture screw (Fig. 3.14) and then connect an exhaust gas analyser in accordance with the maker's instructions.
3 Start the engine, increase its speed momentarily to clear the manifold and then turn the volume screw until the vehicle tachometer indicates the specified idle speed.
4 Now turn the mixture screw until the specified exhaust gas CO level is obtained.
5 Readjust the idle speed and then recheck the CO level.
6 If an exhaust gas analyser is not available, unscrew the mixture screw until the engine idle speed is at its highest then turn the screw in again until the speed just starts to fall. Regard this as a temporary setting until it can be checked using an exhaust gas analyser.

13 Carburettor – removal and refitting

1 Remove the air cleaner.
2 Release the pressure in the cooling system by unscrewing the expansion tank cap.
3 Disconnect the coolant hoses from the automatic choke housing and the carburettor throttle block. If the hoses are tied up as high as possible, coolant loss will be minimal and the need to drain the cooling system avoided.
4 Disconnect the fuel hose, the throttle link rod and the electric leads from the solenoid valves and choke (where fitted).
5 Unscrew the carburettor mounting nuts and lift the unit from the inlet manifold.
6 Refitting is a reversal of removal, always use a new flange gasket.
7 Bleed the choke housing and top up the cooling system through the expansion tank.

Fig. 3.11 Weber DARA carburettor (Sec 11)

130   Chapter 3 Fuel system

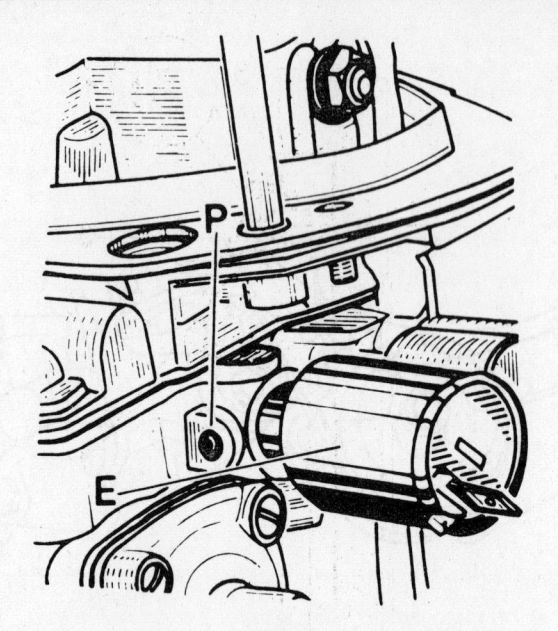

Fig. 3.12 Anti-run-on fuel cut-off solenoid valve (E) (Sec 11)

P Locking screw

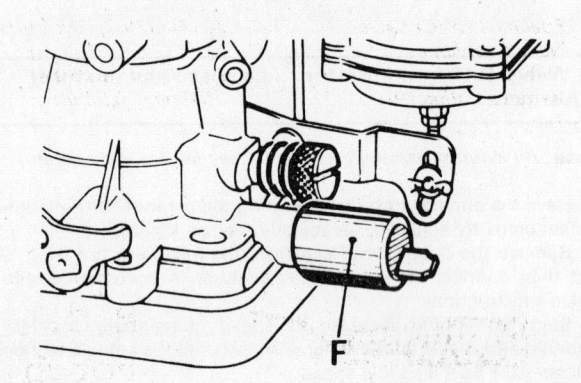

Fig. 3.13 Constant idle circuit cut-off solenoid valve (F) (Sec 11)

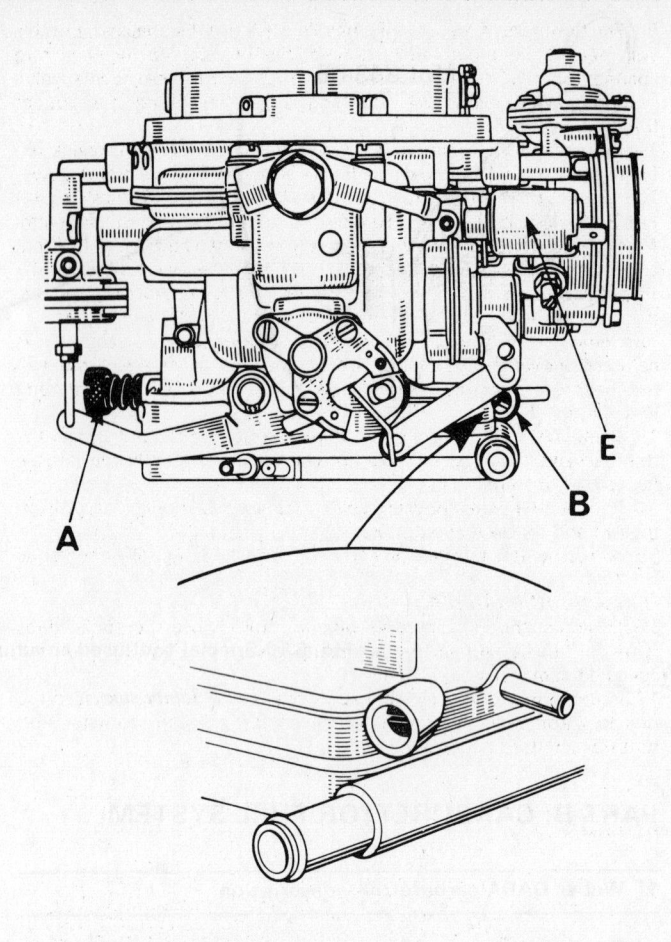

Fig. 3.14 Carburettor adjustment screws (Secs 11 and 12)

A Volume (idle speed) screw   E Fuel cut-off solenoid valve
B Fuel mixture screw

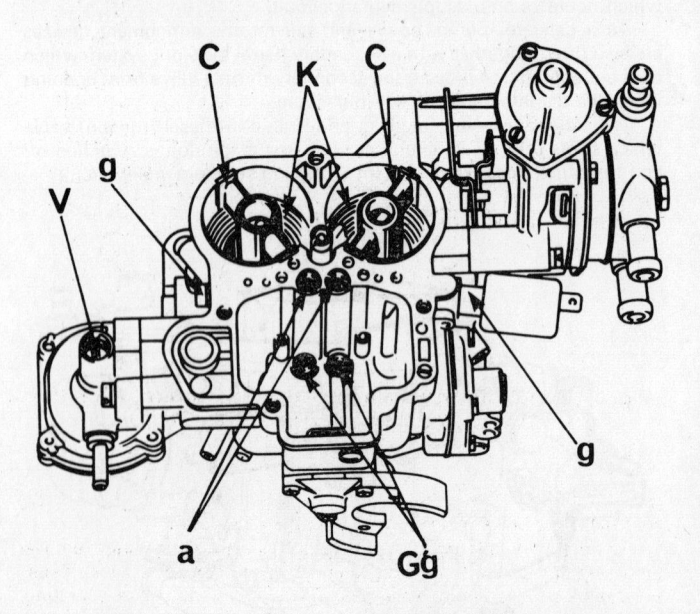

Fig. 3.15 Location of carburettor jets (Sec 14)

a Air compensating jet   Gg Main jet
C Diffuser   K   Venturi
g Idling jet

14 Weber DARA carburettor – overhaul and adjustment

1 Overhaul of a well-worn carburettor is not usually a worthwhile exercise, it is far better to obtain a new or unworn unit.
2 However, if carburettor performance is suspect, it can be worthwhile dismantling the unit and renewing diaphragms, seals and gaskets. Seek advice from a dealer regarding the availability of such parts.
3 The operations described in paragraphs 5 to 10 can be carried out without the need to remove the carburettor from the engine.
4 The adjustments described in this Section are vital to the success of the overhaul.
5 Remove the air cleaner (Section 3), and disconnect the fuel hose from the carburettor.
6 Unscrew the plug just above the fuel inlet nozzle, remove and clean the filter gauze. Refit the gauze and the plug.
7 Extract the screws and take off the top cover. This will be as far as most overhaul work will need to go, as the jets can be removed and blown through with air from a tyre pump – never probe them with wire or their calibration will be ruined. In extreme cases of clogging a nylon bristle may be used to clear a jet.
8 Mop out dirt and sediment from the fuel bowl.

Chapter 3 Fuel system

9 The tightness of the fuel inlet needle valve may be checked, but this will mean driving out the float pivot pin in order to locate a ring spanner on the valve. Fuel seeping out through the needle valve seating washer will cause too high a fuel level and consequent flooding.

10 Before fitting the carburettor cover with a new gasket, check the float setting. Hold the cover vertically with the float hanging down. The nearest point of the float to the surface of the cover gasket (A – Fig. 3.16) should be 7.0 mm (0.28 in). If necessary bend the float arm. Now gently lift the float with the finger. Dimension A + B (float stroke) should be 15.0 mm (0.59 in). If necessary, bend the tab (5).

11 Where it was decided to strip the carburettor completely, remove the unit, as described in Section 13, and clean away external dirt.

12 Remove the top cover and jets as previously described. A carburettor overhaul kit should now be obtained which will contain all the necessary gaskets, diaphragms and seals which will require renewal.

13 Remove the diaphragm units, invert the carburettor and extract the throttle valve block screws. Do not attempt to dismantle the throttle flap plates or spindles.

14 Remove the choke housing cover and housing if necessary. Clean, inspect and renew any worn items.

15 As reassembly progresses, carry out the following adjustments.

Initial throttle opening

16 To adjust the initial throttle opening, fully close the choke flaps with the fingers and set the adjusting screw (1 – Fig. 3.17) on the specified cam step (Fig. 3.18).

17 Using a twist drill or similar gauge, check that the gap between the primary throttle flap and barrel is as given in the Specifications. If not, turn the screw as necessary.

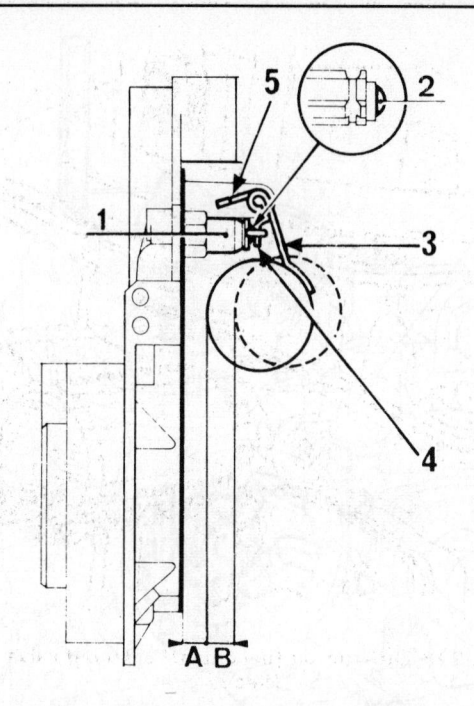

Fig. 3.16 Float setting diagram (Sec 14)

1 Fuel inlet needle valve
2 Needle valve ball
3 Float arm
4 Valve operating tab
5 Tab

A Float to gasket surface = 7.0 mm (0.28 in)
A + B (float stroke) = 15.0 mm (0.59 in)

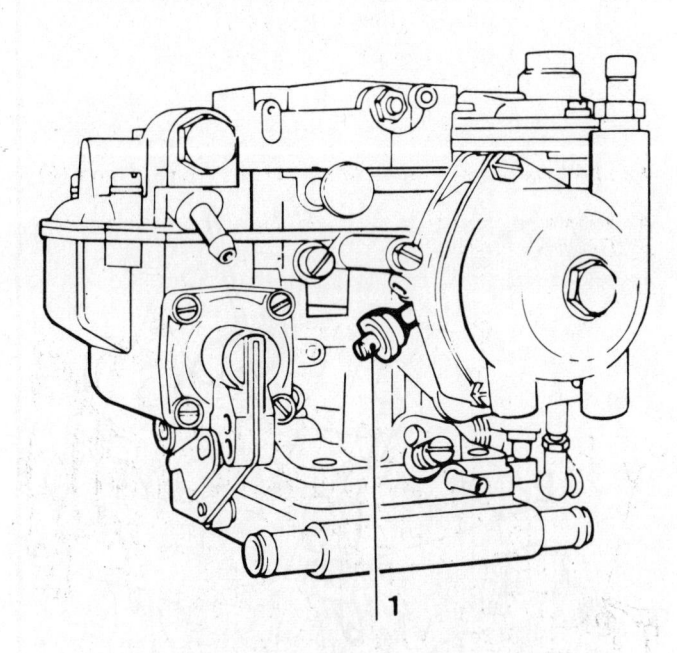

Fig. 3.17 Initial throttle opening adjusting screw (1) on the Weber 32 DARA carburettor (Sec 14)

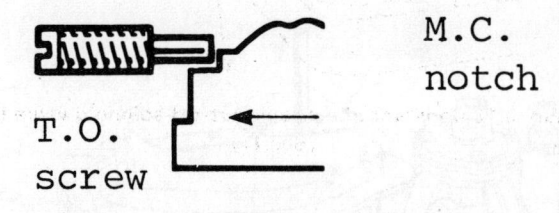

Fig. 3.18 Throttle opening screw on 'medium cold' cam notch (Sec 14)

Automatic choke (vacuum part-open) setting

18 To adjust the automatic choke (vacuum part-open setting) remove the cover and bi-metallic spring then manually close the choke flaps. Fully raise the vacuum capsule pushrod and turn the choke operating lever against it. Using a twist drill check that the gap between the choke flap and barrel is as given in the Specifications. If not, turn the adjustment screw inside the top of the vacuum capsule as necessary. Refit the bi-metallic spring and cover, making sure that the spring engages the lever correctly and the assembly marks on the cover and body are in alignment.

Deflooding mechanism – adjustment

19 To adjust the deflooding mechanism, fully close the choke flaps manually then fully open the throttle lever. Using a twist drill check that the gap between the choke flaps and barrel is as given in the Specifications. If not, turn the adjusting screw as necessary. After making an adjustment check the initial throttle opening, as described in paragraphs 16 and 17.

20 To adjust the float level proceed as described in paragraph 10.

Defuming valve

21 Some carburettors are fitted with a defuming valve which vents the float chamber to atmosphere when idling.

22 To adjust, hold the choke flap open and depress the defuming valve lever (1 – Fig. 3.22). Using a twist drill, measure the throttle flap opening. If it is not as specified (see Specifications) turn the nut (E).

Fig. 3.19 Automatic choke adjustment (Sec 14)

1 Initial throttle opening screw
2 Operating lever
3 Cam
4 Pushrod
5 Adjusting screw

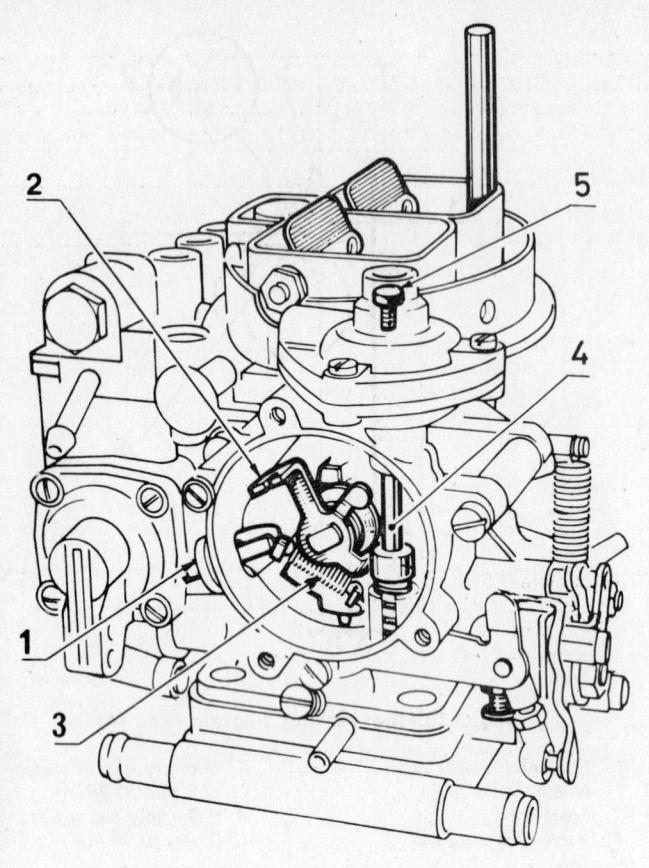

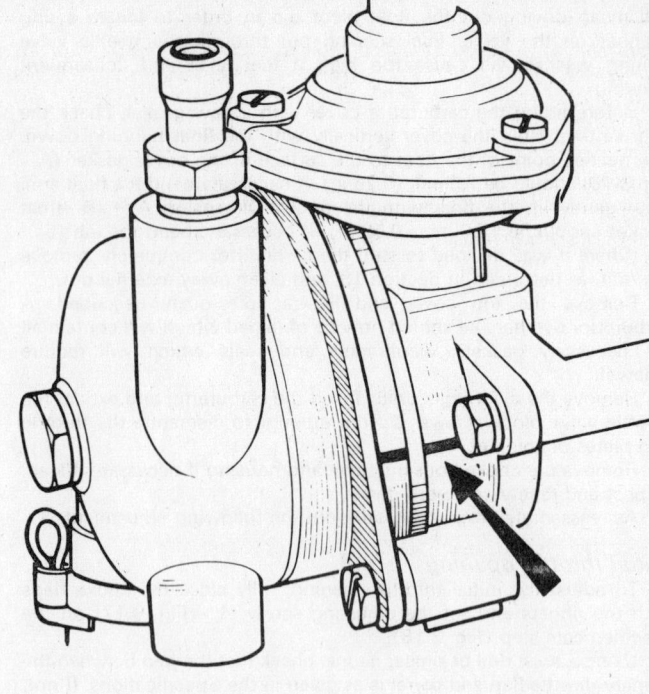

Fig. 3.20 Automatic choke housing alignment marks (Sec 14)

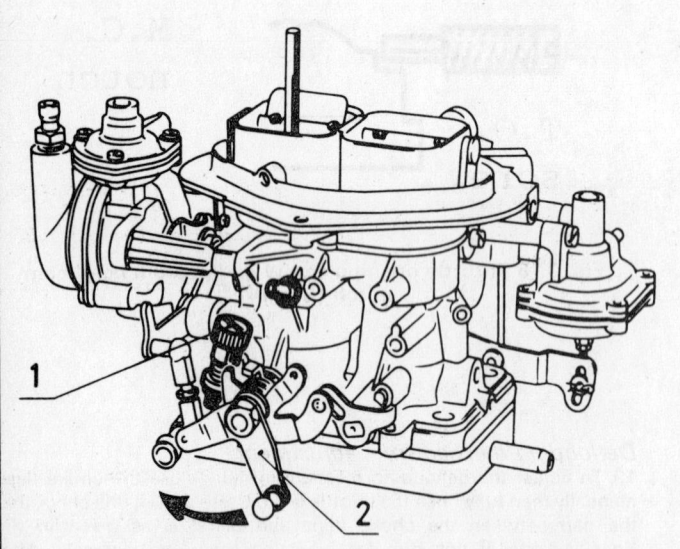

Fig. 3.21 Deflooding mechanism (Sec 14)

1 Adjusting screw
2 Throttle lever

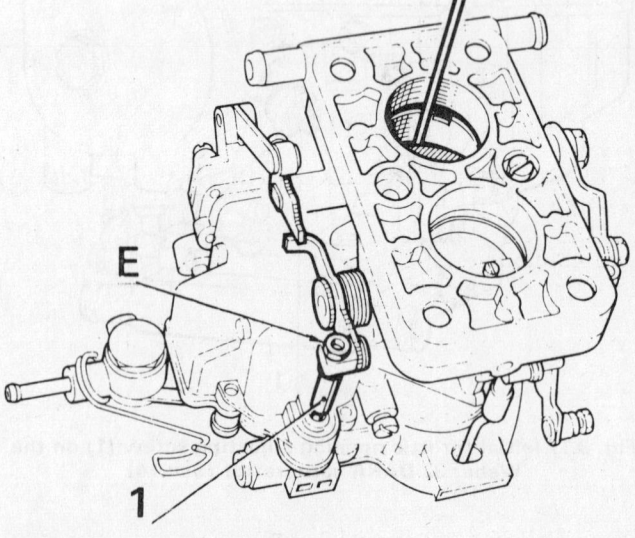

Fig. 3.22 Adjusting defuming valve (Sec 14)

E Nut
1 Valve lever

Chapter 3 Fuel system

15 Fault diagnosis – carburettor fuel system

Unsatisfactory engine performance and excessive fuel consumption are not necessarily the fault of the fuel system. In fact they more commonly occur as a result of ignition and timing faults. Before acting on the following it is necessary to check the ignition system first. Even though a fault may lie in the fuel system it will be difficult to trace unless the ignition is correct. The faults below, therefore, assume that this has been attended to first (where appropriate).

Symptom	Reason(s)
Smell of petrol when engine is stopped	Leaking fuel lines or unions Leaking fuel tank
Smell of petrol when engine is idling	Leaking fuel line unions between pump and carburettor Overflow of fuel from float chamber due to wrong level setting, ineffective needle valve or punctured float
Excessive fuel consumption for reasons not covered by leaks or float chamber faults	Worn jets Over-rich setting Sticking mechanism Dirty air cleaner element Sticking air cleaner thermostatic mechanism
Difficult starting, uneven running, lack of power, cutting out	One or more jets blocked or restricted Float chamber fuel level too low or needle valve sticking Fuel pump not delivering sufficient fuel Faulty solenoid fuel shut-off valve Induction leak
Difficult starting when cold	(Automatic choke maladjusted) Automatic choke not cocked before starting Weak mixture
Difficult starting when hot	Automatic choke malfunction Accelerator pedal pumped before starting Vapour lock (especially in hot weather or at high altitude) Rich mixture
Engine does not respond properly to throttle	Faulty accelerator pump Blocked jet(s) Slack in accelerator cable
Engine idle speed drops when hot	Overheated fuel pump
Engine runs on	Faulty fuel cut-off valve

PART C: TYPE R (RENIX) FUEL INJECTION SYSTEM

16 Description

This fuel system is of the pressure/speed type. The quantity of fuel injected is determined by both the vacuum in the inlet manifold and the engine speed.

The inlet vacuum pressure controls the basic injection time which is then corrected to suit other engine conditions and requirements. These include coolant temperature, air temperature, battery voltage and atmospheric pressure.

The fuel injection system also calculates the ignition advance requirements to suit any particular set of engine conditions.

A computer controls both the fuel injection and ignition systems.

17 Idle speed and mixture – adjustment

Note: *An exhaust gas analyser will be needed for this operation*

1984 models

1 Have the engine at full operating temperature with an exhaust gas analyser connected in accordance with the manufacturer's instructions.

2 Remove the tamperproof cap from the mixture potentiometer screw (B – Fig. 3.24).
3 With the engine idling, turn the air bypass screw (A – Fig. 3.25) until the specified idling speed is obtained.
4 Now turn screw (B) in or out until the correct exhaust gas CO level is obtained. Re-adjust the idle speed.
5 Switch off the engine and disconnect the analyser.

1985 and later models

6 The air bypass screw is no longer fitted to later models; the idling speed cannot be altered as it is controlled by an air regulating solenoid valve monitored by a power module and throttle switch to form an electronic idling system.
7 Main components of this system comprise a computer, mounted with the one for the ignition system; an air regulating valve, located at the front left-hand side of the engine near the alternator; and a three-position throttle switch.
8 The computer incorporates an integrated idling adjustment circuit.
9 The air regulating (idle speed adjusting) valve (1 – Fig. 3.26) consists of two coils which are fed by signals which position the valve between the closed and fully open position according to requirements.
10 The computer controls the idle speed – faster at cold start and then reducing it as the engine warms up.

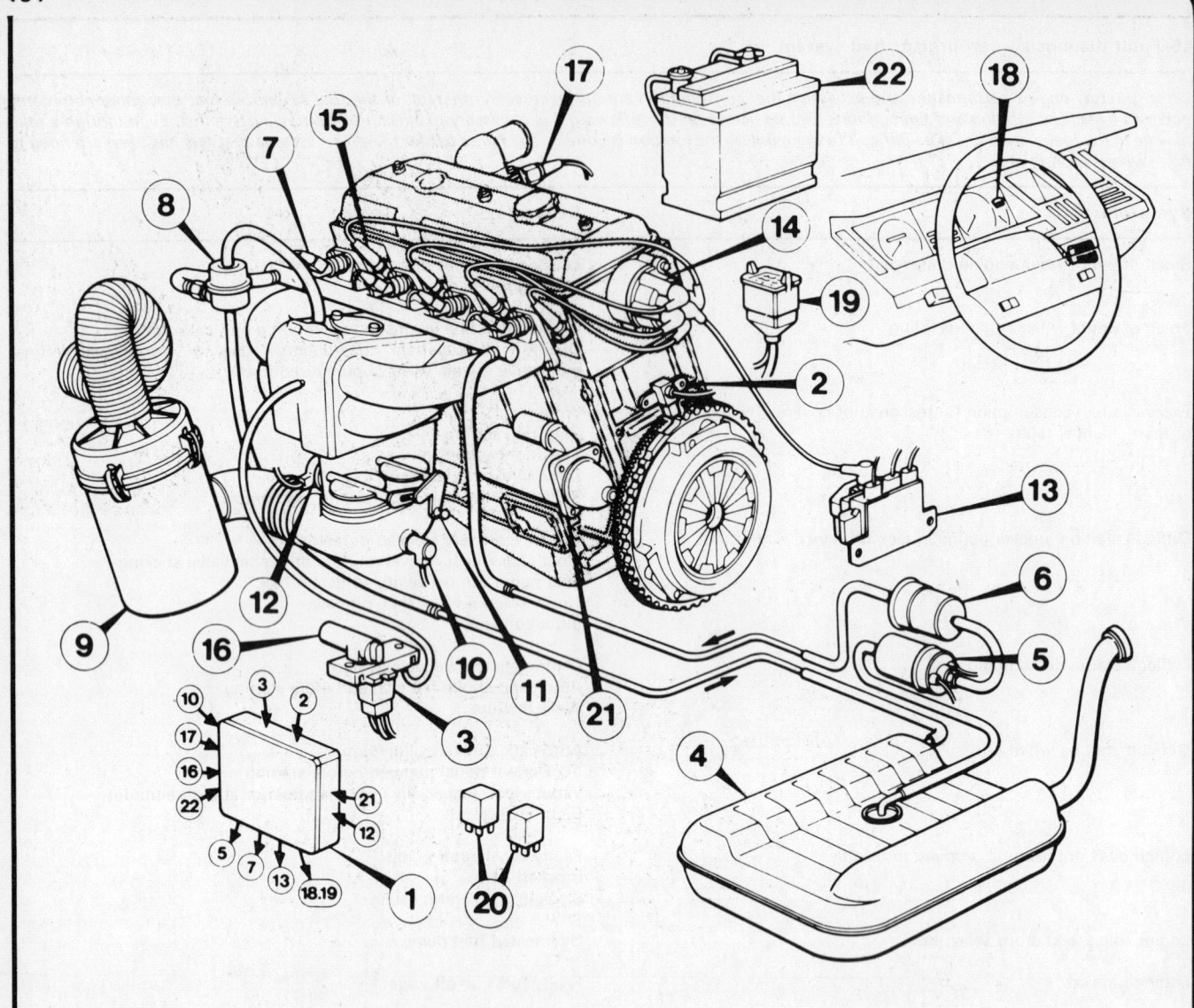

Fig. 3.23 Type R (Renix) fuel injection system components (1984 models) (Sec 16)

1 Fuel injection/ignition computer
2 Speed and position sensor
3 Absolute pressure sensor
4 Fuel tank
5 Electric fuel pump
6 Fuel filter
7 Fuel injectors
8 Fuel pressure regulator
9 Air cleaner
10 Air temperature sensor
11 Throttle casing
12 No load/full load switch
13 Ignition power module
14 Ignition distributor
15 Spark plugs
16 Idle mixture potentiometer
17 Coolant temperature sensor
18 Electronic fault warning lamp
19 Diagnostic plug
20 Relay
21 Starter motor
22 Battery

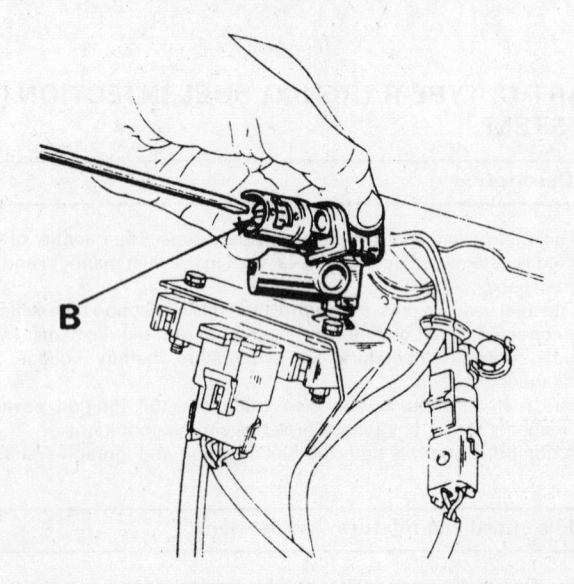

Fig. 3.24 Fuel mixture potentiometer screw (B) (Sec 17)

Chapter 3 Fuel system   135

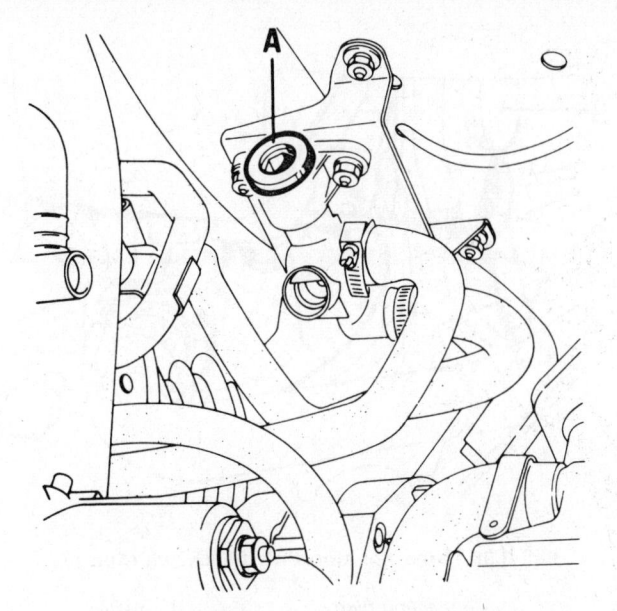

Fig. 3.25 Air bypass screw (A) (Sec 17)

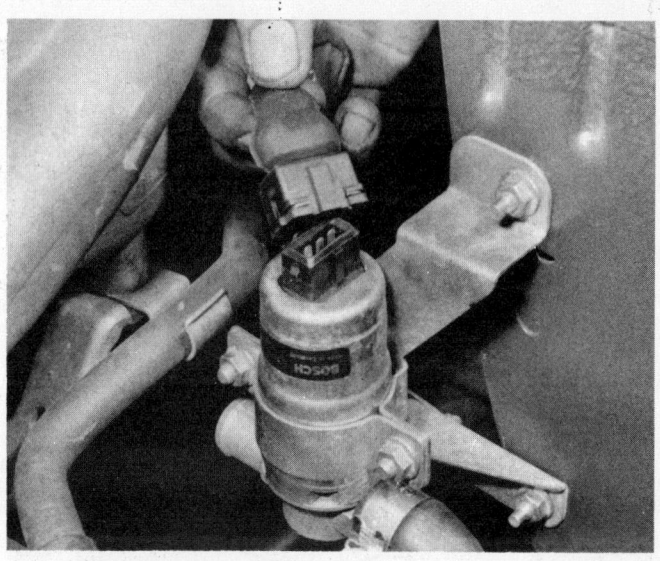

17.11 Air regulating valve (Type R fuel injection)

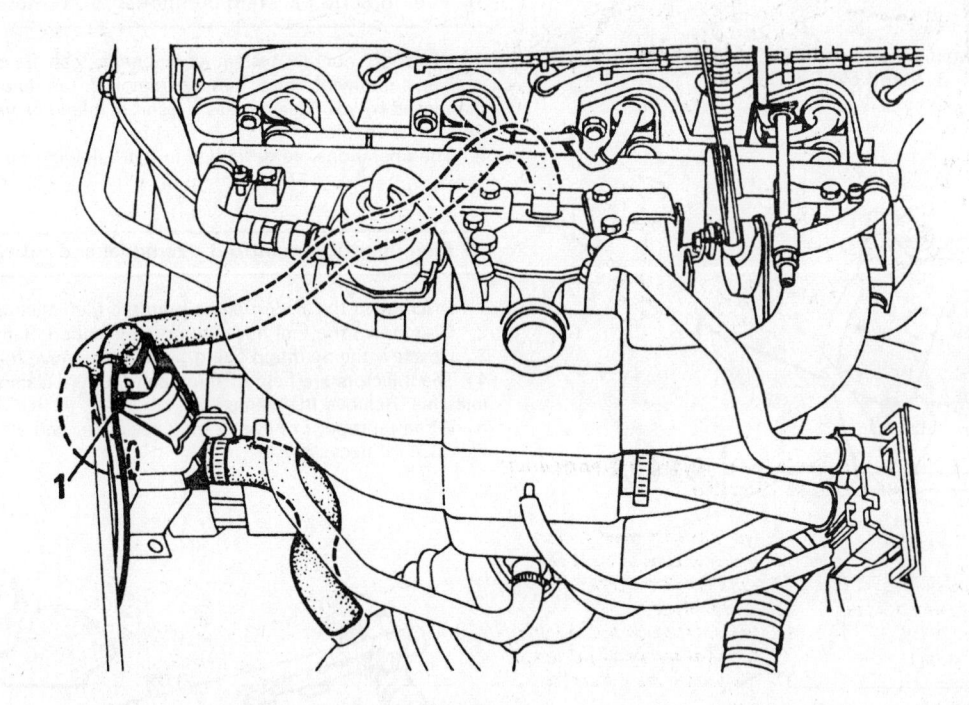

Fig. 3.26 Air regulating (idle speed adjusting) valve (1) on later models (Sec 17)

Air regulating valve
11 This is located on the left-hand headlamp mounting frame (photo).
12 If the valve is removed, take care when refitting it that the arrow on the base of the valve indicates the direction of air flow.
13 The valve can be checked in the following way. Remove the valve and rotate it rapidly in the hand first in one direction and then in the other. The valve should open and close.
14 To test with the valve in the car, pull off the wiring plug and apply battery voltage to terminal 4 (Fig. 3.27).
15 Connect terminal 5 briefly to earth with the engine running. The speed should drop rapidly to below 800 rev/min.
16 Connect terminal 3 briefly to earth and the speed should rise to over 2000 rev/min.

Three-position throttle switch
17 The switch can be removed if the wiring plug is pulled off and the two fixing screws removed.
18 When refitting align the flat on the switch to coincide with the flat on the throttle butterfly valve plate spindle and then move it in the direction of the arrow until the light throttle 'click' is heard.
19 Tighten the fixing screws.

Throttle butterfly valve housing idle stop (automatic transmission) – adjustment
20 On cars equipped with automatic transmission, the idle stop on the swivel lever should be correctly adjusted. With the accelerator pedal released, there should be a clearance of between 0.1 and 0.2 mm (0.0039 and 0.0079 in) as shown in Fig. 3.29.

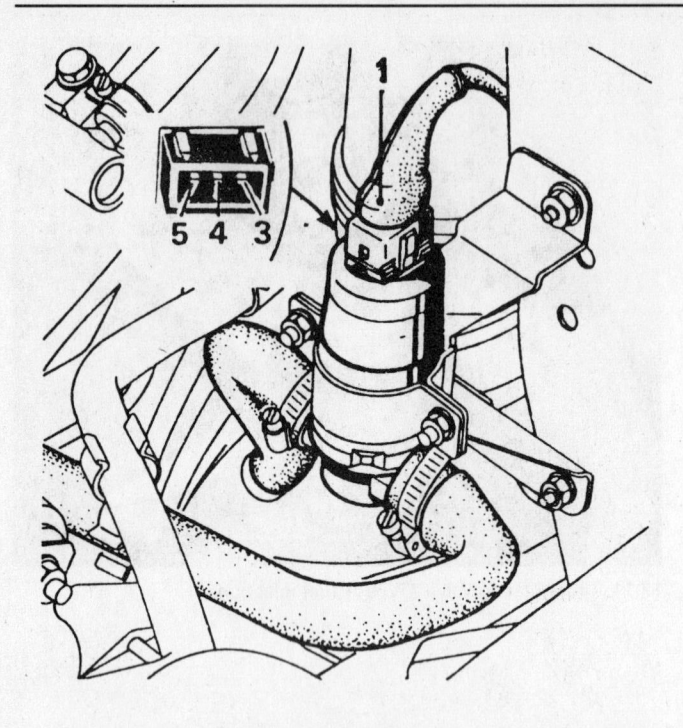

Fig. 3.27 Air regulating valve connector (1) and terminals (3, 4 and 5) (Sec 17)

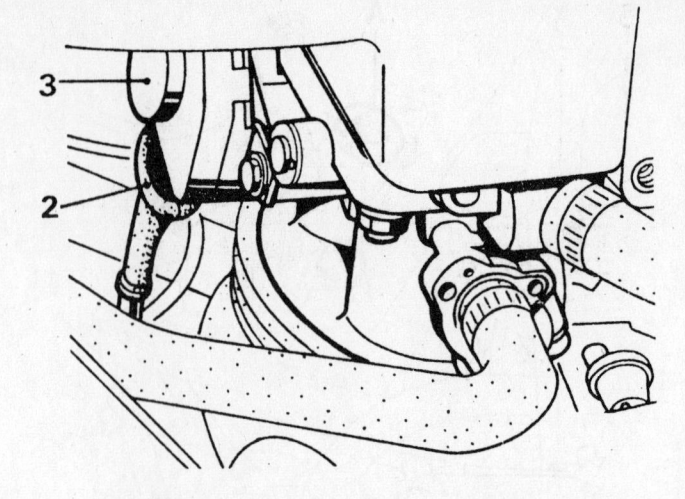

Fig. 3.28 Three-position throttle switch (Sec 17)

2 Connecting plug 3 Switch

18 Fuel injection system components – removal (general)

1 Although not all system components can be checked without special equipment, they can be removed for renewal if a fault is diagnosed by your dealer or as a result of reference to Sections 25 and 26 of this Chapter.
2 The operations are described in the following Sections.

19 Fuel injection manifold – removal and refitting

1 Disconnect the wiring plugs from the fuel injectors.
2 Disconnect the fuel hoses from the injection manifold.
3 Unscrew the manifold fixing bolts and remove the manifold.
4 The injectors are held to the manifold by a clamp for each pair of injectors. Remove the clamps.
5 When refitting, fit new 'O'-ring seals (4), and also new caps (5 – Fig. 3.31) if necessary.

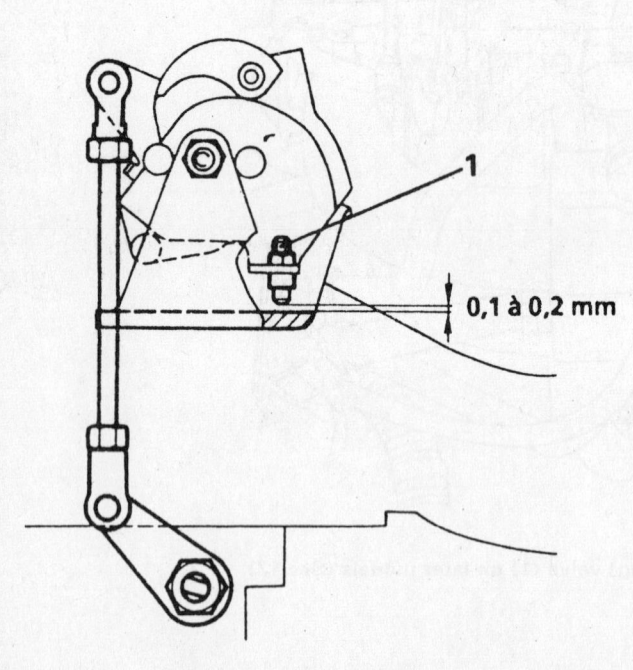

Fig. 3.29 Throttle switch (automatic transmission models) (Sec 17)

1 Adjusting screw and locknuts

21 Too high or too low an idle speed will probably be due to an air leak in the induction system or a fault in, or incorrect adjustment of one of the foregoing components.
Mixture
22 Adjustment of the mixture (CO level) is as described in earlier paragraphs of this Section.

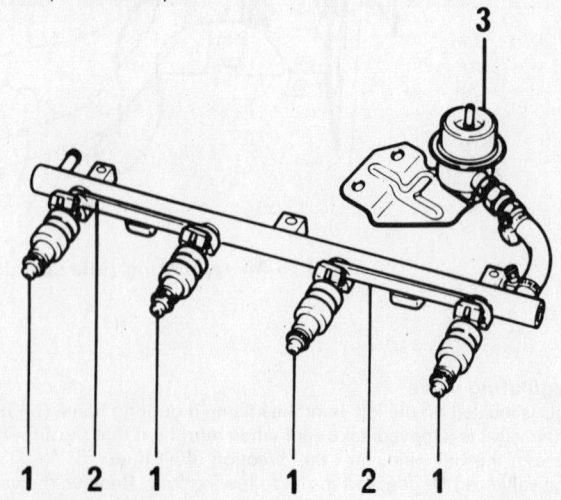

Fig. 3.30 Fuel injection manifold (Sec 19)

1 Injector 3 Fuel pressure regulator
2 Securing clamps

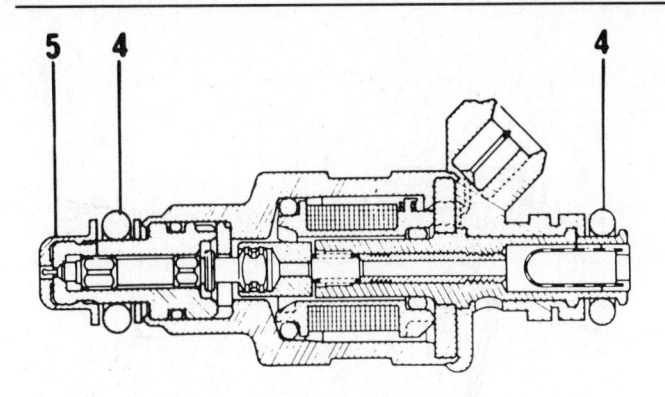

Fig. 3.31 Sectional view of a fuel injector (Sec 19)

4 O-ring   5 Cap

20 Fuel pressure regulator – removal and refitting

1 Disconnect the fuel and vacuum hoses from the pressure regulator.
2 Unbolt the support lug and unscrew the fixing nut and remove the pressure regulator.
3 Refitting is a reversal of removal.

21 Computer – removal and refitting

1 The computer is located on the left-hand wing valance within the engine compartment.
2 Release the clip, open the weatherproof case and remove the computer unit (photos).
3 Trace the routing of the wiring harness and disconnect the electrical connections as necessary.
4 Refitting is a reversal of removal.

22 Sensors – removal and refitting

Coolant temperature sensor
1 This should only be removed when the engine is cold.
2 Disconnect the electrical leads.
3 Remove the expansion tank cap to release any pressure in the cooling system.
4 Unscrew the sensor and quickly plug the hole to minimise coolant loss.
5 Smear the threads with a little gasket cement before screwing in the new sensor.

Air temperature sensor
6 Disconnect the wiring plug from the sensor.
7 Remove the air inlet duct from the throttle housing.
8 Remove the sensor.
9 Refitting is a reversal of removal.

Absolute pressure sensor
10 Disconnect the wiring plug from the sensor.
11 Disconnect the sensor from its mounting plate and the pipe from the inlet manifold.
12 Always remove the sensor by levering it off with a screwdriver, never pull on the connecting pipe. Refitting is a reversal of removal.

Speed sensor
13 Disconnect the wiring plug, unscrew the shouldered bolts and remove the sensor.
14 Refitting is a reversal of removal; no clearance adjustment is required.

21.2A Computer weatherproof case

21.2B Computer with weatherproof case removed

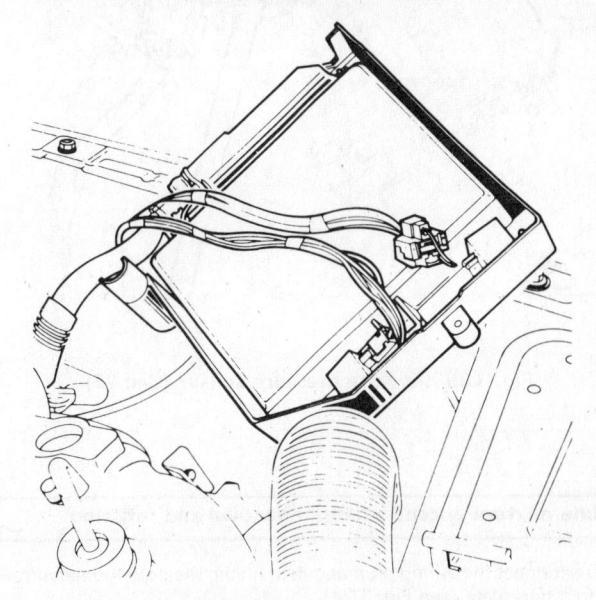

Fig. 3.32 Computer (Sec 21)

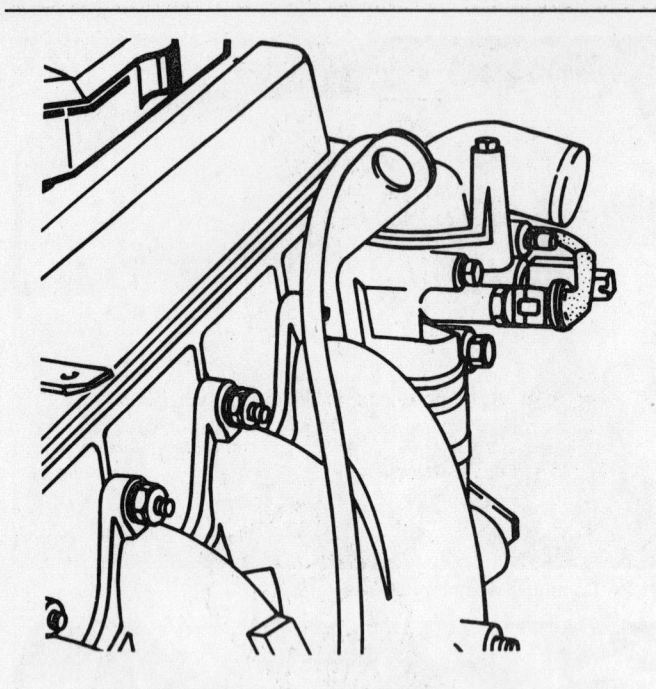

Fig. 3.33 Coolant temperature sensor (Sec 22)

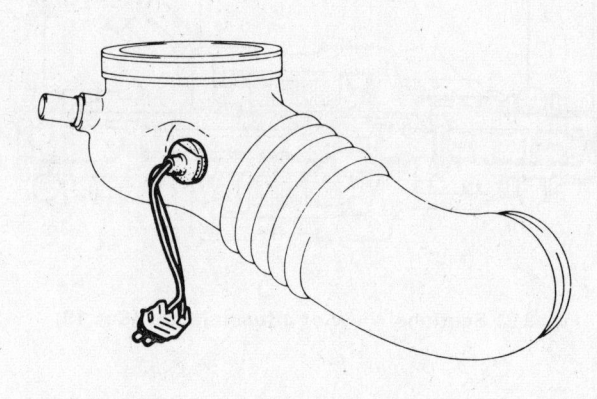

Fig. 3.34 Air temperature sensor (Sec 22)

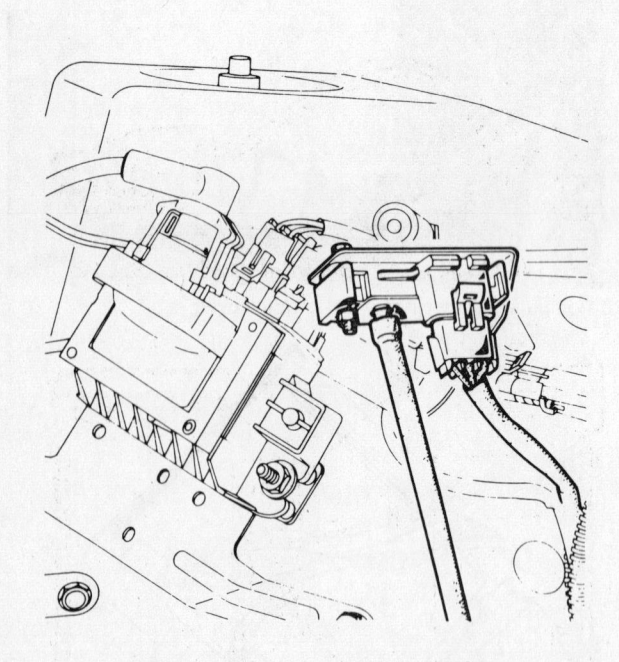

Fig. 3.35 Absolute pressure sensor (Sec 22)

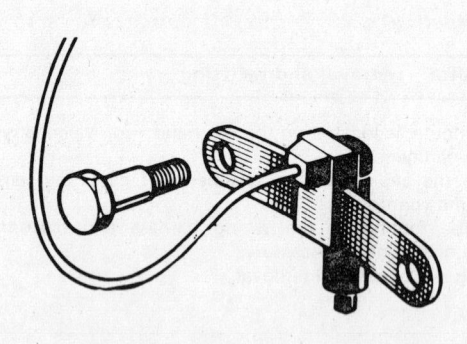

Fig. 3.36 Speed sensor (Sec 22)

23 Idle mixture potentiometer – removal and refitting

1 Disconnect the wiring plug and then unclip the potentiometer from its mounting plate (see Fig. 3.24).
2 After refitting, adjust the idle speed and mixture as described in Section 17.

24 No load/full load switch – removal, refitting and adjustment

Note: *An ohmmeter will be needed for switch adjustment*

1 This is a switch/sensor assembly which signals the computer when the throttle valve is fully open or fully closed.
2 The appropriate signal is sent 10° before the fully open position is reached and 2° before fully closed. This causes the fuel injectors to supply the greater volume of fuel at full throttle (maximum air increase).
3 The 'no load' position provides the deceleration shut-off.
4 To remove the switch, disconnect the wiring plug and drain the cooling system.
5 Unbolt and remove the inlet manifold.
6 Unbolt and remove the throttle housing.
7 Refitting is a reversal of removal, then adjust the switch.
8 Do this by connecting an ohmmeter to the switch terminals.
9 Release the screw (1 – Fig. 3.38) and turn the screw (3) until, with the throttle valve closed, 0 is indicated on the ohmmeter and, when fully open, infinity.

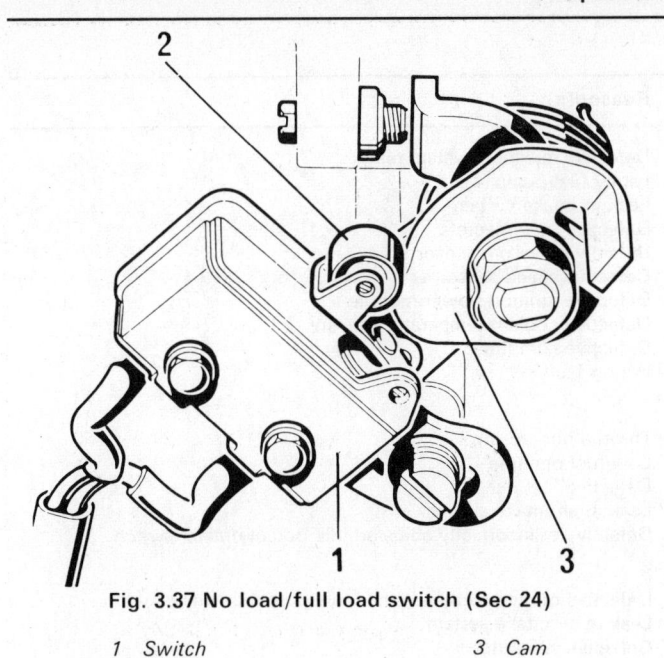

Chapter 3 Fuel system

Fig. 3.37 No load/full load switch (Sec 24)

1 Switch
2 Roller
3 Cam

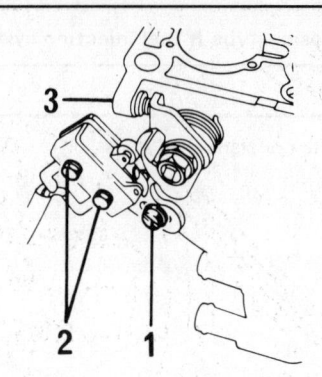

Fig. 3.38 No load/full load switch screws (Sec 24)

1 Clamp screw
2 Switch fixing screws
3 Adjustment screw

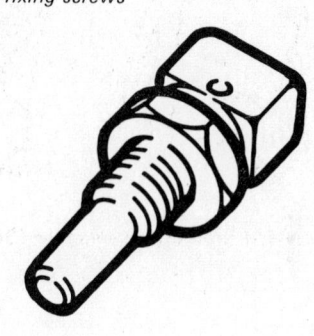

25 Component – testing

Note: *An ohmmeter will be needed for full component testing*

Fuel injectors
1 Disconnect the wiring plugs from the injectors.
2 Remove the injector manifold and place each injector in turn over a small container. Use self-locking grips to supplement the injector clamps.
3 Switch on the ignition; there should be no ejection of fuel. Now apply battery voltage to each injector in turn, fuel should be ejected.
4 Refit the injection manifold.

Coolant temperature sensor
5 Remove the sensor from the engine and leave it to stabilise for ten minutes. Now measure its resistance with an ohmmeter. At a temperature of between 19 and 21°C (66 and 70°F) resistance should be between 2.2 and 2.8 Kohms. At a temperature of between 79 and 81°C (174 and 178°F) resistance should be between 0.28 and 0.37 Kohms.

Air temperature sensor
6 Place an accurate thermometer in the air cleaner air intake as a means of measuring the sensor temperature level.
7 Using an ohmmeter, check the resistance using the following table.

Temperature	Resistance (Kohms)
0°C (32°F)	254 to 266
20°C (68°F)	283 to 297
40°C (104°F)	315 to 329

Absolute pressure sensor
8 First check the vacuum pipe and its connections, also make sure that the calibrated jet is correctly located.
9 Check for continuity between terminal (A) on the sensor connector and terminal (17) on the computer connector (Figs. 3.23 and 3.41).
10 Check the computer earth between terminals 1, 2 and 10 and a good earth.

Speed sensor
11 Measure the resistance at the sensor connector plug. It should be between 142 and 184 ohms.

Idle speed mixture potentiometer
12 Measure the resistance at its connector. It should be between 0 and 10 ohms fully anti-clockwise to 10 Kohms clockwise through a total movement of between 265 and 275°.

Fig. 3.39 Coolant temperature sensor removed (Sec 25)

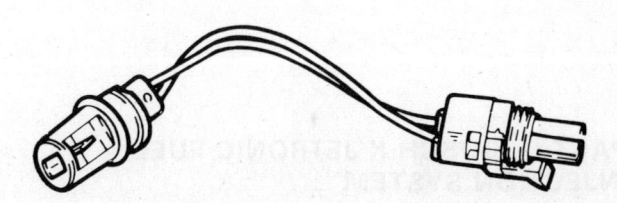

Fig. 3.40 Air temperature sensor removed (Sec 25)

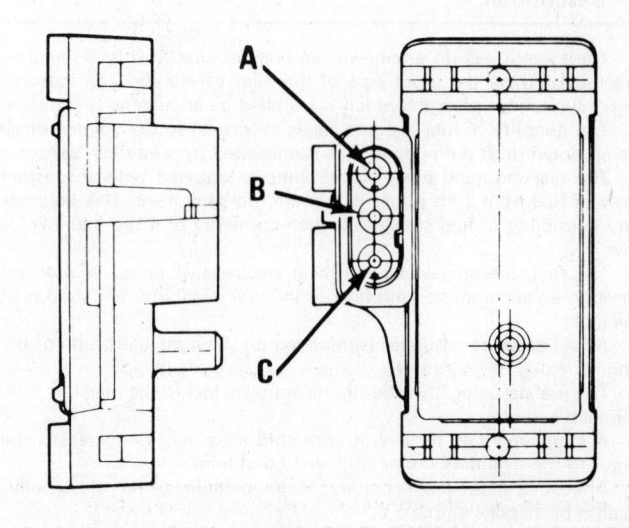

Fig. 3.41 Absolute pressure sensor and terminals (Sec 25)

A Earth
B Output voltage
C +5V

26 Fault diagnosis – type R fuel injection system

Symptom	Reason(s)
Engine will not start or starts then stops	Defective injection system relay Faulty fuel pump Leak in intake system Defective fuel injectors Defective pressure sensor Defective speed sensor Defective ignition power module Defective coolant temperature sensor Defective computer Wiring fault
Uneven idling	Throttle not closing Low fuel pressure Defective fuel injector Leak in air intake system Defective or incorrectly adjusted idle potentiometer switch
Poor acceleration	Defective or incorrectly adjusted idle potentiometer switch Leak in air intake system Defective computer Wiring fault
Misfiring	Poor central earth connection
Excessive fuel usage	Defective injectors High fuel pressure Defective coolant temperature sensor Defective computer Wiring fault

PART D: BOSCH K JETRONIC FUEL INJECTION SYSTEM

27 Description

The system has no engine-driven components. The final injectors are mounted on the input side of the inlet valves and continuously inject fuel under pressure which is supplied by an electric fuel pump.

The quantity of fuel injected varies according to the volume of air being drawn into the engine. This is measured by an airflow sensor.

The rear-mounted electric fuel pump is supplied with a constant flow of fuel from a lift pump within the fuel tank itself. This prevents any possibility of fuel starvation when cornering or if the fuel level is low.

The fuel, pressurised by the main electric fuel pump, is pumped through an accumulator and filter to the fuel distributor and cold start injector.

A fuel pressure regulator is mounted on the right-hand side of the engine.

The fuel metering/distributor unit supplies fuel to the injectors, one for each cylinder.

A supplementary air circuit with cold start injector increases the input to the cylinders under cold start conditions.

Airflow is regulated according to temperature by an electrically-heated bi-metallic strip.

Power for the injection system components is through a tachymetric relay located near the steering column. The relay in turn is connected to the ignition system and cuts off current to the fuel pumps, the control pressure regulator and the supplementary air device when less than one pulse per second is being received.

28 Idle speed and mixture – adjustment

1984 models
Note: *An exhaust gas analyser will be needed for this operation*

1　Have the engine at normal operating temperature.
2　The exhaust manifolds each have a manifold connection hole which is plugged (photo). Remove the plugs and connect pipes and

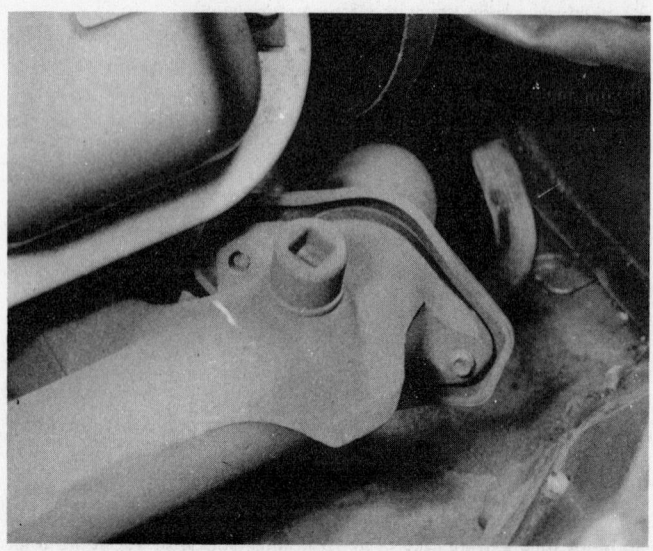

28.2 Exhaust manifold CO test hole plug

141

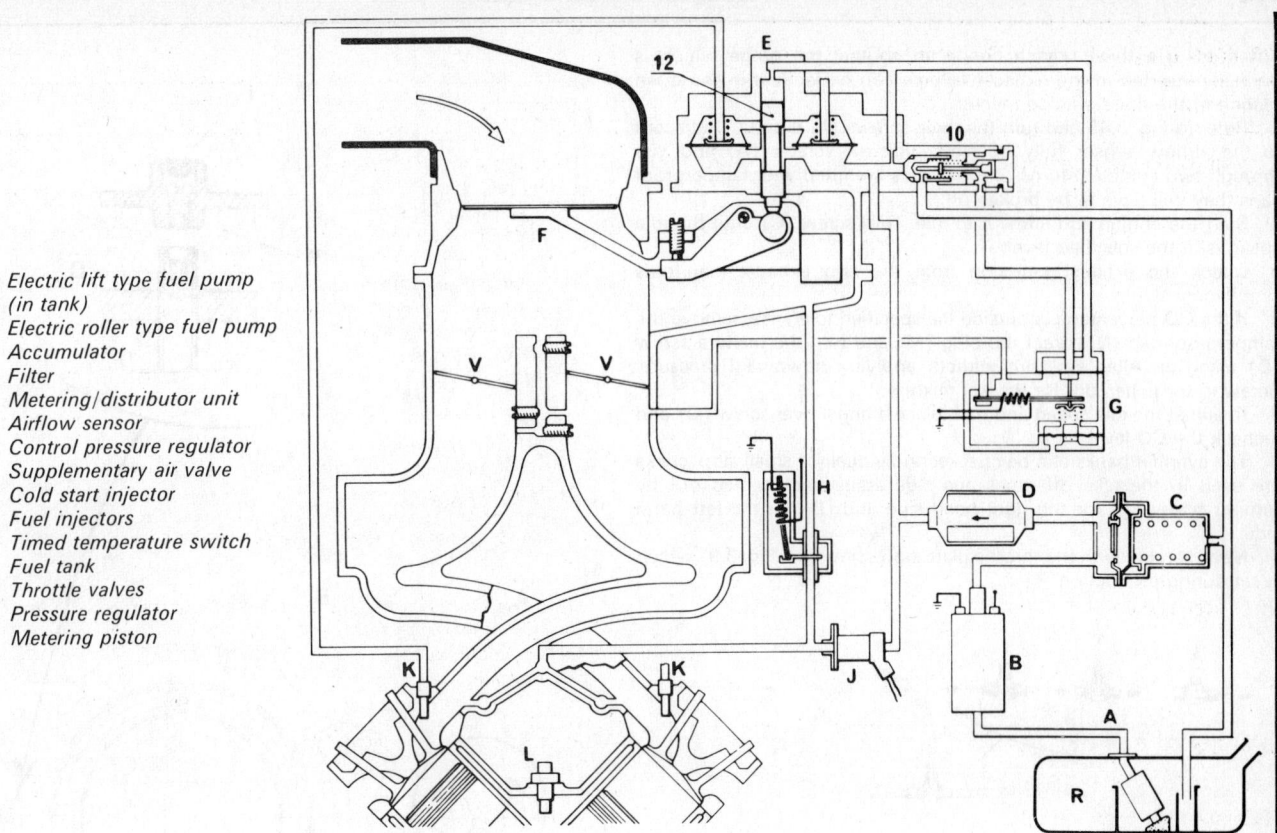

A   Electric lift type fuel pump (in tank)
B   Electric roller type fuel pump
C   Accumulator
D   Filter
E   Metering/distributor unit
F   Airflow sensor
G   Control pressure regulator
H   Supplementary air valve
J   Cold start injector
K   Fuel injectors
L   Timed temperature switch
R   Fuel tank
V   Throttle valves
10 Pressure regulator
12 Metering piston

Fig. 3.42 Bosch K Jetronic fuel injection system (Sec 27)

A   Fuel tank lift pump
B   Electric roller type fuel pump
C   Pressure accumulator
D   Fuel filter
E   Metering/distributor unit
F   Air flow sensor
G   Control pressure regulator
H   Supplementary air device
J   Cold start injector
K   Fuel injectors
L   Timed temperature switch
M   Tachymetric relay

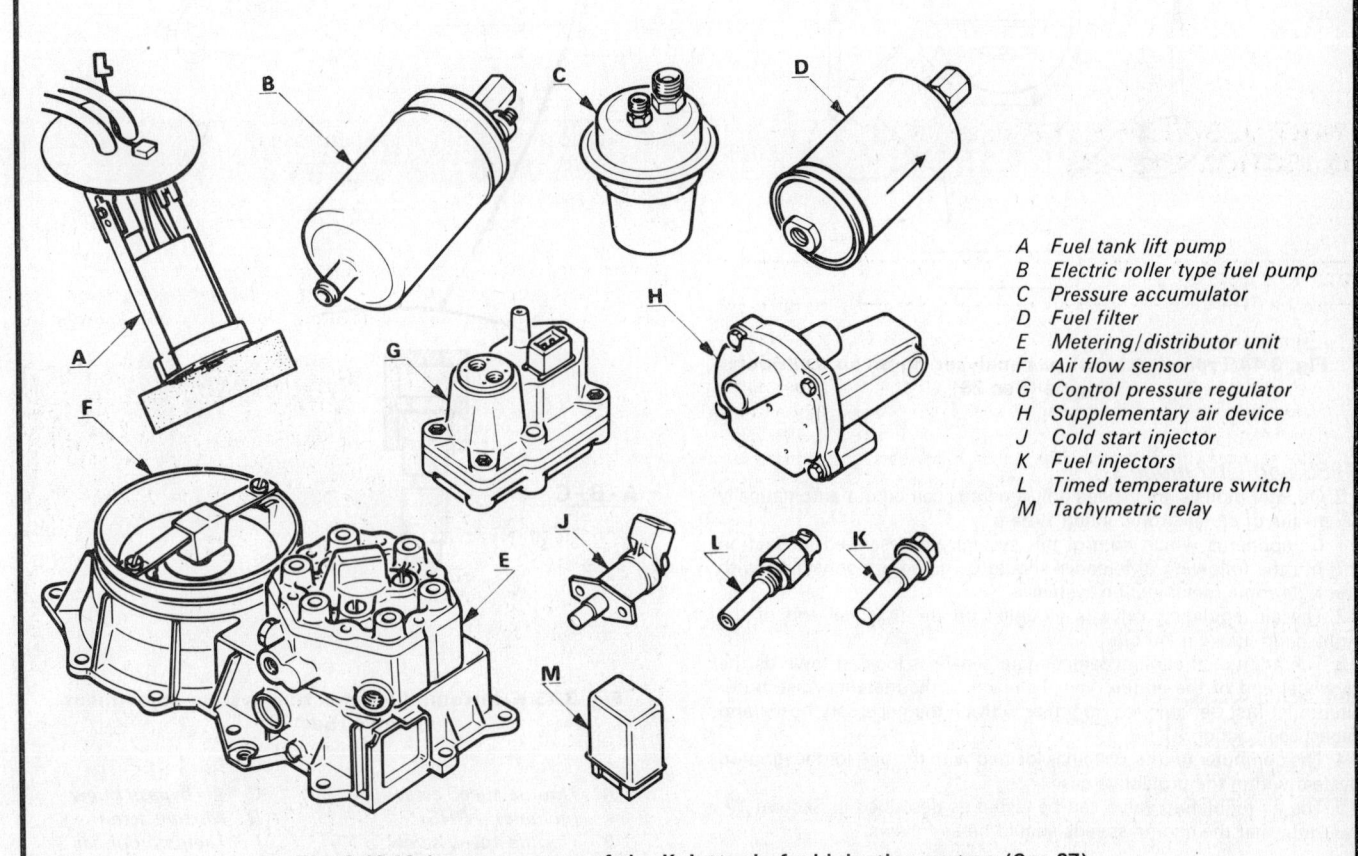

Fig. 3.43 Main components of the K Jetronic fuel injection system (Sec 27)

142　　　　　　　　　　　　　　　　　Chapter 3 Fuel system

hoses using a three branch connector so that the probe which is normally inserted in the exhaust tailpipe can be located in the open branch of the three-way connector.

3 Refer to Fig. 3.45 and turn the three screws (A, B and C) adjacent to the airflow sensor fully in. Now unscrew screws (A) and (B) through two complete turns. If the screws are fitted with tamperproof caps they will have to be broken off.

4 Start the engine and allow it to idle. Turn screw (C) until the idle speed is at the specified level.

5 Check the exhaust emission from the two exhaust manifolds simultaneously.

6 If the CO percentage is outside the specified tolerance, remove the tamperproof cap (I) extract the plug (M) and turn the mixture screw (D) using an Allen key, but without applying downward pressure. Screw in for richer, out for weaker mixture.

7 Readjust the idle speed and then place a finger over screw (D) and recheck the CO level.

8 The cylinder banks can be checked individually if small stop-cocks are used in the take off pipes and side adjustment carried out by turning screw (A) for the right-hand side and (B) for the left-hand side.

9 Never tamper with the throttle plate stop screw (4 – Fig. 3.9) which is set during production.

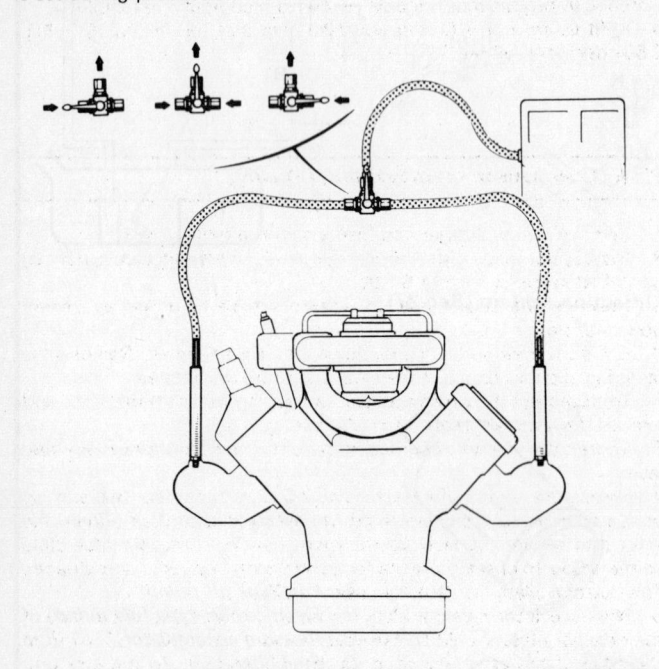

Fig. 3.44 Typical exhaust gas analyser rig on six-cylinder engine (Sec 28)

1985 and later models

10 On later models, idle speed adjustment is carried out automatically by means of an electronic idling system.

11 Components which control this system are described in Section 17, but the following differences should be noted in connection with the K Jetronic fuel injection system.

12 The air regulating valve is mounted on the flywheel end of the right-hand intake manifold.

13 The additional coolant temperature sensor is located towards the flywheel end of the engine Vee. To reach it, the metering/distributor unit must first be removed, together with all the necessary hoses and pipes; see Section 31.

14 The computer for the system is located with the one for the ignition system within the protective case.

15 The air regulating valve can be tested as described in Section 17, but note that the engine speeds should be as follows:

Earthing terminal 5 – speed drops to 825 rev/min (manual transmission) or 750 rev/min (automatic transmission)

Earthing terminal 3 – speed increases to 1500 rev/min or more

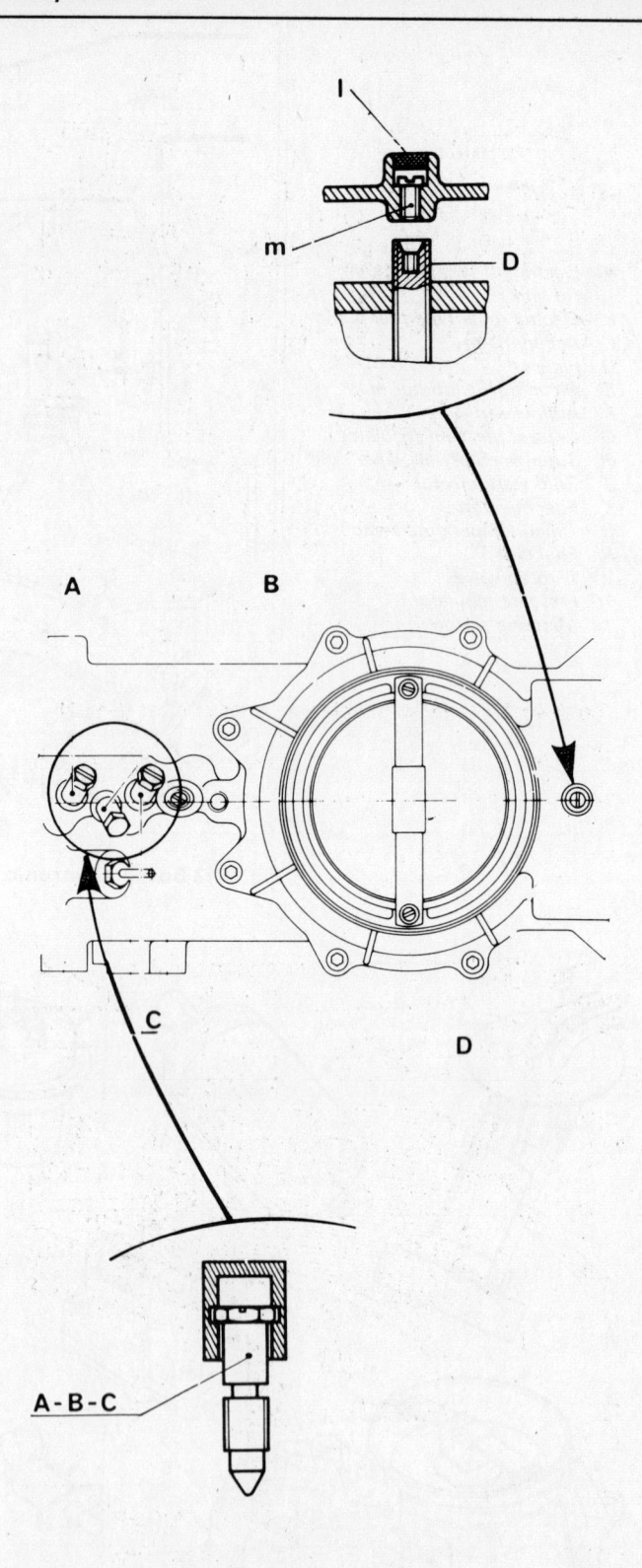

Fig. 3.45 K Jetronic fuel injection system adjustment screws (Sec 28)

A Throttle speed screw (volume) – RHS
B Throttle speed screw (volume) – LHS
C Air bypass screw
D Mixture screw
I Tamperproof cap
M Threaded plug

Chapter 3 Fuel system

29 Fuel injection system components – removal (general)

1 Although not all system components can be checked without special equipment, they can be removed for renewal if a fault is diagnosed by your dealer or as a result of reference to Sections 38 and 39 of this Chapter.
2 The operations are described in the following Sections.

30 Metering/distributor unit – removal and refitting

1 Remove the air cleaner (Section 3).
2 Disconnect the battery.
3 Identify (with numbered tape) the pipe connections to the unit, and unscrew and remove the bolts from the banjo type unions (photo).
4 Unscrew the three fixing bolts and remove the metering/distributor unit from the engine. Remove the O-ring seal.
5 Do not remove the plugs (1 – Fig. 3.49). In fact it is not recommended that the unit is dismantled, as some internal parts are matched in production. If faulty, renew as a complete assembly.
6 Refit using a new O-ring and take care that the piston (Q – Fig. 3.50) does not fall out.

31 Airflow sensor – removal and refitting

1 Remove the air cleaner and disconnect the battery.
2 Disconnect the pipes from the fuel metering/distributor as described in the preceding Section.
3 Unscrew the socket-headed screws which hold the air sensor upper casing.
4 Lift off the air sensor/metering/distributor assembly. Remove the metering/distributor unit if the sensor is to be dismantled.
5 To dismantle the airflow sensor, extract the circlip (Fig. 3.51) and remove the washers, seals, balls and spring.
6 Loosen the screws (17), and remove the pivot pin rocker arm and lever.
7 Renew the seals when reassembling and centralise the arm by wrapping a sheet of paper around the detector plate then tighten the screws to secure the lever to the pivot pin. Turn the detector plate centre screw to check the plate for centralising and remove the paper. The detector plate top face is marked 'TOP'.
8 Now check the detector plate height adjustment. The upper face of the detector plate should be between 0 and 0.1 mm (0.0039 in) from the edge of the air intake cone on the fuel metering/distributor unit side. If the plate is too high, it can be lowered by carefully tapping down the guide pin (4 – Fig. 3.53) using a punch. Do not overdo this as trying to knock it back would not prove satisfactory.
9 Refitting the air flow sensor is a reversal of removal, but renew all seals.

32 Lower air casing – removal and refitting

1 Remove the air cleaner.
2 Disconnect the fuel pipes from the metering/distribution unit, as described in Section 30. Remove the metering/distributor unit.
3 Release the clip on the hose which connects the air intake tube (50 – Fig. 3.54) to the supplementary air valve and free the hose. The other end of the hose connects to the lower air casing.
4 Unscrew the socket-headed screws and the bolt which hold the air casing to the inlet manifold (photo).
5 Disconnect the link rod (6 – Fig. 3.9) from the throttle reel. Note the balljoint clip.
6 Withdraw the lower air casing and disconnect the vacuum hose (photo).
7 Refitting is a reversal of removal, but renew all seals.

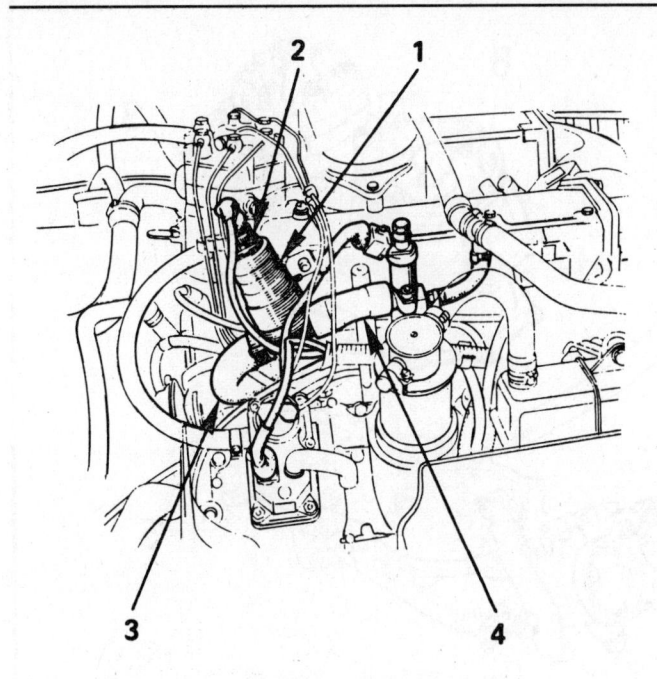

Fig. 3.46 Air regulating valve on later models (Sec 28)

1 Valve
2 Connecting plug
3 Hose
4 Hose

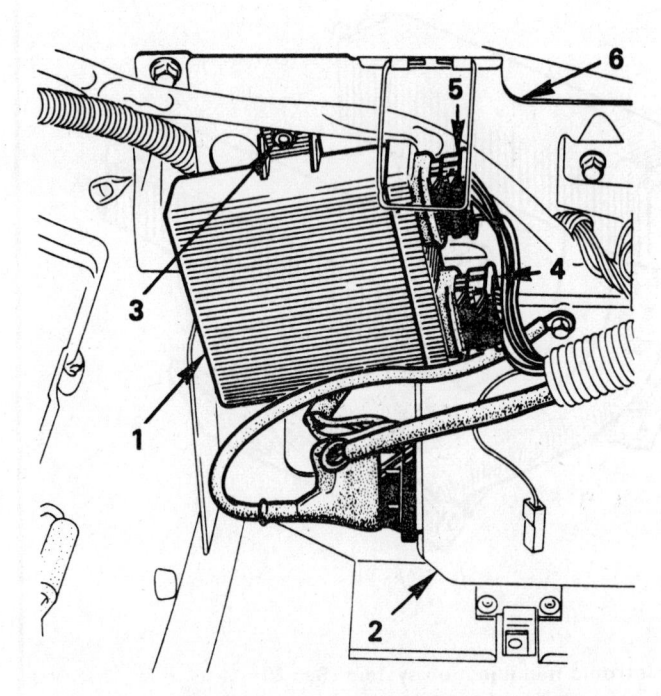

Fig. 3.47 Automatic idle speed adjustment computer on later models (Sec 28)

1 Computer (automatic idling)
2 Computer (ignition)
3 Fixing screw
4 Five-pin connector
5 Six-pin connector
6 Support plate

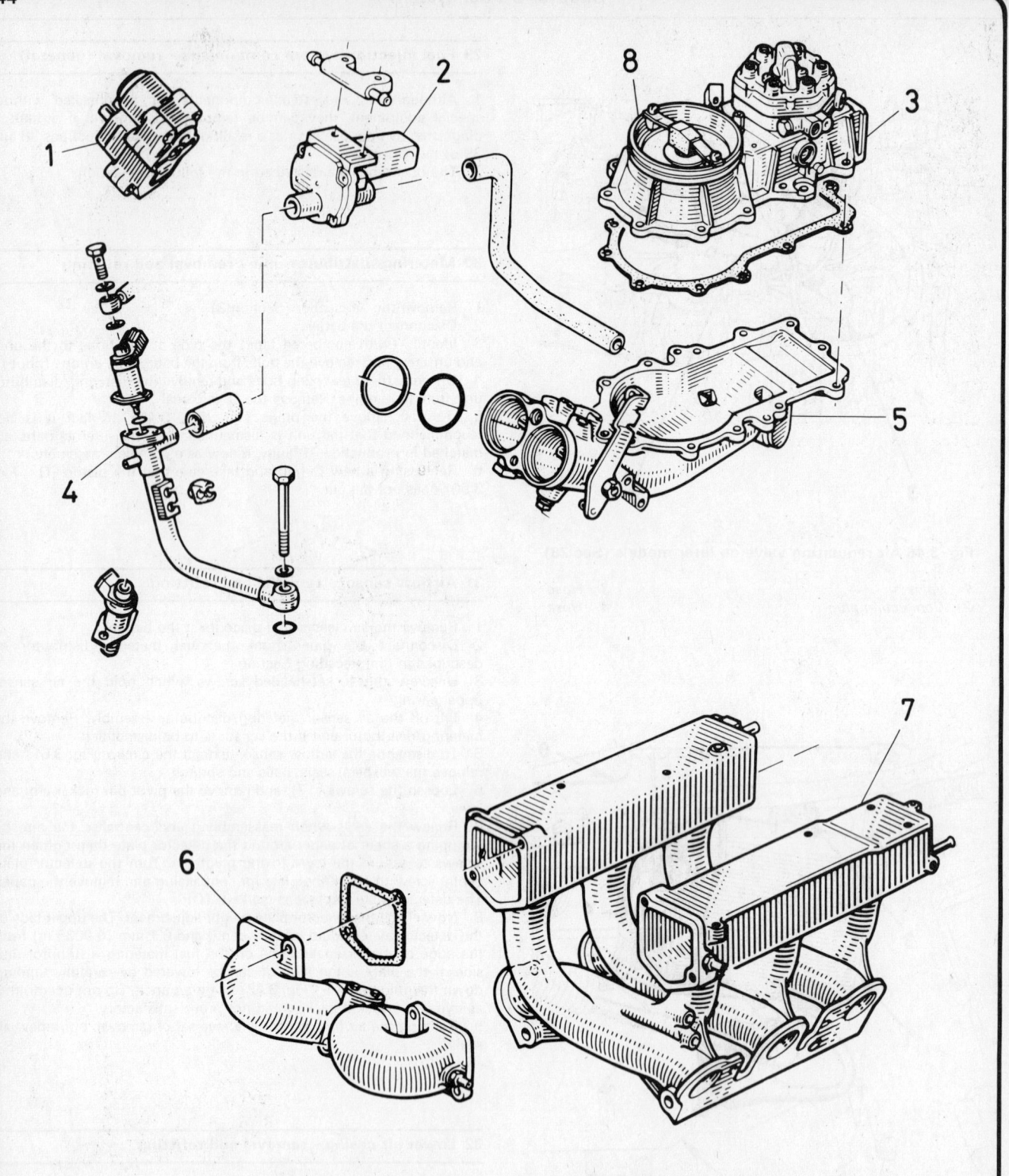

Fig. 3.48 Main engine components of the K Jetronic fuel injection system (Sec 29)

1   Control pressure regulator
2   Supplementary air valve
3   Fuel metering/distributor unit
4   Cold start injector
5   Lower air casing
6   Air distributor front manifold
7   Intake manifold/air distributor
8   Airflow sensor

30.3 Unscrewing banjo type union bolt

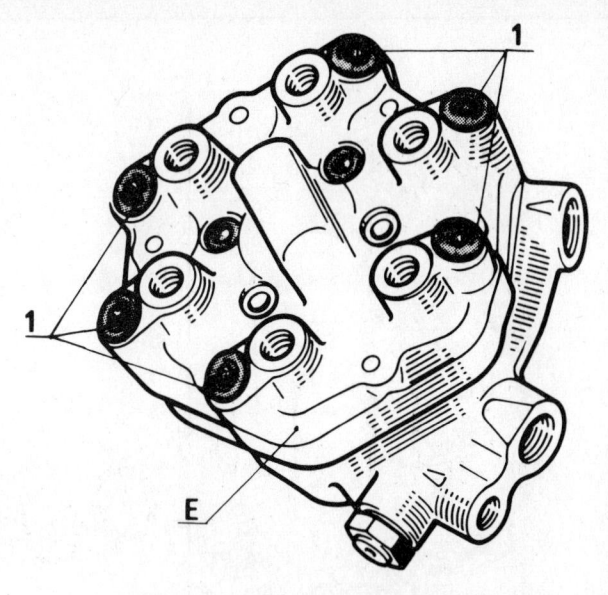

Fig. 3.49 Fuel metering/distributor unit (Sec 30)

1 Plugs (do not remove)   E Unit upper section

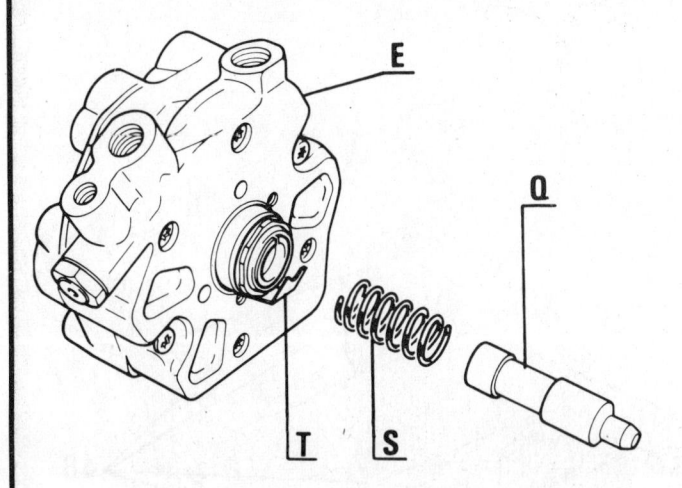

Fig. 3.50 Fuel metering unit components (Sec 30)

E   Unit   S   Coil spring
Q   Metering piston   T   Tab

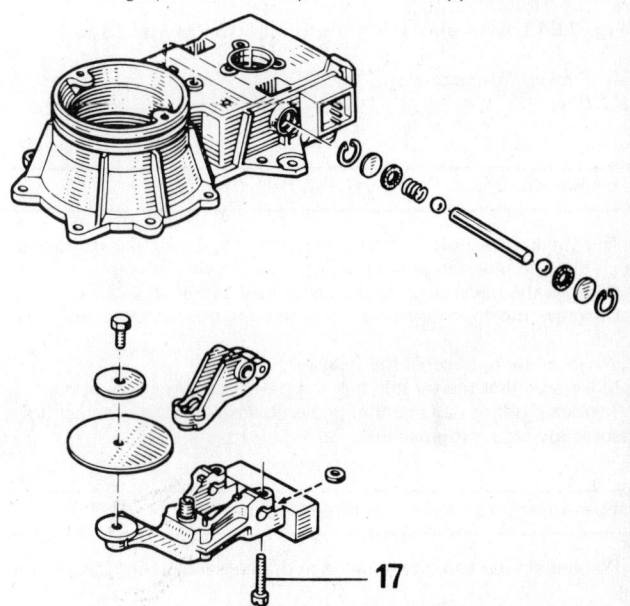

Fig. 3.51 Air flow sensor components (Sec 31)

17   Screw

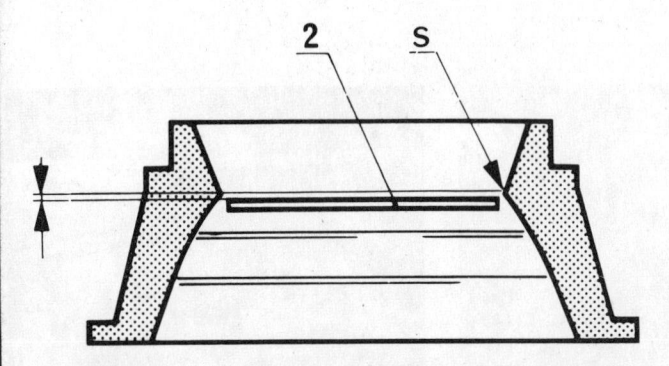

Fig. 3.52 Airflow sensor detector plate height setting diagram (Sec 31)

2   Plate stop   S   Edge of air intake cone

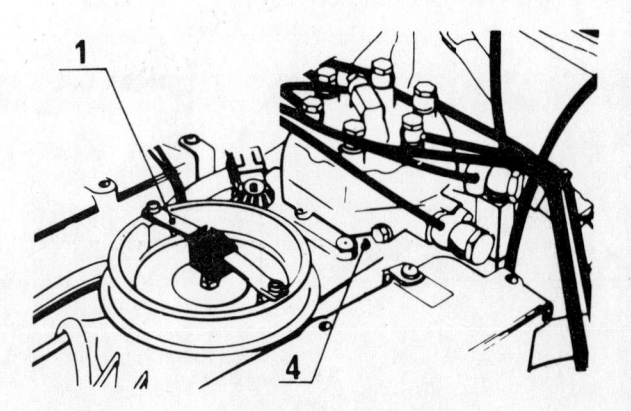

Fig. 3.53 Air flow sensor detector plate support (1) and guide pin (4) (Sec 31)

Chapter 3 Fuel system

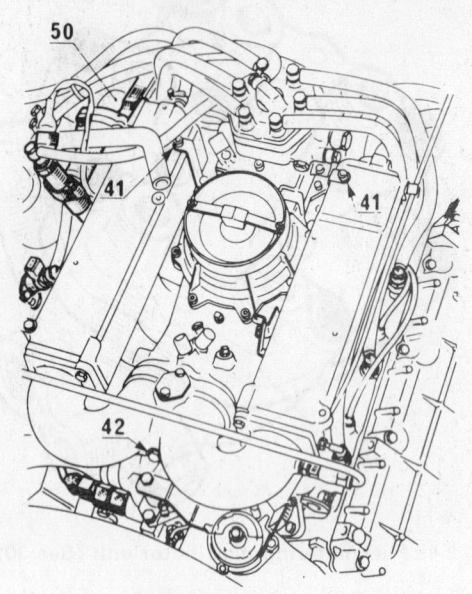

Fig. 3.54 Lower air casing disconnection points (Sec 32)

41 Socket-head screw
42 Bolt
50 Air intake tube

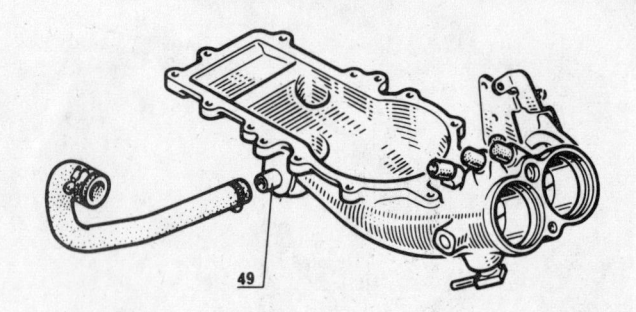

Fig. 3.55 Air intake tube connecting hose at lower air casing (49) (Sec 32)

33 Intake manifold – removal and refitting

1 The intake manifold is combined with the air intake distributor which form an integrated unit supplying both cylinder banks.
2 Remove the lower air casing as described in Section 32.
3 Unscrew the four retaining bolts and lift the manifold from the engine.
4 When refitting, observe the following points.
5 Make sure that the throttle link rod passes **under** the support.
6 Take care not to damage the ignition distributor cap as the airflow sensor is lowered into position.

34 Fuel injectors – removal and refitting

1 Disconnect the banjo type union from the fuel metering/distributor unit.
2 Release the fuel injector clip and union. Use an open-ended spanner to hold the flats on the injector against rotation while the union is unscrewed (photo).
3 Refitting is a reversal of removal.

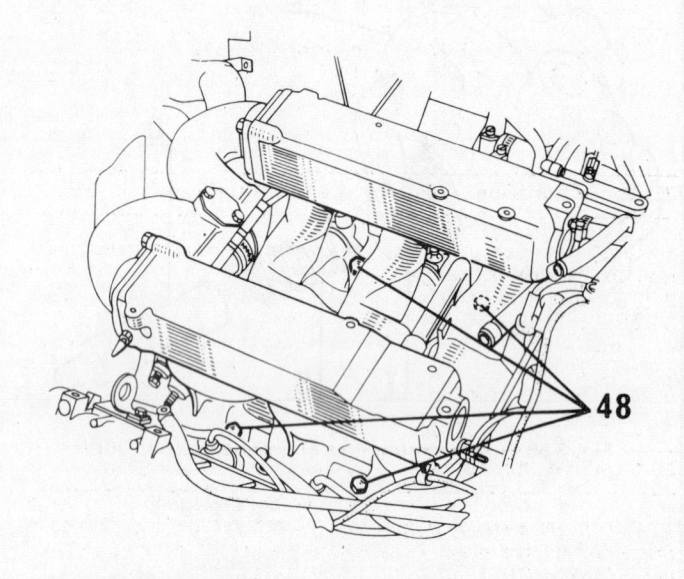

Fig. 3.56 Intake manifold fixing bolts (48) (Sec 33)

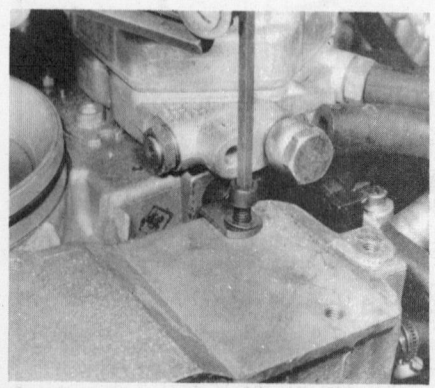

32.4 Unscrewing air casing socket-headed screw

32.6 Lower air casing vacuum hose

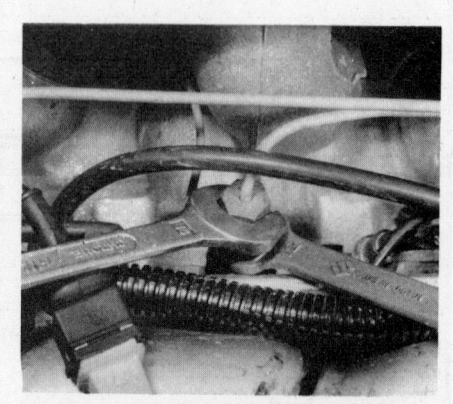

34.2 Releasing a fuel injector

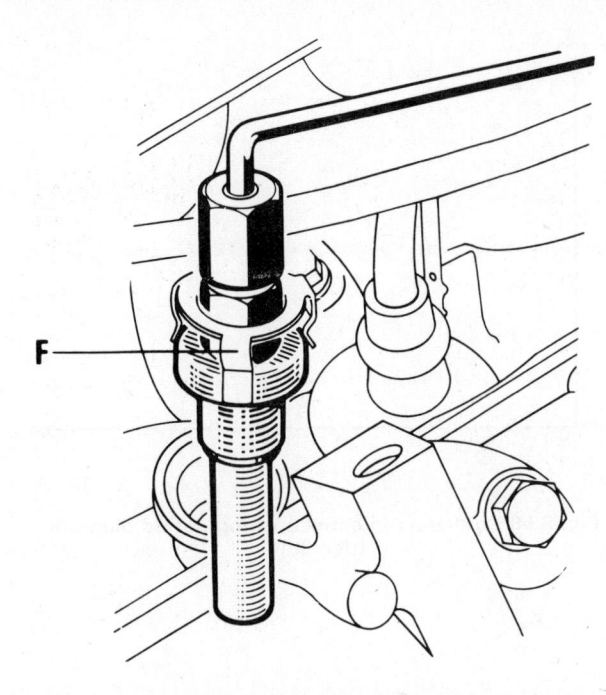

Chapter 3 Fuel system

Fig. 3.57 Fuel injector (Sec 34)

F Clip

35.1 Control pressure regulator

36.1 Cold start injector

36.2 Supplementary air valve

35 Control pressure regulator – removal and refitting

1 Disconnect the fuel pipes by unscrewing the bolts from the banjo type unions (photo).
2 Disconnect the wiring plug.
3 Disconnect the flexible hose.
4 Unscrew the socket-headed mounting bolts, not the flange connecting bolts, and remove the regulator.
5 Refitting is a reversal of removal.

36 Supplementary air valve – removal and refitting

1 Disconnect the wiring plug from the cold start injector, also disconnect the fuel pipe by unscrewing the bolt from the banjo type union (photo).
2 Disconnect the flexible hoses from the supplementary air valve (photo).
3 Remove the valve, together with the cold start injector.
4 Refitting is a reversal of removal.

37 Cold start injector – removal and refitting

1 Disconnect the fuel pipe and wiring plug from the cold start injector.
2 Unscrew the flange socket-headed screws which hold the cold start injector to the supplementary air valve and remove it.
3 Refitting is a reversal of removal, but use new copper washers at the banjo union and a new flange gasket.

38 Fuel injection system components – testing

Note: A test lamp or voltmeter and an ohmmeter will be needed for these tests.

1 Without a pressure gauge and other special equipment the components which can be tested are limited to the following items.

Chapter 3 Fuel system

2 Before testing system components, first check that all system wiring and connecting plugs are secure and that fuel hoses and pipelines unions are tight.

Cold start injector

3 With the engine hot and idling, directly connect the cold start injector terminals to the battery. The engine should start to choke by an excess of fuel.
4 With the engine cold, disconnect the electrical leads from the cold start injector, and connect a test lamp or voltmeter between the injector terminals.
5 Operate the starter motor and a voltage will be indicated, or the test lamp will illuminate and stay on for a period allied to the temperature of the timed temperature switch (Fig. 3.58).
6 To check the timed temperature switch, first remove it and quickly plug the hole to prevent loss of coolant.
7 Place the switch in cold water and connect a test wire between the battery positive terminal through a test lamp to one terminal of the timed temperature switch. Connect another wire between the battery negative terminal and the switch body.
8 The test lamp should illuminate or voltage be indicated on the voltmeter.
9 Heat the water and the test lamp should go out or zero voltage be indicated at between 31 and 39°C (88 and 102°F).
10 The resistance between the terminals of the switch connector and earth can be checked using an ohmmeter. Test first at temperatures below 30°C (86°F) then higher than 40°C (104°F).
11 Record the readings and compare them with the following table.

Temperature	Terminal G to earth	Terminal W to earth	Terminal G to W
Less than 30°C (86°F)	25 to 40 ohms	0 ohms	25 to 40 ohms
More than 40°C (104°F)	50 to 80 ohms	100 to 160 ohms	50 to 80 ohms

Supplementary air valve

12 Without removing the valve, which should be at a temperature of approximately 20°C (68°F), disconnect the electrical plugs and air hoses.
13 Look through the air ducting and observe the diaphragm which should be partially open.
14 Connect battery voltage direct to the valve terminals. After a period of five minutes, the diaphragm should be fully closed. If it is not, check the resistance between the valve terminals which should be between 25 and 35 ohms.
15 Any failure to perform in accordance with the foregoing tests, will mean renewal of the component concerned.

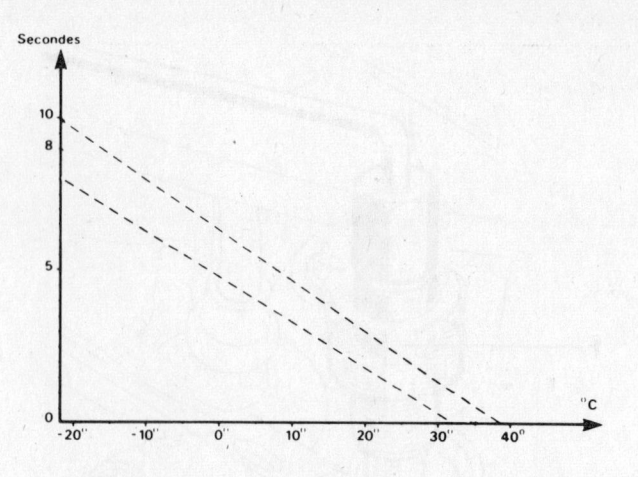

Fig. 3.58 Cold start injector opening period diagram (Sec 38)

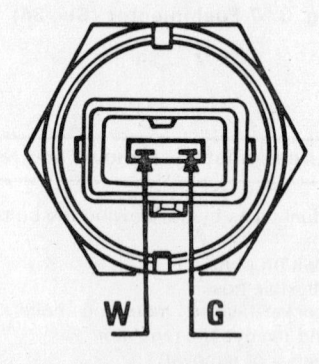

Fig. 3.59 Timed temperature switch connecting plug terminals (Sec 38)

G and W See text

39 Fault diagnosis – Bosch K Jetronic fuel injection system

Symptom	Reason(s)
Poor cold starting	Faulty fuel pump Cold start injector not opening Supplementary air valve not opening Cold start injector leaking Timed temperature switch faulty Leak in fuel system Airflow sensor plate or piston sticking Clogged air cleaner element Air intake system leaking
Poor hot starting	Faulty fuel pump Control pressure too high or too low Air intake system leaking Airflow sensor plate or piston sticking Injector leaking Mixture too rich or too weak

Chapter 3 Fuel system

Sympton	Reason(s)
Rough idling during warm up	Cold start injector leaking Supplementary air valve not closing or opening Defective vacuum advance Clogged air cleaner element Leak in air intake system Injector leaking Leak in fuel system
Rough idling when hot	Cold start injector leaking Supplementary air valve not closing Injector leaking Defective vacuum advance Mixture incorrect Leak in air intake system Airflow sensor plate or piston sticking
Backfiring in inlet manifold	Weak mixture Leak in air intake system Defective vacuum advance circuit
Backfiring in exhaust system	Rich mixture Cold start injector leaking Leak in fuel system
Misfire at normal roadspeeds	Faulty contact in fuel pump Fuel pressure too high or too low Leak in fuel system
Lack of power	Clogged air cleaner Throttle not opening fully Weak mixture Rich mixture Cold start injector leaking Leak in air intake system Airflow sensor plate or piston sticking Control pressure too high or low
'Pinking'	Injector leaking or low opening pressure Airflow sensor plate or piston sticking
Inability to adjust idling speed	Supplementary air valve not closing

PART E: TURBOCHARGER

40 Description

A Garrett T3 turbocharger is used on the 2458 cc V6 engine.
The turbocharger uses the exhaust gas pressure to drive a turbine which pressurises the inlet system, thus providing increased efficiency and power. The turbine rotates at a very high speed and has its own oil supply direct from the engine.
Air is supplied by the turbocharger which is located between the V cylinder banks above the clutch housing. Between the compressor and the intake manifold, the air passes through an intercooler and plenum chamber.
Regulation of boost pressure is achieved by an exhaust system wastegate operated by a pressure sensor in the intake manifold.
A bypass circuit prevents the turbocharger supplying boost during deceleration.
A safety feature is the provision of a pressure sensor to cut the fuel supply off if boost pressure becomes excessive.
Under-bonnet temperature is controlled after switching off the ignition by an extractor fan and associated ducting.
Engine oil is cooled by passing it through a cooler incorporated in the oil filter base.
Although the fuel system used on Turbo versions is the R (Renix) type already described, the location of the pressure regulator should be noted (Fig. 3.62), also the various take-off points and safety components used specifically on this engine and shown in Fig. 3.63.

The air intake system is fitted with special type hose clips and if they ever require renewal use only exact replacements.
The crankcase ventilation system is modified to incorporate an oil separator.

41 Throttle casing – removal and refitting

1 Remove the inlet manifold from its location between the air intercooler and the throttle casing.
2 Disconnect the electrical leads from the throttle switch and the air temperature sensor.
3 Unbolt and remove the throttle casing.
4 Refitting is a reversal of removal.
5 Check the link rod adjustment. If necessary, pass a 5.0 mm diameter rod through the holes in the throttle drum/support and adjust the length of the rod with the throttle held against the idle stop (Fig. 3.66).

42 Air intercooler – removal and refitting

1 Disconnect and remove the battery.
2 Release the braking system anti wheel lock computer and move it aside.
3 Remove the battery support.
4 Disconnect the intercooler ducts, remove the mounting bolts and withdraw the intercooler.
5 Refitting is a reversal of removal.

150

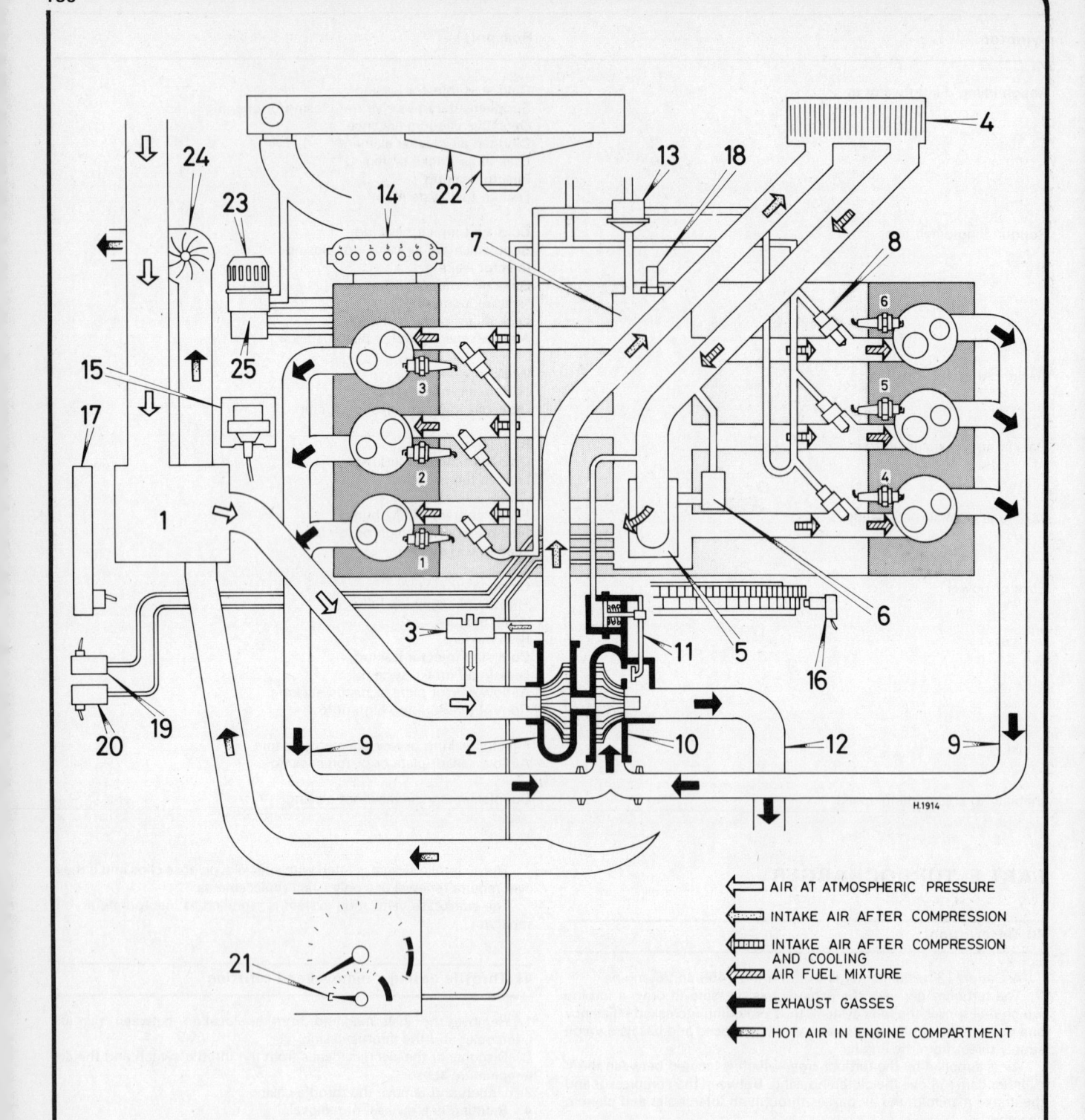

Fig. 3.60 Turbocharger system (Sec 40)

1 Air cleaner
2 Compressor
3 Compressor spill valve
4 Intercooler
5 Plenum chamber
6 Idle speed regulator valve
7 Inlet manifold
8 Injectors
9 Exhaust manifolds
10 Turbine
11 Turbo control (aneroid and wastegate)
12 Exhaust pipe
13 Fuel pressure regulator
14 Ignition distributor
15 Ignition power module and coil
16 Position/speed sensor
17 Injection/ignition computer
18 Knock detector
19 Injection system pressure sensor
20 Safety pressure sensor
21 Boost pressure sensor
22 Radiator with electric cooling fan
23 Oil filter
24 Hot air extractor fan
25 Oil cooler

⇐ AIR AT ATMOSPHERIC PRESSURE
⇐ INTAKE AIR AFTER COMPRESSION
⇐ INTAKE AIR AFTER COMPRESSION AND COOLING
⇐ AIR FUEL MIXTURE
⬅ EXHAUST GASSES
⬅ HOT AIR IN ENGINE COMPARTMENT

151

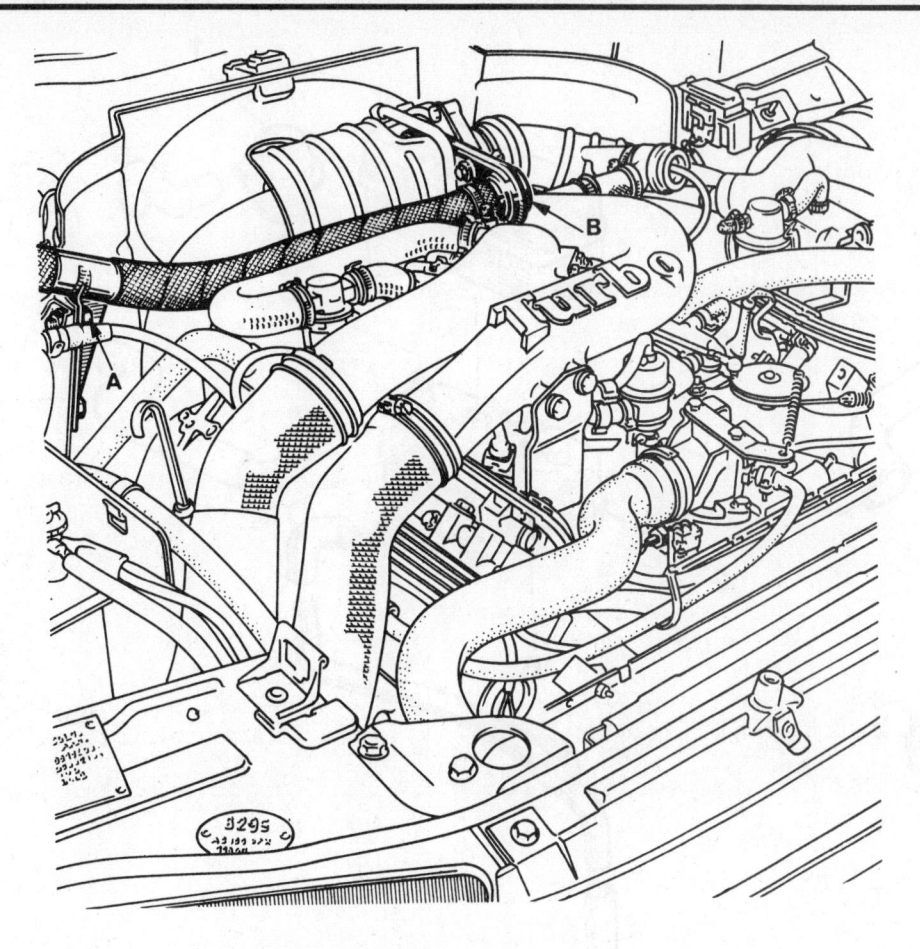

Fig. 3.61 Turbocharger cooling system (Sec 40)

A Air duct   B Scoop fixed to heat shield

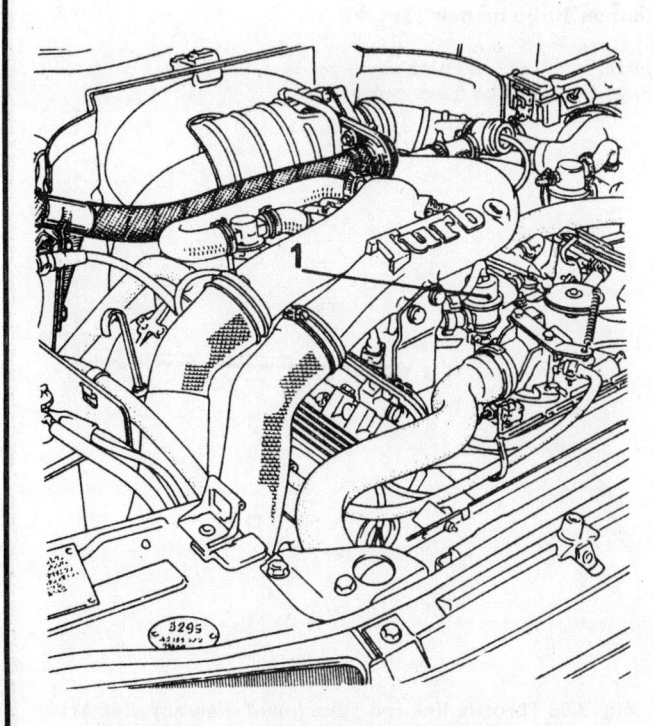

Fig. 3.62 Fuel pressure regulator (1) (Sec 40)

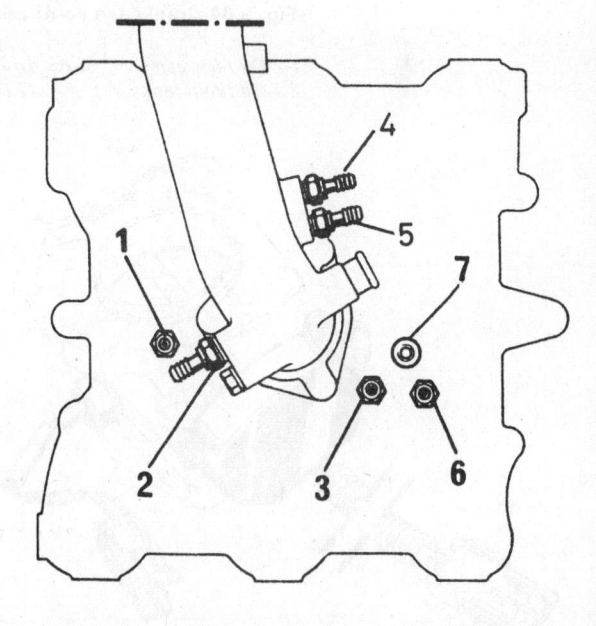

Fig. 3.63 Turbocharger take-off points (Sec 40)

1 Fuel pressure regulator
2 Turbocharger pressure gauge
3 Bypass valve
4 Safety pressure limiter
5 Turbocharger pressure regulator
6 Pressure sensor
7 Brake servo

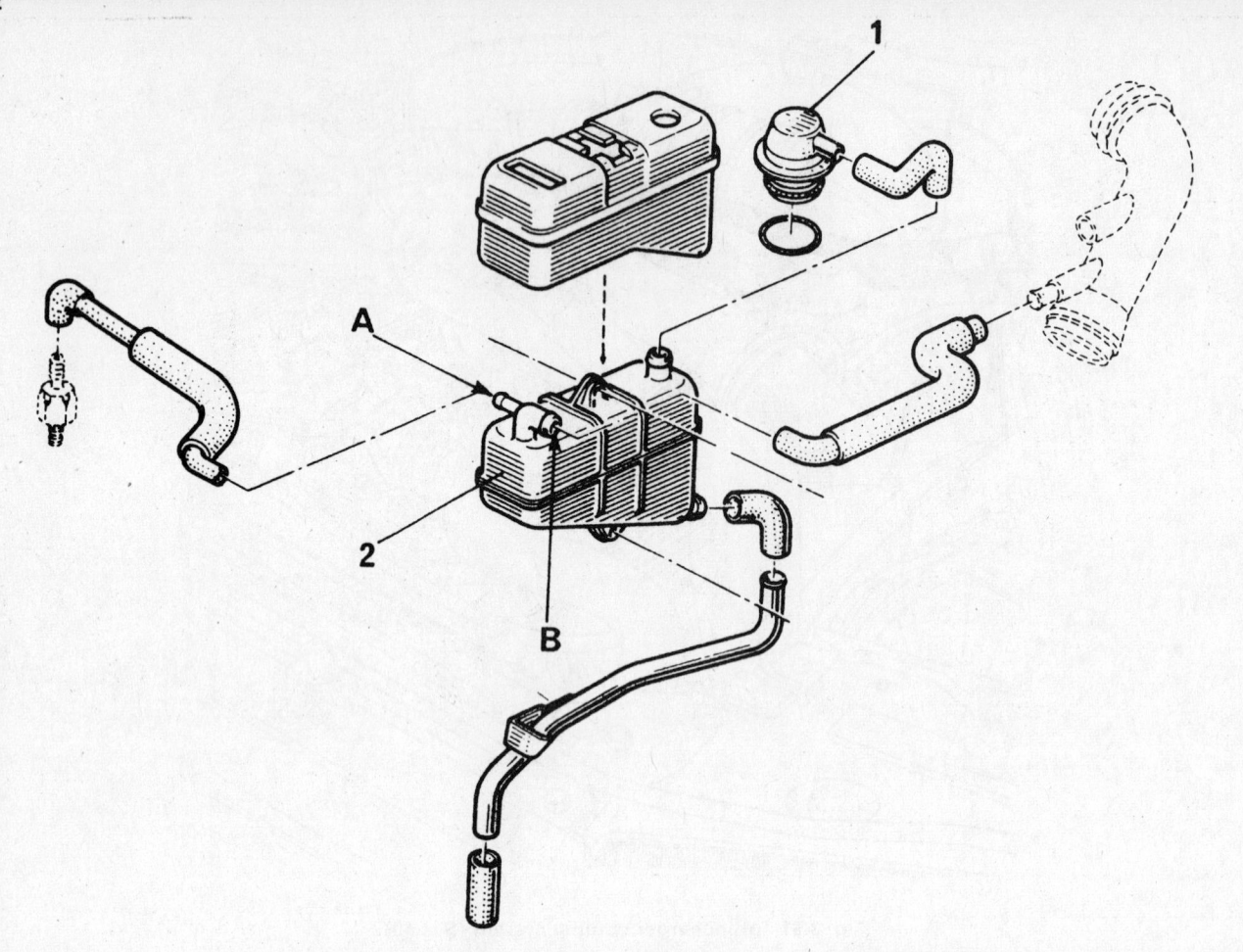

Fig. 3.64 Crankcase ventilation system on Turbo models (Sec 40)

1 Oil filler cap
2 Oil drain tank
A To inlet manifold (2.0 mm orifice)
B To intake duct (8.5 mm orifice)

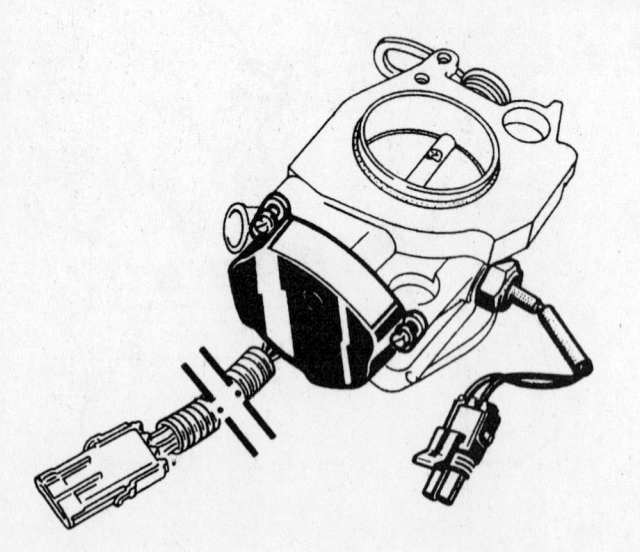

Fig. 3.65 Throttle casing (Sec 41)

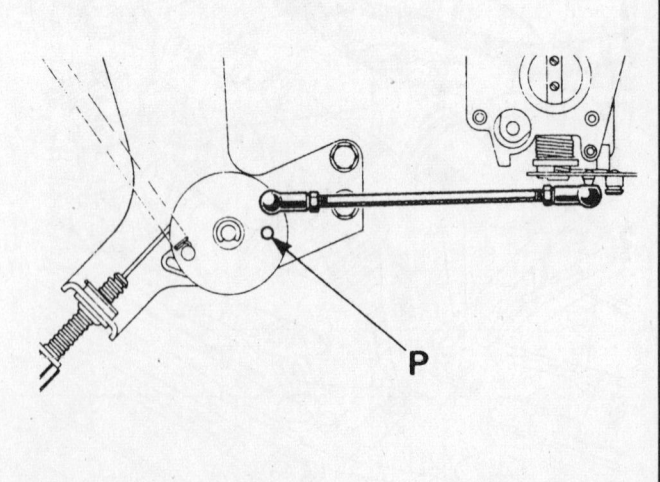

Fig. 3.66 Throttle link rod adjustment diagram (Sec 41)

P Throttle reel hole – 5.0 mm diameter

Chapter 3 Fuel system

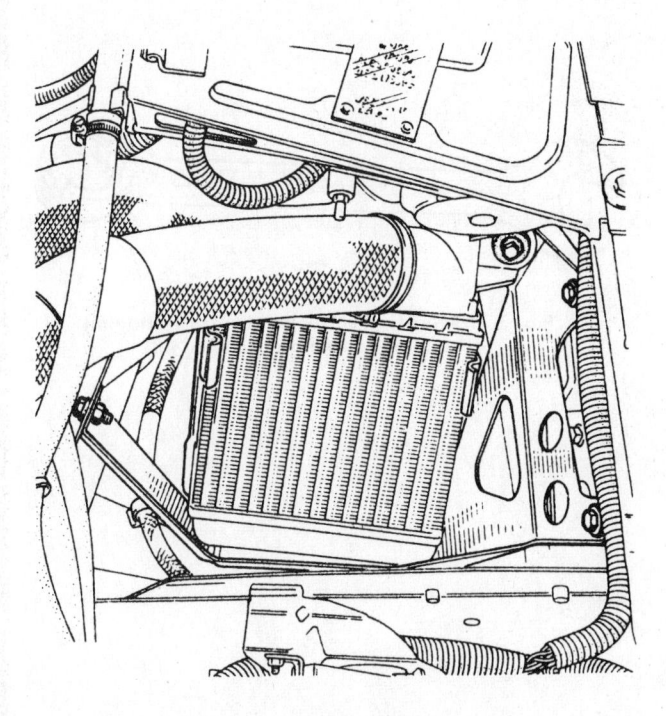

Fig. 3.67 Turbo intercooler (Sec 42)

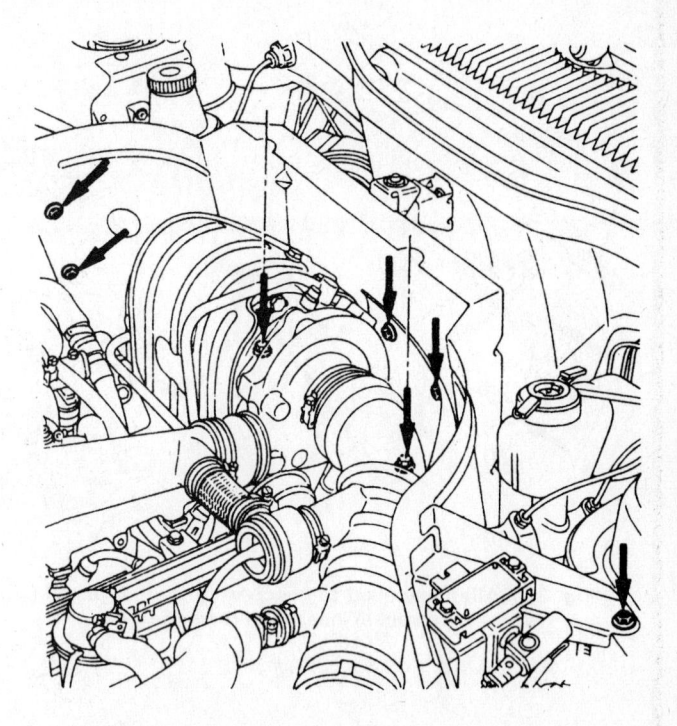

Fig. 3.68 Turbocharger heat shield, intake scoop and air fan ducting screws (arrowed) (Sec 43)

43 Turbocharger – removal and refitting

1 Remove the turbocharger heat shield, the intake scoop and the cooling air fan duct.
2 Disconnect the exhaust pipe (2 – Fig. 3.69) and remove it.
3 Remove the heat shield (1) and the pressurised air pipe (3), then disconnect the oil supply pipe (4). Disconnect the oil return pipe. The use of an Allen key will be required to unscrew the socket-headed fixing screws.
4 To remove the turbocharger fixing bolts modify a 17 mm ring spanner as shown (Fig. 3.71). Remove the bolts and the turbocharger.
5 When refitting the turbocharger use all new gaskets located on clean mating surfaces and renew the nuts which secure the turbocharger to the exhaust manifold.
6 Reconnect the oil return pipe and the exhaust and air intake system.
7 Prime the turbocharger with oil poured in through the hole (4 – Fig. 3.72).
8 Disconnect the three-way connector from the ignition power module and then operate the starter motor until oil starts to run out of the hole. Reconnect the oil input pipe union with a new gasket.
9 Reconnect the three-way connector and then run the engine at idling speed for a few minutes to re-establish the oil flow.
10 Never run the engine with the air intake system disconnected.

44 Turbocharger pressure regulator – adjusting, removal and refitting

1 The pressure regulator, when correctly adjusted, is vital to the correct operation of the turbo charger.
2 Due to the need for special gauges, it is not possible for the home mechanic to adjust it but for information purposes, if the regulator adjusting rod locknut is released and the end fitting screwed further onto the rod, the pressure is increased. If it is unscrewed, the pressure is reduced.
3 To remove the pressure regulator, first disconnect the oil supply and return pipes from the turbocharger as described in the preceding Section.
4 Disconnect the hose from the pressure regulator.

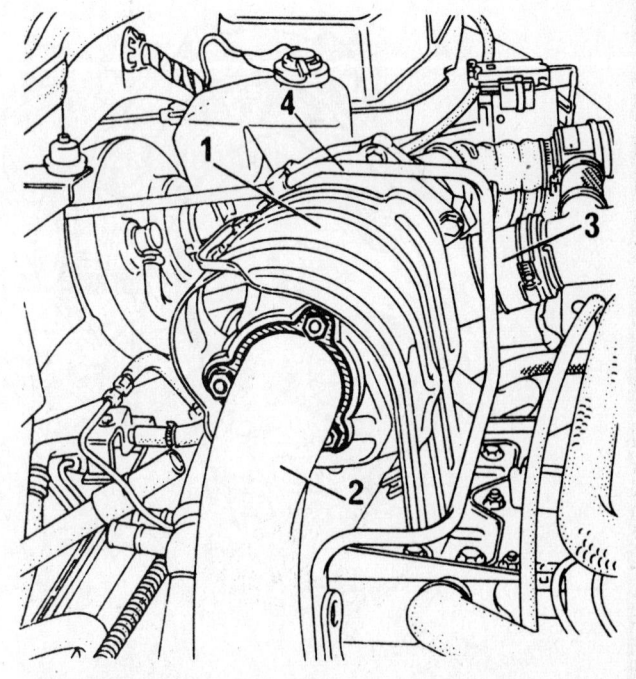

Fig. 3.69 Turbocharger disconnection points (Sec 43)

1 Heat shield
2 Exhaust pipe
3 Pressurised air
4 Oil supply

5 Extract the circlip (2 – Fig. 3.73) and disconnect the adjusting rod (3) from the valve operating arm (6).
6 Unscrew and remove the mounting bolts (4) and withdraw the pressure regulator by rotating it and passing it downwards beside the turbocharger.

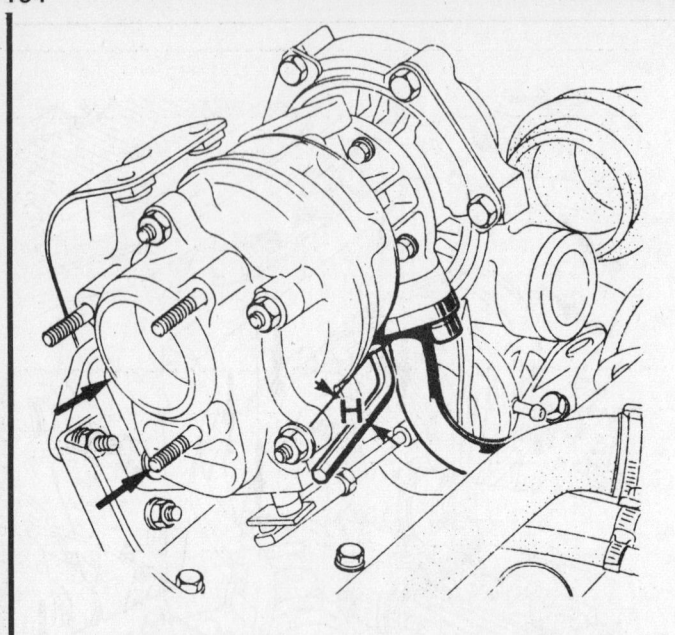

Fig. 3.70 Allen key used to unscrew turbocharger oil return pipe socket-headed screws (Sec 43)

H = 30 mm

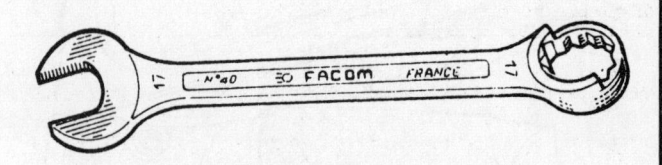

Fig. 3.71 Modified ring spanner for removing turbocharger fixing nuts (Sec 43)

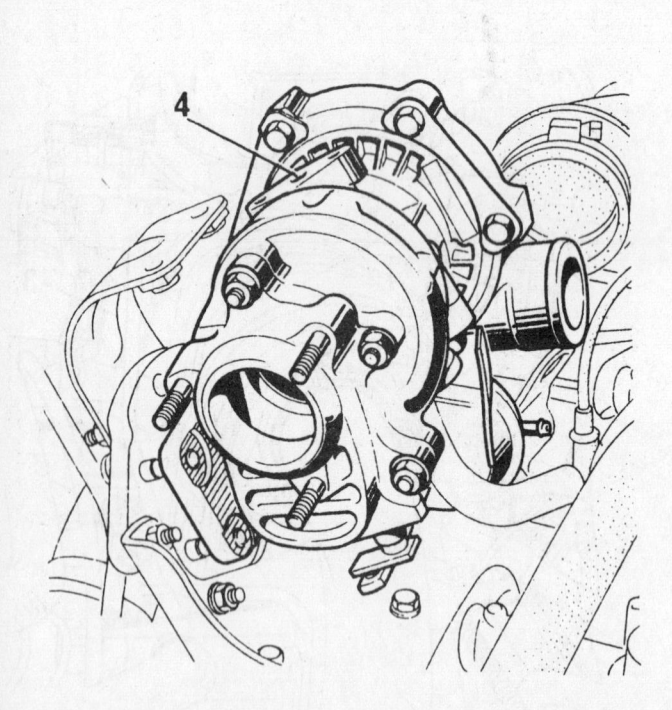

Fig. 3.72 Turbocharger oil input hole (4) (Sec 43)

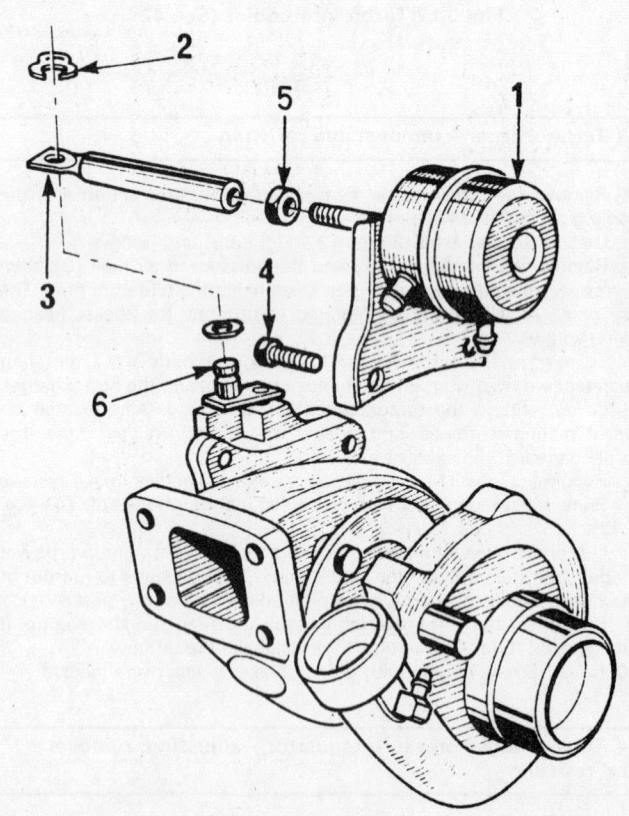

Fig. 3.73 Turbocharger pressure regulator (Sec 44)

1 Regulator unit
2 Circlip
3 Threaded rod
4 Fixing bolt
5 Locknut
6 Valve operating arm

Chapter 3 Fuel system 155

PART F: EMISSION CONTROL, MANIFOLDS AND EXHAUST SYSTEM

45 Emission control system – description

Apart from the fitting of a crankcase ventilation system to all models, certain vehicles operating in some territories are equipped with more sophisticated emission control systems which include exhaust gas recirculation (EGR) and air injection (AI) systems.

46 EGR system (four-cylinder carburettor engine) – description and testing

1 The EGR valve is controlled by the vacuum in the carburettor primary venturi.
2 At pre-determined engine coolant temperature levels, the EGR valve is inoperative.
3 The EGR valve is essentially a diaphragm valve that, according to the inlet manifold pressure, opens or closes to permit recirculation of a certain volume of exhaust gas into the combustion cycle to reduce the generation of oxides of nitrogen (NOx).

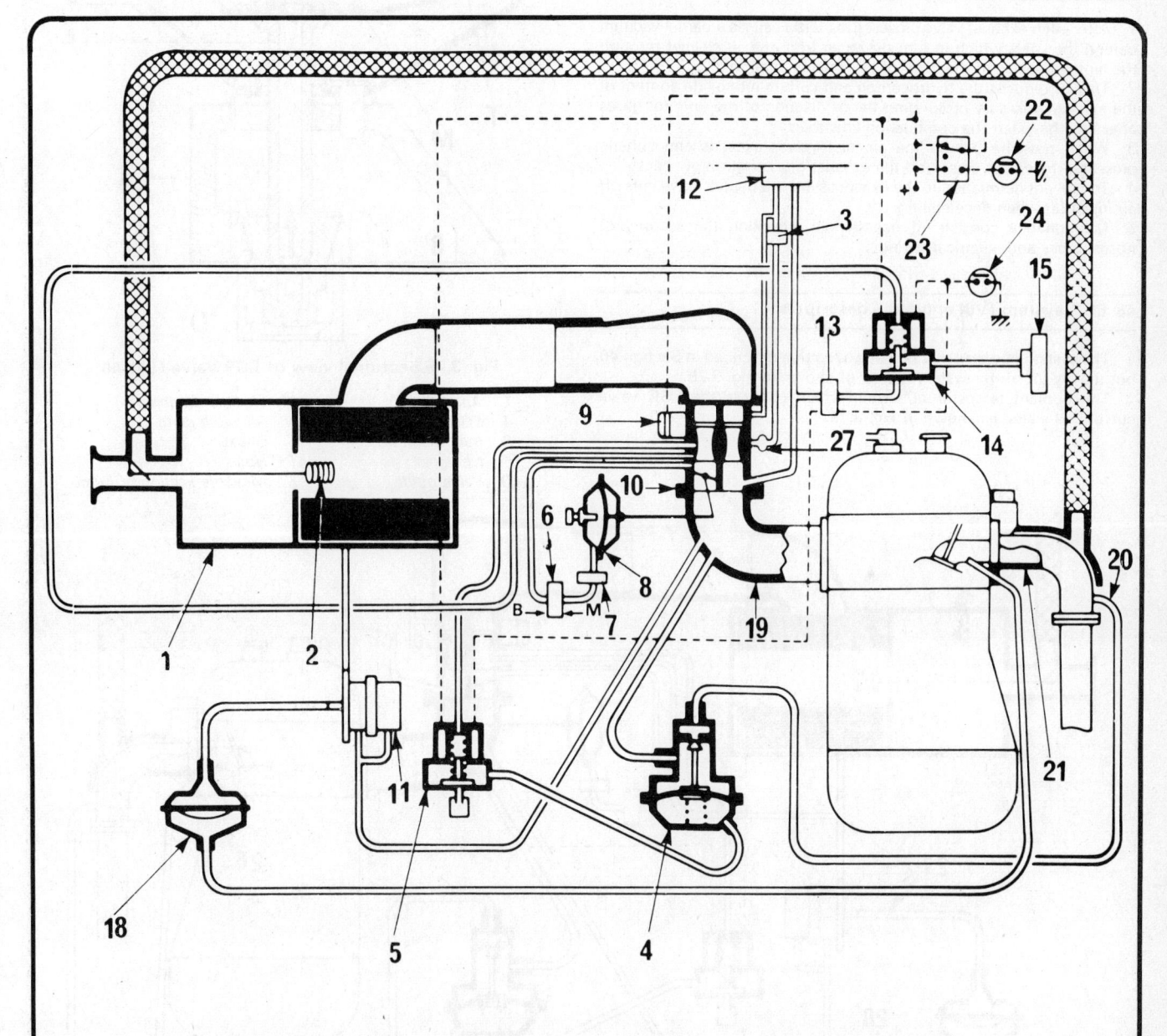

Fig. 3.74 Emission control EGR system on four-cylinder carburettor engine (Sec 46)

1. Thermostatically-controlled air cleaner
2. Wax capsule
3. Delay valve (grey)
4. EGR valve
5. Solenoid valve
6. Delay valve (brown [M]/white [B])
7. Volume chamber
8. Throttle opener
9. Automatic choke
10. Carburettor
11. Bypass valve
12. Thermovalve
13. Delay valve (green)
14. Solenoid valve
15. Ignition
18. Air inlet valve
19. Inlet manifold
20. Calibrated orifice
21. Exhaust manifold
22. Ambient temperature thermal switch
23. Relay
24. Thermal switch (60° – 140°F)
27. Secondary venturi locking device

4 Regular maintenance should be carried out to ensure that all system connections are secure. The exhaust gas injection nozzles at the inlet manifold should be periodically cleaned and the EGR valve renewed at the intervals specified in Routine Maintenance.
5 To check that the EGR valve is working, observe the valve guide (D – Fig. 3.75) while an assistant revs the engine which should be at normal operating temperature. The valve guide should be seen to move.

47 AI system (four-cylinder carburettor engine) – description and testing

1 After each exhaust valve closes, gas inertia causes a partial vacuum behind the valve which in turn draws air from the air cleaner through the air inlet valves to the back of the exhaust manifold.
2 This air reduces the hydrocarbon and carbon monoxide content of the exhaust gases by prolonging the oxidisation of the unburnt gases after they have left the combustion chambers.
3 When pressure behind the air inlet valves exceeds atmospheric pressure, the valves close and the exhaust gas cannot blow back.
4 To prevent detonation in the exhaust system a bypass valve cuts off air injection when decelerating.
5 Maintenance consists of occasionally checking the security of hoses, pipes and electrical wiring.

48 EGR system (V6i engine) – description

The system operates in a similar way to that described in Section 46 but the layout of the system differs, as shown in Fig. 3.78.
The coolant temperature switch which controls the EGR valve cut-off is located as shown in Fig. 3.79.

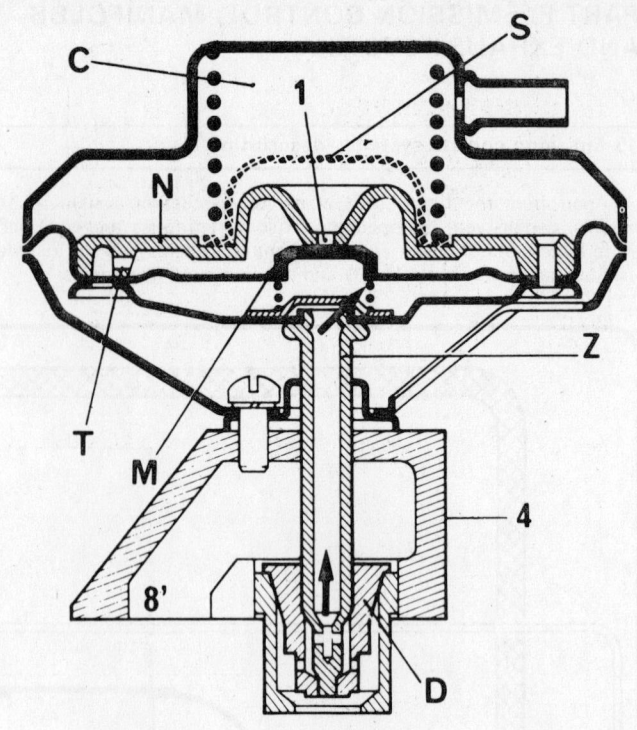

Fig. 3.75 Sectional view of EGR valve (Sec 46)

1	Aperture	M	Flexible diaphragm
4	EGR valve body	N	Movable plate
8	Inlet manifold	S	Gauze
C	Chamber	T	Holes
D	Valve guide	Z	Movable section of guide

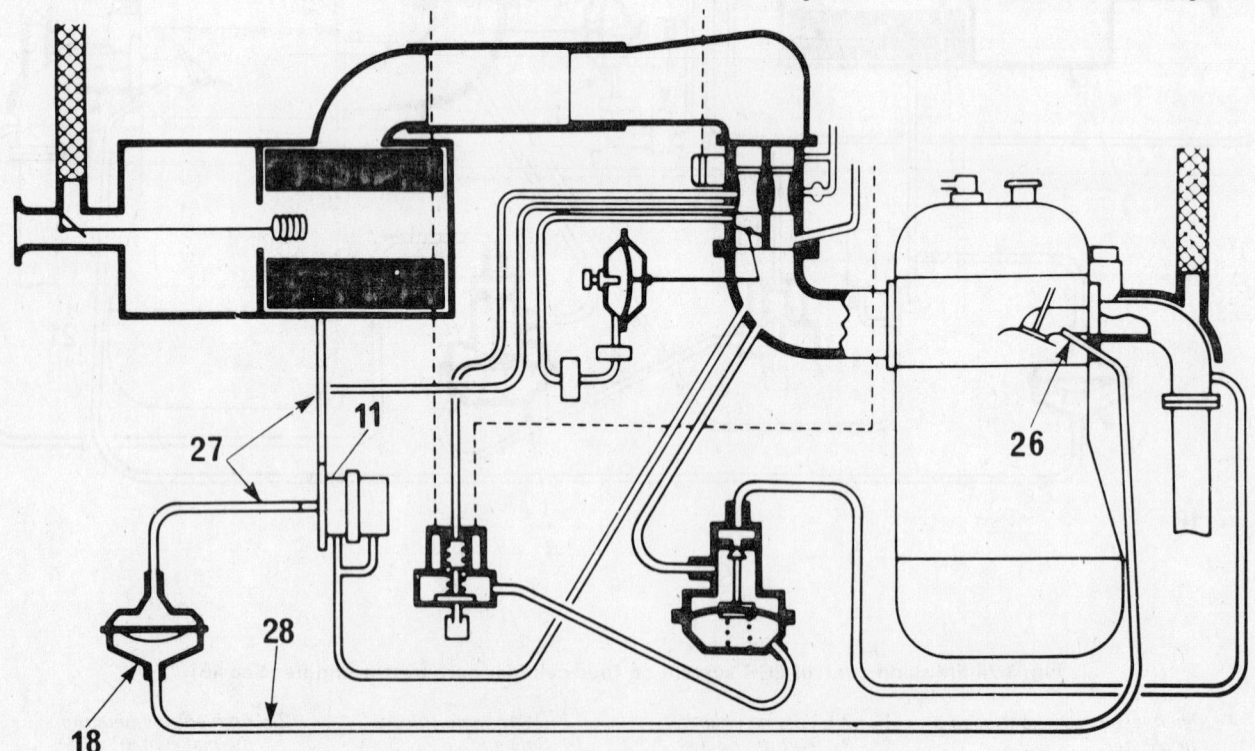

Fig. 3.76 Air injection (AI) system on four-cylinder carburettor engine (Sec 47)

11 Bypass valve
18 Valve
26 Air injector
27 Hose
28 Pipe

Fig. 3.78 Emission control (EGR) system on V6 engine (Sec 48)

- A   Fuel tank lift pump
- B   Main electric fuel pump
- C   Pressure accumulator
- D   Fuel filter
- E   Fuel metering/distributor unit
- F   Airflow sensor unit
- G   Control pressure regulator
- H   Auxiliary air control
- J   Cold start injector
- K   Fuel injectors
- L   Timed thermal switch
- M   To air cleaner
- R   Fuel tank
- 10 Fuel pressure regulator
- 12 Metering piston
- 13 EGR valve
- 14 EGR solenoid valve
- 15 Temperature switch (coolant above 45°C − 13°F)
- 16 Vacuum advance solenoid valve
- 17 Ignition distributor
- 18 Delay valve
- 20 Noise reduction chamber
- 21 Air inlet valves
- 22 Air injector
- 23 Air take-off pipe (air cleaner)
- 24 Exhaust gas take-off pipes
- 26 Exhaust gas recycling pipe to inlet manifold

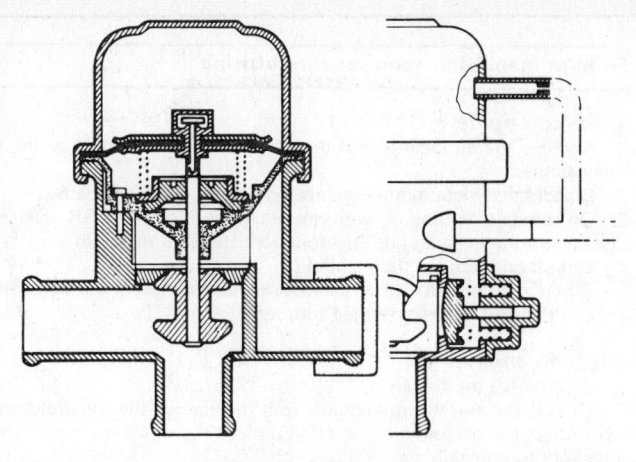

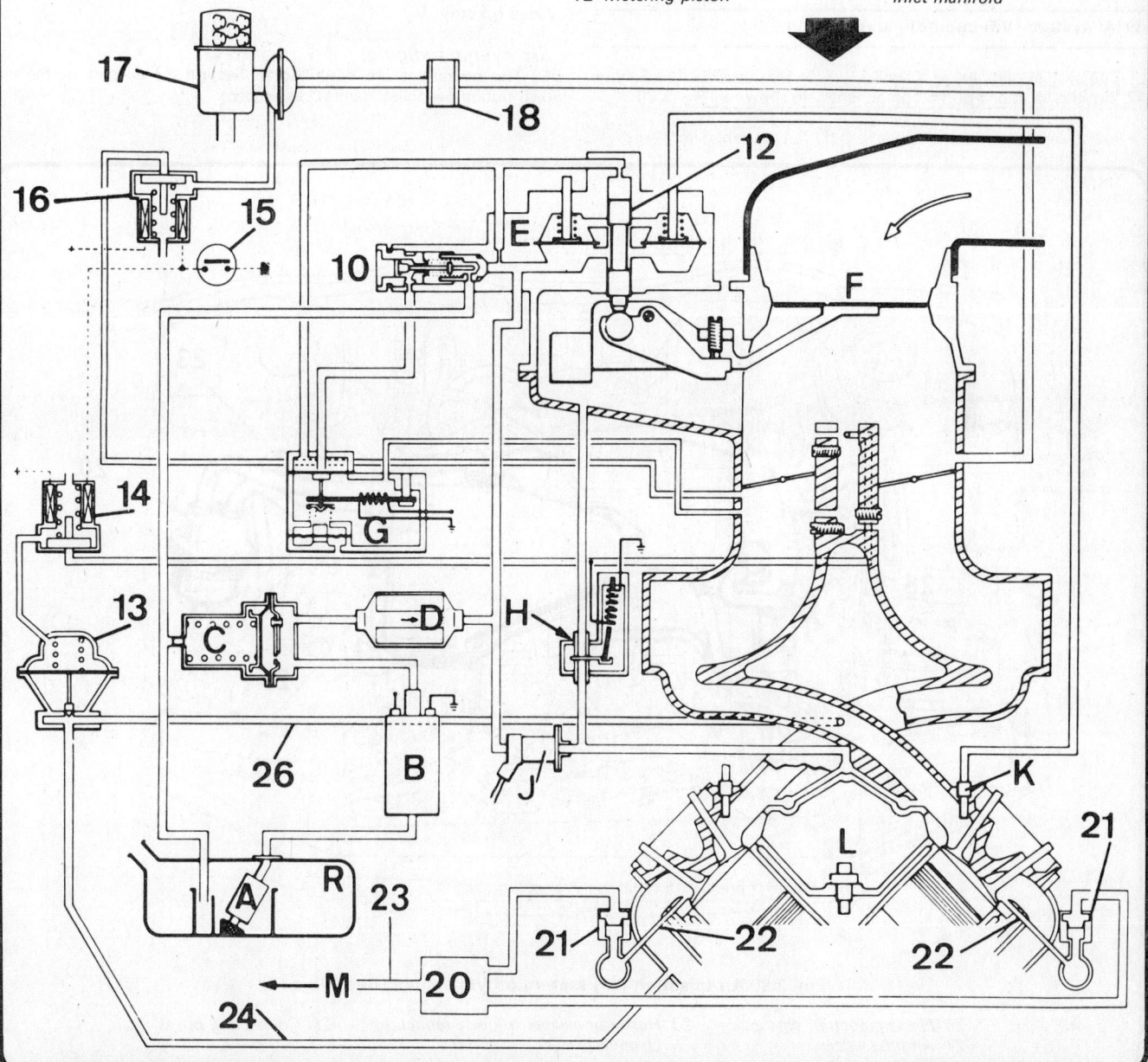

Fig. 3.77 Sectional view of the AI system bypass valve (Sec 47)

157

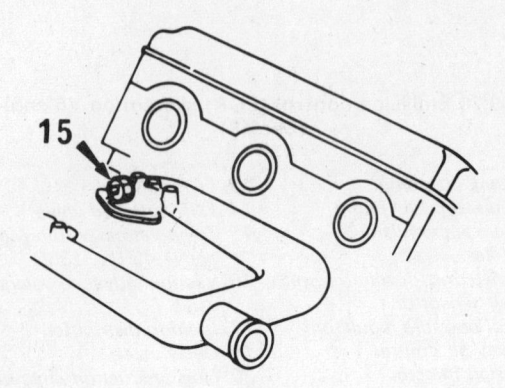

Fig. 3.79 EGR valve cut-off temperature switch (15) (Sec 48)

49 AI system (V6i engine) – description

The system operates in a similar way to that described in Section 47, but the locations of the components are shown in Fig. 3.80.

50 Inlet manifold – removal and refitting

1995 cc engine
1 Remove the air cleaner and the carburettor as described earlier in this Chapter.
2 Disconnect vacuum and coolant hoses and electrical leads.
3 On models equipped with an emission control EGR system, disconnect the exhaust gas injectors from the inlet manifold.
4 Unbolt and remove the manifold.
5 When refitting, use new gaskets on clean mating surfaces and tighten the nuts to the specified torque.

2165 cc engine
6 Disconnect the battery.
7 Unbolt the throttle quandrant from the top of the manifold and disconnect the balljointed link rod and place the quadrant, with cable attached, to one side.
8 Unbolt the fuel injection manifold from the inlet manifold.
9 Remove the air cleaner as described in Section 3.
10 Disconnect all hoses from the inlet manifold, including those for the brake servo, ignition computer and fuel pressure regulator.
11 Unbolt and remove the inlet manifold.
12 Refitting is a reversal of removal, use new gaskets on clean mating faces (photo).

Six-cylinder engines
13 The operations are described in Section 33 as part of the air distribution box/inlet manifold assembly.

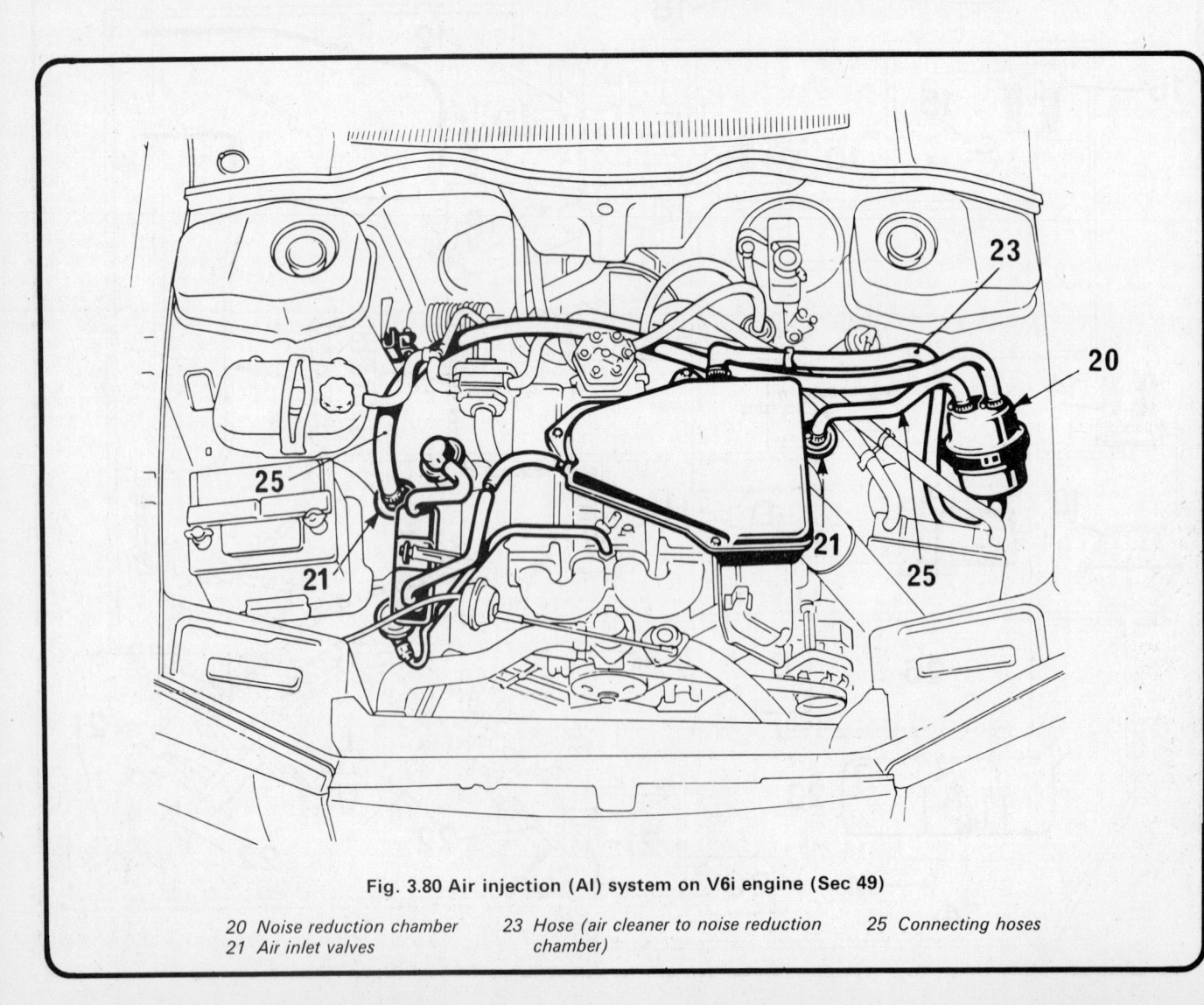

Fig. 3.80 Air injection (AI) system on V6i engine (Sec 49)

20 Noise reduction chamber
21 Air inlet valves
23 Hose (air cleaner to noise reduction chamber)
25 Connecting hoses

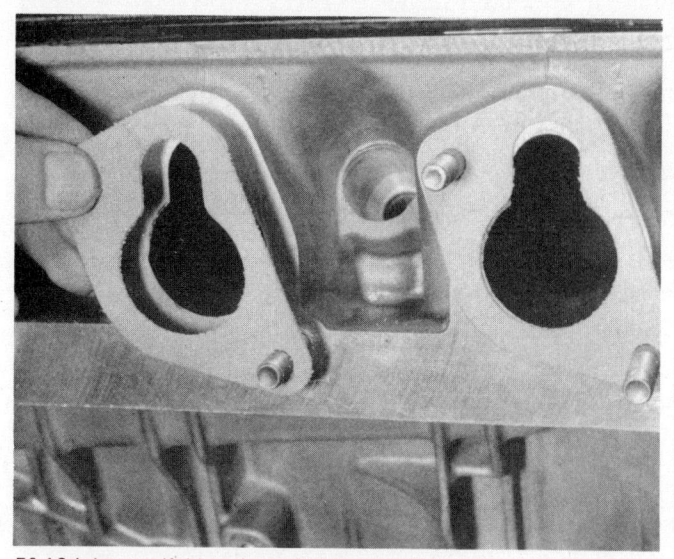

Chapter 3 Fuel system

50.12 Inlet manifold gaskets (2165 cc engine)

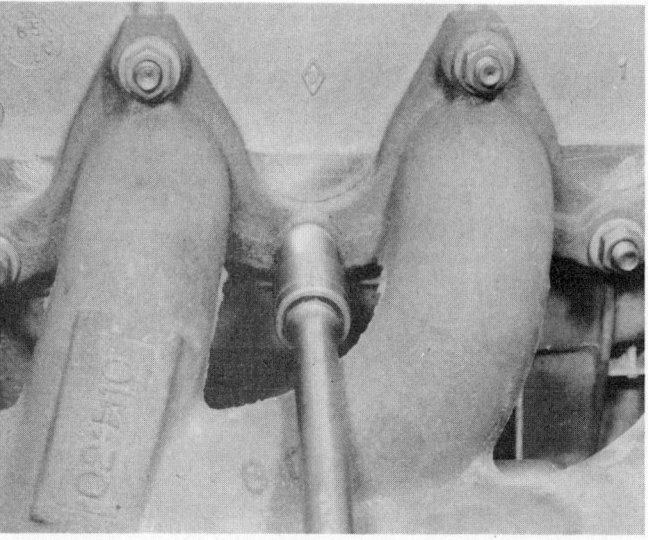

51.4A Unscrewing exhaust manifold nut (2165 cc engine)

51 Exhaust manifold – removal and refitting

Four-cylinder engines
1. Disconnect the exhaust downpipe from the manifold. The retaining nuts are best unscrewed by using a long extension and socket wrench from under the car.
2. On 1995 cc models note the hot air collector plate fixed to the manifold.
3. If an emission control air injection system is fitted, disconnect the injection pipes from the manifold.
4. Unscrew the exhaust manifold nuts and lift the manifold from the cylinder head (photos).
5. Use new gaskets on clean mating surfaces when refitting and tighten the fixing nuts to the specified torque (photo).

Six-cylinder engines (not Turbo)
6. Remove the air cleaner.
7. Unclip wiring and hoses as necessary for access to the manifolds. On the right-hand side of the engine, access will be improved if the crankcase ventilation system hoses and oil separator are first removed.
8. Disconnect both exhaust downpipes from the rear ends of the manifold.
9. Unscrew the exhaust manifold fixing nuts and lift away the manifold (photo).
10. When refitting, use new gaskets on clean mating surfaces and tighten the fixing nuts to specified torque.

51.4B Removing exhaust manifold (2165 cc engine)

51.5 Exhaust manifold gasket (2165 cc engine)

51.9 Exhaust manifold nuts (2664 cc engine)

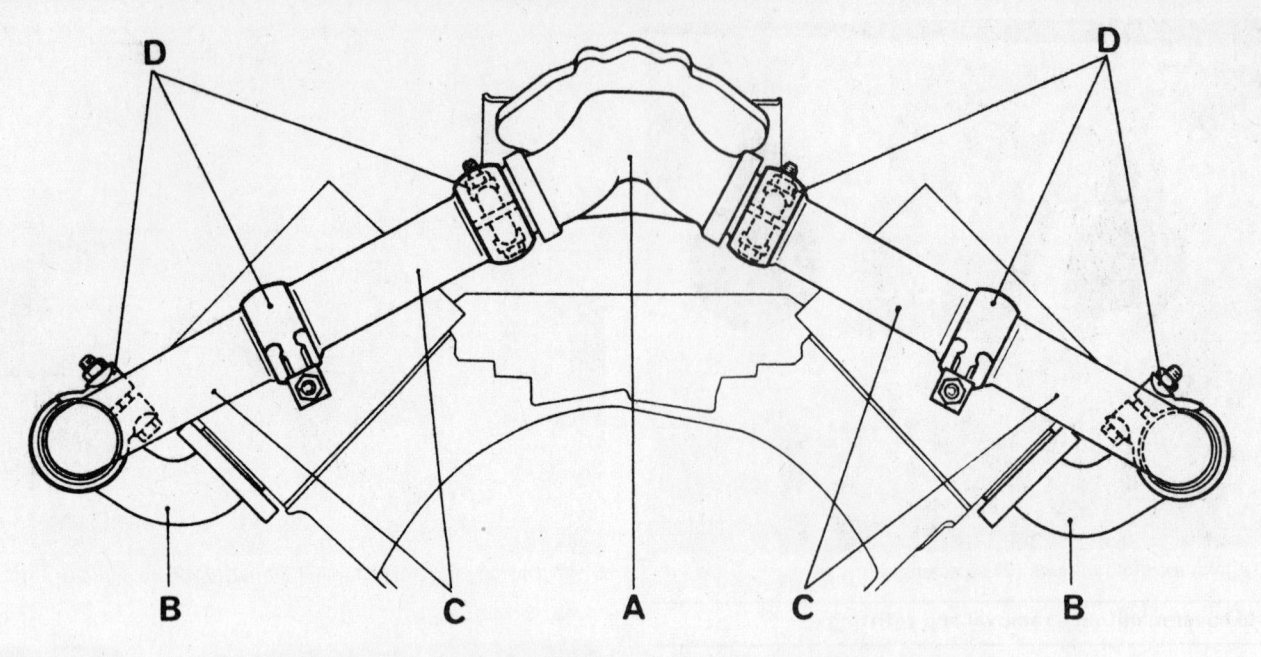

Fig. 3.81 Turbocharger manifold section tightening sequence (Sec 51)

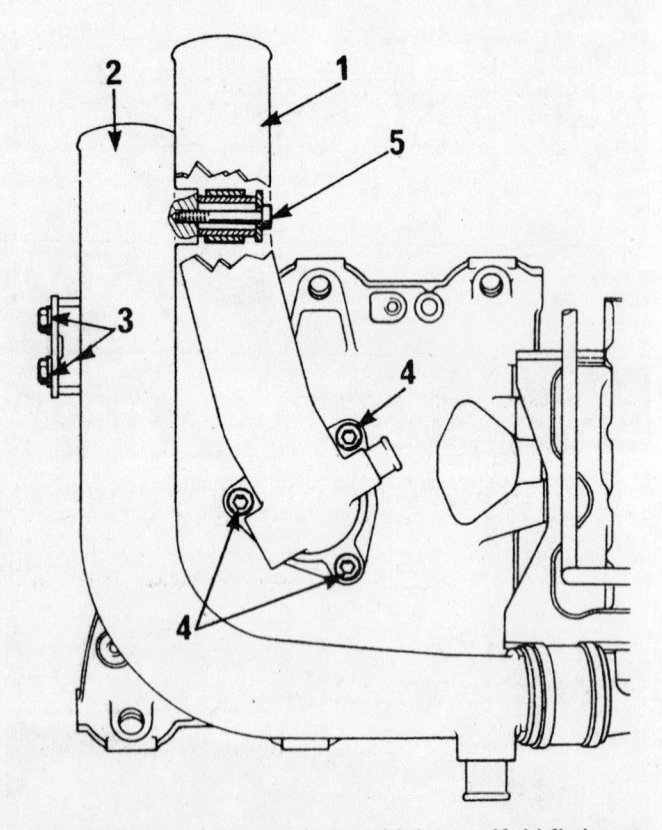

14 Fit the interconnecting pipes (C) with their clamps positioned as shown (D).
15 Tighten all nuts and bolts to the specified torque.
16 Run the engine for fifteen minutes and retighten to the specified torque **without** loosening the nut or bolt first.
17 Whenever a system nut is released for any reason it must be renewed.
18 If the two manifolds shown in fig. 3.82 have been removed, it is important that they are refitted in the following way.
19 Place the two manifolds (1 and 2) in position and fit the fixing bolts only finger tight. Tighten the two bolts (3) then the three bolts (4) and finally the connecting bolt (5).

52 Exhaust system

1 The exhaust system differs in design according to the size of engine used (photo).
2 The complete system or just part of it can be removed for repair or

Fig. 3.82 Turbocharger duct and inlet manifold fitting sequence (Sec 51)

For key see text

Turbo engine
11 On these engines, it is important that the manifold sections are refitted in the following sequence.
12 Refer to Fig. 3.81 and fit the centre section (A).
13 Using new gaskets, fit the outer sections (B) and screw up the bolts finger tight only at this stage.

52.1 Dual expansion boxes (2664 cc engine)

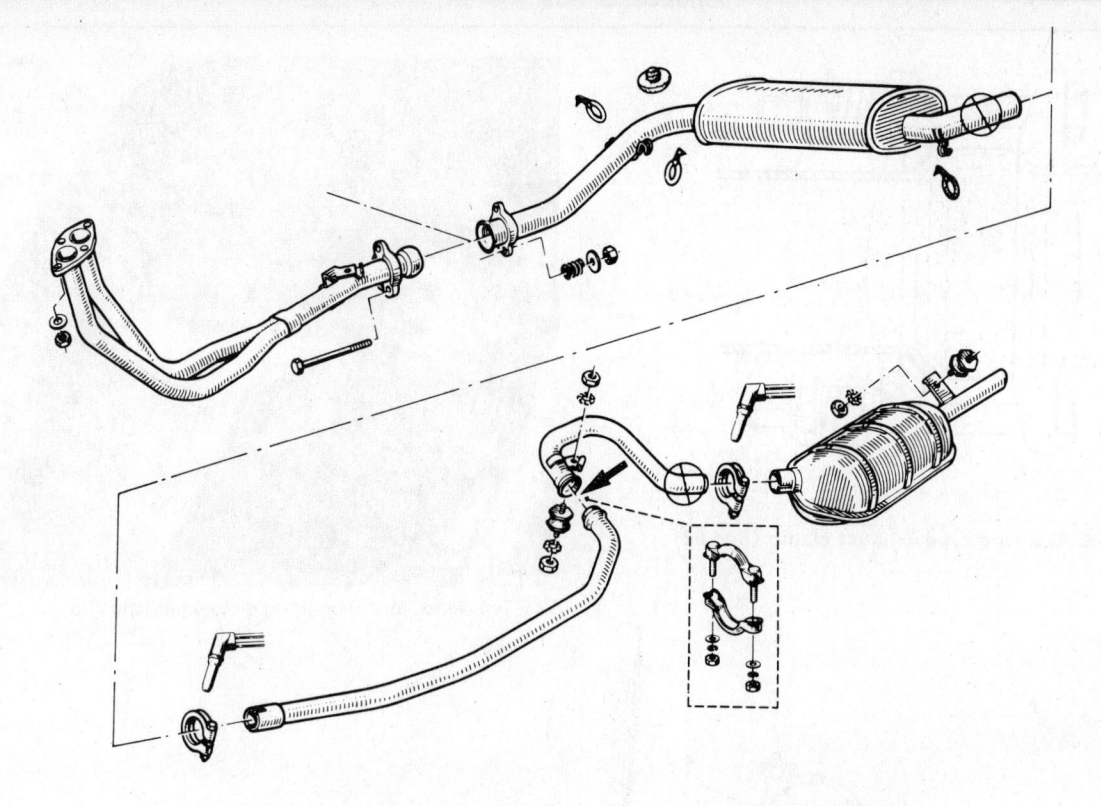

Fig. 3.83 Typical four-cylinder engine exhaust system (Sec 52)

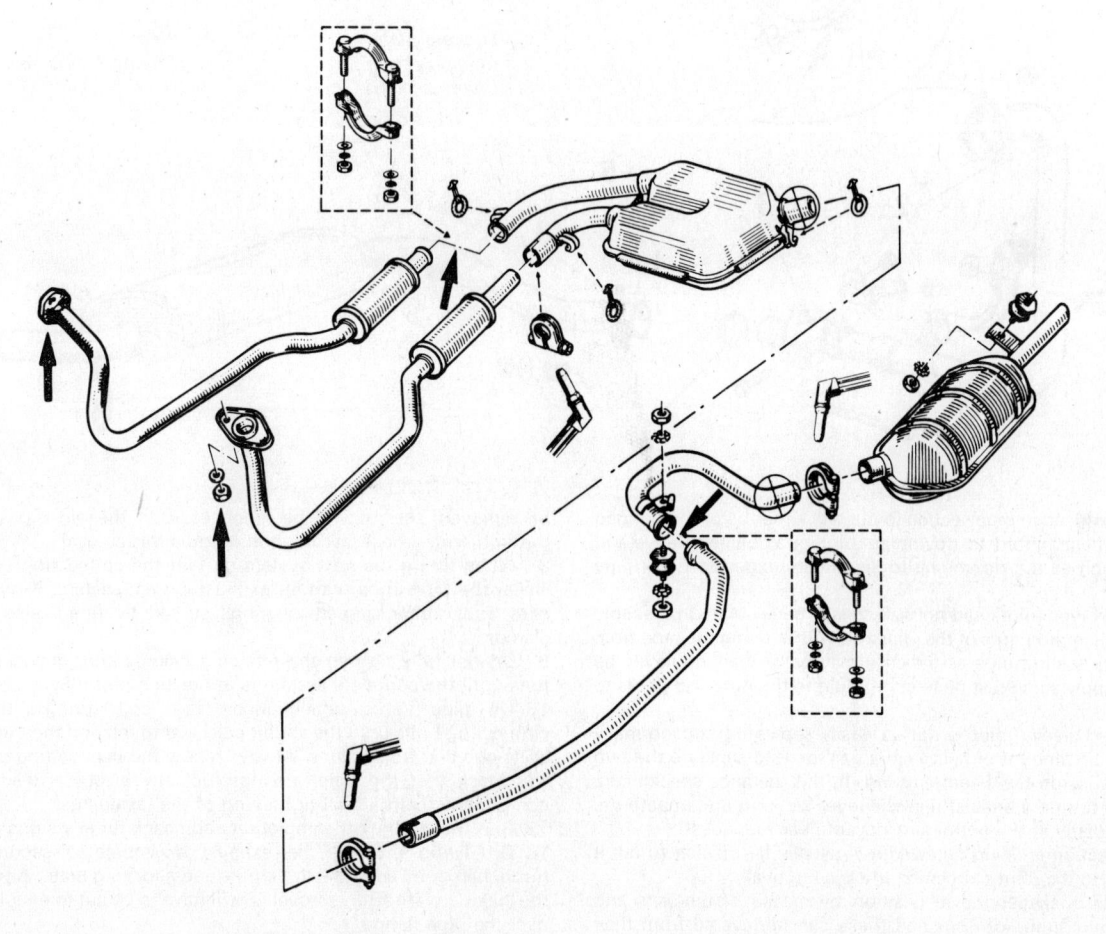

Fig. 3.84 Typical six-cylinder engine exhaust system (Sec 52)

Chapter 3 Fuel system

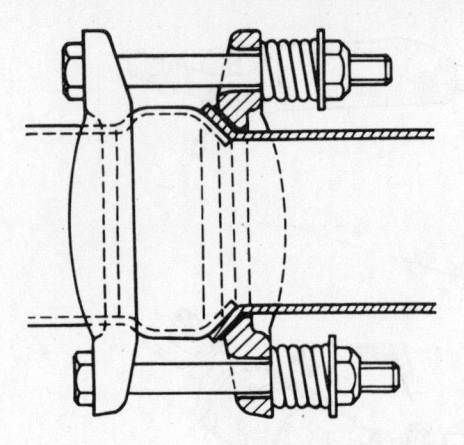

Fig. 3.85 Coil spring type exhaust clamp (Sec 52)

52.6 Typical exhaust pipe flexible mounting and clip

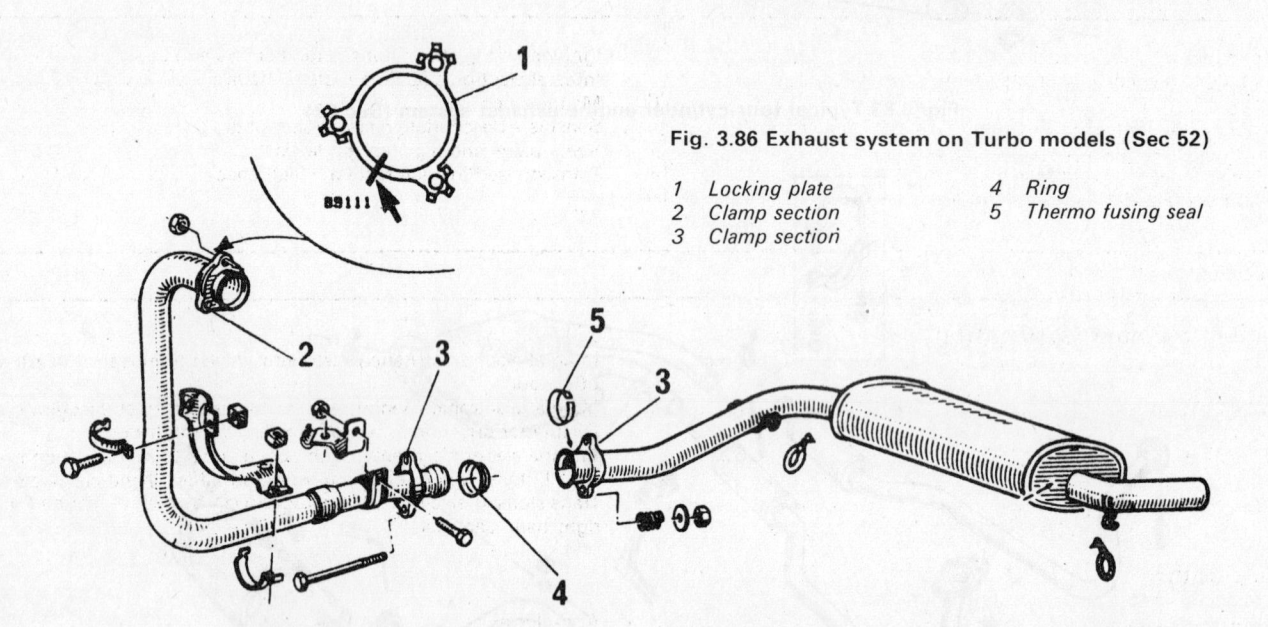

Fig. 3.86 Exhaust system on Turbo models (Sec 52)

1 Locking plate
2 Clamp section
3 Clamp section
4 Ring
5 Thermo fusing seal

renewal. The respective pipe section joints will either be of the flanged type, such as the manifold-to-downpipe joint, or a collared sleeve and clamp joint such as the downpipe-to-secondary expansion box pipe joint.
3 The flanged type joints can normally easily be separated by simply removing the clamping nuts of the joint and withdrawing the pipe from the manifold or mating pipe as the case may be. It may, however, be necessary to apply some rust penetrating fluid to the nuts and joints to ease removal.
4 The collared sleeve joint is not so easily separated and normally requires a certain amount of heat from a gas torch to separate the two pipes after the clamp has been removed. In this instance special care must be taken to guard against fire whenever working underneath the vehicle, particularly at the petrol tank end or near fuel lines!
5 If a pipe section is being renewed it is usually far quicker to cut it through close to the joint concerned to ease removal.
6 The system is suspended in position by means of brackets and rubber mounting straps or rings and these can be levered from their mounting brackets when necessary (photo).
7 The foregoing operations should be followed if only certain sections of the system are to be renewed. If the complete system must be renewed, then it will be much easier if the old pipes are sawn through with a hacksaw in short lengths for removal.
8 When fitting the new system de-burr the connecting sockets and smear the pipe ends with an exhaust system sealant. New mounting clips and rubber suspension rings should be fitted, also new pipe clamps.
9 Do not fully tighten the retaining clamps and joint clamp/flange nuts until the complete system is in position and fully suspended.
10 On pipe joints which incorporate coil springs, tighten the connecting bolts until the spring coils just touch and then unscrew the bolts one and a half turns. Always renew the heat setting washer.
11 Check that the pipes do not foul any electrical leads or other components before final tightening of the fastenings.
12 Run the engine on completion and check for any signs of leakage.
13 On Turbo models, the exhaust downpipe is secured to the turbocharger by a flange, three nuts and a locking plate. Always renew the locking plate after removal. It will have to be cut to enable it to pass over the pipe flange.
14 When reassembling, use a new thermo fusing seal (5 – Fig. 3.86).
15 Make sure that there is a clearance of 20.0 mm (0.79 in) between the downpipe and side-member.

Chapter 4 Ignition system

Contents

Description 1	Ignition timing (transistorised ignition system) 5
Distributor (integral electronic system) – removal and refitting 6	Integral electronic ignition system – fault check 4
	Maintenance 2
Distributor (transistorised electronic system) – removal, overhaul and refitting 7	Sensors – description, removal and refitting 8
	Spark plugs and high tension leads 9
Fault diagnosis – ignition system 11	Transistorised ignition system – fault check 3
Ignition switch – removal and refitting 10	

Specifications

System type and application

B297 (1995 cc)	Integral electronic (Renix) distributor driven from rear of camshaft by offset dog
B29E (2165 cc)	Integral electronic (Renix) with connections to fuel injection system. Distributor driven from rear of camshaft by offset dog
B295 (2458 cc)	Integral electronic (Renix) with connections to fuel injection system. Distributor driven by coupling from front of left-hand camshaft
B298 (2664 cc)	Transistorised electronic, distributor driven by gear at rear end of right-hand camshaft

Distributor

Rotation:
    4-cylinder engines .... Anti-clockwise
    6-cylinder engines .... Clockwise

Firing order (No 1 at flywheel end):
    4-cylinder engines 1–3–4–2
    6-cylinder engines 1–6–3–5–2–4

Ignition timing (dynamic) .... 9 to 11° BTDC at idle speed with vacuum disconnected. Only adjustable on 2664 cc engines

Distributor advance (2664 cc engine):
    Centrifugal advance .... Starts at 1200 rev/min
                                                      16 to 20° at 3000 rev/min
                                                      Maximum advance 26 to 30° at 4860 rev/min
    Vacuum advance .... Starts at 102 mm Hg (4.0 in Hg)
                                                       Ends at 254 mm Hg (10.0 in Hg)
                                                        Maximum advance = 18 to 20°

Spark plugs

	Champion	Electrode gap
B297 (1995 cc)	S7YC	0.6 mm (0.024 in)
B29E (2165 cc)	S7YC	0.9 mm (0.035 in)
B295 (2458 cc)	S6YC	0.6 mm (0.024 in)
B298 (2664 cc)	RBN9GY	0.6 mm (0.024 in)

Torque wrench setting

	Nm	lbf ft
Spark plug	20	15

Chapter 4 Ignition system

1 Description

1 The type of ignition system varies according to vehicle model; see Specifications.
2 Both ignition systems require the minimum of maintenance; even ignition timing is carried out automatically according to engine load and speed requirements with the integral electronic ignition system.

Transistorised ignition system

3 The main components of the system are shown in Fig. 4.1.
4 An impulse goes from the distributor to the control unit. This makes and breaks the current to the ignition coil with the help of impulses from the impulse sender. The dwell angle is regulated electronically. From the ignition coil, the high tension impulse goes as usual to the spark plugs via the distributor rotor arm (photo).
5 The ignition advance is controlled by means of centrifugal weights and a vacuum chamber.
6 The impulse sender opens and closes a magnetic, as opposed to an electrical, circuit. This induces impulses in the magnetic pick-up or coil. The impulse sender is made up of four main parts: the stator, the magnetic pick-up (or coil), the rotor and the magnet. Whilst the stator, pick-up and magnet are connected to the distributor housing, the rotor is connected to, and rotated with, the distributor shaft.
7 The control unit, or electronic module, is of a solid state design employing transistors. It converts and amplifies the impulses from the impulse sender and sends them on to the coil. It also performs a second function, control of the dwell angle.
8 The rotor has six teeth and the stator has three teeth. The magnet creates a magnetic field which passes through the stator. When the pole teeth are opposite each other, the magnetic circuit is closed; when the teeth are apart, the circuit is open. In this way the rotor opens and closes the magnetic field as it rotates; this generates the current pulses in the magnetic pick-up.

Integral electronic (Renix) ignition system

9 The system is fully computerized and the main ignition functions take place within the intregral ignition module. The distributor is considerably reduced in size as it only incorporates the rotor arm.
10 The computer determines the correct ignition timing and dwell by processing signals from the TDC/BDC sensor, the vacuum capsule and knock sensor.
11 Although the vacuum capsule is similar in appearance to the conventional unit, the internal components are different and it **must not** be removed, otherwise a thin wire to the computer will be broken. However, the coil does not incorporate internal connections to the computer and it can therefore be removed separately. Make sure that the coil wires are fitted correctly; the red wire (9) goes to the positive (+) terminal (7), and the black wire (10) goes to the negative (–) terminal (8).

Special precautions

12 Particular care must be taken whenever any work or checks are being performed on the ignition system or associated engine components necessitating the detachment of the HT leads.
13 **Do not** attempt to run the engine with any HT lead disconnected or serious damage to the system could result. If a spark plug is removed for testing with the lead attached, ensure that it is properly earthed.
14 Take care to avoid receiving electric shocks from the HT system. The HT voltage is higher than in conventional ignition systems and could be hazardous to people in poor health.

2 Maintenance

1 At the intervals specified in Routine Maintenance clean or renew the spark plugs and check the high tension (HT) leads; see Section 9.
2 Remove the distributor cap and inspect it for cracks, and the condition of the centre carbon brush.

1.4 Ignition coil (V6i engine)

3 Check the rotor contact end and the cap contacts for erosion. They may be cleaned carefully to remove deposits but, if badly eroded, renew the components.

3 Transistorised ignition system – fault check

Note: *A magnet and an ohmmeter will be needed for these checks.*

1 If the engine will not start carry out the following preliminary check.
2 Remove the distributor cap, pull the coil HT lead from the cap and earth the lead.
3 Disconnect the other end of this lead and hold it close to its socket in the ignition coil. Switch on the ignition.
4 Holding a magnet in the hand, move it up and down near the distributor windings. A spark should jump the gap between the end of the HT lead and the coil socket.
5 If no spark is evident, connect an ohmmeter across the terminals on both resistances (C – Fig. 4.1). Current should flow; resistance 1.0 ohm.
6 To check the ignition coil, connect an ohmmeter between the battery (+) terminal and distributor terminals on the ignition coil. Current should flow.
7 Connect the ohmmeter between the battery (+) terminal and distributor terminals on the ignition coil. Current should flow.
8 Connect an ohmmeter between the battery (+) terminal and HT lead socket on the ignition coil; current should flow.
9 Check the distributor by connecting the ohmmeter between the two terminals in the connector; current should flow.
10 Connect the ohmmeter between one of the two terminals in the distributor wiring connector and the distributor body. Current should not flow.
11 Disconnect the distributor wiring connector and then connect the two leads from the TDC sensor to the two sockets in the connector on the electronic control unit (Fig. 4.3). Take a magnet and pass it quickly towards and then away from the TDC sensor. A spark should jump the gap between the end of the disconnected coil HT lead and the coil socket. If no spark is evident then the distributor is faulty.
12 If the ignition fault occurs while the engine is running, being evident by misfiring or intermittent cutting out, then, provided the spark plugs and HT leads are in good condition, rectification can only be satisfactorily carried out by substitution of new units for the coil and electronic control unit.
13 A diagnostic socket is fitted to enable Renault mechanics to quickly pinpoint any problem within the ignition system and also to time it accurately. Unfortunately, without the necessary associated equipment, the diagnostic socket is of little use to the home mechanic (photo).

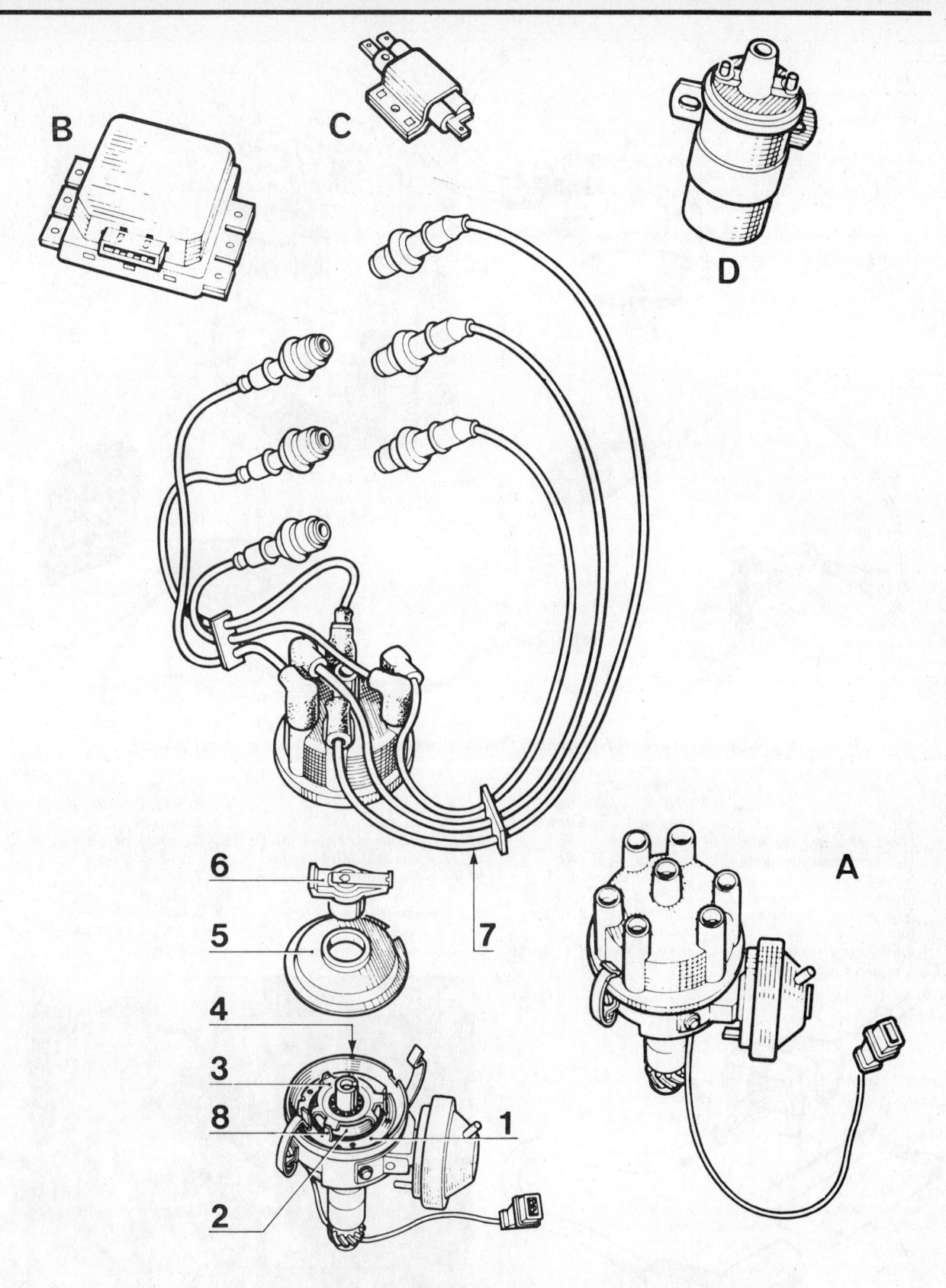

Fig. 4.1 Transistorised ignition system (2664 cc models) (Sec 1)

1 Permanent magnet
2 Pick-up coil
3 Rotor
4 No 1 cylinder firing point
5 Dustproof cover
6 Rotor arm
7 HT lead harness
8 Impulse sender segment

A Distributor (inductive pick-up unit)
B Electronic control unit
C Resistors
D Coil (special)

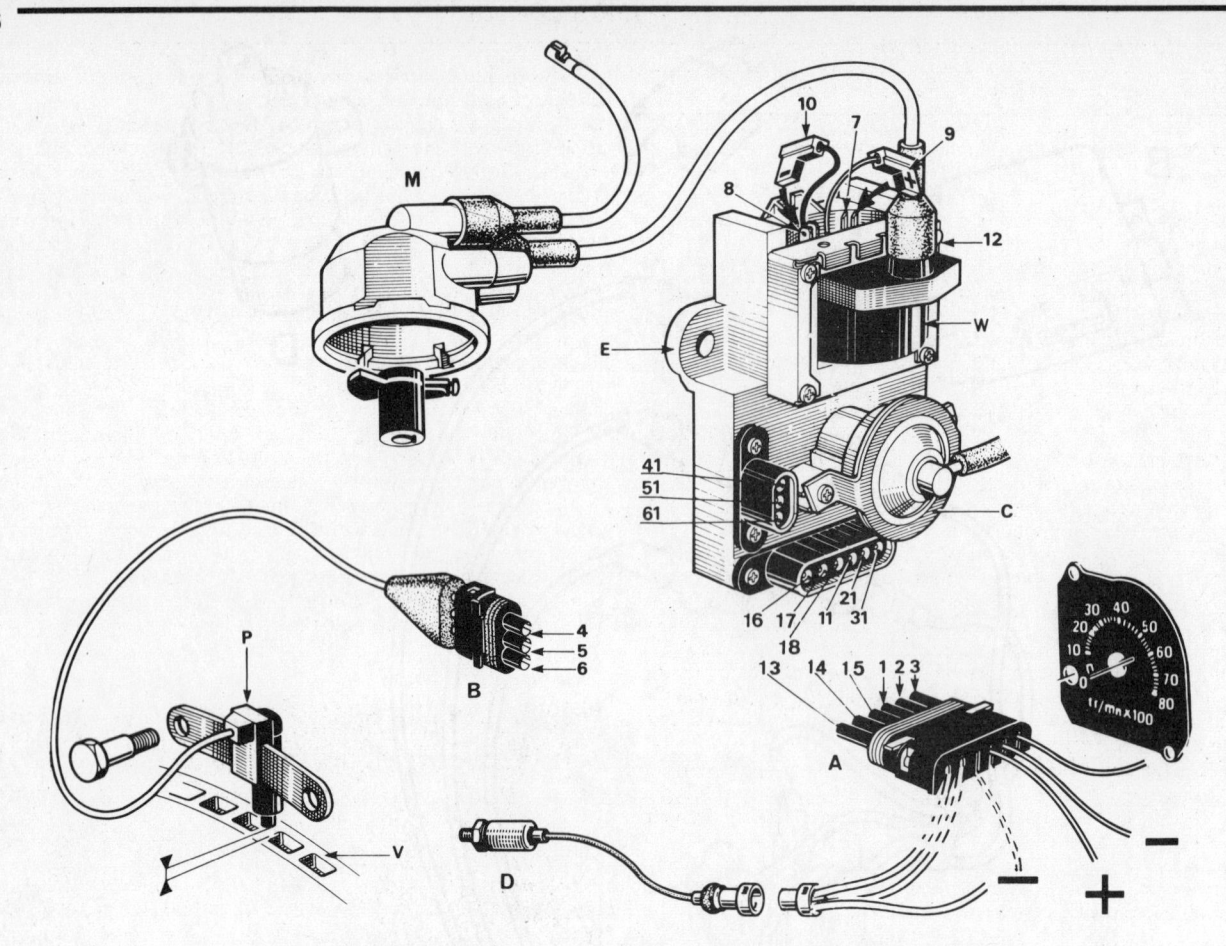

Fig. 4.2 Integral electronic ignition system components (4-cylinder models) (Sec 1)

1 + Feed	12 HT terminal	18 Terminal for +4° offset on flywheel	C Vacuum sensor
2 Earth	13 Detector information (pinking ...) or earth	21 Module earth	D Pinking detector (depending on version)
3 Rev counter	14 Detector information (pinking ...) or earth	31 Rev counter wire terminal	E Computer or module
4 Flywheel sensor winding		41 Flywheel sensor information	M Distributor cap
5 Flywheel sensor winding	15 Detector information (pinking ...) or earth	51 Flywheel sensor information	P TDC sensor
6 Screening		61 Screening	V Flywheel
7 Ignition coil + terminal	16 Terminal for +2° offset on flywheel	A Connection – computer or module feed	W Ignition coil
8 Ignition coil – terminal			Terminals 9 and 11 are connected direct inside.
9 Ignition coil + lead	17 Terminal for –8° offset on flywheel	B Connector – flywheel sensor	
10 Ignition coil – lead			
11 Module + entry			

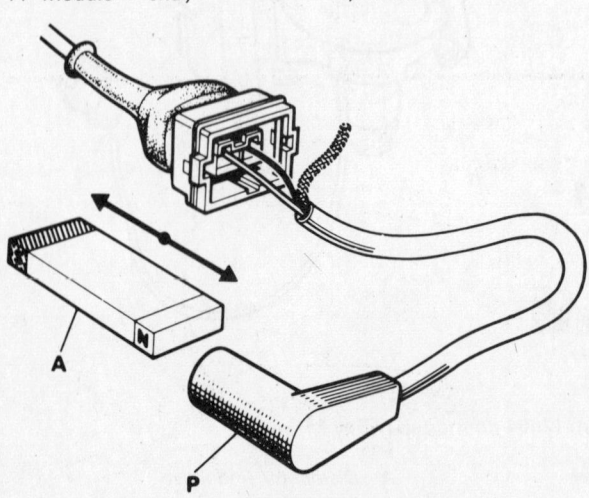

Fig. 4.3 Test circuit (Sec 3)

A Magnet   P TDC sensor

3.13 Diagnostic plug

Chapter 4 Ignition system

4 Integral electronic ignition system – fault check

Note: *A test lamp, a voltmeter and an ohmmeter will be needed for these checks.*

1 First check that all wiring and connections are in good condition and fitted correctly.
2 Connect a voltmeter between the ignition coil terminal 7 (Fig. 4.2) and earth, and check that 9.5 volt is registered with the ignition switched on.
3 Disconnect the computer supply multiplug and, using a voltmeter, check that 9.5 volt is registered on terminal 1 with the ignition on and the starter operating.
4 Switch off the ignition and, using an ohmmeter, check that zero ohms is registered between the multiplug terminal 2 and earth. Also check that there is zero ohms between the module terminals 9 and 11 – if not, the computer is faulty.
5 Refit the multiplug and, using the voltmeter, check that there is 9.5 volt at terminal 9 with the ignition on.
6 Switch off the ignition and disconnect the TDC sensor plug from the module and, using an ohmmeter between terminals 4 and 5, check that the resistance is 100 to 200 ohms. If not, renew the sensor.
7 With the ignition still off, check that the resistance between terminals 5 and 6, and also between terminals 4 and 6, is infinity.
8 Switch on the ignition and then reconnect the TDC sensor plug to the module then disconnect the coil wires (9 and 10) and connect a test bulb between them. Spin the engine on the starter and check that the bulb flashes. Switch off the ignition.
9 With the ignition switched off, connect an ohmmeter between the ignition coil terminals 7 and 12, and check that the resistance is 2500 to 5500 ohm. Also check that the resistance between terminals 7 and 8 is between 0.4 and 0.8 ohm.
10 Disconnect the computer supply multiplug and, using an ohmmeter, check that the resistance between the tachometer terminals 2 and 3 is 20 000 ohm. Reconnect the multiplug.

5 Ignition timing (transistorised ignition system)

Note: *A stroboscopic timing lamp will be needed for this operation*

1 Ignition timing is only required on the transistorised ignition system, not on the integral electronic ignition system.
2 The timing can only be carried out dynamically using a stroboscope.
3 If necessary, highlight the timing marks on the flywheel/driveplate and the relevant graduation on the timing scale (Fig. 4.4) with quick-drying white paint. Typist's correcting fluid is ideal (photo).
4 The engine must be at normal operating temperature. Disconnect and plug the distributor vacuum pipe.
5 Connect a timing light (strobe) to No 1 spark plug HT lead in accordance with the maker's instructions. Some timing lights connect in series with the spark plug; others require additional connections to a 12 volt battery or to an external power supply. Position all leads clear of moving parts, and make sure that you do not get articles of clothing, long hair etc, caught in engine moving parts during subsequent operations.
6 Start the engine and allow it to idle. Point the timing light at the timing marks, they will appear stationary and, if the timing is correct, in alignment.
7 If the marks are not aligned, slacken the distributor clamp bolt and move the distributor body slightly to bring the marks into alignment accordingly. Tighten the clamp bolt and recheck.
8 Accurate checking of the centrifugal and vacuum advance mechanisms is beyond the scope of the DIY mechanic. A rough check may be made by increasing engine speed and checking that the timing mark appears to move away from the pointer on the timing scale. If the vacuum pipe is reconnected, a further advance should be observed. Jerky movement of the timing mark, or flutter, suggests that the automatic advance mechanisms are worn or sticking.
9 When adjustment is complete, switch off the engine, disconnect the timing light and remake the original connections.

5.3 Flywheel housing timing marks (V6 engine)

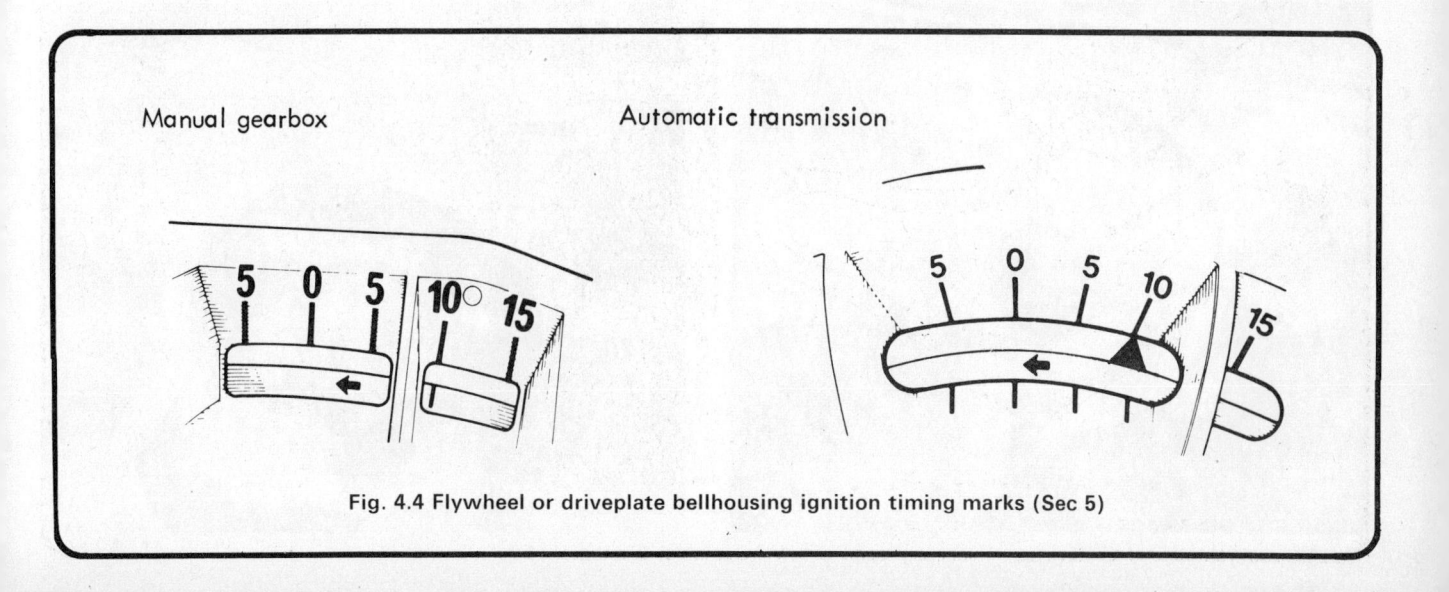

Fig. 4.4 Flywheel or driveplate bellhousing ignition timing marks (Sec 5)

6.1 Distributor off-set drive dog

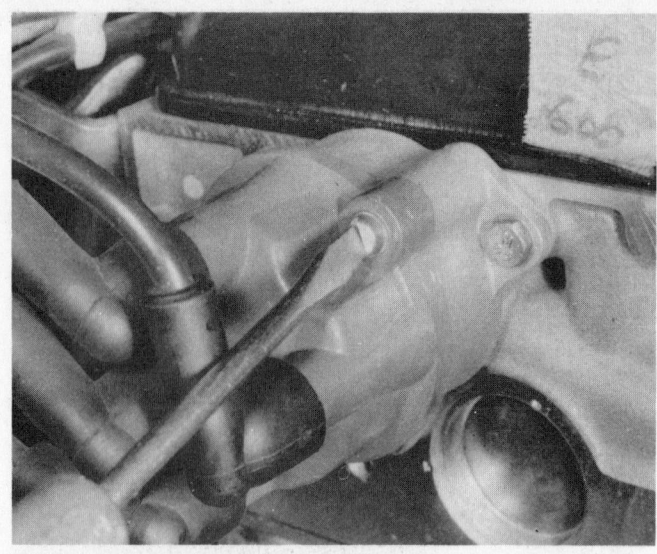

6.2 Distributor cap screw (4-cylinder engine)

6 Distributor (integral electronic system) – removal and refitting

1 The distributor is driven by an offset dog from the rear end of the camshaft on 1995 and 2165 cc models. On 2458 cc Turbo models, the distributor is driven from the front end of the left-hand camshaft (photo).

Non-Turbo models
2 Extract the screws from the distributor cap and place the cap to one side with the HT leads still attached (photo).
3 Remove the rotor and the shield (photos).
4 Unbolt and remove the distributor body. This consists of purely a drive unit for the rotor and no repair or overhaul is possible (photo).

Turbo models
5 Unclip and remove the cover.
6 Unscrew the three fixing screws and take off the distributor cap.
7 Unscrew the three fixing screws and remove the rotor.
8 Take off the insulator and O-ring.
9 The body, drive dog and associated components may be removed after unscrewing the socket-headed retaining bolt with an Allen key.

All models
10 Refitting both types of distributor is a reversal of removal.

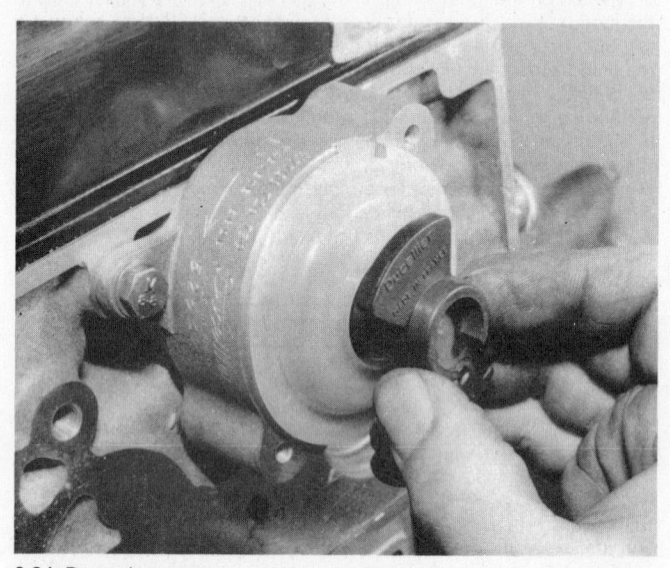

6.3A Removing rotor

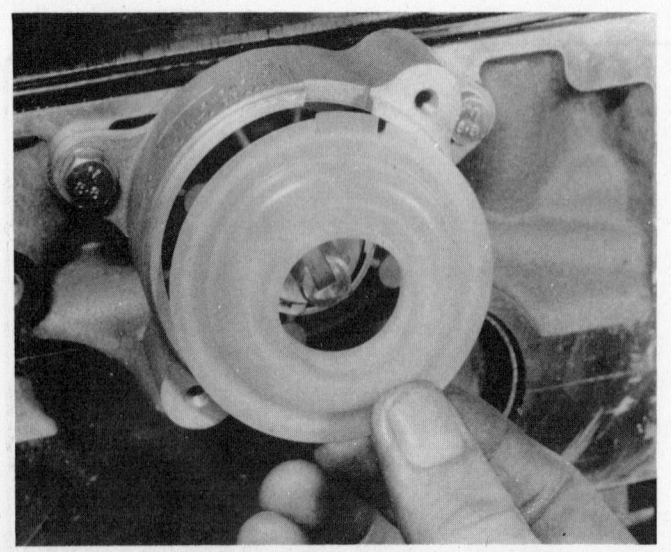

6.3B Removing distributor shield

6.4 Distributor (4-cylinder engine)

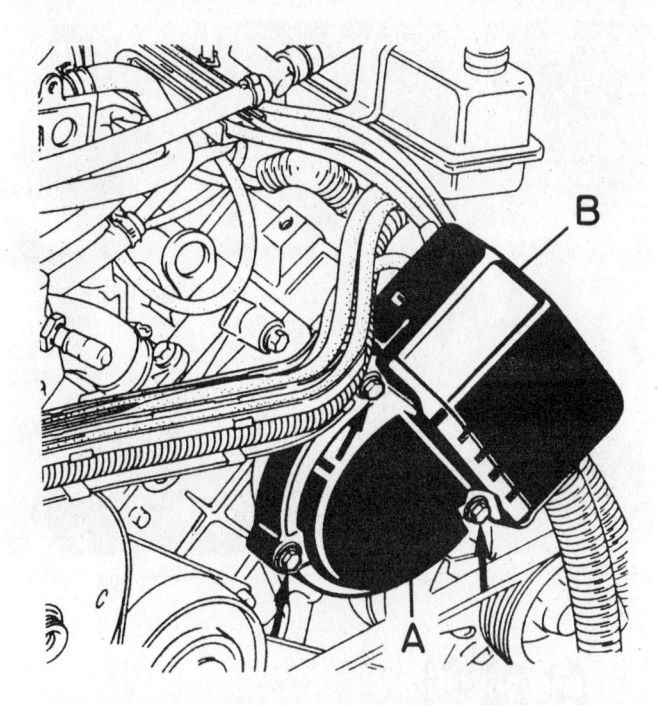

Chapter 4 Ignition system

Fig. 4.5 Distributor on Turbo models (Sec 6)

A Distributor cap   B HT socket cover

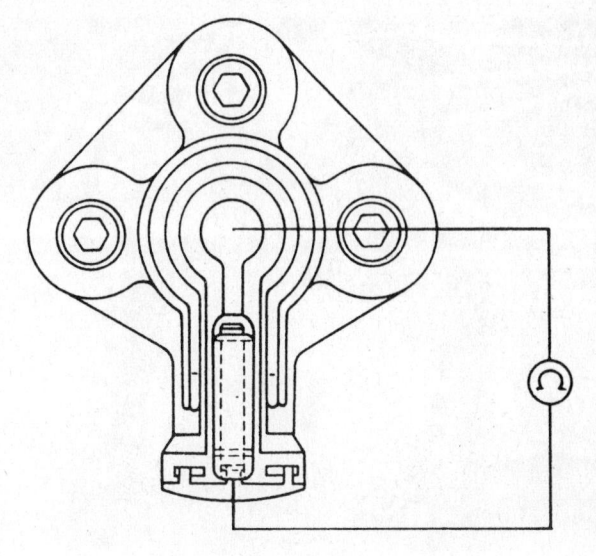

Fig. 4.6 Turbo distributor rotor arm (Sec 6)

Rotor arm resistance 0.8 Kohm to 1.3 Kohm

7 Distributor (transistorised electronic system) – removal, overhaul and refitting

1 The distributor is driven by a gear on the rear of the right-hand camshaft. Unfortunately, its location means that even the simple job of removing the distributor cap turns it into a major operation requiring the prior removal of the airflow meter/sensor and fuel metering/distributor unit of the fuel injection system.
2 Refer to Chapter 3 and remove the air cleaner.
3 Unbolt and remove the double U-shaped manifold from the front of the intake manifold/air distributor unit (photos).
4 Identify and disconnect the fuel pipes from the fuel distributor. This will necessitate unscrewing the banjo union bolts.
5 Extract the socket-headed screws which hold the air floor meter/sensor housing and lift the assembly, complete with fuel distributor, upwards. At the same time pull off the vacuum and fuel hoses (photos).
6 Turn the crankshaft pulley bolt until No 1 piston is at TDC. This will be indicated when the line or notch in the flywheel or driveplate is aligned with the O mark on the timing scale at the bellhousing aperture.
7 Prise back the clips and remove the distributor cap and place it to one side with HT leads still attached (photo).
8 Check that the contact end of the rotor is in fact pointing at the No 1 spark plug contact in the distributor cap (see Fig. 4.7). If not align it by turning the crankshaft pulley bolt.
9 Mark the rim of the distributor body in alignment with the contact end of the rotor if it is not already marked.
10 Mark the position of the distributor clamp plate in relation to the cylinder head.
11 Disconnect the LT wiring plug, unscrew the clamp plate nut and withdraw the distributor.

Overhaul

12 It is recommended that a well worn unit is renewed rather than attempt to replace many individual components.
13 Remove the rotor and dust shield.
14 Extract the circlip (1 – Fig. 4.9) and take out the washer (2).
15 Using two screwdrivers, prise off the reluctor.
16 Extract the second circlip now exposed and extract the three screws which hold the coil assembly. Take care not to lose the rotor arm key.

7.3A Removing U-shaped manifold

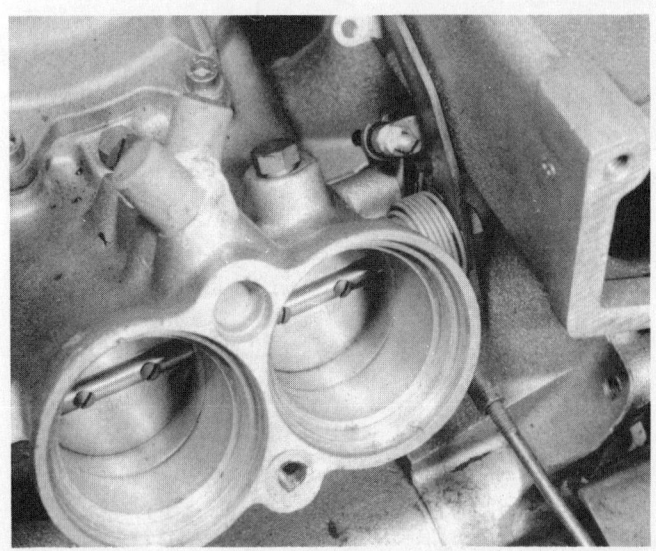

7.3B U-shaped manifold removed

170   Chapter 4 Ignition system

7.5A Air flow sensor/fuel metering distributor unit showing socket-headed fixing screws

7.5B Lifting airflow sensor and fuel metering/distributor unit

7.7 Distributor (V6 engine) with cap removed

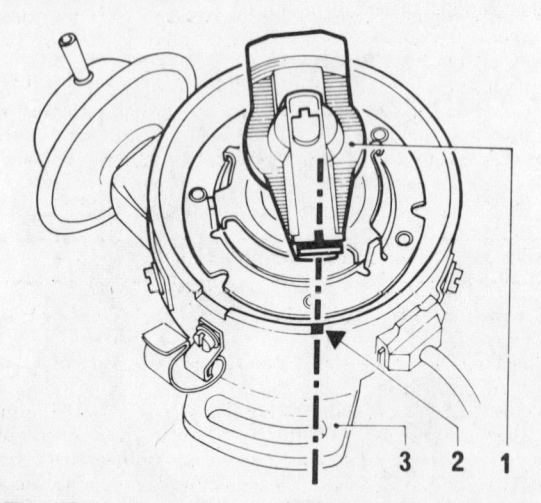

Fig. 4.7 Distributor rotor on 2664 cc engine (Sec 7)

1 Rotor
2 Alignment mark (No 1 at TDC)
3 Distributor clamp plate

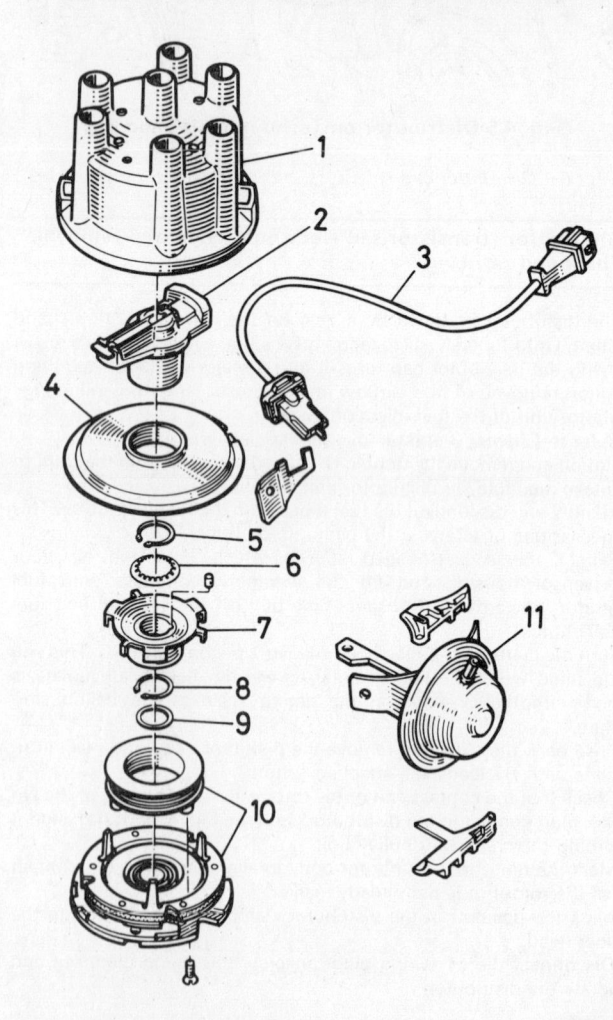

Fig. 4.8 Components of the transistorised ignition system distributor (Sec 7)

1 Distributor cap
2 Rotor arm
3 Connecting lead
4 Dust cover
5 Circlip
6 Washer
7 Reluctor
8 Circlip
9 Washer
10 Magnet/pick-up coil
11 Vacuum unit

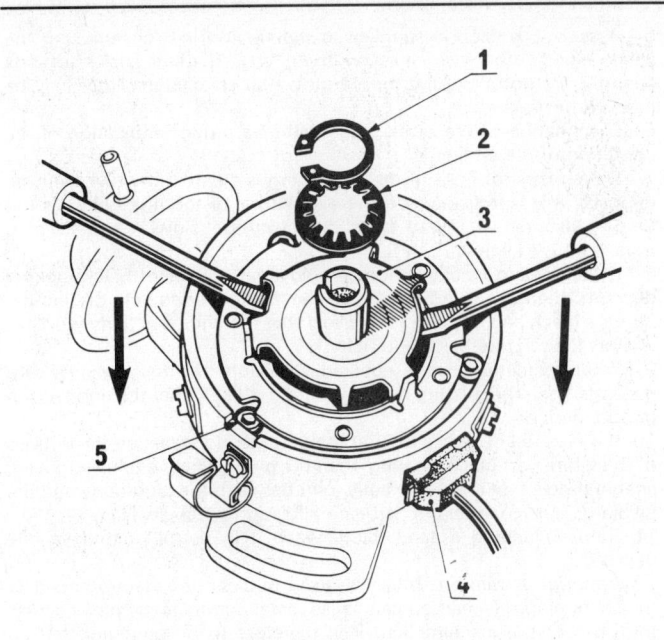

Chapter 4 Ignition system

Fig. 4.9 Removing distributor reluctor (Sec 7)

1 Circlip
2 Washer
3 Reluctor
4 Wiring connector
5 Magnet/pick-up coil

7.21 U-shaped manifold-to-throttle housing O-ring seals

Refitting

17 Reassembly is a reversal of dismantling.
18 Before refitting the distributor, check that No. 1 piston is at TDC.
19 When the distributor is pushed into position, the action of the gears meshing will cause the rotor to turn, so anticipate this by setting the rotor a few degrees out of alignment with the rim mark so that, when the distributor is fully installed, the contact end of the rotor arm will be aligned as was the case before removal.
20 Align the distributor body mark with the one on the cylinder head and tighten the clamp nut. Fit the distributor cap.
21 Refitting the fuel injection components is a reversal of removal, but renew all seals, including the O-rings at the air meter U-shaped manifold (photo).
22 Check and adjust the ignition timing (Section 5).

8 Sensors – description, removal and refitting

1 Where an integral electronic ignition system is fitted, an angular position sensor is used to monitor TDC and BDC from the magnet blocks attached to the flywheel or driveplate periphery. The sensor is secured by special shouldered bolts to the bellhousing and is preset in production; no adjustment being required.
2 On Turbo models, a knock sensor is located on the induction manifold. Ignition is then retarded on the affected cylinder by 7° and on other cylinders by 1°.
3 In the absence of further knocking, the 7° retard alters to 1° after a few seconds and the 1° retard is removed from all cylinders after a period of several minutes.
4 When removing and refitting a sensor, take care not to damage or bend it.

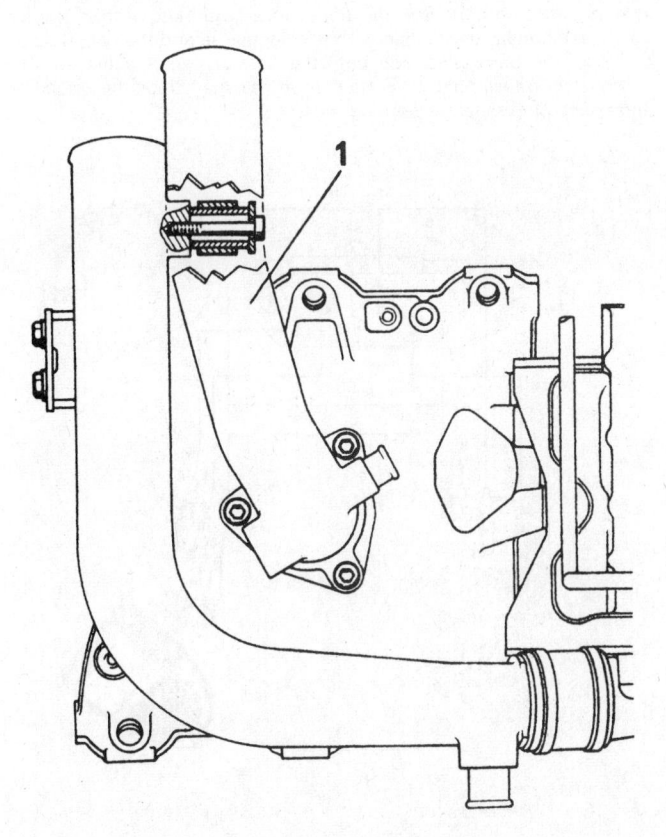

9 Spark plugs and high tension leads

1 Correctly functioning spark plugs are essential for efficient engine operation.
2 At the intervals specified in Routine Maintenance, the plugs should be removed, cleaned and regapped.
3 To remove the plugs, open the bonnet, and pull the HT leads from them. Grip the rubber end fitting, not the lead, otherwise the connection to the end fitting may fracture. On Turbo models, the inlet manifold (1 – Fig. 4.10) must first be removed to gain access to the spark plugs on the right-hand cylinder bank (see Chapter 3, Section 51).
4 Brush or vacuum out accumulated dirt or grit from the spark plug recesses in the cylinder head otherwise it may drop into the combustion chamber when the plug is removed.

Fig. 4.10 Inlet manifold (1) on Turbo models obscuring spark plugs (Sec 9)

Chapter 4 Ignition system

5 A special wrench is supplied in the car tool kit for removing the spark plugs, otherwise unscrew them with a deep socket or box spanner. Do not allow the tool to tilt or the ceramic insulator may be cracked or broken.

6 Examination of the spark plugs will give a good indication of the condition of the engine.

7 If the insulator nose of the spark plug is clean and white, with no deposits, this is indicative of a weak mixture, or too hot a plug (a hot plug transfers heat away from the electrode slowly, a cold plug transfers heat away quickly).

8 If the tip and insulator nose are covered with hard black-looking deposits, then this is indicative that the mixture is too rich. Should the plug be black and oily, then it is likely that the engine is fairly worn, as well as the mixture being too rich.

9 If the insulator nose is covered with light tan to greyish brown deposits, then the mixture is correct and it is likely that the engine is in good condition.

10 If there are any traces of long brown tapering stains on the outside of the white portion of the plug, then the plug will have to be renewed, as this shows that there is a faulty joint between the plug body and the insulator, and compression is being allowed to leak away.

11 Before cleaning a spark plug, wash it in petrol to remove oily deposits.

12 Although a wire brush can be used to clean the electrode end of the spark plug this method can cause metal conductance paths across the nose of the insulator and it is therefore to be preferred that an abrasive powder cleaning machine is used. Such machines are available quite cheaply from motor accessory stores or you may prefer to take the plugs to your dealer who will not only be able to clean them but also to check the sparking efficiency of each plug under compression.

13 The spark plug gap is of considerable importance, as, if it is too large or too small, the size of the spark and its efficiency will be seriously impaired. For the best results the spark plug gap should be set in accordance with the Specifications at the beginning of this Chapter.

14 To set it, measure the gap with a feeler gauge and then bend open, or close, the outer electrode until the correct gap is achieved. The centre electrode should never be bent as this may crack the insulation and cause plug failure if nothing worse.

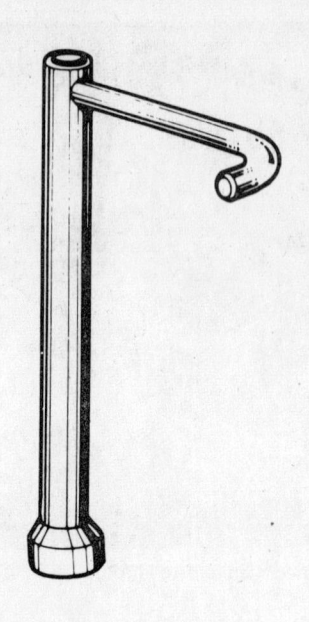

Fig. 4.11 Special spark plug wrench (Sec 9)

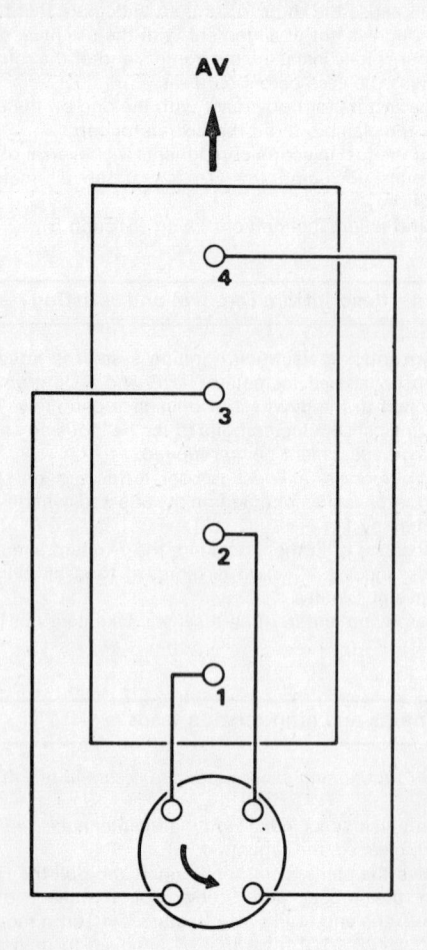

Fig. 4.12 Spark plug lead connections on 4-cylinder engine (Sec 9)

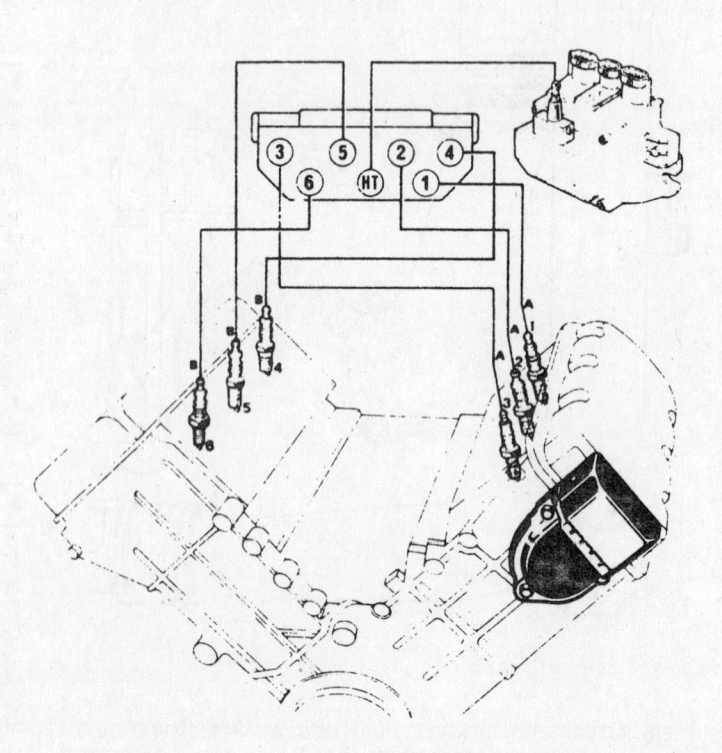

Fig. 4.13 Spark plug lead connections on 6-cylinder Turbo engine (Sec 9)

Chapter 4 Ignition system

15 Special spark plug electrode gap adjusting tools are available from most motor accessory stores.
16 Before refitting the spark plugs, wash each one thoroughly again using clean fuel in order to remove all trace of abrasive powder and then apply a smear of grease to the plug threads.
17 Screw each plug in by hand. This will ensure that there is no chance of cross-threading.
18 Tighten to the specified torque. If a torque wrench is not available, just lightly tighten each plug. It is better to undertighten than strip the threads from the light alloy cylinder head. Also with the taper seat type plugs used (no sealing washers) overtightening can make them very difficult to unscrew.
19 When connecting the spark plug leads, make sure that they are connected in their correct order according to engine type (see Figs. 4.12, 4.13, 4.14). No. 1 cylinder is at the flywheel (driveplate) end of the engine.
20 The spark plug leads require no routine attention other than being kept clean by wiping them regularly.
21 In order to minimise corrosion in the distributor cap lead sockets, smear the HT cable end fittings with a light coating of petroleum jelly.

10 Ignition switch – removal and refitting

1 The operations are described in Chapter 10, Section 29, in conjunction with the steering column lock.

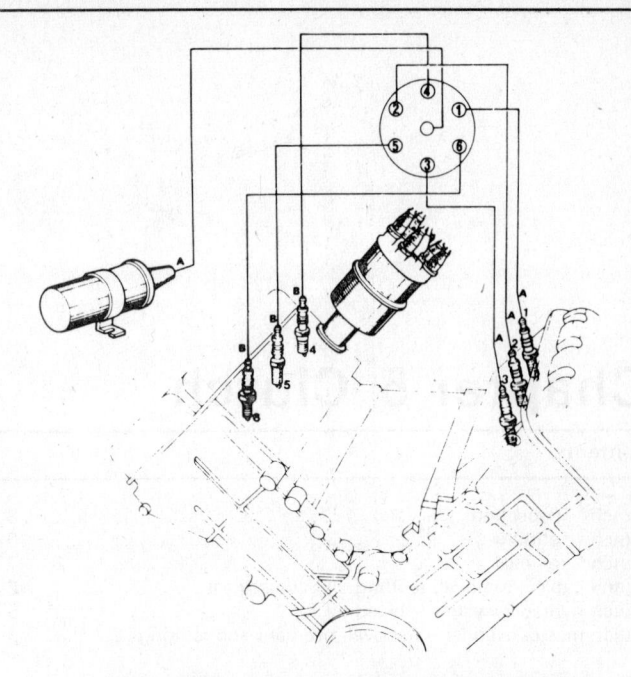

Fig. 4.14 Spark plug lead connections on 6-cylinder engine with transistorised electronic ignition (Sec 9)

11 Fault diagnosis – ignition system

The following gives general indications only. For more detailed system checks, refer to Section 3 or 4.

Symptom	Reason(s)
Engine fails to start	Faulty or disconnected leads Faulty spark plug Fault in ignition coil Fault in pick-up or control unit
Engine starts and runs but misfires	Incorrect timing Fouled spark plug Incorrectly connected HT leads Check in distributor cap or rotor Poor battery, engine and earth connections
Engine overheats, lacks power	Incorrect timing* Fault in vacuum unit or mechanical advance*
Engine 'pinks'	Timing too advanced* Advance mechanism stuck in advance position* Low fuel octane Upper cylinder oil used in fuel Excessive oil vapour from crankcase ventilation system (worn piston rings) Automatic transmission fluid seeping into inlet manifold caused by faulty transmission vacuum unit

*Transistorised electronic system only

Chapter 5 Clutch

Contents

Clutch – inspection	8	Clutch pedal – removal and refitting	6
Clutch – refitting	10	Clutch release mechanism	9
Clutch – removal	7	Clutch slave cylinder – removal, overhaul and refitting	4
Clutch cable – removal, refitting and adjustment	2	Description	1
Clutch hydraulic system – bleeding	5	Fault diagnosis – clutch	11
Clutch master cylinder – removal, overhaul and refitting	3		

Specifications

General
Type .... Single dry plate, with diaphragm spring and sealed ball release bearing in constant contact with diaphragm spring fingers

Actuator:
  B297 (1995 cc) .... Cable
  B29E (2165 cc), B295 (2458 cc) and B298 (2664 cc) .... Hydraulic

Driven plate diameter
Models B297 and B29E .... 215.0 mm (8.5 in)
Models B295 and B298 .... 235.0 mm (9.3 in)

Torque wrench setting
	Nm	lbf ft
Pressure plate cover to flywheel	24	18

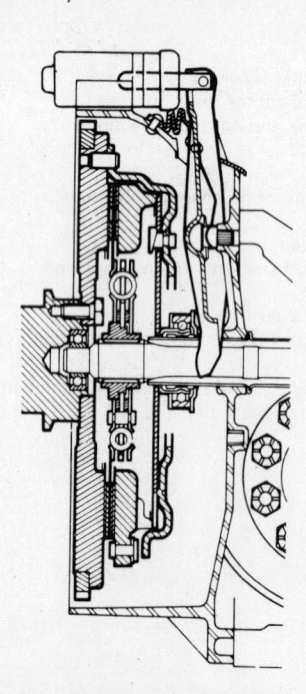

Fig. 5.1 Sectional view of hydraulically-operated clutch (Sec 1)

1 Description

The clutch is of single dry plate type with a diaphragm spring pressure plate.

The driven plate incorporates torsion springs to give resilience to the hub.

The release bearing is of sealed ball type, self-centering and in constant contact with the diaphragm spring fingers.

Clutch actuation may be hydraulic or by cable according to model; refer to Specifications. The master cylinder is fed by fluid from the brake fluid reservoir.

2 Clutch cable – removal, refitting and adjustment

1 Release the locknuts on the cable end fitting at the clutch release lever.
2 Disconnect the cable from the cut-out in the release lever.
3 Working under the facia panel, disconnect the cable from the clutch pedal arm link.
4 Withdraw the cable into the engine compartment.
5 Fit the new cable by reversing the removal operations then adjust in the following way.
6 Support the clutch pedal in the fully raised position.
7 Adjust the locknuts on the cable end fitting until there is a free movement (G) at the end of the damping block (Fig. 5.4) of between 3.0 and 4.0 mm (0.12 and 0.16 in).
8 Apply grease to the cable end fitting cut-out in the release lever.

Chapter 5 Clutch

3 Clutch master cylinder – removal, overhaul and refitting

1 On left-hand drive cars, remove the front left-hand roadwheel and the plastic cover (P – FIg.. 5.5) from under the wheel arch.
2 The master cylinder is supplied with fluid from the brake fluid reservoir.
3 Disconnect the fluid supply hose from the master cylinder also the pipeline union which serves the slave cylinder. Catch any fluid in a suitable container (photo).
4 Working under the facia panel, disconnect the clutch pedal pushrod from the pedal arm.
5 Unbolt and remove the master cylinder, the bolts being accessible from inside the car.
6 Clean away external dirt, extract the circlip from the end of the cylinder and withdraw the pushrod and piston assembly.
7 Examine the surfaces of the piston and cylinder. If they are scored or corroded, renew the master cylinder complete.
8 If the components are in good condition, clean them in methylated spirit or hydraulic fluid – nothing else – and discard the piston seal.
9 Obtain a repair kit which will contain all the necessary renewable items.
10 Fit the new seal, manipulating it into position with the fingers. Dip the piston into clean hydraulic fluid and push it into the cylinder. Do not trap the seal lips.
11 Hold the piston depressed with the pushrod and fit the circlip.
12 Refit the master cylinder and then check the pushrod setting.
13 The pushrod length (L – Fig. 5.7) should be beteeen 129.0 and 130.0 mm (5.08 and 5.12 in).
14 There should be an endfloat (J – Fig. 5.8) of between 0.2 and 0.5 mm (0.008 and 0.020 in) which will be indicated by a pedal free movement of between 1.0 and 2.5 mm (0.04 and 0.10 in).
15 Adjust as necessary by releasing the locknut, disconnecting the clevis fork from the pedal arm and turning the fork in or out. The clevis fork is fixed to the pedal arm by a clevis pin and split pin.
16 Bleed the system, as described in Section 5.

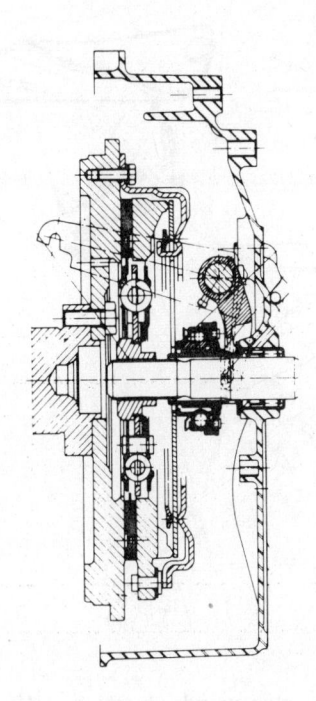

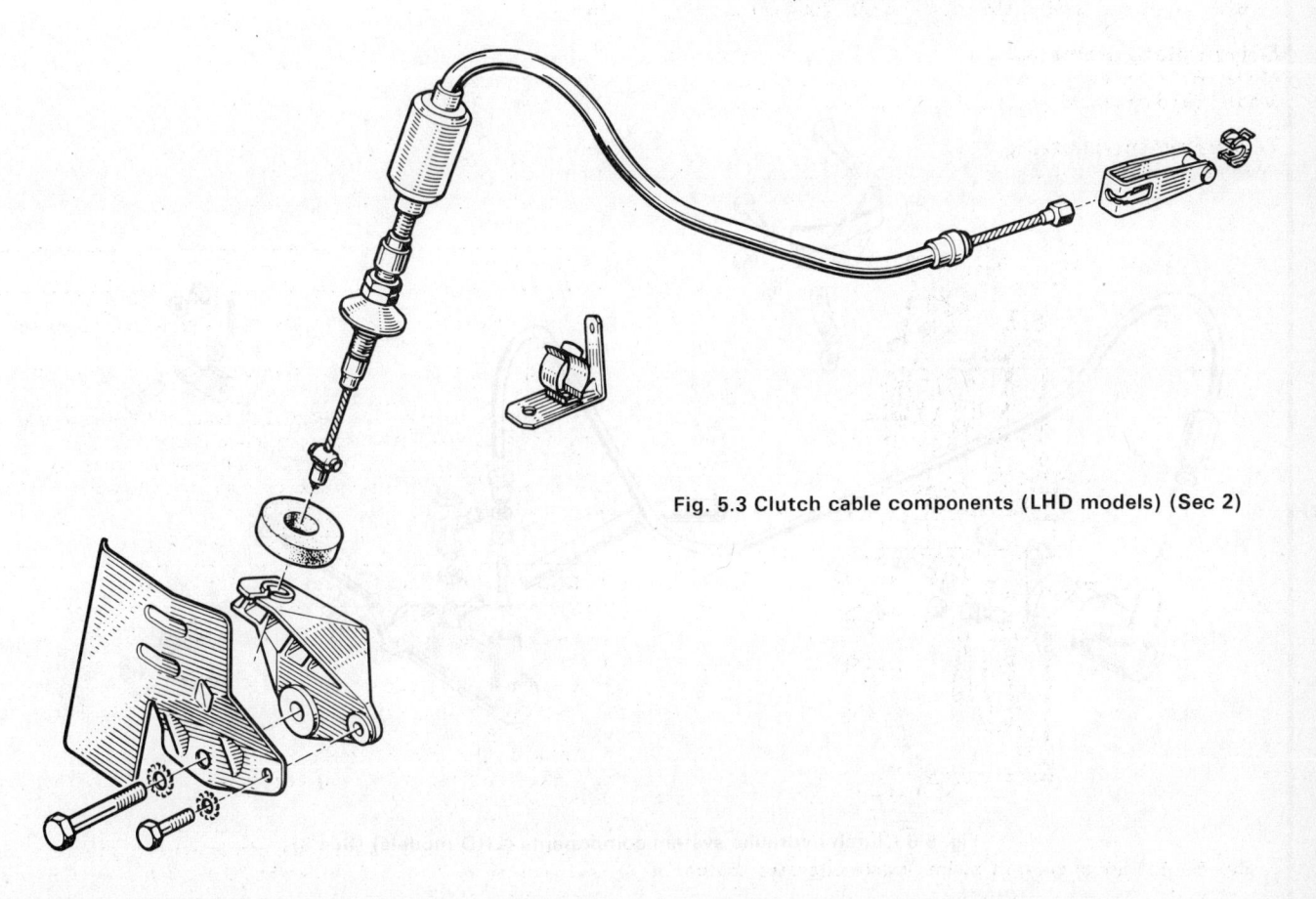

Fig. 5.2 Sectional view of cable-operated clutch (Sec 1)

Fig. 5.3 Clutch cable components (LHD models) (Sec 2)

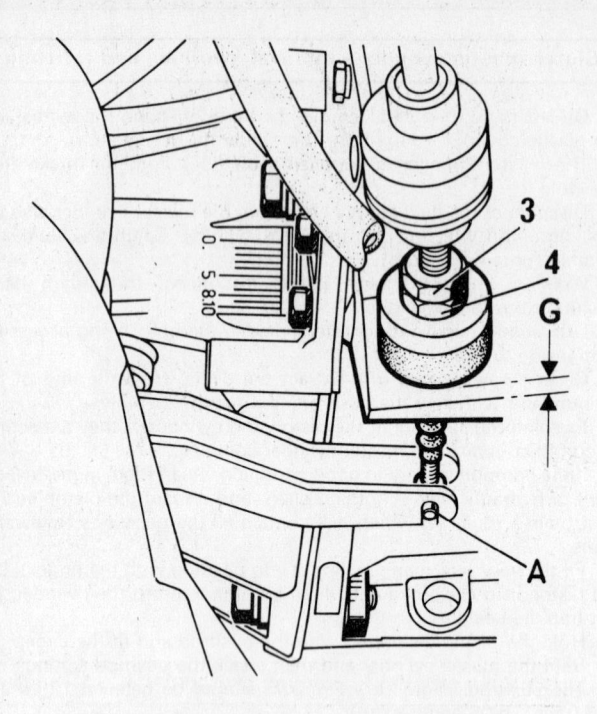

Fig. 5.4 Clutch cable at release lever (Sec 2)

A Grease application point
G Clearance at damping block
3 Locknut
4 Locknut

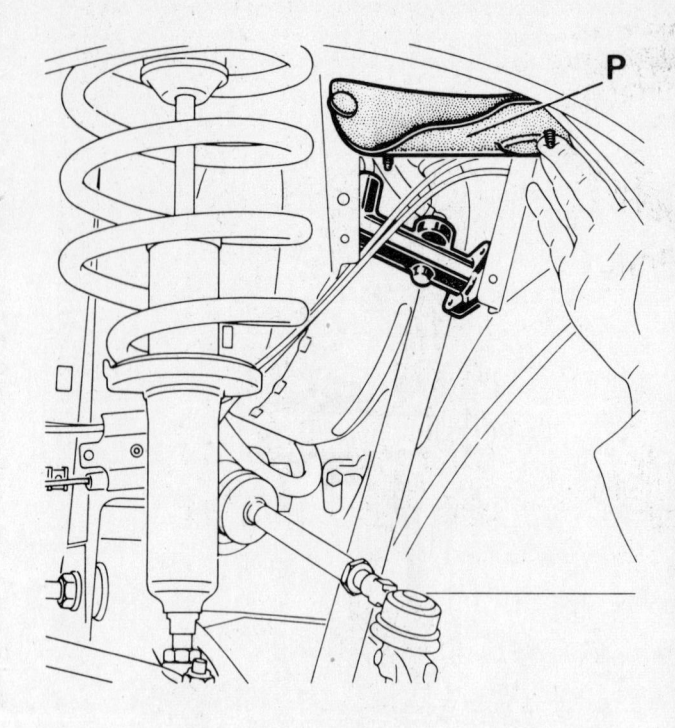

Fig. 5.5 Clutch master cylinder location (LHD models) (Sec 3)

P Plastic cover

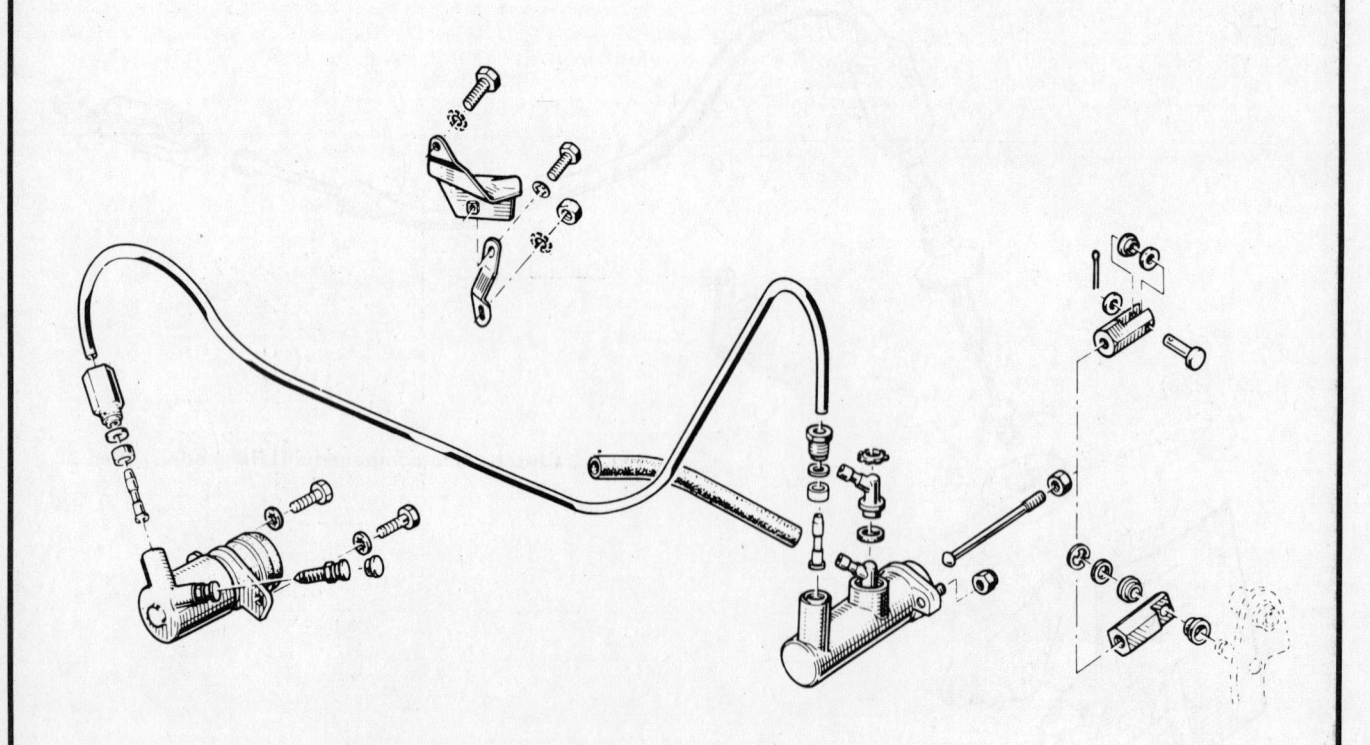

Fig. 5.6 Clutch hydraulic system components (LHD models) (Sec 3)

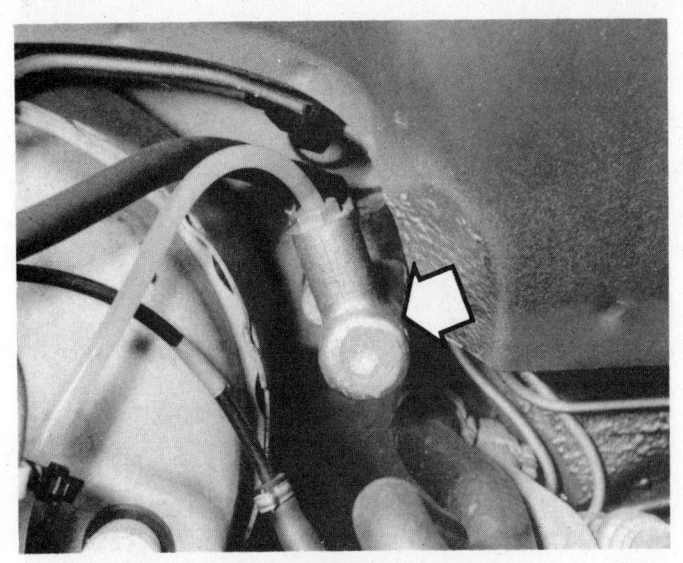

Chapter 5 Clutch 177

3.3 Clutch master cylinder (arrowed)

4.2A Slave cylinder (2165 cc) with fluid inlet arrowed

4.2B Clutch slave cylinder (V6) – arrowed

4 Clutch slave cylinder – removal, overhaul and refitting

Six-cylinder models
1 Unbolt the power steering pump and move it to one side of the engine compartment.

All models
2 Disconnect the fluid feed pipe from the slave cylinder. Allow the fluid to drain into a suitable container (photos).
3 Unscrew the two fixing bolts and draw the cylinder off the pushrod which will remain attached to the clutch release fork. **Note:** On Turbo models, the slave cylinder is shielded from the turbocharger by baffle panels which must be removed for access. On four cylinder (2165 cc) models an exhaust heat shield is used (photo).

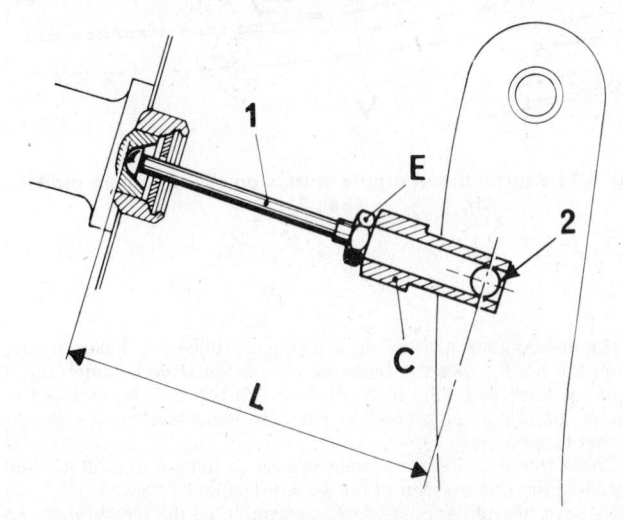

Fig. 5.7 Clutch pushrod setting diagram (Sec 3)

1 Pushrod
2 Clevis pin
C Clevis fork
E Locknut
L = 129.0 to 130.0 mm
    (5.08 to 5.12 in)

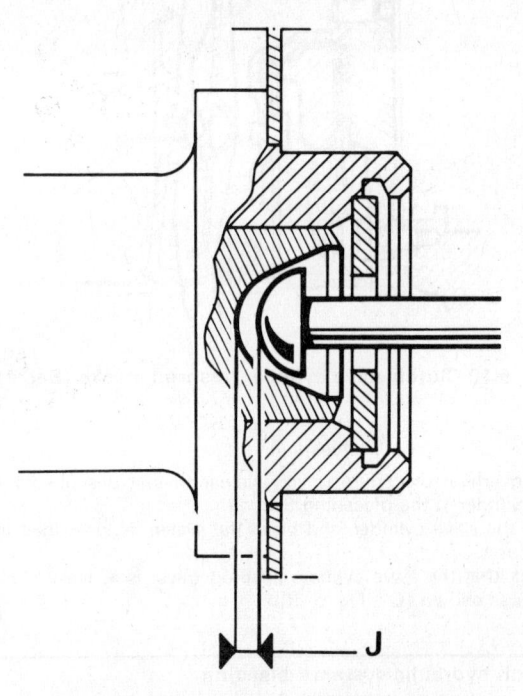

Fig. 5.8 Clutch pushrod endfloat (Sec 3)

J = 0.2 to 0.5 mm
   (0.008 to 0.020 in)

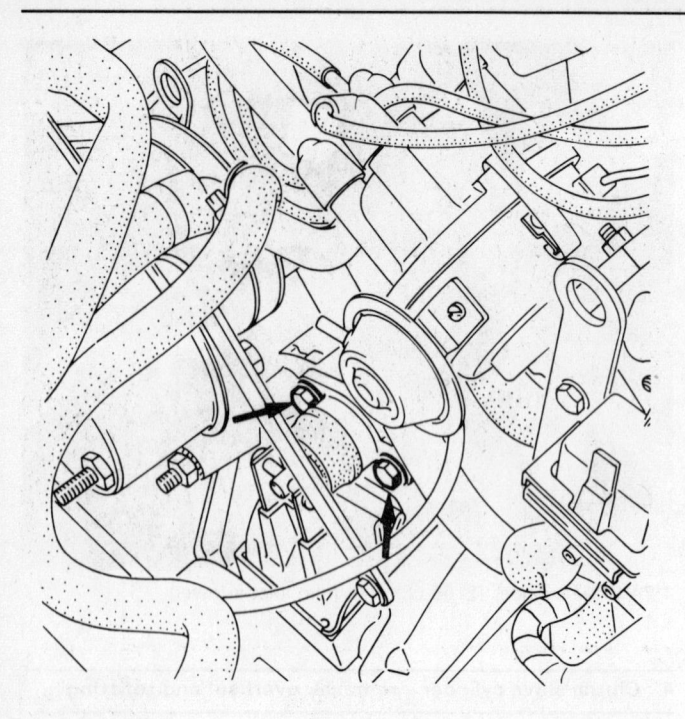

Chapter 5 Clutch

Fig. 5.9 Clutch slave cylinder fixing bolts (arrowed) (Sec 4)

4.3 Clutch slave cylinder heat shield

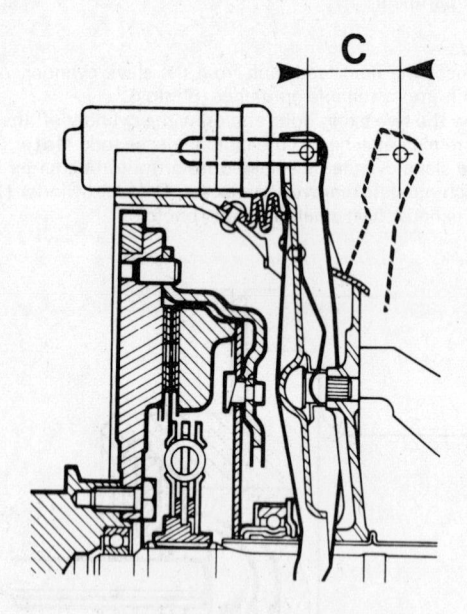

Fig. 5.10 Clutch slave cylinder pushrod stroke (Sec 4)

C = 11.0 mm (0.43 in) minimum

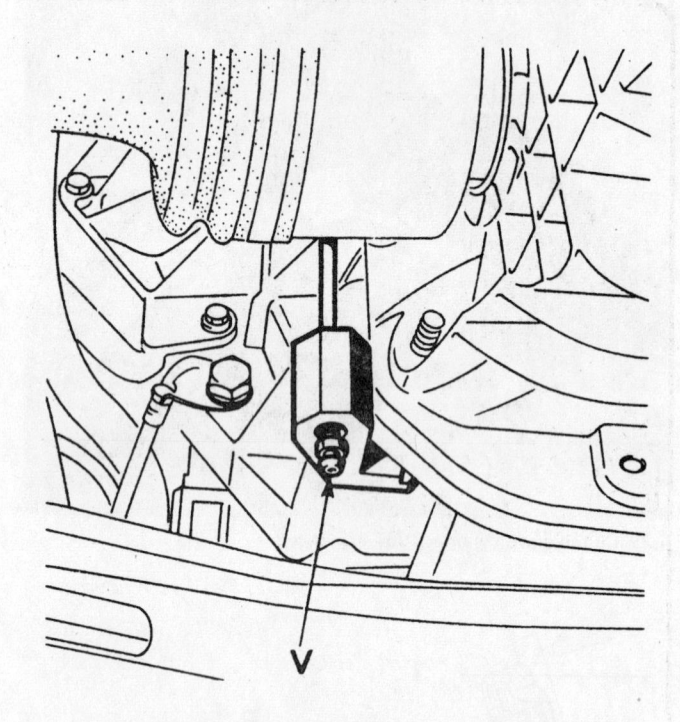

Fig. 5.11 Clutch bleed nipple extension (V) on Turbo models (Sec 5)

4 The overhaul procedure is very similar to that described for the master cylinder in the preceding Section.
5 Refit the slave cylinder, and bleed the system as described in the next Section.
6 Check that the slave cylinder pushrod travel is at least 11.0 mm (0.43 in) as shown (C – Fig. 5-20).

5 Clutch hydraulic system – bleeding

Bleeding the hydraulic system is necessary whenever air has entered the system; after disconnecting any component or as a result of leakage (which should first be rectified.)

1 Gather together a clean jar, a length of rubber of plastic tubing which fits tightly over the bleed nipple on the slave cylinder and a supply of fresh hydraulic fluid. An assistant will also be required to operate the clutch pedal and to keep the fluid level in the master cylinder topped up.
2 Check that the master cylinder reservoir is full – if not, fill it – and also cover the bottom inch of the jar with hydraulic fluid.
3 Remove the rubber dust cap (if present) from the bleed nipple on the slave cylinder and place one end of the tube securely over it. On Turbo models, the bleed nipple is located on an extension bolted to the corner of the gearbox (Fig. 5.11). Place the other end of the tube in the jar, ensuring that the tube orifice is below the level of the fluid.
4 Using a ring or open-ended spanner, unscrew the bleed nipple

Chapter 5 Clutch 179

approximately one turn and then have your assistant slowly depress the clutch pedal.
5 Tighten the bleed nipple while the pedal is held in the fully depressed position.
6 Have your assistant release the clutch pedal slowly, allowing it to return fully. After waiting four seconds to allow the master cylinder to recuperate, repeat the above procedure.
7 Keep the master cylinder reservoir topped up throughout the bleeding operation, otherwise further air will be introduced into the system.
8 When clean hydraulic fluid free from air bubbles can be seen coming from the end of the tube, tighten the bleed nipple, remove the rubber tube and refit the dust cap.
9 Finally top up the master cylinder reservoir and refit the cap. Discard the old hydraulic fluid as it is contaminated and must not be re-used in the system.
10 Reference should be made to Chapter 9, Section 15 for details of bleeding using one-way valve or pressure bleeding equipment which is also suitable for the clutch hydraulic system.

6 Clutch pedal – removal and refitting

1 Disconnect the clutch cable or pushrod from the pedal arm (photo). Disconnect the pedal springs.

6.1 Clutch pedal pushrod

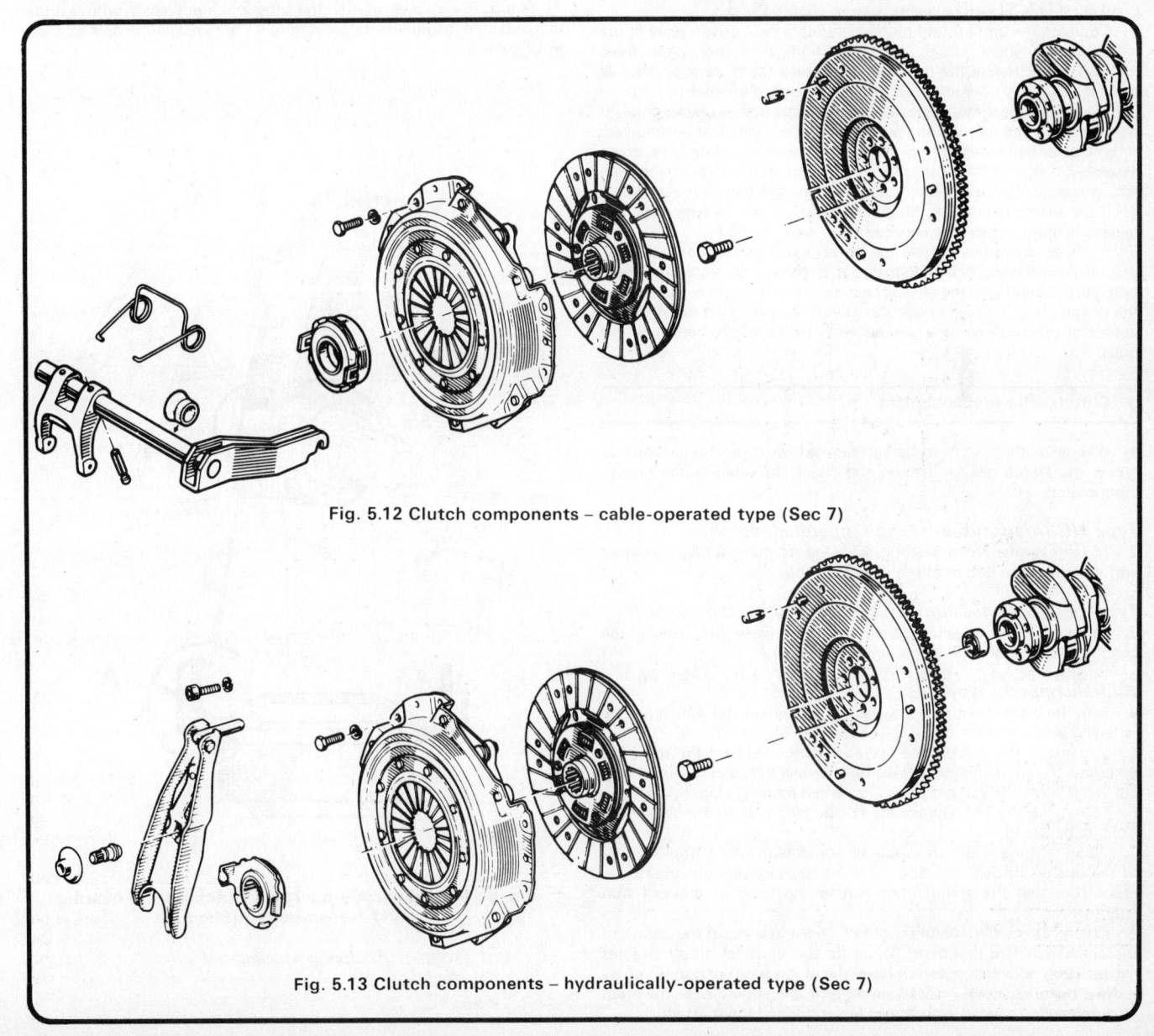

Fig. 5.12 Clutch components – cable-operated type (Sec 7)

Fig. 5.13 Clutch components – hydraulically-operated type (Sec 7)

2 Extract the spring clip from the end of the pedal pivot shaft, push the cross-shaft out of the pedal bushes until the pedal can be removed.
3 The pedal bushes can be renewed if worn.
4 Refitting is a reversal of removal but check the pushrod setting as described in Section 3.

7 Clutch – removal

1 Access to the clutch is obtained by removing the transmission (Chapter 6).
2 Unscrew the clutch cover fixing bolts evenly and progressively until the diaphragm spring pressure is relieved and then lift the pressure plate assembly from its dowels and catch the driven plate (friction disc) as it drops out.

8 Clutch – inspection

1 Examine the driven plate friction linings for wear and loose rivets, and also for rim distortion, cracks, broken hub springs and worn splines. The surface of the friction linings may be highly glazed but as long as the friction material can be clearly seen this is satisfactory. If the friction material is dark in appearance, further investigation is necessary as it is a sign of oil contamination caused by a leaking crankshaft rear oil seal or gearbox input shaft oil seal.
2 Compare the amount of lining wear with a new driven plate at the stores in your local garage, and if the linings are more than three quarters worn, renew the driven plate. Renew the plate regardless of wear if it is badly contaminated or mechanically damaged.
3 Check the machined faces of the flywheel and pressure plate. If deep grooves or heavy scoring are apparent the units must be renewed. It will also be necessary to renew the pressure plate and cover assembly if it is cracked or split or if the pressure of the diaphragm spring is suspect. Do not attempt to machine the flywheel, renew it.
4 If the diaphragm spring fingers are stepped due to wear from the release bearing, renew the pressure plate cover assembly.
5 If either the driven plate or cover assembly require renewal, practical experience has shown that it is always advisable to renew both components and the release bearing at the same time. Renewal of the driven plate or cover assembly individually can often cause clutch judder or vibration as new components do not easily bed in to old ones.

9 Clutch release mechanism

1 Whenever the clutch is being renewed, take the opportunity to renew the clutch release bearing and check the other clutch release components.

Type NG3 transmission (cable-operated clutch)
2 To remove the release bearing, prise the retaining spring legs apart and withdraw the bearing from its guide tube.

Type UN1 transmission (hydraulically-operated clutch)
3 To remove the release bearing, tilt the release fork, unclip the bearing and slide it from its guide tube (photo).

All transmission types
4 Turn the bearing with the fingers. If it is rough or shaky it should be renewed without question.
5 To remove the fork (cable-operated clutch) extract the roll pins, withdraw the shaft and take off the fork. Renew the shaft bushes if they are worn. Note the roll pin projection when refitting (Fig. 5.15).
6 To remove the fork (hydraulically-operated clutch) simply slide it from its ball-stud.
7 When refitting a fork (hydraulically-operated clutch), apply grease to the bushes or ball-stud and, with the hydraulically-operated type, make sure that the spring goes behind the head of the ball-stud (photo).
8 Finally, check the condition of the spigot bearing in the centre of the crankshaft rear mounting flange for the flywheel on all engines except 1995 cc which does not have one. If it is worn or rattles, renew it. New bearings are of greased sealed type and should have the outer circumference smeared with thread locking fluid before fitting.

10 Clutch – refitting

Note: *An alignment tool will be needed for 'hydraulic clutch' models.*

1 Remove any oil, dirt or protective coating from the flywheel, pressure plate and bellhousing interior.
2 Locate the driven plate on the flywheel so that its greater projecting hub is away from the flywheel (photo).
3 Hold it against the flywheel and locate the pressure plate cover assembly on its dowels. Screw in the retaining bolts finger tight only (photo).
4 Where a cable-operated clutch is fitted move the splined hub of the driven plate until it is centralised, within the tips of the diaphragm spring fingers. Check the alignment visually.
5 On hydraulically-operated clutches, centralise the driven plate using a clutch alignment tool. Such a tool is obtainable from many motor parts stores (photo) or one can be fabricated from a piece of dowel, rod or tubing so that the end locates in the spigot bearing and the centre of the tool is a sliding fit in the splined hub of the driven plate. Winding insulating tape around the tool is one method of obtaining the required stepped diameters.
6 Once the driven plate has been centralised, tighten the clutch cover bolts progressively in diametrically opposite sequence to the specified torque (photo). Remove the alignment tool, where used.
7 Smear the splines of the transmission input shaft with a little molybdenum disulphide grease and refit the transmission as described in Chapter 6.

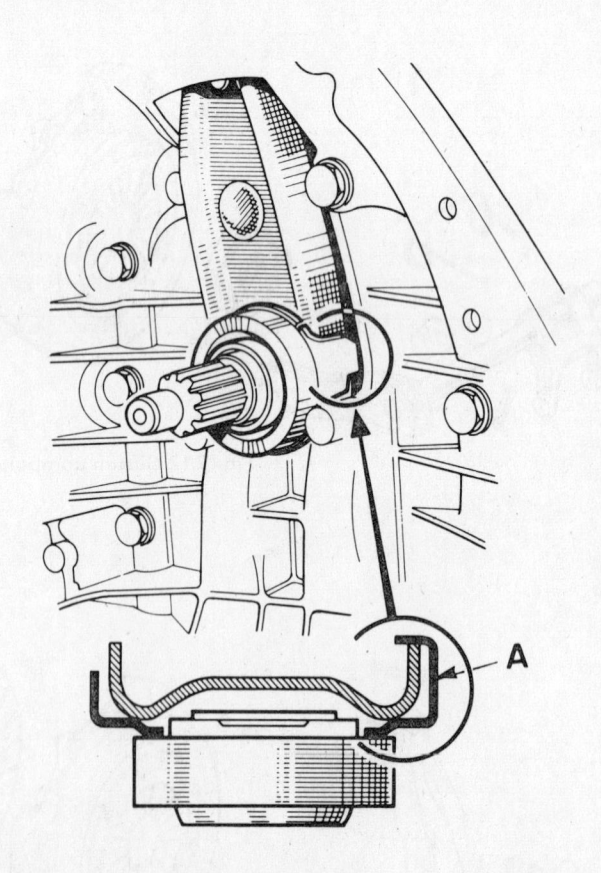

Fig. 5.14 Hydraulically-operated clutch release bearing (UN1 transmission) (Sec 9)

A Spring retaining clip

181

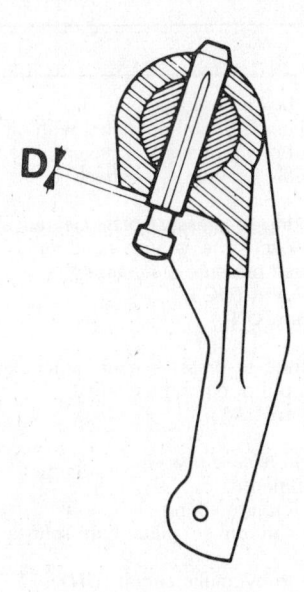

Fig. 5.15 Clutch release fork rollpin projection – cable-operated clutch (Sec 9)

D = 1.0 mm (0.039 in)

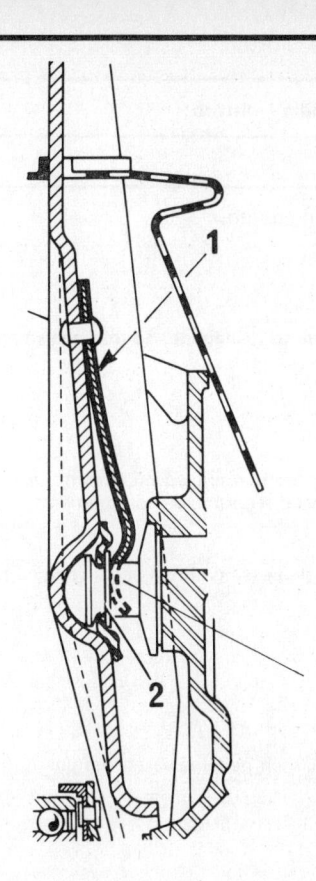

Fig. 5.16 Clutch release fork spring (1) and ball-stud (2) – hydraulically-operated clutch (Sec 9)

9.3 Clutch release bearing (hook arrowed)

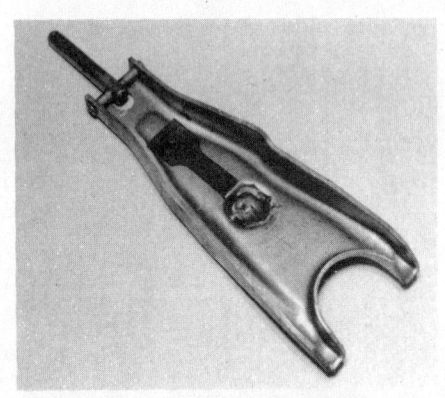

9.7 Rear side of clutch release arm

10.2 Offering clutch driven plate to flywheel

10.3 Offering up clutch pressure plate/cover

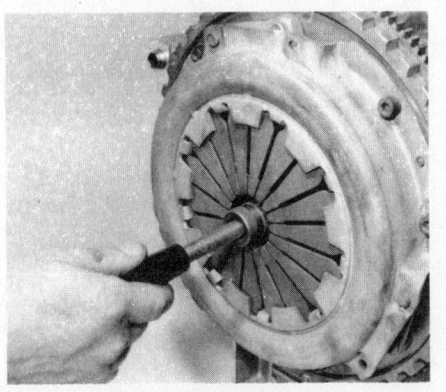

10.5 Centralising clutch driven plate

10.6 Tightening a clutch cover socket-headed bolt

11 Fault diagnosis – clutch

Symptom	Reason(s)
Judder when taking up drive	Loose engine or gearbox mountings Badly worn friction linings or contaminated with oil Worn splines on gearbox input shaft or driven plate hub Worn input shaft needle bearing in flywheel
Clutch spin (failure to disengage) so that gears cannot be meshed*	Incorrect release bearing-to-pressure plate clearance Rust on splines (may occur after vehicle standing idle for long periods) Damaged or misaligned pressure plate assembly Cable stretched or broken (NG3) Air in hydraulic system (UN1)
Clutch slip (increase in engine speed does not result in increase in vehicle road speed – particularly on gradients)	Incorrect release bearing-to-pressure plate finger clearance Friction linings worn out or oil contaminated Air in hydraulic system (UN1)
Noise evident on depressing clutch pedal (engine running)	Dry, worn or damaged release bearing Incorrect pedal adjustment Weak or broken pedal return spring Excessive play between driven plate hub splines and input shaft splines Low fluid level or air in hydraulic circuit (UN1) Incorrectly adjusted cable (NG3)
Noise evident as clutch pedal released (engine running)	Distorted driven plate Broken or weak driven plate torsion springs Incorrect pedal adjustment Distorted or worn input shaft Release bearing loose on retainer hub

*This condition may also be due to the driven plate being rusted to the flywheel or pressure plate. It is possible to free it by applying the handbrake, depressing the clutch pedal, engaging top gear and operating the starter motor. If really badly corroded, then the engine will not turn over, but in the majority of cases the driven plate will free. Once the engine starts, rev it up and slip the clutch several times to clear the rust deposits.

Chapter 6 Manual transmission

Contents

Part A – Type NG3 transmission
Description	1
Differential/final drive – overhaul	13
Differential oil seals – removal without removing transmission	15
Gearchange linkage – removal and refitting	3
Maintenance	2
Primary shaft – overhaul	7
Rear cover – dismantling and reassembly	11
Reverse idle shaft and gear – dismantling and reassembly	9
Secondary shaft – overhaul	8
Selector forks and shaft – dismantling and reassembly	10
Transmission – overhaul (general)	5
Transmission – reassembly	14
Transmission – removal and refitting	4
Transmission – removal of major assemblies	6
Transmission housing – inspection	12

Part B – Type UN1 transmission
Description and maintenance	16
Differential/final drive – overhaul	26
Gearchange linkage	17
Primary shaft – overhaul	20
Rear cover – dismantling and reassembly	24
Reverse idler shaft and gear – dismantling and reassembly	22
Secondary shaft – overhaul	21
Selector forks and shafts – dismantling and reassembly	23
Transmission – reassembly	21
Transmission – removal and refitting	18
Transmission – removal of major assemblies	19
Transmission housing – inspection	25

Part C – All transmissions
Fault diagnosis – manual transmission	28

Specifications

General
Type .... Five forward speeds (all synchromesh) and reverse
Application:
   Type NG3 .... B297 (1995 cc) – 4-cylinder
   Type UN1 .... B29E (2165 cc) – 4-cylinder
                                               B295 (2458 cc) – 6-cylinder
                                               B298 (2664 cc) – 6-cylinder

Ratios
	B297	B29E	B295 and B298
1st	4.09:1	3.36:1	3.36:1
2nd	2.18:1	2.06:1	2.06:1
3rd	1.41:1	1.38:1	1.38:1
4th	1.03:1	1.04:1	0.96:1
5th	0.78:1	0.82:1	0.76:1
Reverse	3.55:1	3.55:1	3.55:1
Final drive	3.56:1	3.89:1	3.89:1

Differential bearing preload
NG3 transmission:
   Used bearings .... Nil
   New bearings .... 3.0 to 9.5 kgf (7.0 to 21.0 lbf)
UN1 transmission:
   Used bearings .... Nil
   New bearings .... 14.5 to 21.0 kgf (32.0 to 47.0 lbf)

Pinion-to-crownwheel backlash
NG3 transmission .... 0.12 to 0.25 mm (0.005 to 0.010 in)
UN1 transmission .... Non-adjustable (one ring nut for bearing preload only)

Oil capacity
NG3 transmission .... 2.0 litre (3.5 pints)
UN1 transmission .... 3.4 litre (6.0 pints)

Torque wrench settings
	Nm	lbf ft
NG3 transmission		
Casing flange bolts	19	14
Casing centre bolts	30	22
Primary shaft nut	128	94
Secondary shaft nut	147	108
Crownwheel bolts	122	90
Reverse swivel bolt	24	18
Rear cover bolts	15	11
Bellhousing-to-engine bolts	54	40
UN1 transmission		
Casing bolts:		
8.0 mm	24	18
10.0 mm	49	36
Crownwheel bolts	122	90
Primary shaft nut	132	97
Secondary shaft nut	195	144
Reverse swivel bolt	24	18
Rear cover bolts	24	18
Clutch housing bolts (to gearcase)	49	36
Spacer plate bolts	49	36
Bellhousing-to-engine bolts	54	40

PART A: TYPE NG3 TRANSMISSION

1 Description

The transmission is located behind the engine in a similar manner to rear-wheel-drive cars but, of course, power is transmitted to the front roadwheels through driveshafts. The driveshafts are splined to the side gears of the differential/final drive which is integrated into the transmission casing.

Five forward synchromesh gears are provided with a reverse gear. The gearchange control is of floor-mounted type.

The transmission casing is of four section type: the clutch bellhousing, the rear cover which houses the selector finger and 5th gear, and the two main casing sections which are split longitudinally.

2 Maintenance

1 At the specified intervals, check the oil level, preferably when the transmission is cold. The car should be standing on a level surface.
2 Unscrew and remove the combined filler/level plug from the side of the transmission (photo). The oil should just start to dribble out. If not, top up with the specified type of oil. Refit the plug.
3 To change the oil at the specified intervals (see Routine Maintenance), drain the oil hot by removing the filler/level and the

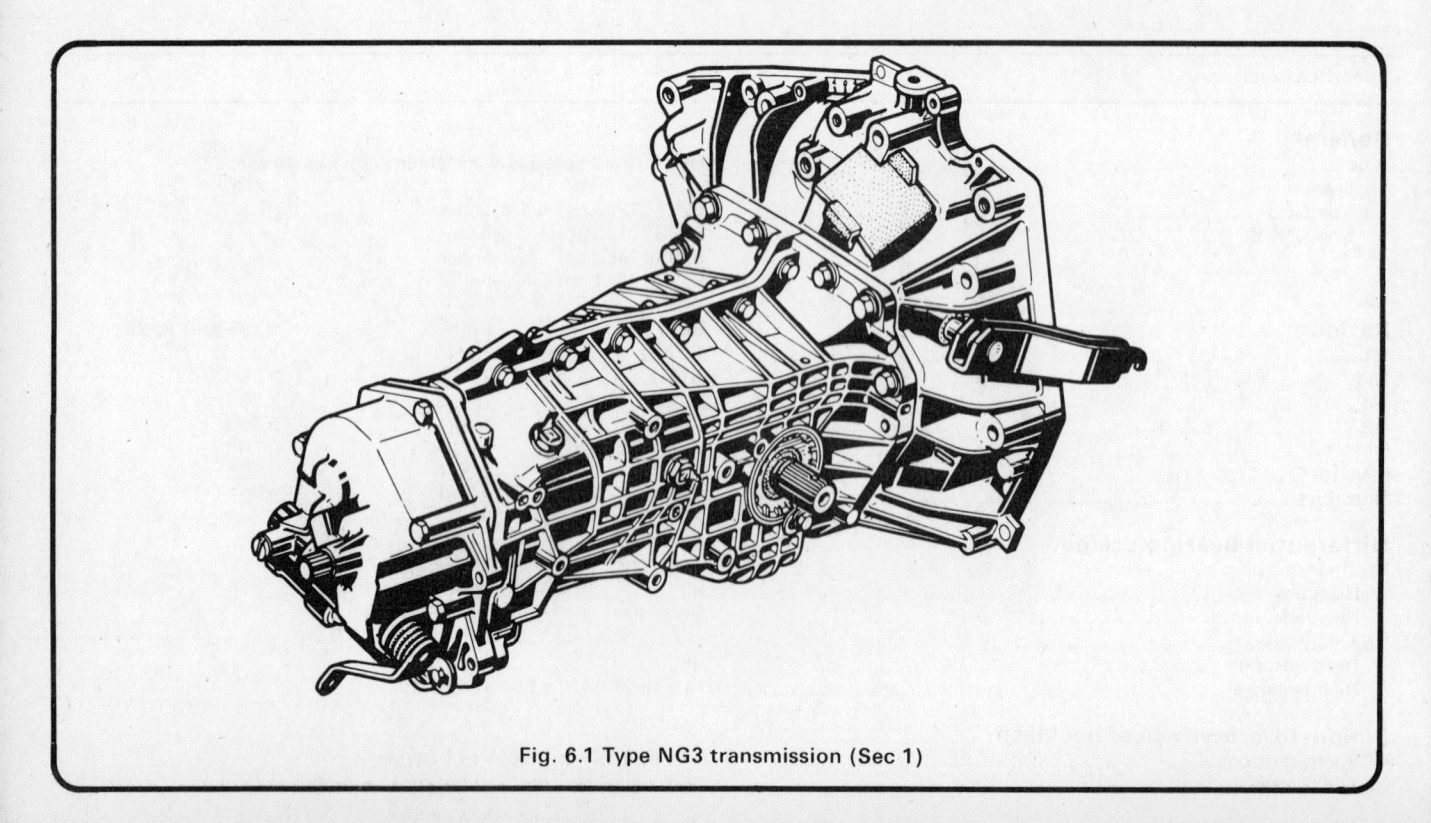

Fig. 6.1 Type NG3 transmission (Sec 1)

Chapter 6 Manual transmission

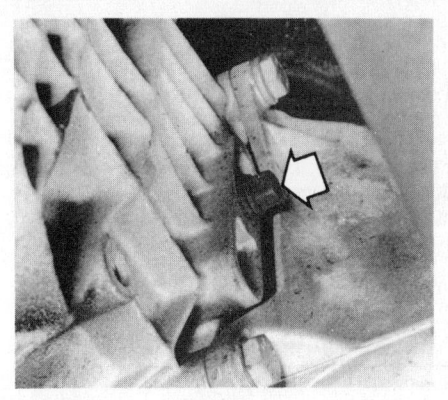

2.2 Transmission filler/level plug (arrowed)

2.3 Transmission drain plug

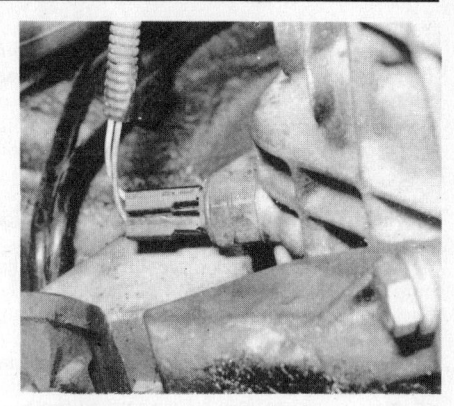

4.6 Reversing lamp switch leads

drain plugs (photo); having placed a suitable container under the transmission.
4 Refit the drain plug then refill with the correct grade and quantity of oil.
5 When the oil just starts to dribble out of the filler/level plug hole, refit the plug.

3 Gearchange linkage – removal and refitting

1 The system consists of two remote control rods and a cable to release a locking finger when the ring on the gear lever is lifted during selection of reverse gear. This is part of an interlock device to prevent engagement of reverse gear when shifting from 3rd to 2nd.
2 To disconnect the controls from the transmission, first unscrew the cable retaining union nut. Its location differs according to transmission type (see Fig. 6.3).
3 Unscrew the nut and disconnect the remote control rod from the selector arm on the transmission. The rod will slide out of the slotted arm.
4 Prise the remaining rod off its flexible bush to separate the rod from the other selector arm.
5 Disconnect both rods from the base of the gearchange lever. One rod is secured by a nut, the other by a ball socket which is simply prised apart.
6 The cable can be released after unclipping its stop and removing the sleeve.
7 Working inside the car, remove the centre console (Chapter 11).
8 Unscrew and remove the gear lever knob.
9 Extract the four screws from the bellows retainer and withdraw the bellows.
10 Withdraw the gear lever.
11 Refitting is a reversal of removal, but smear the threads of the cable union with gasket cement before screwing it into the transmission casing.

4 Transmission – removal and refitting

Note: *A trolley jack will be useful for this operation*
1 The transmission may be removed on its own, as described in this Section, or together with the engine; refer to Chapter 1.
2 Drain in the transmission oil and disconnect the battery.
3 Place the car over an inspection pit or raise its front end sufficiently high for the clutch bellhousing to pass under the car when the transmission is withdrawn. Remove the front roadwheels.
4 Disconnect both driveshafts (as described in Chapter 8) from the transmission.

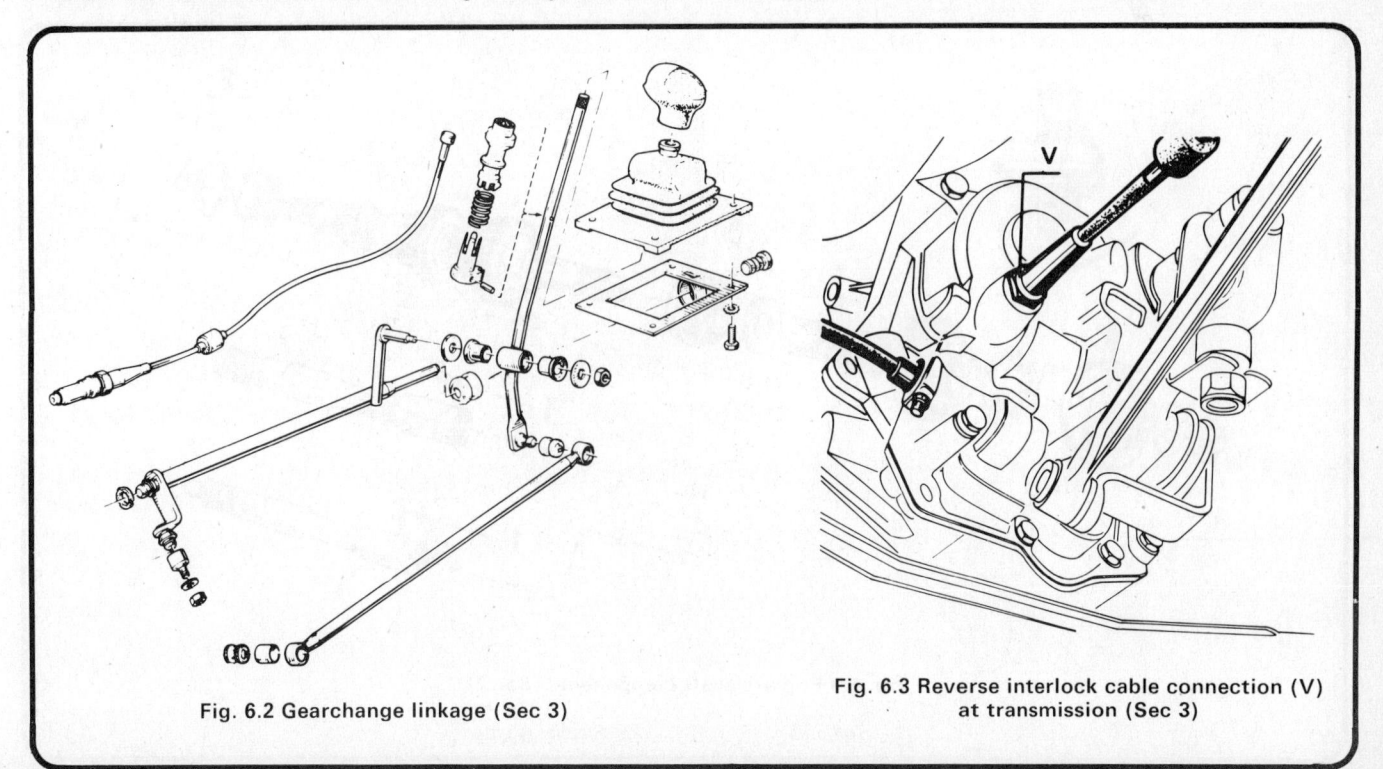

Fig. 6.2 Gearchange linkage (Sec 3)

Fig. 6.3 Reverse interlock cable connection (V) at transmission (Sec 3)

Chapter 6 Manual transmission

5 Disconnect the sensor for the speedometer.
6 Disconnect the leads from the reversing lamp (photo).
7 Disconnect the gearchange linkage, as described in the preceding Section.
8 Unbolt the TDC sensor and tie it out of the way.
9 Disconnect the clutch cable from its bracket and from the release lever; see Chapter 5.
10 Disconnect the exhaust downpipe.
11 Remove the cover plate from the lower front face of the flywheel housing.
12 Unscrew the starter motor mounting bolts to release the starter from the flywheel housing.
13 Unscrew and remove the bolts which connect the transmission to the engine.
14 Support the engine using a jack or hoist and place a second jack, preferably of trolley type, under the transmission.
15 Disconnect the transmission flexible mountings and brackets.
16 Raise the transmission slightly to clear the crossmember during withdrawal to the rear.
17 Remove the transmission from under the car.
18 Refitting is a reversal of removal, but observe the following points.
19 Make sure that the clutch driven plate has been centralised, as described in Chapter 5.
20 Tighten all nuts and bolts to the specified torque and renew those of self-locking type.
21 Adjust the clutch cable (Chapter 5).
22 Fill the transmission with oil of the specified type.

5 Transmission – overhaul (general)

1 Before stripping the transmission, assess the general condition. If the unit is noisy and gearchanging is notchy or far from silent then it may be more economic to obtain a new or factory-reconditioned transmission.

6 Transmission – removal of major assemblies

1 With the transmission removed, clean away external dirt and grease using paraffin and a stiff brush, or a water-soluble solvent.
2 Place the unit on a bench or, if it must be dismantled on the floor, rest it on a sheet of hardboard.
3 Remove the clutch release bearing and lever.
4 Unbolt and remove the clutch bellhousing.
5 Select 3rd or 4th gear, unscrew the 5th detent plug and extract the spring and ball.
6 Unbolt and remove the rear cover. Tilt the cover as it is withdrawn to disengage the selector finger.
7 Unscrew the casing bolts, then position the transmission so that the crownwheel teeth point downwards and lift off the casing half-section.
8 Remove the differential/final drive.
9 Remove the primary and secondary geartrains.

7 Primary shaft – overhaul

Note: *A bearing puller will be needed for this operation*

1 Drive out the roll pin and separate the clutch shaft from the primary shaft. Remove the washers.
2 Support the shaft vertically in a vice fitted with jaw protectors. Release the nut staking, and remove the nut.
3 Using a bearing puller, claws placed behind 5th gear, draw off 5th gear and the synchro-hub together.
4 Extract the shaft bearings, again using the puller.
5 Examine all the components carefully and renew as necessary.
6 Commence reasssembly by pressing on the new bearings with a press or by driving them on using a piece of tubing placed against the bearing inner track. A double ball type bearing is used at the 5th gear end of the shaft and a roller bearing at the clutch shaft end. The bearings should be fitted so that their engraved marks are visible from the ends of the shaft.
7 Two types of 5th gear synchro-hub could be fitted. On the early type with continuous inner splines apply a little locking fluid to the splines before fitting the unit. On the later type with short splines, support the 4th gear and press on the synchro-hub with a load of not less than 100 kgf (220 lbf) or more than 1500 kgf (3300 lbf).
8 Finally, fit the washer, then apply a little locking fluid to the threads of a new nut before fitting it and tightening it to the specified torque. Lock the nut by staking.
9 Locate the special washer then fit the clutch shaft to the primary shaft and drive in the roll pin to secure.

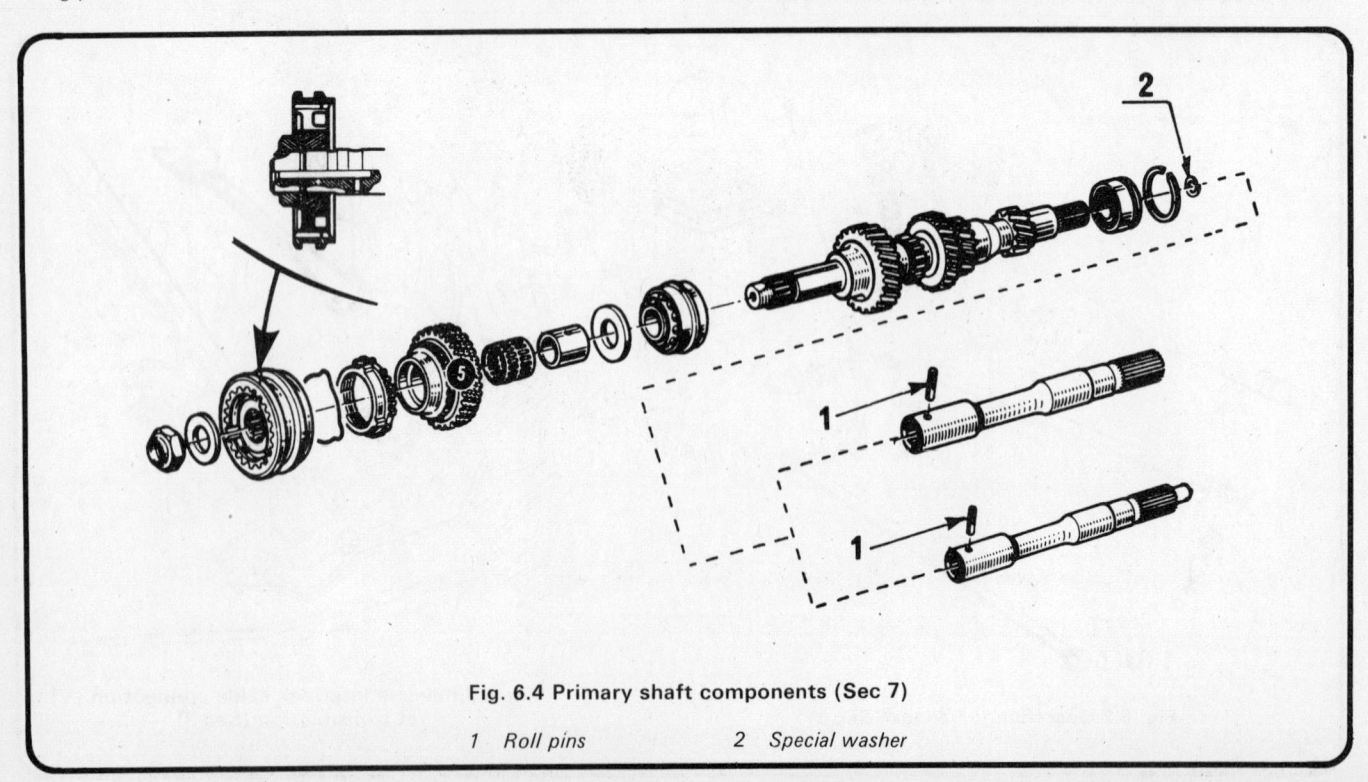

Fig. 6.4 Primary shaft components (Sec 7)

1 Roll pins   2 Special washer

Chapter 6 Manual transmission

8 Secondary shaft – overhaul

Note: *A bearing puller will be needed for this operation*

1 Grip the shaft by 1st gear in a vice fitted with jaw protectors so that the shaft is vertical.
2 Select 1st gear by moving the synchro sleeve incorporating reverse gear teeth downwards.
3 Unscrew the speedometer drive gear nut.
4 Using the puller, remove 5th gear.
5 Remove the secondary shaft components in the order shown in Fig. 6.6. Extract the circlips before attempting removal. Mark the synchro-hubs and sleeves in relation to each other to ensure correct refitting. Note that the roller bearing at the pinion end does not have an inner track, and to prevent the bearing falling apart a clip can be fitted over it.
6 Renew any worn or damaged components also the circlips and shaft nut.
7 Reassemble the components in reverse order to removal. The synchro-hubs are a sliding fit on the shaft splines, however, if the hub becomes tight, remove it, turn it, and engage it with different splines. Note the correct fitted positions of the synchro units shown in Fig. 6.9. Alternative types of 3rd/4th synchro units are shown in Fig. 6.10.
8 On this gearbox, two alternative types of 5th gear are fitted. With the first type having continuous inner splines apply a small amount of locking fluid to the splines before fitting the gear. The second type gear is free turning for three-quarters of its splines and the final splines must be pressed into position to give the correct preload to the double taper roller bearing. Position the assembly in the press as shown in Fig. 6.11 with a spring balance and cord around the bearing outer track. Press on the gear until the preload is between 1.5 and 4.0 kgf (3.4 and 9.0 lbf) and note that the press loading must not be less than 100 kgf (220 lbf) 1500 kgf (3300 lbf).
9 Finally select 1st then fit the speedo drive nut using a little locking fluid, tighten and lock it.

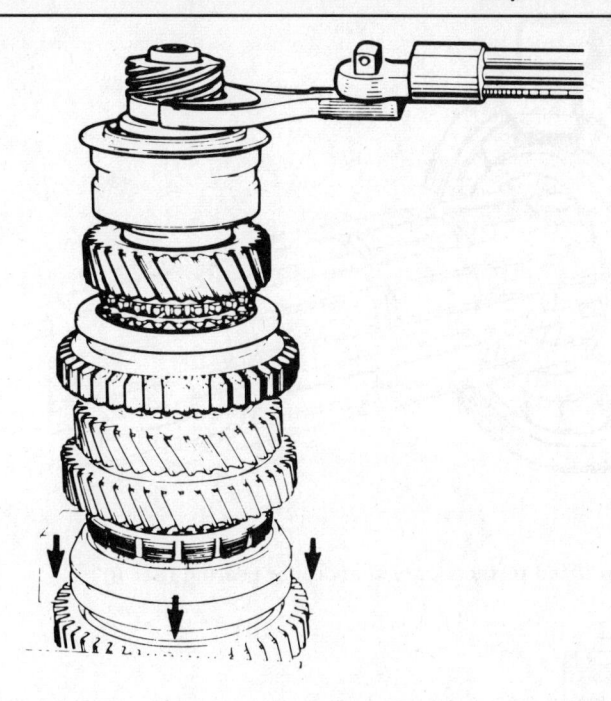

Fig. 6.5 Unscrewing secondary shaft speedometer drive gear nut (Sec 8)

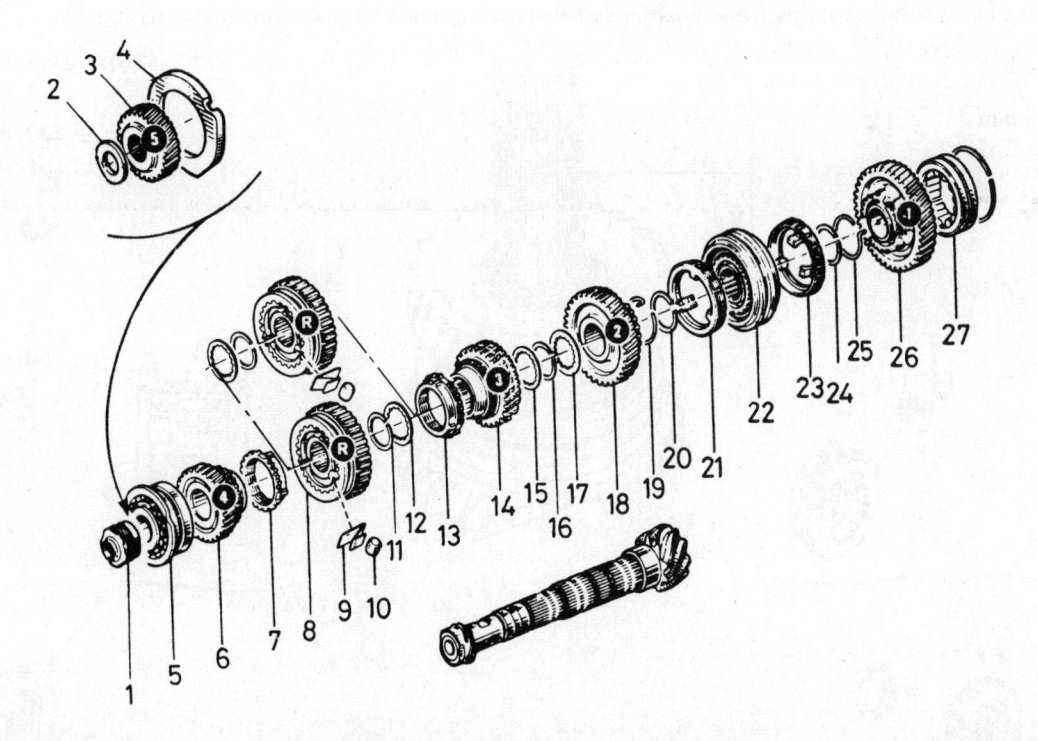

Fig. 6.6 Secondary shaft components (Sec 8)

1 Speedometer drivegear	8 3rd/4th synchro unit and reverse gear	14 3rd gear	21 Baulk ring
2 Washer	9 Spring clip	15 Washer	22 1st/2nd synchro unit
3 5th gear	10 Roller	16 Circlip	23 Baulk ring
4 Thrust washer	11 Circlip	17 Washer	24 Circlip
5 Bearing	12 Washer	18 2nd gear	25 Clip
6 4th gear	13 Baulk ring	19 Clip	26 1st gear
7 Baulk ring		20 Circlip	27 Bearing

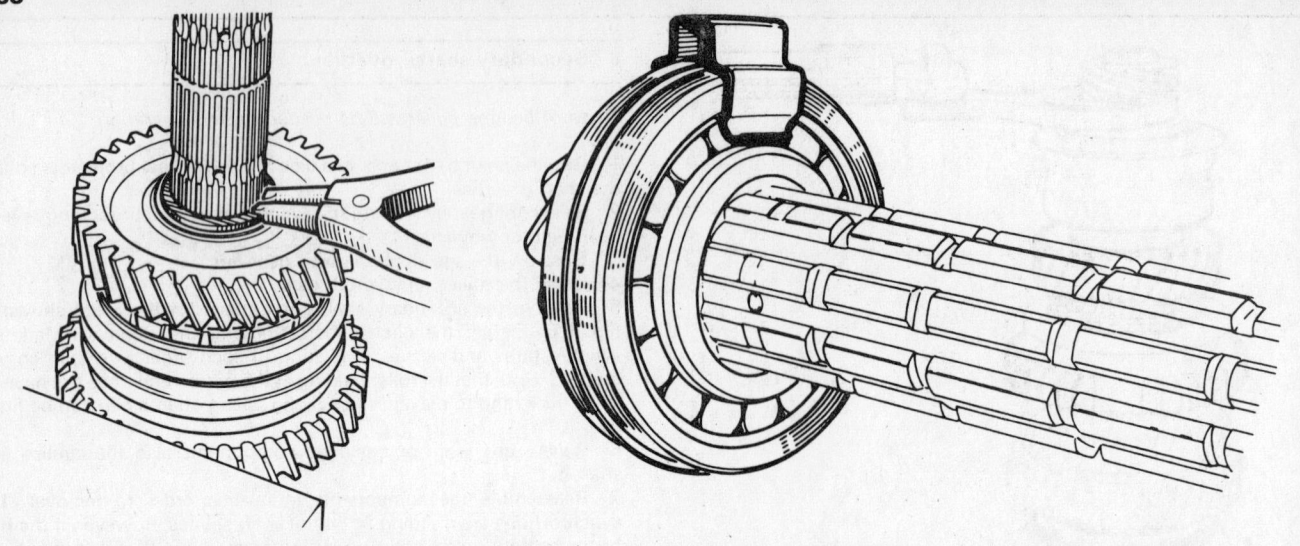

Fig. 6.7 Extracting secondary shaft circlip (Sec 8)

Fig. 6.8 Clip fitted to secondary shaft roller bearing (Sec 8)

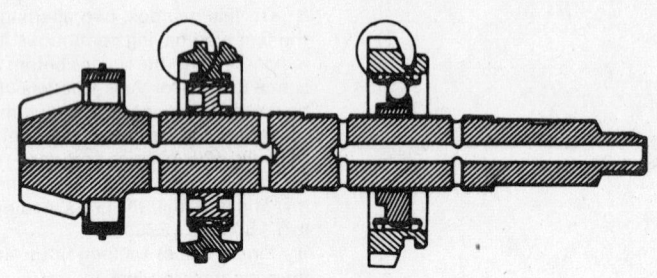

Fig. 6.9 The correctly fitted positions of the synchro units on the secondary shaft (Sec 8)

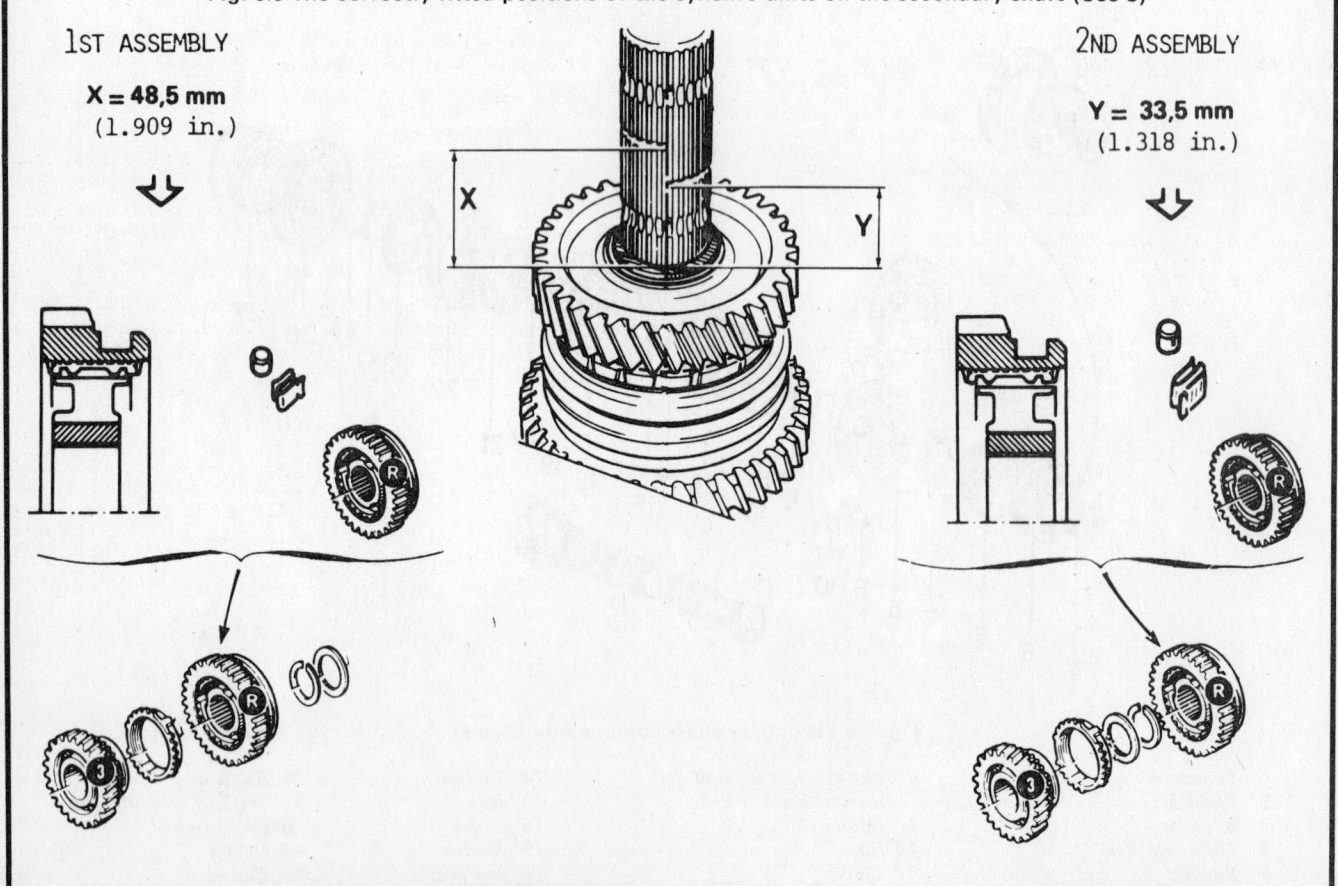

1ST ASSEMBLY

X = 48,5 mm
(1.909 in.)

2ND ASSEMBLY

Y = 33,5 mm
(1.318 in.)

Fig. 6.10 Two types of 3rd/4th synchro unit fitted to the secondary shaft (Sec 8)

Chapter 6 Manual transmission

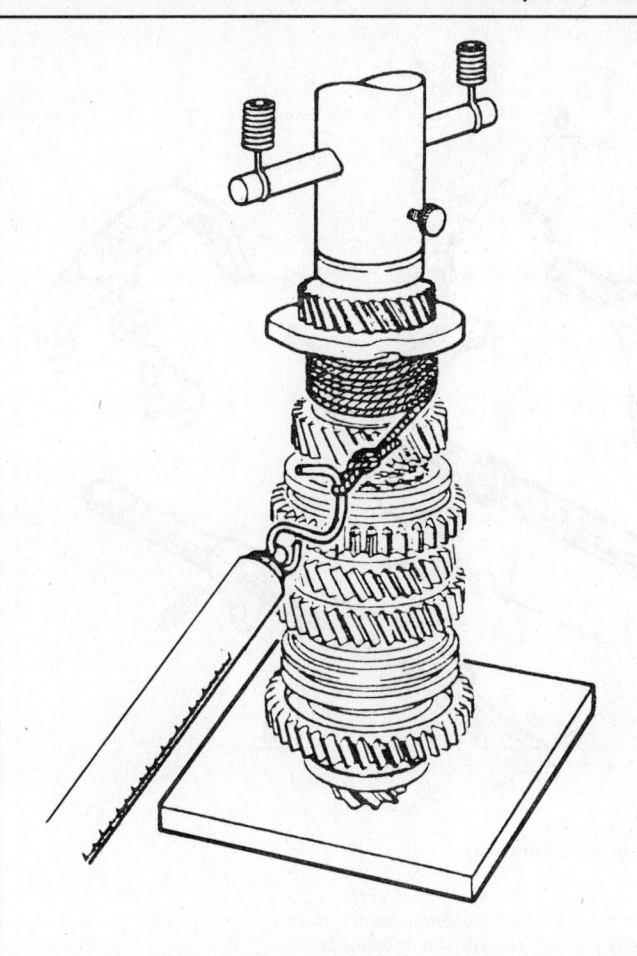

Fig. 6.11 Pressing 5th gear onto the secondary shaft (Sec 8)

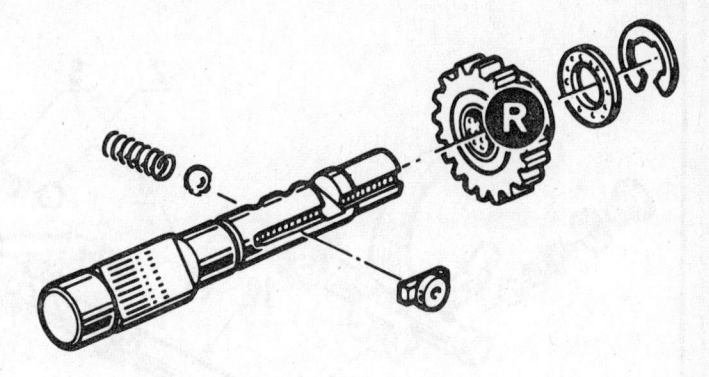

Fig. 6.12 Reverse idler shaft components (Sec 9)

9 Reverse idler shaft and gear – dismantling and reassembly

1 Extract the circlip then withdraw the shaft followed by the gear, friction washer and guide. Recover the interlocking ball and spring.
2 To refit, locate the spring and ball in the casing.
3 Insert the shaft and locate the gear on it with the hub facing the differential end.
4 Fit the friction washer with the bronze face toward the gear.
5 Locate the guide in the bore then push in the shaft and fit the circlip.

10 Selector forks and shafts – dismantling and reassembly

1 Select neutral and then pull out 5th gear selector shaft. Recover the locking ball.
2 Drive out the roll pin securing the selector forks.
3 Pull out the 3rd/4th selector shaft, remove the fork, and recover the detent ball and spring.
4 Remove the interlock disc and pull out the 1st/2nd selector shaft. Recover the detent ball and spring.
5 Unscrew the bolt and remove the reverse selector lever.
6 Pull out the reverse selector shaft.
7 When refitting the shaft, note that the slots on the roll pins must face the rear casing. The refitting procedure is a reversal of removal. After fitting the 5th gear shaft select 3rd or 4th to prevent the shaft moving when the rear cover is fitted.
8 **Note:** From 1986, a spring and ball are incorporated in the 1st/2nd selector shaft. When reassembling the fork to the shaft; make sure that the roll pin hole in the fork is nearest the rear cover. The slits in the roll pins must also be towards the rear cover.

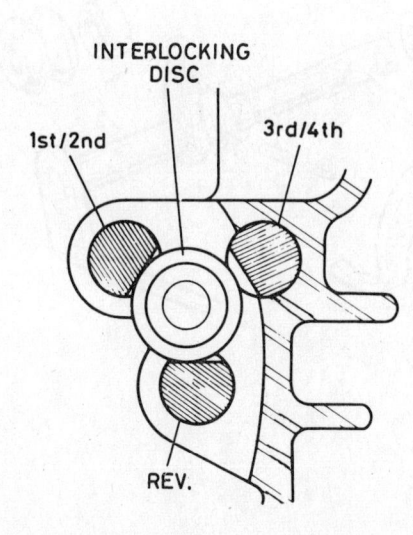

Fig. 6.13 Selector shaft interlock disc (Sec 10)

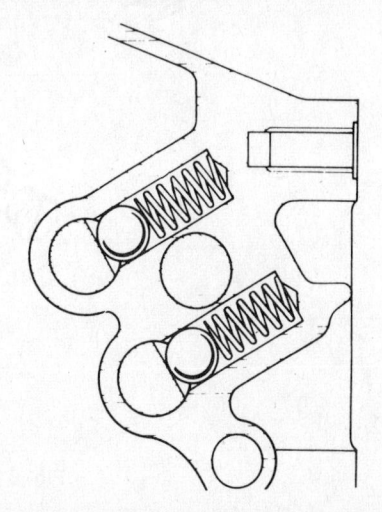

Fig. 6.14 Location of detent balls and springs (Sec 10)

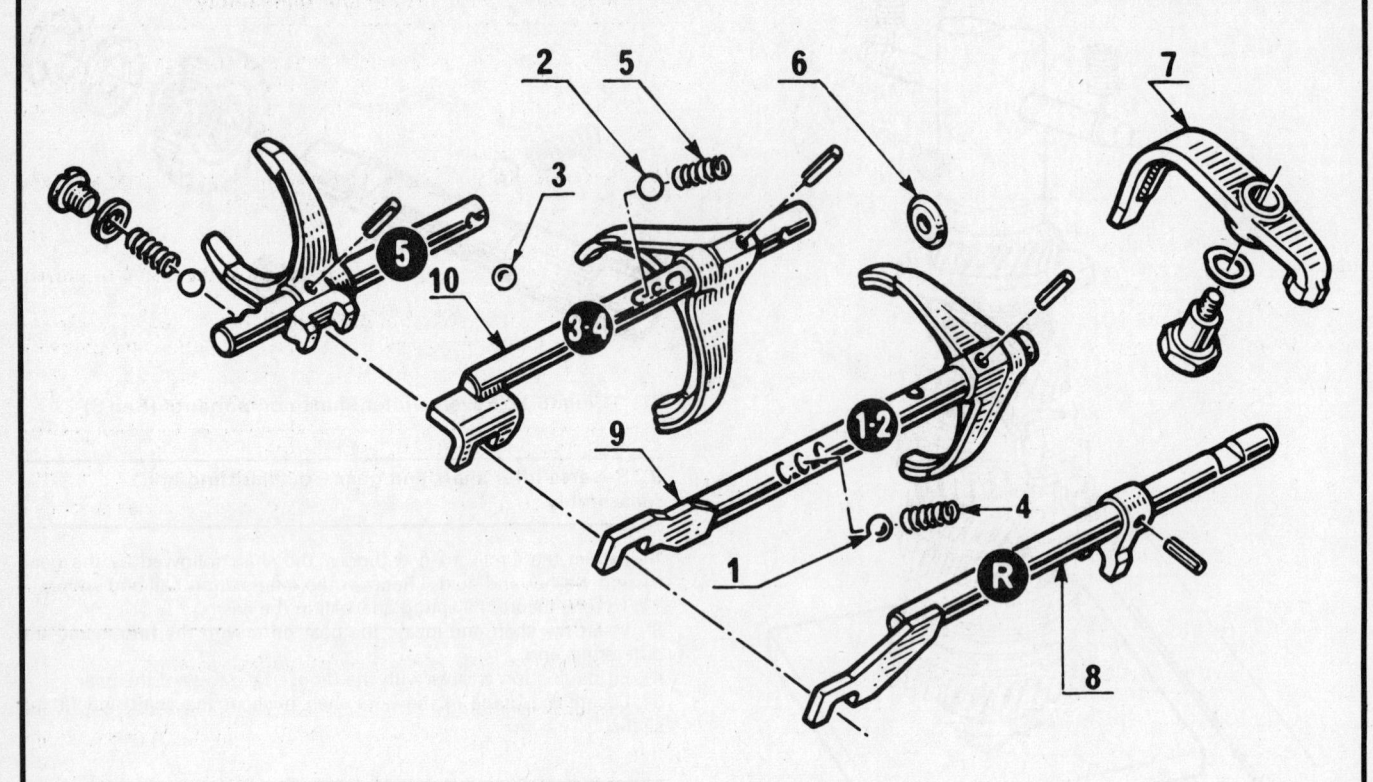

Fig. 6.15 Selector components (Sec 10)

1 Detent ball
2 Detent ball
3 Detent ball
4 Spring
5 Spring
6 Interlock disc
7 Reverse swivel arm
8 Reverse selector shaft
9 1st/2nd selector shaft
10 3rd/4th selector shaft

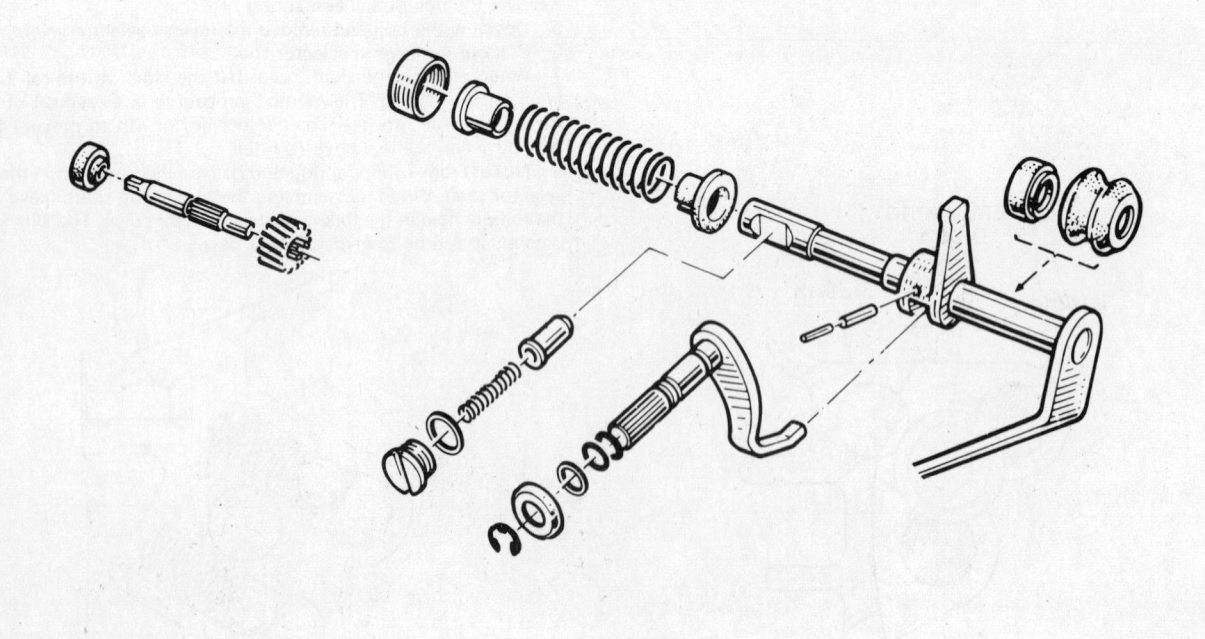

Fig. 6.16 Rear cover components (Sec 11)

Chapter 6 Manual transmission

11 Rear cover – dismantling and reassembly

1 Drive out the roll pin(s) using a suitable punch.
2 Extract the circlip and prise the bush from the selector lever shaft.
3 Unscrew the plug and extract the reverse shaft stop plunger and spring.
4 Remove the selector components.
5 To remove the speedometer drivegear, prise up the short arms and withdraw the shaft, followed by the gear. Note that the gear must be renewed after removal.
6 Extract the oil seals.
7 Remove any sharp edges from the ends of the shafts to prevent damage to the oil seals.
8 Reassembly is a reversal of dismantling, but fit new oil seals and make sure that the speedometer drivegear arms fully enter the groove in the shaft.

12 Transmission housing – inspection

1 Examine all housing sections for cracks, and renew oil seals as a matter of routine.

13 Differential/final drive – overhaul

Note: *A bearing puller will be needed for this operation*

1 It is unlikely that the differential will require overhaul, but if the bearings, or other components must be renewed, proceed in the following way.
2 Using a suitable puller, remove the tapered roller bearings.
3 Unbolt and remove the crownwheel. If the crownwheel is renewed, then the secondary shaft must be renewed as well to ensure that the pinion gear is matched to the crownwheel (supplied as a matching pair).
4 Remove the sun wheels and planet gears and separate the differential components. The collar will be destroyed, so renew it.
5 Now turn your attention to the ring nuts in the transmission casing. Unscrew the lockbolt and remove the locktab.

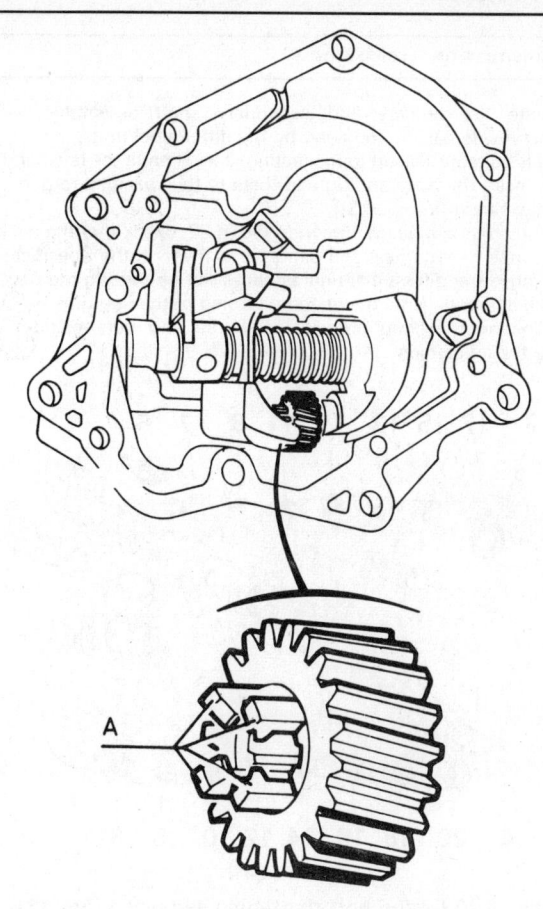

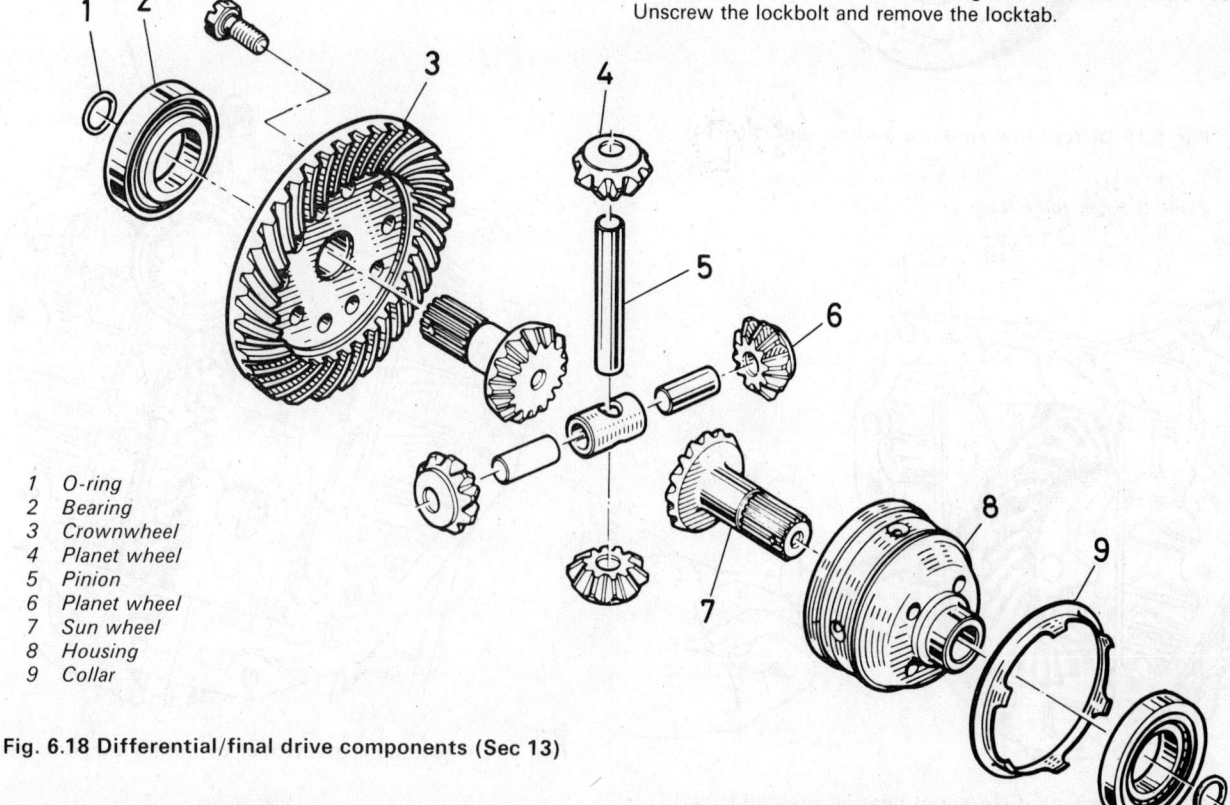

Fig. 6.17 Speedometer driven gear retaining arms (A) (Sec 11)

1 O-ring
2 Bearing
3 Crownwheel
4 Planet wheel
5 Pinion
6 Planet wheel
7 Sun wheel
8 Housing
9 Collar

Fig. 6.18 Differential/final drive components (Sec 13)

6 Make up a suitable tool to unscrew the ring nuts and remove them.
7 Remove the bearing outer tracks from the casing sections and fit the new ones.
8 Fit new oil seals and O-rings to the ring nuts. The oil seals are inserted from the rear face of the ring nut and pressed in until the seal face (2) is tight against the ring nut face (3 – Fig. 6.19).
9 Reassemble the differential and press on the new bearings. Always use new self-locking nuts on the crownwheel and tighten to the specified torque. Note that the bearing on the crownwheel side has a smaller diameter inner track than the opposite bearing.
10 If new differential bearings have been fitted the casing ring nuts must be adjusted to obtain the correct preload. First fit the differential in the casings without the primary or secondary shafts and tighten the casing bolts to the correct torque and in the correct sequence (Fig. 6.20).
11 Screw in the ring nuts turning the one on the differential housing side slightly more than the opposite side. Continue turning until the differential unit becomes hard to move.
12 Using a spring balance and cord (Fig. 6.21) check that the differential turns within the specified loadings, and adjust the ring nuts accordingly.
13 Separate the casings and remove the differential pending reassembly of the transmission.

14 Transmission – reassembly

1 Locate the primary and secondary shaft assemblies in the right-hand side casing, followed by the differential unit.
2 Apply suitable sealant to the mating faces then fit the left-hand side casing, insert the bolts and tighten them to the specified torque in the sequence shown in Fig. 6.20.
3 Using a dial gauge, as shown in Fig. 6.22, check that the backlash between the crownwheel and pinion is as given in the Specifications. If not, move the differential unit as necessary by turning the ring nuts by equal amounts. Note the unequal turning of the ring nuts will result in an incorrect bearing preload. Mark the ring nut positions after making the adjustment.

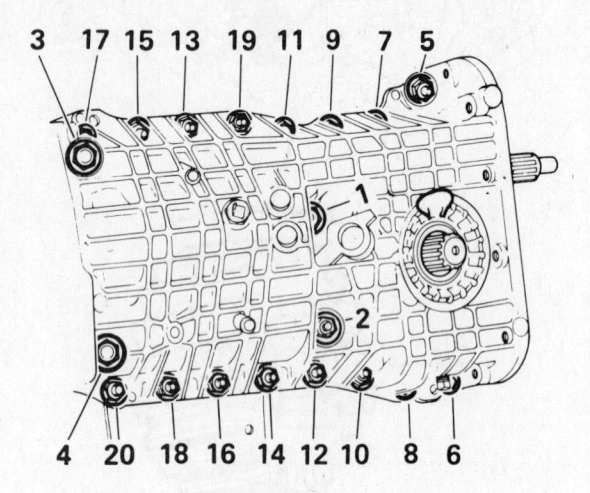

Fig. 6.20 Casing bolt tightening sequence (Sec 13)

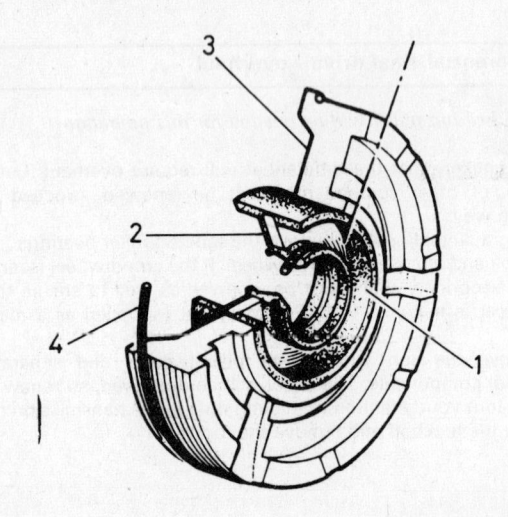

Fig. 6.19 Differential ring nut and oil seal (Sec 13)

1   Oil seal
2 and 3   Faces to be flush
4   O-ring

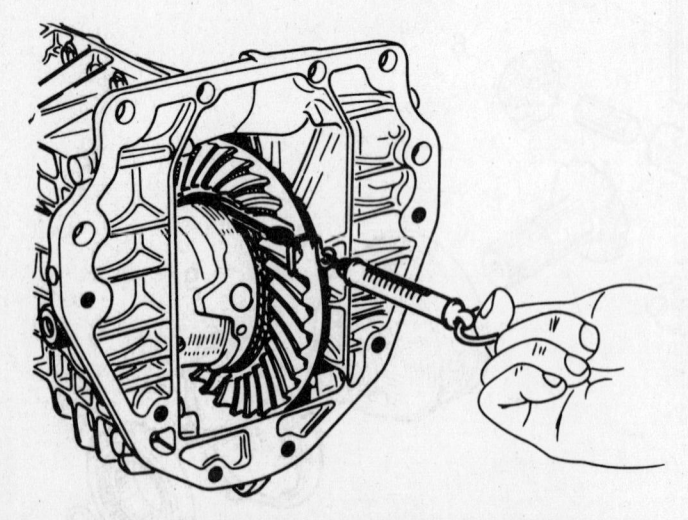

Fig. 6.21 Checking differential bearing preload (Sec 13)

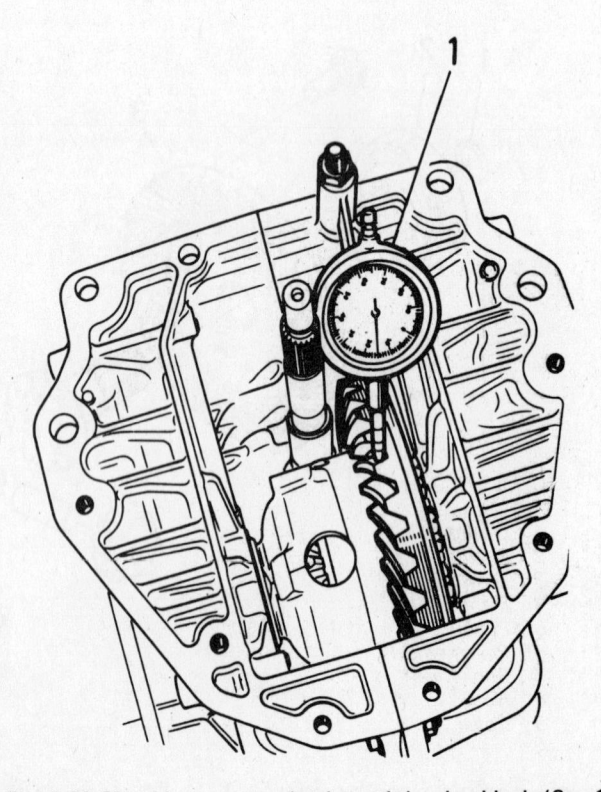

Fig. 6.22 Checking crownwheel-to-pinion backlash (Sec 15)

1   Dial gauge

Chapter 6 Manual transmission

4 Fit the rear cover, together with a new gasket, then fit the bolts and tighten them to the specified torque.
5 Insert the 5th gear detent ball and spring then tighten the plug into the cover.
6 Wrap adhesive tape lightly over the clutch shaft splines and smear with a little grease.
7 Refit the clutch housing using a new gasket and tighten the bolts. Remove the tape.
8 Check that the ring nuts are positioned correctly then fit the lockplates and tighten the bolts.

15 Differential oil seals – renewal without removing transmission

1 The differential oil seals may be renewed without having to remove the transmission from the car.
2 Disconnect the driveshaft (as described in Chapter 8) from the transmission.
3 To renew the seals, mark the ring nuts in relation to the casings then unbolt the lockplates and unscrew the ring nuts, noting the exact number of turns necessary to remove them.
4 Remove the old seals and O-rings then press in the new seals and locate the O-rings in the grooves as described in Section 13.
5 Wrap adhesive tape over the shaft splines and lubricate the seals and O-rings with a little grease.
6 Screw the ring nuts into their original positions and fit the lockplates. Remove the tape.
7 Reconnect the driveshaft, fill the transmission with oil.

PART B – TYPE UN1 TRANSMISSION

16 Description and maintenance

Refer to Sections 1 and 2, noting the slight design differences shown in Fig. 6.23.

17 Gearchange linkage

Refer to Section 3, but note the different location of the cable union (Fig. 6.24).

18 Transmission – removal and refitting

1 The operations are similar to those described in Section 4 but the following differences must be taken into account.
2 The clutch slave cylinder must be unbolted and tied up out of the way. There is no need to disconnect the hydraulic connections. The bolt (V – Fig. 6.25) cannot be removed.
3 Unbolt the power steering pump, disconnect the drivebelt and move the pump and reservoir to one side.
4 On 2664 cc (B298) models, remove the engine damper (M – Fig. 6.26) from the left-hand side.
5 The rear end of the transmission will have to be lowered in order to be able to remove the right-hand mounting.
6 When refitting the transmission, make sure that the bolt shown in Fig. 6.25 is in position before the transmission is offered up.

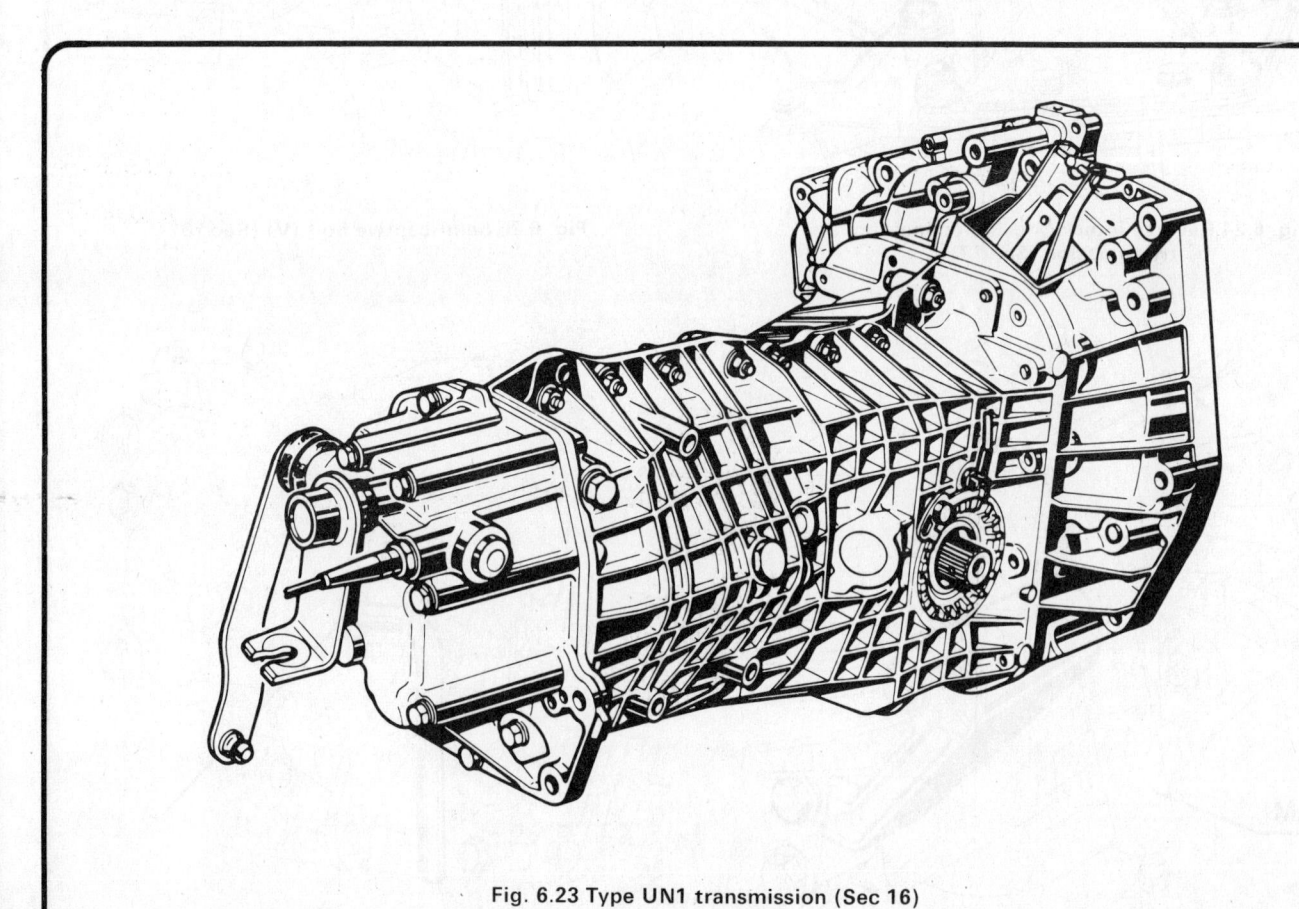

Fig. 6.23 Type UN1 transmission (Sec 16)

Chapter 6 Manual transmission

19 Transmission – removal of major assemblies

Note: *A puller will be needed for this operation*

1 With the transmission removed from the car, clean away external dirt and grease using paraffin and a stiff brush, or a water-soluble solvent.
2 Place the unit on a bench or, if it must be dismantled on the floor, rest it on a sheet of hardboard.
3 Remove the clutch release bearing and arm.
4 Unbolt and remove the bellhousing. Twelve bolts are used to hold it in position, two of them being longer and located at the positioning dowels.
5 Tap the bellhousing free if necessary using a soft-faced mallet. Peel off the old gasket.
6 Drive out the double roll pins and remove the selector arm and dust excluder from the selector rod.
7 From the side of the rear cover, unscrew the plug and extract 5th gear detent spring and ball. Use a pencil magnet if necessary to remove the ball.
8 Unscrew the rear cover fixing bolts.

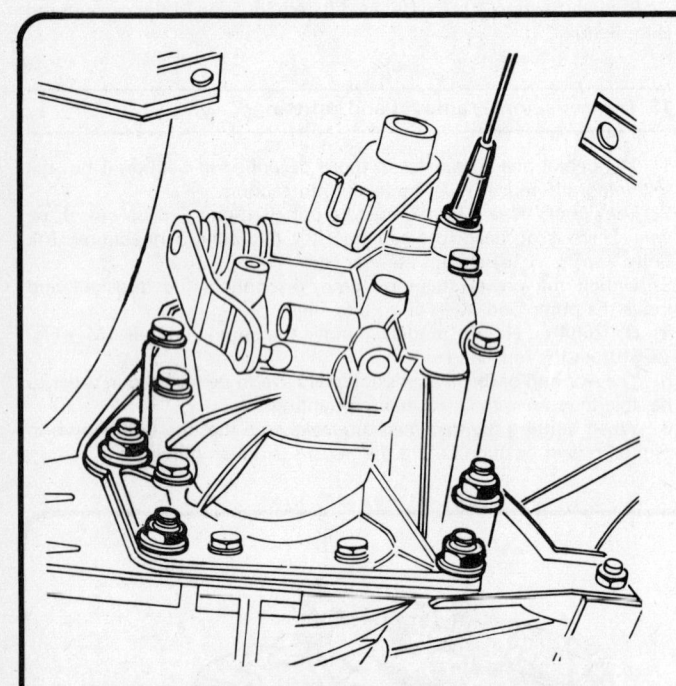

Fig. 6.24 Reverse interlock cable connection at transmission (Sec 17)

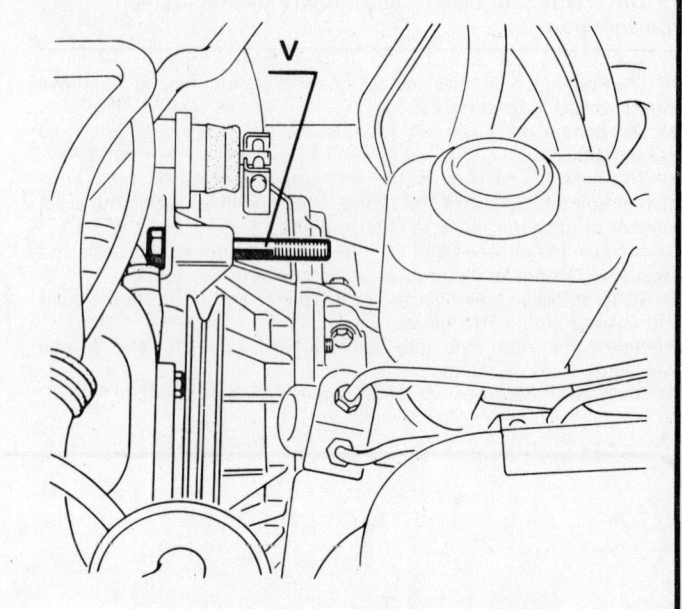

Fig. 6.25 Semi-captive bolt (V) (Sec 18)

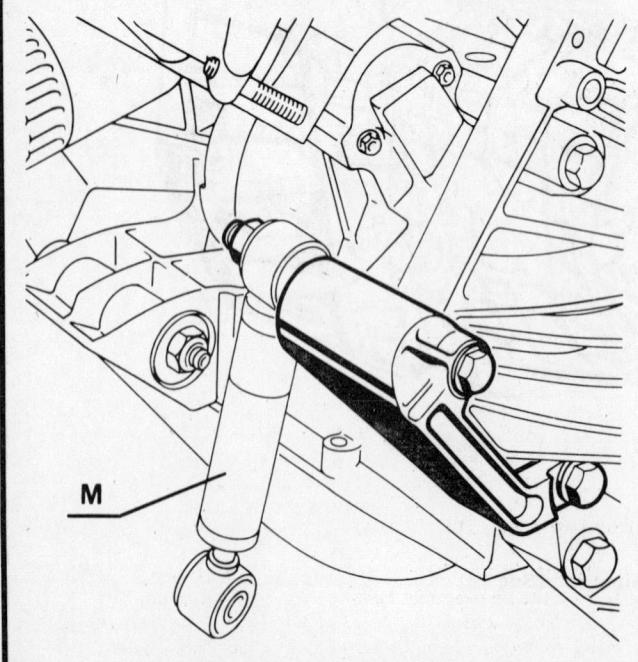

Fig. 6.26 Engine damper (M) (Sec 18)

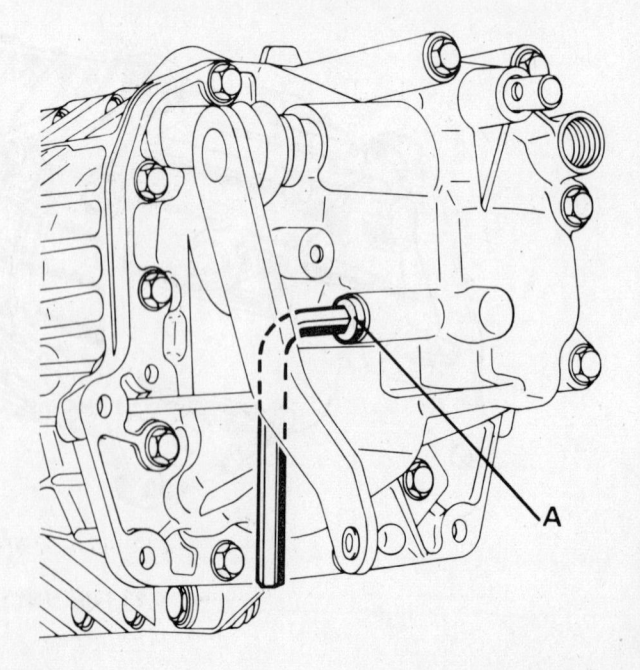

Fig. 6.27 5th gear selector shaft detent plug (A) (Sec 19)

Chapter 6 Manual transmission

9 Pull the cover off the transmission casing, at the same time pushing in the selector rod to free the cover.
10 Before separating the gearcase halves, consideration should be given as to whether or not the primary and secondary shaft assemblies are likely to require attention. If it is decided they will require attention, the respective primary and secondary shaft rear locknuts must be loosened. This is best done at this stage as the shafts can be prevented from turning by tapping the selector rods to 'in gear' positions to lock the shafts.
11 Once the shafts are locked, relieve the nut staking from the secondary and primary shafts. Get an assistant to steady the gearbox whilst you loosen off the primary, and secondary shaft locknuts.
12 Reset the selectors in the neutral mode.
13 From the primary shaft, take off the shaft nut and the dished washer (concave side to synchro unit).
14 Using a puller with thin claws, draw the locking gear from the primary shaft.
15 Take off the baulk ring and 5th gear synchro unit, complete with selector fork.
16 Remove 5th sliding gear, the split needle roller bearing, the bush and the thrust washer.
17 From the secondary shaft, remove the nut, dished washer (concave side to gear).
18 Using a puller, remove 5th gear. If it is difficult to engage the available puller behind the gear, leave it until the shaft is dismantled after removal; see paragraph 1, Section 21.
19 Unbolt the spacer plate.
20 Unscrew and remove the reverse lamp switch.
21 Unscrew all the bolts which hold the split casing sections together, noting that the centre two have their nuts to the reverse lamp switch side.
22 Position the transmission so that the teeth on the crownwheel are pointing down and remove the half-casing upwards.
23 Lift out the differential/final drive.
24 Remove the geartrains.

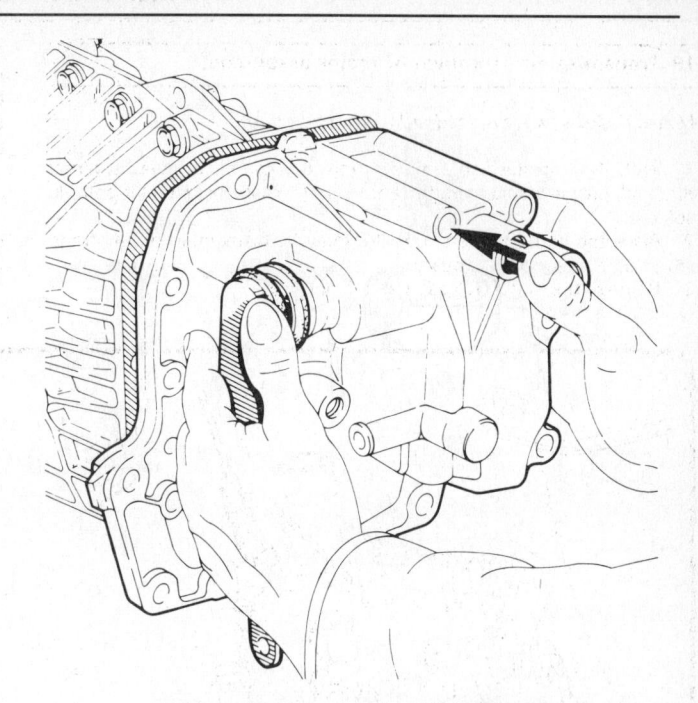

Fig. 6.28 Removing rear cover (Sec 19)

Push in the selector rod

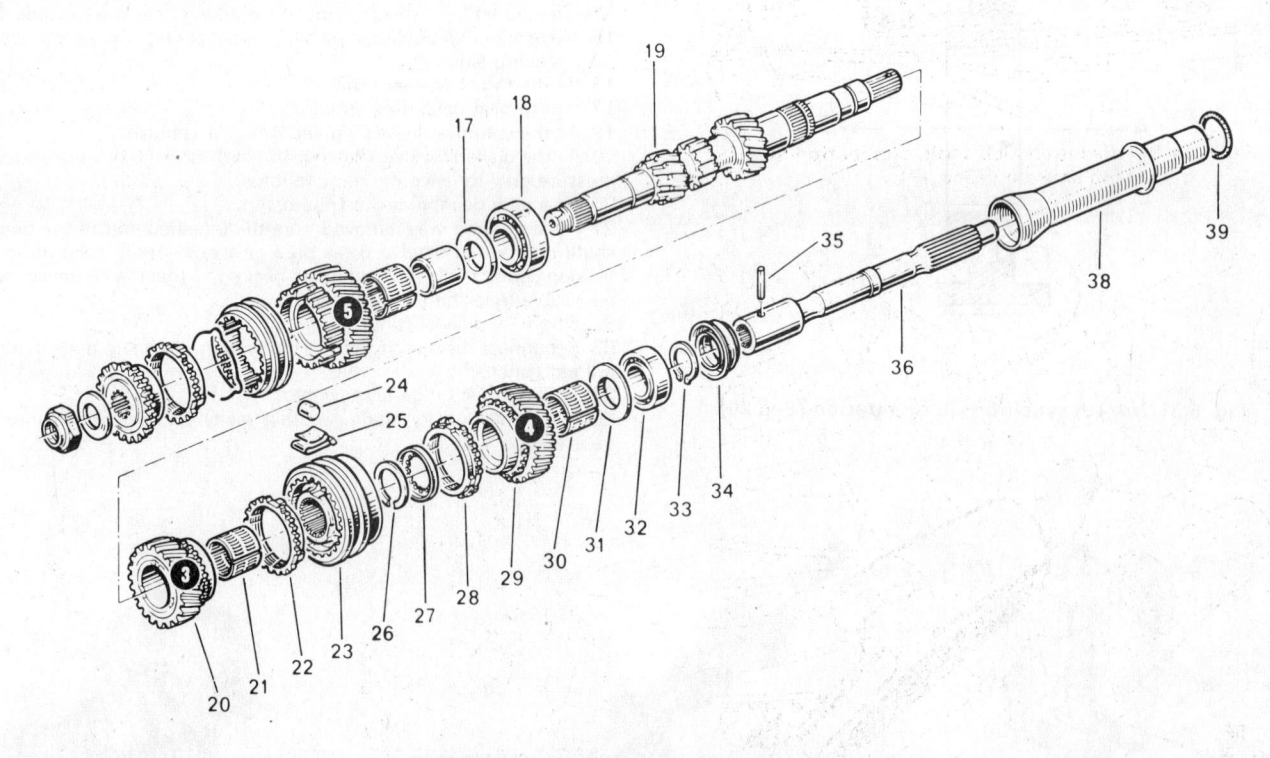

Fig. 6.29 Primary shaft components (Sec 20)

17 Thrust washer	23 3rd/4th synchro unit	29 4th gear	34 Oil seal
18 Bearing	24 Roller	30 Split needle roller bearing	35 Roll pin
19 Primary shaft	25 Roller spring clip	31 Thrust washer	36 Clutch shaft
20 3rd gear	26 Circlip	32 Bearing	38 Guide tube
21 Split needle roller bearing	27 Washer	33 Circlip	39 O-ring
22 3rd gear baulk ring	28 4th gear baulk ring		

Chapter 6 Manual transmission

25 Remove the casing magnet.
26 Remove the reverse idler shaft and gear; see Section 22.
27 If necessary dismantle the selector components, as described in Section 23.

20 Primary shaft – overhaul

Note: *A puller will be needed for this operation*

1 Pull off the thrust guide tube.
2 Drive out the roll pin and separate the clutch shaft from the primary shaft.
3 Using circlip pliers, extract the circlip.
4 Place the claws of a puller behind 4th gear and pull the gear and the bearing from the shaft.
5 Note the concave side of the dished washer is against the gear.
6 Remove the split needle bearing.
7 Take off the baulk ring, the 3rd/4th synchro sleeve and three rollers.
8 Remove the thrust washer and extract the synchro-hub circlip.
9 Remove the synchro-hub and 3rd gear, with the baulk rings and split needle bearing. Remove the bearing from the threaded end of the shaft if it is worn or damaged (photos).
10 Clean and check all components for wear or damage and renew them as necessary.
11 If there has been a history of noisy gearchanging, certainly renew the baulk rings and the synchro assembly if the splines or teeth show signs of wear.
12 Commence reassembly by fitting 3rd gear and the split needle roller bearing to the non-threaded end of the primary shaft (photo). Apply oil as reassembly progresses.
13 Locate the baulk ring on the shaft and then fit 3rd/4th synchro-hub (photo), making sure that it is positioned on the shaft as shown in Fig. 6.31. Fit a new hub circlip (photo).
14 Fit the synchro sleeve, the three rollers and spring clips (photos). Make sure that the extended tips of the spring clips are towards 4th.
15 Make sure that the large teeth of the baulk ring engage correctly in their synchro slots.
16 Fit the thrust washer (photo).
17 Fit 4th gear baulk ring (photo).
18 Fit the split needle bearing and 4th gear (photo).
19 Fit the dished washer (flat side to gear) then the bearing, applying pressure only to its centre track (photos).
20 Fit a new bearing circlip (photo).
21 If the bearing was removed from the threaded end of the primary shaft, now is the time to press on a new one. Apply pressure to the bearing centre track only and fit the bearing so that the engraved marks are visible from the end of the shaft.
22 Fit a new oil seal (photo).
23 Reconnect the clutch shaft to the primary shaft and drive in a new roll pin (photo).
24 Fit the thrust guide tube (photo).
25 Place the primary shaft geartrain ready for assembling into the transmission casing.

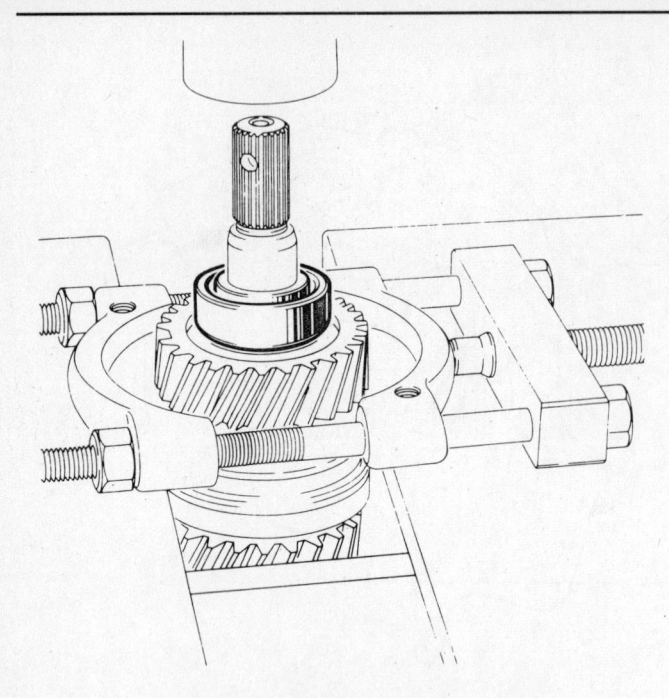

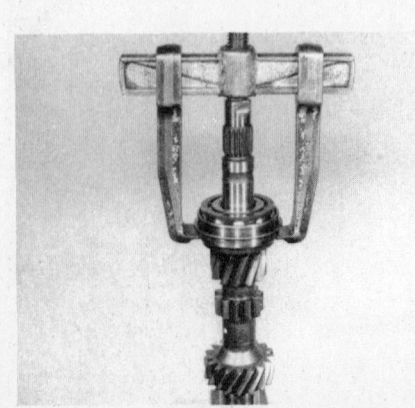

Fig. 6.30 Pulling off 4th gear (Sec 20)

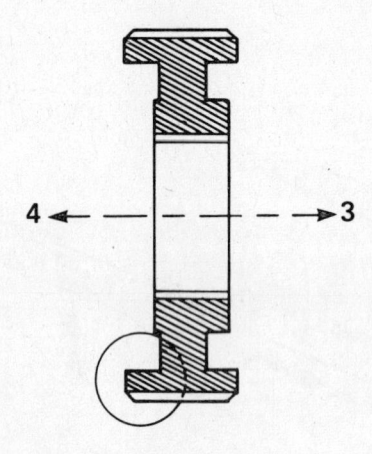

Fig. 6.31 3rd/4th synchro-hub orientation (Sec 20)

20.9A Removing a primary shaft bearing

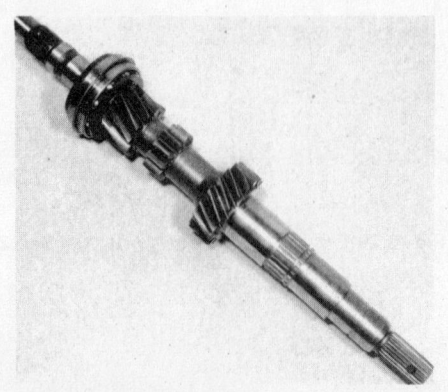

20.9B Primary shaft stripped

20.12 Fitting 3rd gear to the primary shaft

20.13A Fitting 3rd gear baulk ring

20.13B 3rd/4th synchro-hub

20.13C Fitting hub circlip

20.14A Fitting 3rd/4th synchro sleeve

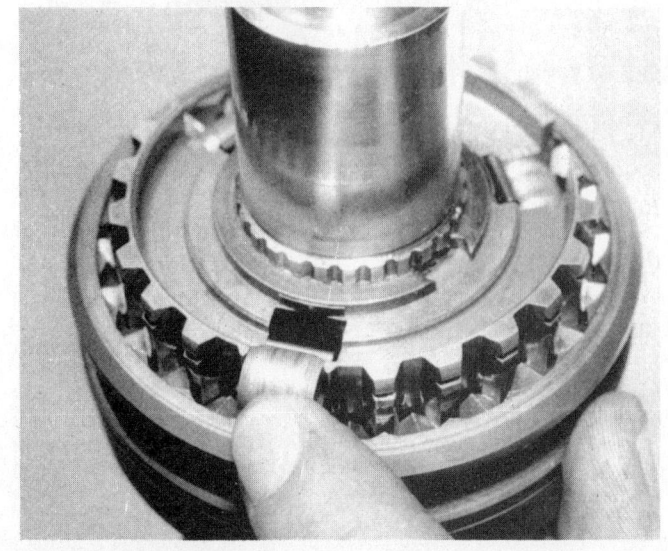

20.14B Fitting synchro rollers

20.14C Fitting synchro roller spring clips

Chapter 6 Manual transmission

20.16 Fitting the thrust washer

20.17 4th gear baulk ring

20.18 Split needle bearing and 4th gear

20.19A Dished washer

20.19B Fitting primary shaft roller bearing

20.20 Fitting bearing circlip

20.22 Fitting the oil seal

20.23 Connecting clutch shaft to primary shaft

20.24 Thrust guide tube

21 Secondary shaft – overhaul

Note: A puller will be needed for this operation

1 If 5th gear was not removed earlier, place the claws of a puller behind the spacer plate and remove the gear from the shaft.
2 Using the puller, remove the double tapered roller bearing.
3 Remove 1st gear and the split needle roller bearing. Note the spring at the rear of the gear.
4 Remove the baulk ring.
5 Remove 1st/2nd synchro sleeve.
6 Extract the circlip and then remove 1st/2nd synchro-hub, together with 2nd gear, by placing the claws of the puller behind the gear (photo).
7 Separate the hub, baulk ring and spring ring (inside 2nd gear).
8 Remove the split needle roller bearing.
9 The remaining gears (3rd and 4th) are a shrink fit and cannot be removed from the shaft (photo).
10 Refer to Section 20, paragraphs 10 and 11.
11 Commence reassembly by fitting the synchro spring inside 2nd gear so that the three slots are covered (photo).
12 Fit the split needle roller bearing and 2nd gear to the shaft (photos).
13 Locate the baulk ring (photo).
14 Warm 1st/2nd synchro-hub in an oven or boiling water and press it or drive it onto the shaft using a length of tubing; orientate as shown in Fig. 6.34. Hold the baulk ring so that its lugs are not damaged during fitting of the hub (photos).
15 Fit the hub circlip (photo).
16 Fit the 1st/2nd synchro sleeve with reverse gear (photo).
17 Fit the baulk ring (photo).

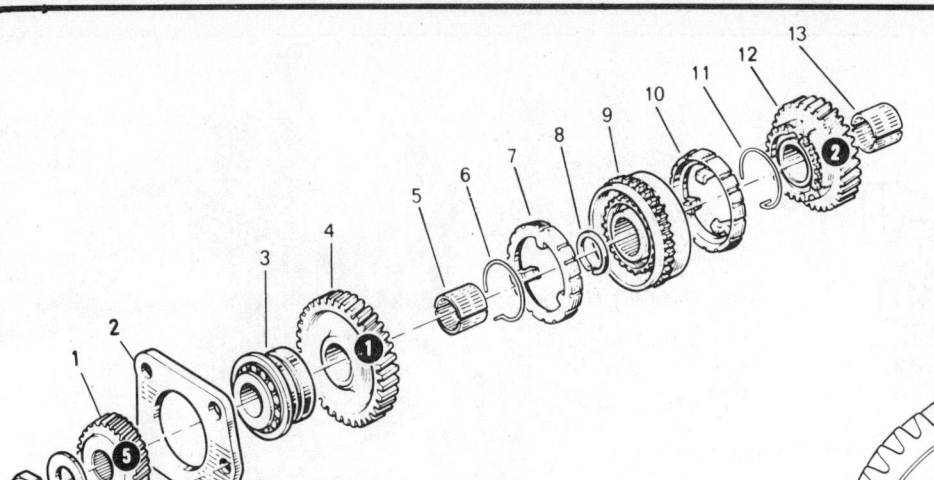

Fig. 6.32 Secondary shaft components (Sec 21)

1. 5th speed gear
2. Spacer plate
3. Double taper roller bearing
4. 1st gear
5. Split needle roller bearing
6. Synchro spring
7. 1st gear baulk ring
8. Circlip
9. 1st/2nd synchro unit
10. 2nd gear baulk ring
11. Synchro spring
12. 2nd gear
13. Split needle roller bearing

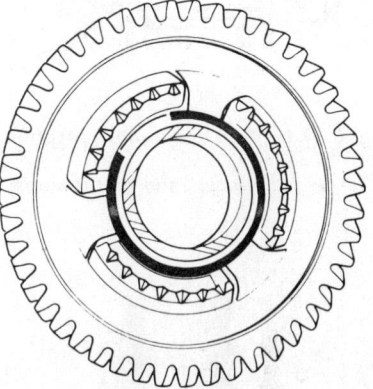

Fig. 6.33 Synchro spring correctly positioned in gear (Sec 21)

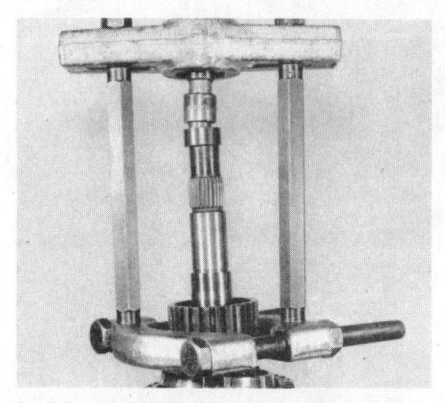

21.6 Removing 1st/2nd synchro-hub from secondary shaft

21.9 Secondary shaft stripped

21.11 2nd gear synchro spring (arrowed)

21.12A 2nd gear split needle roller bearing

21.12B 2nd gear

21.13 2nd gear baulk ring

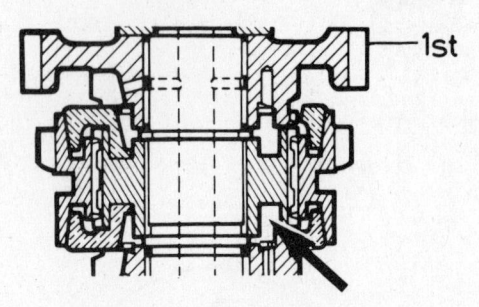

Fig. 6.34 1st/2nd synchro orientation (Sec 21)

Deeper hub groove (arrowed) towards 2nd gear

21.14A 1st/2nd synchro-hub

21.14B 1st/2nd synchro-hub fitted

21.15 Fitting synchro-hub circlip

21.16 1st/2nd synchro sleeve with reverse gear

21.17 Baulk ring

Chapter 6 Manual transmission

18 Fit the split needle roller bearing and 1st gear (photos); making sure that the sycnhro spring is in place inside it covering the three slots.
19 Fit the double tapered roller bearing so that the flange is towards the end of the shaft. Apply pressure only to the bearing centre track. Make sure that the spacer is between the races (photos).
20 Place the secondary shaft geartrain ready for assembling into the transmission casing.

22 Reverse idler shaft and gear – dismantling and reassembly

1 Remove the reverse idler shaft, gear and thrust washers.
2 The casing bushes cannot be renewed.
3 When refitting, note the location of the thinner and thicker thrust washers and make sure that the shaft roll pin locates in its cut-out (photo).

23 Selector forks and shafts – dismantling and reassembly

1 Move the selector forks and shafts as necessary to be able to drive out the fork securing roll pins. Withdraw the shafts and forks, noting carefully which way round the forks are fitted to the shafts. Take care to retrieve the detent balls and springs.
2 Unscrew the pivot bolt and remove the reverse selector swivel and shaft.
3 Inspect all components and pay particular attention to the selector forks for wear at their tips.
4 Reassemble 3rd/4th fork and shaft then 1st/2nd followed by reverse. Apply thread locking fluid to the threads of the reverse swivel pivot bolt and tighten to the specified torque (photo).
5 Make sure that the slits in all roll pins face the rear cover, and note the different sizes of roll pins used on the 1st/2nd fork (Fig. 6.39).
6 The 5th gear selector fork was removed during withdrawal of 5th gear from the primary shaft, see Section 19.

21.18A 1st gear split needle roller bearing

21.18B 1st gear

21.18C 1st gear fitted

21.19A Tapered roller bearing race

21.19B Using a piece of tubing to fit roller bearing race

21.19C Bearing double track

21.19D Bearing race spacer

21.19E Second tapered roller bearing race

21.19F Double roller bearing assembled and fitted

22.3A Reverse idler shaft and gear

22.3B Reverse idler shaft locking roll pin (arrowed)

Fig. 6.35 Reverse idler components (Sec 22)

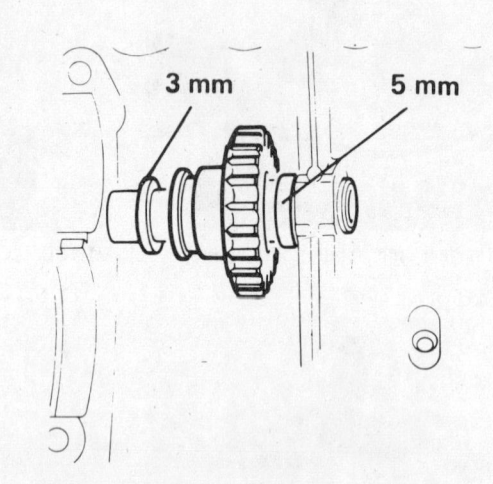

Fig. 6.36 Reverse idler shaft thrust washers (Sec 22)

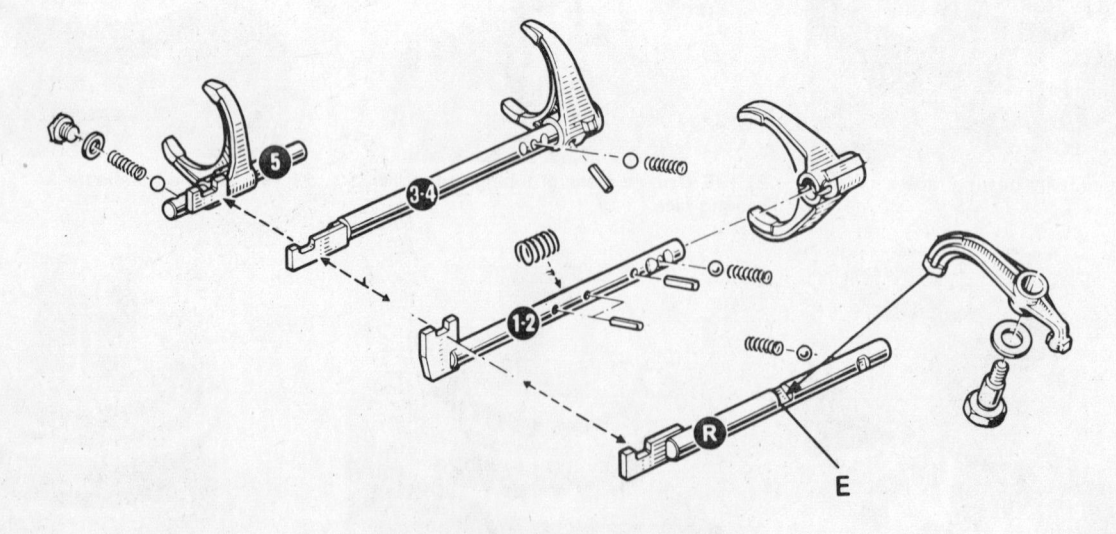

Fig. 6.37 Selector components (Sec 23)

E Notch for reverse swivel arm

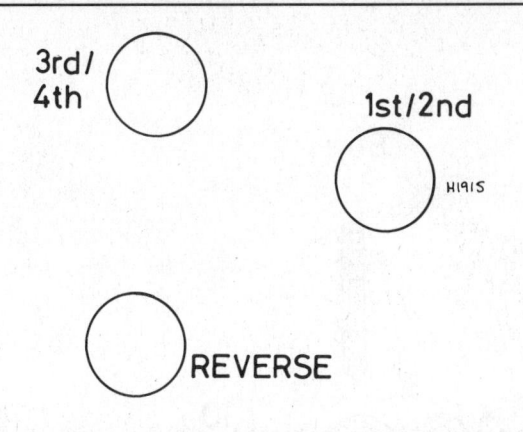

Chapter 6 Manual transmission

Fig. 6.38 Selector shaft identification diagram (Sec 23)

Viewed from differential end

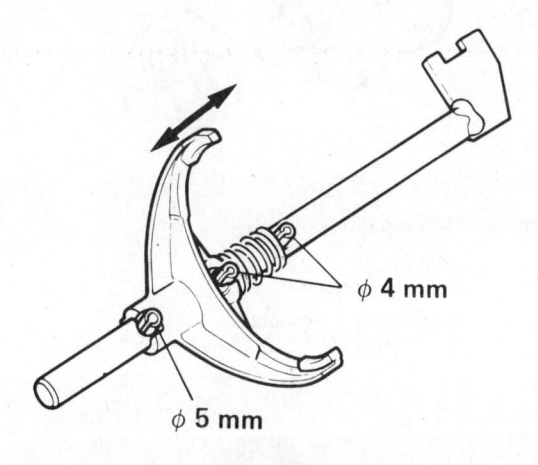

23.4 Selector shaft and fork arrangement

A 3/4
B 1/2
C Reverse swivel arm

Fig. 6.39 1st/2nd selector fork roll pins (Sec 23)

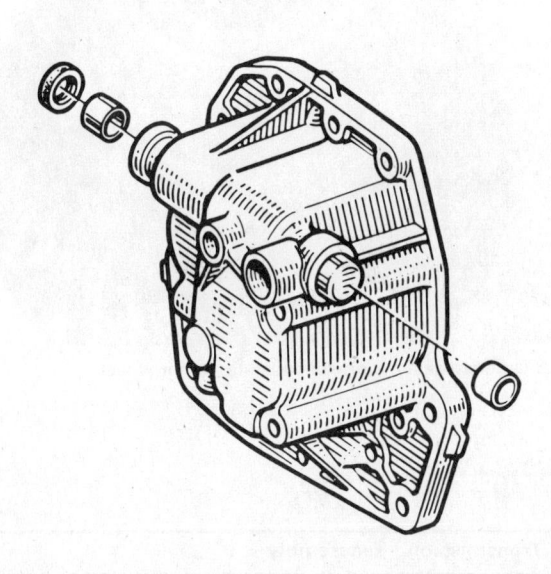

24 Rear cover – dismantling and reassembly

1 Drive out the double roll pins from the swivel arm.
2 Using a forked tool, compress the coil spring, withdraw the shaft and remove the swivel arm.
3 Reassembly is a reversal of dismantling.

25 Transmission housing – inspection

Refer to Section 12.

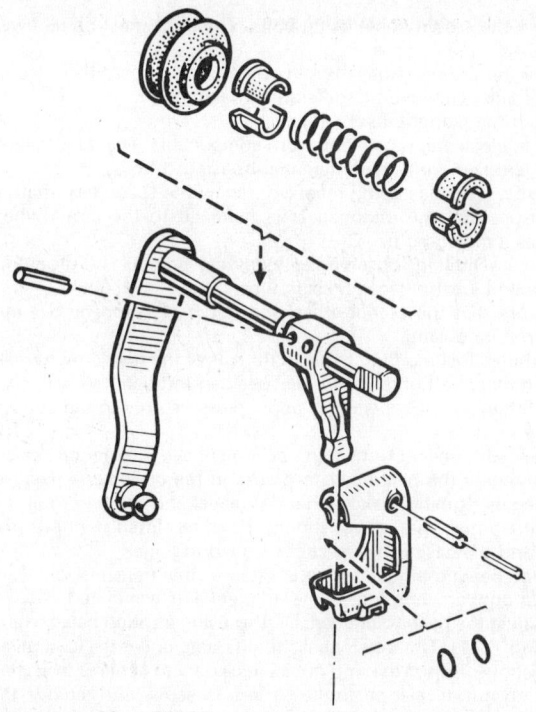

Fig. 6.40 Rear cover components (Sec 24)

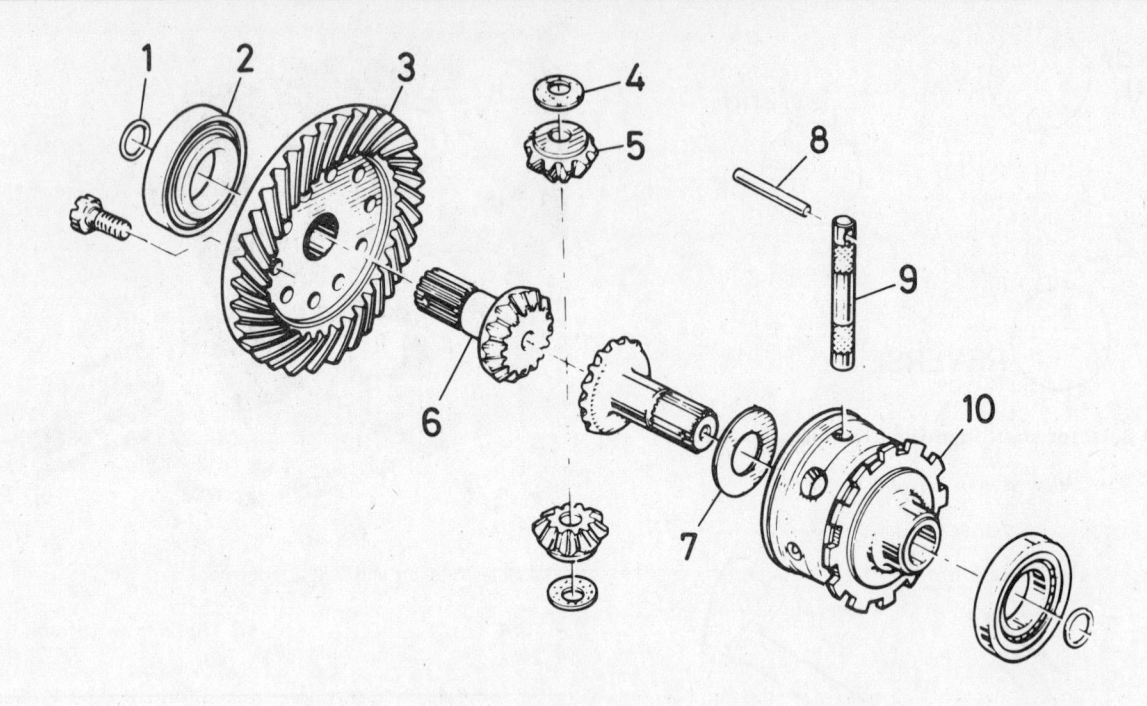

Fig. 6.41 Differential/final drive components (Sec 26)

1 O-ring	5 Planet wheel	8 Roll pin
2 Bearing	6 Sun wheel	9 Pinion shaft
3 Crownwheel	7 Anti-friction washer	10 Differential housing
4 Dished washer		

26 Differential/final drive – overhaul

Note: *A puller will be needed for this operation*

1 Unscrew the crownwheel fixing bolts. The bolts must be renewed at reassembly.
2 Remove the O-rings from the sun wheels and discard them.
3 Using a puller, remove the differential bearings.
4 Drive out the planet wheel roll pin.
5 Separate, clean and renew any worn components. The target wheel cannot be removed from the differential housing (photo).
6 If the crownwheel is being renewed, renew the secondary shaft as well to ensure that the pinion gear is matched to the crownwheel (supplied as a matching pair).
7 When reassembling, observe the following points. The groove in the anti-friction washer faces the sun wheel.
8 Make sure that the planet wheel roll pin is recessed by 5.0 mm (0.20 in) in the housing.
9 Apply thread locking fluid to clean threads of the new crownwheel bolts which must be tightened to the specified torque.
10 When fitting the new bearings, apply pressure only to the bearing centre track.
11 Use new sun wheel O-rings, the bearings have integral oil seals.
12 Always renew the bearing outer tracks in the casings. A piece of suitable diameter tubing makes a good removal and refitting tool.
13 The differential/final drive should now be fitted without the geartrains and the casing half-sections bolted together.
14 Only one bearing adjusting nut is fitted to this transmission. Turn the ring nut until any endfloat in the differential is eliminated.
15 Now adjust the bearing preload. Do this using a spring balance and cord (see Fig. 6.21). The differential should rotate under the load given in Specifications. Adjust the ring nut as necessary to achieve this, then mark the ring nut in relation to the casing. Unscrew and remove the ring nut, counting the number of turns required to remove it.
16 Separate the casings and remove the differential pending reassembly of the transmission.

26.5 Differential showing speedometer target wheel

27 Transmission – reassembly

1 Have the casings and components clean. Lubricate the parts as reassembly proceeds.
2 The reverse idle and selector components will have been reassembled as described in Section 22 and 23.

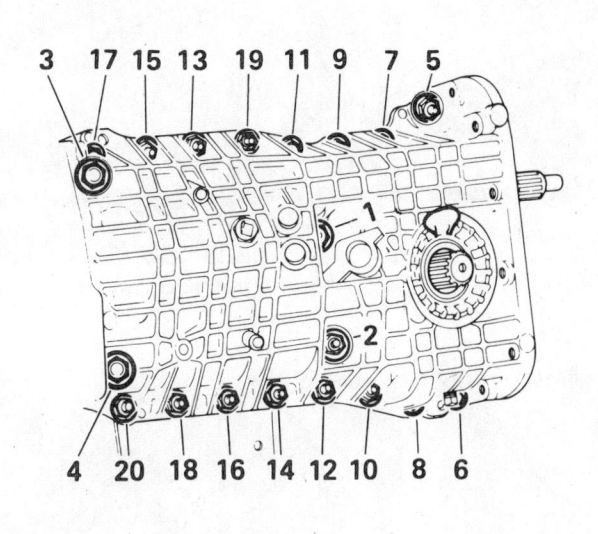

Fig. 6.42 Casing bolt tightening sequence (Sec 26)

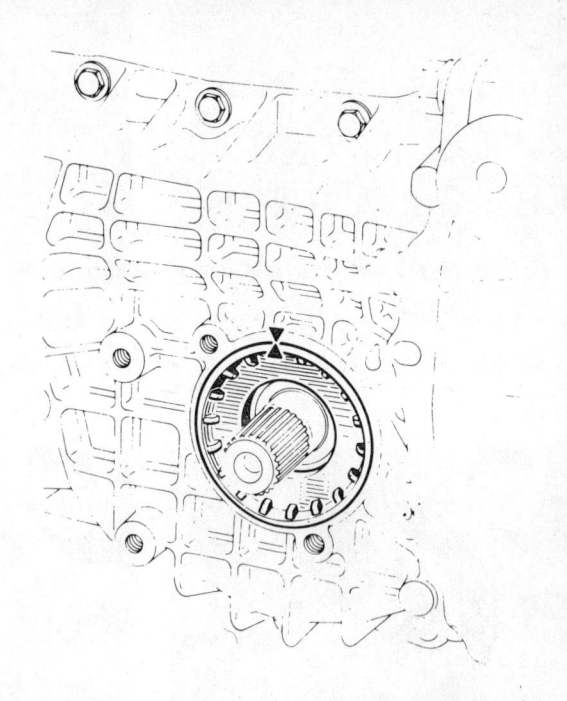

Fig. 6.43 Ring nut-to-casing alignment mark (Sec 26)

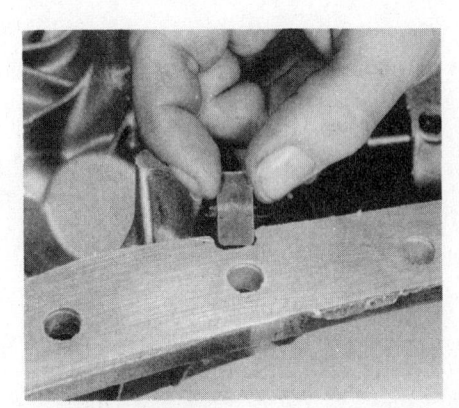

27.3 Casing magnet

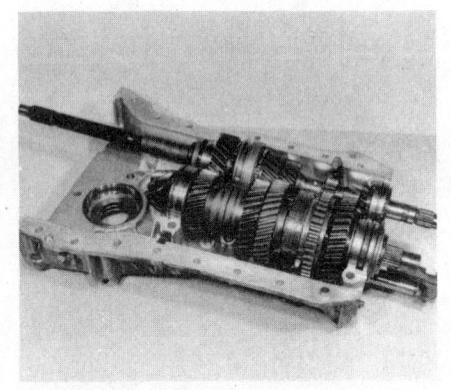

27.4 Geartrains installed

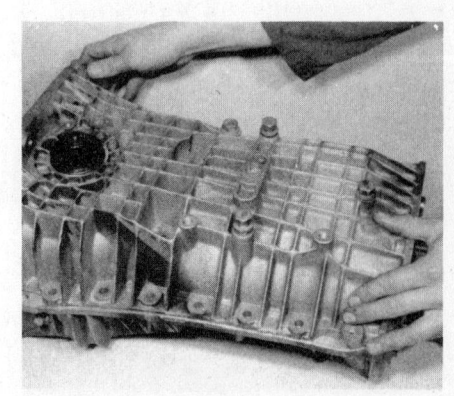

27.6A Connecting casing sections

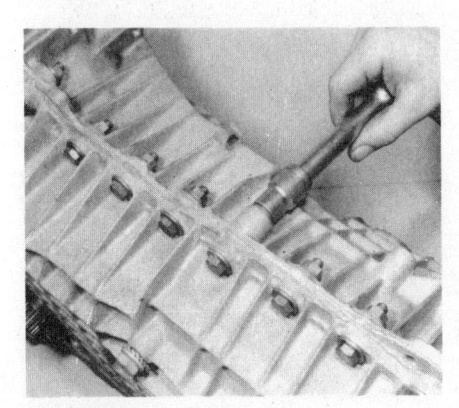

27.6B Tightening casing bolts

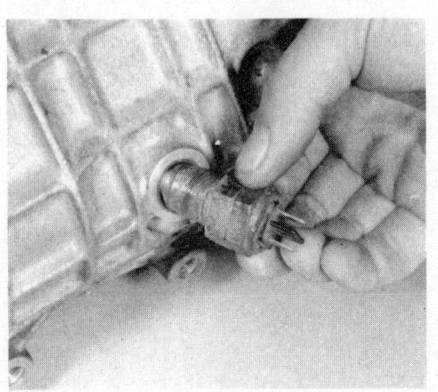

27.7 Fitting reversing lamp switch

27.8 Secondary shaft spacer plate

27.9 5th gear on secondary shaft

27.10 Dished washer on secondary shaft

27.11A Primary shaft thrust washer

27.11B 5th gear bush on primary shaft

27.11C 5th gear split needle roller bearing on primary shaft

27.11D 5th sliding gear on primary shaft

27.12 5th gear synchro sleeve, fork and shaft

27.13 5th gear baulk ring and locking gear

27.15A Dished washer

27.15B Primary shaft nut

27.17A Tightening shaft nut

27.17B Staking shaft nut

Chapter 6 Manual transmission

3 Fit the casing magnet (photo).
4 Refit the geartrains, meshed together, into the half-casing which has the adjustable differential bearing ring nut (photo).
5 Lower the differential into position (crownwheel teeth pointing downward).
6 Smear the casing mating flanges with jointing compound and bolt them together (photos). Tighten to torque in the sequence shown in Fig. 6.42. The bolts (1 and 2) should be liberally smeared with jointing compound.
7 Smear the threads of the reverse lamp switch with jointing compound and screw it into position (photo).
8 Bolt the secondary shaft spacer plate into position (photo).
9 Fit 5th gear to the secondary shaft (photo).
10 Fit the dished washer and screw on a new nut, having applied thread locking fluid to its threads (photo).
11 To the primary shaft, fit the thrust washer, the bush, the needle roller bearing and 5th sliding gear (photos).
12 Fit 5th gear synchro sleeve complete with selector fork and shaft (photo).
13 Fit the baulk ring (photo).
14 Fit the locking gear.
15 Fit the dished washer (concave side to gear) and screw on a new nut having applied thread locking fluid to its threads (photos).
16 Move the 5th and 2nd selector rods to engage two gears at once and so lock up the geartrains.
17 Tighten the two shift nuts to the specified torque and lock them by staking. Reselect neutral (photos).
18 Fit the rear cover, having smeared the mating surfaces with sealant. It is important that the cover is tilted when offering it up to ensure that the selector arm engages positively in the selector shaft dog notches. Once the cover is fitted, tighten the bolts to the specified torque and check that all gears can be obtained. Fit the dust excluder, selector arm and double roll pins to secure it (photos).
19 Refit 5th gear interlock ball and spring. Apply sealant to the plug threads and screw it into position (photos).
20 Fit the dust excluder and selector arm using a new roll pin.
21 Using a new gasket, bolt the bell housing into position (photos).
22 Fit the clutch release bearing and arm.
23 Refill the transmission with oil after it has been refitted to the car.

27.18A Fitting the rear cover

27.18B Tightening rear cover bolts

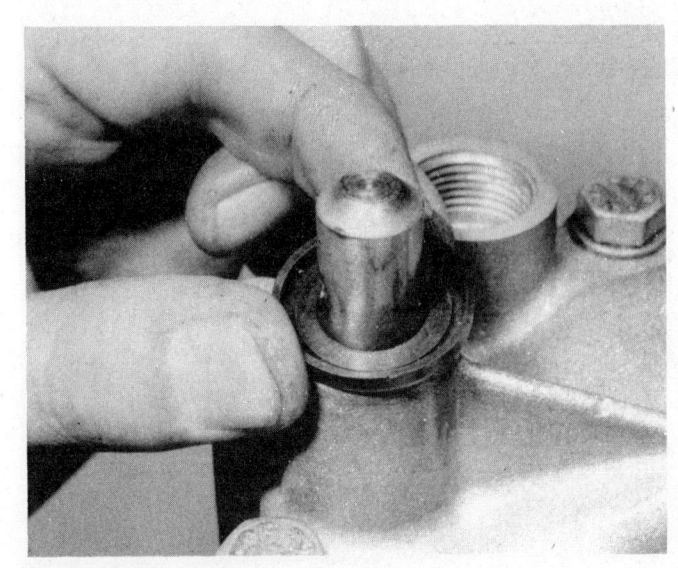

27.18C Selector arm dust excluder

27.18D Fitting selector arm

27.18E Fitting selector arm roll pin

27.19A 5th gear interlock ball

27.19B Interlock spring

27.19C Interlock ball/spring plug

27.21A Fitting bellhousing to transmission casing

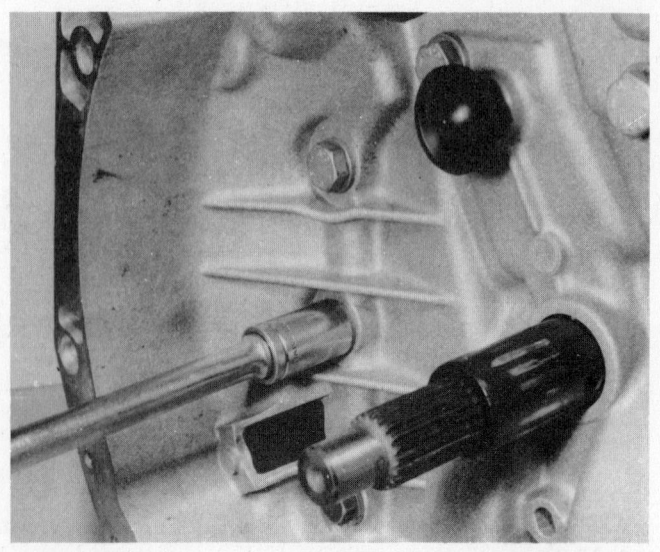

27.21B Tightening a bellhousing bolt

Chapter 6 Manual transmission

PART C – ALL TRANSMISSIONS

28 Fault diagnosis – manual transmission

Symptom	Reason(s)
Weak or ineffective synchromesh	Synchro bulk rings worn, split or damaged
	Synchromesh units worn, or damaged
Jumps out of gear	Gearchange mechanism worn
	Synchromesh units badly worn
	Selector fork badly worn
Excessive noise	Incorrect grade of oil in gearbox or oil level too low
	Gearteeth excessively worn or damaged
	Intermediate gear thrust washers or spacers worn allowing excessive end play
	Worn bearings
Difficulty in engaging gears	Clutch pedal adjustment incorrect
Noise when cornering	Wheel bearing or driveshaft fault
	Differential fault

Note: *It is sometimes difficult to decide whether it is worthwhile removing and dismantling the gearbox for a fault which may be nothing more than a minor irritant. Gearboxes which howl, or where the synchromesh can be beaten by a quick gearchange, may continue to perform for a long time in this state. A worn gearbox usually needs a complete rebuild to eliminate noise because the various gears, if re-aligned on new bearings, will continue to howl when different bearing surfaces are presented to each other. The decision to overhaul, therefore, must be considered with regard to time and money available, relative to the degree of noise or malfunction that the driver has to suffer.*

Chapter 7 Automatic transmission

Contents

Automatic transmission (Type MJ3) – removal and refitting 11	Maintenance 2
Automatic transmission (Type 4141) – removal and refitting 12	Selector control lever and cable (Type MJ3 transmission) –
Description 1	removal, refitting and adjustment 7
Fault diagnosis – automatic transmission 13	Selector control lever and cable (Type 4141 transmission)
Fluid filter – renewal 3	removal, refitting and adjustment 8
Governor control cable (Type 4141 transmission) – adjustment 6	Transmission oil seals – renewal 9
Kickdown switch (Type MJ3 transmission) – testing and	Vacuum unit – removal and refitting 10
adjustment 5	
Kickdown switch (Type 4141 transmission) – testing and	
adjustment 4	

Specifications

General

Type	MJ3 or 4141, three-speed automatic, computer-controlled	
Application:		
Four-cylinder models	Type MJ3	
Six-cylinder models	Type 4141	
Ratios:	**Type MJ3**	**Type 4141**
1st	2.50 : 1	2.44 : 1
2nd	1.52 : 1	1.47 : 1
3rd	1.00 : 1	1.00 : 1
Reverse	2.00 : 1	1.92 : 1
Final drive	3.56 : 1	3.44 : 1

Fluid capacity

From dry	6.0 litre (10.6 pints)
At fluid renewal	2.5 litre (4.4 pints)
Final drive (Type 4141 only)	1.6 litre (2.8 pints)

Torque wrench settings

	Nm	lbf ft
Driveplate-to-converter bolts	30	22
Fluid cooler pipe unions	19	14
Sump pan fixing screws	6	4
Differential cover plate bolts	19	14
Torque converter housing-to-engine connecting bolts	54	40

Chapter 7 Automatic transmission

1 Description

One of the two types of automatic transmission may be fitted, according to whether the car is four-cylinder or six-cylinder, see Specifications.

Both types of transmission are similar in operating characteristics, but vary slightly in design detail and maintenance requirements.

Type 4141

Fully automatic gearchanging is provided without the use of a clutch, but override selection is still available to the driver.

The automatic transmission consists of three main assemblies: the torque converter, the final drive and the gearbox. A cutaway view of the transmission unit, with the three main assemblies sub-divided, is shown in Fig. 7.1.

The converter takes the place of the conventional clutch and transmits the drive automatically from the engine to the gearbox, providing increased torque when starting off.

The converter receives its lubricant from the pump mounted on the rear of the gearcase, and is driven directly by the engine. This pump also distributes fluid to the respective gears, clutch and brake assemblies within the gearbox.

The gearbox comprises an epicyclic geartrain giving three forward and one reverse gear; selection of which is dependent on the hydraulic pressure supplied to the respective clutches and brakes. The hydraulic pressure is regulated by the hydraulic distributor, and gear selection is determined by two solenoid valves. These are actuated by the electrically-operated governor/computer. The exact hydraulic pressure is regulated by a vacuum capsule and pilot valve operating according to engine loading.

The clutches (E1 and E2) and brakes (F1 and F2) are multi-disc oil bath type and, according to the hydraulic loading, engage or release the epicyclic geartrain components.

The governor is, in effect, a low output alternator which provides variable current to the computer. It is driven by a worm gear on the final drive pinion and its output depends on the vehicle speed and engine loading.

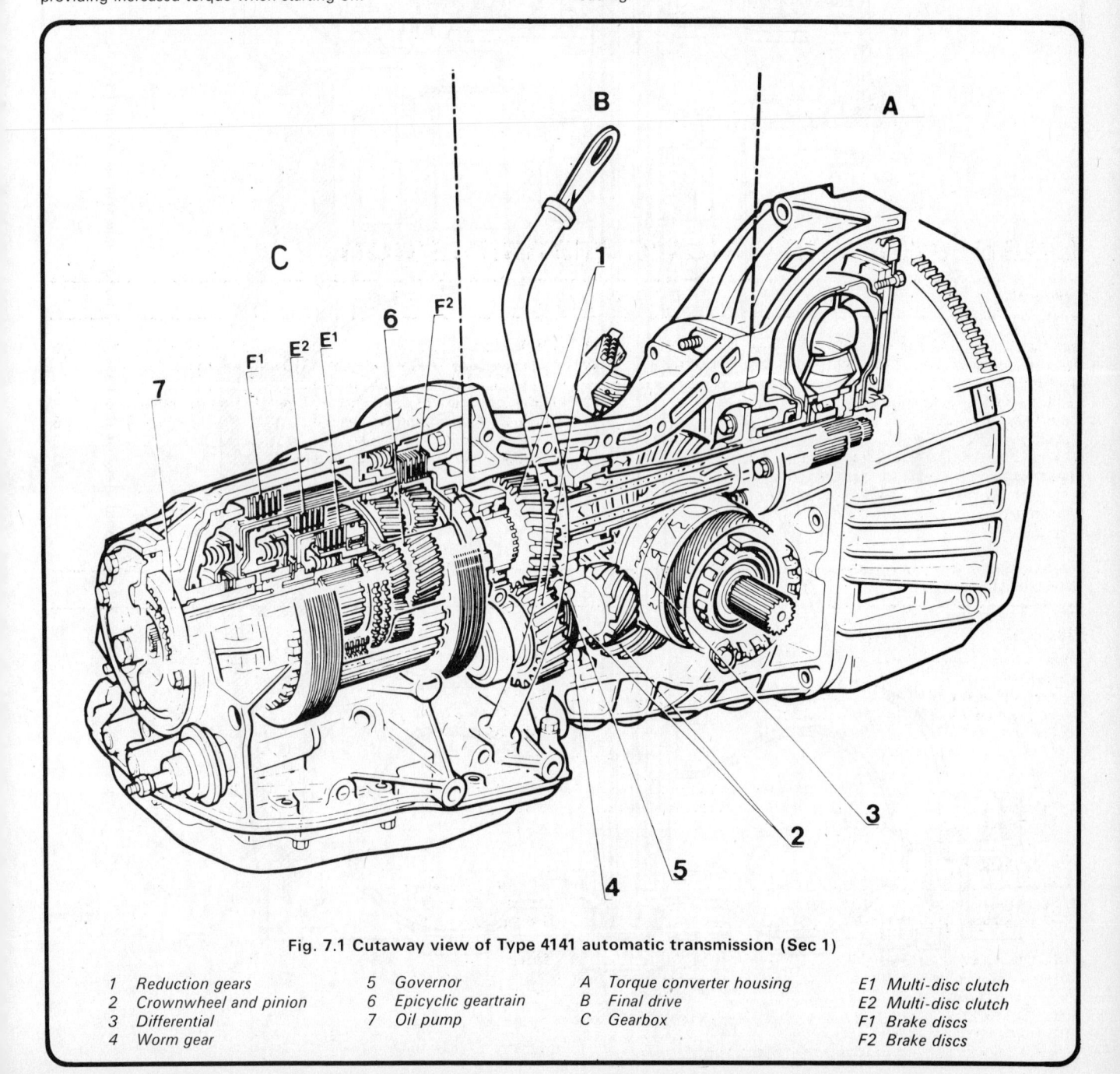

Fig. 7.1 Cutaway view of Type 4141 automatic transmission (Sec 1)

1 Reduction gears	5 Governor	A Torque converter housing	E1 Multi-disc clutch
2 Crownwheel and pinion	6 Epicyclic geartrain	B Final drive	E2 Multi-disc clutch
3 Differential	7 Oil pump	C Gearbox	F1 Brake discs
4 Worm gear			F2 Brake discs

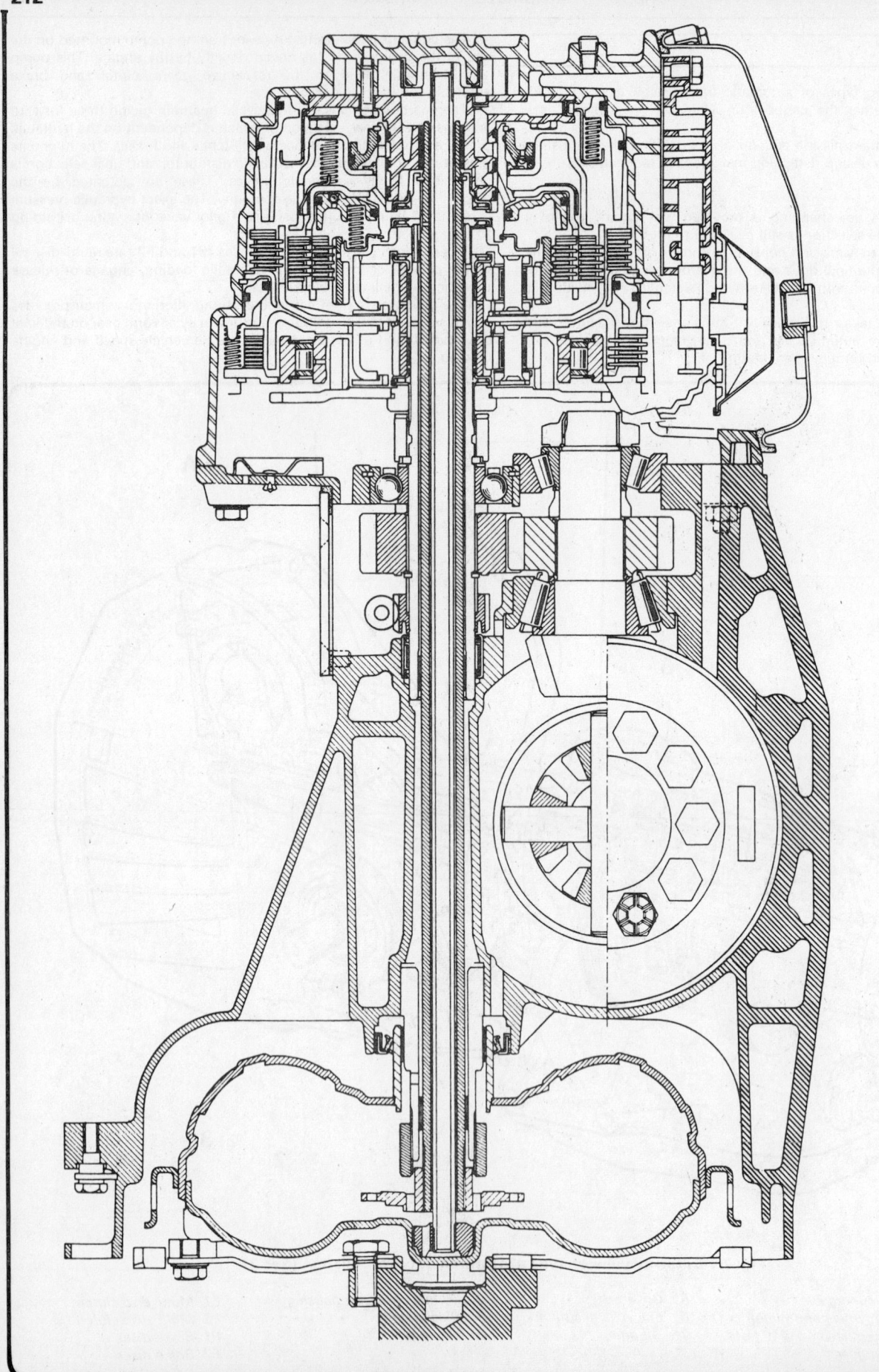

Fig. 7.2 Sectional view of the Type MJ3 automatic transmission (Sec 1)

213

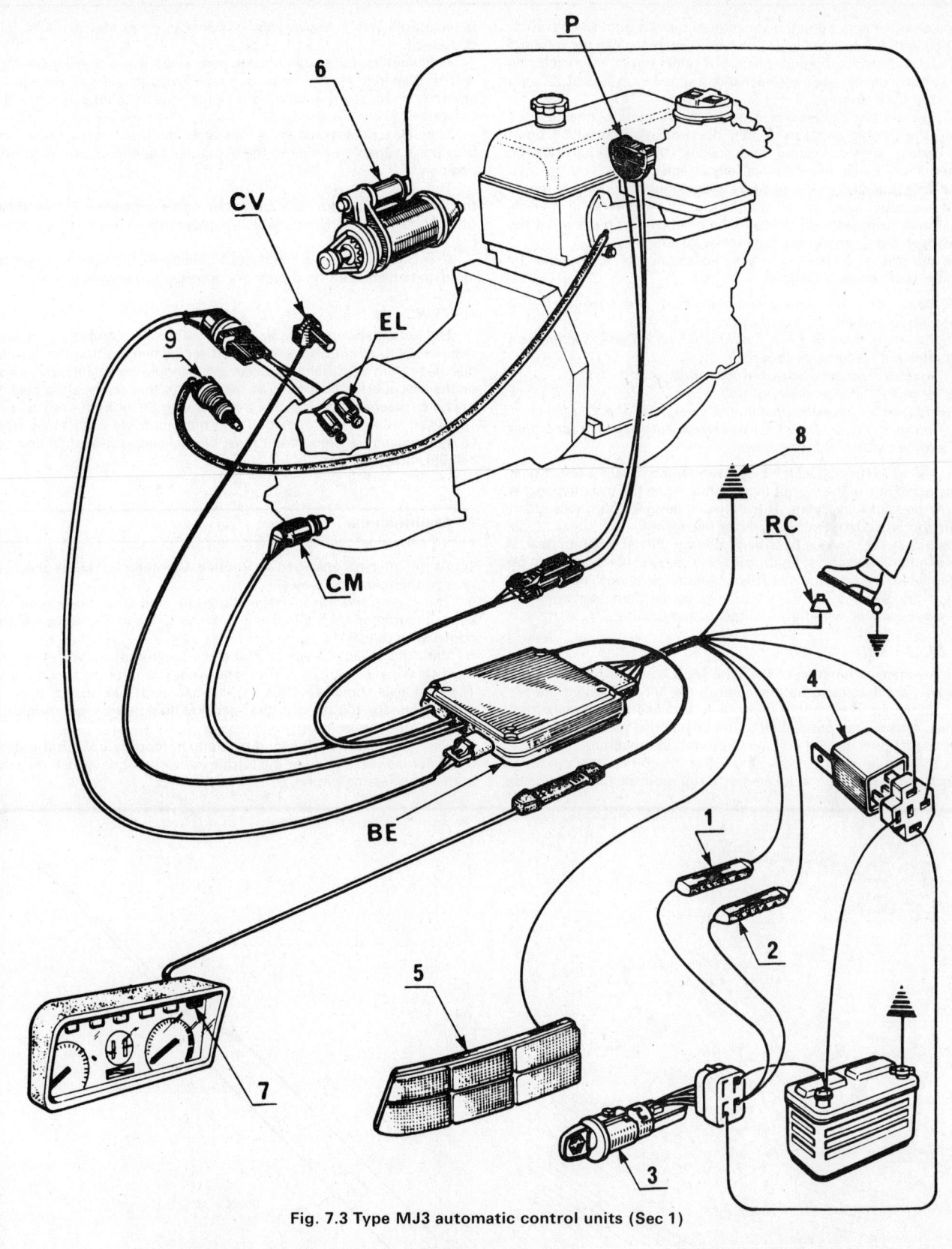

Fig. 7.3 Type MJ3 automatic control units (Sec 1)

1 Reversing light fuse (5A)
2 Supply fuse (1.5A)
3 Starter switch
4 Starter relay
5 Reversing lights
6 Starter
7 Instrument panel (warning light)
8 Automatic transmission earth
9 Vacuum capsule
BE Computer or module
CM Multi-purpose switch
CV Speed sensor
EL Pilot solenoid valves
P Load potentiometer
RC Kick-down switch

The computer acts upon the variation of current from the governor combined with the selected lever position to open or close the solenoid valves accordingly. In addition it acts as a safety device to prevent the 1st gear 'hold' position being selected at a speed in excess of 22 mph (35 kph) at light throttle.

The system also incorporates a kickdown switch, operated by pressing the throttle pedal to its fully open position, at which point under certain engine loading and speeds, the computer will be activated and a lower gear automatically selected.

The drive selected in the gearbox is transferred to the differential unit via stepdown gears, which compensate for the difference in levels between the main gear assemblies in the gearbox and the level of the crownwheel and pinion in the differential housing.

The selector lever is centrally situated within the car and has six positional alternatives, as follows:

P (Park): With the lever in this position, the transmission is neutralised and the drive wheels are locked
R (Reverse): Reverse gear position, which when selected also actuates the reversing light switch
N (Neutral): The transmission is in neutral
1 (1st gear): 1st gear hold position
2 (2nd gear): Automatic operation between 1st and 2nd gears
A (Automatic) or D (Drive): Gears engage automatically according to engine loading and car speed

In addition to the above, the kickdown switch causes a lower gear to be selected at a higher speed than normal when the throttle pedal is suddenly pressed fully down. This device is designed to give sudden acceleration when required, such as for overtaking.

Because of the obvious hazards of starting the car when in gear, a starter inhibitor switch is fitted and only allows the starter to be operated when the selector is in Park or Neutral position. The inhibitor switch is fitted below the transmission governor/computer units. Its removal necessitates withdrawal of the oil sump plate.

Type MJ3

The mechanical function of the Type MJ3 is similar to that of the Type 4141 just described, but the controls are fully computerised.

The computer or module is continually supplied with signals from the speed sensor, load potentiometer, multi-purpose switch, kickdown switch and its own control unit, and from this information it transmits signals to the pilot solenoid valves to select the correct gear range.

The kickdown switch is located at the bottom of the accelerator pedal travel and a lower gear is selected when the pedal is fully depressed.

The multi-purpose switch is located on the rear of the transmission and is operated by the range selector lever. It controls the engine starting circuit, the reversing light circuit and the pilot solenoid valve circuit.

The load potentiometer is operated by the throttle valve and provides a variable voltage to the computer, dependent on the throttle position.

The speed sensor is mounted on the left-hand side of the transmission and it provides an output signal dependent on the speed of the parking pawl wheel, which is proportional to the speed of the car.

A vacuum capsule is connected directly to the inlet manifold to regulate transmission fluid pressure according to engine loading.

All types

The automatic transmission is a relatively complex unit and therefore, should problems occur, it is recommended that the fault be discussed with your Renault dealer, who should be able to advise you on the best course of action to be taken. Items that can be attempted by the home mechanic are given in the following Sections in this Chapter. To obtain trouble-free operation and maximum life expectancy from your automatic transmission, it must be serviced as described and not be subjected to abuse.

2 Maintenance

1 At the intervals specified in Routine Maintenance, check the fluid level in the following way.
2 The engine and transmission should be cold. Start the engine and allow it to run for a few minutes in order to fill the torque converter and cooler with fluid.
3 With the selector lever in P and the engine idling, withdraw the dipstick, wipe it clean, re-insert it and withdraw it for the second time. The fluid level should be between the 'min' and 'max' marks.
4 If necessary, top up with the specified fluid poured in through the dipstick guide tube.
5 The transmission will very rarely require topping-up, but if it does check for leaks probably at the torque converter or driveshaft oil seals in the transmission casing.

Fig. 7.4 Automatic transmission fluid dipstick location (Type MJ3) (Sec 2)

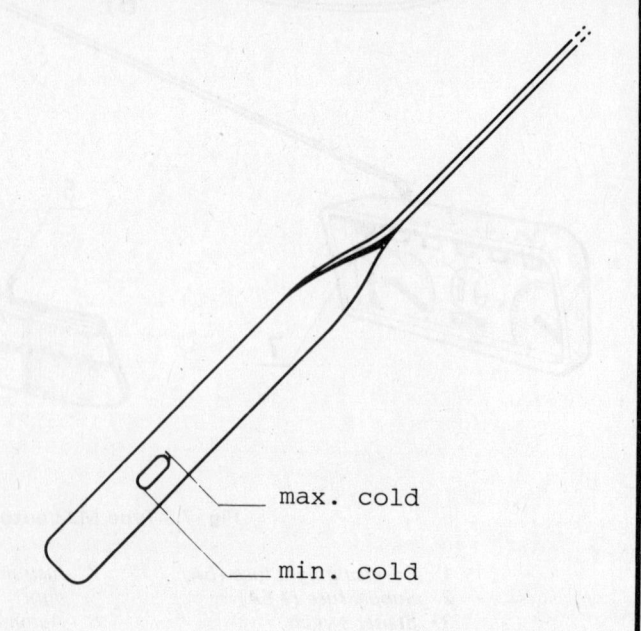

Fig. 7.5 Dipstick marking (Sec 2)

Chapter 7 Automatic transmission

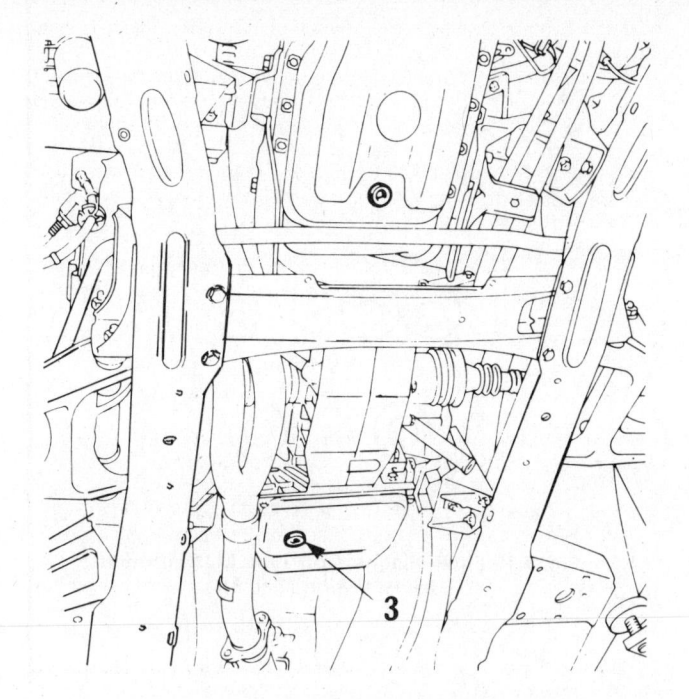

Fig. 7.6 Drain plug (3) (Type MJ3) (Sec 2)

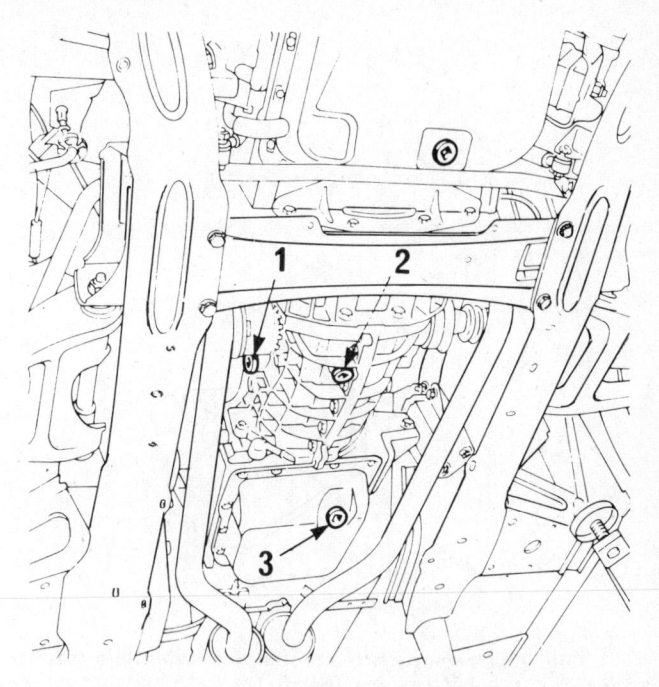

Fig. 7.7 Drain and filler/level plugs (Type 4141) (Sec 2)

1 Final drive filler/level plug
2 Final drive drain plug
3 Transmission fluid drain plug

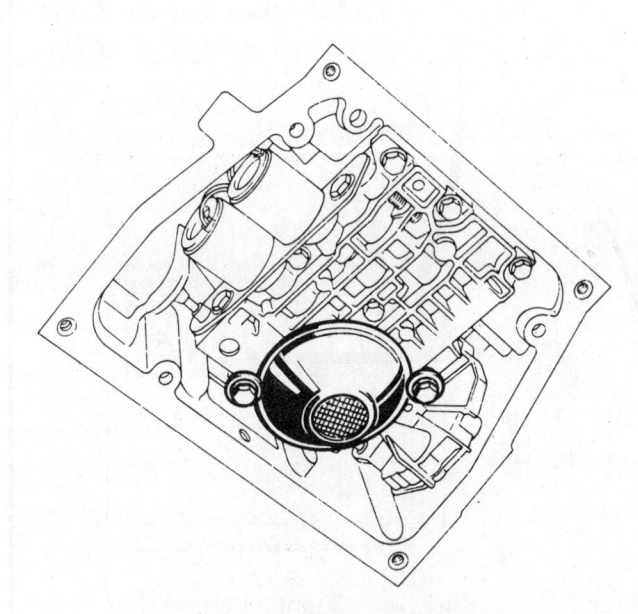

Fig. 7.8 Circular type fluid filter (Sec 3)

6 The transmission fluid should be drained, preferably hot, at the intervals specified in Routine Maintenance. Place a suitable container under the drain plug and then remove the plug, taking care that the extremely hot fluid does not scald you.
7 On Type MJ3 transmissions the fluid is common to the geartrains and integral final drive. On Type 4141 transmissions, separate lubricant is used in the final drive and this should be drained at the same time.
8 Refit the drain plugs, fill the transmission through the dipstick guide tube, and the separate final drive (Type 4141) through the combined filler/level plug hole. Remember that different lubricants are used in the transmission and Type 4141 final drive; refer to lubricant details in the introductory Section of this Manual.

3 Fluid filter – renewal

1 This is not a routine service operation, but may be required if the transmission fluid has become contaminated or regular renewal has been neglected.
2 Drain the fluid as previously described, unbolt and remove the sump pan and gasket.
3 Unbolt the gauze type filter and remove it. With the rectangular type of filter, note particularly which way round it is fitted as it can be installed incorrectly.
4 Fit the new filter using a new sealing ring.
5 Before offering up the sump pan (with a new gasket) check that the magnets are located as shown in Fig. 7.10 on Type MJ3 units. If only one magnet is used originally, additional ones should be purchased and fitted.
6 Refill the transmission with clean fluid.

4 Kickdown switch (Type 4141 transmission) – testing and adjustment

1 Disconnect the lead from the kickdown switch and then connect a test lamp between the switch and the battery (+) terminal.
2 Switch on the ignition and have an assistant depress the accelerator pedal fully. The test bulb should illuminate.
3 If it does not, adjust in the following way. Mark the kickdown cable sleeve with a line (T – Fig. 7.11) 3.0 mm (0.12 in) from the switch cover.
4 With the accelerator fully depressed, the line should be flush with the switch cover. If it is not, reposition the clip (F).
5 Recheck the operation using the test lamp.

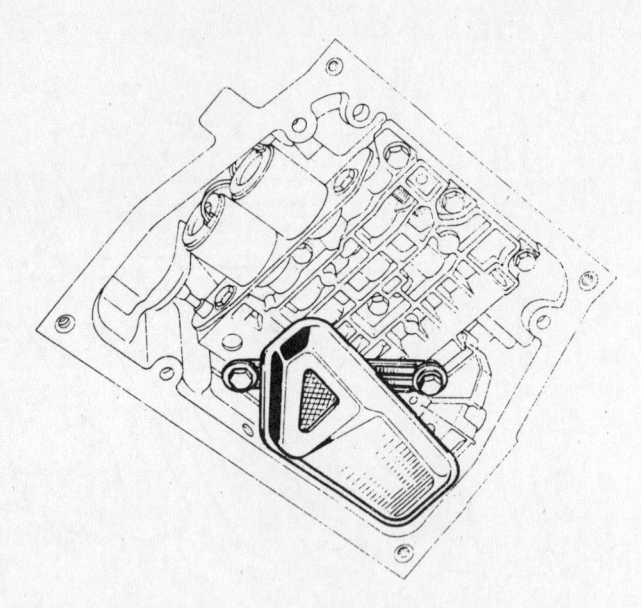

Fig. 7.9 Correctly fitted rectangular type fluid filter (Sec 3)

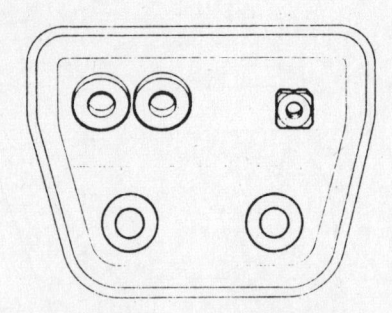

Fig. 7.10 Filter magnets on Type MJ3 automatic transmission (Sec 3)

Kick-down switch wire (white with red end)

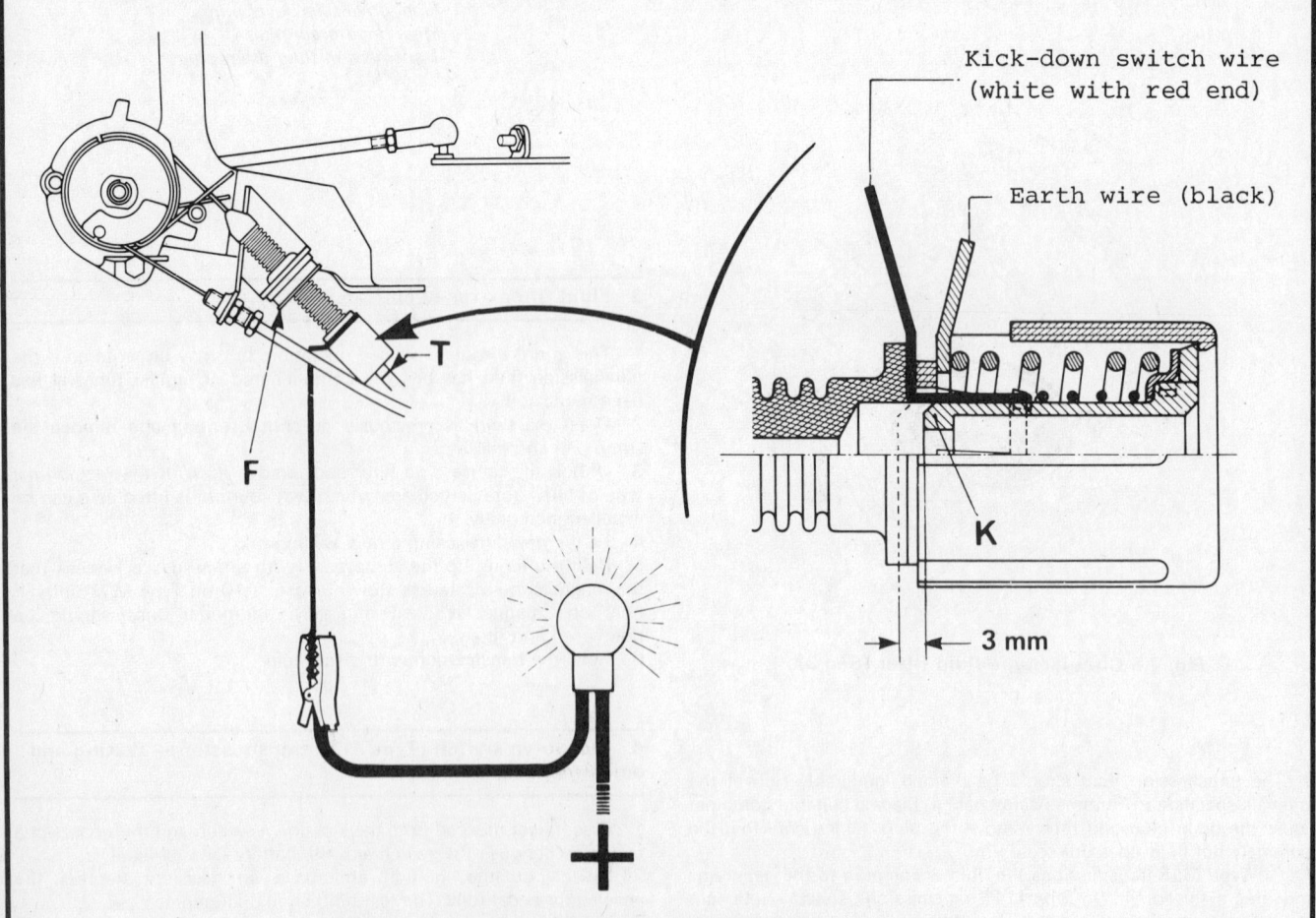

Earth wire (black)

3 mm

Fig. 7.11 Kickdown switch and test circuit (Type 4141) (Sec 4)

F Clip   T Alignment mark
K Stop

Chapter 7 Automatic transmission

5 Kickdown switch (Type MJ3 transmission) – testing and adjustment

1 The switch is incorporated in the load potentiometer in the computer.
2 Slightly loosen the potentiometer fixing screws. On 2165 cc engined models, the screws are accessible through holes in the quadrant if an assistant depresses the accelerator hard.
3 Disconnect the potentiometer wiring plug and connect a battery and test lamp.
4 With the accelerator pedal held fully down, turn the potentiometer until the test lamp illuminates. Tighten the two screws.
5 If the test lamp does not come on at all, and all the wiring is satisfactory, renew the potentiometer/kickdown switch.

6 Governor control cable (Type 4141 transmission) – adjustment

1 The throttle linkage and kickdown switch must be correctly adjusted.
2 Screw the cable sleeve (6 – Fig. 7.13) in as far as it will go.
3 Have an assistant depress the accelerator pedal fully and then tension the cable to just eliminate all slackness by turning the sleeve (F – Fig. 7.14). Tighten the locknut.
4 Adjust the cable sleeve (G – Fig. 7.15) to obtain a clearance (J) of between 0.3 and 0.5 mm (0.012 and 0.020 in).

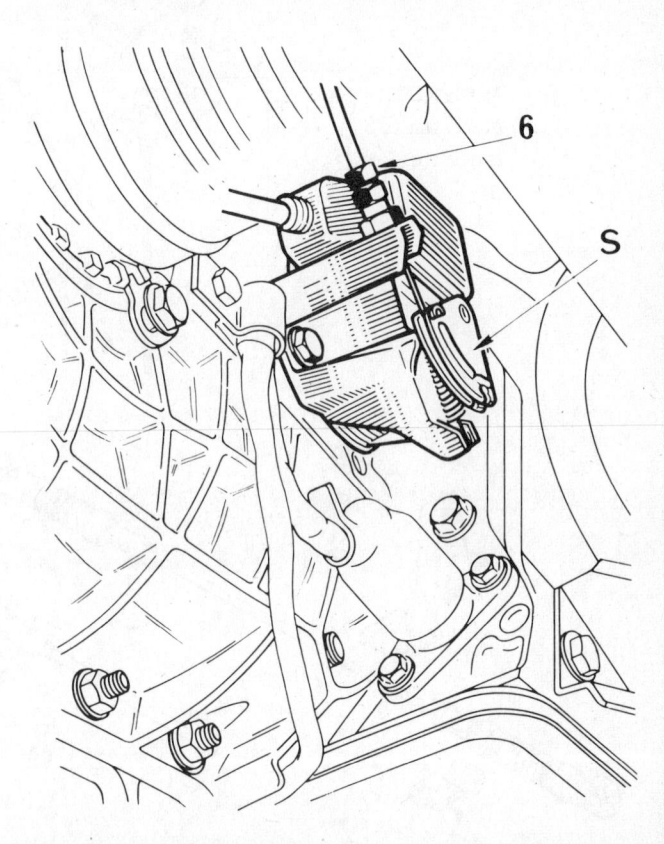

Fig. 7.13 Governor cable sleeve (6) and quadrant (S) (Sec 6)

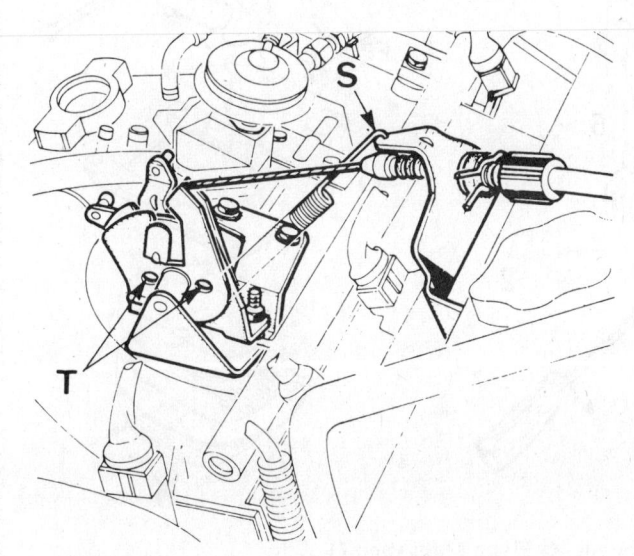

Fig. 7.12 Kickdown switch (load potentiometer) (Type MJ3) (Sec 5)

S Return spring
T Potentiometer fixing screws

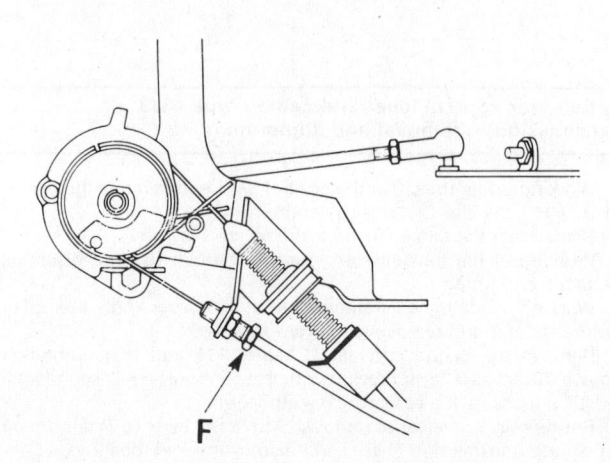

Fig. 7.14 Governor cable at throttle housing (Sec 6)

F Cable sleeve

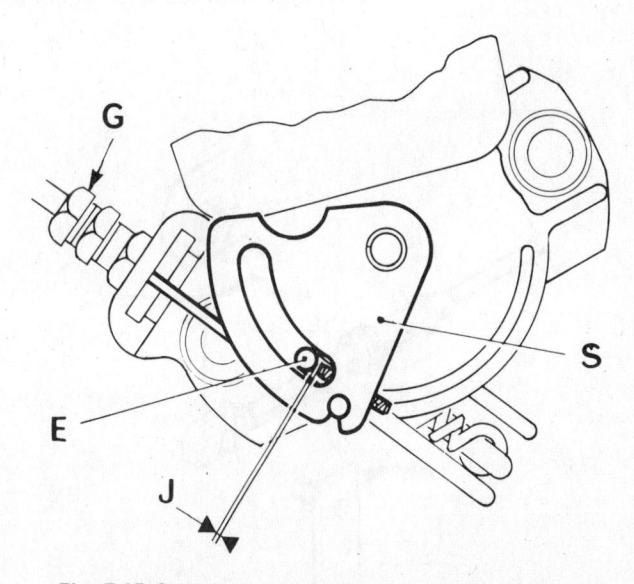

Fig. 7.15 Governor cable setting at quadrant (Sec 6)

E Stop pin
G Cable sleeve
J Clearance = 0.3 to 0.5 mm (0.012 to 0.020 in)
S Quadrant

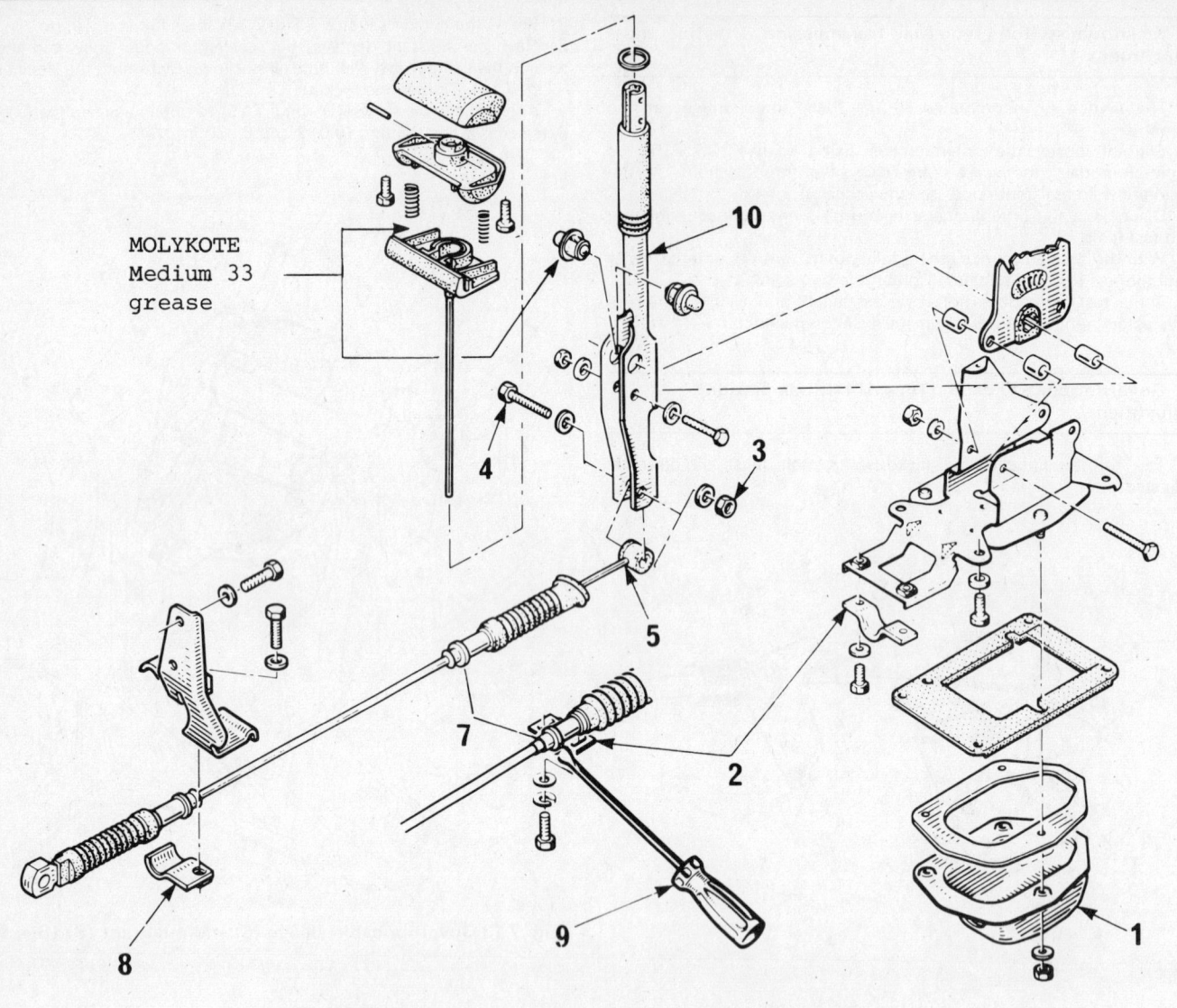

Fig. 7.16 Speed selector control components (Type MJ3) (Sec 7)

1 Cover	4 Pin	8 Clamp
2 Clamp	5 Cable	9 Screwdriver
3 Nut	7 Sleeve stop	10 Lever

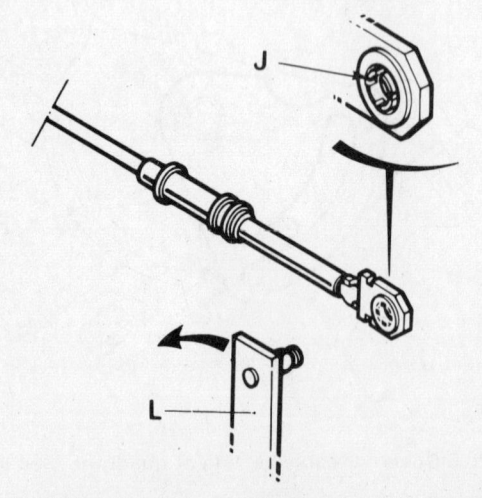

Fig. 7.17 Speed selector cable notches (J) and arm (L) on transmission (Sec 7)

7 Selector control lever and cable (Type MJ3 transmission) – removal and adjustment

1 Working under the car at the control lever end, remove the cover (1 – Fig. 7.16) the clip (2), nut (3) and pin (4).
2 Disconnect the cable (5) from the lever (10).
3 Working at the transmission, remove the clip (8) and disconnect the cable end fitting.
4 Working inside the car, remove the selector lever knob after driving out the roll pin and removing the two screws.
5 Remove the centre console (Chapter 11) and then unbolt and remove the selector lever bracket with the control lever. Disconnect the electrical leads as the assembly is withdrawn.
6 Refitting is a reversal of removal. Move the lever to P and the arm (L) on the transmission (Fig. 7.17) also to the Park position.
7 Connect the selector cable so that the notches (J) are towards the balljoint.
8 Using a screwdriver as a lever, prise the cable to preload it before tightening the clip bolts (2 – Fig. 7.16).

Chapter 7 Automatic transmission

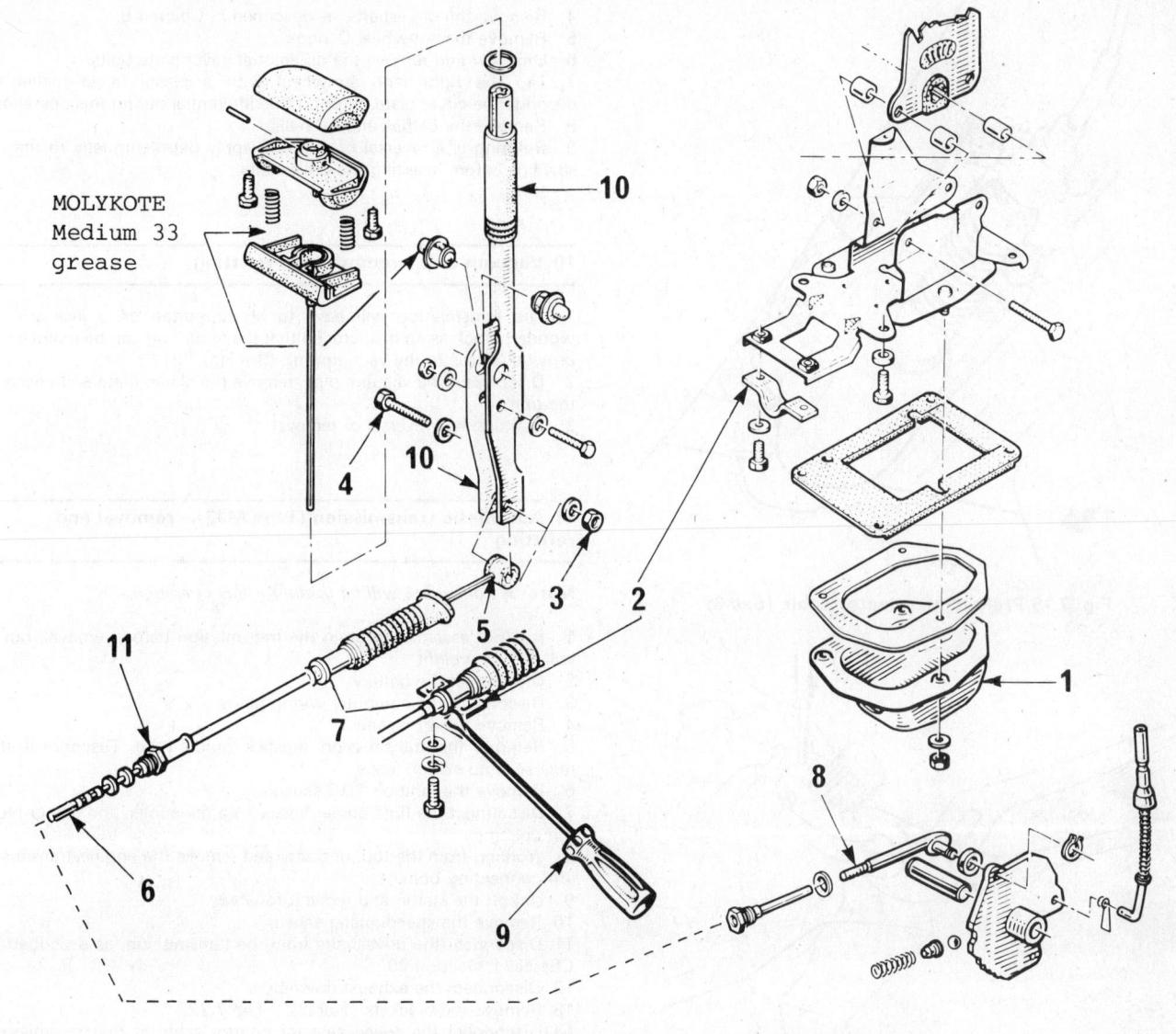

Fig. 7.18 Speed selector control components (Type 4141) (Sec 8)

1 Cover
2 Clamp
3 Nut
4 Pin
5 Cable
6 Cable end fitting
7 Cable stop
8 Quadrant link
9 Screwdriver
10 Lever
11 Union nut

MOLYKOTE Medium 33 grease

8 Selector control lever and cable (Type 4141 transmission) – removal, refitting and adjustment

1 Removal and refitting operations for the selector lever are similar to those described in the preceding Section. Removal and refitting of the control cable, however, is entirely different.
2 Place the selector lever in N.
3 Working underneath the car, remove the cover plate (1 – Fig. 7.18), clip (2), nut (3), and the pin (4). Disconnect the cable from the lever.
4 Working at the transmission casing, release the cable union nut (11). Without the special tool (B Vi. 868) make up a suitable spanner (such as a box spanner slotted lengthwise) which will have to be used in order to reach it. Once the union is released, unscrew the cable and then jack it hard to free the sealing washers and withdraw it from the transmission casing.
5 When fitting the cable, set the lever in P.
6 Screw in the cable union and tighten it.
7 Pull on the cable to select P and connect it to the hand control lever.
8 Preload the cable using a screwdriver as shown before tightening the cable clamp bolts (Fig. 7.19).
9 Check gear selection, and that starter actuation only occurs in N and P positions.
10 Should the cable renewal be due to the old one breaking, then the sump pan will have to be removed in order to extract the cable end fitting and sealing washers.

9 Transmission oil seals – renewal

Torque converter oil seal

1 This is accessible after having removed the transmission (or engine) and the torque converter.
2 Tap the front face of the old seal at one point with a punch so that it tilts and can be extracted.
3 Tap the new seal squarely into position having smeared its lips with petroleum jelly.

Chapter 7 Automatic transmission

Driveshaft oil seals

4 Remove the driveshafts, as described in Chapter 8.
5 Remove the sunwheel O-rings.
6 Unscrew and remove the differential cover plate bolts.
7 Tap the right-hand sunwheel with a plastic-faced mallet to dislodge the cover plate. Support the differential during this operation.
8 Remove the baffles and oil seals.
9 Refitting is a reversal of removal; apply petroleum jelly to the oil seal lips before inserting the driveshafts.

10 Vacuum unit – removal and refitting

1 The transmission will have to be supported on a jack with a wooden block as an insulator so that the mounting can be removed to provide access to the vacuum unit (9 – Fig. 7.3).
2 Disconnect the vacuum pipe, remove the clamp plate and unscrew the unit.
3 Refitting is a reversal of removal.

11 Automatic transmission (Type MJ3) – removal and refitting

Note: A trolley jack will be useful for this operation.

1 It is not essential to drain the transmission before removal, but it will reduce weight.
2 Disconnect the battery.
3 Disconnect the computer wiring plugs.
4 Remove the air cleaner.
5 Remove the transmission dipstick guide tube. Disconnect the reverse lamp switch leads.
6 Remove the ignition TDC sensor.
7 Disconnect the fluid cooler hoses from the cooler, and cap or plug the openings.
8 Working from the top, unscrew and remove the engine/transmission connecting bolts.
9 Unbolt the starter and move it forward.
10 Remove the speedometer sensor.
11 Disconnect the driveshafts from the transmission, as described in Chapter 1, Section 20.
12 Disconnect the exhaust downpipe.
13 Remove the brackets (1 and 2 – Fig. 7.22).
14 Disconnect the speed selector control cable at the transmission end.

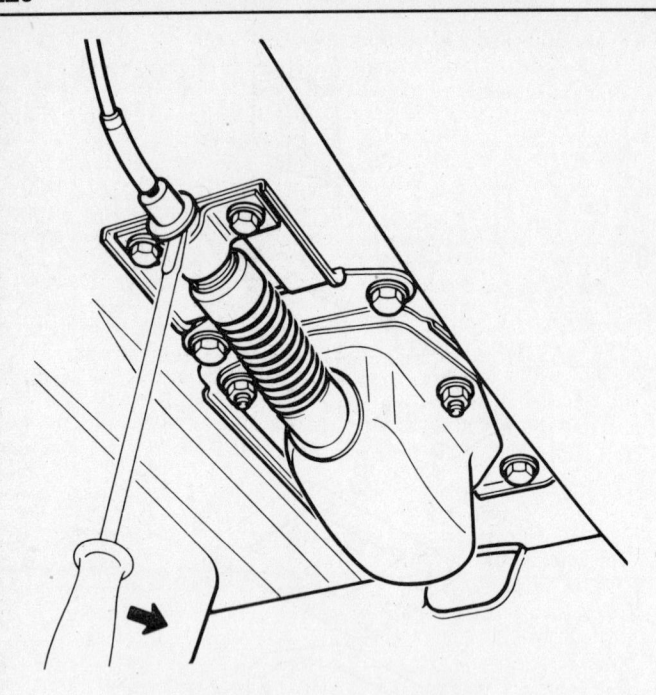

Fig. 7.19 Preloading selector cable (Sec 8)

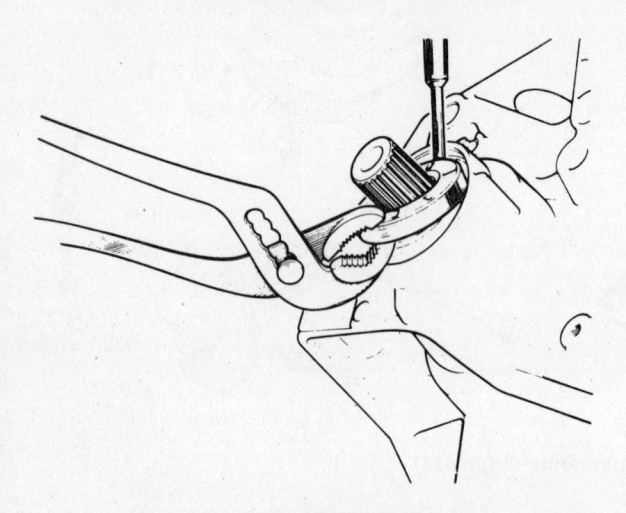

Fig. 7.20 Removing a differential oil seal (Sec 9)

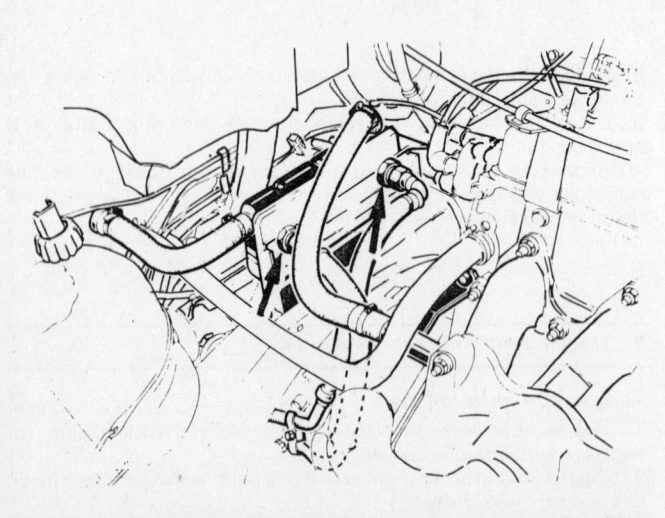

Fig. 7.21 Type MJ3 fluid cooler hoses – arrowed (Sec 11)

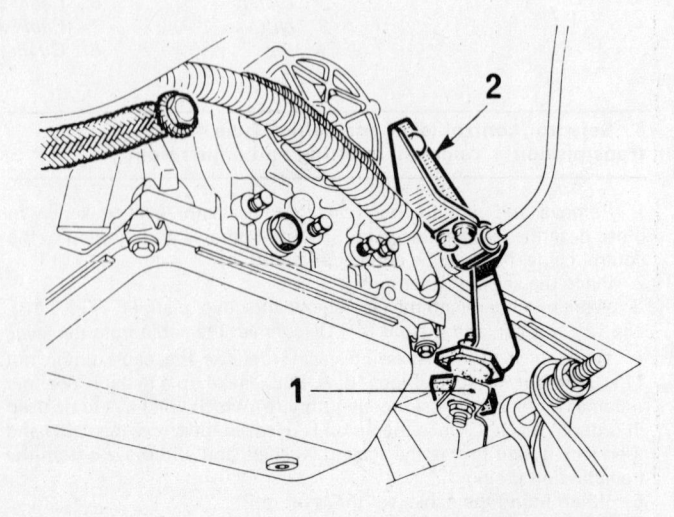

Fig. 7.22 Exhaust bracket (1) and control cable bracket (2) (Sec 11)

Chapter 7 Automatic transmission   221

15 Disconnect the transmission earth strap.
16 On 1995 cc engined models, unbolt the cover plate from the lower front face of the torque converter housing.
17 Working through the starter motor aperture, disconnect the driveplate from the torque converter, as described in Chapter 1, Section 21.
18 Unless the car is over an inspection pit, or up on a car lift, the rear end must now be raised and securely supported so that enough clearance will be provided for the transmission to be withdrawn rearwards and removed from under the car.
19 Support the transmission on a trolley jack and remove the left-hand mounting, complete with brackets.
20 Remove the vacuum capsule pipe.
21 Remove the right-hand mounting, complete with brackets.
22 Unscrew and remove the remaining lower engine/transmission connecting bolts.
23 Lower the trolley jack and withdraw the transmission. Guide the computer out with the transmission. The engine should be supported safely on the front crossmember.
24 Fit a retaining plate to hold the torque converter in full engagement with the oil seal, as described in Chapter 1, Section 21.
25 Refitting is a reversal of removal, but observe the following points.
26 Check that the positioning dowels are in place before offering the transmission to the engine, also grease the recess in the crankshaft which accepts the torque converter.
27 Align the converter in relation to the driveplate, as shown in Fig. 7.25.
28 Tighten the driveplate bolts to the specified torque.
29 Tighten the exhaust flange bolts until the spring coils touch then unscrew the bolts through $1^{1}/_{2}$ turns.
30 Reconnect the driveshafts.
31 Check and top up the transmission fluid.

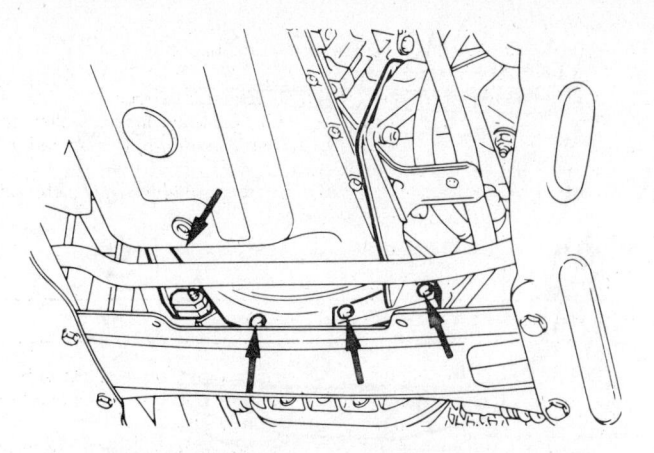

Fig. 7.23 Torque converter cover plate bolts on 1995 cc models – arrowed (Sec 11)

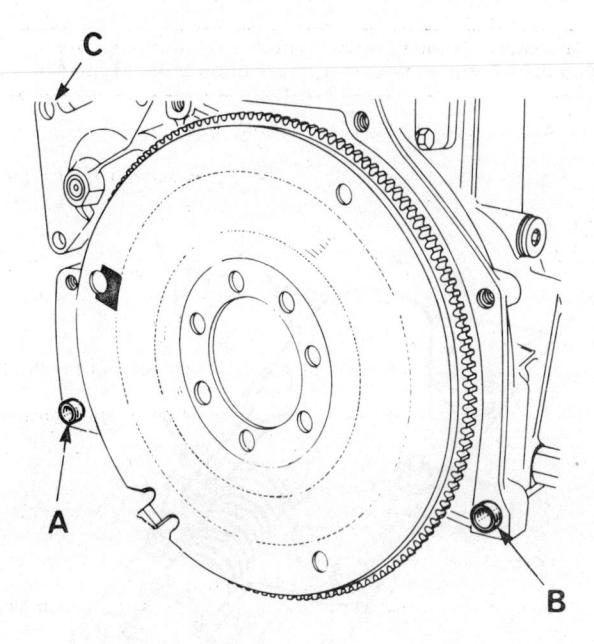

Fig. 7.24 Locating dowels (Sec 11)

A Engine locating dowel   C Starter motor locating dowel
B Engine locating dowel

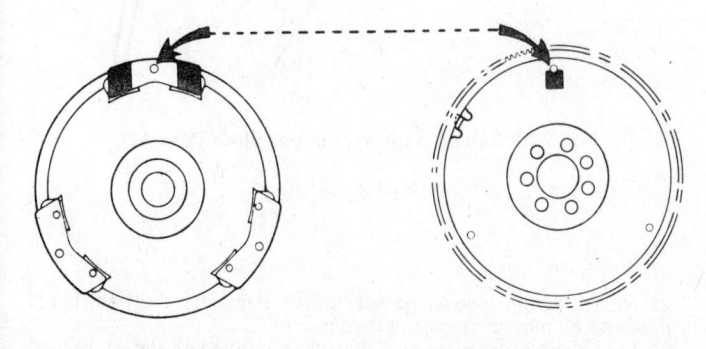

Fig. 7.25 Converter-to-driveplate alignment (Type MJ3) (Sec 11)

12 Automatic transmission (Type 4141) – removal and refitting

Note: *A trolley jack will be needed for this operation*

1 It is not essential to drain the transmission before removal, but it will reduce weight.
2 Drain the separate final drive oil.
3 Disconnect the battery.
4 Release the computer and speedometer wiring from the body clips.
5 Disconnect the fluid cooler pipes from the cooler.
6 Disconnect the vacuum pipe from the inlet manifold.
7 Disconnect the scuttle drain pipes.
8 Unbolt the power steering pump and tie it to the side of the engine compartment.
9 Disconnect the exhaust downpipes from the manifolds.
10 Remove the transmission dipstick guide tube.
11 Unscrew and remove the torque converter housing-to-engine upper connecting bolts.
12 Remove the speedometer sensor.
13 Unbolt the starter motor and move it forward.
14 Disconnect the driveshafts from the transmission, as described in Chapter 1, Section 20.
15 Remove the engine damper and its bracket. Disconnect the governor cable.
16 Unbolt and remove the cover plate from the lower face of the torque converter housing.
17 Remove the TDC sensor and bracket.
18 Disconnect the speed selector control cable at the transmission end.
19 Disconnect the earth strap from the left-hand transmission mounting.
20 Disconnect the driveplate from the torque converter, as described in Chapter 1, Section 21.
21 Unless the car is over an inspection pit, or up on a car lift, the rear end must now be raised and securely supported so that enough clearance will be provided for the transmission to be withdrawn rearwards and removed from under the car.
22 Support the transmission on a trolley jack, if available, and remove both mountings from the transmission.
23 Unscrew and remove the remaining lower engine/transmission connecting bolts.

Chapter 7 Automatic transmission

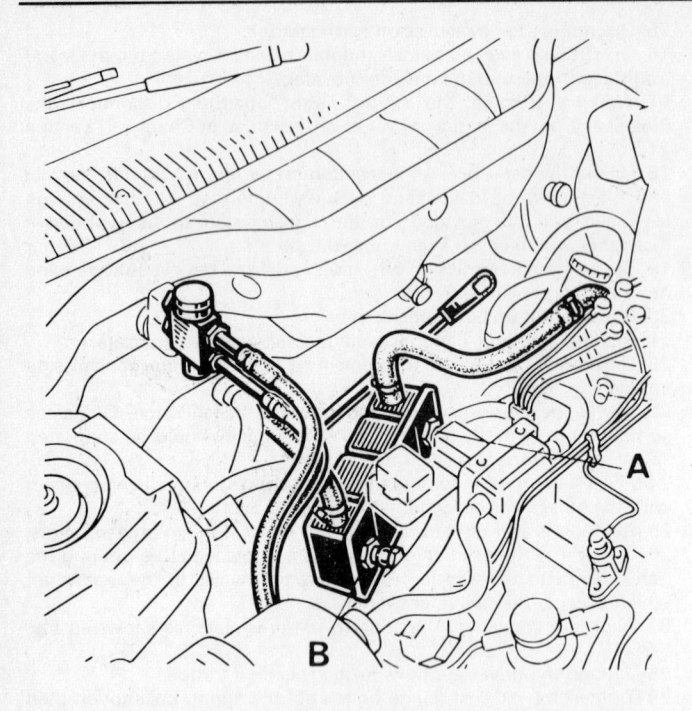

Fig. 7.26 Type 4141 fluid cooler hoses (A and B) (Sec 12)

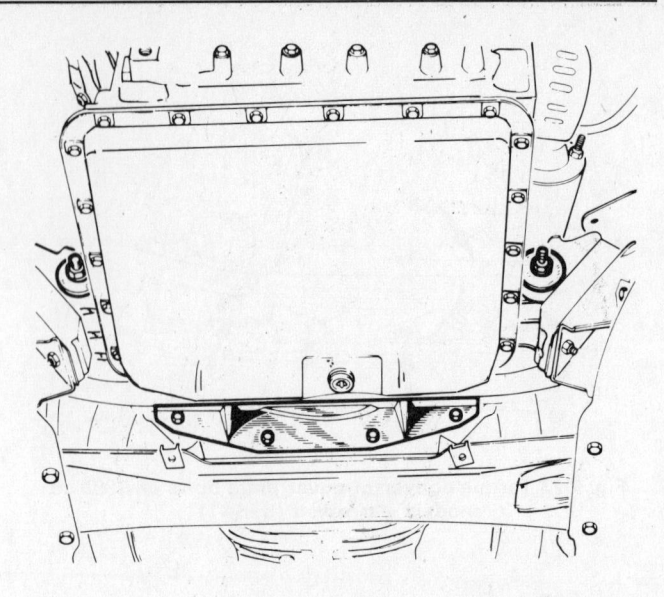

Fig. 7.27 Torque converter cover plate bolts (Type 4141) (Sec 12)

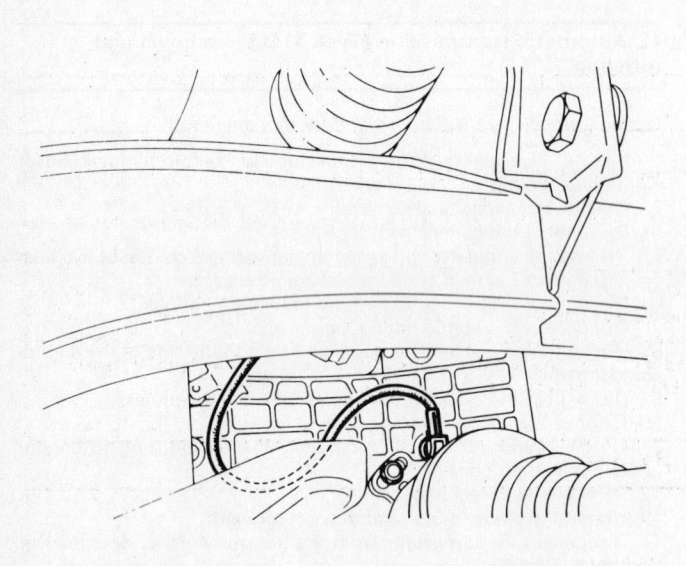

Fig. 7.28 TDC sensor and bracket (Sec 12)

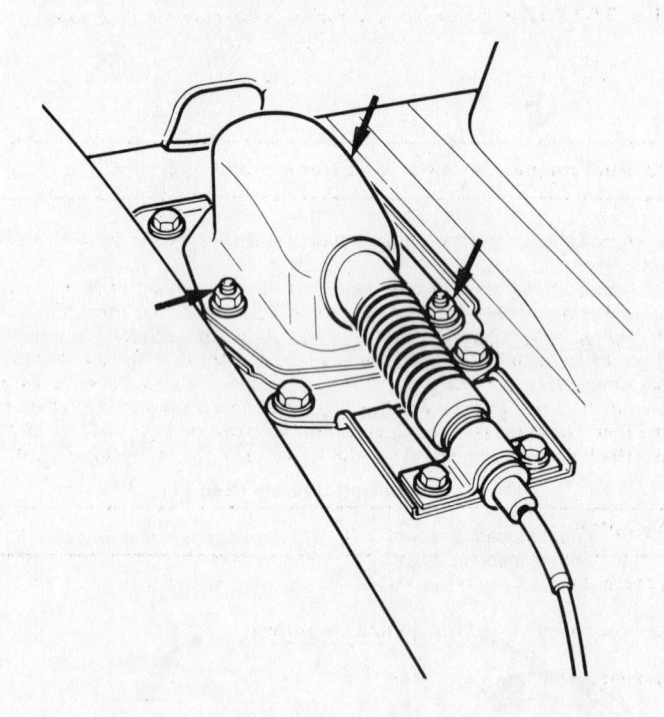

Fig. 7.29 Selector control cover plate (Sec 12)

Nuts arrowed

24 Lower the trolley jack and withdraw the transmission. Guide the computer out with the transmission. The engine should rest safely on the front crossmember.
25 Fit a retaining plate to hold the torque converter in full engagement with the oil seal, as described in Chapter 1, Section 21.
26 Refitting is a reversal of removal, but observe the following points.
27 Make sure that the transmission locating dowels are in position on the engine mating face and that the power steering pump mounting bolt is in place in the transmission before the transmission is offered to the engine, as it cannot be fitted later.
28 When connecting the driveplate to the torque converter, no particular alignment is required but tighten the bolts to the specified torque.

29 Note that the longest socket-headed screw for the right-hand transmission mounting goes at the top.
30 Locate the cover plate over the starter positioning dowel, noting the position of the shortest bolt (E – Fig. 7.31).
31 Tension the power steering pump drivebelt.
32 Adjust the governor cable and selector control, as described in earlier Sections.
33 Refill the final drive and the gearbox (if drained).

Chapter 7 Automatic transmission

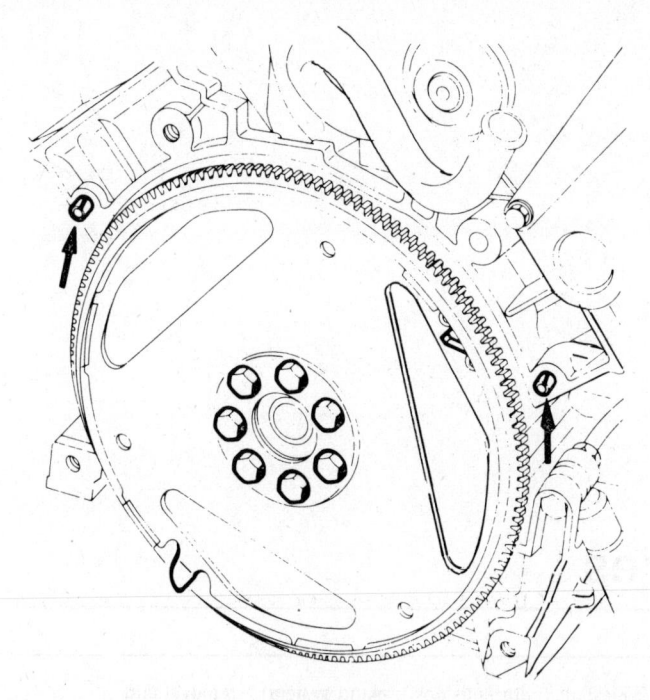

Fig. 7.30 Transmission-to-engine locating dowels – arrowed (Sec 12)

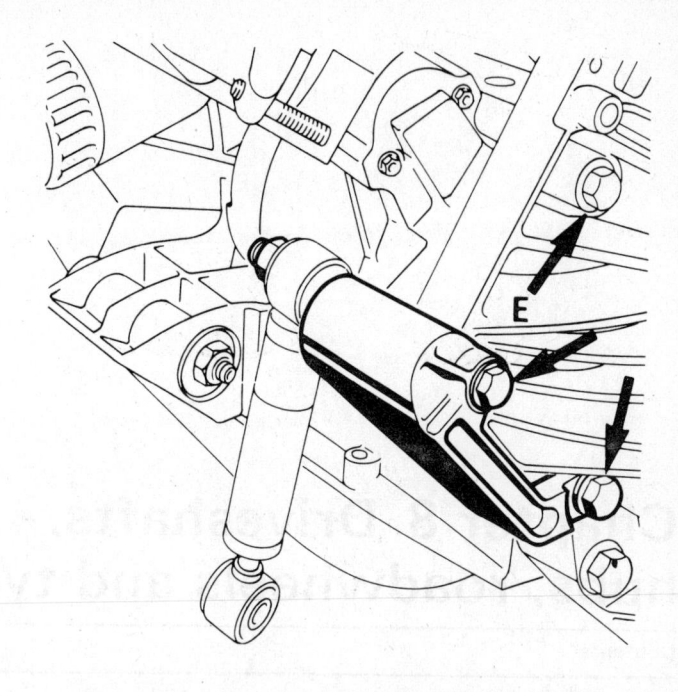

Fig. 7.31 Location of cover plate bolts (Sec 12)

E Shortest bolt

13 Fault diagnosis – automatic transmission

1 Automatic transmission faults are almost always the result of low fluid level or incorrect adjustment of the selector linkage or governor control cable.
2 If these items are checked and found to be in order, the fault is probably internal and should be diagnosed by your Renault dealer who is specially equipped to pinpoint the problem and effect any necessary repairs.
3 Do not allow any defect in the operation of the automatic transmission to go unchecked – it could prove expensive!
4 If the starter fails to function at any time, it is possible that the starter inhibitor switch is at fault but first check that, (a) the selector control lever adjustment is correct, and (b) the transmission wiring harness plugs and socket connections are secure. Check the starter circuit wiring for continuity to the switch plug. The inhibitor switch is situated beneath the transmission governor/computer unit and within the hydraulic section of the unit. The sump plate would have to be removed to gain access to the switch and this is therefore best entrusted to your Renault dealer.
5 The following guide to simple faults should be regarded as being the limit of fault diagnosis for the home mechanic.

Symptom	Reason(s)
Engine stalling or rough idling	Leaking vacuum pipe
Loss of transmission fluid without visible leak	Faulty vacuum unit
Violent gearchanges	Faulty vacuum unit
'Creep' in N	Selector linkage requires adjustment
Slip in D or R	Low fluid level
Incorrect gearchange speeds	Throttle, kickdown switch or load potentiometer requires adjustment
Jerk at speed selection	Fault in vacuum unit or pipe
No drive or slipping during speed range changes	Clogged fluid filter

Chapter 8 Driveshafts, hubs, roadwheels and tyres

Contents

Description and maintenance	1
Driveshaft bellows and joints – general	4
Driveshaft bellows (inboard end, Type GI 76) – renewal	5
Driveshaft bellows (inboard end, Type GI 82) – renewal	6
Driveshaft bellows (outboard end, Type GE 86) – renewal	7
Driveshaft bellows (outboard end, Type RF 95 six-ball) – renewal	8
Driveshaft coupling joints – renewal	9
Driveshaft (with anti-lock braking system) – removal and refitting	3
Driveshaft (without anti-lock braking system) – removal and refitting	2
Fault diagnosis – drivebelts, hubs, roadwheels and tyres	13
Front hub bearings – renewal	10
Rear hub bearings – renewal	11
Wheels and tyres – general care and maintenance	12

Specifications

Driveshafts
Type	Tubular with CV joint at each end
Inboard joint	GI 76 tripod or GI 82 tripod
Outboard joint	GE 86 tripod or RF 95 six-ball
Lubricant capacity:	
Inboard joint	150 g (5.3 oz)
Outboard joint	140 g (4.9 oz)

Hubs
Front	Renewable split track ball-bearing
Rear	Double ball sealed, renewable only complete with hub

Roadwheels
Size:
Four-cylinder models	5^{1}/$_{2}$J x 14
Six-cylinder models	6J x 15
Type	Pressed steel or light alloy

Tyres
Size:
Four-cylinder models	165 R 14 T
	165 HR 14
	185/70 R 14 T
	185/70 HR 14
Six-cylinder models	195/60 HR 15
	205/60 VR 15
Type	Tubeless radial ply

Pressures (cold)*

	Front	Rear
Four-cylinder models	2.1 bar (30 lbf/in^2)	2.2 bar (32 lbf/in^2)
Six-cylinder models:		
Non-turbo	2.0 bar (29 lbf/in^2)	2.2 bar (32 lbf/in^2)
Turbo	2.5 bar (36 lbf/in^2)	2.3 bar (33 lbf/in^2)

* Cars with automatic transmission increase front tyre pressure by 0.1 bar (1.5 lbf/in^2)

Chapter 8 Driveshafts, hubs, roadwheels and tyres

Torque wrench settings	Nm	lbf ft
Driveshaft nut	245	181
Suspension arm upper balljoint nut	64	47
Steering tie-rod end balljoint nut	39	29
Front caliper guide bolt	34	25
Front hub bearing retaining plate screws	15	11
Rear hub nut	156	115
Rear caliper bracket bolts	64	47
Roadwheel bolts:		
4-bolt fixing	88	65
5-bolt fixing	98	72

1 Description and maintenance

1 The driveshafts which transmit power from the transmission final drive to the front roadwheels incorporate a constant velocity (CV) joint at each end.
2 The inboard joint is of tripod type, while the outboard joint may be of tripod or six-ball type, depending upon car model.
3 Maintenance consists of visual inspection of the joint bellows. Any small splits will cause rapid loss of lubricant so faulty bellows must be renewed immediately.

2 Driveshaft (without anti-lock braking system) – removal and refitting

Note: *A balljoint splitter will be needed for this operation.*

1 Raise the front of the car and remove the roadwheel from the side from which the driveshaft is to be removed. Drain the transmission oil.
2 Have an assistant apply the footbrake hard while the hub/driveshaft unit is unscrewed. Alternatively, hold the hub against rotation by making up a lever similar to the one shown in Fig. 8.1
3 Refer to Chapter 9 and unbolt the disc caliper. Tie it up out of the way.
4 Unscrew, but do not remove, the nuts from the suspension arm upper and lower balljoints.
5 Place a jack under the lower suspension arm and then, using a suitable extractor, disconnect the upper balljoint from the eye in the hub carrier.
6 Using the same tool, disconnect the tie-rod end balljoint from the steering arm.
7 Tilt the hub carrier downwards and release the splined end of the driveshaft from it (photo).
8 Using a suitable drift, drive the double roll pins out of the driveshaft at the transmission end.
9 Withdraw the driveshaft (photo).
10 Refitting is a reversal of removal, but observe the following points.
11 Smear the inboard shaft splines with molybdenum disulphide grease.
12 Use new roll pins (photo) and seal their ends with suitable sealant (hard setting type).
13 Apply thread locking fluid to the threads (clean) for the caliper fixing bolts.
14 Tighten all nuts and bolts to the specified torque. Renew the driveshaft/hub self-locking nut.
15 Apply the footbrake two or three times to position the disc pads against the disc.
16 Refill the transmission with oil.

3 Driveshaft (with anti-lock braking system) – removal and refitting

Note: *A balljoint splitter and a driveshaft extractor (T.Av 1050) will be needed for this operation.*

1 On some Turbo models, the driveshafts are bonded into the hubs and their removal requires the use of a special tool. It may be possible to use a large three-legged puller as a substitute.
2 Clean the splines before refitting the driveshaft and apply Loctite Scelbloc.

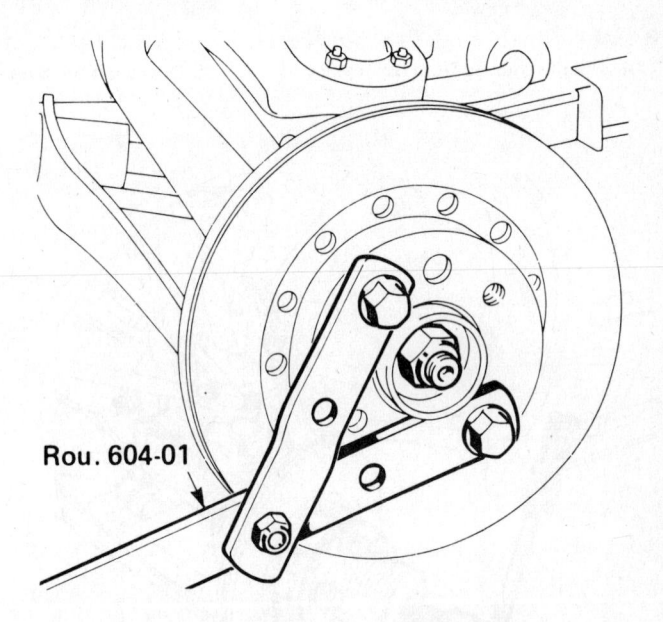

Fig. 8.1 Tool for holding front hub against rotation (Sec 2)

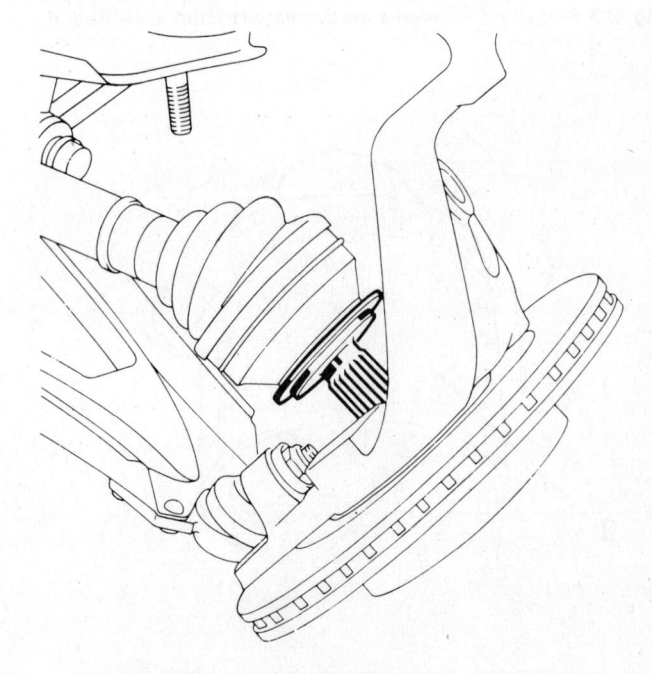

Fig. 8.2 Disconnecting driveshaft from hub carrier (Sec 2)

226 Chapter 8 Driveshafts, hubs, roadwheels and tyres

2.7 Driveshaft removed from hub carrier

2.9 Driveshaft removed from transmission

2.12 Driving in new driveshaft roll pins

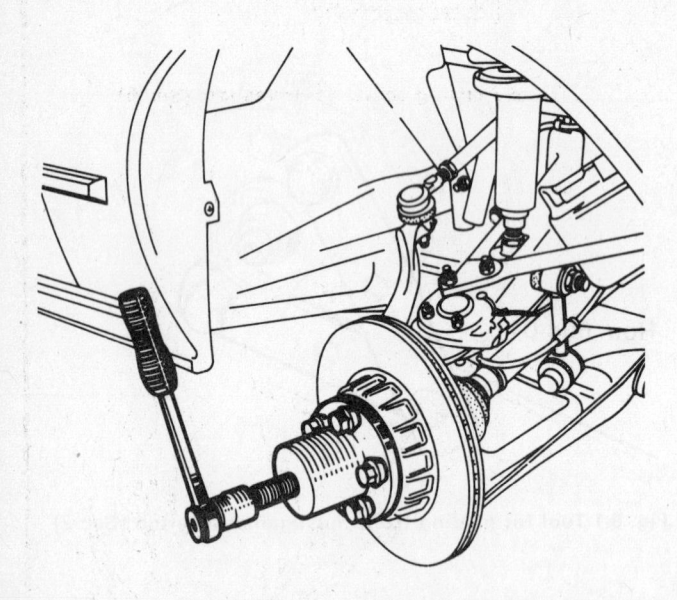

Fig. 8.3 Releasing bonded type driveshaft from hub (Sec 3)

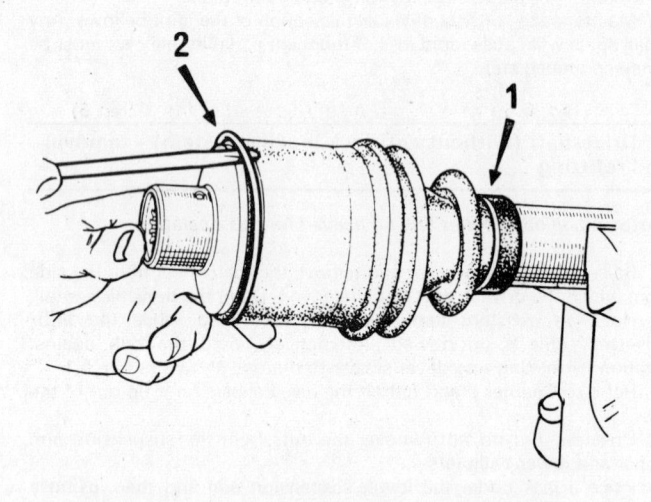

Fig. 8.4 Driveshaft bellows ring (1) and spring clip (2) (Sec 5)

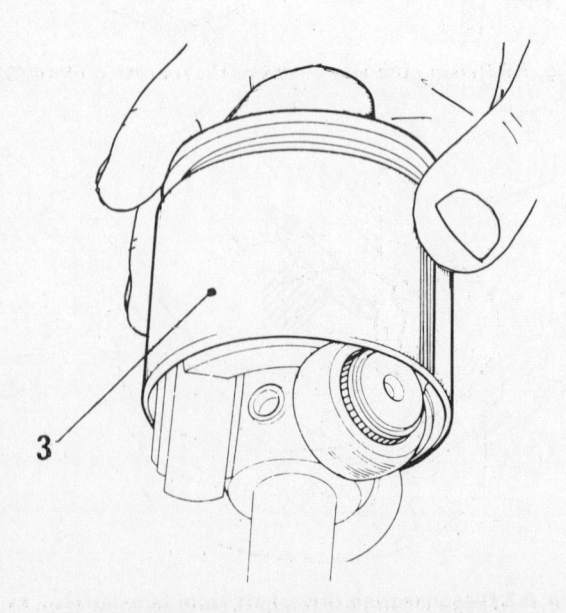

Fig. 8.5 Removing yoke (3) from spider (Sec 5)

3 On later models the driveshaft incorporates helical splines; bonding no longer being required.
4 The operations are otherwise as described in Section 2.

4 Driveshaft bellows and joints – general

1 The driveshaft joint bellows vary in size depending upon the particular joint which they protect.
2 The original bellows should therefore be removed and taken as a pattern to the supplier so that the replacement can be matched with it.
3 The different types of driveshaft joints are fitted in a random manner during production of the vehicle. Inspection after dismantling will identify the particular type fitted.
4 Always check the availability of individual driveshaft joints before dismantling. Apart from the spider on Type GI 76 and GI 82 couplings, Renault supply only complete driveshafts as replacements. It is possible to obtain pattern parts as replacements for the other joints.

5 Driveshaft bellows (inboard end, Type GI 76) – renewal

1 Remove the driveshaft, as described in an earlier Section.
2 Grip the driveshaft in the jaws of a vice and remove the bellows securing rings.
3 Cut the bellows along their length, and remove and discard them.
4 Withdraw the yoke from the spider.

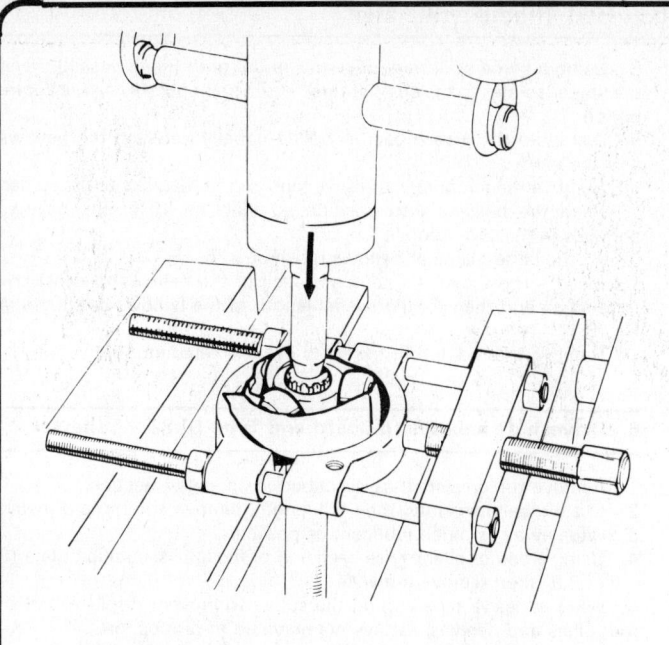

Fig. 8.6 Pressing driveshaft out of spider (Sec 5)

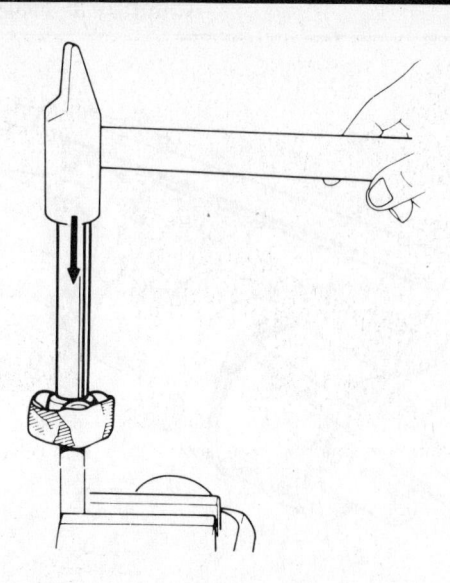

Fig. 8.7 Fitting spider to driveshaft (Sec 5)

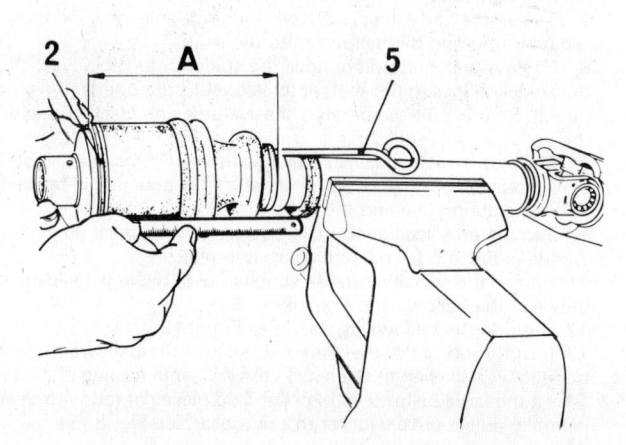

Fig. 8.8 Driveshaft gaiter setting diagram (Type GI 76) (Sec 5)

2 Bellows clip
5 Rod to release air

A = 156.0 mm (6.14 in)

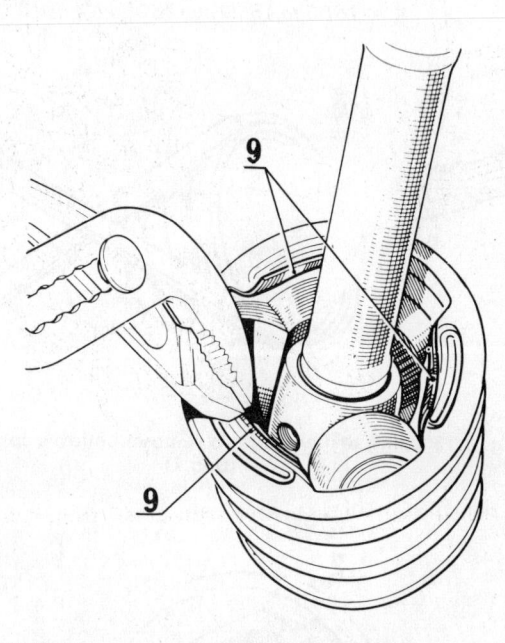

Fig. 8.9 Raising the end of the anti-separation plate (9) (Sec 6)

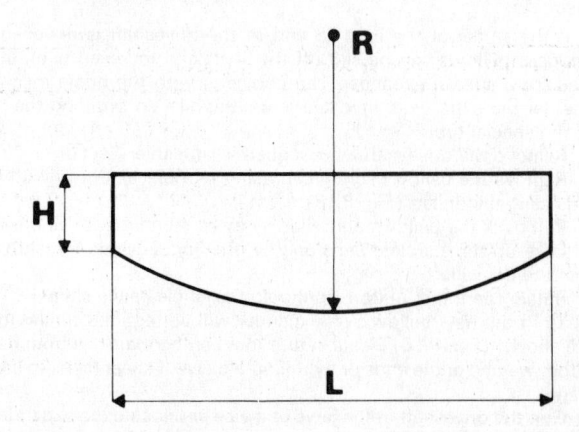

Fig. 8.10 Driveshaft wedge fabrication diagram (Sec 6)

H = 6.0 mm (0.24 in) R = 45.0 mm (1.8 in)
L = 40.0 mm (1.6 in)

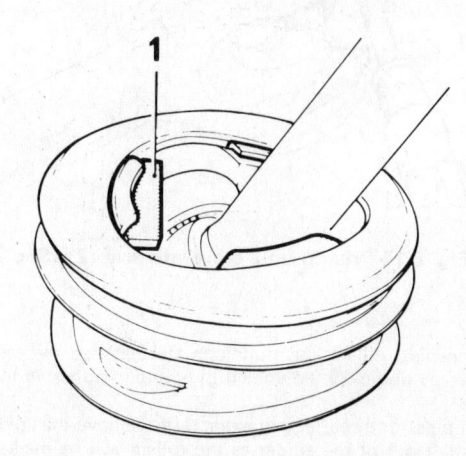

Fig. 8.11 Wedge (1) in position in driveshaft joint (Sec 6)

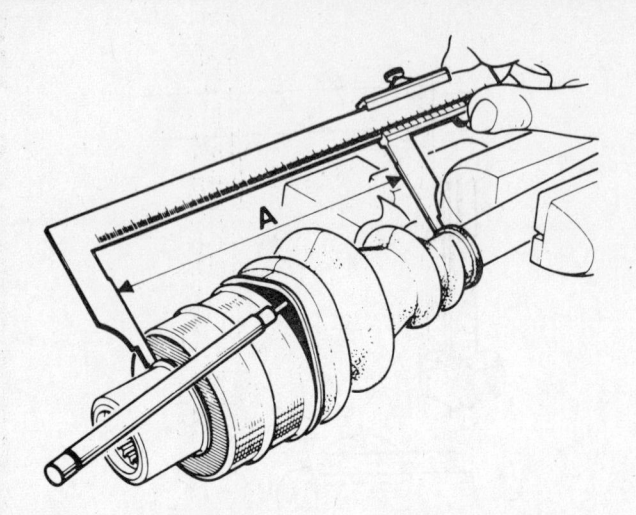

Chapter 8 Driveshafts, hubs, roadwheels and tyres

Fig. 8.12 Bellows positioning diagram (Type GI 82) (Sec 6)

A = 161.0 to 163.0 mm (6.3 to 6.4 in)

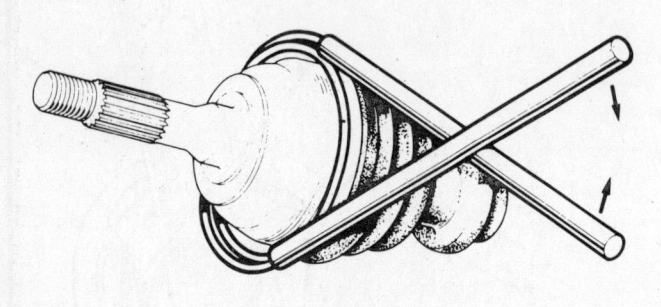

Fig. 8.13 Using drilled rods to remove bellows spring clip (Sec 7)

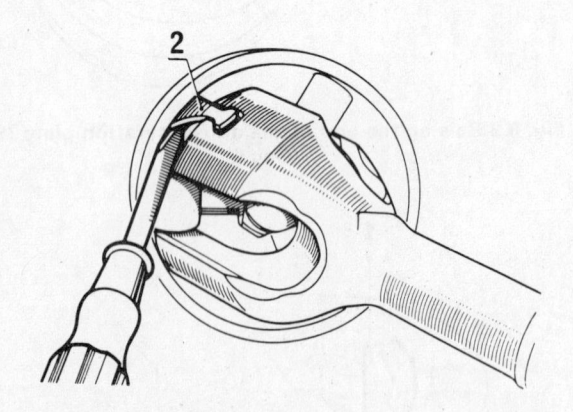

Fig. 8.14 Prising up a starplate arm (2) (Sec 7)

5 The needles, rollers and trunnions are matched in production so keep them as originally assembled by winding adhesive tape around the spider.
6 Use a press or three-legged extractor to remove the spider. Do not tap the shaft out of the spider as the rollers will be marked and the needles damaged.
7 Smear the driveshaft with oil and slide the new bellows small securing ring and bellows onto the shaft.
8 Using a piece of tubing, drive the spider onto the driveshaft. Peen the spider splines to the shaft at three equidistant points using a centre punch.
9 Distribute the grease pack supplied equally between the bellows and the yoke.
10 Remove the temporary adhesive tape and fit the yoke to the spider.
11 Push the bellows into position so that the lip of the bellows engages in the yoke groove.
12 Fit the large diameter bellows retaining clip.
13 Insert a rod under the narrower neck of the bellows to expel any trapped air and then set the overall length of the bellows as shown in Fig. 8.8.
14 Remove the rod and fit the small bellows retaining ring.

6 Driveshaft bellows (inboard end Type GI 82) – renewal

1 Remove the driveshaft as described in an earlier Section.
2 Cut the bellows along their full length, remove and discard them.
3 Wipe away as much lubricant as possible.
4 Using a pair of pliers, raise each end of the anti-separation plate (9 – Fig. 8.9) then remove the yoke.
5 Wind adhesive tape around the spider to prevent displacement of the rollers and needles, as they are matched in production.
6 Use a press or a three-legged extractor to remove the spider. Do not tap the shaft out of the spider as the rollers will be marked and the needles damaged.
7 Smear the driveshaft with oil and slide the new bellows small securing ring and the bellows onto the shaft.
8 Using a piece of tubing, drive the spider onto the driveshaft. Peen the spider splines to the shaft at three equidistant points using a centre punch. Remove the temporary adhesive tape and insert the spider into the yoke.
9 Make up a wedge in accordance with the dimensions shown in Fig. 8.10 from 2.5 mm (0.098 in) sheet metal and tap it between the anti-separation plate and the yoke.
10 Tap the anti-separation plate into its original position, remove the wedge and use it for the remaining two plates.
11 Spread the sachet of grease supplied with the new bellows equally between the bellows and the yoke.
12 Position the bellows as shown in Fig. 8.12.
13 Insert a rod under the narrower neck of the bellows to expel any trapped air and then fit the small bellows retaining ring.
14 Fit the large bellows spring clip. Two pieces of rod with their ends centrally drilled are useful for this purpose (see Fig. 8.13).

7 Driveshaft bellows (outboard end, Type GE 86) – renewal

Note: *A bellows expander (T.Av 537-02) will be needed for this operation.*

If the spider at the inboard end of the driveshaft is removed the outboard bellows can be slid off the shaft and renewed without the need for a special expander. The choice is with the home mechanic whether the extra work involved is worthwhile, so avoiding the need for the special tool.

1 Remove the driveshaft as described in an earlier Section.
2 Remove the bellows spring clip using two pieces of rod with their ends centrally drilled (Fig. 8.13).
3 Pull back the bellows and wipe away as much grease as possible.
4 Prise up the starplate arms one by one and separate the stub axle from the driveshaft.
5 Retain the thrust button, spring and axial clearance shim.
6 To fit the new bellows, an expander will be required similar to the one shown (Fig. 8.16). A substitute tool can be made up from pieces of rod, welded at the ends or by rolling a piece of sheet metal to a cone shape.
7 Grip the driveshaft in the jaws of a vice and locate the expander on the end of the yoke.
8 Lubricate the expander and inside of the bellows and slide the bellows up the expander, making sure that the first fold is fully extended. Pull the bellows half way up the expander two or three times to stretch the rubber and then pull them right over the shaft yoke.
9 If the driveshaft has an intermediate bush, locate the bush in the shaft grooves.

229

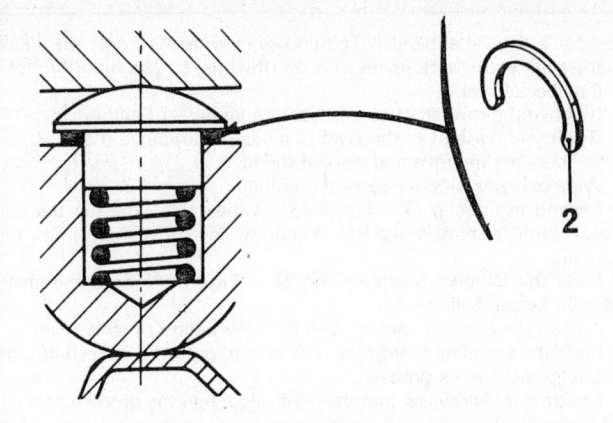

Fig. 8.15 Type GE 86 driveshaft joint thrust button, spring and shim (2) (Sec 7)

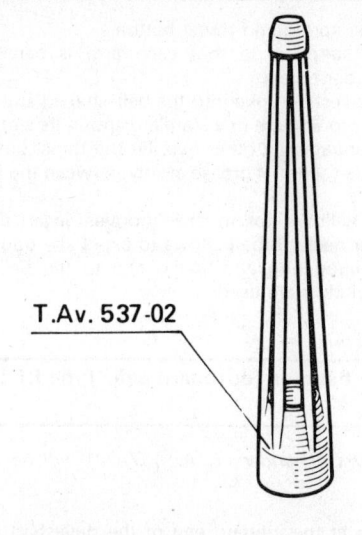

Fig. 8.16 Bellows expander fitting tool (Sec 7)

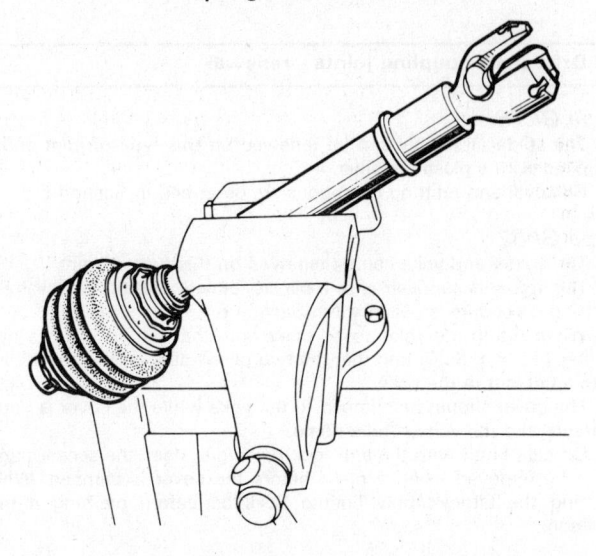

T.Av. 537-02

Fig. 8.17 Driveshaft secured in vice (Sec 7)

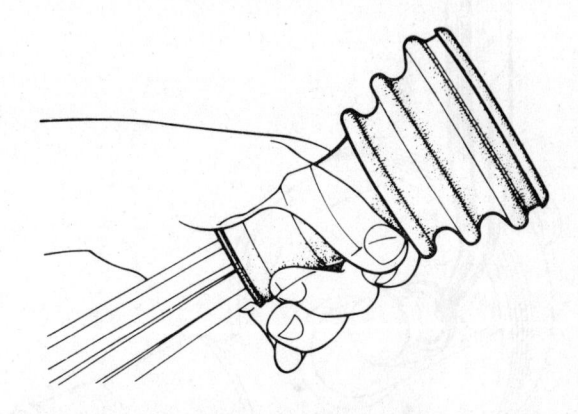

Fig. 8.18 Working bellows up expander tool (Sec 7)

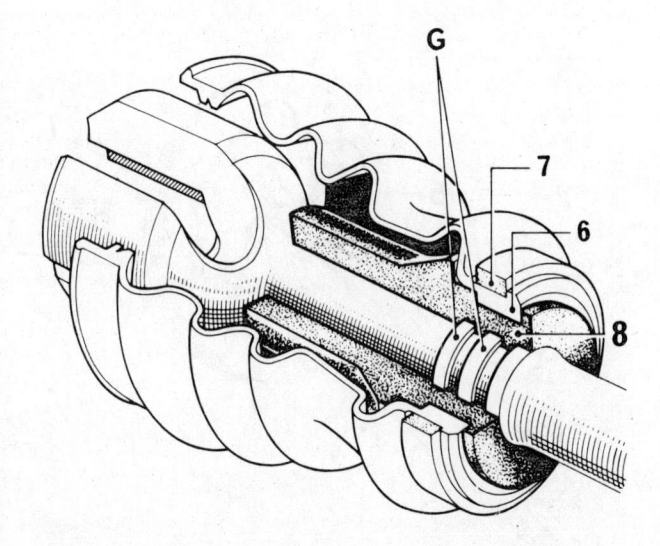

Fig. 8.19 Driveshaft intermediate bush (Sec 7)

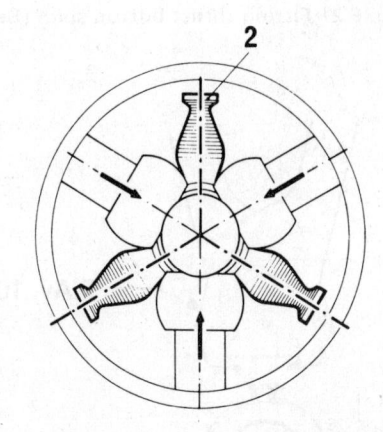

Fig. 8.20 Starplate (2) correctly located (Sec 7)

G   Shaft ribs
6   Bellows
7   Bellows retaining ring
8   Intermediate bush

Chapter 8 Driveshafts, hubs, roadwheels and tyres

10 Fit the spider spring and thrust button.
11 Locate the starplate so that each arm is centrally positioned between the spider rollers.
12 Insert the driveshaft yoke into the bell-shaped stub axle.
13 Tilt the shaft to engage one starplate arm in its slot and then use a screwdriver to engage the other two. Fit the thrust button shim.
14 Distribute the sachet of grease evenly between the bellows and the stub axle.
15 Locate the bellows lips in their grooves, insert a rod under the smaller diameter neck of the bellows to expel any trapped air.
16 Fit the retaining ring and spring clip to the bellows. On some models, crimped clips are used.

8 Driveshaft bellows (outboard end, Type RF 95 six-ball) – renewal

Note: A bellows expander (T. Av 537-02) will be needed for this operation.

If the spider at the inboard end of the driveshaft is removed, the outboard bellows can be slid off the shaft and renewed without the need for a special expander. The choice is with the home mechanic whether the extra work involved is worthwhile, so avoiding the need for the special tool.

1 Remove the driveshaft as described in an earlier Section.
2 Grip the driveshaft in the jaws of a vice and remove the clips, cut the bellows and remove and discard them.
3 Wipe away as much grease as possible.
4 Expand the circlip (7 – Fig. 8.23) at the same time tap the ball hub(s) with a plastic mallet. Withdraw the driveshaft from the coupling.
5 Slide the bellows retaining ring (A – Fig. 8.24) onto the shaft, followed by the bellows (3).
6 Spread the grease supplied with the new bellows inside them.
7 Slide the coupling, complete with circlip, onto the driveshaft until the circlip locks in its groove.
8 Locate the bellows so that their lips engage in the grooves of shaft and stub axle.
9 Fit the bellows retaining ring and the spring clip using drilled rods (shown in Fig. 8.13).

9 Driveshaft coupling joints – renewal

Type GI 76
1 The spider assembly can be renewed on this type of joint and is supplied with a plastic retainer.
2 Removal and refitting operations are described in Section 5.

Type GI 82
3 The spider and yoke can be renewed on this type of joint.
4 The spider is supplied with a plastic retainer and the removal and refitting procedure is described in Section 6.
5 When fitting the yoke cover, make sure that the yoke has a new O-ring (3 – Fig. 8.26) and that the two raised pips on the cover align with a cut-out in the yoke.
6 The cover should be crimped to the yoke while the cover is under pressure and the yoke is fully home.
7 On cars fitted with the anti-lock braking system, the sensor target must be removed using a press before the cover is crimped. When refitting the target, apply Loctite Scelbloc before pressing it into position.

Type GE 86
8 The joint components may be renewed after dismantling as described in Section 7.

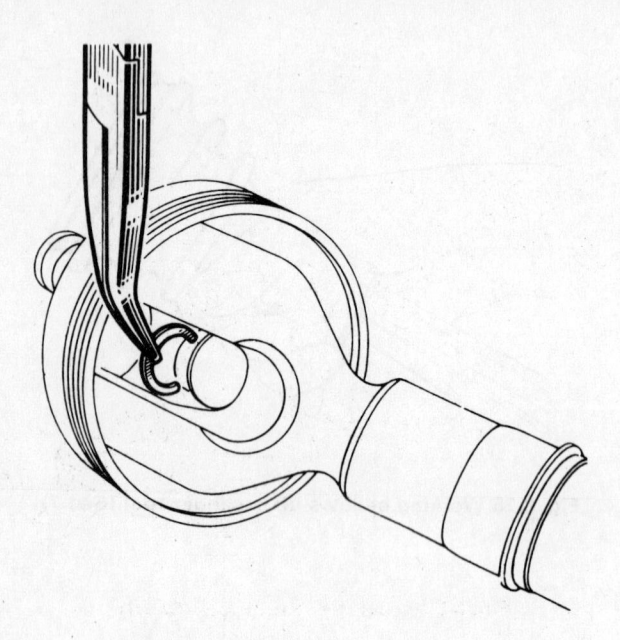

Fig. 8.21 Fitting thrust button shim (Sec 7)

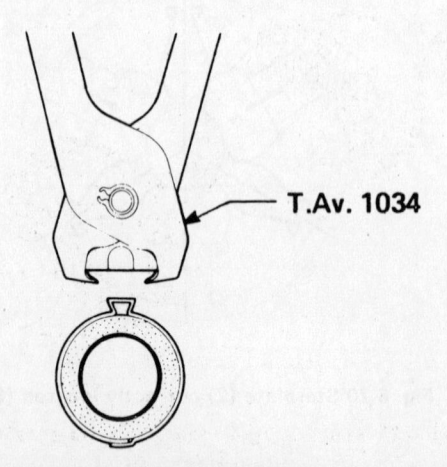

Fig. 8.22 Crimped type driveshaft bellows clip and securing tool (Sec 7)

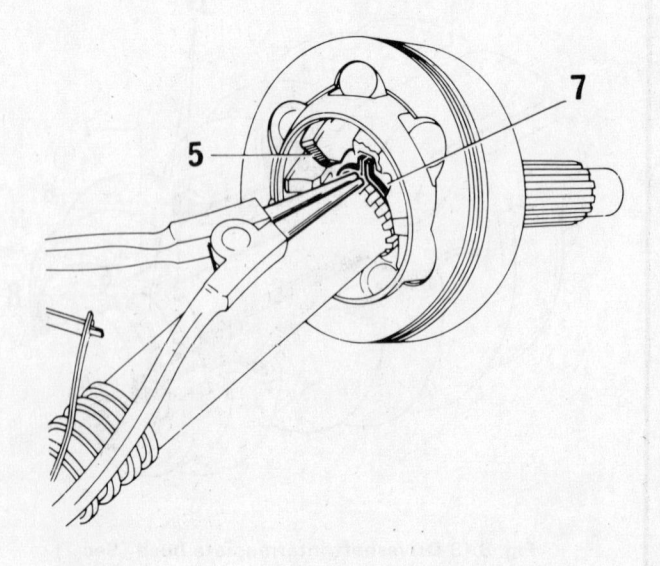

Fig. 8.23 Expanding circlip (7) in six-ball type coupling (Sec 8)

5 Ball hub face

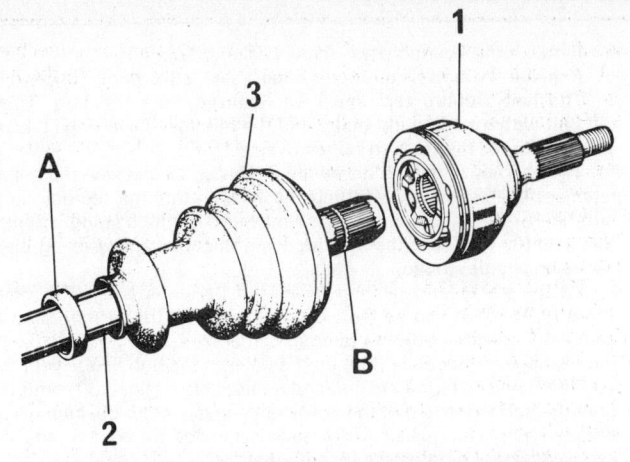

Fig. 8.24 Six-ball type driveshaft coupling (Sec 8)

1 Six-ball coupling
2 Driveshaft
3 Bellows
A Bellows rubber retaining ring
B Circlip groove

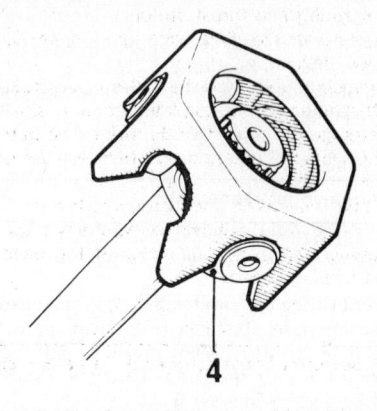

Fig. 8.25 Spider retainer (4) on Type GI 76 driveshaft joint (Sec 9)

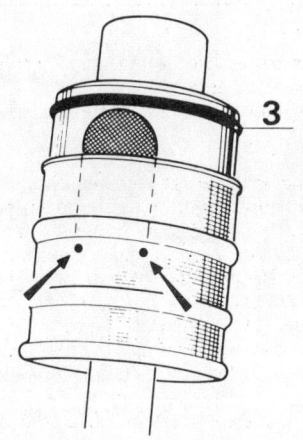

Fig. 8.26 Yoke cover on Type GI 82 joint (Sec 9)

3 O-ring seal
Align the pips (arrowed) with the cut-out

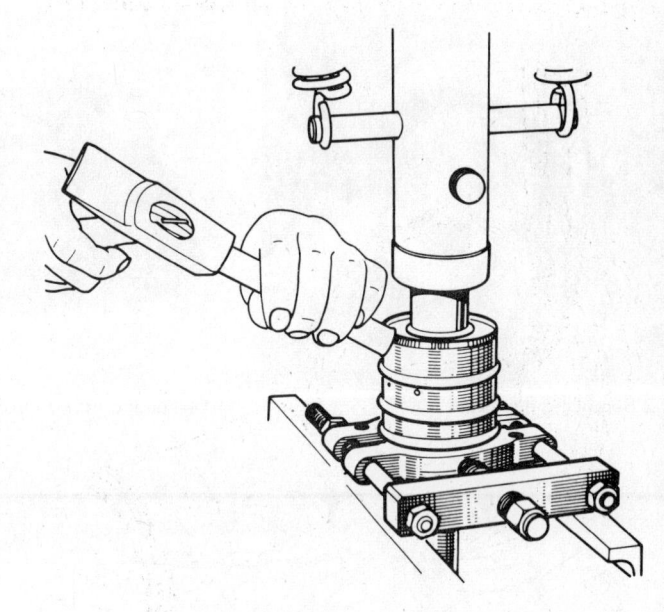

Fig. 8.27 Crimping joint cover (Sec 9)

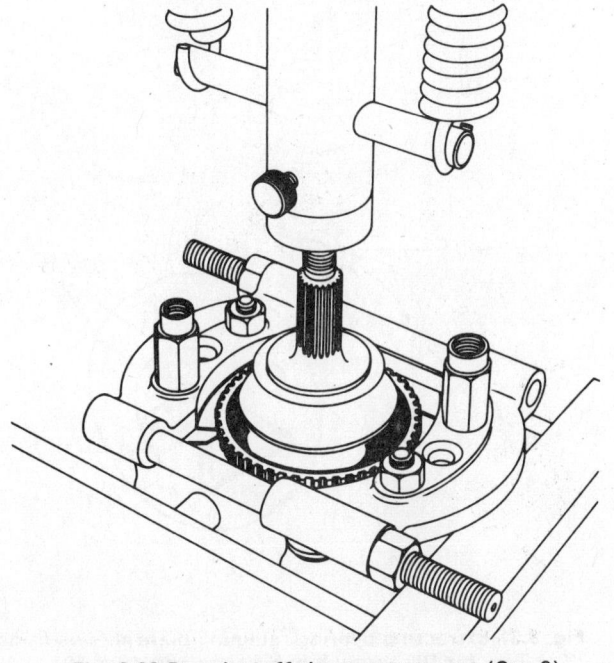

Fig. 8.28 Pressing off the sensor target (Sec 9)

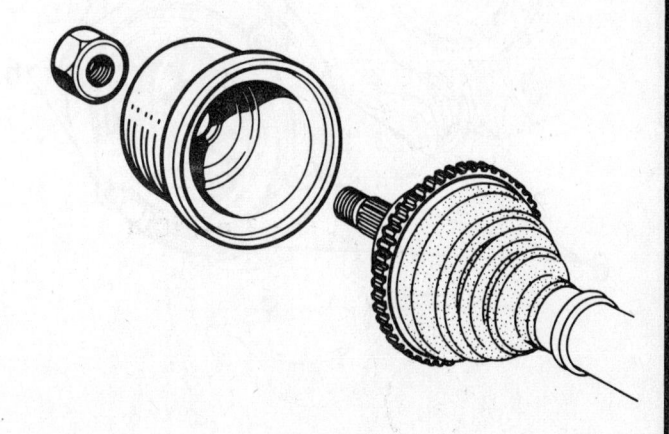

Fig. 8.29 Sensor target located on driveshaft joint (Sec 9)

Chapter 8 Driveshafts, hubs, roadwheels and tyres

Type RF 95 (six-ball)

9 The driveshaft or the stub axle/coupling may be renewed after separation, as described in Section 8.
10 Dismantling of the ball cage serves no purpose as individual spare parts are not available.

10 Front hub bearings – renewal

Note: *A suitable-sized Torx bit will be needed for this operation.*

1 Raise the front of the car and remove the roadwheel.
2 Have an assistant apply the brake pedal hard and then unscrew the driveshaft/hub nut (photo). Alternatively, hold the hub against rotation by bolting a lever to the hub as shown in Fig. 8.1. Remove the brake disc as described in Chapter 9.
3 One of two methods may be used to dismantle the hub.
4 Using two roadwheel bolts and two steel packing pieces (photo), screw in the bolts gradually and evenly to force the hub from the bearing retaining plate.
5 Remove the bearing plate screws using a Torx bit.
6 Remove the hub bearing and inner track half-section (photo).
7 The half-section track must be removed from the hub. This is difficult without a bearing puller and the hub may have to be taken to your dealer for the track to be removed.
8 The alternative way to dismantle the hub is to unscrew the bearing retainer plate screws by inserting the Torx bit through the hole in the hub flange. The hub can then be removed with the bearing assembly. Separate the hub from the bearing using the bolts and packing pieces as previously described.
9 Fit the tracks and bearing races and pack the balls with grease (photo). The hub can be tapped or pressed into the bearing, and the complete assembly fitted to the hub carrier by inserting the screws for the bearing retainer plate through the holes in the hub flange (photos).
10 Alternatively, fit the bearing and retainer plate (fig. 8.30), and then offer the hub to the end of the driveshaft and tap it into position using a wooden or plastic mallet. Once several threads of the driveshaft are exposed, screw on a new hub/driveshaft nut.
11 Refit the disc and caliper (Chapter 9). Continue screwing on the self-locking nut and tighten to the specified torque.
12 Refit the roadwheel and lower the car.

10.2 Removing driveshaft nut

10.4 Forcing off hub from bearing retainer

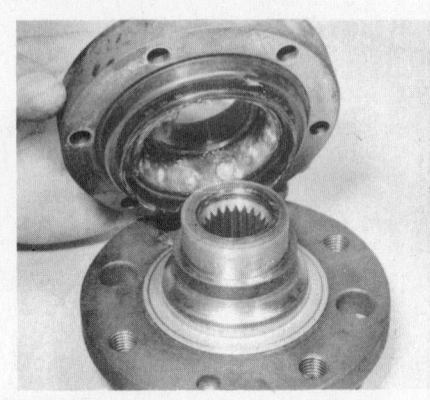

10.6 Bearing and hub separated

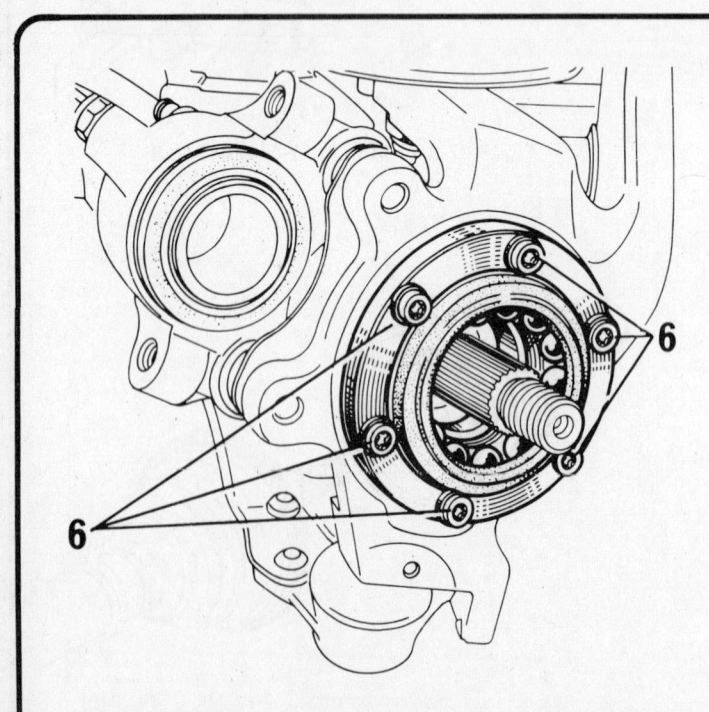

Fig. 8.30 Bearing retaining plate screws (6) (Sec 10)

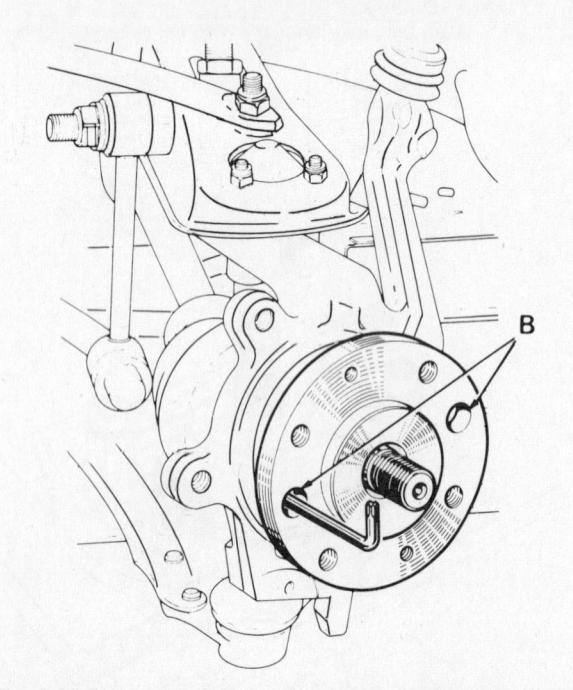

Fig. 8.31 Extracting bearing retaining plate screws using Torx bit (B) through hub flange hole (Sec 10)

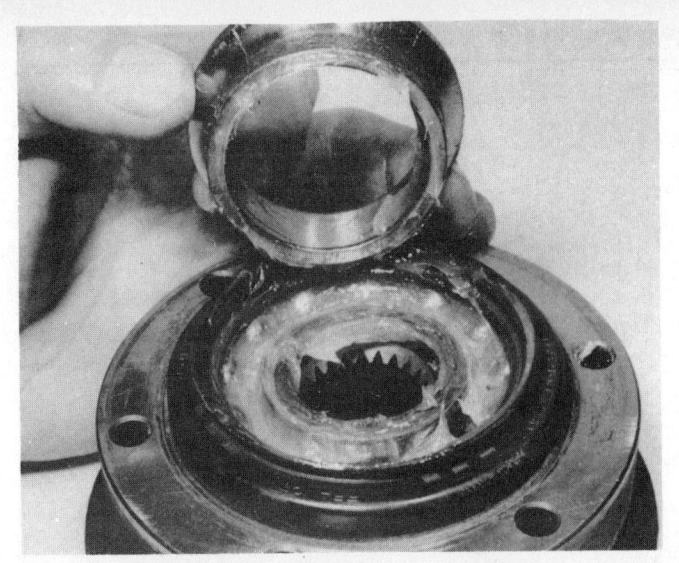

10.9A New bearing inner track

10.9B Tapping hub and bearing together

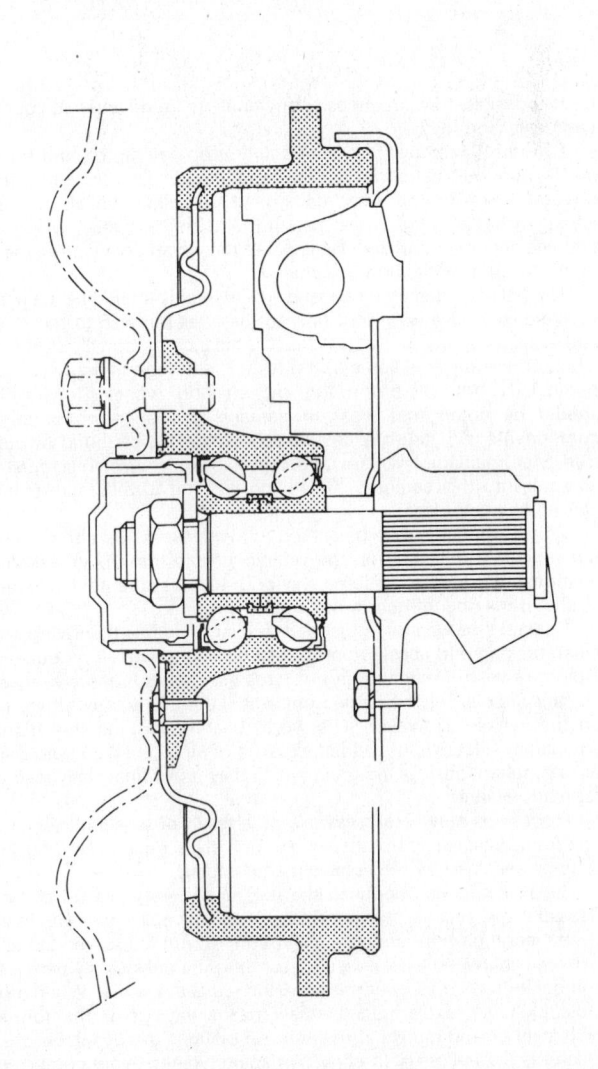

Fig. 8.32 Sectional view of rear hub bearings on four-cylinder models (Sec 11)

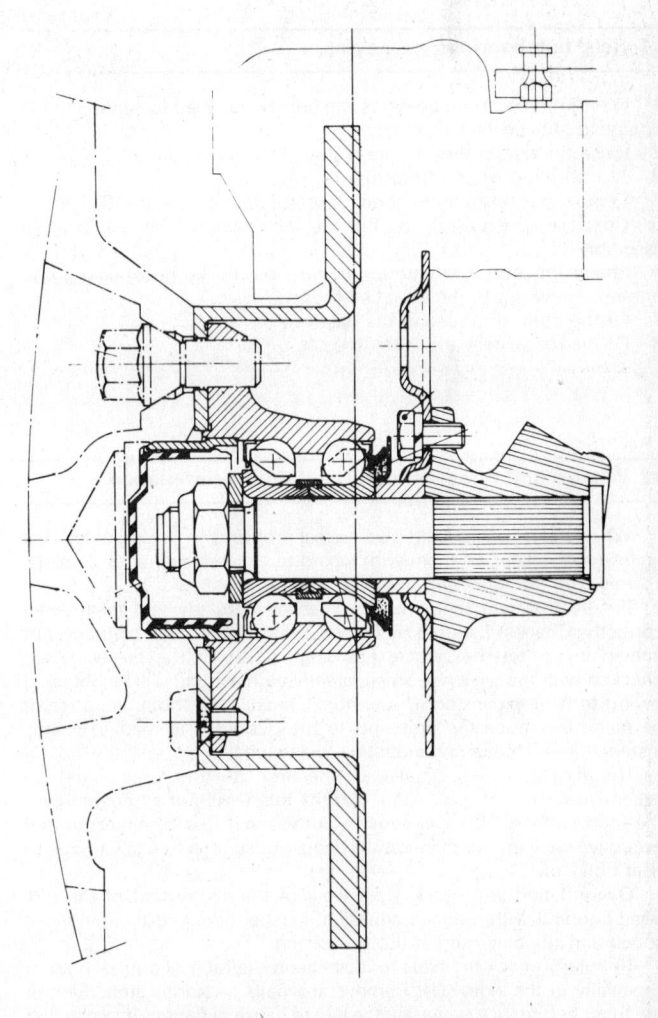

Fig. 8.33 Sectional view of rear hub bearings on six-cylinder models (Sec 11)

234 Chapter 8 Driveshafts, hubs, roadwheels and tyres

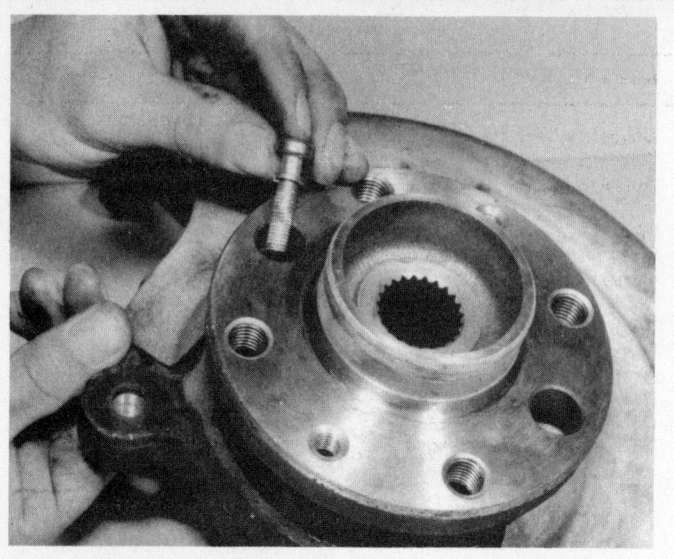

10.9C Fitting bearing retainer plate screw

11.3 Rear hub dust cap

11 Rear hub bearings – renewal

1 Wear in the rear hub bearings can only be rectified by renewing the complete hub/bearing assembly.
2 Raise the rear of the car and remove the roadwheel.
3 Tap off the dust cap (photo).
4 Remove the brake drum or disc, according to model, see Chapter 9.
5 Unscrew and remove the hub nut and remove the hub/bearing assembly.
6 Smear the stub axle with a little gear oil, fit the hub/bearing and tighten a new nut to the specified torque.
7 Fit the drum or disc, and the dust cap.
8 Fit the roadwheel and lower the car.

12 Wheels and tyres – general care and maintenance

Wheels and tyres should give no real problems in use provided that a close eye is kept on them with regard to excessive wear or damage. To this end, the following points should be noted.

Ensure that tyre pressures are checked regularly and maintained correctly. Checking should be carried out with the tyres cold and not immediately after the vehicle has been in use. If the pressures are checked with the tyres hot, an apparently high reading will be obtained owing to heat expansion. Under no circumstances should an attempt be made to reduce the pressures to the quoted cold reading in this instance, or effective underinflation will result.

Underinflation will cause overheating of the tyre owing to excessive flexing of the casing, and the tread will not sit correctly on the road surface. This will cause a consequent loss of adhesion and excessive wear, not to mention the danger of sudden tyre failure due to heat build-up.

Overinflation will cause rapid wear of the centre part of the tyre tread coupled with reduced adhesion, harsher ride, and the danger of shock damage occurring in the tyre casing.

Regularly check the tyres for damage in the form of cuts or bulges, especially in the sidewalls. Remove any nails or stones embedded in the tread before they penetrate the tyre to cause deflation. If removal of a nail *does* reveal that the tyre has been punctured, refit the nail so that its point of penetration is marked. Then immediately change the wheel and have the tyre repaired by a tyre dealer. Do *not* drive on a tyre in such a condition. In many cases a puncture can be simply repaired by the use of an inner tube of the correct size and type. If in any doubt as to the possible consequences of any damage found, consult your local tyre dealer for advice.

Periodically remove the wheels and clean any dirt or mud from the inside and outside surfaces. Examine the wheel rims for signs of rusting, corrosion or other damage. Light alloy wheels are easily damaged by 'kerbing' whilst parking, and similarly steel wheels may become dented or buckled. Renewal of the wheel is very often the only course of remedial action possible.

The balance of each wheel and tyre assembly should be maintained to avoid excessive wear, not only to the tyres but also to the steering and suspension components. Wheel imbalance is normally signified by vibration through the vehicle's bodyshell, although in many cases it is particularly noticeable through the steering wheel. Conversely, it should be noted that wear or damage in suspension or steering components may cause excessive tyre wear. Out-of-round or out-of-true tyres, damaged wheels and wheel bearing wear/maladjustment also fall into this category. Balancing will not usually cure vibration caused by such wear.

Wheel balancing may be carried out with the wheel either on or off the vehicle. If balanced on the vehicle, ensure that the wheel-to-hub relationship is marked in some way prior to subsequent wheel removal so that it may be refitted in its original position.

General tyre wear is influenced to a large degree by driving style – harsh braking and acceleration or fast cornering will all produce more rapid tyre wear. Interchanging of tyres may result in more even wear, but this should only be carried out where there is no mix of tyre types on the vehicle. However, it is worth bearing in mind that if this is completely effective, the added expense of replacing a complete set of tyres simultaneously is incurred, which may prove financially restrictive for many owners.

Front tyres may wear unevenly as a result of wheel misalignment. The front wheels should always be correctly aligned according to the settings specified by the vehicle manufacturer.

Legal restrictions apply to the mixing of tyre types on a vehicle. Basically this means that a vehicle must not have tyres of differing construction on the same axle. Although it is not recommended to mix tyre types between front axle and rear axle, the only legally permissible combination is crossply at the front and radial at the rear. When mixing radial ply tyres, textile braced radials must always go on the front axle, with steel braced radials at the rear. An obvious disadvantage of such mixing is the necessity to carry two spare tyres to avoid contravening the law in the event of a puncture.

In the UK, the Motor Vehicles Construction and Use Regulations apply to many aspects of tyre fitting and usage. It is suggested that a copy of these regulations is obtained from your local police if in doubt as to the current legal requirements with regard to tyre condition, minimum tread depth, etc.

13 Fault diagnosis – driveshafts, hubs, roadwheels and tyres

Symptom	Reason(s)
Vibration	Driveshaft bent Worn CV joints Out of balance roadwheels
'Clonk' on taking up drive or on the over run	Worn CV joints Worn splines on driveshaft, hub or differential side gears Loose driveshaft nut or wheel bolts
Noise or roar when cornering	Worn hub bearings Loose driveshaft/hub vent
Rapid or uneven wear of tyre tread	Out of balance roadwheels Distorted roadwheel Incorrect front wheel alignment

Chapter 9 Braking system

Contents

Anti-lock braking system – description	23	Maintenance and inspection	2
Anti-lock braking system – maintenance and testing	24	Master cylinder – removal and refitting	12
Brake disc – inspection, removal and refitting	7	Pressure regulating valve	13
Brake drum – inspection and renovation	11	Rear brake shoes (Bendix type) – inspection and renewal	8
Brake hydraulic system – bleeding	15	Rear brake shoes (Girling type) – inspection and renewal	9
Brake pedal – removal and refitting	22	Rear disc brake caliper – removal, overhaul and refitting	6
Description	1	Rear disc pads – inspection and renewal	4
Fault diagnosis – braking system	26	Rear wheel cylinder – removal, overhaul and refitting	10
Front disc brake caliper – removal, overhaul and refitting	5	Stop-lamp switch-adjustment	25
Front disc pads – inspection and removal	3	Vacuum servo air filter – renewal	17
Handbrake – adjustment	20	Vacuum servo non-return valve – renewal	18
Handbrake cables – renewal	21	Vacuum servo unit-description and testing	16
Hydraulic pipes and hoses – inspection, removal and refitting	14	Vacuum servo unit – removal and refitting	19

Specifications

System type
Four wheel hydraulic, dual circuit. Front disc, rear drum on four-cylinder models. Four wheel disc on six-cylinder models. Servo assistance and load-sensitive pressure regulating valve. Handbrake cable operated to rear wheels. Anti-locking system as factory-fitted option; standard on Turbo versions

Front disc brakes
Type .... Ventilated

	Four-cylinder models	Six-cylinder models
Diameter	259.1 mm (10.2 in)	279.4 mm (11.0 in)
Thickness	20.1 mm (0.79 in)	22.1 mm (0.87 in)
Maximum disc run-out	0.07 mm (0.003 in)	0.07 mm (0.003 in)
Minimum pad wear thickness (including backing)	9.0 mm (0.35 in)	9.0 mm (0.35 in)

Rear disc brakes
Type .... Solid
Diameter .... 254.0 mm (10.0 in)
Thickness .... 11.9 mm (0.47 in)
Maximum disc run-out .... 0.07 mm (0.003 in)
Minimum pad wear thickness (including backing):
    Four-cylinder models .... 5.0 mm (0.20 in)
    Six-cylinder models .... 7.0 mm (0.28 in)

Chapter 9 Braking system

Rear drum brakes
Type .... Leading and trailing shoe with automatic adjuster
Diameter .... 228.6 mm (9.0 in)
Maximum internal diameter after refinishing .... 229.6 mm (9.04 in)

Hydraulic units
Front caliper cylinder diameter:
    Four-cylinder models .... 54.0 mm (2.13 in)
    Six-cylinder models .... 60.0 mm (2.4 in)
Rear wheel cylinder diameter .... 22.0 mm (0.87 in)
Rear caliper cylinder diameter .... 36.0 mm (1.42 in)

Vacuum servo unit
Diameter .... 203.2 mm (8.0 in)

Torque wrench settings

	Nm	lbf ft
Master cylinder-to-servo bolts	12	9
Servo unit to bulkhead	19	14
Caliper bracket bolts	64	47
Caliper guide bolt	34	25
Hose and pipeline unions	12	9
Anti-lock brake sensor screws:		
Front	30	22
Rear	15	11

1 Description

The braking system is of four wheel dual circuit hydraulic type. The circuits are split diagonally; one front wheel and one diagonally opposite rear wheel.

On four-cylinder models, the front brakes are of ventilated disc type with automatically adjusted drum brakes at the rear.

On six-cylinder models four wheel disc brakes are fitted, the front ones being of ventilated type with solid ones at the rear.

All models have servo assistance and a load-sensitive pressure regulating valve to control rear wheel braking.

The handbrake is cable-operated to the rear wheels.

An anti-lock braking system is available as a factory-fitted option on certain models and is standard on Turbo versions (see Section 23).

2 Maintenance and inspection

1 Regularly check the fluid level in the brake master cylinder reservoir. The need for topping-up should be very infrequent and then only a small quantity should be required to compensate for friction lining wear (photo).
2 The fluid level should never fall below the 'Danger' mark on the translucent reservoir. Any large loss of fluid will be due to a leak in the hydraulic system of either the brake or clutch systems as both use the common reservoir.
3 If a leak cannot be observed, check the area around the master cylinder pushrods inside the car. A leak here will indicate faulty piston seals.
4 When topping-up use only clean fluid which has been stored in an air tight container.
5 At the intervals specified in Routine Maintenance, inspect the disc pads and shoe linings for wear.

3 Front disc pads – inspection and renewal

1 Raise the front of the car and support it securely.
2 Remove the front roadwheels.
3 Inspect the thickness of the friction material on the disc pads (photo). If it has worn below the minimum specified, renew the pads in the following way.
4 Unscrew the caliper guide bolts using a ring spanner and an open-ended spanner to lock the guide pins.
5 Lift off the caliper and support it or tie it up out of the way.
6 Disconnect the sensor wire. Remove the disc pads and discard them.

2.1 Brake fluid reservoir cap and level switch

3.3 Front disc caliper pad inspection hole

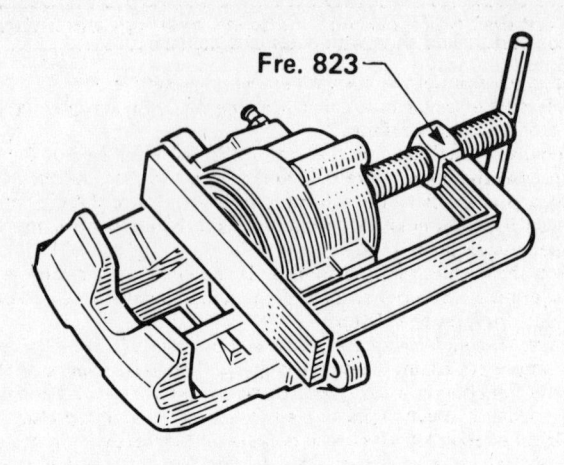

Fig. 9.1 Using clamp to depress caliper piston (Sec 3)

7 Brush away all dust and dirt, but *avoid inhaling it as it is injurious to health*.
8 The piston should now be pushed fully into its cylinder. Do this using a flat bar or a clamp similar to the one shown (Fig. 9.1). As the piston is depressed, the fluid will rise in the reservoir, so anticipate this by syphoning some out beforehand. An old battery hydrometer or poultry baster is useful for this, but make sure that they are clean. Do not syphon out fluid by mouth.
9 Fit the new pads; the inboard one being the one with the sensor wire.
10 Locate the springs correctly, then the caliper.
11 Fit the guide bolts, having applied thread locking fluid to their threads.
12 Tighten the bottom guide bolt and then the upper one to the specified torque.
13 Renew the pads on the opposite wheel in a similar way.
14 Apply the footbrake several times to position the pads against the discs.
15 Fit the roadwheels and lower the car to the floor.
16 Top up the master cylinder reservoir to the correct level.

4 Rear disc pads – inspection and renewal

1 Raise the rear of the car, support it securely and remove the rear roadwheels.
2 Check the wear of the friction material on the pads. If it has worn below the minimum specified, renew the pads in the following way.
3 Disconnect the handbrake cable from the lever on the caliper, refer to Section 21.
4 Extract the two lockpins and then tap out the two sliding keys (photos).
5 Lift the caliper from the disc and tie it up out of the way (photo).
6 Remove the disc pads and springs.
7 The piston must now be fully retracted into the cylinder. Do this by turning the piston with the square section shaft of a screwdriver until the piston continues to turn but will not go in any further. As the piston goes in, the fluid level in the master cylinder reservoir will rise, anticipate this by syphoning out some fluid (see Section 3, paragraph 8). Finally set the piston so that the line (R – Fig. 9.5) is nearest to the bleed screw.
8 Locate the springs under the pads and fit the pads (photo).
9 Fit the caliper by engaging one end of the caliper between the spring clip and the keyway on the bracket, compress the springs and engage the opposite end.
10 Insert the first key then insert a screwdriver in the second key slot and use it as a lever until the slot will accept the key.
11 Fit new key retaining clips.
12 Renew the pads on the opposite wheel.
13 Reconnect the handbrake cable.
14 Apply the footbrake several times to position the pads against the discs.
15 Fit the roadwheels and lower the car to the floor.
16 Top up the master cylinder reservoir to the correct level.

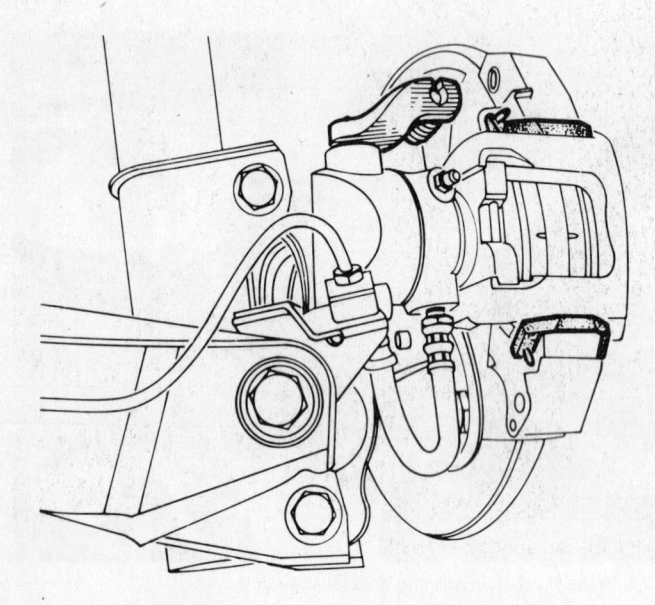

Fig. 9.2 Rear caliper sliding keys (Sec 4)

4.4A Sliding key lockpin (arrowed)

4.4B Sliding key withdrawn

4.5 Lifting caliper from disc

Chapter 9 Braking system

5 Front disc brake caliper – removal, overhaul and refitting

1 Raise the front of the car and remove the roadwheel.
2 Release the flexible brake hose from the caliper by unscrewing it no more than a quarter of a turn.
3 Remove the caliper from the disc, as described in Section 3.
4 Unscrew the caliper from the flexible hose and then cap the end of the hose or use a brake hose clamp to prevent loss of fluid (photo).
5 Clean away external dirt from the caliper and remove the dust excluder and retaining ring.
6 Eject the piston from the cylinder by applying air pressure to the fluid inlet hole. Only low air pressure is required such as is generated by a foot-operated tyre pump.
7 Inspect the surfaces of the piston and cylinder. If there is evidence of scoring or corrosion, renew the caliper cylinder assembly complete.
8 If the components are in good condition, remove the seal and wash them in clean hydraulic fluid or methylated spirit – nothing else.
9 Obtain a repair kit which will contain all the necessary renewable items. Fit the new seal; using the fingers to manipulate it into its groove.
10 Dip the piston in clean hydraulic fluid and insert the piston into the cylinder.
11 Fit the dust excluder and retaining ring.
12 Screw the caliper onto the flexible hose, fit the caliper to its bracket

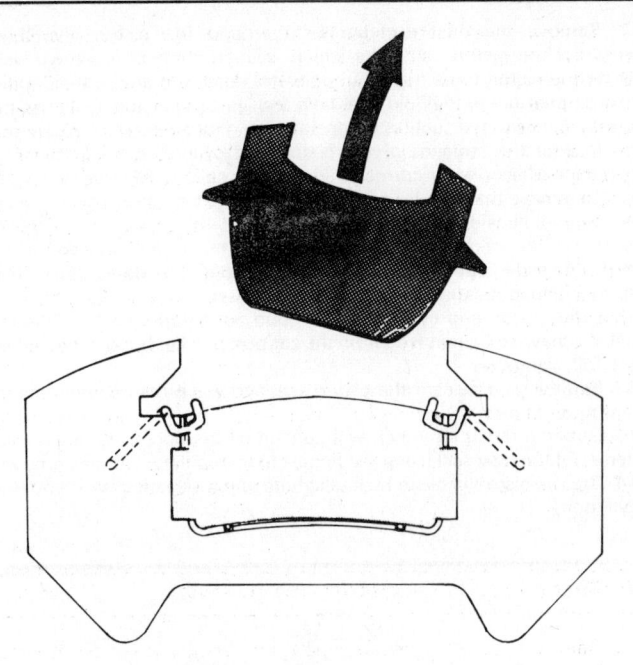

Fig. 9.3 Removing rear disc pad (Sec 4)

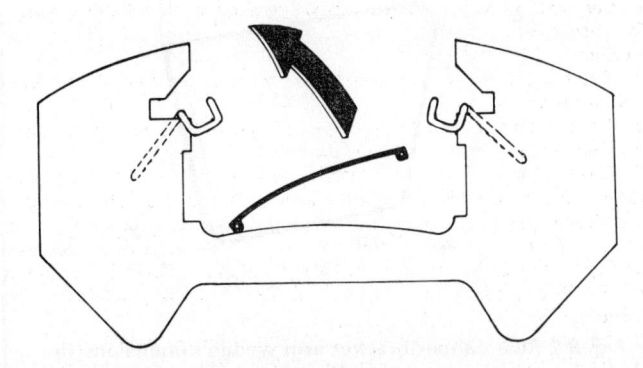

Fig. 9.4 Rear disc pad spring (Sec 4)

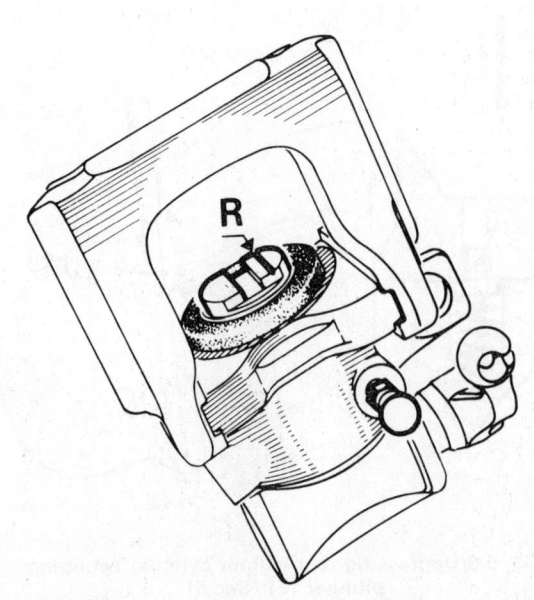

Fig. 9.5 Rear caliper piston alignment (Sec 4)

Line R to be nearest bleed screw

4.8 Disc pad spring (arrowed)

5.4 Brake hose clamp

making sure that the piston is fully depressed into its cylinder.
13 Tighten the flexible hose union and bleed the hydraulic circuit (Section 15).
14 Refit the roadwheel and lower the car.
15 Apply the footbrake two or three times.

6 Rear disc brake caliper – removal, overhaul and refitting

1 Raise the rear of the car and remove the roadwheel.
2 Disconnect the handbrake cable from the lever on the caliper (see Section 21).
3 Release the flexible brake hose from the caliper by unscrewing it by no more than a quarter of a turn.
4 Remove the caliper from the disc as described in Section 4.
5 Unscrew the caliper from the hose and cap the end of the hose or use a brake hose clamp to prevent loss of fluid.
6 Clean away external dirt from the caliper and grip it in the jaws of a vice fitted with jaw protectors.
7 Remove the dust excluder from around the piston and then unscrew the piston using the square section shaft of a screwdriver. Once the piston turns freely but does not come out any further, apply low air pressure to the fluid inlet hole and eject the piston. Only low air pressure is required such as is generated by a foot-operated tyre pump.
8 Inspect the surfaces of the piston and cylinder bore. If there is evidence of scoring or corrosion, renew the caliper cylinder.
9 To remove the cylinder a wedge will have to be made in accordance with the dimensions shown in Fig. 9.7.
10 Drive the wedge in to slightly separate the support bracket arms and so slide out the cylinder which can be done once the spring-loaded locating pin has been depressed.
11 If the piston and cylinder are in good condition then the cylinder will not have to be removed from the support bracket but overhauled in the following way:
12 Remove and discard the piston seal and wash the components in methylated spirit.
13 Obtain a repair kit which will contain all the necessary renewable items. Fit the new seal using the fingers to manipulate it into its groove.
14 Dip the piston in clean hydraulic fluid and insert the piston into the cylinder.

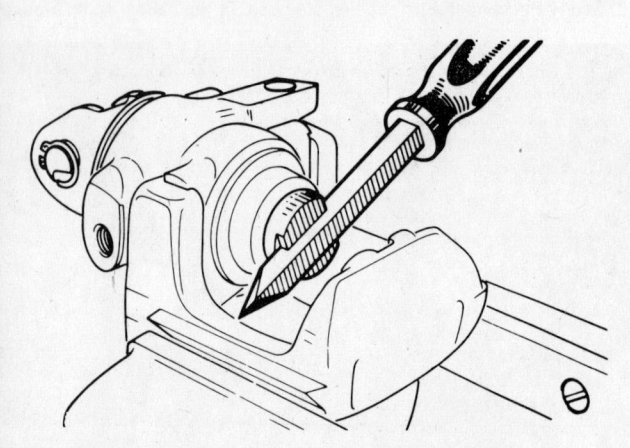

Fig. 9.6 Unscrewing rear caliper piston (Sec 6)

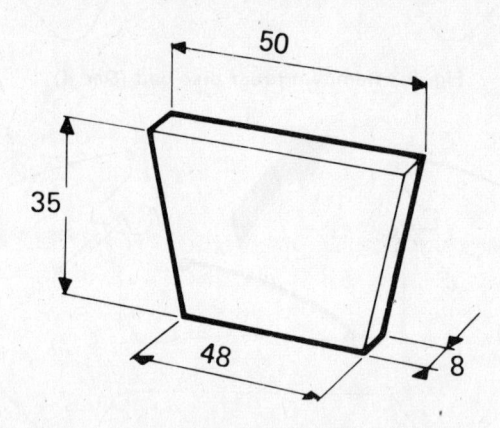

Fig. 9.7 Rear caliper bracket arm wedge dimensions (in mm) (Sec 6)

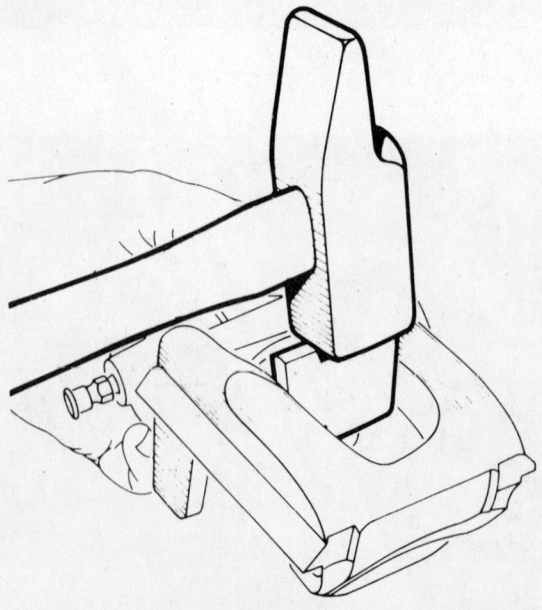

Fig. 9.8 Driving in wedge to spread caliper bracket arms (Sec 6)

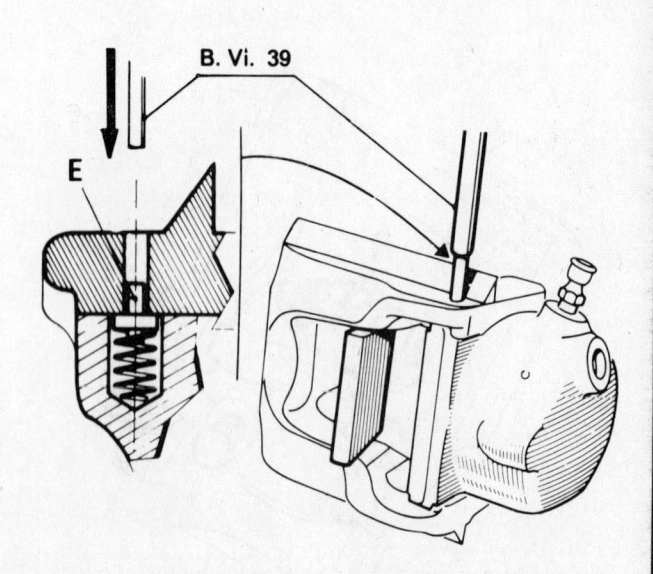

Fig. 9.9 Depressing rear caliper cylinder retaining plunger (E) (Sec 6)

Chapter 9 Braking system

15 Using the square section shaft of a screwdriver, turn the piston until it turns but will not go in any further. Set the piston so that the line (R – Fig. 9.5) is nearest the bleed screw.
16 Fit a new dust excluder.
17 If the handbrake operating mechanism is worn or faulty, dismantle it in the following way before renewing the piston seal.
18 Grip the caliper in a vice fitted with jaw protectors.
19 Refer to Fig. 9.10 and remove the dust excluder (1), the piston (2), dust cover (3), and the circlip (4).

20 Compress the spring washers (5) and pull out the shaft (6).
21 Remove the plunger cam (7), spring (8), adjusting screw (9), plain washer (10) and spring washer (5).
22 Refer to Fig. 9.11 and drive out the sleeve (12) using a drift, and remove the sealing ring (11).
23 Clean all components and renew any that are worn.
24 Reassembly is a reversal of removal but observe the following points.
25 Drive in the sleeve (12) until it is flush with face (A) (Fig. 9.12).
26 Make sure that the spring washers are fitted as shown; convex face to convex face.
27 Align the piston as previously described.
28 Refit the caliper after screwing it onto the hydraulic hose. Tighten the hose union.
29 Reconnect the handbrake cable.
30 Bleed the hydraulic circuit (Section 15).
31 Fit the roadwheel and lower the car.
32 Apply the footbrake several times and adjust the handbrake cable (Section 20).

7 Brake disc – inspection, removal and refitting

Note: *A suitably-sized Torx bit will be needed for this operation.*

Inspection
1 Whenever the disc pads are being checked for wear, take the opportunity to inspect the disc for deep grooving, scoring or cracks. Light scoring is normal.
2 Rust will normally build up on the edge of a disc and this should be removed periodically by holding a sharp tool against the disc while it is rotated (photo).
3 If there has been a history of brake judder, check the discs for run-out. Do this using a dial gauge or feeler blades between the disc and a fixed point while the disc is rotated. Run-out should be within specified limit (see Specifications).
4 If the disc is grooved, cracked or distorted it must be renewed in the following way; refinishing of a disc must not be carried out as this can cause the caliper piston to over extend when the brakes are applied.

Front disc renewal
5 Raise the front of the car and remove the roadwheel.
6 Unscrew and remove the two caliper bracket bolts (A – Fig. 9.13).
7 Lift the caliper off the disc and tie it up out of the way.
8 Unscrew the disc fixing screws, these are of Torx type so a special bit will be required.
9 Remove the disc.
10 Clean any protective coating from the new disc and locate it on the hub.
11 Screw in and tighten the fixing screws.

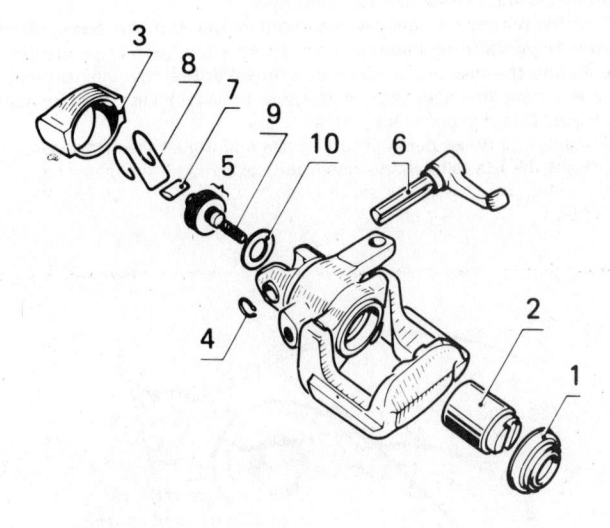

Fig. 9.10 Exploded view of rear caliper (Sec 6)

1 Dust excluder
2 Piston
3 Rear dust cover
4 Circlip
5 Spring washers
6 Shaft
7 Plunger cam
8 Spring
9 Adjusting screw
10 Washer

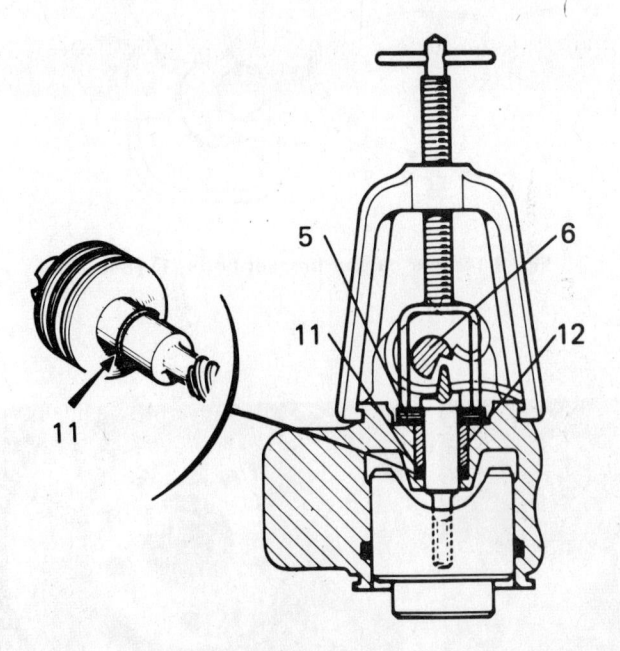

Fig. 9.11 Using a puller to compress spring washers in rear caliper. Inset O-ring (11) on adjusting screw (Sec 6)

5 Spring washer
6 Shaft
11 O-ring
12 Sleeve

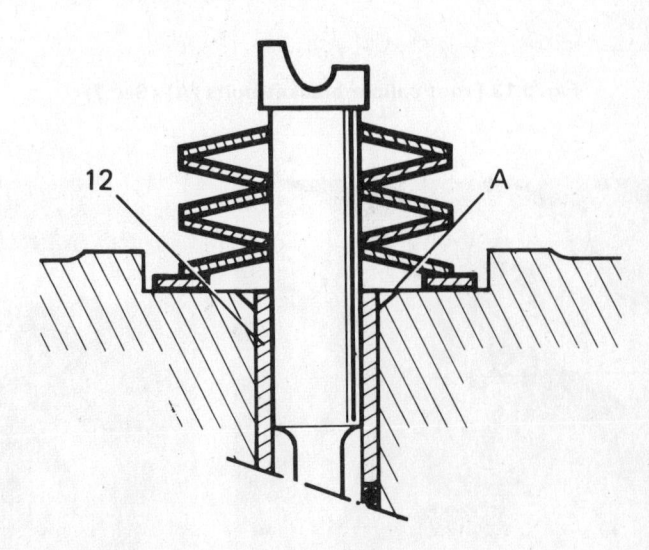

Fig. 9.12 Rear caliper sleeve (12) fitted flush (A) (Sec 6)

Chapter 9 Braking system

12 Clean the threads of the caliper bracket fixing bolts, apply thread locking fluid to them, locate the caliper and screw in the bolts; tightening them to the specified torque.
13 Apply the brake pedal several times to position the pads up against the disc, fit the roadwheel and lower the car.

Rear disc renewal
14 Raise the rear of the car and remove the roadwheel.
15 Refer to Section 4 and remove the disc pads and disconnect the handbrake cable. Tie the caliper up out of the way.
16 Unscrew and remove the two caliper bracket bolts (1 – Fig. 9.14) and lift off the caliper bracket (photo).
17 Using a Torx bit, unscrew and remove the two disc fixing screws and remove the disc (photos).
18 Fitting the disc is a reversal of removal but remove any protective coating from the new disc, and apply thread locking fluid to clean threads of the caliper bracket bolts.
19 Apply the brake pedal several times and adjust the handbrake.
20 Refit the roadwheel and lower the car.

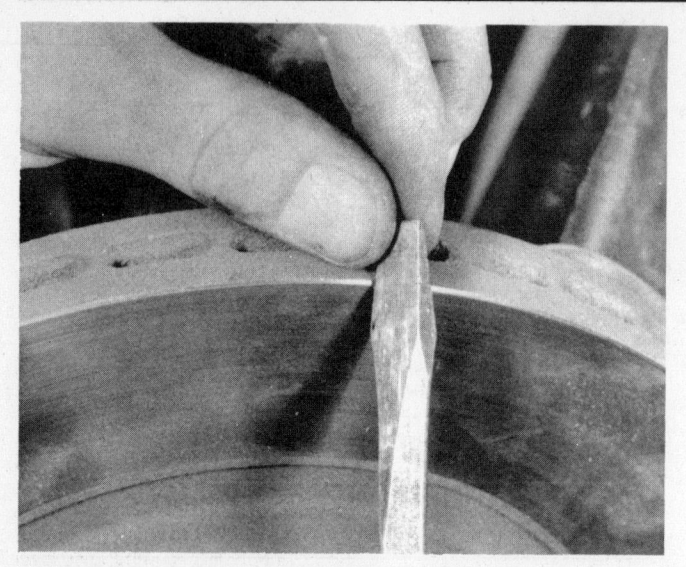

7.2 Removing rust deposits from a disc

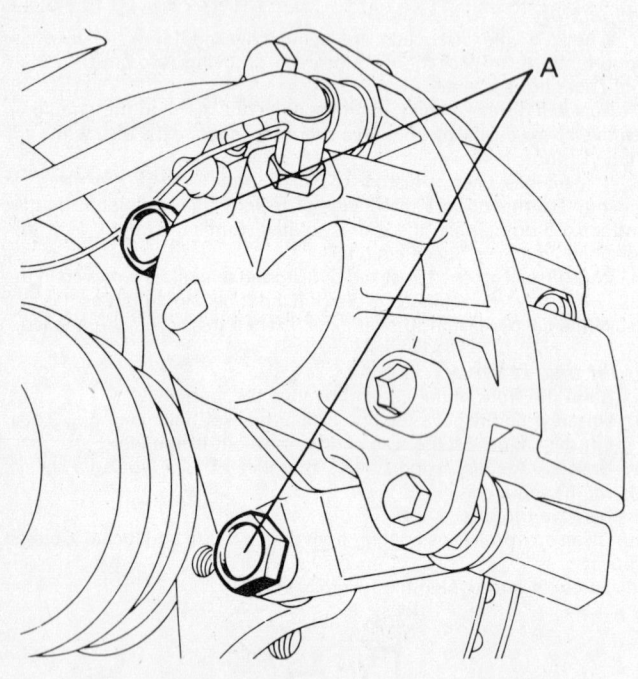

Fig. 9.13 Front caliper bracket bolts (A) (Sec 7)

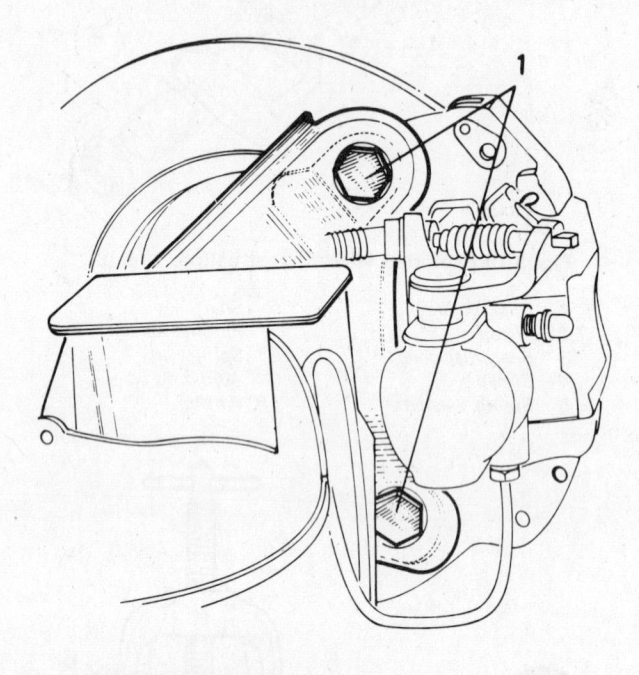

Fig. 9.14 Rear caliper bracket bolts (1) (Sec 7)

7.16 Removing a rear caliper bracket

7.17A Extracting a rear disc Torx screw

7.17B Removing a rear disc

8 Rear brake shoes (Bendix type) – inspection and renewal

Note: *A suitably-sized Torx bit will be needed for this operation*

1   Raise the car and remove the roadwheels.
2   To inspect the shoes, take out the two Torx brake drum securing screws and pull off the drum. If it is tight, try gently tapping with a plastic or copper-faced hammer. If this is not successful, prise out the blanking plug from the brake backplate, insert a screwdriver and depress the toothed segment of the automatic adjuster so that the toothed adjuster lever can be moved towards the stub axle.
3   Brush away dust and dirt, *taking care not to inhale it as it is injurious to health.*
4   If the shoe friction linings have worn down to or below the specified minimum, then the shoes should be renewed.
5   Obtain factory relined shoes, do not attempt to reline the old shoes as this seldom proves satisfactory. Tap off the hub dust cap, unscrew the hub nut and remove the hub assembly.
6   Note carefully the position of the leading and trailing shoes and how the friction lining exposes more of the shoe web at one end than the other.
7   Note also into which holes in the shoe web the return springs engage.
8   Disconnect the upper return spring.
9   Disconnect the handbrake cable from the trailing shoe.
10 Remove the two shoe hold-down springs. Do this by inserting a screwdriver into the centre of the spring and rotating it.
11 Move the toothed automatic adjuster lever (C – Fig. 9.15) as far as possible in the direction of the stub axle.
12 Pull both shoes away from the backplate and then disconnect the link (B – Fig. 9.16) from the leading shoe.
13 Move the toothed adjuster lever back to its original position. Rotate the toothed segment to do this.
14 Twist the leading shoe at right-angles to the backplate and disconnect the lower shoe return spring.
15 Remove both shoes. While the shoes are removed, do not touch the brake pedal, or the wheel cylinder pistons will be ejected.
16 Place the old shoes on the bench and transfer the handbrake lever and automatic adjuster lever to the new shoes. The levers are retained by spring clips.
17 Release the spring (E – Fig. 9.17) and transfer the link to the new trailing shoe.
18 Fit the new shoes by reversing the removal operations. Apply a smear of high melting-point grease to the shoe-rubbing high spots on the brake backplate, to the shoe lower anchor block recesses and to the end faces of the wheel cylinder pistons. Set the toothed automatic adjuster lever fully towards the stub axle.
19 The self-adjuster must now be set. To do this, have the handbrake lever hard up against the trailing shoe then measure the gap (H – Fig. 9.18). It should be approximately 1.0 mm (0.039 in). If it is not, then renew the link tension spring and the shoe upper and lower return springs.
20 Clean the interior of the drum and fit it, together with the hub, taking care not to damage the oil seal. Tighten the hub nut to the specified torque (Chapter 8).
21 Refit the dust cap over the nut.
22 Refit the backplate plug, if removed.
23 Repeat the operations on the opposite wheel.
24 Adjust the handbrake cable (Section 20).
25 Refit the roadwheels and lower the car.

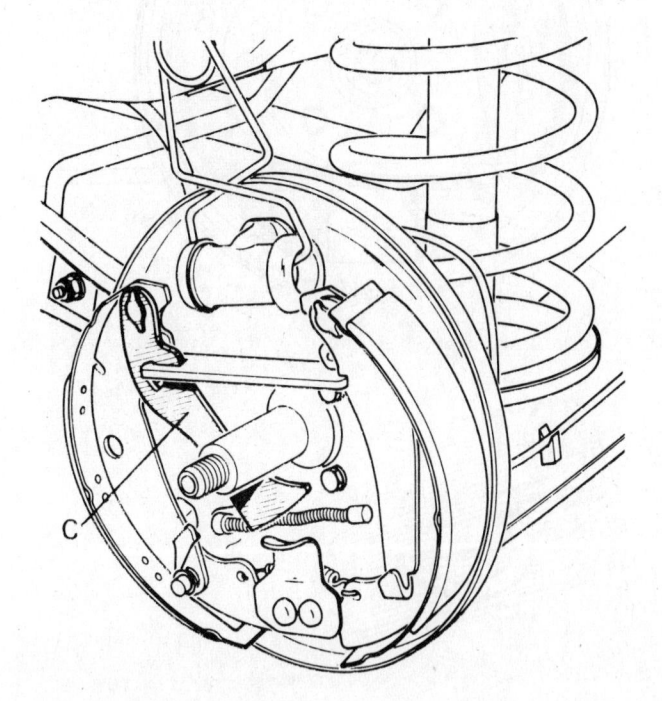

Fig. 9.15 Left-hand rear drum brake – Bendix type (Sec 8)

C   *Automatic adjuster lever*

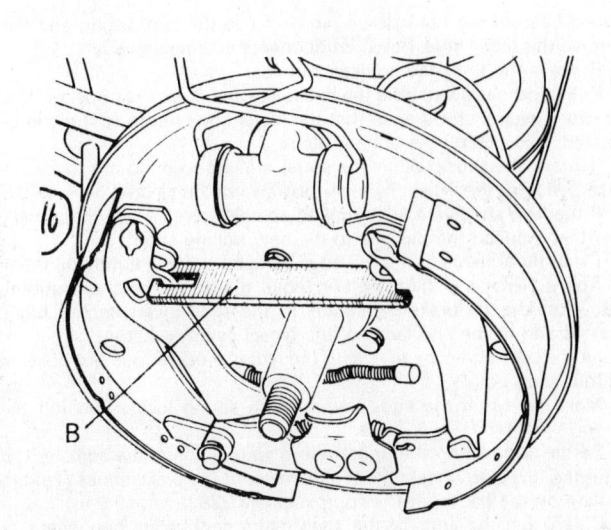

Fig. 9.16 Shoe link (B) on left-hand drum brake (Sec 8)

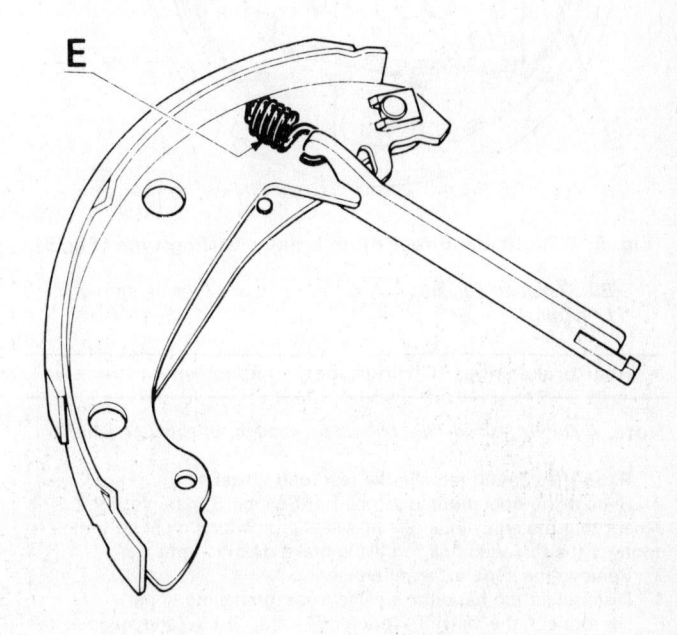

Fig. 9.17 Shoe strut tension spring (E) (Sec 8)

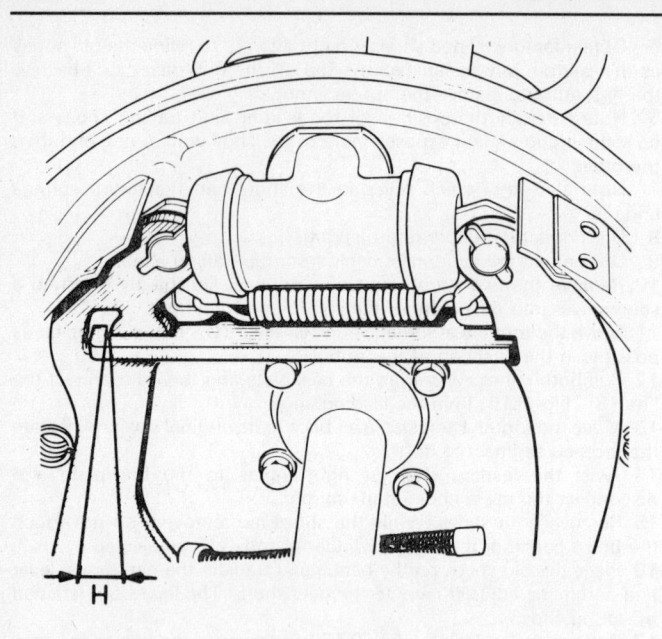

Fig. 9.18 Self-adjuster lever-to-strut gap (Sec 8)

$H = 1.0$ mm (0.039 in)

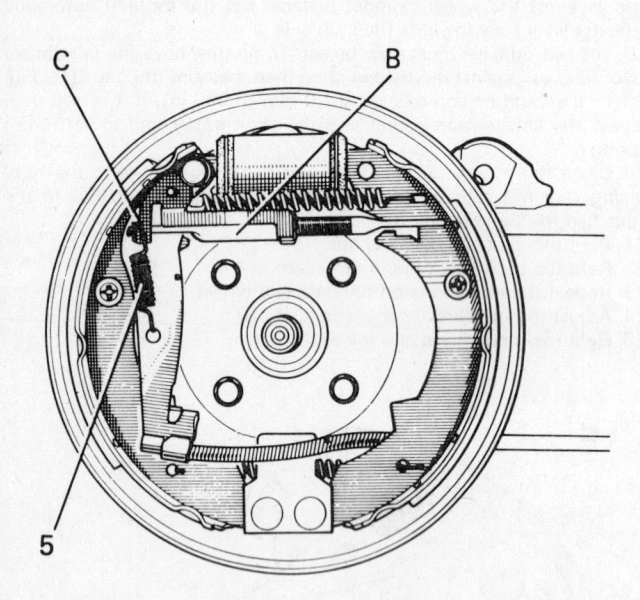

Fig. 9.19 Right-hand rear drum brake – Girling type (Sec 9)

- B Threaded adjuster rod
- C Adjuster lever
- 5 Tension spring

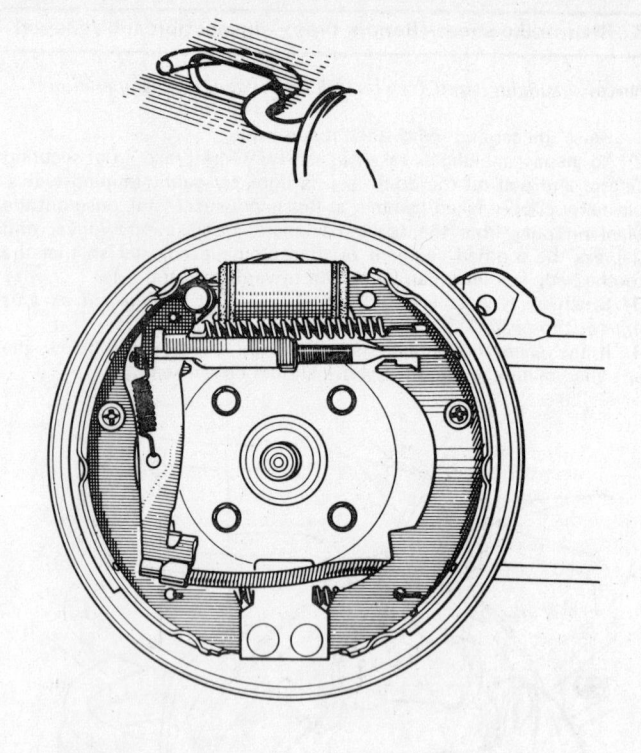

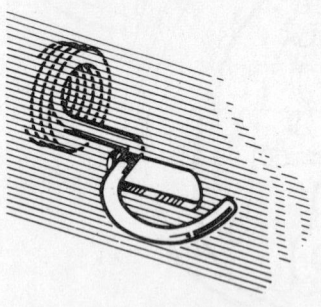

Fig. 9.20 Shoe return spring engagement in hole in shoe (Sec 9)

9 Rear brake shoes (Girling type) – inspection and renewal

Note: A suitably-sized Torx bit will be needed for this operation

1 Raise the car and remove the rear road wheel.
2 Repeat the operations described in Section 8, paragraphs 2 to 7, except that the adjuster is of star wheel type which must be turned to contract the threaded link rod if the brake drum is tight.
3 Remove the shoe upper return spring.
4 Disconnect the handbrake cable from the trailing shoe.
5 Disconnect the tension spring (5 – Fig. 9.19) and remove the adjuster lever (C).
6 Grip one of the shoe hold-down spring caps with a pair of pliers, depress the cup and turn it through 90° so that it will pass over the tee shaped head of the hold-down pin. Remove the cup, spring and pin. Remove the other hold-down components in a similar way.
7 Remove the threaded adjuster rod (B).
8 Pull the shoes away from the backplate and cross the upper ends of the shoes over each other so that the lower shoe return spring can be released from behind the anchor block.
9 Remove the shoes. While the shoes are removed, do not touch the brake pedal, or the wheel cylinder pistons will be ejected.
10 If the new shoes are not supplied complete with a new handbrake lever then transfer the old one to the new trailing shoe.
11 Clean the threads of the threaded adjuster rod and lightly oil them.
12 Apply a smear of high melting-point grease to the shoe-rubbing high spots on the brake backplate, to the shoe lower anchor block recesses and to the end faces of the wheel cylinder pistons.
13 Fit the new shoes by reversing the removal operations but observe the following points.
14 Make sure that the shoe lower return spring locates behind the anchor block tab.
15 Before connecting the upper return spring, adjust the adjuster rod by turning the star wheel until the diameter of the brake shoes (held in position on the backplate) is approximately 228.0 mm (9.0 in).
16 Check that the ends of the shoe return springs are connected as shown in Fig. 9.20.
17 Carry out the operations described in paragraphs 20 to 25 of the preceding Section.

Chapter 9 Braking system

10 Rear wheel cylinder – removal, overhaul and refitting

Note: *A suitably-sized Torx bit will be needed for this operation*

1 Raise the rear of the car and remove the roadwheel.
2 Remove the brake drum, as described in Section 8 or 9.
3 Prise and chock the upper ends of the brake shoes so that they are away from the wheel cylinder pistons.
4 Disconnect the flexible brake hose from the rigid pipeline at the support bracket. Use two spanners to do this and cap the end of the pipeline to prevent fluid loss. A rubber bleed nipple cap is useful for this purpose.
5 Extract the spring clip to free the hose from the bracket.
6 Unscrew the hose from the wheel cylinder.
7 Unscrew the wheel cylinder mounting bolts (A – Fig. 9.21) and remove the cylinder.
8 Clean away external dirt, remove the dust excluders and shake out the pistons, seals and spring.
9 Examine the condition of the piston and cylinder bore surfaces. If they are scored or corroded, renew the cylinder complete.
10 If the pistons and cylinder are in good condition then discard the seals and wash the components in methylated spirit.
11 Obtain a repair kit which will contain all the necessary renewable items.
12 Reassemble after dipping each component in clean hydraulic fluid.
13 Refit the wheel cylinder, connect the hose and pipelines.
14 Locate the brake shoes on the pistons.
15 Fit the brake drum, tighten the Torx screws, fit the roadwheel, lower the car and bleed the hydraulic circuit (Section 15).

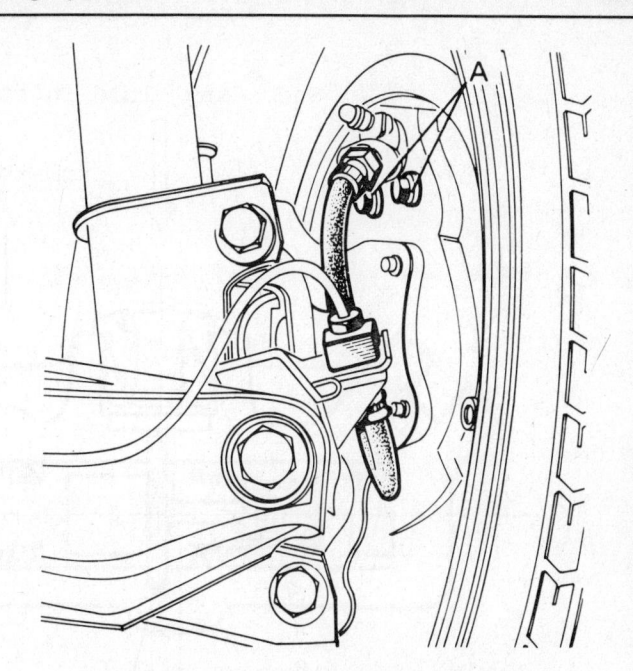

Fig. 9.21 Wheel cylinder fixing bolts (A) (Sec 10)

11 Brake drum – inspection and renovation

1 Whenever the brake drums are removed to examine the shoe linings for wear, take the opportunity to inspect the interior of the drums for grooving, scoring or cracks.
2 The drums may be machined to renovate them provided both drums are done at the same time and the internal diameter will not be increased to more than the specified maximum.

12 Master cylinder – removal and refitting

1 Syphon the fluid from the master cylinder reservoir. Use a syringe, a clean battery hydrometer or a poultry baster to do this, **never** use the mouth to suck the fluid out through a tube.
2 Disconnect the leads from the low level switch and pull the reservoir upwards out of the sealing grommets.
3 Note the locations of the hydraulic pipes and then disconnect them

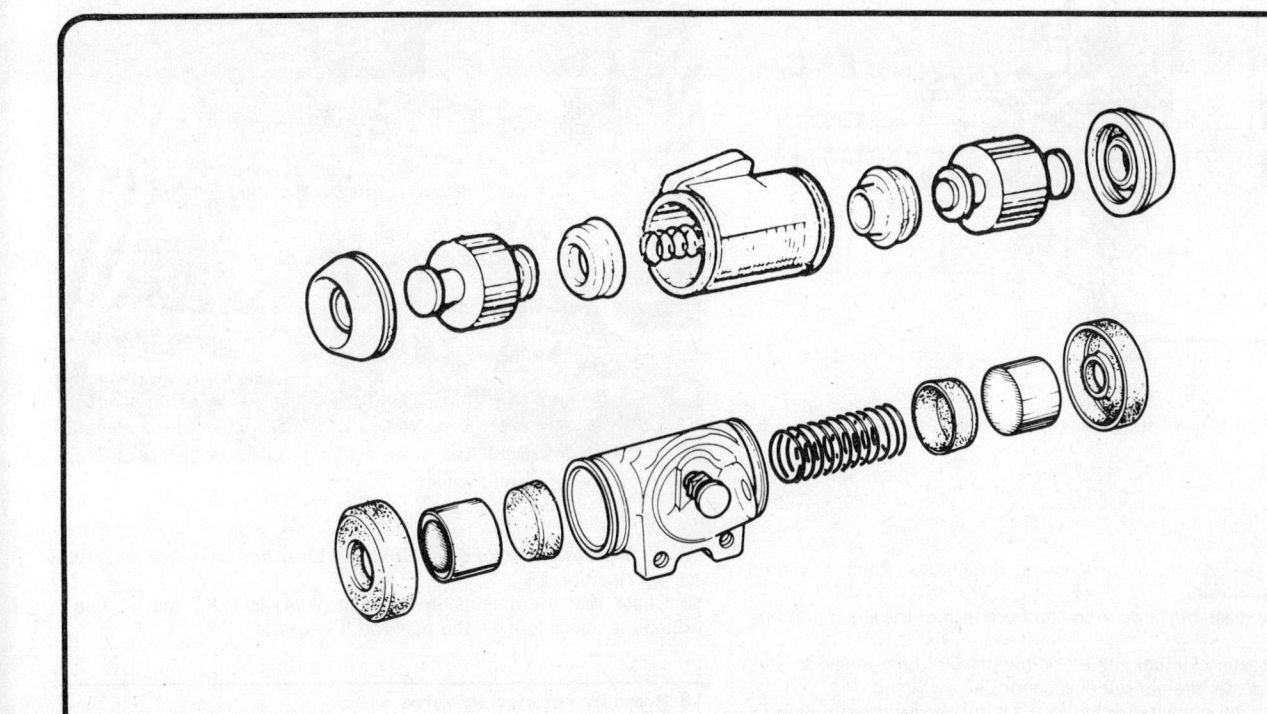

Fig. 9.22 Types of wheel cylinder (Sec 10)

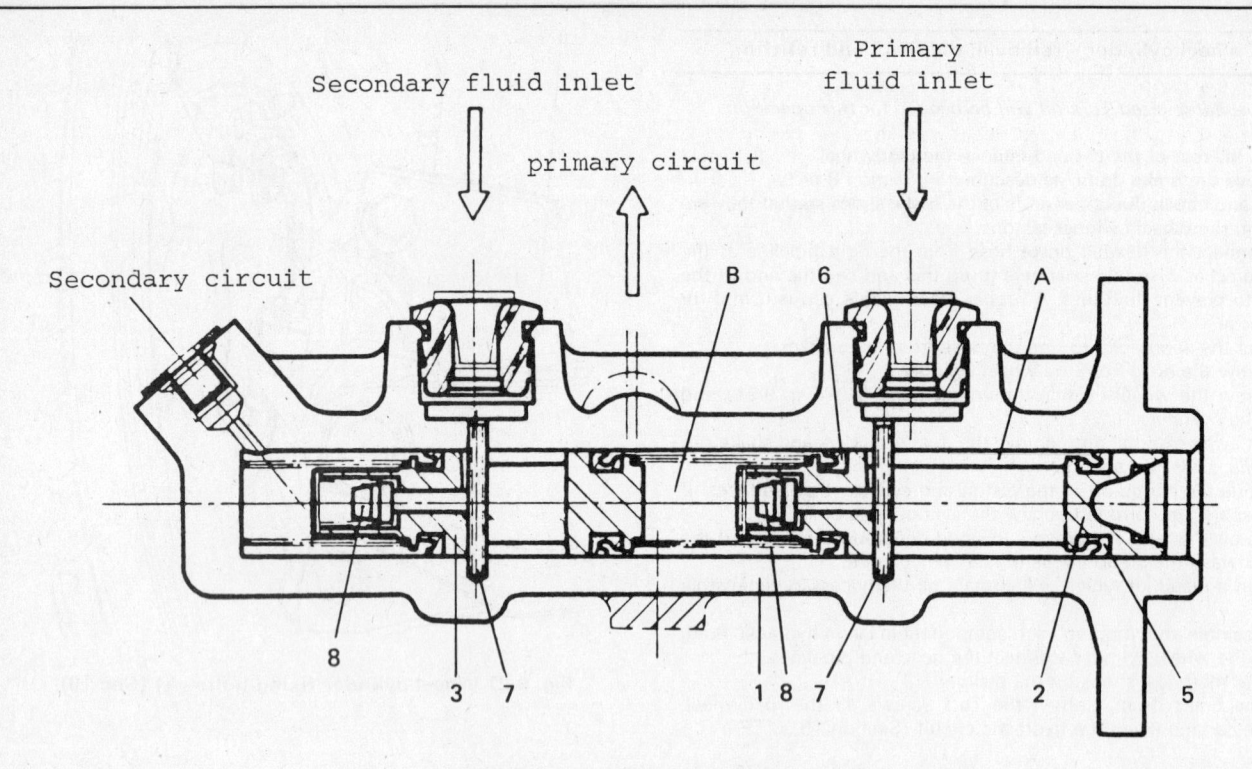

Fig. 9.23 Sectional view of brake master cylinder (Sec 12)

A Primary piston assembly	1 Cylinder body	3 Secondary piston
B Secondary piston assembly	2 Primary piston	5 Seal
		6 Seal
		7 Roll pins
		8 Valve

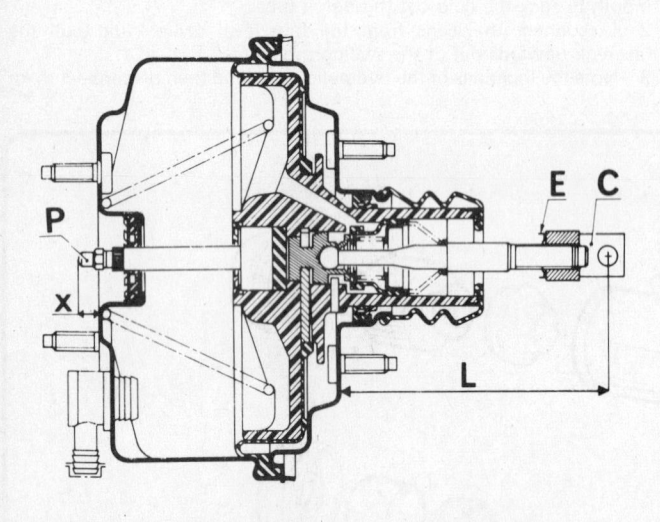

Fig. 9.24 Servo pushrod setting diagram (Sec 12)

C Clevis fork
E Locknut
L = 128.5 mm (5.1 in)
P Pushrod nut
X = 9.0 mm (0.35 in)

13.1 Pressure regulating valve

from the master cylinder by unscrewing the unions. Catch any fluid which may drain out.
4 Unbolt the master cylinder from the front face of the servo unit and remove it.
5 A faulty master cylinder cannot be overhauled, only renewed, due to the fact that the internal roll pins cannot be extracted.
6 Before fitting the master cylinder, check the pushrod protrusion (X – Fig. 9.24). This should be 9.0 mm (0.35 in). Adjust if necessary by means of the pushrod nut.
7 Bolt the master cylinder into position and connect the fluid pipes.
8 Push the reservoir firmly into its grommets.

9 Fill the reservoir with clean fluid and bleed the complete hydraulic system (Section 15).
10 Check that the dimension (L – Fig. 9.24) is 128.5 mm (5.1 in). Adjust the clevis fork on the pushrod if necessary.

13 Pressure regulating valve

1 Checking the operation of the pressure regulating valve (photo) is not within the scope of the home mechanic due to the need for special hydraulic pressure gauges.

Chapter 9 Braking system

2 However, an obviously faulty unit can be renewed and the new unit given an initial setting in the following way.
3 Remove the clip (G – Fig. 9.25).
4 Remove the suspension arm link rod (H).
5 Identify and disconnect the hydraulic pipes from the valve block.
6 Unscrew and remove the three fixing bolts and remove the regulating valve from its bracket.
7 The new pressure regulating valve will be supplied preset with two clips fitted as shown in Fig. 9.26.
8 To fit the unit, first slacken nut (A).

9 Locate the rod (B) in the suspension arm link rod (H in Fig. 9.25) and refit the clip (G).
10 Tighten nut (A) while preventing the sleeve (C) from turning (Fig. 9.26).
11 Remove the temporary clips (1) and (2).
12 Check the initial setting. With the arm (E – Fig. 9.27) being pressed in the direction of the arrow, dimension (H) should be 32.0 mm (1.26 in). Adjust if necessary by releasing nut (A) and turning the sleeve (C – Fig. 9.28). On no account disturb the nut (F – Fig. 9.27).
13 Bleed the hydraulic circuit (Section 15).
14 Have the valve pressure checked by your Renault dealer as soon as possible.

14 Hydraulic pipes and hoses – inspection, removal and refitting

1 At intervals given in Routine Maintenance carefully examine all brake pipes, hoses, hose connections and pipe unions.
2 First check for signs of leakage at the pipe unions. Then examine the flexible hoses for signs of cracking, chafing and fraying.
3 The brake pipes must be examined carefully and methodically. they

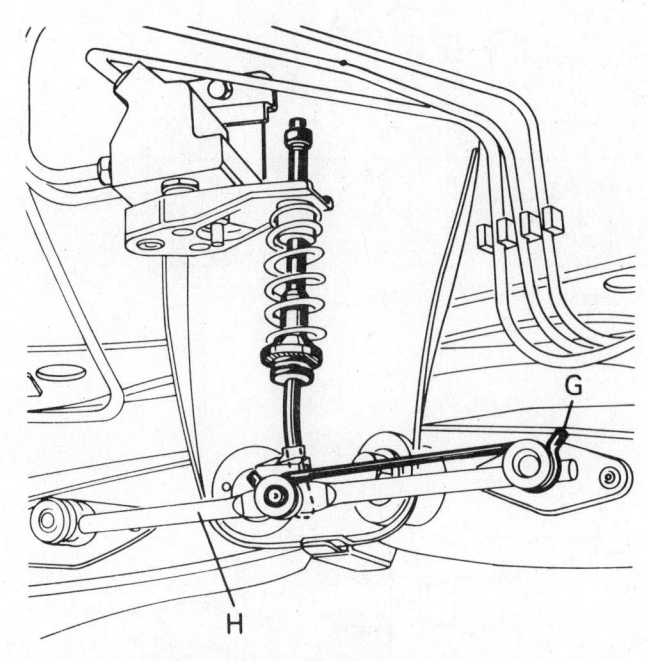

Fig. 9.25 Pressure regulating valve (Sec 13)

G Clip   H Suspension arm link rod

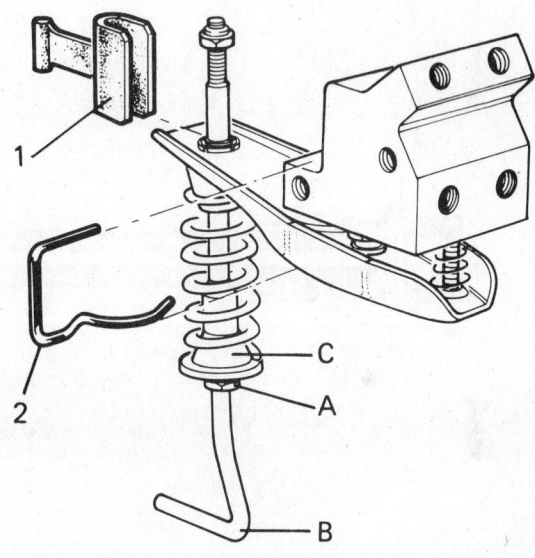

Fig. 9.26 Clips fitted to new pressure regulating valve (Sec 13)

1 Clip   A Locknut
2 Clip   B Rod
                    C Sleeve

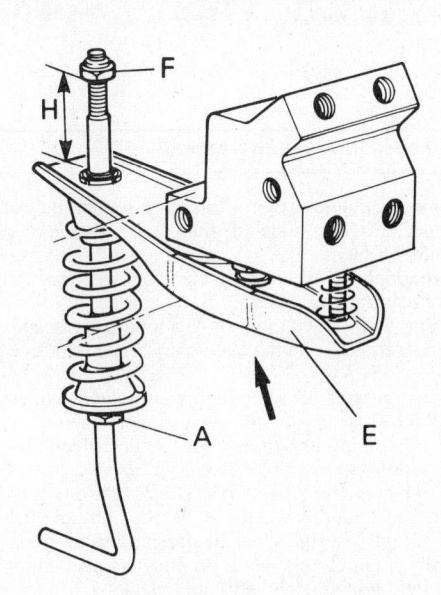

Fig. 9.27 Pressure regulating valve initial setting diagram (Sec 13)

A Locknut   F Nut (do not move)
E Arm   H = 32.0 mm (1.26 in)

Press arm in direction of arrow

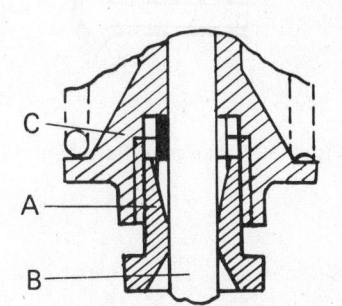

Fig. 9.28 Sectional view of pressure regulating valve lower end (Sec 13)

A Locknut   C Sleeve
B Rod

248 Chapter 9 Braking system

must be cleaned off and checked for signs of dents, corrosion or other damage. Corrosion should be scraped off and, if the depth of pitting is significant, the pipes renewed. This is particularly likely in those areas underneath the vehicle body where the pipes are exposed and unprotected.

4 If any section of pipe or hose is to be removed, first unscrew the master cylinder reservoir filler cap and place a piece of polythene over the filter neck. Secure the polythene with an elastic band ensuring that an airtight seal is obtained. This will minimise brake fluid loss when the pipe or hose is removed.

5 Brake pipe removal is usually quite straightforward. The union nuts at each end are undone, the pipe and union pulled out and the centre section of the pipe removed from the body clips. Where the union nuts are exposed to the full force of the weather they can sometimes be quite tight. As only a open-ended spanner can be used, burring of the flats on the nuts is not uncommon when attempting to undo them. For this reason a self-locking wrench or special brake union split ring spanner should be used (photo).

6 To remove a flexible hose, wipe the unions and brackets free of dirt and undo the union nut from the brake pipe end.

7 Next extract the hose retaining clip, or unscrew the nut, and lift the end of the hose out of its bracket. If a front hose is being removed, it can now be unscrewed from the brake caliper.

8 Brake pipes can be obtained individually, or in sets, from most accessory shops or garages with the end flares and union nuts in place.

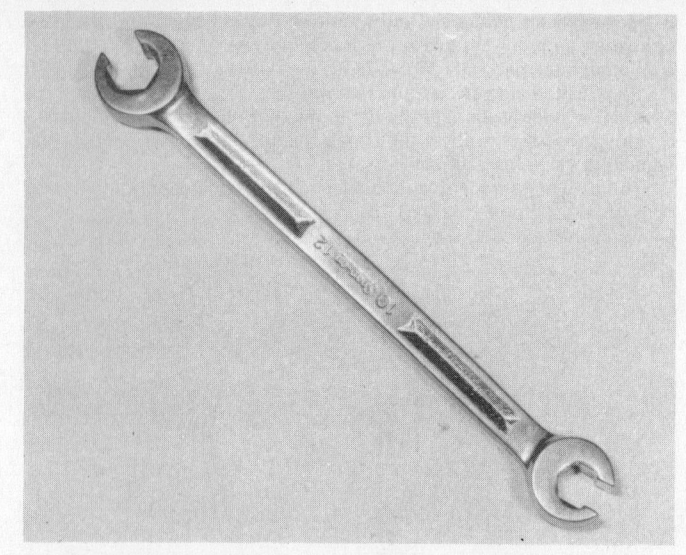

14.5 Brake pipe union spanner

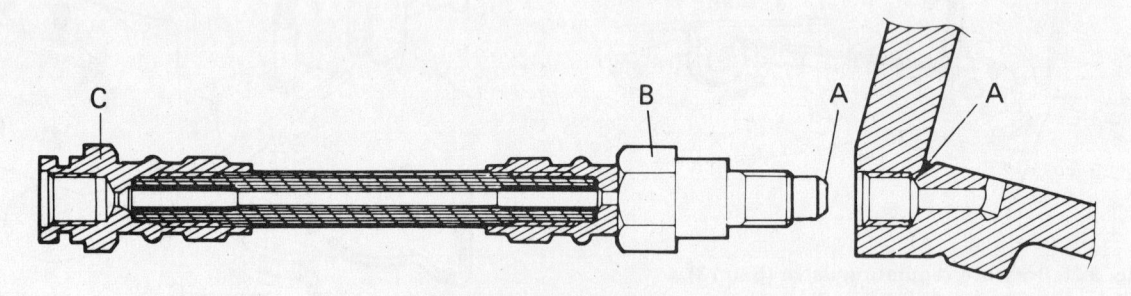

Fig. 9.29 Flexible brake hose (Sec 14)

A Taper seat sealing (no washer used)
B Union nut
C End fitting

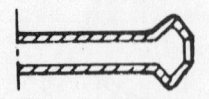

Fig. 9.30 Brake pipe flare (Sec 14)

The pipe is then bent to shape, using the old pipe as a guide, and is ready for fitting to the car.

9 Refitting the pipes and hoses is a reverse of the removal procedure. Make sure that the hoses are not linked when in position and also make sure that the brake pipes are securely supported in their clips. After refitting, remove the polythene from the reservoir and bleed the brake hydraulic system, as described in the next Section.

15 Brake hydraulic system – bleeding

1 If the master cylinder or the pressure regulating valve has been disconnected and reconnected then the complete system (both circuits) must be bled.

2 If a component of one circuit has been disturbed then only that particular circuit need be bled.

3 Bleed one rear brake and its diagonally opposite front brake. Repeat in this sequence on the remaining circuit if the complete system is to be bled.

4 Unless the pressure bleeding method is being used, do not forget to keep the fluid level in the master cylinder reservoir topped up to prevent air from being drawn into the system which would make any work done worthless.

5 Before commencing operations, check that all system hoses and pipes are in good condition with all unions tight and free from leaks.

6 Take great care not to allow hydraulic fluid to come into contact with the vehicle paintwork as it is an effective paint stripper. Wash off any spilled fluid immediately with cold water.

7 As the system incorporates a vacuum servo, destroy the vacuum by giving several applications of the brake pedal in quick succession.

8 On cars equipped with an anti-lock braking system, it is essential during bleeding to avoid operating the hydraulic unit as this could actuate the pump and create air pockets in the hydraulic unit.

9 If the hydraulic unit is being renewed, the new one will be supplied already primed with brake fluid to obviate air locks.

Chapter 9 Braking system

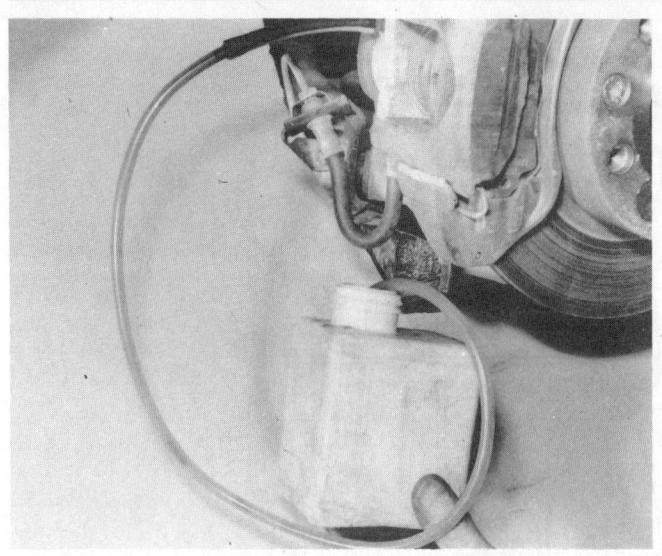

15.12 Bleed tube and container

16.10 Brake servo vacuum hoses and one-way valve/angled connector

Bleeding – two man method
10 Gather together a clean jar and a length of rubber or plastic tubing which will be a tight fit on the brake bleed screws.
11 Engage the help of an assistant.
12 Push one end of the bleed tube onto the first bleed screw and immerse the other end in the jar which should contain enough hydraulic fluid to cover the end of the tube (photo).
13 Open the bleed screw one half a turn and have your assistant depress the brake pedal fully then slowly release it. Tighten the bleed screw at the end of each pedal downstroke to obviate any chance of air or fluid being drawn back into the system.
14 Repeat this operation until clean hydraulic fluid, free from air bubbles, can be seen coming through into the jar. Tighten the bleed screw at the end of a pedal downstroke and remove the bleed tube.
15 Bleed the remaining screws in a similar way.

Bleeding – using one way valve kit
16 There is a number of one-man, one-way brake bleeding kits available from motor accessory shops. It is recommended that one of these kits is used wherever possible as it will greatly simplify the bleeding operation and also reduce the risk of air or fluid being drawn back into the system, quite apart from being able to do the work without the help of an assistant.
17 To use the kit, connect the tube to the bleed screw and open the screw one half a turn.
18 Depress the brake pedal fully and slowly release it. The one-way valve in the kit will prevent expelled air from returning at the end of each pedal downstroke. Repeat this operation several times to be sure of ejecting all air from the system. Some kits include a translucent container which can be positioned so that the air bubbles can actually be seen being ejected from the system.
19 Tighten the bleed screw, remove the tube and repeat the operations on the remaining brakes.
20 On completion, depress the brake pedal. If it still feels spongy repeat the bleeding operations as air must still be trapped in the system.

Bleeding – using a pressure bleeding kit
21 These kits too are available from motor accessory shops and are usually operated by air pressure from the spare tyre.
22 By connecting a pressurised container to the master cylinder fluid reservoir, bleeding is then carried out by simply opening each bleed screw in turn and allowing the fluid to run out, rather like turning on a tap, until no air is visible in the expelled fluid.
23 By using this method, the large reserve of hydraulic fluid provides a safeguard against air being drawn into the master cylinder during bleeding which often occurs if the fluid level in the reservoir is not maintained.
24 Pressure bleeding is particulary effective when bleeding 'difficult' systems or when bleeding the complete system at time of routine fluid renewal.

All methods
25 It may also be necessary to bleed the clutch hydraulic circuit (Chapter 5).
26 When bleeding is completed, check and top up the fluid level in the master cylinder reservoir.
27 Check the feel of the pedal. If it feels at all spongy, air must still be present in the system and further bleeding is indicated. Failure to bleed satisfactorily after a reasonable period of the bleeding operation, may be due to worn master cylinder seals.
28 Discard brake fluid which has been expelled. It is almost certain to be contaminated with moisture, air and dirt making it unsuitable for further use. Clean fluid should always be stored in an airtight container as it absorbs moisture readily (hygroscopic) which lowers its boiling point and could affect braking performance under severe conditions.

16 Vacuum servo unit – description and testing

1 A vacuum servo unit is fitted to the master cylinder, to provide power assistance to the driver when the brake pedal is depressed.
2 The unit operates by vacuum obtained from the induction manifold and comprises, basically, a booster diaphragm and a non-return valve.
3 The servo unit piston rod acts as the master cylinder pushrod. The driver's braking effort is transmitted through another pushrod to the servo unit piston and its built-in control system. The servo unit piston does not fit tightly into the cylinder, but has a strong diaphragm to keep its edges in constant contact with the cylinder wall, so assuring an airtight seal between the two parts. The forward chamber is held under vacuum conditions created in the inlet manifold of the engine and, during periods when the brake pedal is not in use, the controls open a passage to the rear chamber so placing it under vacuum. When the brake pedal is depressed, the vacuum passage to the rear chamber is cut off and the chamber opened to atmospheric pressure. The consequent rush of air pushes the servo piston forward in the vacuum chamber and operates the main pushrod to the master cylinder. The controls are designed so that assistance is given under all conditions and, when the brakes are not required, vacuum in the rear chamber is established when the brake pedal is released. Air from the atmosphere entering the rear chamber is passed through a small air filter.

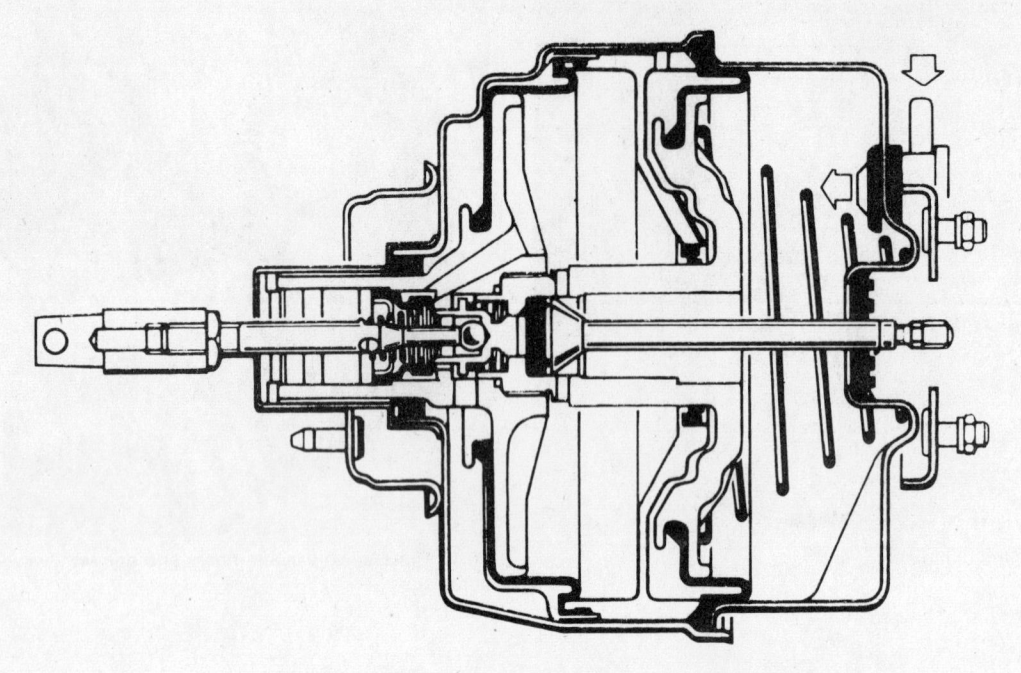

Fig. 9.31 Sectional view of typical vacuum servo unit (Sec 16)

4 On some models, the servo is of two piston type which produces a greater servo effect for the same unit diameter.
5 Operation of the servo can be checked in the following way.
6 With the engine off, depress the brake pedal several times. The pedal travel should remain the same.
7 Depress the brake pedal fully and hold it down. Start the engine and feel that the pedal moves down slightly.
8 Hold the pedal depressed with the engine running. Switch off the engine, holding the pedal depressed. The pedal should not rise nor fall.
9 Start the engine and run it for at least a minute. Switch off, then depress the brake pedal several times. The pedal travel should decrease with each application.
10 If the foregoing tests do not prove satisfactory, check the servo vacuum hose and non-return valve for security and leakage at the valve grommet (photo).
11 If the brake servo operates properly in the test, but still gives less effective service on the road, the air filter through which air flows into the servo should be inspected. A dirty filter will limit the formation of a difference in pressure across the servo diaphragm.
12 The servo unit itself cannot be repaired and therefore a complete renewal is necessary if the measures described are not effective.

17 Vacuum servo air filter – removal

1 Working under the facia panel, pull the dust excluder off the servo and slide it up the pushrod.
2 Using a scriber or similar tool, pick out the filter and cut it to remove it.
3 Cut the new filter as shown in Fig. 9.32 and push it neatly into position. Refit the dust excluding boot.

18 Vacuum servo non-return valve – renewal

1 Disconnect the vacuum hose from the connector on the front face of the servo unit.
2 Pull and twist the non-return valve elbow from its grommet.
3 Refitting is a reversal of removal; use a new grommet. A smear of rubber grease on the elbow spigot will ease its entry into the grommet. Do not force the elbow when fitting as it is possible to push the grommet into the interior of the servo.

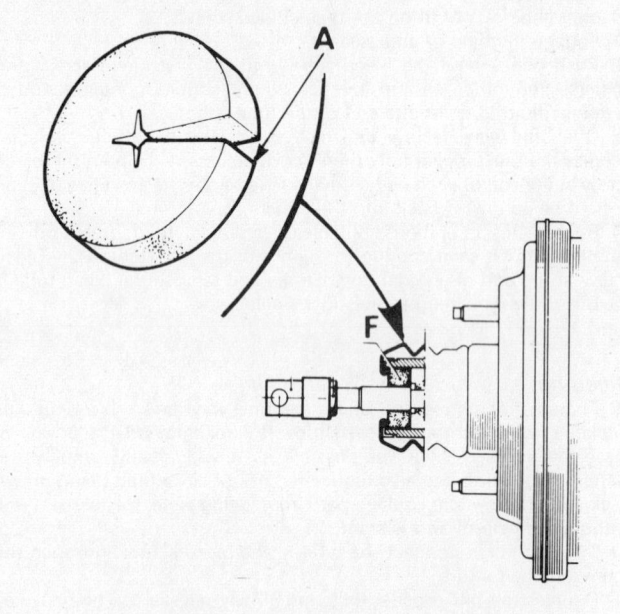

Fig. 9.32 Vacuum servo filter (Sec 17)

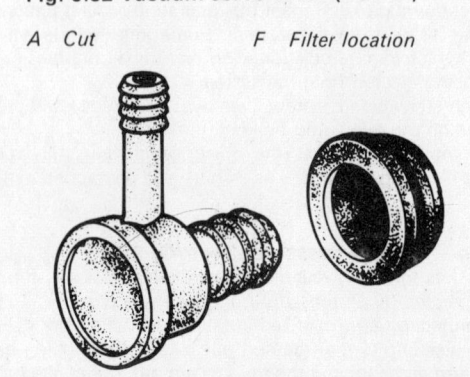

A Cut   F Filter location

Fig. 9.33 Vacuum servo non-return valve and sealing grommet (Sec 18)

Chapter 9 Braking system

19 Vacuum servo unit – removal and refitting

1 Remove the master cylinder, as described in Section 12.
2 Disconnect the vacuum hose from the servo unit.
3 Disconnect the pushrod from the brake pedal by extracting the split pin and the clevis pin.
4 Working inside the car unscrew the brake servo mounting nuts.
5 Withdraw the servo from the engine compartment.
6 When refitting the servo unit and the master cylinder, check the pushrod dimensions (X and L) as described in Section 12.

20 Handbrake – adjustment

All models
1 The handbrake will normally be kept in adjustment by the action of the drum brake automatic adjuster or the self-adjusting action of the rear disc calipers.
2 However, the cable may require periodic adjustment to compensate for cable stretch.
3 Raise the rear of the car, support securely and fully release the handbrake.
4 Remove the handbrake primary rod protective cover from under the car.
5 Loosen the locknuts (3) on the sleeve (1) (Fig. 9.35).

Drum brakes
6 Turn the sleeve until, with the rear roadwheels being turned, the shoes can just be heard making contact with the drum.
7 Unscrew the sleeve until the shoes no longer drag.
8 When correctly adjusted, the handbrake lever should fully apply the brakes when pulled over 9 notches.

Disc brakes
9 Turn the sleeve to slacken the cables until there is a clearance (J – Fig. 9.36) of 5.0 mm (0.2 in) between the cable end nipple and the caliper lever. Pull the nipple to check this on both calipers.

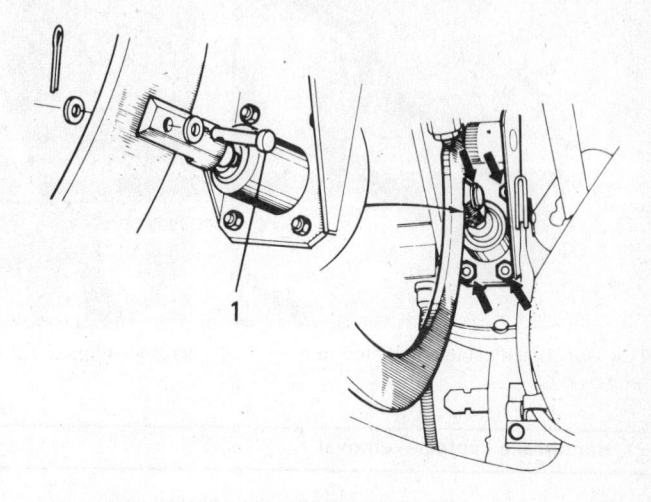

Fig. 9.34 Servo push-rod clevis pin (1) and mounting nuts (arrowed) (Sec 19)

10 Now turn the adjuster sleeve until any clearance between the nipple and the lever is just eliminated.
11 If the adjustment has been correctly carried out the brakes will be fully applied when the handbrake lever has been pulled over 10 to 12 notches. Never over adjust, a shorter handbrake lever travel will cause rapid wear of the disc pads.

All models
12 Tighten the locknuts on the sleeve, refit the cover and lower the car to the floor.

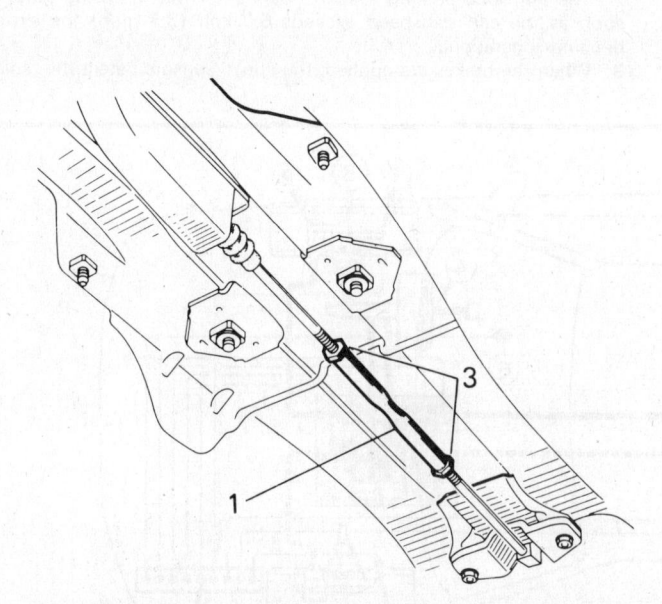

Fig. 9.35 Handbrake adjuster sleeve (1) and locknuts (3) (Sec 20)

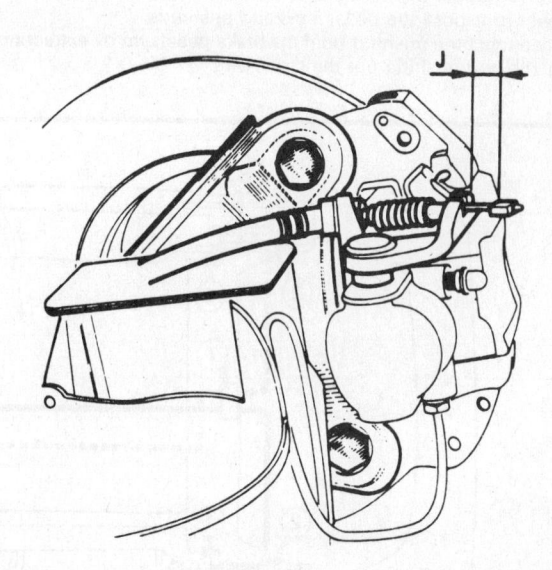

Fig. 9.36 Handbrake cable clearance at rear caliper (Sec 20)

J = 5.0 mm (0.2 in)

252   Chapter 9 Braking system

21.4 Handbrake cable connected to rear caliper lever

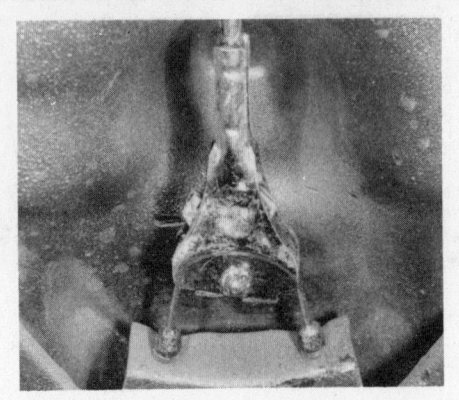

21.6 Handbrake cable equaliser

21.7 Handbrake cable clip

21 Handbrake cables – removal

1 Raise the rear of the car and remove the rear roadwheels.
2 Fully release the handbrake.
3 On models with drum brakes, remove the drums and disconnect the cable(s) from the shoe lever(s).
4 On models with rear disc brakes, disconnect the cable(s) from the caliper lever(s) (photo).
5 If there is any difficulty in releasing the cable end fittings, slacken the adjuster sleeve on the handbrake primary rod right off.
6 Working at the cable equaliser, extract the clip, remove the clevis pin and disconnect the equaliser from the primary rod. Slip the cable out of the equaliser groove (photo).
7 Disconnect the cables from their body clips (photo).
8 Renew the cable by reversing the removal operations. Apply grease to the equaliser groove. Adjust as described in Section 20.

22 Brake pedal – removal and refitting

1 Working under the facia panel, extract the two screws and remove the cover to expose the pedal arms and pushrods.
2 Disconnect the pushrod from the brake pedal arm by extracting the spring clip and pushing out the clevis pin.
3 Disconnect the return and tension springs and remove the spring clip from the end of the pedal cross-shaft.
4 Push the cross-shaft out of the pedal bracket until the pedal can be removed.
5 Refitting is a reversal of removal.

23 Anti-lock braking system – description

1 The hydraulic components used in this system are the same as those used in a conventional braking system but the following additional items are fitted. Refer to Fig. 9.37.

    A speed sensor (1) on each wheel
    Four targets (8) mounted at the front of the driveshafts and at the rear of the rear hubs
    An electronic computer (5) comprising a self-monitoring system, a hydraulic unit (3) which incorporates a pressure regulating valve, and a high pressure pump
    A warning lamp is fitted to the instrument panel

2 The anti-lock braking system operates in the following way. As soon as the car roadspeed exceeds 6.0 kph (3.7 mph) the system becomes operational.
3 When the brakes are applied, the speed sensors detect the rapidly

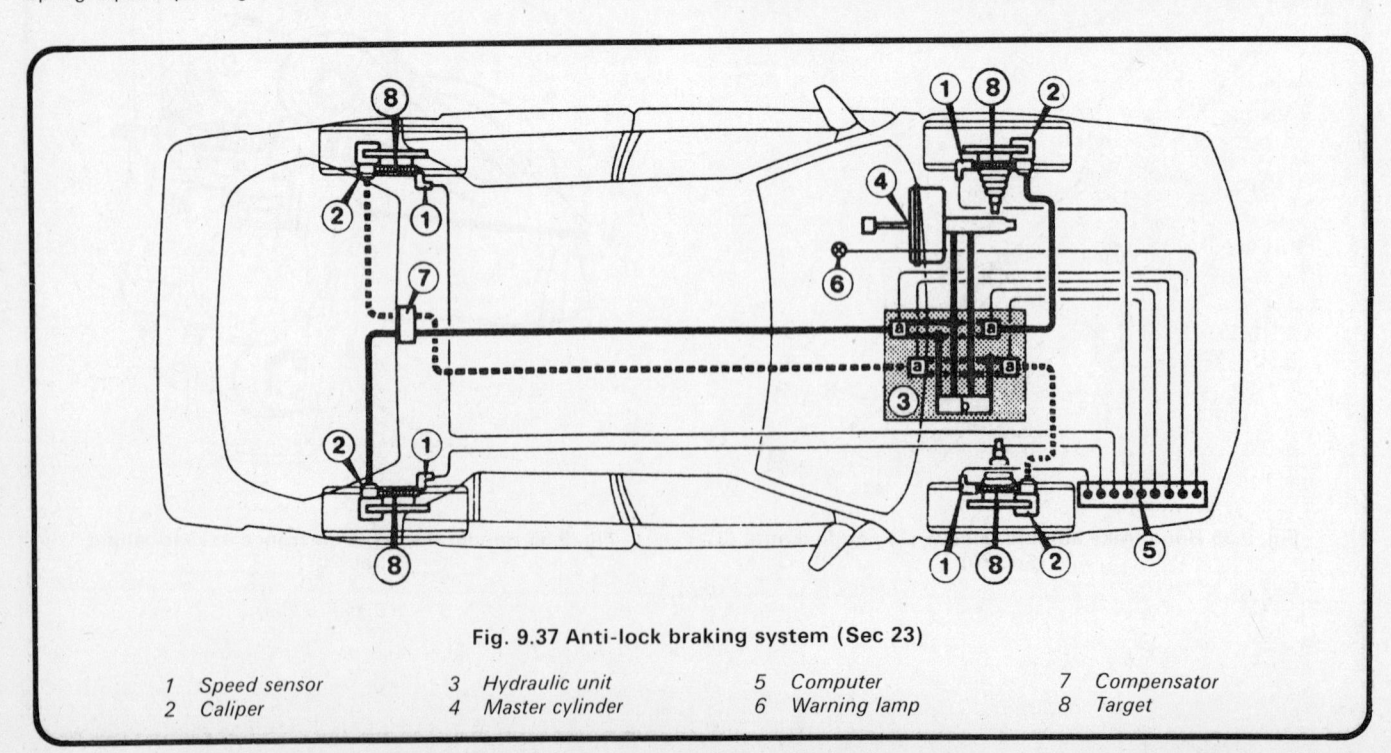

Fig. 9.37 Anti-lock braking system (Sec 23)

| 1 Speed sensor | 3 Hydraulic unit | 5 Computer | 7 Compensator |
| 2 Caliper | 4 Master cylinder | 6 Warning lamp | 8 Target |

Chapter 9 Braking system

falling speed of any roadwheel which happens if it starts to lock. The computer operates a regulator valve to prevent hydraulic braking pressure rising and to reduce it until the particular roadwheel deceleration ceases.

4 As soon as this deceleration ceases, a reverse phase commences to again raise the hydraulic pressure to normal.

5 The anti-lock cycle can be repeated up to ten times per second.

6 In order to maintain even braking, one rear wheel locking will subject the opposite brake to the same pressure variation cycle.

7 The warning lamp indicates to the driver the fact that the system is non-operational. Should this happen, the normal braking system remains fully operational.

24 Anti-lock braking system – maintenance and testing

Note: *A 35 contact junction block (MS 1048), a voltmeter and an ohmmeter (or multimeter) will be needed for these operations*

1 It is important that the speed sensor securing screws are kept tightened to the specified torque (see Specifications).

2 The most likely reason for a fault occurring in the system is a poor or corroded electrical contact. Inspect all leads and terminals, clean if necessary. If the fault remains, disconnect the computer from its base and substitute the junction block MS1048.

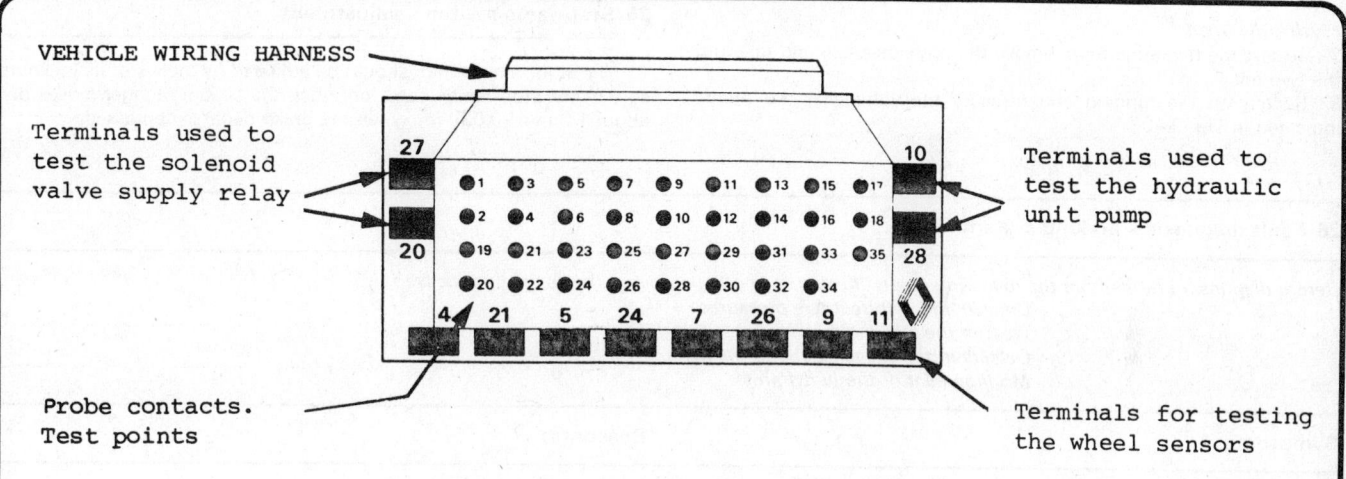

Fig. 9.38 Anti-lock braking test junction block (Sec 24)

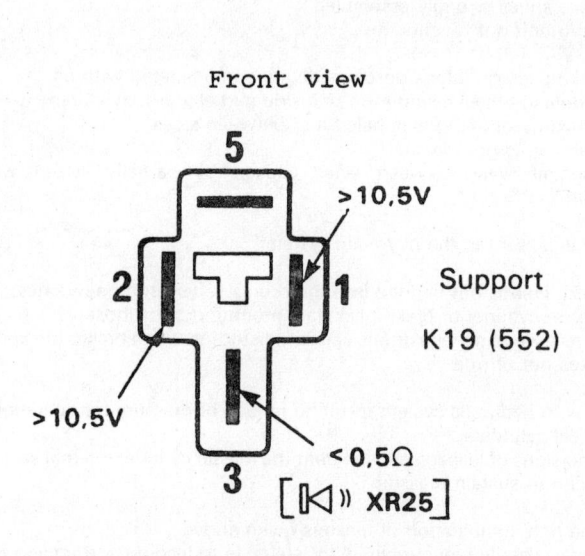

Fig. 9.39 Anti-lock braking system supply relay (Sec 24)

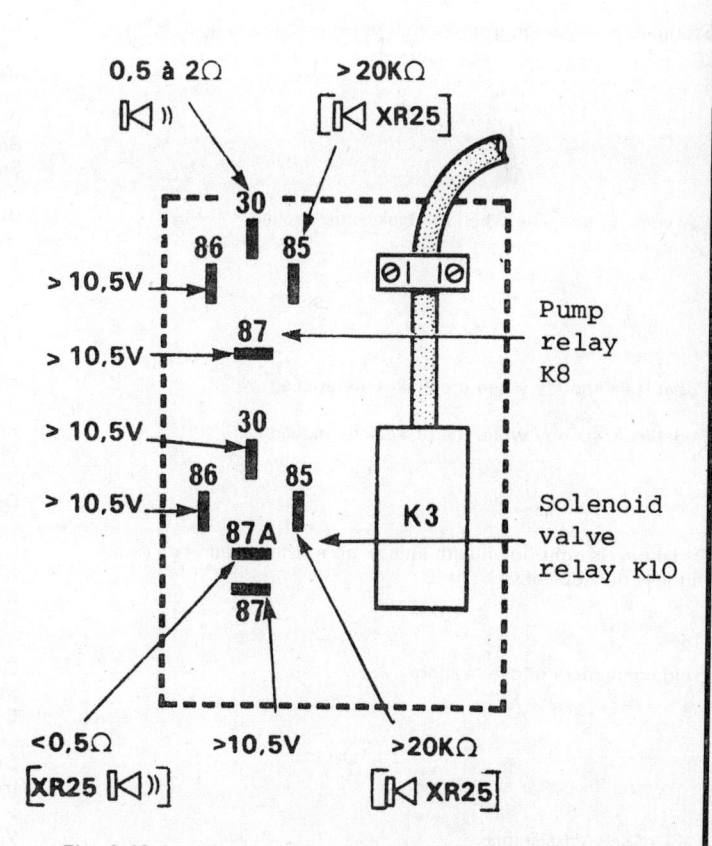

Fig. 9.40 Anti-lock braking system hydraulic unit relay sockets (Sec 24)

Chapter 9 Braking system

Hydraulic pump
3 The hydraulic pump can be tested by connecting terminals 10 and 28 on the junction block (ignition on). Listen for the pump running; limit the test to a maximum of two seconds.

Supply relay
4 The supply relay, located under the expansion bottle for the cooling system, can be tested if it is first disconnected.
5 Switch on the ignition and measure the voltages shown in Fig. 9.39.
6 Switch off the ignition, and measure the resistance at the points shown.

Hydraulic unit
7 To test the hydraulic unit, remove the protective cap and take out the two relays.
8 Switch on the ignition and measure the voltage at the points indicated in Fig. 9.40.
9 Switch off the ignition and measure the resistances at the points shown.

Solenoid valve relay
10 Connect the terminals 20 and 27 on the junction block (ignition on). The solenoid valve should be heard to operate.

Wheel sensors
11 Refer to Fig. 9.38 and connect the voltmeter to terminal 11. Turn each roadwheel sharply, and a voltage exceeding 0.1 volts should be indicated. If it is not, check the lead connections to the sensors and for play in the wheel bearings.

25 Stop-lamp switch – adjustment

1 The stop-lamp switch should be adjusted by means of its locknuts so that the stop-lamps come on when the switch plunger moves out about 1.0 mm (0.039 in) when the brake pedal is depressed.

26 Fault diagnosis – braking system

Before diagnosing faults from the following chart, check that any braking irregularities are not caused by:
- Uneven and incorrect tyre pressures
- Wear in the steering mechanism
- Defects in the suspension and dampers
- Misalignment of the bodyframe

Symptom	Reason(s)
Pedal travels a long way before the brakes operate	Incorrect pedal adjustment Brake shoes set too far from the drums
Stopping ability poor, even though pedal pressure is firm	Linings, discs or drums badly worn or scored One or more wheel hydraulic cylinders seized, resulting in some brake shoes not pressing against the drums (or pads against disc) Brake linings contaminated with oil Wrong type of linings fitted (too hard) Brake shoes wrongly assembled Servo unit not functioning
Car veers to one side when the brakes are applied	Brake pads or linings on one side are contaminated with oil Hydraulic wheel cylinder on one side partially or fully seized A mixture of lining materials fitted between sides Brake discs not matched Unequal wear between sides caused by partially seized wheel cylinders
Pedal feels spongy when the brakes are applied	Air is present in the hydraulic system
Pedal feels springy when the brakes are applied	Brake linings not bedded into the drums (after fitting new ones) Master cylinder or brake backplate mounting bolts loose Severe wear in brake drums causing distortion when brakes are applied Discs out of true
Pedal travels right down with little or no resistance and brakes are virtually non-operative	Leak in hydraulic system resulting in lack of pressure for operating wheel cylinders If no signs of leakage are apparent the master cylinder internal seals are failing to sustain pressure
Binding, juddering, overheating	One or a combination of reasons given above Shoes installed incorrectly with reference to leading and trailing ends Broken shoe return spring Disc out-of-round Drum distorted Incorrect pedal or handbrake adjustment
Lack of servo assistance	Vacuum hose disconnected or leaking Non-return valve defective or incorrectly fitted Servo internal defect

Chapter 10 Suspension and steering

Contents

Description	1
Fault diagnosis – suspension and steering	30
Front anti-roll bar – removal and refitting	3
Front shock absorber and coil spring – removal and refitting	4
Front suspension hub carrier – removal and refitting	8
Front suspension lower track control arm – removal and refitting	7
Front suspension upper track control arm – removal and refitting	5
Front suspension upper track control arm balljoint – renewal	6
Maintenance	2
Power-assisted steering pump – removal and refitting	23
Power steering circuit – bleeding	25
Power steering gear and pump – overhaul	24
Rear anti-roll bar – removal and refitting	11
Rear coil spring – removal and refitting	10
Rear shock absorber – removal and refitting	9
Rear suspension radius rod – removal and refitting	12
Rear suspension track control arm – removal and refitting	13
Rear wheel alignment	27
Steering angles and front wheel alignment	26
Steering column – removal, overhaul and refitting	19
Steering column lock/ignition switch – removal and refitting	29
Steering gear (manual) – removal and refitting	21
Steering gear (power-assisted) – removal and refitting	22
Steering rack bellows – renewal	15
Steering rack slipper – adjustment	20
Steering tie-rod end balljoint – renewal	14
Steering universally-jointed shaft (manual steering) – removal and refitting	17
Steering universally-jointed shaft (power-assisted steering) – removal and refitting	18
Steering wheel – removal and refitting	16
Steering wheel height adjuster	28

Specifications

Front suspension
Type .... Independent, double wishbones, coil springs, telescopic shock absorbers. Anti-roll bar

Rear suspension
Type .... Independent; lower control arms, telescopic struts with coil springs. Anti-roll bar

Camber (non-adjustable):
- V6 Turbo .... 1°15′ ± 30′
- All other models .... −1°15′ ± 30′

Toe-in:
- V6 Turbo .... 1 mm ± 1 mm (0.04 in ± 0.04 in)
- All other models .... 0 ± 1 mm (0 ± 0.04 in)

Steering
Type .... Rack and pinion with safety column and rake adjuster. Power-assistance on all models except TS

Power steering fluid capacity:
- With integral reservoir .... 0.7 litre (1.2 pints)
- With remote reservoir .... 1.1 litre (1.9 pints)

Caster:
Suspension setting (H5 – H2) – see Section 26

	Power-assisted steering	Manual steering
10.0 mm (0.39 in)	4° 0′	2° 0′
35.0 mm (1.38 in)	3° 30′	1° 30′
60.0 mm (2.36 in)	3° 0′	1° 0′
85.0 mm (3.35 in)	2° 30′	0° 30′
110 mm (4.33 in)	2° 0′	0°
Maximum side-to-side difference	1° 0′	1° 0′

One complete turn of the radius rod will alter the castor 0° 15′

Chapter 10 Suspension and steering

Camber (non-adjustable)	0° 30'
Maximum side-to-side difference	1° 0'
Steering axis inclination (non-adjustable)	12° 30' ± 30'
Maximum side-to-side difference	1° 0'
Toe-out	3 mm ± 1 mm (0.12 in ± 0.04)
Number of turns of steering wheel:	
TS	4.5
All other models	3.0
Turning circle (between kerbs)	10.8 m (35.4 ft)

Torque wrench settings

	Nm	lbf ft
Front suspension		
Shock absorber top mounting nut	15	11
Upper arm inboard pivot nut	92	68
Upper arm balljoint taper pin nut	64	47
Upper arm balljoint fixing nuts	21	15
Radius rod to upper arm nuts	68	50
Shock absorber bottom mounting nut	58	43
Shock absorber bottom mounting pivot	79	58
Lower arm inboard pivot nut	128	94
Lower arm balljoint nut	64	47
Anti-roll bar clamp bolts	15	11
Driveshaft nut	245	181
Rear suspension		
Shock absorber upper mounting nut	30	22
Shock absorber bottom mounting bolts	98	72
Radius rod bolts	34	25
Anti-roll bar and bolt	16	12
Track control arm inboard pivot bolt	68	50
Track control arm to hub carrier bolt	88	65
Hub self-locking nut	156	115
Anti-roll bar clamp bolts	15	11
Steering		
Tie-rod end balljoint taper pin nut	39	29
Tie-rod end balljoint locknut	34	25
Steering wheel fixing screws	15	11
Roadwheel bolts:		
4-bolt fixing	88	65
5-bolt fixing	98	72

1 Description

The front suspension is of independent, double wishbone type with coil springs and telescopic hydraulic shock absorbers. An anti-roll bar is fitted. The upper wishbones are positively located by radius rods which are adjustable to alter the castor angle.

The rear suspension is also of independent type, having lower control arms, telescopic struts with coil springs and an anti-roll bar. Lateral control arms (radius rods) are used to positively locate the lower, transverse, control arms.

The steering gear is of power-assisted rack and pinion type on all models except the TS.

The steering column is of safety, collapsible type.

2 Maintenance

1 At the specified intervals, check all suspension flexible bushes for wear.
2 Occasionally check the condition and security of all steering and suspension nuts, bolts and components.
3 Inspect the struts and shock absorbers for signs of fluid leakage. If anything more than a slight weep from the top gland is evident, then the unit must be renewed.
4 If the car tends to roll on bends or dip under heavy braking, check the action of the struts and shock absorbers by pressing the corner of the car downwards and then releasing it. The up and down momentum of the car should be damped out immediately. If the car oscillates up and down several times, the condition of the particular unit should be checked after removal from the car.
5 Inspect the steering rack bellows for splits and loss of lubricant. Look particularly closely at the bottom of the bellows pleats when the steering is at full lock. Splits here can often pass unnoticed.
6 With the help of an assistant, check for wear in the steering tie-rod end balljoints. Move the steering wheel quickly a few degrees in each direction and observe the balljoints for shake or lost movement. If evident renew the tie-rod ends.
7 Check the front suspension lower arm balljoint by inserting a lever carefully between the arm and the hub carrier and checking for vertical movement.
8 At the specified intervals, check the front wheel alignment.
9 On models with power-assisted steering, check the fluid level in the reservoir at the intervals specified in Routine Maintenance.
10 On four-cylinder models with a remotely-sited fluid reservoir, keep the level just above the bottom of the gauze strainer (photo).
11 On six-cylinder models with the fluid reservoir integral with the pump, keep the level up to 1 (Fig. 10.5) if the fluid is hot, or to 2 if cold.
12 Use only specified fluid for topping-up.

3 Front anti-roll bar – removal and refitting

1 Unbolt the anti-roll bar links from the upper wishbone arms (photo).
2 Unbolt the insulator clamps which hold the anti-roll bar to the body member (photo).
3 Remove the anti-roll bar from the car.
4 The links can be removed from the anti-roll bar after extracting the circlip and removing the spacer.
5 Renew the rubber insulators and the links if their flexible bushes are worn.
6 Refitting is a reversal of removal, but tighten the nuts to the specified torque when the weight of the car is on the roadwheels.

257

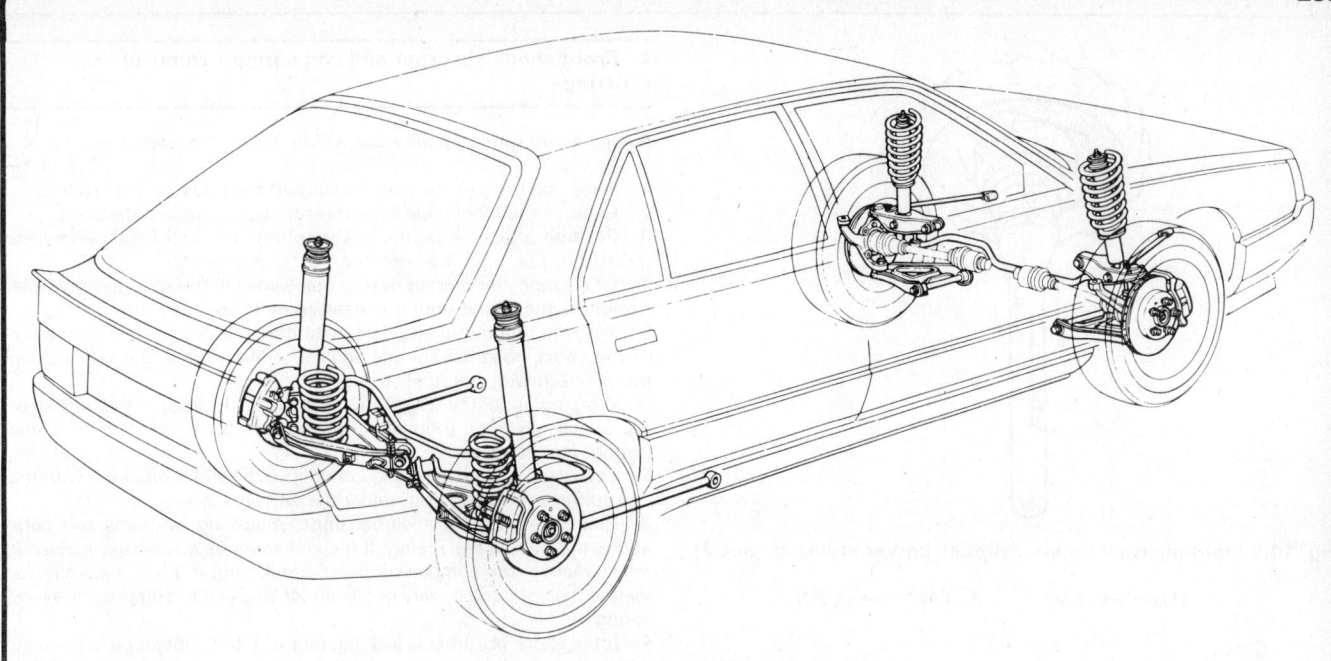

Fig. 10.1 Suspension layout (Sec 1)

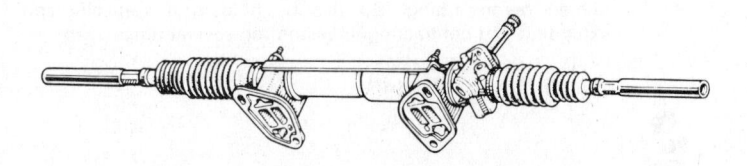

Fig. 10.2 Manual steering gear (LHD shown) (Sec 1)

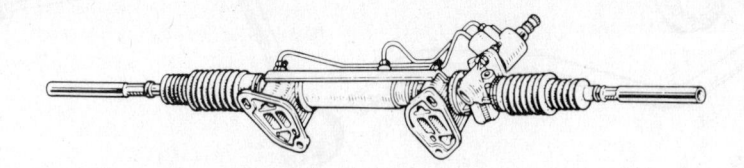

Fig. 10.3 Power-assisted steering gear (LHD shown) (Sec 1)

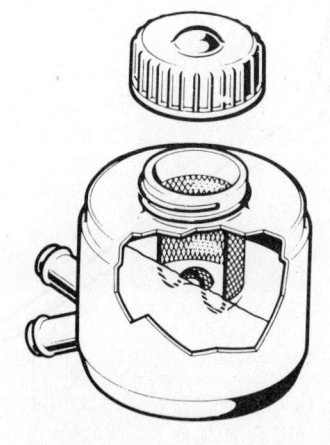

Fig. 10.4 Fluid reservoir on early four-cylinder power-assisted steering (Sec 2)

2.10 Later type power steering fluid reservoir on four-cylinder models

3.1 Front anti-roll bar and link

3.2 Front anti-roll bar insulator and clamp

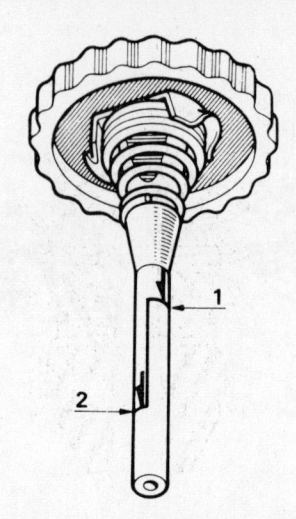

Chapter 10 Suspension and steering

Fig. 10.5 Fluid dipstick on six-cylinder power steering (Sec 2)

1 Max mark (hot)   2 Max mark (cold)

4 Front shock absorber and coil spring – removal and refitting

Note: *A coil spring compressor will be needed for this operation.*

1 Raise the front of the car and support the body on axle stands.
2 Remove the roadwheel from the side being worked upon.
3 Position a jack under the lower control arm and gently compress the suspension until the control arms are horizontal.
4 Fit a spring compressor over as many coils of the spring as possible. Compress the spring until it is freed from its locating cups.
5 Working within the engine compartment, hold the flats on the top of the shock absorber spindle against rotation while the self-locking nut is unscrewed (photos).
6 Working under the wheel arch, release the locknut at the bottom of the shock absorber (photo), unscrew the shock absorber from the bottom mounting pin.
7 Lower the suspension jack and remove the shock absorber with the coil spring (clamped) from under the wheel arch.
8 Remove the shock absorber upper mounting cushions and cups and withdraw the coil spring. If the coil spring is not being renewed it may remain in the compressor pending refitting. If it is to be renewed, release the compressor very gently and then use it to compress the new spring.
9 If the shock absorber is leaking, renew it. If its operation is suspect, grip its lower mounting in a vice fitted with jaw protectors so that the unit is vertical.
10 Fully extend and contract the shock absorber six times. Any evidence of jerky operation, lack of resistance or seizure will indicate the need for a new unit.
11 The new unit will probably have been stored flat so it must be primed before fitting. Do this by holding it vertically and fully extending and contracting its piston rod several times.

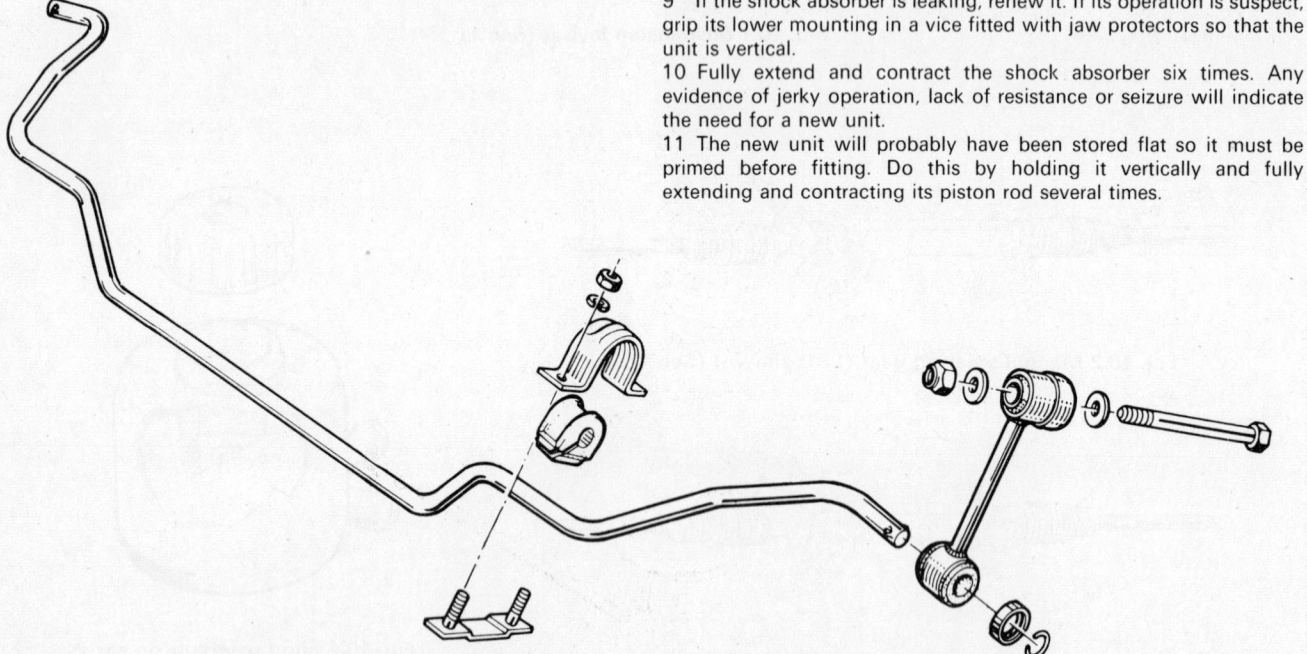

Fig. 10.6 Front anti-roll bar (Sec 3)

4.5A Front shock absorber upper mounting spindle cap

4.5B Unscrewing front shock absorber upper mounting spindle nut

4.6 Front shock absorber lower mounting locknut (arrowed)

Chapter 10 Suspension and steering

12 Refitting is a reversal of removal, but observe the following points.
13 Make sure that the end of the bottom spring coil is up against the stop in its lower seat.
14 Make sure that the upper mounting cups and cushions are correctly located.
15 Tighten all nuts to the specified torque after the weight of the car has been lowered onto its roadwheels.

5 Front suspension upper track control arm – removal and refitting

Note: *A balljoint splitter will be needed for this operation.*

1 Raise the front of the car, support the body on axle stands and remove the roadwheel.
2 Place a jack under the suspension lower track control arm.
3 Release the locknut on the shock absorber lower mounting.
4 Disconnect the radius rod from the suspension upper track control arm.
5 Unscrew the nut on the upper balljoint taper pin and then, using a suitable tool, disconnect the balljoint from the hub carrier (photo).
6 Remove the anti-roll bar as described in Section 3.
7 Remove the shock absorber lower mounting pivot pin, raise the control arm and unscrew the shock absorber lower mounting.
8 Unscrew the track control arm inboard pivot bolt and withdraw the track control arm (photo).

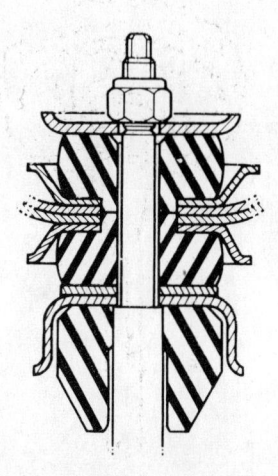

Fig. 10.7 Front shock absorber upper mounting (Sec 4)

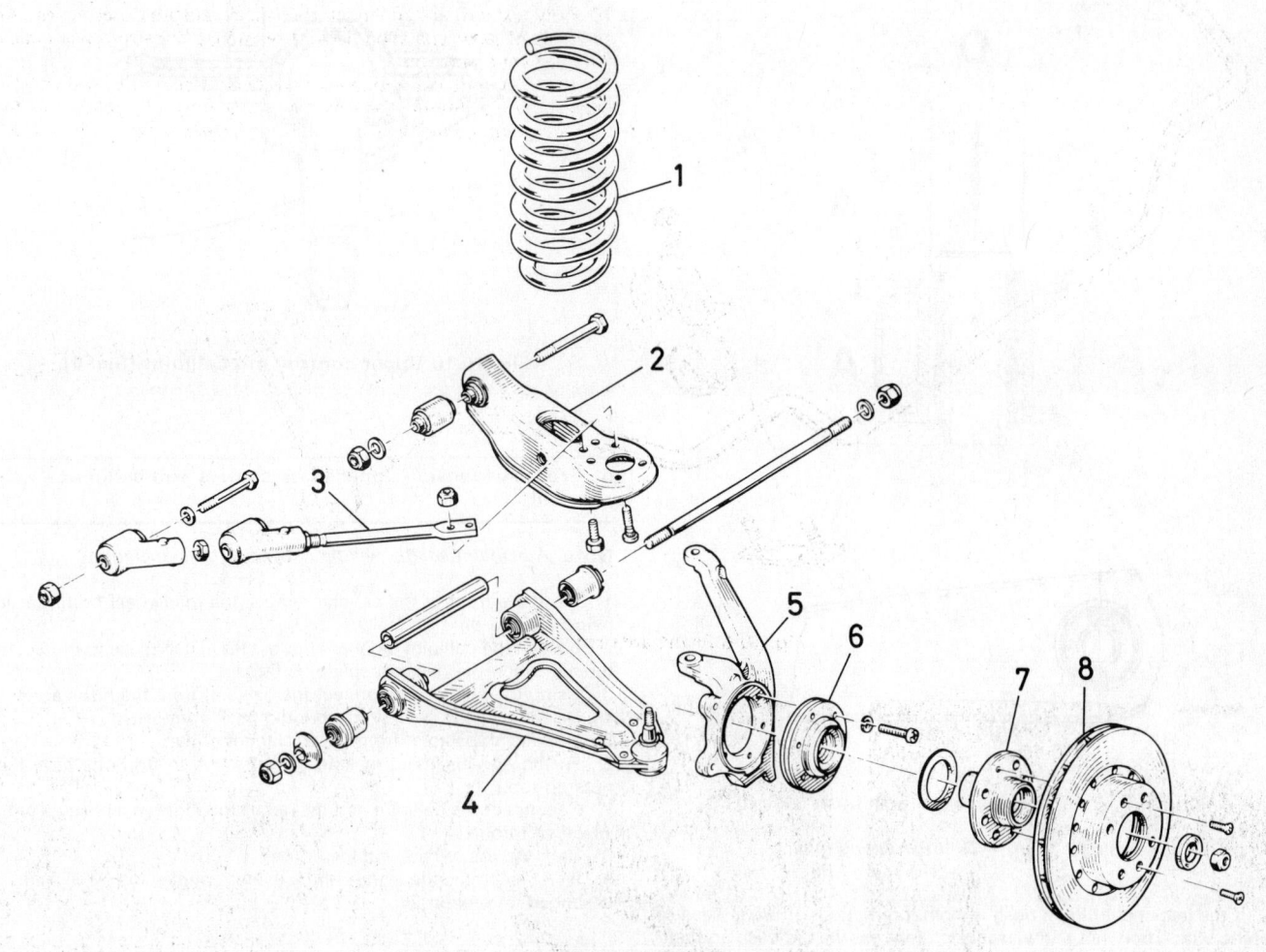

Fig. 10.8 Front suspension components (Sec 5)

1 Coil spring
2 Upper track control arm (wishbone)
3 Radius rod
4 Lower track control arm (wishbone)
5 Hub carrier
6 Hub
7 Bearing retaining plate
8 Brake disc

Chapter 10 Suspension and steering

5.5 Front suspension upper arm balljoint taper pin

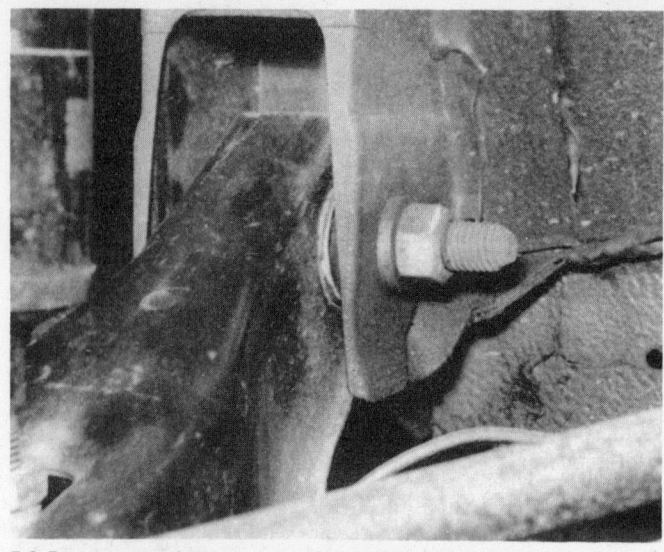

5.8 Front suspension upper arm inboard pivot bolt

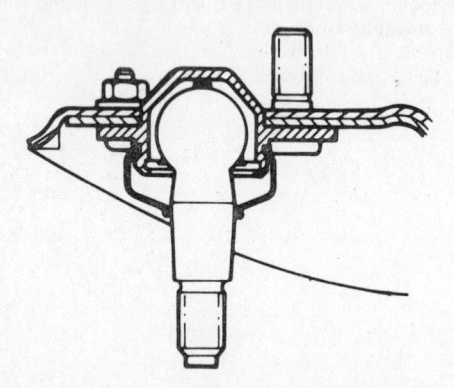

Fig. 10.10 Upper control arm balljoint (Sec 6)

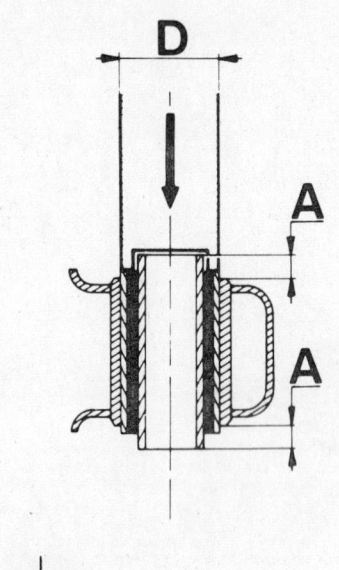

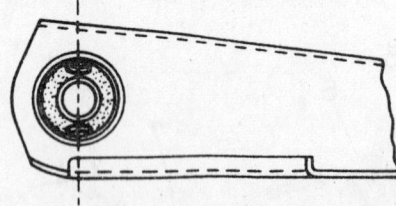

Fig. 10.9 Upper track control arm bush (Sec 5)

A = 7.5 mm (0.30 in)   D Bush fitting tube = 34.5 mm (1.36 in) o.d.

9 If the track control arm bush is worn, press it out or use a piece of tubing with a bolt, nut and washers to draw out the old bush and refit the new one. Make sure that the new bush is centralised as shown in Fig. 10.9.
10 The balljoint may be renewed as described in the next Section.
11 Refitting is a reversal of removal. Do not fully tighten to the specified torque. The track control arm pivot bolt and the shock absorber bottom mounting pivot pin until the weight of the car is on the roadwheels.

6 Front suspension upper track control arm balljoint – renewal

Note: *A balljoint splitter will be needed for this operation.*

1 Raise the front of the car and remove the roadwheel from the side being worked on.
2 Loosen the balljoint taper pin nut but leave it in place to protect the threads when the balljoint splitter is fitted.
3 Using the splitter, disconnect the balljoint from the hub carrier.
4 Disconnect the radius rod from the track control arm.
5 Unbolt the balljoint from the track control arm.
6 Fit the new balljoint, making sure that the fixing bolts have their nuts on top.
7 Reconnect the balljoint and the radius rod, tighten all fixings to the specified torque.
8 Refit the roadwheel and lower the car.
9 It is recommended that the steering angles are checked, as described in Section 26.

7 Front suspension lower track control arm – removal and refitting

1 Raise the front of the car, support it securely and remove the roadwheel.
2 Unscrew the lower track control arm balljoint nut until it makes

Chapter 10 Suspension and steering

contact with the driveshaft joint. Continue unscrewing it so that it acts as an extractor to force the balljoint taper pin out of the eye of the hub carrier. Renew the nut ready for reassembly (photo).
3 Unscrew the track control arm inboard pivot pin nuts and withdraw the pin towards the front of the car (photo). Note the location of the castor control shim.
4 Worn flexible bushes can be renewed by pressing them out or by using a distance piece, bolt, nut and washers. Fit the new bushes as shown in Fig. 10.13.
5 If the balljoint is worn, drill out the securing rivets and remove it. The new balljoint will be supplied complete with fixing bolts; make sure that the bolt heads are uppermost (photo).
6 Refitting is a reversal of removal, but grease the pivot pin before inserting it and make sure that the castor control shim is correctly located.
7 Tighten all nuts to the specified torque, but leave the pivot pin ones until the weight of the car is on the roadwheels.

8 Front suspension hub carrier – removal and refitting

Note: *A balljoint splitter and suitably-sized Torx bit will be needed for this operation.*

1 Raise the front of the car and remove the roadwheel from the side being worked upon.
2 Have an assistant apply the brakes fully and unscrew the driveshaft/hub nut. Alternatively, use a long lever between two roadwheel bolts to prevent rotation.

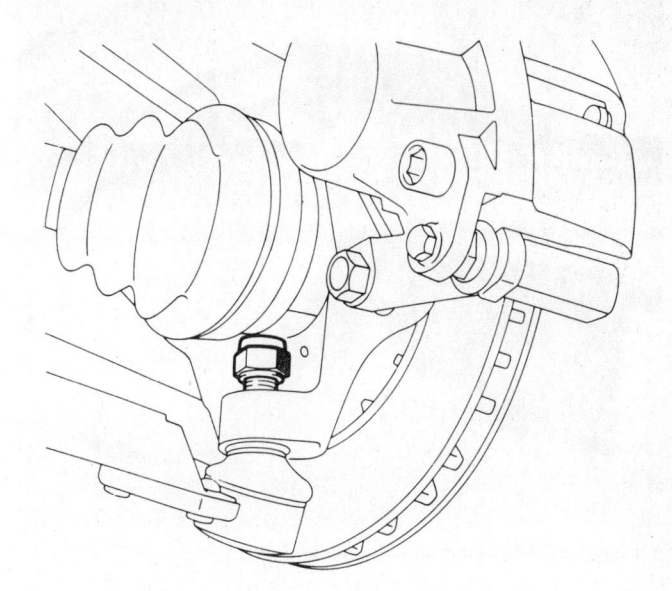

Fig. 10.11 Using lower balljoint nut to disconnect balljoint from hub carrier (Sec 7)

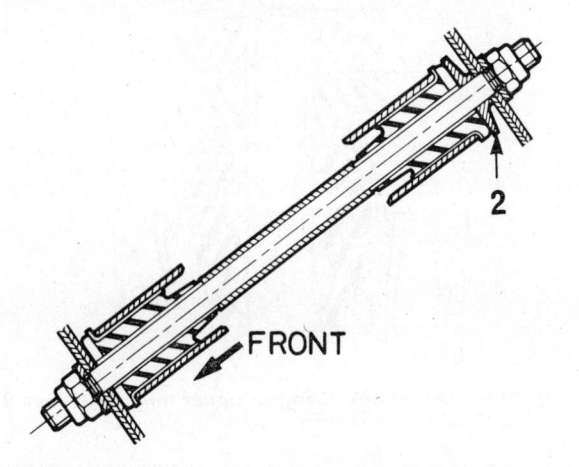

Fig. 10.12 Lower track control arm inboard pivot (Sec 7)

2 Castor shim

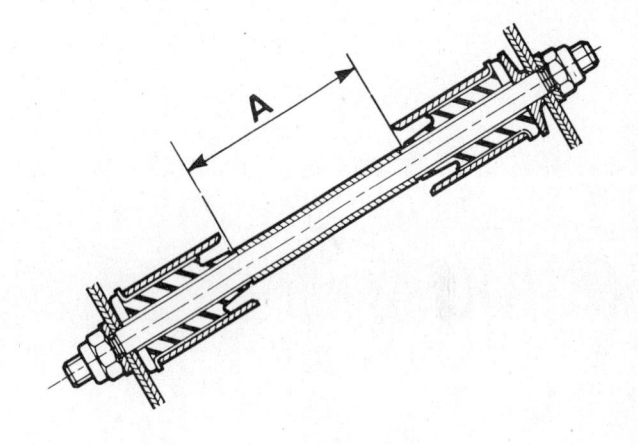

Fig. 10.13 Lower track control arm bush setting diagram (Sec 7)

$A = 112.6$ mm (4.4 in)

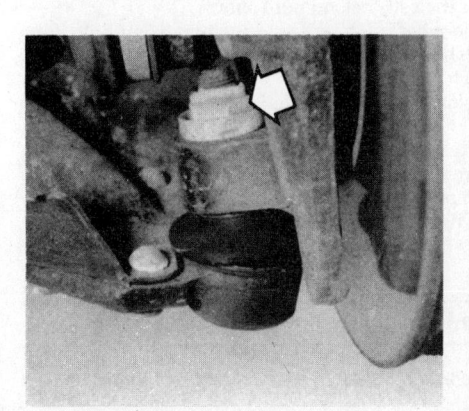

7.2 Front suspension lower arm balljoint nut (arrowed)

7.3 Front suspension lower arm inboard pivot nut

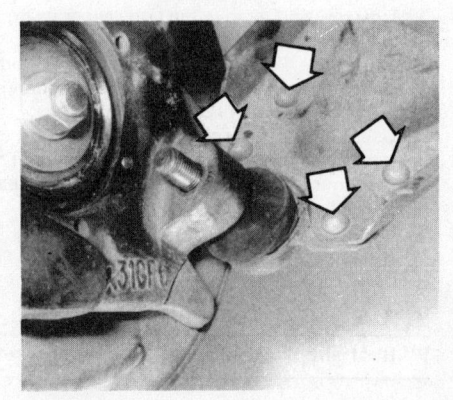

7.5 Front suspension upper arm balljoint rivets (arrowed)

Chapter 10 Suspension and steering

9.5 Rear shock absorber upper mounting

9.6 Rear shock absorber lower mounting (arrowed)

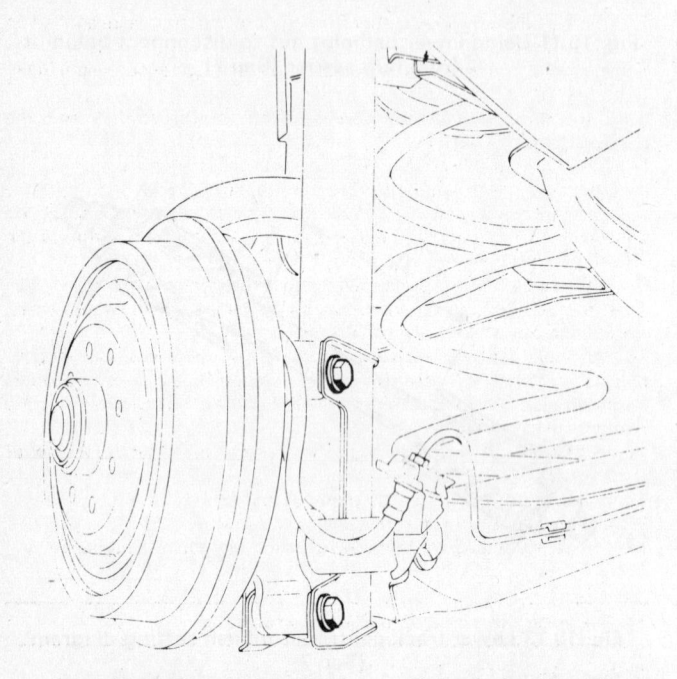

Fig. 10.14 Shock absorber lower fixing bolts (Sec 9)

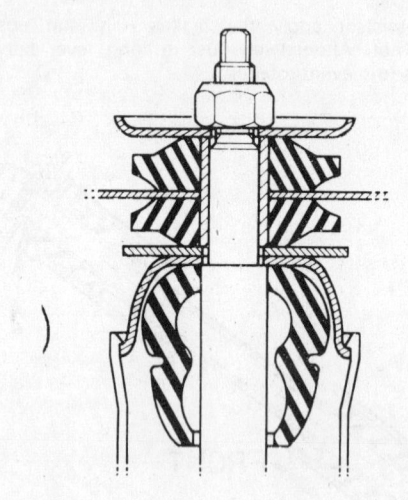

Fig. 10.15 Rear shock absorber upper mounting (Sec 9)

3 Unscrew and remove the two caliper fixing bolts and tie the caliper up out of the way.
4 Extract the Torx screws and take off the brake disc.
5 Using a Torx key, unscrew the hub bearing retaining plate screws.
6 Remove the hub/bearing assembly.
7 Using a balljoint splitter, disconnect the upper track control arm balljoint and the steering tie-rod end balljoint from the hub carrier.
8 Disconnect the lower track control arm balljoint, as described in Section 7.
9 Lift the hub carrier from the car.
10 Refitting is a reversal of removal; tighten all nuts and bolts to the specified torque.

9 Rear shock absorber – removal and refitting

1 Raise the rear of the car and remove the roadwheel from the side being worked upon.

2 Support the suspension arm using a jack or axle stand.
3 Working inside the luggage area, remove the parcel shelf and side shelf from the side concerned (see Chapter 11).
4 Pull off the rubber cap from the shock absorber top mounting.
5 Hold the flats on the shock absorber spindle with an open-ended spanner and unscrew the self-locking nut (photo).
6 Working under the wheel arch, unscrew the two fixing bolts from the bottom of the shock absorber (photo). Compress the shock absorber and remove it.
7 Test the shock absorber as described in Section 4, paragraphs 9, 10 and 11.
8 Refit the shock absorber, making sure that the upper mounting cushions and cups are correctly located (Fig. 10.15).
9 Smear the lower mounting bolts with grease before fitting and only tighten them and the upper mounting nut after the weight of the car is on the roadwheels.

10 Rear coil spring – removal and refitting

Note: *A spring compressor will be needed for this operation.*

1 Raise the rear of the car and remove the roadwheel from the side being worked upon.

Chapter 10 Suspension and steering

11.2 Rear anti-roll bar end link

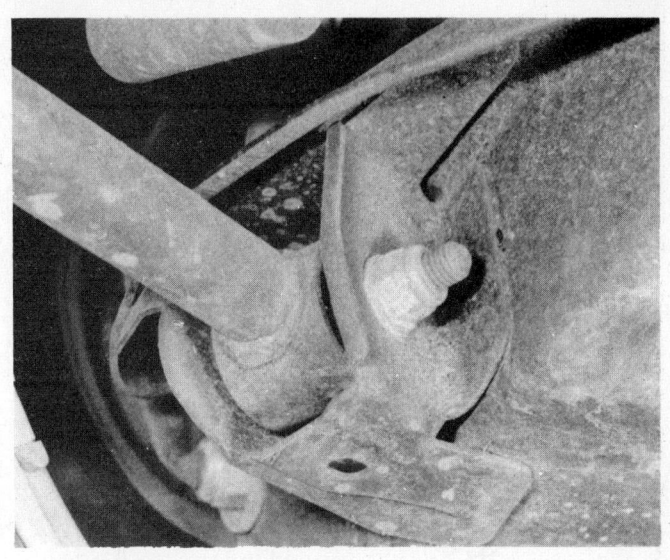

12.2 Rear radius rod attachment

2 Remove the shock absorber as described in the preceding Section.
3 Using a suitable coil spring compressor, slightly compress the spring to relieve the tension in the anti-roll bar.
4 A simple spring compressor may be made up from a length of studding, secured in the shock absorber top mounting hole and with a plate across the lower spring coils held to the studding by a nut.
5 Disconnect the anti-roll bar end link.
6 Disconnect the brake pressure regulator valve link rod from the suspension arm.
7 Lower the suspension arm until the spring can be removed complete with clamp type compressor or, if the studding type compressor is used, release it very gently until the coil spring is free of tension. Do not stretch the brake hose.
8 Refitting is a reversal of removal; make sure that the end of the top coil of the spring butts against the stop in the cup.
9 Compress the spring so that the anti-roll bar and the pressure regulator link rod can be connected. Do not tighten the fixings to the specified torque until the weight of the car is again on its roadwheels.

11 Rear anti-roll bar – removal and refitting

1 It is unlikely that the rear anti-roll bar will ever require removal as the clamp insulators and end link cushions can be renewed without removing it.
2 However, if it must be removed, then the fuel tank will have to be withdrawn first (Chapter 3), and the anti-roll bar disconnected and manoeuvred out of the body member cut-outs after the end-links and clamps have been unbolted (photo).

12 Rear suspension radius rod – removal and refitting

1 Raise the rear of the car and remove the roadwheel and the reaction members which run underneath the fuel tank.
2 Unbolt the radius rod from the suspension arm and from the body bracket and remove the rod (photo).
3 Rebush the rod if necessary and refit.
4 Tighten the bolts to the specified torque only after the weight of the car is again on the roadwheels, and use new self-locking nuts.

13 Rear suspension track control arm – removal and refitting

Note: *A spring compressor will be needed for this operation*

1 Raise the rear of the car and remove the roadwheel from the side being worked upon.

2 Remove the shock absorber and the coil spring, as described in earlier Sections.
3 Disconnect the handbrake cable from the shoe or caliper lever (Chapter 9).
4 Disconnect the brake pressure regulator control arm link from the suspension arm.
5 Disconnect the flexible brake hose and cap its end.
6 Disconnect the radius rod from the suspension arm.
7 Unscrew and remove the pivot bolt from the inboard end of the control arm and withdraw the arm, carefully noting the location of the wheel alignment shims.
8 The hub carrier may be unbolted from the control arm.
9 If the flexible bushes are worn in the control arm or hub carrier, press them out and press new ones in.
10 Refit the control arm, taking care to replace the shims in their original locations. If a new control arm is being fitted, place two shims on either side of the arm then relocate them later when checking the rear wheel alignment.
11 Tighten all nuts and bolts to the specified torque after the weight of the car is on its roadwheels.
12 Bleed the brake hydraulic system (Chapter 9).
13 Check the rear wheel alignment (Section 27).
14 Check the setting of the brake pressure regulator (Chapter 9).

14 Steering tie-rod and balljoint – renewal

Note: *A balljoint splitter will be needed for this operation*

1 Raise the front of the car and remove the roadwheel from the side being worked upon.
2 Hold the flats on the balljoint with an open-ended spanner and release the locknut (photo).
3 Unscrew the balljoint taper pin nut but do not remove it so that it will protect the threads when the splitter is fitted.
4 Fit the splitter and disconnect the balljoint from the eye of the steering arms (photo).
5 Count the number of exposed threads and then hold the tie-rod still while unscrewing the balljoint from it. Remove the locknut.
6 Grease the threads of the new balljoint and screw on the locknut.
7 Screw the balljoint into the tie-rod until the same number of threads are exposed as was originally recorded.
8 Connect the balljoint to the steering arm. Never grease the taper pin as it will rotate when the nut is tightened. Should this happen when tightening any balljoint nut, apply pressure with a lever or grips to the socket housing to force the taper pin further into its conical eye while the nut is being tightened (photo).
9 Tighten the taper pin nut to the specified torque.
10 Set the balljoint at the centre of its arc of travel, hold it still with an open-ended spanner and tighten the locknut.
11 Refit the roadwheel, lower the car.

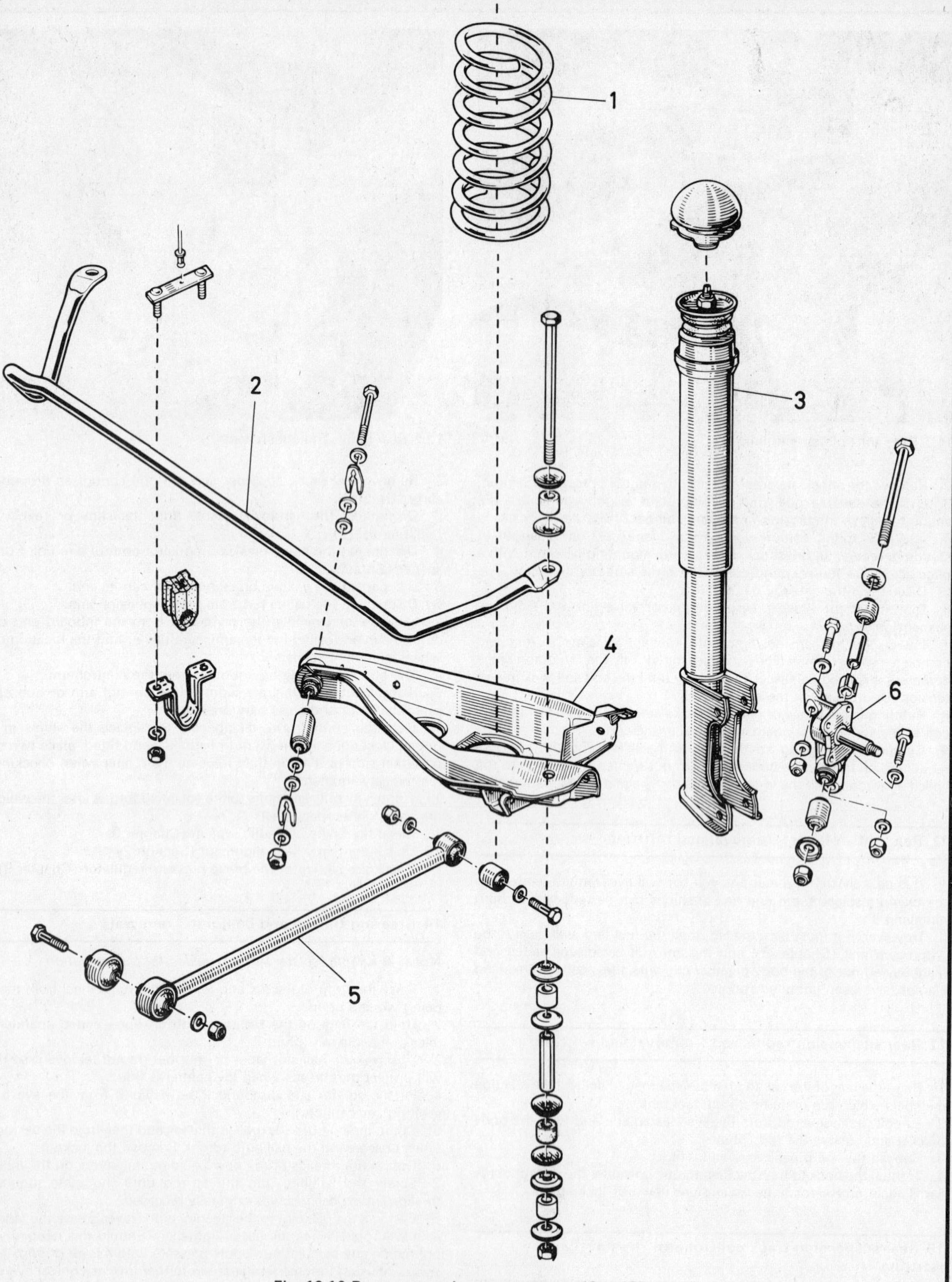

Fig. 10.16 Rear suspension components (Sec 12)

1 Coil spring	3 Rear shock absorber	5 Radius rod
2 Anti-roll bar	4 Track control arm	6 Hub carrier

Chapter 10 Suspension and steering

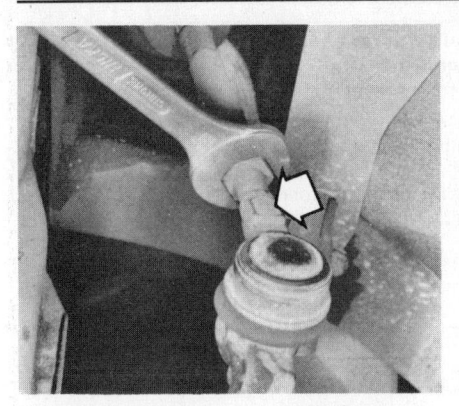

14.2 Steering tie-rod end balljoint flats (arrowed)

14.4 Typical balljoint splitter

14.8 One method of preventing a balljoint taper pin from rotating

16.2 Removing steering wheel hub cover

16.4 Extracting steering shaft safety circlip

16.5A Steering wheel hub and Torx screws

12 However precisely the new balljoint may have been fitted, due to the manufacturing differences alone, the front wheel alignment should be checked as described in Section 26.

15 Steering rack bellows – renewal

Note: *A balljoint splitter will be needed for this operation.*

1 Remove the tie-rod end balljoint as described in the preceding Section.
2 Release the bellows clip and slide the bellows off the end of the tie-rod.
3 If there is no evidence of dirt or water contamination, the new bellows may be fitted and clipped in position. If dirt or water are found to have contaminated the rack lubricant, extend the rack and wipe away as much old lubricant as possible. Pack some specified grease inside the new bellows and smear it liberally on the exposed rack teeth. Fit the new bellows.
4 Refit the balljoint and roadwheel, check the wheel alignment as described in Section 26.

16 Steering wheel – removal and refitting

Note: *A suitably-sized Torx bit will be needed for this operation.*

1 Set the front roadwheels in the straight-ahead attitude.
2 Pull off the cover from the steering wheel hub (photo).
3 Mark the hub-to-shaft relationship.
4 Extract the safety circlip from the top of steering column shaft (photo).
5 Extract the Torx screws and then screw them into the alternative threaded holes. As the screws are tightened evenly, the hub will be drawn off (photo). Remove the hub and steering wheel (photo).
6 Refit the steering wheel and hub to the shaft – aligning the previously-made marks. Hold the steering wheel in position on the shaft, and tighten the hub Torx screws. Refit the circlip.
7 Road test the car to check the steering wheel position. If necessary, slacken the Torx screws and move the wheel; the elongated screws will allow fine adjustment.
8 Retighten the Torx screws and refit the hub cover.

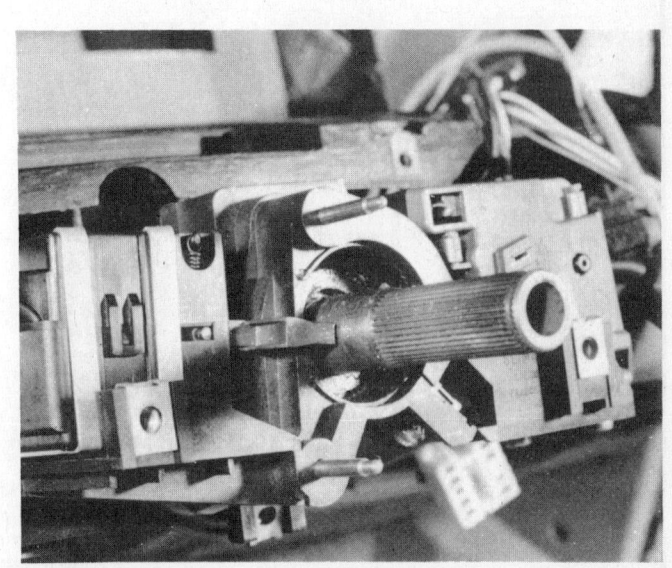

16.5B Steering shaft with steering wheel removed

Chapter 10 Suspension and steering

17 Steering universally-jointed shaft (manual steering) – removal and refitting

1 Working within the engine compartment, prise out the plastic clips which secure the rubber boot to the bulkhead. Push the boot into the car interior and remove the split plastic bush.
2 Unscrew the pinch-bolt which holds the universally-jointed shaft to the pinion on the steering rack housing.
3 Working under the facia panel, remove the key and pinch-bolts from the coupling at the bottom of the steering column tube. Remove the shaft. Check its length and renew it if it is outside specified tolerance (Fig. 10.18). This will be due to impact damage as the shaft is collapsible.
4 Refitting is a reversal of removal; tighten all bolts. The bolts at the upper coupling must only be tightened when they are aligned as shown (A and B – Fig. 10.19).

18 Steering universally-jointed shaft (power assisted steering) – removal and refitting

1 The operations are similar to those described in the preceding Section except that only two pinch-bolts must be removed, both accessible from inside the car.
2 Check the length of the shaft and renew it if it is outside specified tolerance. This will be due to impact damage as the shaft is collapsible (Fig. 10.21).

19 Steering column – removal, overhaul and refitting

1 Remove the steering wheel as described in Section 16.
2 Remove the universally-jointed shaft as described in Section 17 or 18.

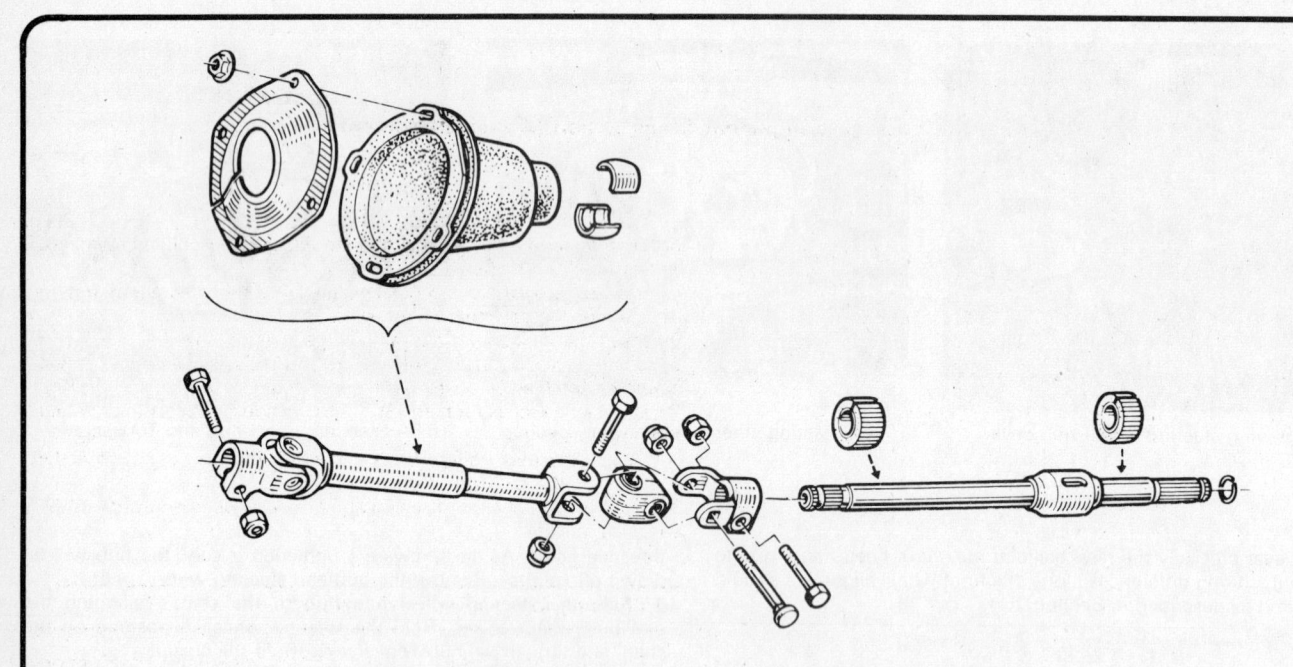

Fig. 10.17 Steering shaft components (manual) (Sec 17)

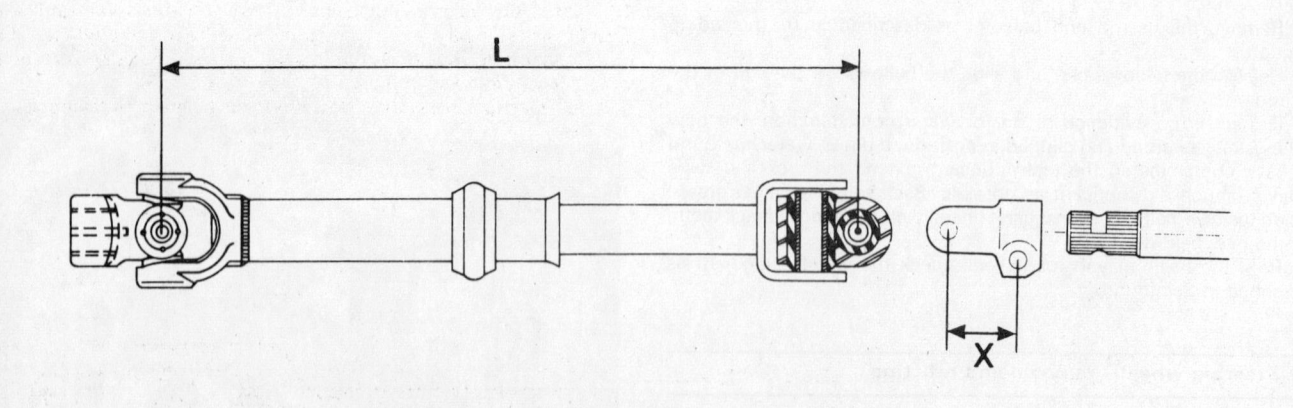

Fig. 10.18 Alternative types of manual steering shafts (Sec 17)

Where   X = 30.0 mm (1.18 in)
         L = 306.0 mm ± 1 mm (12.05 in ± 0.04 in)
Where   X = 33.0 mm (1.30 in)
         L = 303.0 mm ± 1 mm (11.93 in ± 0.04 in)

Chapter 10 Suspension and steering

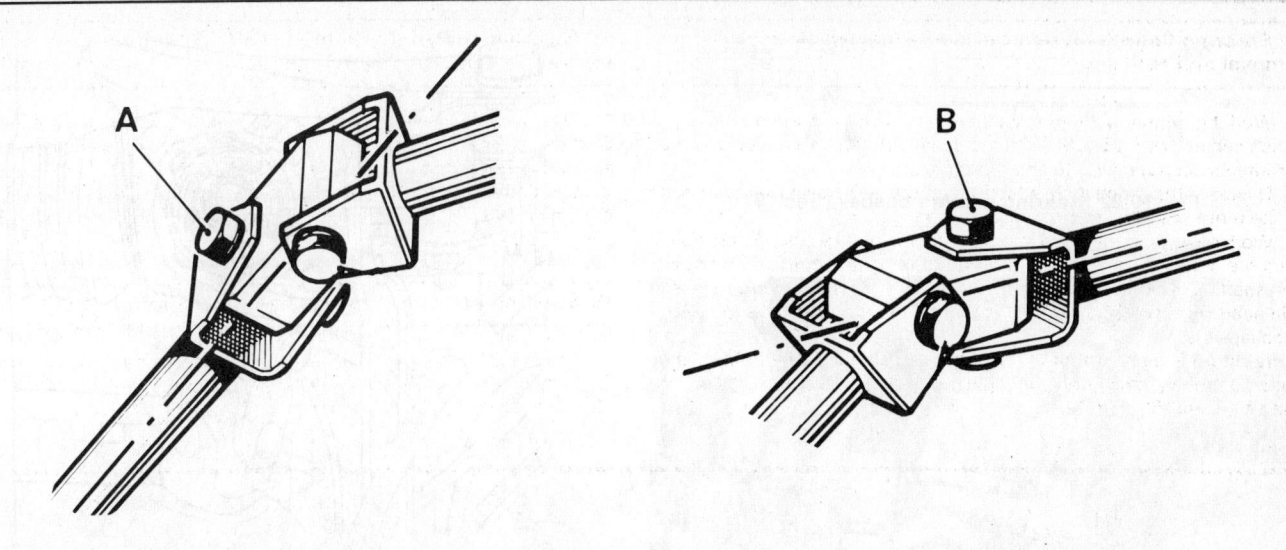

Fig. 10.19 Steering shaft coupling alignment for pinch bolt (A and B) tightening (Sec 17)

3 Disconnect the battery and remove the steering column switches, as described in Chapter 12.
4 Working under the facia panel, unscrew the column mounting bolts, lower the column and remove it (photo).
5 Insert the ignition key and unlock the steering shaft.
6 Using a plastic-faced hammer, tap the shaft upwards out of the column tube. The top bearing will come with the shaft.
7 Use a long piece of tubing (35.0 mm outside diameter) to drive out the bottom bearing.
8 Grease the new bottom bush and tap it into position. Insert the shaft and fit the new top bush (greased).
9 Refitting is a reversal of removal, tighten all bolts and nuts to the specified torque.

20 Steering rack slipper – adjustment

1 Raise the front of the car so that the roadwheels hang free. Support securely with axle stands.
2 Remove the front roadwheels.
3 Unlock the rack slipper adjuster screw by bandaging the lockplate tab.
4 Tighten the adjuster nut using a 10.0 mm Allen key until the steering wheel is hard to turn, then unscrew the nut one quarter of a turn. The steering should now be easily turned without any hard spots when moved from lock to lock.
5 Lock the nut by bending down two lockplate tabs at opposite notches.

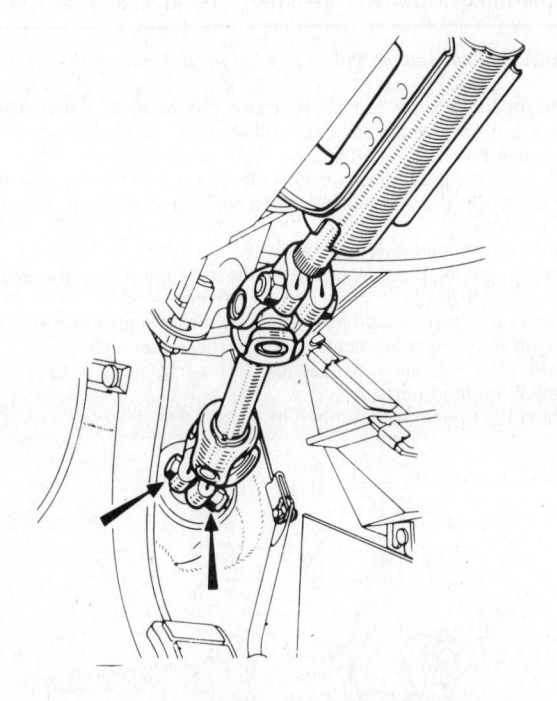

Fig. 10.20 Steering shaft pinch-bolts (power-assisted steering) (Sec 18)

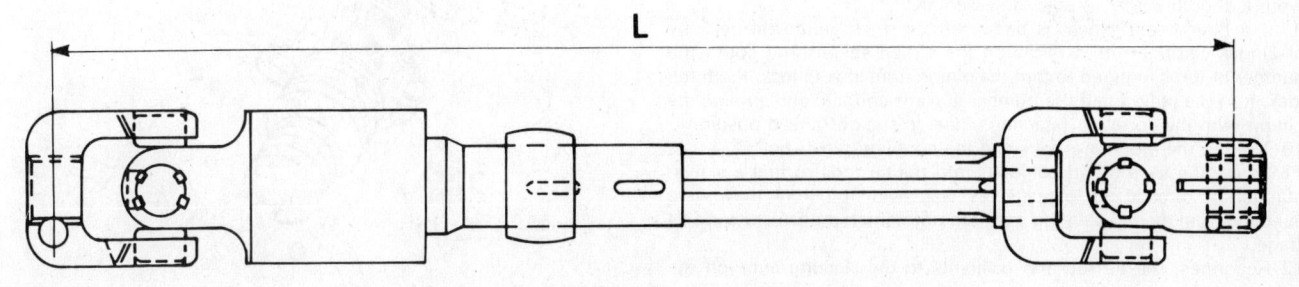

Fig. 10.21 Power-assisted steering shaft (Sec 18)

L = 341.0 mm ± 1 mm (13.43 in ± 0.04 in)

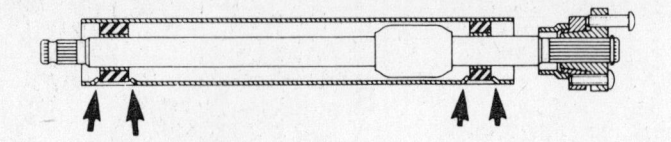

Fig. 10.22 Steering column bushes (Sec 19)

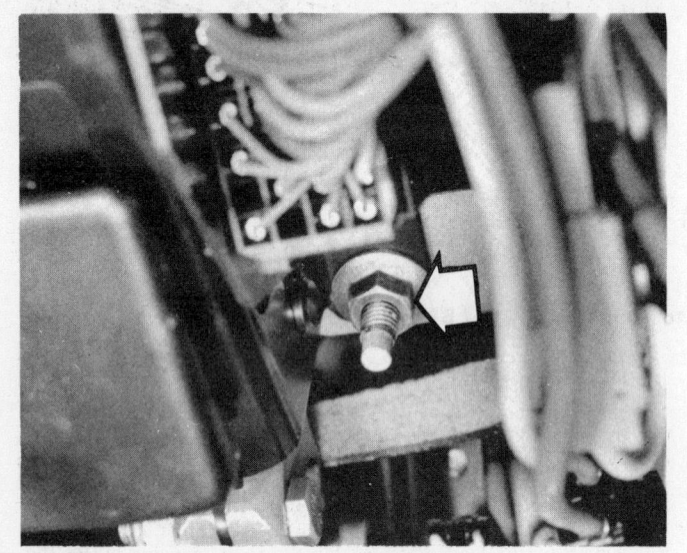

19.4 Steering column mounting nut (arrowed)

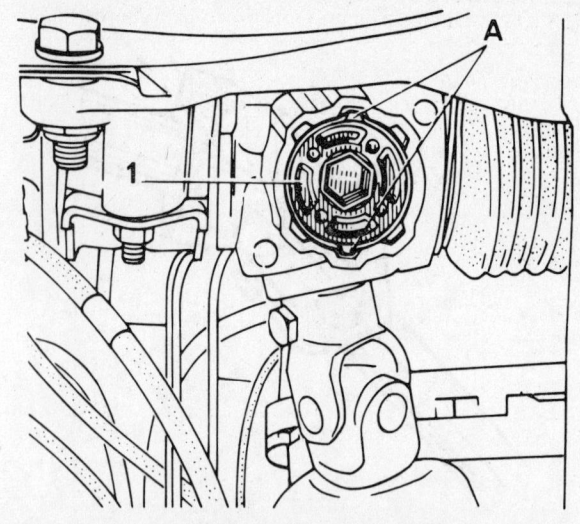

Fig. 10.23 Steering rack slipper adjuster nut (1) and lockplate tabs (A) (Sec 20)

21 Steering gear (manual) – removal and refitting

Note: *A balljoint splitter will be needed for this operation.*

1 Raise the front of the car and support securely. Remove the roadwheel; right-hand wheel for left-hand drive and left-hand wheel for right-hand drive.
2 Centre the steering. Do this by turning the steering to full lock and marking a point on its rim in relation to a similar mark on the facia panel.
3 Turn the steering wheel to full opposite lock and count the number of turns. Now divide the number of turns by two and turn the steering wheel from full lock by this number.
4 Disconnect the tie-rod end balljoints from the steering arms on the hub carriers.
5 Unscrew the pinch-bolt from the shaft coupling at the steering gear pinion. Mark the relationship of the coupling to the pinion shaft.
6 Unscrew the four bolts which hold the steering gear to the crossmember.
7 Pull the steering gear downwards to release it from the steering column shaft and then withdraw it sideways through the aperture under the wheel arch.
8 It is not recommended that the steering gear is overhauled, but if worn it should be renewed as an assembly.
9 If a new steering gear is being refitted, then centre the rack by placing a cardboard disc firmly on the pinion splines and count the number of turns required to turn the pinion from lock to lock. From full lock, turn the pinion half the number of turns counted and engage the pinion with the coupling (steering wheel in straight-ahead position).
10 Tighten the mounting bolts and the coupling pinch-bolt.
11 Screw the tie-rod end balljoints into the tie-rods so that a similar number of threads are exposed as was evident before they were removed from the old steering gear. Do not tighten the locknuts at this stage.
12 Reconnect the tie-rod and balljoints to the steering arms on the hub carriers.
13 Refit the roadwheels, lower the car to the floor.
14 Adjust the front wheel alignment and tighten the tie-rod end balljoint locknuts.

22 Steering gear (power-assisted) – removal and refitting

Note: *A balljoint splitter will be needed for this operation.*

1 The operations are similar to those described in the preceding Section, but observe the following differences.
2 Remove both front roadwheels.
3 On some models, a heat shield is fitted over the steering gear and a cover plate over the universally-jointed steering shaft. Remove both of these if fitted.
4 Remove the fluid pipe bracket.
5 Disconnect the pipes from the fluid reservoir and from the steering pump. Allow the fluid to drain into a container.
6 Unbolt the steering gear from the body crossmember and withdraw it through the inner wing aperture under the wheel arch.
7 Refit after reference to Section 21, paragraphs 9 to 14 and reconnect the fluid pipes.
8 Bleed the system as described in Section 25.

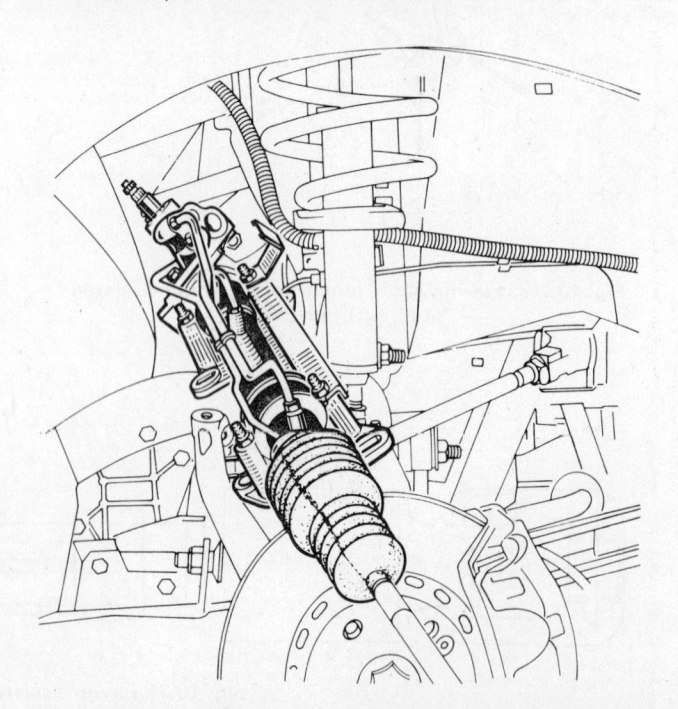

Fig. 10.24 Removing power-assisted steering gear (Sec 22)

Chapter 10 Suspension and steering

23 Power-assisted steering pump – removal and refitting

Four-cylinder model
1 Disconnect the fluid hoses and allow the fluid to drain into a container.
2 Release the mounting and tensioner bolts and remove the pump drivebelt.
3 Unbolt and remove the pump.

Six-cylinder models
4 Disconnect the fluid hoses and allow the fluid to drain into a container (photo).
5 Release the belt tensioner pulley and take off the drivebelt.
6 Unbolt the pump and remove it.

Refitting
7 This is a reversal of removal, tension the drivebelt (Chapter 2) and bleed the system as described in Section 25.

24 Power steering gear and pump – overhaul

1 In view of the special tools and skills required, it is not recommended that the steering gear or pump are dismantled or overhauled, but renewal by obtaining a new or factory reconditioned unit.

25 Power steering circuit – bleeding

1 This will only be required if any part of the hydraulic system has been disconnected.
2 Fill the reservoir to the brim with the specified fluid.
3 Turn the steering to full lock in both directions. Top up if required.
4 Start the engine and turn the steering slowly from lock to lock. Switch off and top up to the correct level.

23.4 Power steering pump (six-cylinder models)

26 Steering angles and front wheel alignment

1 Accurate front wheel alignment is essential to provide good steering and roadholding characteristics and to ensure slow and even tyre wear. Before considering the steering angles, check that the tyres are correctly inflated, that the front wheels are not buckled, the hub bearings are not worn or incorrectly adjusted and that the steering linkage is in good order, without slackness or wear at the joints.
2 Wheel alignment consists of four factors:
Camber, is the angle at which the road wheels are set from the vertical when viewed from the front or rear of the vehicle. Positive camber is the angle (in degrees) that the wheels are tilted outwards at the top from the vertical.
Castor, is the angle between the steering axis and a vertical when viewed from each side of the vehicle. Positive castor is indicated when

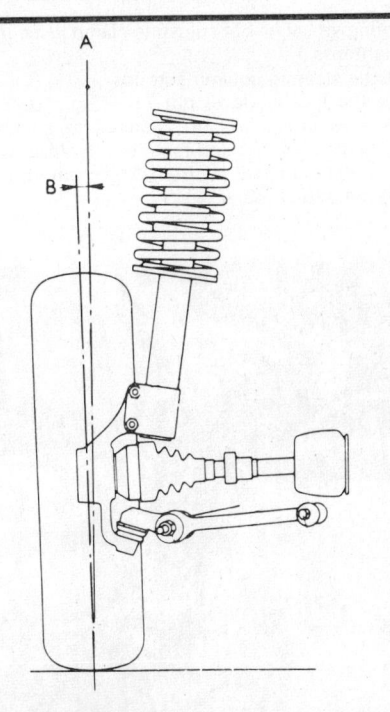

Fig. 10.25 Camber angle diagram (Sec 26)

A Vertical   B Camber angle (positive)

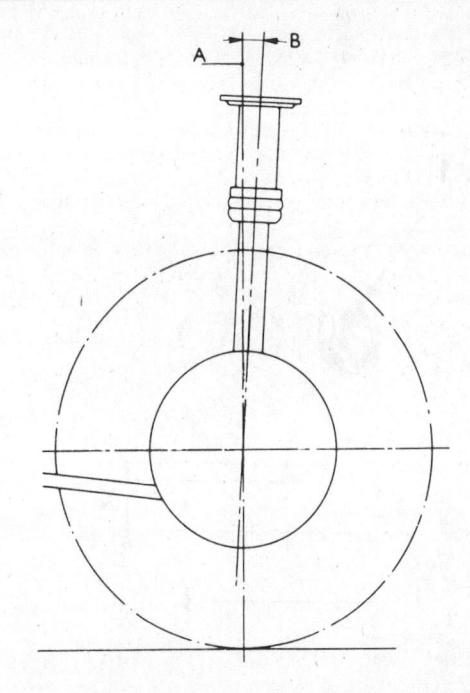

Fig. 10.26 Castor angle diagram (LH wheel) (Sec 26)

A Vertical   B Castor angle (positive)

the steering axis is inclined towards the rear of the vehicle at its upper end.

Steering axis inclination, is the angle when viewed from the front or rear of the vehicle between vertical and an imaginary line drawn between the top and bottom strut mountings.

Toe, is the amount by which the distance between the front inside edges of the roadwheel rims differs from that between the rear inside edges. If the distance between the front edges is less than that at the rear, the wheels are said to toe-in. If the distance between the front inside edges is greater than that at the rear, the wheels toe-out.

3 Owing to the need for precision gauges to measure the small angles of the steering and suspension settings, it is preferable that measuring of camber and castor is left to a service station having the necessary equipment.

4 The camber and steering axis inclination angles are set in production and cannot be adjusted. Where they differ from those specified, suspect collision damage or gross wear in the steering or suspension components.

5 The castor angle is adjustable by shortening or lengthening the radius rods. The castor angle will depend upon the suspension height measurement. Measure from the floor to the centre of the front suspension arm inboard pivot bolt (H2). Measure from the floor to the centre of the rear suspension arm inboard pivot bolt (H5). Subtract H2 from H5 and then refer to the table in Specfications.

6 To check the front wheel alignment, first make sure that the lengths of both tie-rods are equal when the steering is in the straight-ahead position. Adjust if necessary by releasing the tie-rod end locknuts and turning the tie-rods until the lengths of the exposed threads are equal on each side.

7 Obtain a tracking gauge. These are available in various forms from accessory stores or one can be fabricated from a length of steel tubing suitably cranked to clear the sump and bellhousing and having a setscrew and locknut at one end.

8 With the gauge, measure the distance between the two wheel inner rims (at hub height) at the rear of the wheel. Push the vehicle forward to rotate the wheel through 180° (half a turn) and measure the distance between the wheel inner rims, again at hub height, at the front of the wheel. This last measurement should differ from the first by the appropriate toe-out according to specification (see Specifications).

9 Where the toe-out is found to be incorrect, release the tie-rod balljoint locknuts and turn the tie-rods equally. Only turn them a quarter of a turn at a time before re-checking the alignment.

10 Turn each tie-rod in the same direction when viewed from the centre line of the car otherwise the rods will become unequal in length. This would cause the steering wheel spoke position to alter and cause problems on turns with tyre scrubbing. On completion, tighten the tie-rod locknuts without disturbing their setting, check that the balljoint is at the centre of its arc of travel.

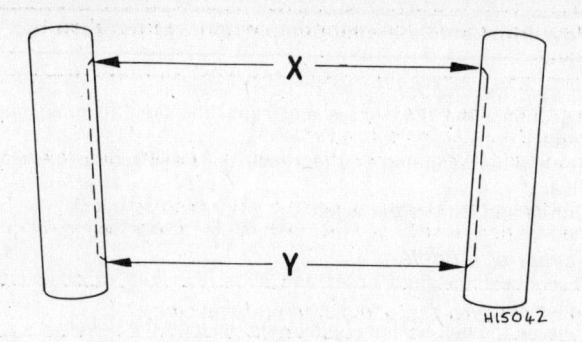

Fig. 10.27 Front wheel alignment diagram showing toe-out (Sec 26)

$X - Y = $ toe-out

27 Rear wheel alignment

1 The camber angle is set in production and is not adjustable.
2 The roadwheels toe setting is adjustable by altering the number of shims which are located at the suspension control arm pivot bolt.
3 The rear wheel alignment is checked in a similar way to that described for the front wheels in the preceding Section.

28 Steering wheel height adjuster

1 The steering wheel height adjusting clamp lever can be adjusted for degree of tightness.
2 Remove the steering column shrouds.
3 Unscrew the locking lever nut (1 – Fig. 10.28) and lower the steering wheel as far as it will go. Remove the clamp lever.
4 Tighten the ring nut (2) and then refit the lever to its splined shaft so that (X) is 30.0 mm (1.2 in) from the column bracket.
5 Refit the nut and shrouds.

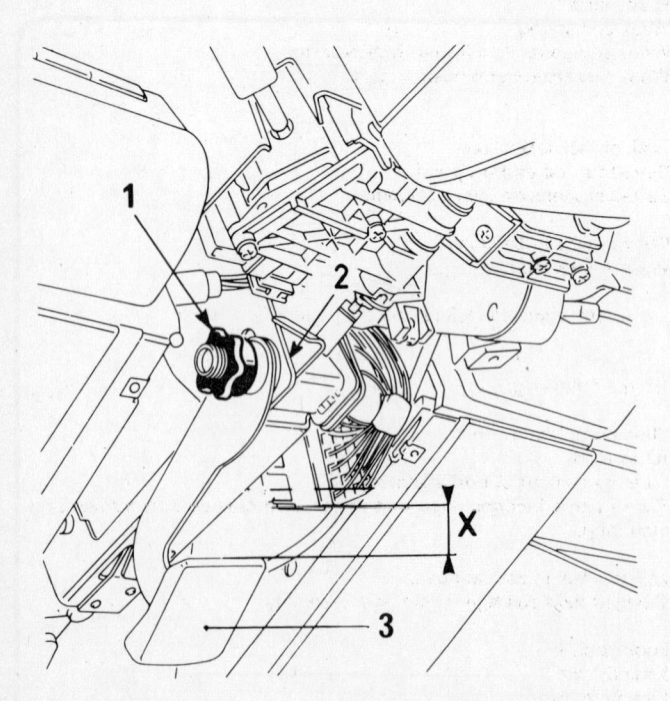

Fig. 10.28 Steering wheel locking lever setting diagram (Sec 28)

1 Nut
2 Ring nut
3 Lever
X = 30.0 mm (1.2 in)

29.5 Extracting ignition switch screw

Chapter 10 Suspension and steering

29 Steering column lock/ignition switch – removal and refitting

1 Disconnect the battery.
2 Remove the steering column shroud.
3 Unbolt the steering column mounting brackets and lower the column.
4 Turn the ignition key to unlock the steering column lock.
5 The ignition switch can be withdrawn if the wiring plug is disconnected and the switch locking screw unscrewed from the switch housing (photo).
6 The column lock is held by shear bolts which will have to be drilled so that an 'easy out' type of screw extractor can be used to withdraw them.
7 When refitting do not fully tighten the new shear-head bolts until the operation of the column lock has been checked using the ignition key, then fully tighten the bolts until their heads break off.
8 Relocate the steering column, refit the column shrouds and reconnect the battery.

30 Fault diagnosis – suspension and steering

Symptom	Reason(s)
Front suspension	
Vehicle wanders	Incorrect wheel alignment
	Worn front control arm balljoints
Heavy or stiff steering	Incorrect front wheel alignment
	Incorrect tyre pressures
Wheel wobble or vibration	Roadwheels out of balance
	Roadwheel buckled
	Incorrect front wheel alignment
	Faulty strut
	Weak coil spring
Excessive pitching or rolling on corners or during braking	Faulty strut
	Weak or broken coil spring
Tyre squeal when cornering	Incorrect front wheel alignment
	Incorrect tyre pressures
Abnormal tyre wear	Incorrect tyre pressures
	Incorrect front wheel alignment
	Worn hub bearing
Rear suspension	
Poor roadholding and wander	Faulty strut
	Weak coil spring
	Worn or incorrectly adjusted hub bearing
	Worn suspension arm bush
Manual stearing gear	
Stiff action	Lack of rack lubrication
	Seized tie-rod end balljoints
	Seized suspension lower balljoint
Free movement at steering wheel	Wear in tie-rod balljoints
	Wear in rack teeth
Knocking when traversing uneven surface	Incorrectly adjusted rack slipper
Power-assisted steering	
The symptoms and reasons applicable to manual steering gear will apply plus the following:	
Stiff action or no return action	Slipping pump drivebelt
	Air in fluid
	Steering column out of alignment
	Castor angle incorrect due to damage or gross wear in bushes and mountings
Steering effort on both locks unequal	Leaking seal in steering gear
	Clogged fluid passage within gear assembly
Noisy pump	Loose pulley
	Kinked hose
	Low fluid level

Chapter 11 Bodywork and fittings

Contents

Bonnet – removal and refitting	8	Major body damage – repair	5
Centre consoles – removal and refitting	18	Minor body damage – repair	4
Description	1	Radiator grille – removal and refitting	7
Door – removal and refitting	14	Rear bumper – removal and refitting	28
Facia panel – removal and refitting	23	Rear door – dismantling and reassembly	13
Front door – dismantling and reassembly	11	Rear door trim panel – removal and refitting	12
Front door trim panel – removal and refitting	10	Rear interior quarter panel – removal and refitting	17
Front seat – removal and refitting	24	Rear parcel shelf	19
Front spoiler/bumper – removal and refitting	6	Rear seat – removal and refitting	25
Front wing – removal and refitting	9	Rear view mirrors	26
Headrests	20	Seat belts – maintenance, removal and refitting	21
Interior grab handles	22	Sunroof – dismantling and reassembly	27
Maintenance – bodywork and underframe	2	Tailgate – removal and refitting	15
Maintenance – upholstery and carpets	3	Windscreen and heated rear screen – renewal	16

Specifications

Type .... Five-door hatchback; all-steel, welded unitary construction

Body dimensions, vehicle weights and other information are given in the introductory Section of this Manual

1 Description

The Renault 25 is a five-door hatchback; lavishly equipped and spaciously designed.

It is of all-steel, welded construction incorporating progressive crumple zones at front and rear, with a rigid crushproof passenger compartment.

Long term protection of the bodywork is afforded by the now generally accepted dipping, underbody coating and wax injection processes.

Note: *Many body components and trim items are secured with Torx screws; therefore a set of Torx bits will be needed.*

2 Maintenance – bodywork and underframe

The general condition of a vehicle's bodywork is the one thing that significantly affects its value. Maintenance is easy but needs to be regular. Neglect, particularly after minor damage, can lead quickly to further deterioration and costly repair bills. It is important also to keep watch on those parts of the vehicle not immediately visible, for instance the underside, inside all the wheel arches and the lower part of the engine compartment.

The basic maintenance routine for the bodywork is washing – preferably with a lot of water, from a hose. This will remove all the loose solids which may have stuck to the vehicle. It is important to flush these off in such a way as to prevent grit from scratching the finish. The wheel arches and underframe need washing in the same way to remove any accumulated mud which will retain moisture and tend to encourage rust. Paradoxically enough, the best time to clean the underframe and wheel arches is in wet weather when the mud is thoroughly wet and soft. In very wet weather the underframe is usually cleaned of large accumulations automatically and this is a good time for inspection.

Periodically, except on vehicles with a wax-based underbody protective coating, it is a good idea to have the whole of the underframe of the vehicle steam cleaned, engine compartment included, so that a thorough inspection can be carried out to see what minor repairs and renovations are necessary. Steam cleaning is available at many garages and is necessary for removal of the accumulation of oily grime which sometimes is allowed to become thick in certain areas. If steam cleaning facilities are not available, there are one or two excellent grease solvents available which can be brush applied. The dirt can then be simply hosed off. Note that these

Chapter 11 Bodywork and fittings

methods should not be used on vehicles with wax-based underbody protective coating or the coating will be removed. Such vehicles should be inspected annually, preferably just prior to winter, when the underbody should be washed down and any damage to the wax coating repaired. Ideally, a completely fresh coat should be applied. It would also be worth considering the use of such wax-based protection for injection into door panels, sills, box sections, etc, as an additional safeguard against rust damage where such protection is not provided by the vehicle manufacturer.

After washing paintwork, wipe off with a chamois leather to give an unspotted clear finish. A coat of clear protective wax polish will give added protection against chemical pollutants in the air. If the paintwork sheen has dulled or oxidised, use a cleaner/polisher combination to restore the brilliance of the shine. This requires a little effort, but such dulling is usually caused because regular washing has been neglected. Care needs to be taken with metallic paintwork, as special non-abrasive cleaner/polisher is required to avoid damage to the finish. Always check that the door and ventilator opening drain holes and pipes are completely clear so that water can be drained out. Bright work should be treated in the same way as paint work. Windscreens and windows can be kept clear of the smeary film which often appears by the use of a proprietary glass cleaner. Never use any form of wax or other body or chromium polish on glass.

3 Maintenance – upholstery and carpets

Mats and carpets should be brushed or vacuum cleaned regularly to keep them free of grit. If they are badly stained remove them from the vehicle for scrubbing or sponging and make quite sure they are dry before refitting. Seats and interior trim panels can be kept clean by wiping with a damp cloth. If they do become stained (which can be more apparent on light coloured upholstery) use a little liquid detergent and a soft nail brush to scour the grime out of the grain of the material. Do not forget to keep the headlining clean in the same way as the upholstery. When using liquid cleaners inside the vehicle do not over-wet the surfaces being cleaned. Excessive damp could get into the seams and padded interior causing stains, offensive odours or even rot. If the inside of the vehicle gets wet accidentally it is worthwhile taking some trouble to dry it out properly, particularly where carpets are involved. *Do not leave oil or electric heaters inside the vehicle for this purpose.*

4 Minor body damage – repair

The colour bodywork repair photographic sequences between pages 32 and 33 illustrate the operations detailed in the following sub-sections.
Note: *For more detailed information about bodywork repair, the Haynes Publishing Group publish a book by Lindsay Porter called The Car Bodywork Repair Manual. This incorporates information on such aspects as rust treatment, painting and glass fibre repairs, as well as details on more ambitious repairs involving welding and panel beating.*

Repair of minor scratches in bodywork

If the scratch is very superficial, and does not penetrate to the metal of the bodywork, repair is very simple. Lightly rub the area of the scratch with a paintwork renovator, or a very fine cutting paste, to remove loose paint from the scratch and to clear the surrounding bodywork of wax polish. Rinse the area with clean water.

Apply touch-up paint to the scratch using a fine paint brush; continue to apply fine layers of paint until the surface of the paint in the scratch is level with the surrounding paintwork. Allow the new paint at least two weeks to harden; then blend it into the surrounding paintwork by rubbing the scratch area with a paintwork renovator or a very fine cutting paste. Finally, apply wax polish.

Where the scratch has penetrated right through to the metal of the bodywork, causing the metal to rust, a different repair technique is required. Remove any loose rust from the bottom of the scratch with a penknife, then apply rust inhibiting paint to prevent the formation of rust in the future. Using a rubber or nylon applicator fill the scratch with bodystopper paste. If required, this paste can be mixed with cellulose thinners to provide a very thin paste which is ideal for filling narrow scratches. Before the stopper-paste in the scratch hardens, wrap a piece of smooth cotton rag around the top of a finger. Dip the finger in cellulose thinners and then quickly sweep it across the surface of the stopper-paste in the scratch; this will ensure that the surface of the stopper-paste is slightly hollowed. The scratch can now be painted over as described earlier in this Section.

Repair of dents in bodywork

When deep denting of the vehicle's bodywork has taken place, the first task is to pull the dent out, until the affected bodywork almost attains its original shape. There is little point in trying to restore the original shape completely, as the metal in the damaged area will have stretched on impact and cannot be reshaped fully to its original contour. It is better to bring the level of the dent up to a point which is about 1/8 in (3 mm) below the level of the surrounding bodywork. In cases where the dent is very shallow anyway, it is not worth trying to pull it out at all. If the underside of the dent is accessible, it can be hammered out gently from behind, using a mallet with a wooden or plastic head. Whilst doing this, hold a suitable block of wood firmly against the outside of the panel to absorb the impact from the hammer blows and thus prevent a large area of the bodywork from being 'belled-out'.

Should the dent be in a section of the bodywork which has a double skin or some other factor making it inaccessible from behind, a different technique is called for. Drill several small holes through the metal inside the area – particularly in the deeper section. Then screw long self-tapping screws into the holes just sufficiently for them to gain a good purchase in the metal. Now the dent can be pulled out by pulling on the protruding heads of the screws with a pair of pliers.

The next stage of the repair is the removal of the paint from the damaged area, and from an inch or so of the surrounding 'sound' bodywork. This is accomplished most easily by using a wire brush or abrasive pad on a power drill, although it can be done just as effectively by hand using sheets of abrasive paper. To complete the preparation for filling, score the surface of the bare metal with a screwdriver or the tang of a file, or alternatively, drill small holes in the affected area. This will provide a really good 'key' for the filler paste.

To complete the repair see the Section on filling and re-spraying.

Repair of rust holes or gashes in bodywork

Remove all paint from the affected area and from an inch or so of the surrounding 'sound' bodywork, using an abrasive pad or a wire brush on a power drill. If these are not available a few sheets of abrasive paper will do the job just as effectively. With the paint removed you will be able to gauge the severity of the corrosion and therefore decide whether to renew the whole panel (if this is possible) or to repair the affected area. New body panels are not as expensive as most people think and it is often quicker and more satisfactory to fit a new panel than to attempt to repair large areas of corrosion.

Remove all fittings from the affected area except those which will act as a guide to the original shape of the damaged bodywork (eg headlamp shells etc). Then, using tin snips or a hacksaw blade, remove all loose metal and any other metal badly affected by corrosion. Hammer the edges of the hole inwards in order to create a slight depression for the filler paste.

Wire brush the affected area to remove the powdery rust from the surface of the remaining metal. Paint the affected area with rust inhibiting paint; if the back of the rusted area is accessible treat this also.

Before filling can take place it will be necessary to block the hole in some way. This can be achieved by the use of aluminium or plastic mesh, or aluminium tape.

Aluminium or plastic mesh is probably the best material to use for a large hole. Cut a piece to the approximate size and shape of the hole to be filled, then position it in the hole so that its edges are below the level of the surrounding bodywork. It can be retained in position by several blobs of filler paste around its periphery.

Aluminium tape should be used for small or very narrow holes. Pull a piece off the roll and trim it to the approximate size and shape required, then pull off the backing paper (if used) and stick the tape over the hole; it can be overlapped if the thickness of one piece is insufficient. Burnish down the edges of the tape with the handle of a screwdriver or similar, to ensure that the tape is securely attached to the metal underneath.

Bodywork repairs – filling and re-spraying

Before using this Section, see the Sections on dent, deep scratch, rust holes and gash repairs.

Many types of bodyfiller are available, but generally speaking those proprietary kits which contain a tin of filler paste and a tube of resin hardener are best for this type of repair. A wide, flexible plastic or nylon applicator will be found invaluable for imparting a smooth and well contoured finish to the surface of the filler.

Mix up a little filler on a clean piece of card or board – measure the hardener carefully (follow the maker's instructions on the pack) otherwise the filler will set too rapidly or too slowly. Using the applicator apply the filler paste to the prepared area; draw the applicator across the surface of the filler to achieve the correct contour and to level the filler surface. As soon as a contour that approximates to the correct one is achieved, stop working the paste – if you carry on too long the paste will become sticky and begin to 'pick up' on the applicator. Continue to add thin layers of filler paste at twenty-minute intervals until the level of the filler is just proud of the surrounding bodywork.

Once the filler has hardened, excess can be removed using a metal plane or file. From then on, progressively finer grades of abrasive paper should be used, starting with a 40 grade production paper and finishing with 400 grade wet-and-dry paper. Always wrap the abrasive paper around a flat rubber, cork, or wooden block – otherwise the surface of the filler will not be completely flat. During the smoothing of the filler surface the wet-and-dry paper should be periodically rinsed in water. This will ensure that a very smooth finish is imparted to the filler at the final stage.

At this stage the 'dent' should be surrounded by a ring of bare metal, which in turn should be encircled by the finely 'feathered' edge of the good paintwork. Rinse the repair area with clean water, until all of the dust produced by the rubbing-down operation has gone.

Spray the whole repair area with a light coat of primer – this will show up any imperfections in the surface of the filler. Repair these imperfections with fresh filler paste or bodystopper, and once more smooth the surface with abrasive paper. If bodystopper is used, it can be mixed with cellulose thinners to form a really thin paste which is ideal for filling small holes. Repeat this spray and repair procedure until you are satisfied that the surface of the filler, and the feathered edge of the paintwork are perfect. Clean the repair area with clean water and allow to dry fully.

The repair area is now ready for final spraying. Paint spraying must be carried out in a warm, dry, windless and dust free atmosphere. This condition can be created artificially if you have access to a large indoor working area, but if you are forced to work in the open, you will have to pick your day very carefully. If you are working indoors, dousing the floor in the work area with water will help to settle the dust which would otherwise be in the atmosphere. If the repair area is confined to one body panel, mask off the surrounding panels; this will help to minimise the effects of a slight mis-match in paint colours. Bodywork fittings (eg chrome strips, door handles etc) will also need to be masked off. Use genuine masking tape and several thicknesses of newspaper for the masking operations.

Before commencing to spray, agitate the aerosol can thoroughly, then spray a test area (an old tin, or similar) until the technique is mastered. Cover the repair area with a thick coat of primer; the thickness should be built up using several thin layers of paint rather than one thick one. Using 400 grade wet-and-dry paper, rub down the surface of the primer until it is really smooth. While doing this, the work area should be thoroughly doused with water, and the wet-and-dry paper periodically rinsed in water. Allow to dry before spraying on more paint.

Spray on the top coat, again building up the thickness by using several thin layers of paint. Start spraying in the centre of the repair area and then, using a circular motion, work outwards until the whole repair area and about 2 inches of the surrounding original paintwork is covered. Remove all masking material 10 to 15 minutes after spraying on the final coat of paint.

Allow the new paint at least two weeks to harden, then, using a paintwork renovator or a very fine cutting paste, blend the edges of the paint into the existing paintwork. Finally, apply wax polish.

Plastic components

With the use of more and more plastic body components by the vehicle manufacturers (eg bumpers, spoilers, and in some cases major body panels), rectification of damage to such items has become a matter of either entrusting repair work to a specialist in this field, or renewing complete components. Repair by the DIY owner is not really feasible owing to the cost of the equipment and materials required for effecting such repairs. The basic technique involves making a groove along the line of the crack in the plastic using a rotary burr in a power drill. The damaged part is then welded back together by using a hot air gun to heat up and fuse a plastic filler rod into the groove. Any excess plastic is then removed and the area rubbed down to a smooth finish. It is important that a filler rod of the correct plastic is used, as body components can be made of a variety of different types (eg polycarbonate, ABS, polypropylene).

If the owner is renewing a complete component himself, he will be left with the problem of finding a suitable paint for finishing which is compatible with the type of plastic used. At one time the use of a universal paint was not possible owing to the complex range of plastics encountered in body component applications. Standard paints, generally speaking, will not bond to plastic or rubber satisfactorily. However, it is now possible to obtain a plastic body parts finishing kit which consists of a pre-primer treatment, a primer and coloured top coat. Full instructions are normally supplied with a kit, but basically the method of use is to first apply the pre-primer to the component concerned and allow it to dry for up to 30 minutes. Then the primer is applied and left to dry for about an hour before finally applying the special coloured top coat. The result is a correctly coloured component where the paint will flex with the plastic or rubber, a property that standard paint does not normally possess.

5 Major body damage – repair

1 This sort of work should be left to your Renault dealer or specialist body repair works.
2 It is essential to have the body aligned on special jigs to ensure that the specified steering and suspension settings are maintained during repair.
3 This is of course beyond the scope of the home mechanic and if not carried out correctly will give rise to unroadworthy behaviour and severe tyre wear.

6 Front spoiler/bumper – removal and refitting

1 Disconnect the battery.
2 Disconnect the front direction indicator lamps (if bumper-mounted).

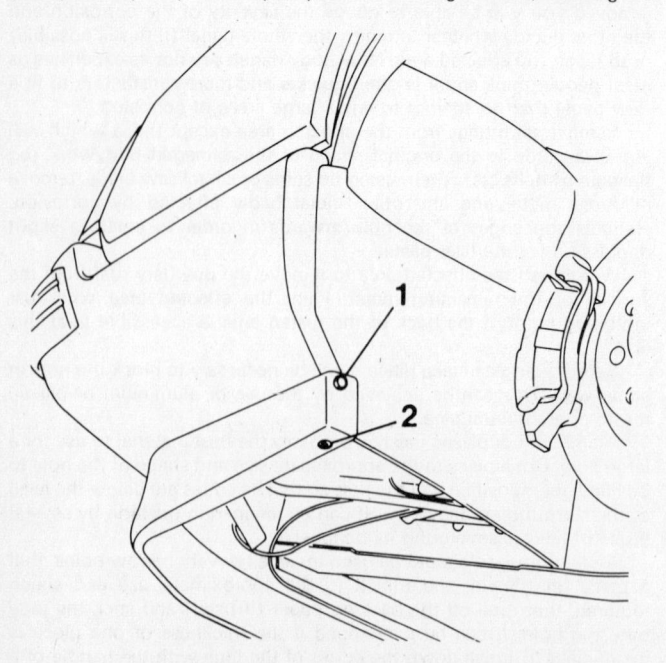

Fig. 11.1 Spoiler end bolt (1) and screw (2) (Sec 6)

Chapter 11 Bodywork and fittings

3 Unscrew the bolts (1 and 2 – Fig. 11.1) from both sides of the car.
4 Prise out the blanking plates (4 – Fig. 11.2) and remove the screws (3).
5 The end fixing bolts (5 – Fig. 11.3) must now be unscrewed but, before this can be done, the following preliminary work must be carried out.
6 On four-cylinder models, unclip the computer from its bracket on the left-hand side of the engine compartment then unscrew the three bracket bolts.
7 On six-cylinder models, unbolt the ignition module and carefully move its mounting bracket to one side.
8 On Turbo versions, unbolt, but do not disconnect, the intercooler and its bracket, and move them aside.
9 Remove the spoiler/bumper which can be dismantled to renew a damaged section by removing the bolts and drilling out the pop rivets as necessary (photo).
10 Refitting is a reversal of removal.

7 Radiator grille – removal and refitting

1 Open the bonnet.
2 The radiator grille is secured by two bolts and four Torx screws.
3 Remove the bolts and screws and withdraw the grille (photos).
4 As mentioned in Chapter 2, side deflectors are fitted to the radiator which are held in place by Torx screws. An upper seal is fitted to the radiator (photo) and certain models have an air blanking plate.
5 Refitting is a reversal of removal.

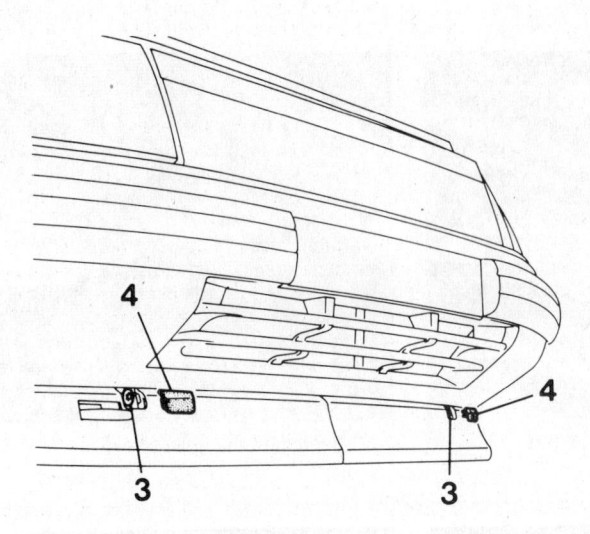

Fig. 11.2 Spoiler blanking plates (4) and screws (3) (Sec 6)

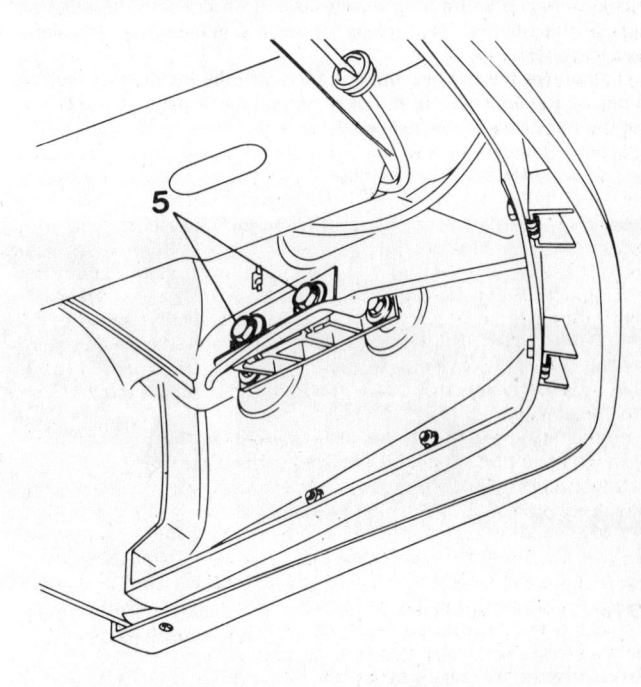

Fig. 11.3 Spoiler/bumper end fixing bolts (5) (Sec 6)

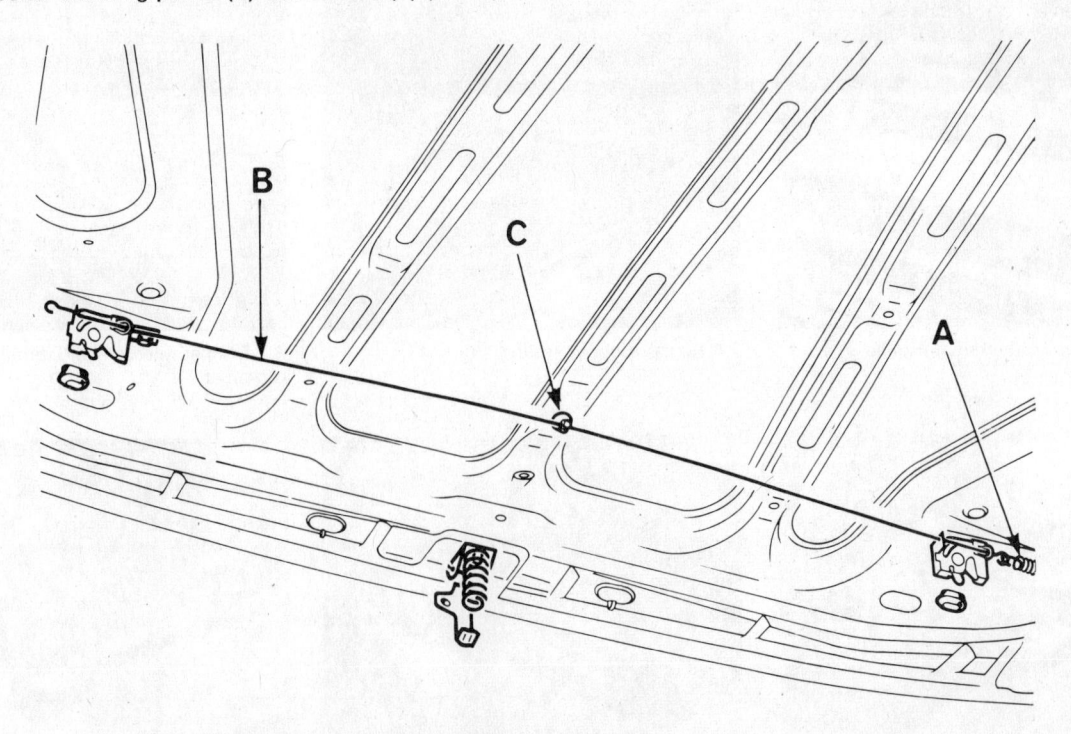

Fig. 11.4 Bonnet lock release cable (Sec 8)

A Tension spring   B Rod   C Clip

8 Bonnet removal and refitting

1 The bonnet is of front opening type; having two side catches operated by a cable release which is routed through the bonnet cavity (photo).
2 The bonnet is supported in the open position by two rear-mounted struts.
3 Should the release cable break, the bonnet can be opened by reaching up through the engine compartment from under the car using a socket wrench and a long extension and unscrewing the nut from each of the bonnet lock strikers. These are mounted on headlamp support panels.
4 To remove the bonnet, open it and mark the location of the rear hinges on the underside of it. Use masking tape or pencil to do this so that the paintwork is not damaged.
5 Have an assistant support the bonnet while the release cable is disconnected from the two latches (photo).
6 The sound deadening panel should be removed from the bonnet underside so that the cable can be unclipped and withdrawn.
7 Disconnect the electrical lead from the 'bonnet closed' switch.
8 There is no need to disconnect the bonnet support struts (photo).
9 Unscrew the hinge bolts and lift the bonnet from the car.
10 When refitting the bonnet, do not fully tighten the hinge bolts until the bonnet has been closed gently and its alignment checked. The gap on each side should be equal. If not, move the bonnet hinges within the tolerance of their elongated bolt holes.
11 The bonnet closure height may be adjusted by altering the number or thickness of plates (B – Fig. 11.5).
12 To prevent bonnet rattle when it is closed, turn the rubber bump stops to alter their height (photo).

6.9 Front spoiler joint details

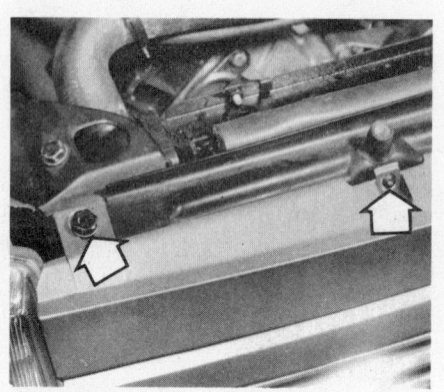

7.3A Radiator grille screw and bolt (arrowed)

7.3B Extracting the grille screw

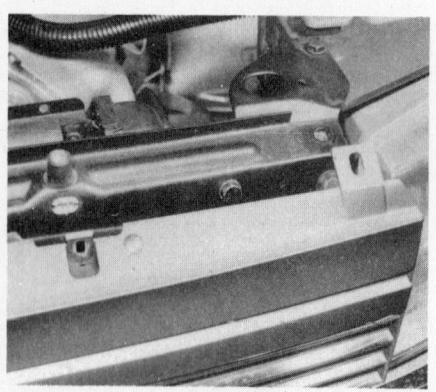

7.3C Withdrawing the radiator grille

7.4 Radiator upper sealing strip

8.1 Bonnet release cable through bonnet lid grommet

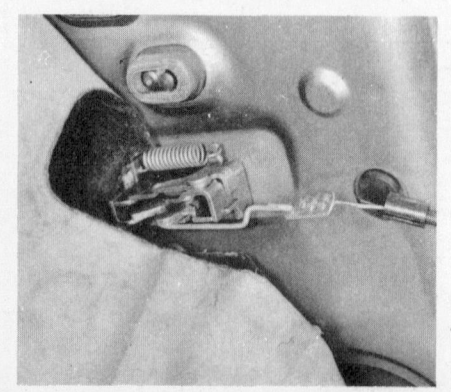

8.5 Bonnet release cable and lock

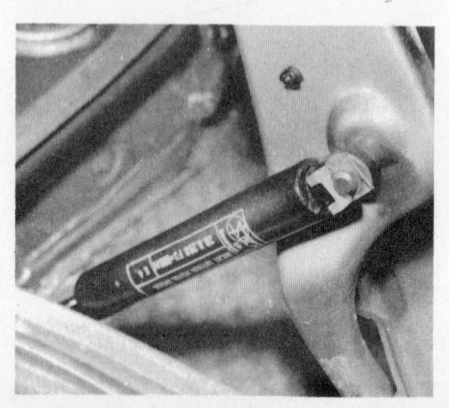

8.8 Bonnet hinge support strut

8.12 Bonnet rubber bump stop

Chapter 11 Bodywork and fittings

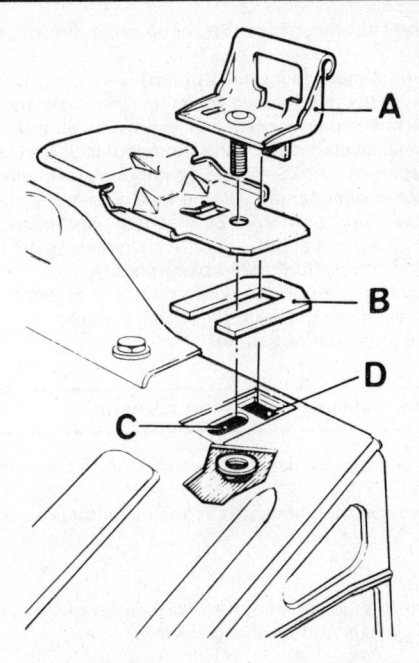

Fig. 11.5 Bonnet lock striker (Sec 8)

- A Striker plate
- B Height adjusting plate
- C Tolerance hole for adjustment
- D Tolerance hole for adjustment

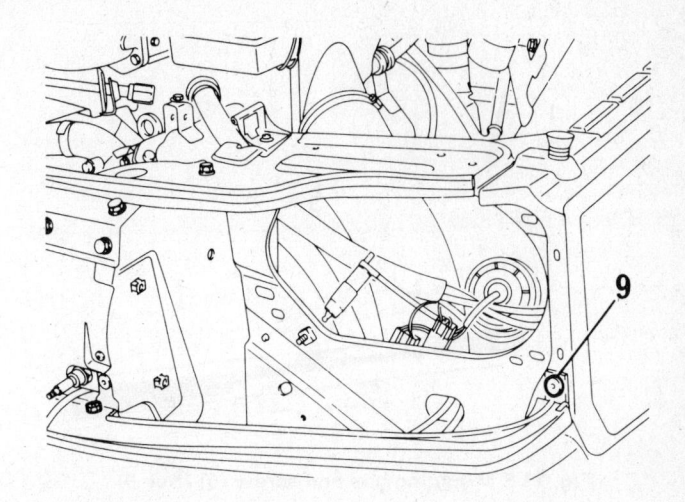

Fig. 11.6 Front end panel moulding rivet (9) (Sec 9)

9 Front wing – removal and refitting

1 Remove the headlamp, as described in Chapter 12, on twin headlamp models, or the direction indicator on single headlamp models.
2 Raise and securely support the front of the car and remove the roadwheel.
3 Remove the bumper end bolts (1) and spoiler bolts (Figs. 11.1 and 11.3).
4 Drill out the rivet (9 – Fig. 11.6) which holds the front panel moulding.
5 Prise out the blanking piece from the side of the spoiler being worked upon and remove the panel screw and clip (Fig. 11.2).
6 Tilt the spoiler/bumper and remove the bolt (6 – Fig. 11.7).
7 Unscrew the line of bolts (7) from the top of the wing. Unscrew bolt (8) from inside the rear of the wheel arch.
8 Unclip the front end of the bodysill moulding, remove the screw and clip (5 – Fig. 11.8).
9 Extract the screws and remove the under wing protective shield (photos).
10 Reaching under the wing, remove the two bolts that hold the wing to the body pillar.
11 The wing is now ready for removal, but its sealing mastic will either have to be cut with a sharp knife or softened using a hot air gun or similar device before the wing can be released.
12 Fitting the new wing is a reversal of removal, but observe the following points.
13 Apply a thick bead of sealing mastic to clean mating flanges before offering the wing into position.
14 Apply protective coating to the underside of the wing and refinish the outer surface to match the body colour.

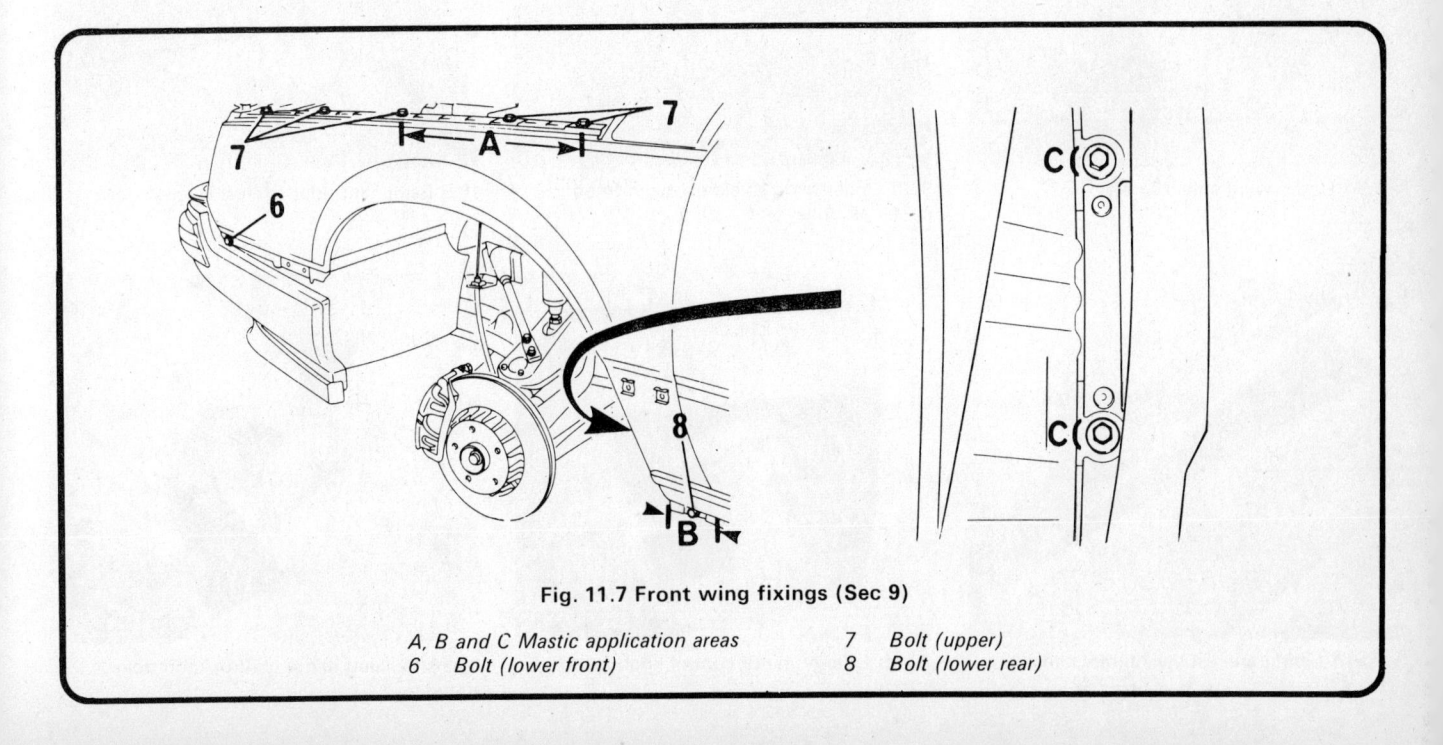

Fig. 11.7 Front wing fixings (Sec 9)

- A, B and C Mastic application areas
- 6 Bolt (lower front)
- 7 Bolt (upper)
- 8 Bolt (lower rear)

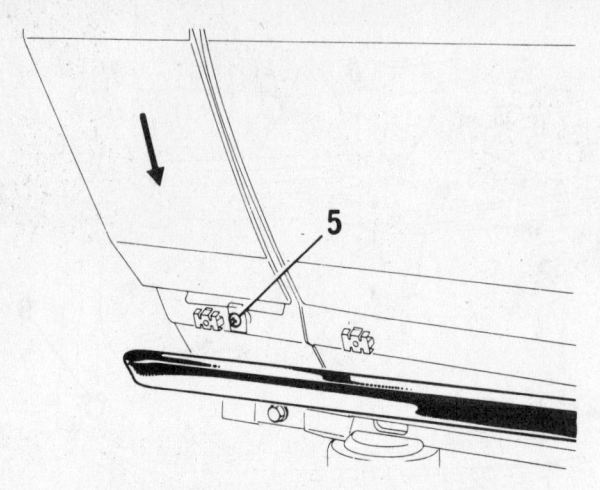

Chapter 11 Bodywork and fittings

Fig. 11.8 Moulding clip and screw (5) (Sec 9)

Remove the moulding in the direction of the arrow

mounting plate and disconnect the wiring plugs from its reverse side (photos).
6 Unscrew the armrest fixing nut (photo).
7 Extract the screw and remove the mirror trim plate (photo).
8 Rotate the radio speaker grille and remove it (photo).
9 Using a small screwdriver, depress the door lock knob tab and pull off the knob (photo). On TS models (without electric windows) prise off the window winder handle using a forked tool.
10 Remove the door trim panel by inserting the fingers or a broad blade between the panel and the door. A sharp jerking action is best to release the plastic clips from their holes (photo).
11 Feed the wiring through the holes in the panel (photo).
12 Carefully peel away the waterproof sheet (photo).
13 Refitting is a reversal of removal.

11 Front door – dismantling and reassembly

1 Remove the door trim panel as described in the preceding Section (photo).
2 The following components can be removed from the door aperture as required.

Radio speaker
3 Disconnect the leads, remove the speaker fixing screws and withdraw the speaker and its shield (photos).

Central door locking
4 The solenoids can be removed after disconnecting the lock link rods (photo) and wiring plugs. On models equipped with this system, door locking and unlocking is carried out by an infra red remote control device which is described in Chapter 12.

Exterior handle and locks
5 The exterior handle is secured by a nut accessible from within the door cavity (photo). Disconnect the link rods, unscrew the nut and remove the handle.

10 Front door trim panel – removal and refitting

1 Lower the window fully and open the door wide.
2 Pull the hinged cover of the door armrest upwards and extract the two upper and two lower Torx fixing screws.
3 There is a small screw at the front end and one inside a small cut-out (photo).
4 Lift the switch mounting plate from the front end of the armrest and release the mirror control knob fixing screw (photos).
5 Push the mirror control knob downwards out of the switch

9.9A Under wing shield screw

9.9B Under wing shield showing turbo cooler air grille

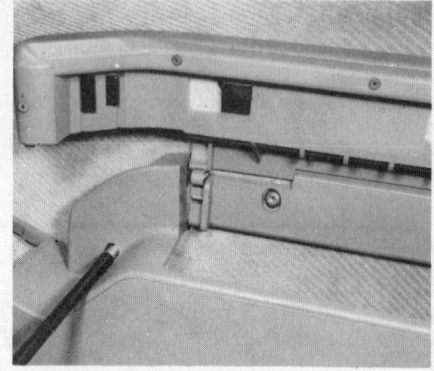

10.3 Extracting door armrest screw

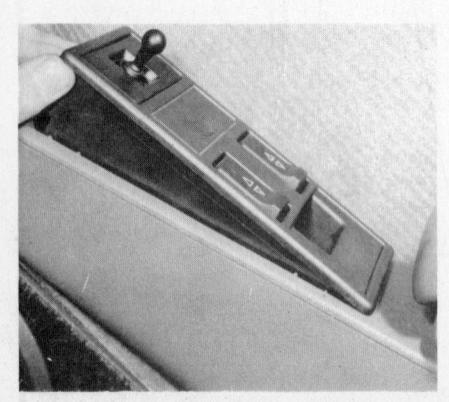

10.4A Lifting armrest switch mounting plate

10.4B Exterior mirror control knob fixing screw

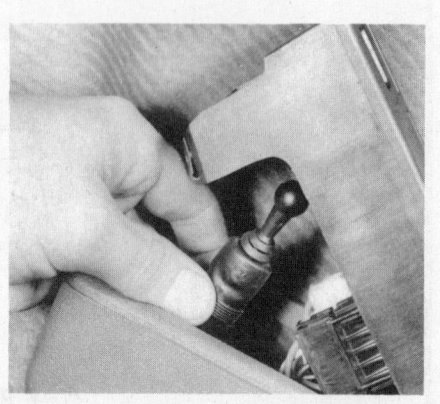

10.5A Pushing mirror control knob from armrest

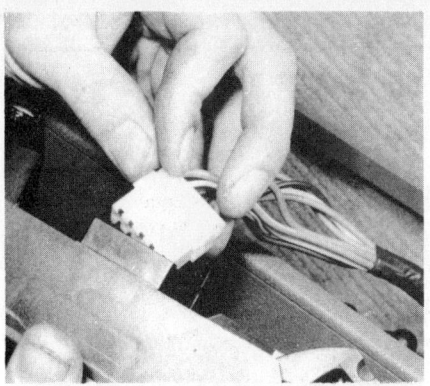

10.5B Disconnecting armrest wiring plug

10.6 Armrest fixing nut

10.7 Extracting exterior mirror escutcheon plate screw

10.8 Removing the radio speaker grille

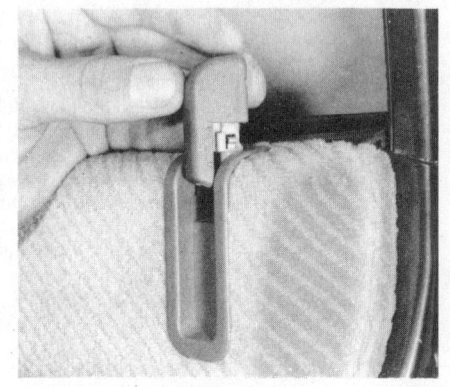

10.9 Removing door lock knob

10.10 Rear trim panel clips

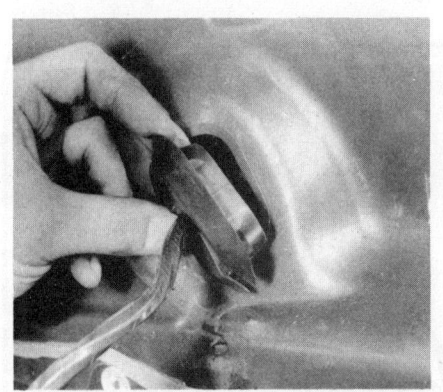

10.11 Door wiring harness and grommet

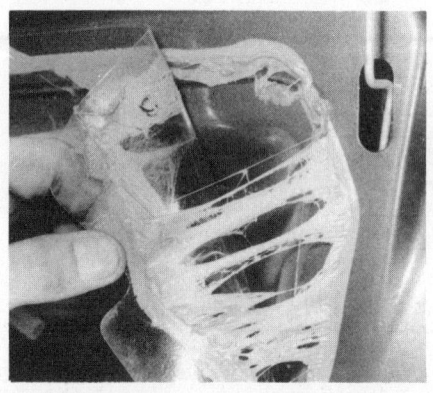

10.12 Pulling off waterproof sheet

11.1 Door with trim panel removed

11.3A Disconnecting radio speaker leads

11.3B Removing door speaker shield

11.4 Door lock solenoid and link rods

280   Chapter 11 Bodywork and fittings

11.5 Door exterior handle nut

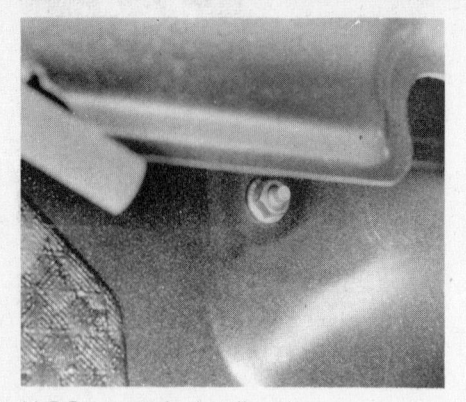

11.7 Door lock

11.10 Electric window control module

6 On TS versions, the cylinder lock can be detached from the door handle after pulling out the retaining clip.
7 On other models, the central door locking system does not require a cylinder lock (see earlier paragraph 4). The door lock is secured to the door edge with three Torx screws (photo).

Window winder mechanism
8 On TS models with a mechanically-operated window regulator, unbolt the lifting bracket from the glass bottom channel.
9 Unscrew the nuts and bolts which hold the regulator to the door panel and withdraw the mechanism through the aperture in the door panel.
10 On other models which have electrically-operated windows, disconnect the wiring plugs from the control module and the motor (photo). Unscrew the six fixing bolts and withdraw the mechanism and motor through the large aperture in the door panel.

Window glass
11 With the glass fully down, pull off the inner glass weatherstrip. No clips are used to retain it.
12 Raise the glass fully and unscrew the two bolts from the glass bottom channel, accessible through the holes (3 – Fig. 11.10).
13 Lower the mechanism (manual or power operated), at the same time holding the glass in the raised position.
14 Swivel the glass and withdraw it towards the outside of the door.

All components
15 Reassembly is a reversal of removal, but observe the following points.
16 If the glass is removed from its bottom channel, reset it in the channel in accordance with the dimension given in Fig. 11.11.
17 Check the glass for smooth operation in its guides. The bottom of the guides can be moved if necessary to 'centre' the glass after releasing the fixing screws.

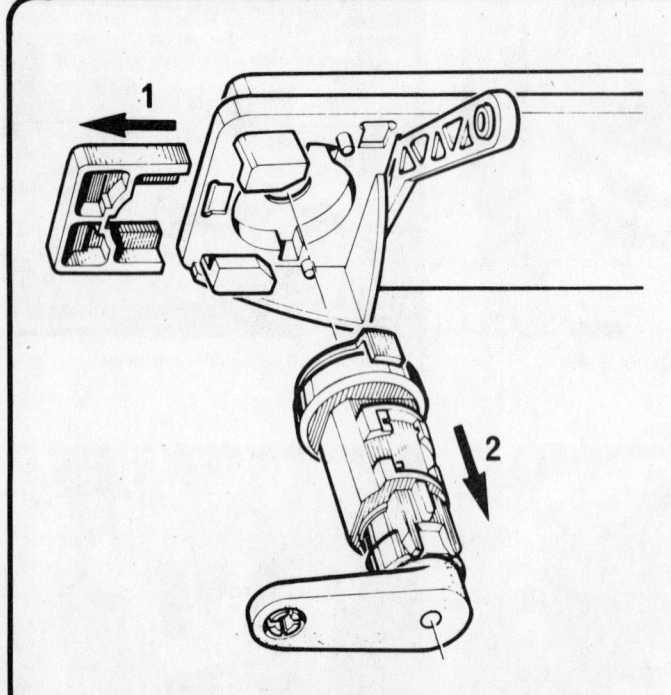

Fig. 11.9 Door cylinder lock (Sec 11)

1 Retaining clip   2 Direction of removal of cylinder

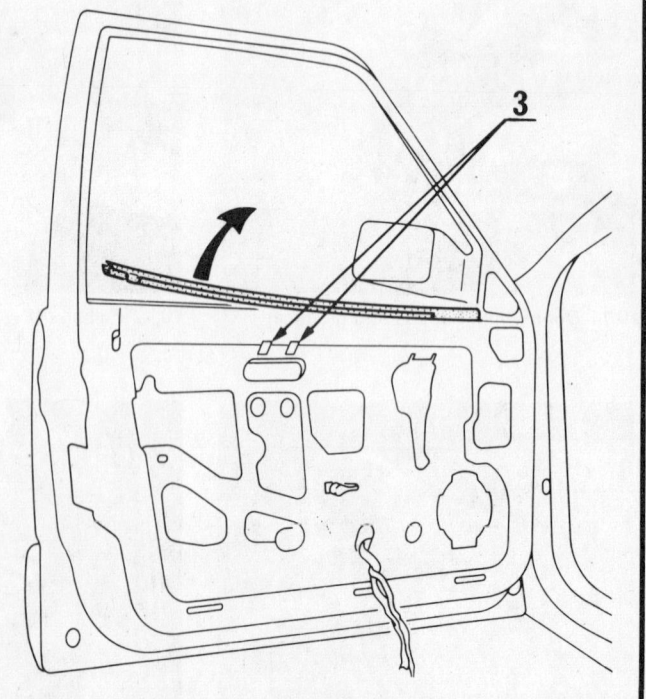

Fig. 11.10 Glass bottom channel bolt access holes (3) (Sec 11)

Pull off the inner glass weatherstrip

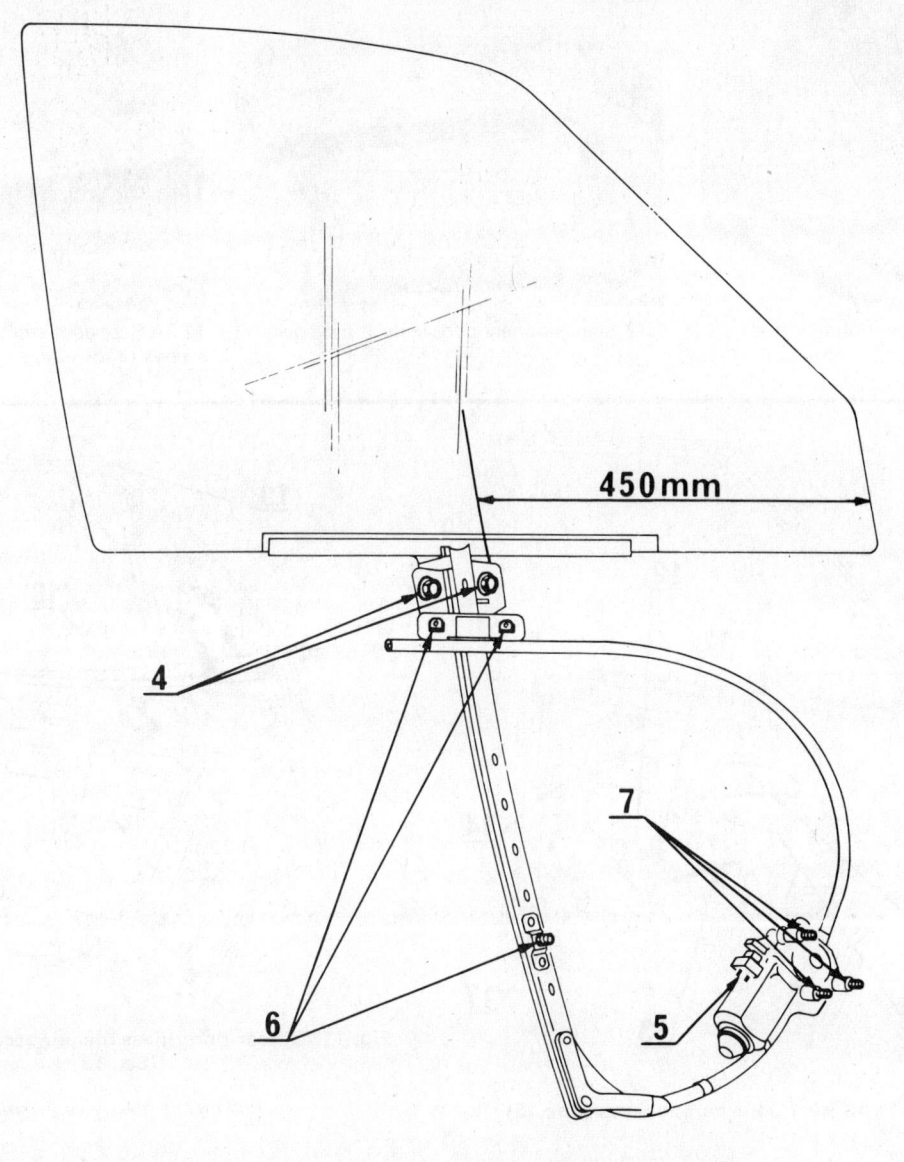

Fig. 11.11 Front door glass bottom channel fixing diagram (Sec 11)

4 Fixing bolts (bottom channel)
5 Motor wiring plug
6 Fixing bolts (cable guide)
7 Motor mounting bolts

12 Rear door trim panel – removal and refitting

1 Open the door wide and extract the screw from the remote control door lock handle (photo).
2 Tap the top cover of the armrest towards the front of the car to release its fixing clips (photo).
3 Extract the top and bottom fixing screws and remove the armrest (photos).
4 On six-cylinder models equipped with rear electric windows, the wiring plugs will have to be disconnected from the armrest switches as the armrest is withdrawn.
5 On models without electric rear windows, prise off the window regulator handle using a forked tool.
6 Remove the trim panel and waterproof sheet as described in Section 10.
7 Refitting is a reversal of removal.

13 Rear door – dismantling and reassembly

1 The operations are very similar to those described in Section 11, but note the location of the components (Fig. 11.12).
2 The door is fitted with a fixed quarter light and this will make removal of the door glass a more complicated operation, as described in the following paragraphs.
3 Remove the window regulator mechanism and then lower the main door glass carefully to its lowest point.
4 Pull out the glass inner and outer weatherstrips and remove the glass divider screws (18 and 19 – Fig. 11.13).
5 Pull out the glass channel and then tilt the divider and remove it.
6 Remove the quarterlight glass with its rubber surround.
7 Pull the main door glass out of the door; withdrawing it towards the outside.
8 Refitting is a reversal of removal.

12.1 Rear door remote control handle screw

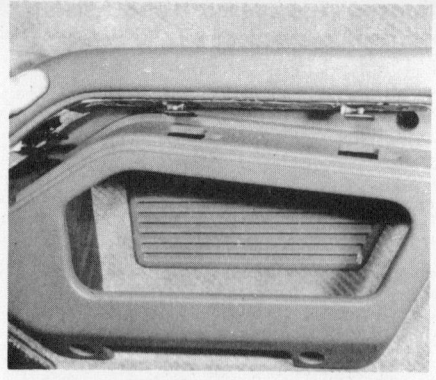

12.2 Removing rear door armrest top cover

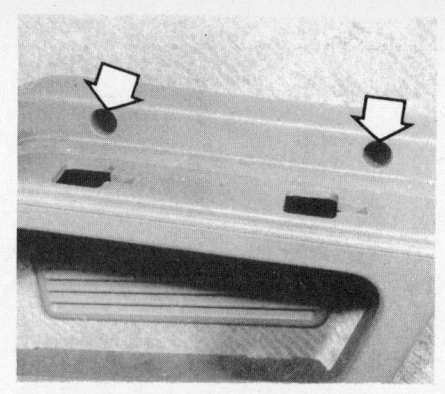

12.3A Rear door armrest upper fixing screws (arrowed)

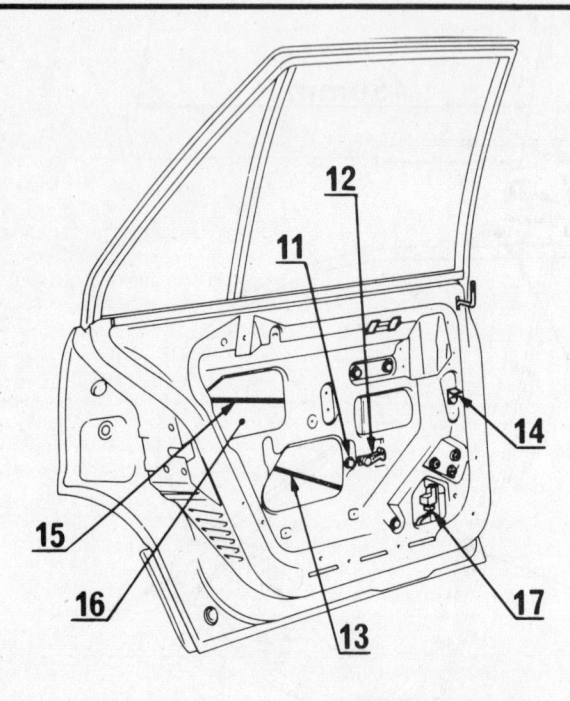

Fig. 11.12 Rear door components (Sec 13)

11 Bolt
12 Remote control lever
13 Link rod
14 Clip
15 Link rod
16 Door aperture
17 Window lift motor

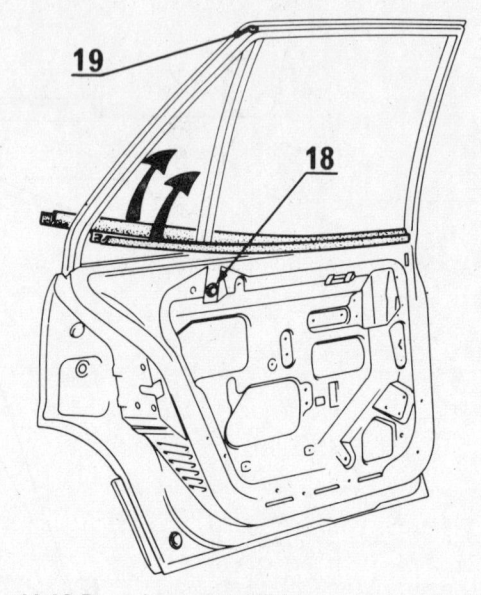

Fig. 11.13 Rear door glass divider screws (18 and 19) (Sec 13)

Pull off the glass weatherstrips

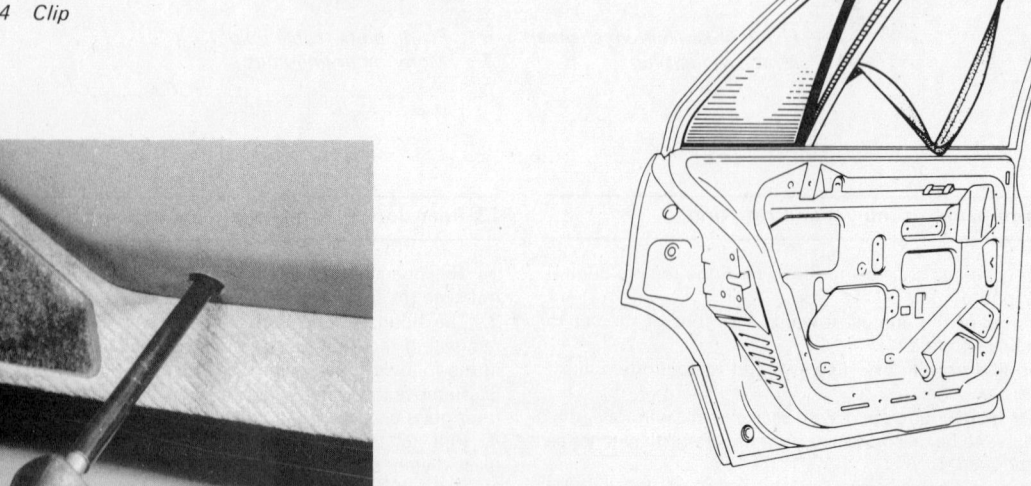

12.3B Rear door armrest lower fixing screw

Fig. 11.14 Removing the door glass channel and divider (Sec 13)

Chapter 11 Bodywork and fittings

bolts which should be unscrewed to remove a door (photo). Never drive out the hinge pivot pin in an attempt to remove a door. If this is done, the hinge pin and its two bushes would have to be renewed.
7 The hinges are welded to the doors.
8 If a front door is being removed, remove the plastic protective liner from under the front wing and remove the panel (23 – Fig. 11.15). The door hinge nuts are now accessible and should be unscrewed.
9 When refitting the door, do not fully tighten the hinge bolts until the door alignment has been checked. Slight adjustment may be carried out by moving the door within the limits of the hinge bolt holes.
10 Check door closure. This should be smooth and without rattle. If necessary, release the striker using a Torx wrench and move it to achieve the desired result (photo).

15 Tailgate – removal and refitting

1 Open the tailgate and extract the nine fixing screws which hold the trim panel to its underside (photo).
2 Disconnect the wiring plugs (photo) and withdraw the wiring harness from the interior of the tailgate. Before withdrawing the wiring, tie a length of string to it. This should be left in place to pull the harness back into the tailgate. Disconnect the washer fluid tube.
3 Remove the rear cover panel from the headlining to expose the tailgate hinge bolts. The hinges are welded to the tailgate and bolted to the bodyshell.
4 With an assistant supporting the tailgate, disconnect the gas support struts. To do this, prise out the locking clip with a small screwdriver and pull the socket from the ball.
5 Unbolt the hinges and lift the tailgate from the car (photo).
6 When refitting the tailgate, do not fully tighten the hinge bolts until the tailgate has been gently closed and its alignment within the body aperture checked. Move the tailgate slightly if necessary within the tolerance of the hinge bolt holes and then fully tighten the bolts.
7 The closure of the tailgate can be adjusted by moving the position of the lock striker.
8 The wiring harness can be drawn into the tailgate using a length of string attached to its end. Certain models have the tailgate lock as part of the central locking system.

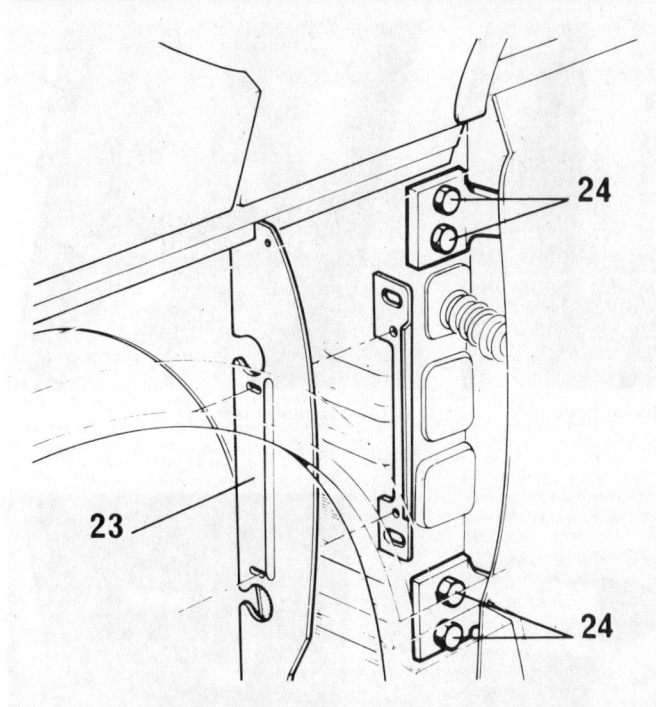

Fig. 11.15 Front door hinge attachment (Sec 14)

23 Under wing cover panel   24 Hinge bolts

16 Windscreen and heated rear screen – renewal

1 Removal and refitting of the front and rear screens should be left to a professional screen replacement specialist.
2 The glass is flush-mounted, and removal requires the use of a hot knife or wire to release the sealant plus the use of special bonding agents when refitting.

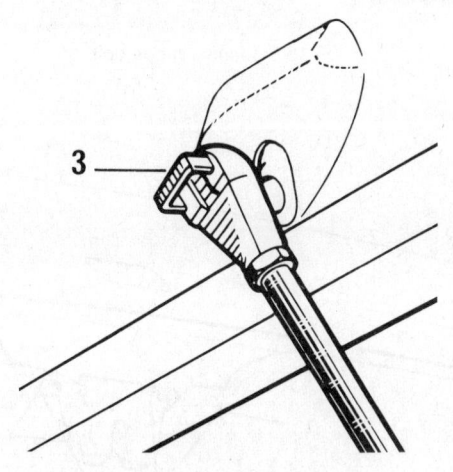

Fig. 11.16 Tailgate strut (Sec 15)

3 Clip

17 Rear interior quarter panel – removal and refitting

1 Pull off the headlining rear cover panel. This is clipped in position.
2 Fold the rear seat back down.
3 Remove the rear parcel shelf (Section 19).
4 Unclip the quarter panel and withdraw it, at the same time feed the seat belts through the cut-outs and disconnect the leads from the map reading lamp.
5 Refitting is a reversal of removal.

14 Door – removal and refitting

1 Disconnect the battery.
2 Open the door wide and disconnect the check link by unscrewing the pillar plate bolts (photo).
3 Remove the door trim panel, as previously described, and disconnect all wiring from the door interior components. Feed the wiring through the protective grommet in the door hinge edge.
4 Support the lower edge of the door on blocks or jacks suitably covered with pads of rag or foam.
5 The position of the door hinges should be marked on the body pillar using masking tape to avoid scratching the paint.
6 The door hinges are bolted to the body pillar and it is the hinge

18 Centre consoles – removal and refitting

Front console
1 Refer to Chapter 12 and remove the radio and cassette (if fitted). Disconnect the battery.
2 Pull up the gear lever gaiter or remove the index plate (autotrans) and remove the front and rear console fixing screws (photos).

Chapter 11 Bodywork and fittings

14.2 Door check link and courtesy lamp/door closed sensor switch (arrowed)

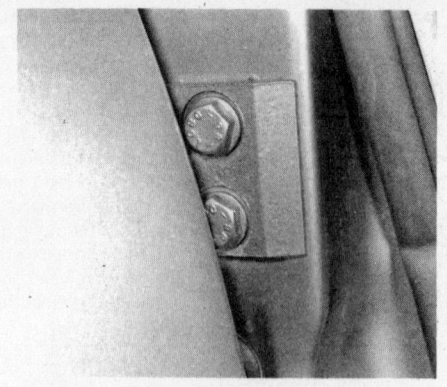

14.6 Rear door hinge bolts on body pillar

14.10 Door lock striker

15.1 Extracting tailgate trim panel screw

15.2 Tailgate wiring harness plug

15.5 Tailgate hinge bolt

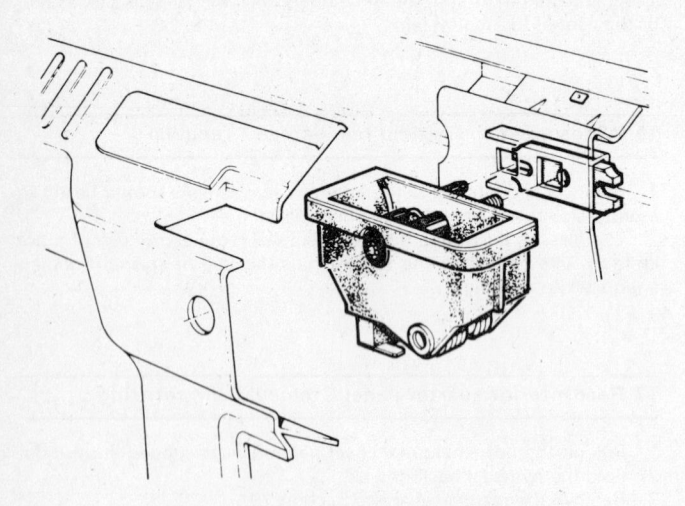

Fig. 11.17 Tailgate lock striker (Sec 15)

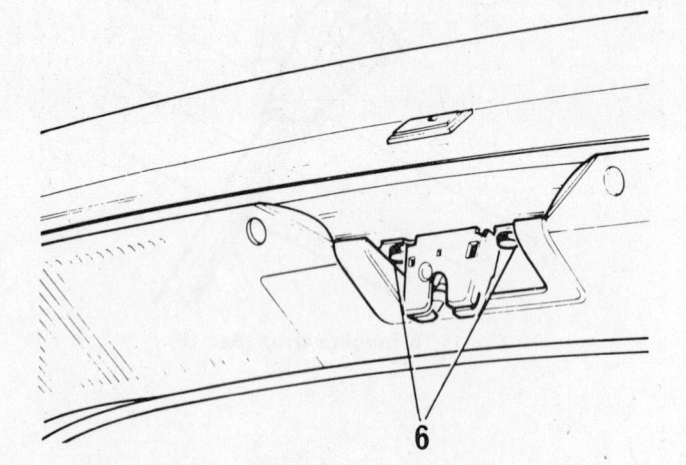

Fig. 11.18 Manually-operated tailgate lock (Sec 15)

6　Fixing bolts

3　Withdraw the console until the wiring plugs can be disconnected from the console switches then lift the console away.

Rear console
4　The operations are similar to those described for the front console but refer to Chapter 2, Section 13, for details of disconnecting the heater air distribution duct to the rear passenger compartment.

All consoles
5　To refit, reverse the removal operations.

19 Rear parcel shelf

1　The rear parcel shelf comprises a detachable centre section and two fixed side sections which incorporate the rear radio speakers.
2　The main shelf may be removed after unhooking the lift straps and pulling it directly backwards.
3　The side sections are held by screws and they may be removed once the speaker leads are disconnected and the screws extracted.
4　When refitting the main shelf, offer it to its fixings and then give it a sharp blow to lock it into its hinge clips.

Chapter 11 Bodywork and fittings

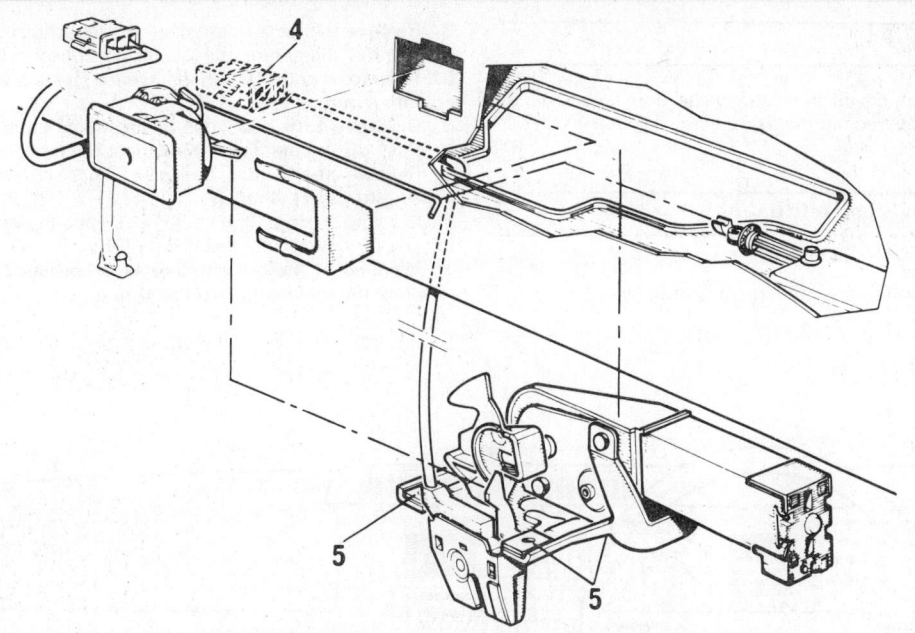

Fig. 11.19 Electrically-operated tailgate lock (Sec 15)

4 Wiring plug 5 Fixing bolts

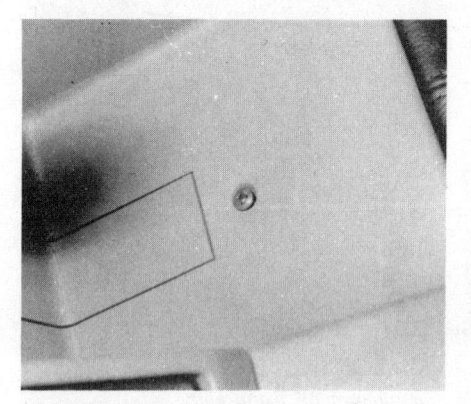

18.2A Extracting front centre console screw (rear)

18.2B Extracting front centre console screw (front)

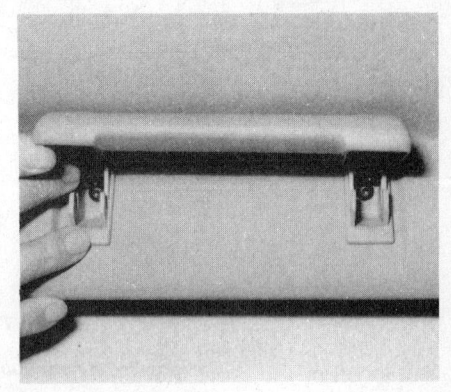

22.1 Grab handle

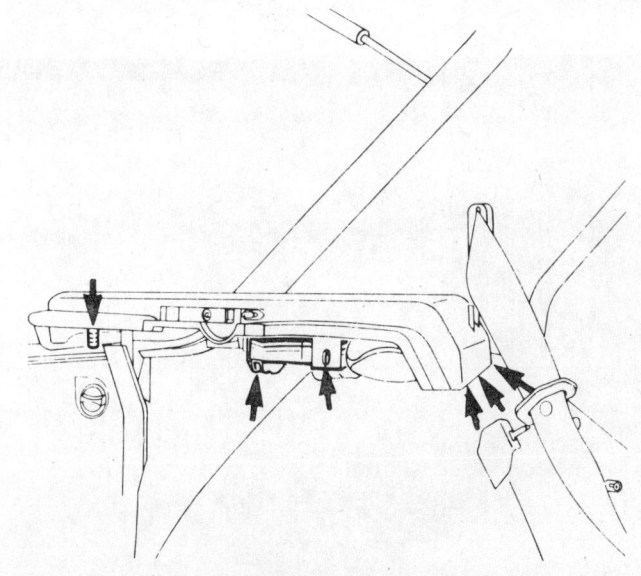

Fig. 11.20 Rear parcel shelf (side section) fixing screws (Sec 19)

20 Headrests

1 The headrests are adjustable for height simply by moving them up and down.
2 To remove them, twist the bevelled plinth at the base of the right-hand headrest stem while pulling the headrest upwards.

21 Seat belts – maintenance, removal and refitting

1 Periodically inspect the seat belts for fraying or other damage. If evident, renew the belt.
2 The belts may be cleaned using warm water and liquid detergent. Do not use solvents of any kind.
3 If the car is involved in a front end collision and the belts have restrained the front or rear seat occupants, renew the belts.
4 When refitting, always maintain the original fitted sequence of washers and spacers.
5 The rear seat belt lower anchor bolts are accessible after removing the seat back. The upper anchor bolts can be reached after removing the rear quarter trim panel (Section 17).
6 On some models, a buzzer warning system is fitted with a timer relay to prevent driving off with the seat belts unfastened.

Chapter 11 Bodywork and fittings

22 Interior grab handles

1 The handles can be removed from above the door apertures by lifting the hand grip and extracting the fixing screws (photo).

23 Facia panel – removal and refitting

1 Disconnect the battery.
2 Refer to Chapter 12 and remove the instrument panel.
3 Remove the facia under cover panel (photo).
4 Remove the centre console, as described in Section 18.
5 Refer to Chapter 10 and remove the steering wheel and the steering column mountings.
6 Disconnect the headlamp beam setting knob and setting plate.
7 Prise off the trim strip and extract the heater control panel screws.
8 Remove the centre console fixing screws and withdraw the console to clear the facia.
9 Working within the engine compartment under the scuttle, unscrew the two facia fixing nuts (photo).
10 Working at the bottom corners of the facia panel inside the car, unscrew the remaining two fixing nuts.

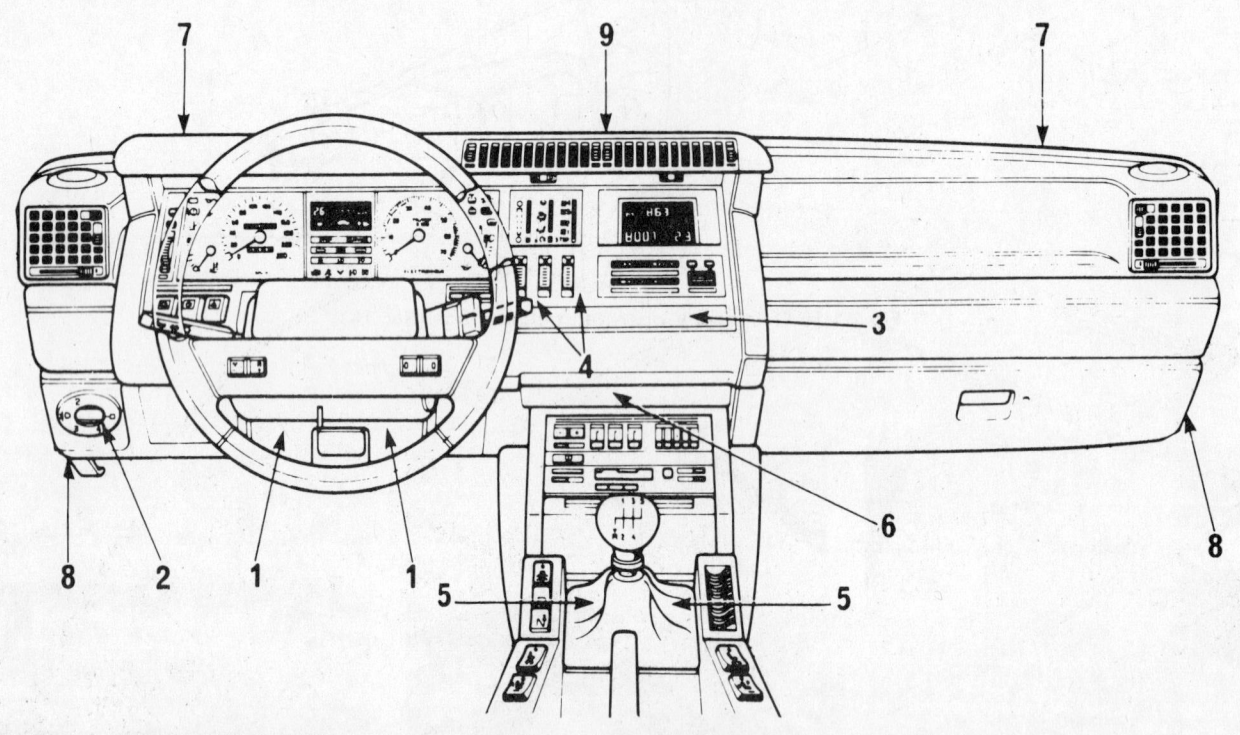

Fig. 11.21 Facia fixing points – LH drive shown (Sec 23)

1 Steering column mounting nuts
2 Headlamp beam adjuster (load level)
3 Trim strip
4 Heater control panel fixing screws
5 Centre console screws
6 Control panel screw
7 Bulkhead (engine compartment) fixing nuts
8 Facia corner fixing nuts
9 Clip

23.3 Facia under cover panel

23.9 Facia fixing nut in engine compartment

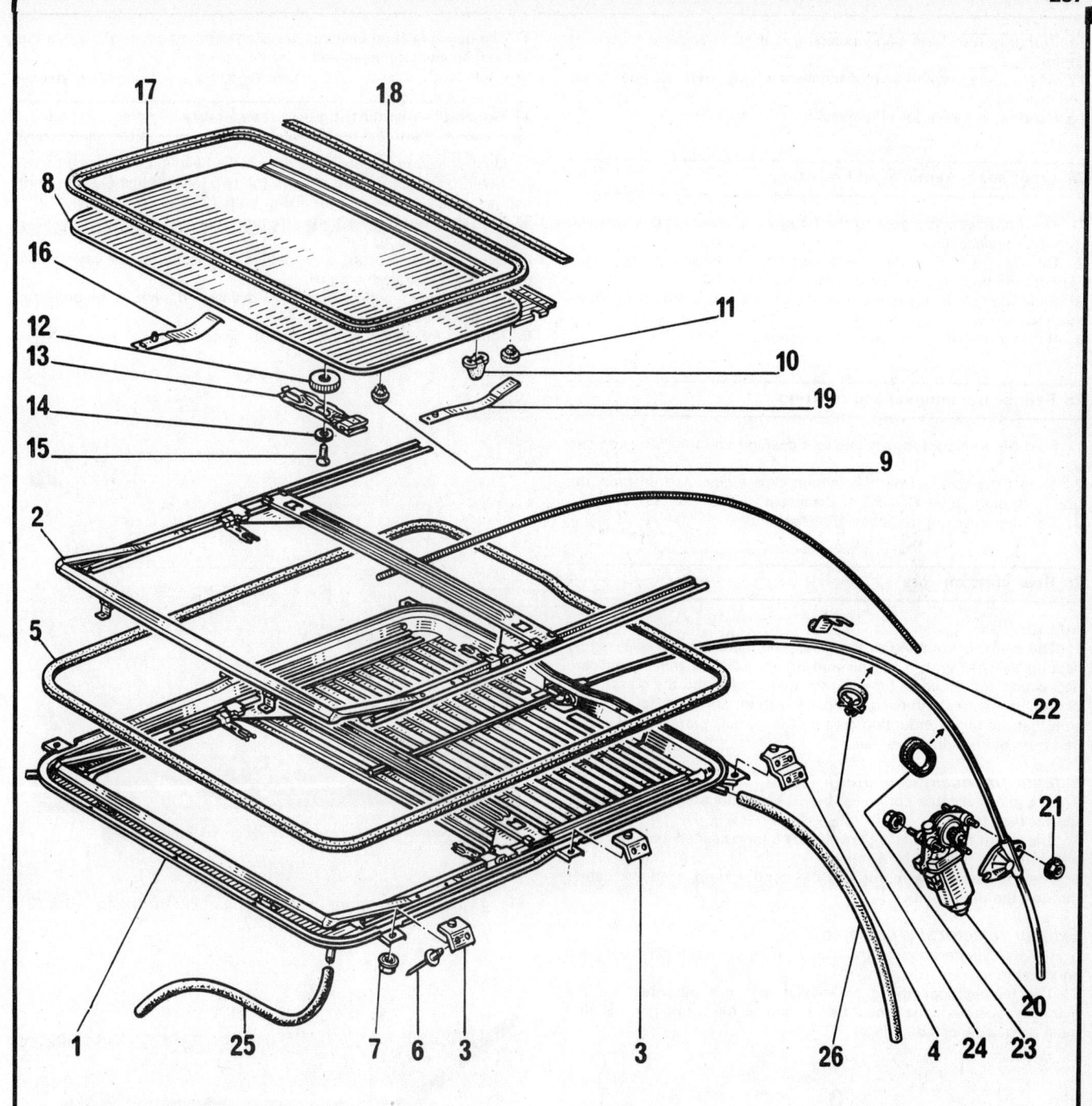

Fig. 11.22 Sunroof components (Sec 27)

1 Gutter	8 Sliding panel	15 Runner guide bolt	21 Motor mounting nuts
2 Sliding panel control frame	9 Deflector stop	16 Guide	22 Cable clip
3 Gutter mounting bracket	10 Plastic slipper	17 Panel frame	23 Grommet
4 Gutter mounting bracket	11 Endstop	18 Sealing strip	24 Water drain pipe clip
5 Seal	12 Adjuster nuts	19 Leaf spring	25 Water drain pipe
6 Rivet	13 Front runner guide	20 Operating motor	26 Water drain pipe
7 Nuts	14 Washer		

288   Chapter 11 Bodywork and fittings

11 Withdraw the facia panel, keeping it level to release if from its centre clip.
12 As the facia is withdrawn, disconnect wiring plugs, vacuum hoses and cables.
13 Refitting is a reversal of removal.

24 Front seat – removal and refitting

1 The front seats are held to the floorpan by Torx screws screwed through the seat slides.
2 On six-cylinder models with electrically-adjustable seats, disconnect the wiring plugs before removing the seats.
3 Slide the seat fixing screw then slide it fully rearwards and extract the front screw.
4 Refit by reversing the removal operations.

25 Rear seat – removal and refitting

1 Fold the seat back down, the seat cushion will fold forwards and upwards.
2 Extract the screws from the scissors type hinges and unscrew the seat back pivot bolts. Remove the rear seat.
3 Refit by reversing the removal operations.

26 Rear view mirrors

Interior
1 The mirror is bounded to the glass. The mirror can be released by heating its base with a hot air gun or by pulling a thin nylon cord backwards and forwards between the mirror base and the glass.
2 A new mirror is supplied complete with an adhesive patch. Make sure that the mirror mounting area on the glass is perfectly clean and free from old bonding material.

Exterior (manually-operated)
3 The mirror is cable-controlled from a knob mounted in the armrest, see Section 10.
4 The mirror fixing nuts are covered by a triangular shaped trim plate which itself is secured by a screw.
5 After releasing the mirror and the control knob, feed the cables through the door cavity.

Exterior (electrically-operated)
6 On six-cylinder models, the exterior mirror is electrically-operated and heated.
7 The removal operations are similar to those described for the manually-operated type except that, of course, the wiring harness plug must be disconnected.

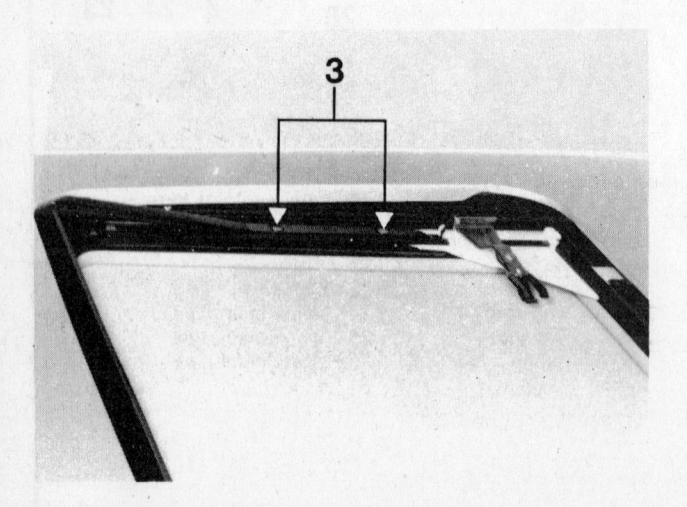

8 The mirror heating element is switched on automatically when the rear screen switch is operated.

27 Sunroof – dismantling and reassembly

1 Open the sunroof panel and detach the headlining from the panel.
2 Refer to Fig. 11.23 and remove the two rear springs (1), the front and rear runner guide and the fixing bolts (2).
3 Turn the front runner guides through a quarter of a turn to release them from the sliding panel.
4 Slightly raise the roof panel which will withdraw the rear runner guides, then remove the panel.
5 If the control mechanism is to be dismantled, remove the deflector pivot bracket bolts (3 – Fig. 11.24).
6 Remove the runner fixing bolts (4 – Fig. 11.25).

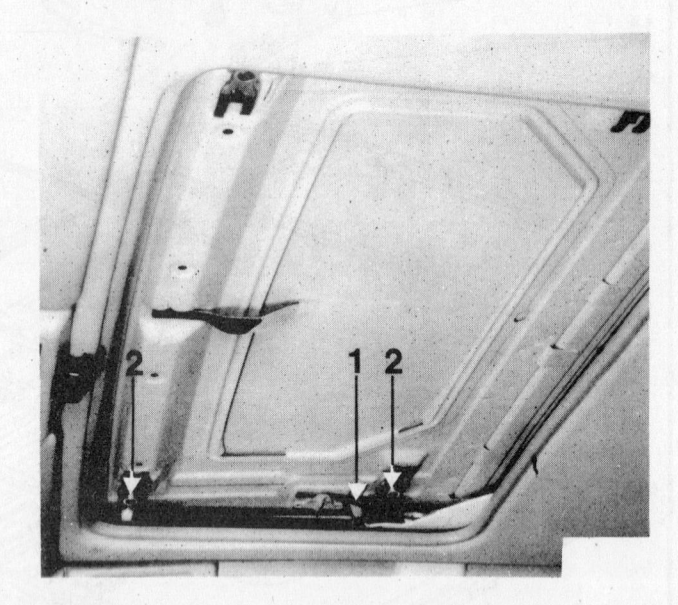

Fig. 11.23 Sunroof panel rear springs (1) and fixing bolts (2) (Sec 27)

Fig. 11.25 Sunroof runner fixing bolt (4) (Sec 27)

Fig. 11.24 Sunroof deflector pivot bracket bolts (3) (Sec 27)

Chapter 11 Bodywork and fittings

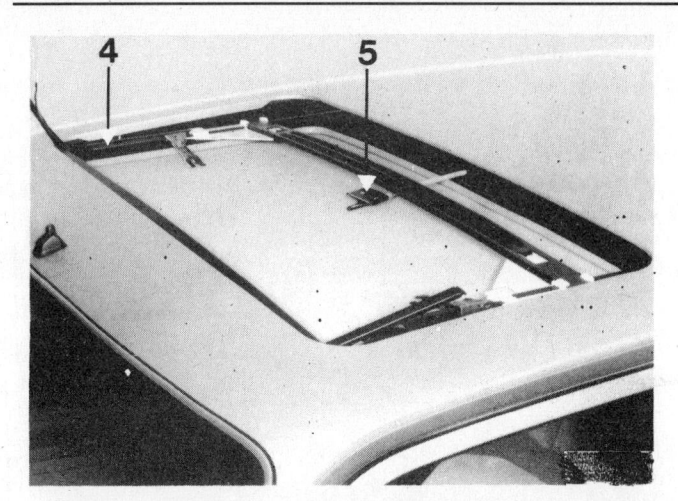

Fig. 11.26 Sunroof control cable clamp bolts (4 and 5) (Sec 27)

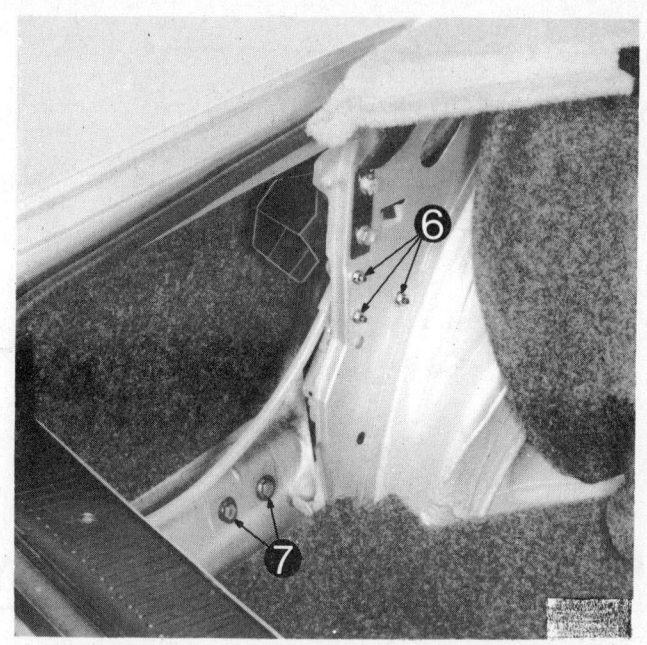

Fig. 11.27 Sunroof motor mounting nuts (6) and rear bumper side mounting bolts (7) (Sec 27)

Fig. 11.28 Sunroof panel control cable routing (Sec 27)

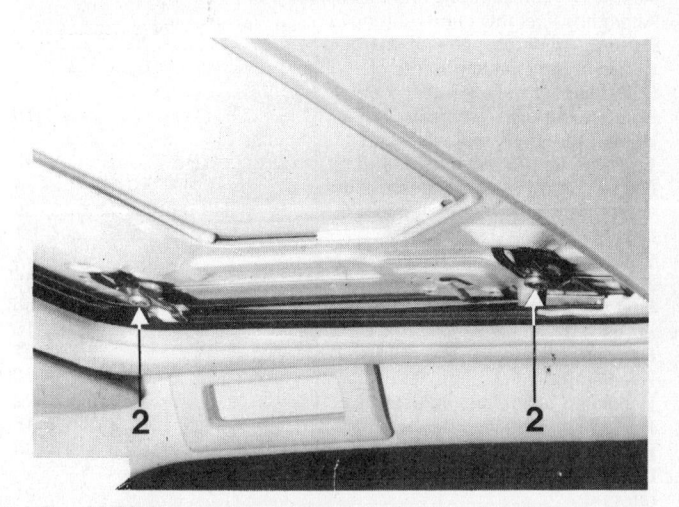

Fig. 11.29 Sunroof panel height adjusting knurled screws (2) (Sec 27)

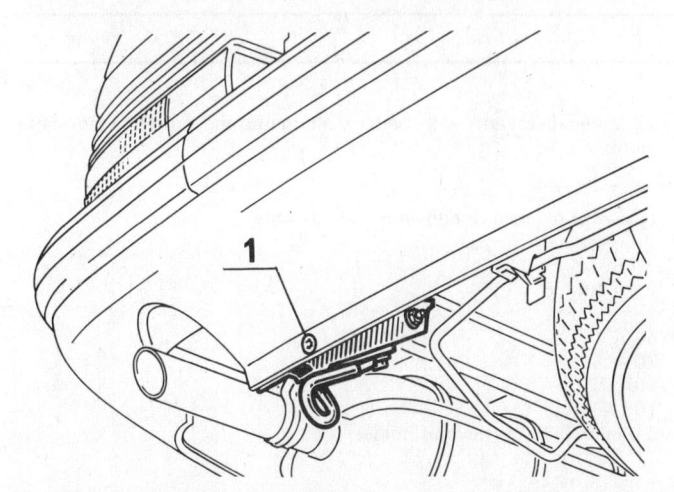

Fig. 11.30 Rear bumper bracket fixing bolt (1) (Sec 28)

7   Remove the control cable clamp bolts (4 and 5 – Fig. 11.26).
8   Withdraw the control mechanism out towards the front of the car.
9   The electric motor can be removed after unscrewing the mounting nuts (6 – Fig. 11.27).
10 Release the cable clamp nuts and pull the cable out towards the front of the car.
11 When reassembling, note the routing of the operating cable under the rear interior quarter trim panel (Fig. 11.28).
12 Assemble the sunroof panel by first inserting the rear runner guides in their frames with the leaf springs under the guides.
13 Adjust the panel by means of the knurled screws until it is flush (when closed) with the roof panel.

28 Rear bumper – removal and refitting

1   The rear bumper is mounted on brackets bolted to the body. One of the brackets has the towing eye welded to it.
2   The ends of the bumper are secured by screws which enter the lower rear wing panel and are accessible from within the luggage area (see Fig. 11.27).

Chapter 12 Electrical system

Contents

Alternator – overhaul	6
Alternator – precautions and maintenance	4
Alternator – removal and refitting	5
Battery – maintenance and charging	2
Battery – removal and refitting	3
Bulb filament defect detection unit	32
Bulbs (exterior) – renewal	19
Bulbs (interior) – renewal	20
Cigar lighter – removal and refitting	43
Cruise control system – description	34
Cruise control system components – removal and refitting	35
Description	1
Electrically-controlled seats	26
Fault diagnosis – electrical system	45
Fuel consumption module	36
Fuses	10
Headlamp – removal and refitting	21
Headlamp beam – alignment	23
Headlamp beam load adjuster	22
Heated rear screen	15
Horns	24
Instrument panel – removal, dismantling, reassembly and refitting	31
Power operated windows	18
Radio aerial – removal and refitting	41
Radio/cassette player – general	38
Radio/cassette player – fitting	40
Radio/cassette player (Renault type) – removal and refitting	39
Radio interference and CB equipment	42
Relays	11
Starter motor – description and testing	7
Starter motor – overhaul	9
Starter motor – removal and refitting	8
Steering column switches – removal and refitting	12
Strip lights	25
Switches – general	13
Tailgate assisted-closure system	16
Tailgate wiper motor – removal and refitting	29
Time and outside temperature display unit	37
Timers – general	14
Vehicle door locking – alternative systems	17
Voice synthesizer	33
Warning buzzers	44
Windscreen/tailgate washer system	30
Windscreen wiper motor and linkage – removal and refitting	28
Wiper blades and arms – removal and refitting	27

Specifications

System type .... 12 V negative earth with battery, alternator and pre-engaged starter motor

Battery
Type and capacity .... Low maintenance or conventional, 50 Ah

Alternator
Make .... Paris-Rhone
Output:
    TS and GTS .... 60 A
    GTX .... 70 A
    V6 .... 90 A
    Models with air conditioning .... 105 A
Minimum brush length (wear limit) .... 5.0 mm (0.20 in) beyond holder

Starter motor
Make and type .... Paris-Rhone, pre-engaged
Maximum brush length (wear limit) .... 10.5 mm (0.39 in)

Chapter 12 Electrical system

Bulbs

Lamp unit	Bulb (W)
Headlamp	Halogen BH1 or H4
Front direction indicator	21
Front parking	3
Tail/stop	5/21
Rear direction indicator	21
Rear fog warning	21
Reversing	21
Rear number plate	5
Radio illumination	0.65
Warning and instrument panel	1.4 (wedge base)
Interior	5

Fuses (typical)

Fuse number	Circuit protected	Rating (A)
1	Heated tailgate screen	20
2	Windscreen wiper 'park'	15
3	Cigar lighter, interior lamp	15
4	Central door locking	25
5	Spare	
6	Spare	
7	Automatic transmission	2
8	Heater controls illumination	5
9	Spare	
10	Spare	
11	Left-hand front seat	30
12	Right-hand front seat	30
13	Heater/air conditioner	25
14	Radio	10
15	Rear foglamp	10
16	Direction indicator flasher	10
17	Left-hand front parking lamp	5
18	Right-hand front parking lamp	5
19	Windscreen wiper	10
20	Reversing lamps	5
21	Instrument panel	3
22	Stop-lamp switch/cruise control	10
23	Left-hand electric windows	30
24	Right-hand electric window/sunroof	30

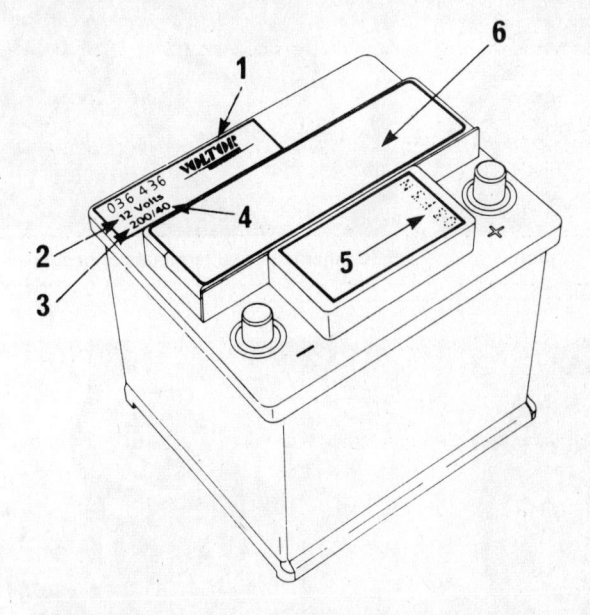

Fig. 12.1 Typical low maintenance battery (Sec 2)

1 Make
2 Voltage
3 Maximum current on starting
4 Capacity (Ah)
5 Date of manufacture
6 Warranty label

1 Description

The electrical system on all models is of 12 volt negative earth type. Main components of the system are a lead-acid battery, an alternator and a pre-engaged starter motor.

All models have a high level of electrical equipment, and higher range versions are equipped with a voice synthesiser, a cruise control system and stereo radio system.

The Renault 25 makes extensive use of electronics both in the engine management system (ignition and fuel injection systems), and for driver information and comfort purposes.

2 Battery – maintenance and charging

1 The battery may be of conventional or low maintenance type, depending upon the date of production of the car.

2 With conventional batteries, check the electrolyte level regularly and add purified water to the cells so that the electrolyte level is 12.5 mm (0.5 in) above the tops of the plates.

3 With low maintenance batteries, the electrolyte level should only require checking after the first four years and every two years thereafter.

4 Never attempt to add acid to a battery, this could prove dangerous. If electrolyte must be added to make up for spillage caused by careless removal of the battery, always have the job done by your dealer or battery specialist.

5 Keep the battery terminals smeared with petroleum jelly to protect them against corrosion.

Chapter 12 Electrical system

6 Any corrosion occurring on the battery platform or surrounding bodywork should be treated immediately to neutralise it. Apply sodium bicarbonate or household ammonia, wash off and then paint the affected areas.
7 The battery will not normally require charging from the mains supply, but if only very short journeys are made with much use being made of the starter and electrical accessories then a regular charge from an outside source may be required.
8 Before charging the battery, remove it from the car as described in the next Section.
9 Connect the leads correctly and make sure that no sparks or naked flame are allowed near the battery during charging as the hydrogen being produced by the battery could cause an explosion.

3 Battery – removal and refitting

1 Open the bonnet and disconnect the battery negative (–) lead then the positive (+) lead in that order.
2 Unbolt the clamp from the lip at the base of the battery casing and lift the battery from its mounting platform.
3 Refit by reversing the removal operations, connect the positive (+) lead and the negative (–) lead in that order.
4 On Renault original equipment batteries, the battery can be quickly disconnected simply by unscrewing the negative terminal thumbscrew through a few turns. There is no need to disconnect the cable from the battery terminal.

4 Alternator – precautions and maintenance

1 The alternator can be damaged if the following precautions are not observed.
2 Never connect the battery leads incorrectly.
3 Never run the engine with the alternator leads disconnected.
4 Do not pull off a battery lead as a means of stopping the engine.
5 When charging the battery from the mains, disconnect it from the car's electrical system or remove it.
6 When using electric welding equipment on the car, always disconnect both battery leads and the alternator leads.
7 Do not operate the starter motor if the engine earth lead is not connected.
8 Regularly inspect the condition of the alternator drivebelt and if frayed or cut, renew it, as described in Chapter 2.
9 The alternator has an integral voltage regulator.

5 Alternator – removal and refitting

1 Disconnect the battery.
2 Remove the air cleaner intake duct to provide greater accessibility.
3 Disconnect the leads from the terminals on the rear cover of the alternator (photos).
4 Remove the drivebelt as described in Chapter 2.
5 Unscrew and remove the mounting and adjuster link bolts and lift the alternator from its mounting bracket (photo).
6 Refit by reversing the removal operations.
7 Tension the drivebelt as described in Chapter 2.

6 Alternator – overhaul

1 Renewal of the brushes should be the limit of overhaul work to the alternator. If the unit has covered a high mileage, exchange it for a new or factory-reconditioned assembly. Renewal of several internal components will prove more expensive than the price of a complete alternator.
2 With the alternator removed from the car, take off its rear plastic cover (photo).
3 Unscrew the two brush holder fixing screws and remove the brush holder (photo).

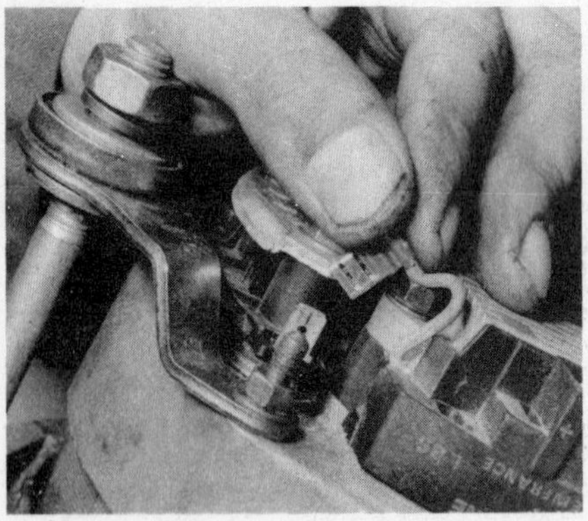

5.3A Disconnecting alternator leads

5.3B Alternator lead connecting plug

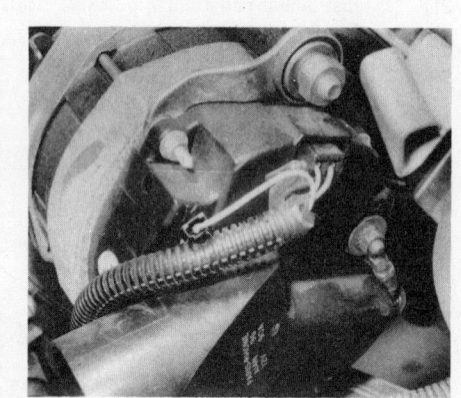

5.3C Alternator rear terminal connections

5.5 Unscrewing alternator mounting bolt

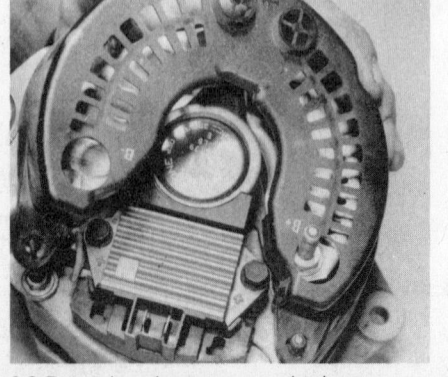

6.2 Removing alternator rear plastic cover

6.3 Removing alternator brush holder

Chapter 12 Electrical system

4 If the brushes are worn down to the stage where they protrude no more than 5.0 mm (0.20 in) beyond their holders, then the complete brush holder should be renewed: separate brushes are not supplied (photo).
5 While the brush holder is removed, take the opportunity to inspect the slip rings. Clean them with a fuel-soaked rag or fine glass paper if they are severely discoloured (photo).
6 Reassembly is a reversal of dismantling.

7 Starter motor – description and testing

1 With a pre-engaged type of starter motor, the drive pinion is brought into mesh with the starter ring gear on the flywheel before the main current is applied.
2 When the starter switch is operated, current flows from the battery to the solenoid which is mounted on the starter body. The plunger in the solenoid moves inwards, so causing a centrally pivoted lever to push the drive pinion into mesh with the starter ring gear. When the solenoid plunger reaches the end of its travel, it closes an internal contact and full starting current flows to the starter field coils. The armature is then able to rotate the crankshaft, so starting the engine.
3 A special freewheel clutch is fitted to the starter drive pinion so that as soon as the engine fires and starts to operate on its own it does not drive the starter motor.
4 When the starter switch is released, the solenoid is de-energised and a spring moves the plunger back to its rest position. This operates the pivoted lever to withdraw the drive pinion from engagement with the starter ring.
5 If the starter motor fails to turn the engine when the switch is operated there are five possible causes:

(a) The battery is faulty
(b) The electrical connections between the switch, solenoid, battery and starter motor are somewhere failing to pass the necessary current from the battery through the starter to earth
(c) The solenoid switch is faulty
(d) The starter motor is mechanically or electrically defective
(e) The starter motor pinion and/or flywheel ring gear is badly worn and in need of replacement

6 To check the battery, switch on the headlights. If they dim after a few seconds the battery is in a discharged state. If the lights glow brightly, operate the starter switch and see what happens to the lights. If they dim then you know that power is reaching the starter motor but failing to turn it. If the starter turns slowly when switched on, proceed to the next check.
7 If, when the starter switch is operated the lights stay bright, then insufficient power is reaching the motor. Remove the battery connections, starter/solenoid power connections and the engine earth strap and thoroughly clean them and refit them. Smear petroleum jelly around the battery connections to prevent corrosion. Corroded connections are the most frequent cause of electric system malfunctions.
8 When the above checks and cleaning tasks have been carried out but without success, you will possibly have heard a clicking noise each time the starter switch was operated. This was the solenoid switch operating, but it does not necessarily follow that the main contacts were closing properly (if no clicking has been heard from the solenoid, it is certainly defective). The solenoid contact can be checked by putting a voltmeter or bulb across the main cable connection on the starter side of the solenoid and earth. When the switch is operated, there should be a reading or lighted bulb. If there is no reading or lighted bulb, the solenoid unit is faulty and should be renewed.
9 If the starter motor operates but doesn't turn the engine over then it is most probable that the starter pinion and/or flywheel ring gear are badly worn, in which case the starter motor will normally be noisy in operation.
10 Finally, if it is established that the solenoid is not faulty and 12 volts are getting to the starter, then the motor is faulty and should be removed for inspection.

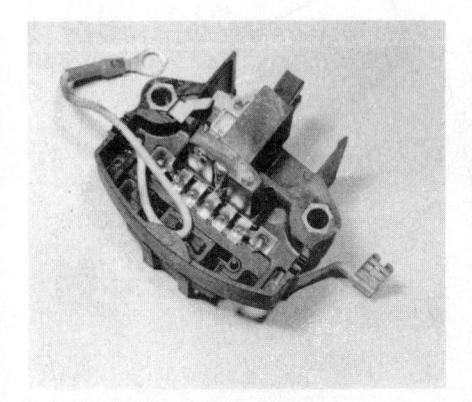

6.4 Alternator brush holder removed

6.5 Alternator slip rings

8.1 Starter motor connections and front mounting bracket (upper)

8.2 Starter motor front mounting bracket (lower)

8.7 Starter motor (six-cylinder)

Chapter 12 Electrical system

8 Starter motor – removal and refitting

Four-cylinder models
Note: *Access for removal of the starter motor from 2165 cc engines will be improved if the inlet manifold is removed first (Chapter 3).*
1 Disconnect the battery and starter motor leads (photo).
2 Disconnect the starter motor front bracket mounting bolts (photo) then unscrew the three bolts which hold the starter to the bellhousing.
3 Withdraw the starter motor upwards.
4 Before refitting the starter motor, first slacken the starter-to-mounting bracket bolts.
5 Bolt the starter to the bellhousing and then fit and tighten the front bracket-to-crankcase bolts. Finally tighten the starter-to-mounting bracket bolts.

Six-cylinder models
6 The starter is easily removed from underneath the car, but on non-Turbo models the oil filter cartridge must first be unscrewed and removed (see Chapter 1).
7 Disconnect the battery and the leads from the starter motor (photo).
8 To reach the starter motor bolts, place the car over an inspection pit, or raise it on jacks and axle stands or ramps.
9 Use a socket wrench on the end of a long extension inserted from the rear and at the side of the transmission to unscrew the starter motor bolts.
10 Remove the starter forward and close to the crankcase.
11 When refitting the starter motor, locate the small cover plate on its dowel before offering the starter into position.
12 When screwing in the starter motor bolts, note the location of the shorter one (A – Fig. 12.3).

Turbo models
13 On these models, some additional work must be carried out before the starter motor can be removed.
14 Disconnect the engine oil cooler pipe support bracket by removing the bolt (C – Fig. 12.14).
15 Disconnect the hose from the union (D – Fig. 12.5) and unscrew and remove the union. Expect some loss of oil.

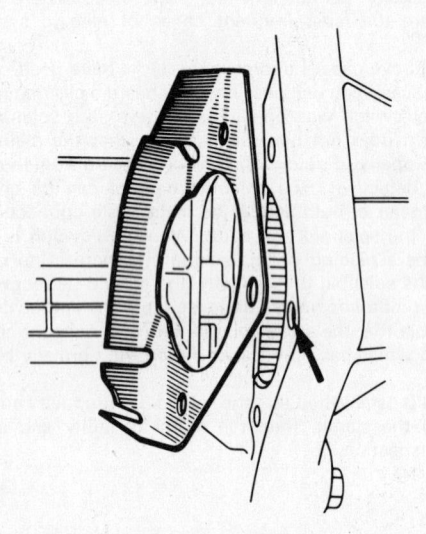

Fig. 12.2 Starter motor mounting plate (six-cylinder models) (Sec 8)

Arrow indicates dowel

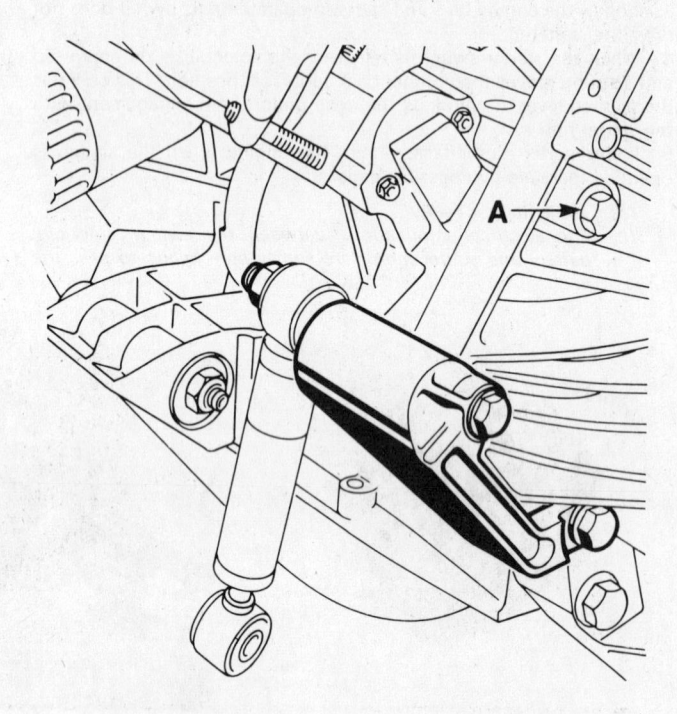

Fig. 12.3 Shorter starter motor bolt (A) (Sec 8)

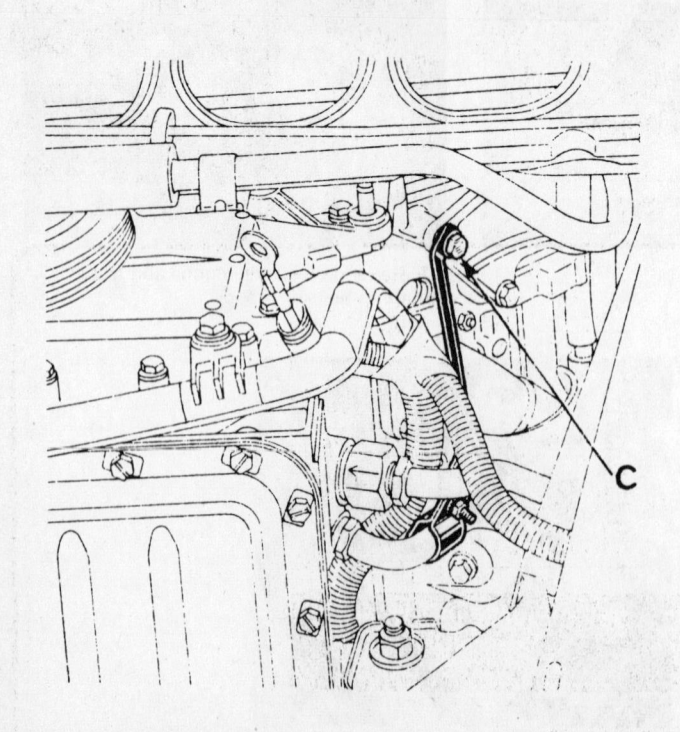

Fig. 12.4 Engine oil cooler pipe bracket and bolt (C) (Sec 8)

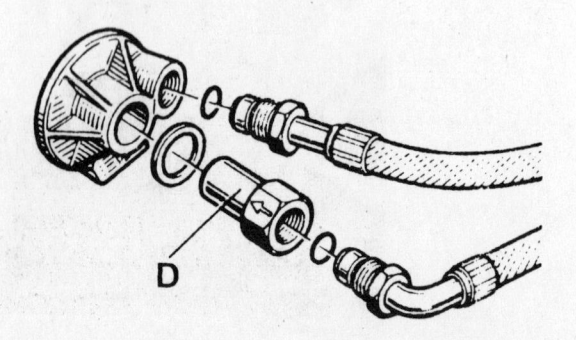

Fig. 12.5 Oil cooler hose and union (D) (Sec 8)

Chapter 12 Electrical system

16 Support the weight of the engine either by attaching a hoist or by placing a jack under the sump pan. Disconnect the left-hand engine mounting and raise the engine slightly to provide clearance for removal of the starter motor.

All models
17 Refitting is a reversal of the removal operations, but top up the engine oil to make up for any lost.

9 Starter motor – overhaul

1 Renewal of the starter motor brushes should be regarded as the limit of overhaul operations.
2 If the starter motor has been in service for a long period and is generally worn, it is more economical to change the original unit for a new or reconditioned one.
3 To check the brushes for wear, first remove the starter from the car as described in the preceding Section.
4 Remove the centre cap screws from the rear cover.
5 Remove the centre cover (photo).
6 Lock the drive pinion teeth with a large screwdriver and then unscrew and remove the rear cover centre bolt. Retain the washer and shims (photo).
7 Unscrew the tie-bolt nuts and take off the rear cover (photos). this is spring-loaded to keep the brushes in contact with the face type commutator. Check the brush wear (see Specifications).
8 The earthed brush leads may be unsoldered and the new brushes resoldered (photo).
9 The field coil brush leads should be cut (photo) and the new ones soldered to the original stubs. Make good the brush lead insulation.

9.5 Starter motor centre cover

9.6 Removing rear cover centre bolt, washer and shims

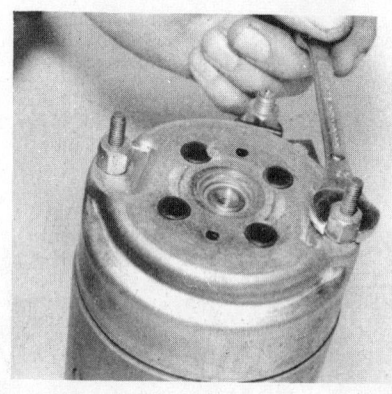

9.7A Unscrewing tie-bolt nut

9.7B Removing starter rear cover

9.8 Unsoldering a brush lead

9.9 Cutting a brush lead

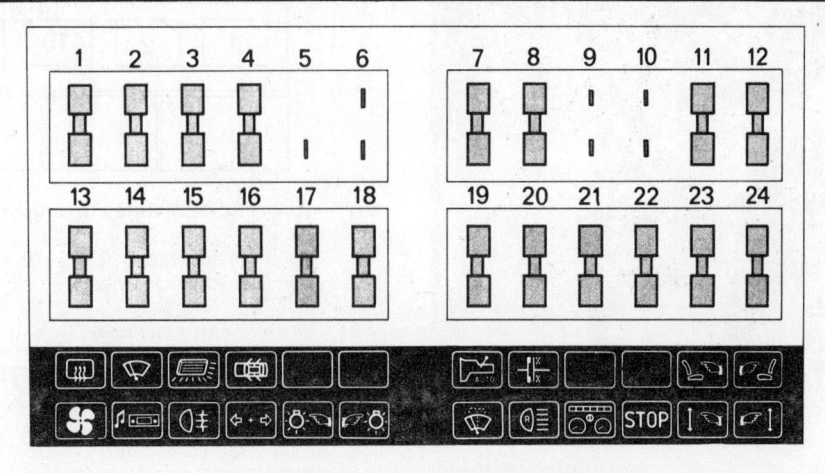

Fig. 12.6 Fuse arrangement (Sec 10)

Chapter 12 Electrical system

10 Clean the commutator with a fuel-soaked cloth, or if very discoloured, very fine glass paper.
11 Fit the rear cover and tie-bolt nuts.
12 Fit the centre bolt, washer and shims. In order to tighten the centre bolt, the armature shaft will have to be gripped close to the drive pinion. Make sure that the shaft is well protected with several layers of tape before applying self-locking grips.
13 Refit the centre cap.

10 Fuses

1 The fuse block is located below the facia panel. Access to the fuses is obtained by releasing the two catches and tilting the lid down (photo).
2 The circuits protected are indicated below the fuses by means of symbols but a typical layout is given in the Specifications.
3 A blown fuse can be removed using the tweezers supplied. Make sure that the new fuse is of the same rating as the old one, and never substitute a piece of wire or other metal object for the proper fuse.
4 If a new fuse blows immediately when the component or system is switched on do not renew it again until the fault has been traced and rectified.

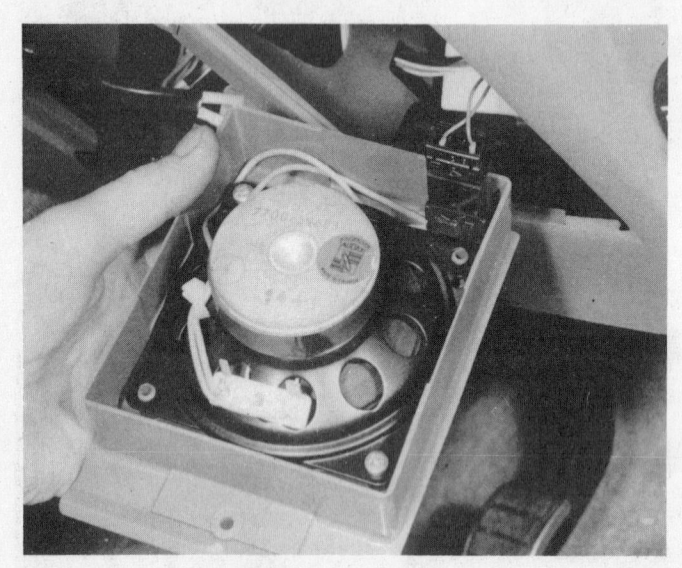

11 Relays

1 The main concentration of relays is located above the fuse block behind the lower facia voice synthesizer (photos). Typical relays are shown in Fig. 12.7.
2 Relays for the radiator electric cooling fans and thermostatic switches are located low down on the left-hand side of the radiator.

12 Steering column switches – removal and refitting

1 Disconnect the battery.
2 Refer to Chapter 10 and remove the steering wheel.

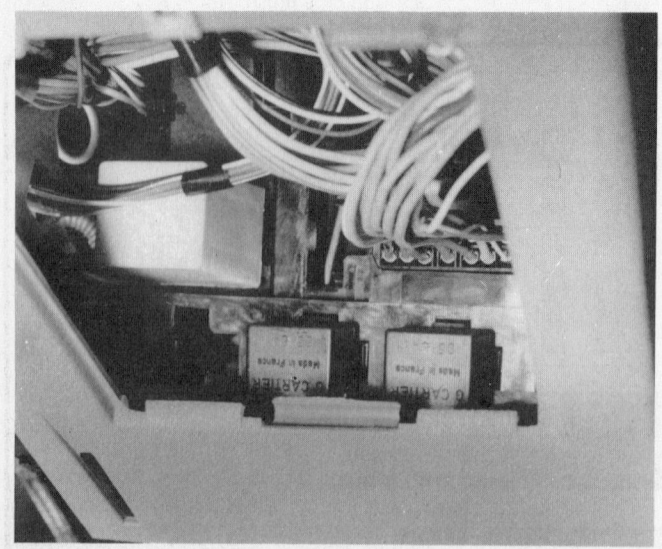

10.1 Fuses in fusebox

11.1A Voice synthesizer speaker

11.1B Relays

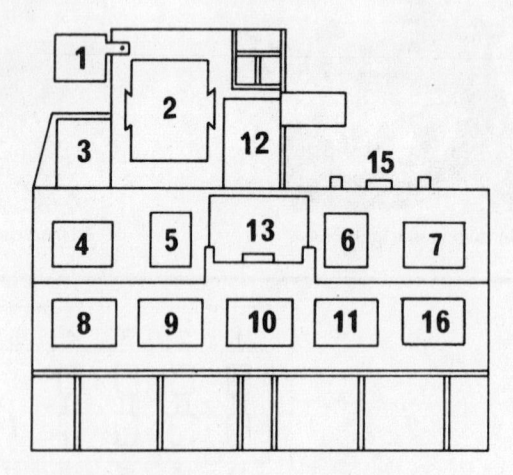

Fig. 12.7 Relay arrangement (Sec 11)

1 Speedometer relay
2 Windscreen wiper timer relay
3 Interior lights timer relay
4 Door locking timer relay
5 Lights 'left on' relay (white)
6 Flasher unit (brown)
7 Headlamp relay (blue)
8 Spare
9 Fuel pump relay (grey)
10 Rear screen demister relay (yellow)
11 Starter relay (red) not 2664 cc engine
12 Driver's door harness
13 Driver's door harness (white)
15 Main feed
16 Spare

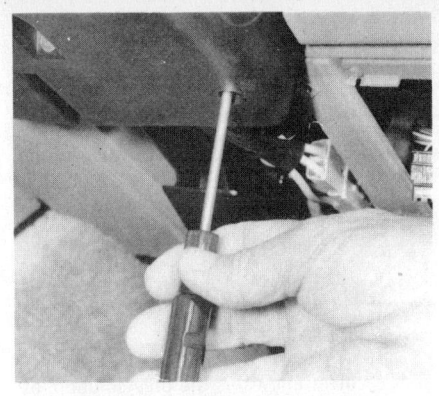

12.3A Steering column lower shroud screw

12.3B Steering column lower shroud screw

12.3C Removing lower shroud

12.3D Removing upper shroud

12.5 Extracting a switch fixing screw

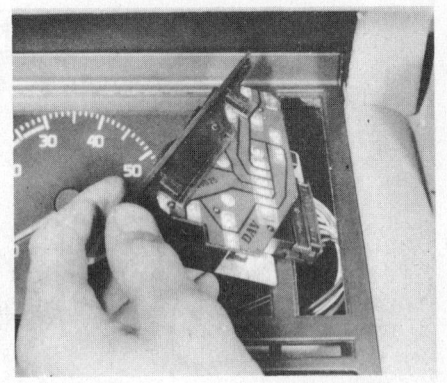

13.2 Removing switch assembly

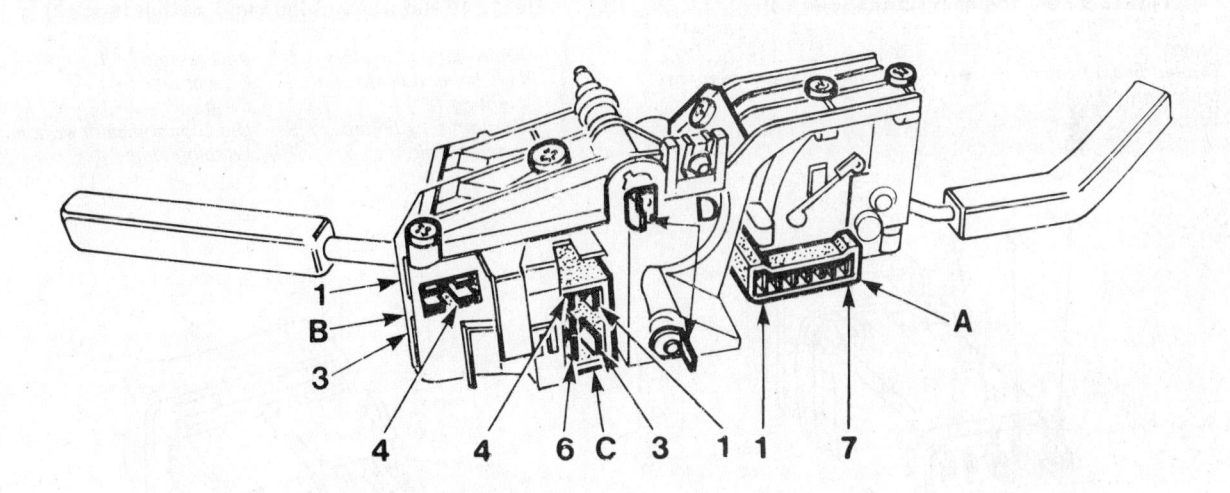

Fig. 12.8 Steering column switches (Sec 12)

Spade connectors at A
1. Timer input
2. Positive (+) after ignition switch
3. Fast wiping speed
4. Normal wiping speed
5. Park
6. Positive (+) after ignition switch
7. Windscreen washer
A. Windscreen wiper plug connector
B. Lighting plug connector

Spade connectors at B
1. Headlamp main beam
2. Headlamp dipped beam
3. Positive (+) before ignition switch
4. Front, rear and rear number plate lamps
C. Direction indicators and horn plug connector

Spade connectors at C
1. Horn
2. Positive (+) before ignition switch
4. Right-hand indicators
5. Flasher unit
6. Left-hand indicators
D. Cruise control plug connector

Chapter 12 Electrical system

3 Remove the steering column upper and lower shrouds (photos). The upper shroud has a switch attached.
4 Disconnect the column switch wiring harness plug.
5 Extract the switch fixing screws and remove the switch (photo).
6 Refitting is a reversal of removal.

13 Switches – general

1 Switches and switch panels are held in position by plastic anchor tabs. If possible, reach behind the switch assembly and compress the tabs before prising the switch or plate from its location.
2 Withdraw the switch until the connecting plug can be disconnected (photo).

14 Timers – general

1 Timing relays are fitted to control the duration of operation of the following:

Windscreen wiper (delay relay)
Electric door locks
Interior lights
Seat belt (buzzer) warning system (certain operating territories only)

2 Should a fault develop in a timer, renewal is the only solution, as repair is not possible.

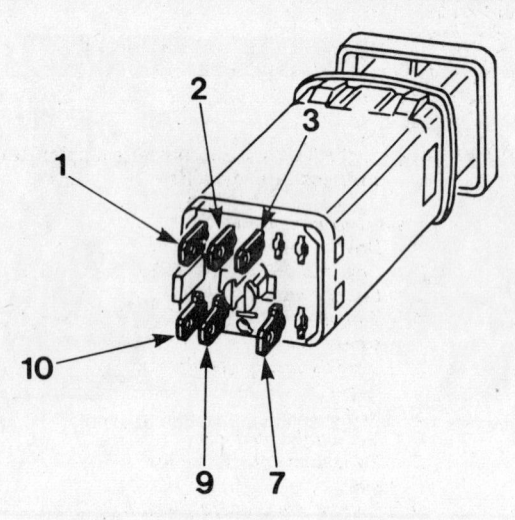

Fig. 12.9 Rear foglamp switch (Sec 13)

1 Spare
2 Dipped beam headlamp
3 Lighting (+)
7 Earth
9 Spare
10 Rear foglamp 'on' warning lamp

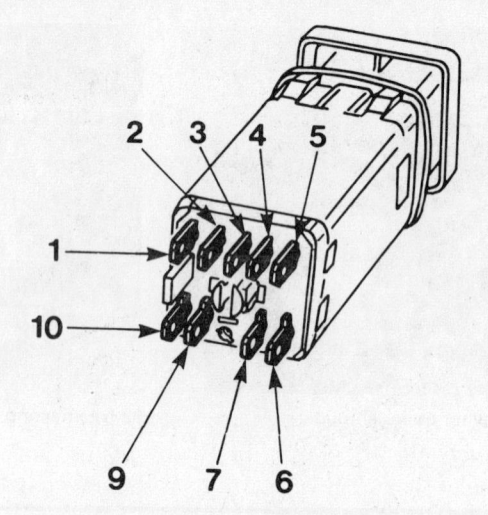

Fig. 12.10 Hazard warning lamp switch (Sec 13)

1 Right-hand indicators
2 Left-hand indicators
3 Lighting (+)
4 Position (+) before ignition switch
5 Accessories (+)
6 Flasher unit (+)
7 Earth
9 Direction indicator switch
10 Warning lamp

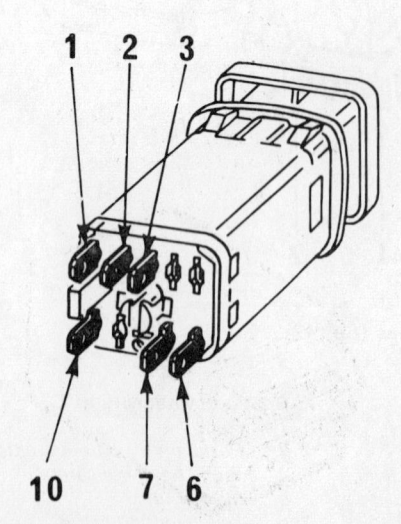

Fig. 12.11 Heated rear screen switch (Sec 13)

1 Positive (+) after ignition switch
2 Relay control
3 Lighting (+)
6 Spare
7 Earth
10 Warning lamp

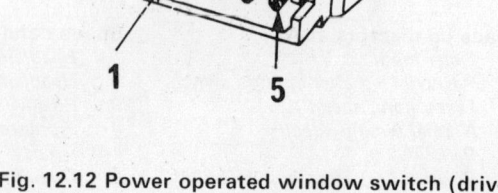

Fig. 12.12 Power operated window switch (driver's) (Sec 13)

1 Motor
2 Positive (+) after ignition switch
3 Earth
4 Switch illumination
5 Motor

299

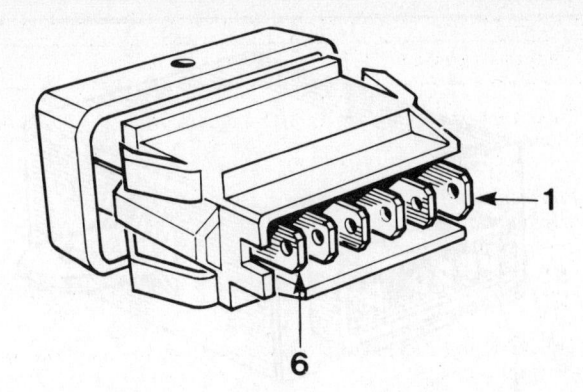

Fig. 12.13 Power operated window switch (passenger's) (Sec 13)

1 Switch illumination
2 Motor
3 Driver's switch
4 Earth
5 Driver's switch
6 Motor

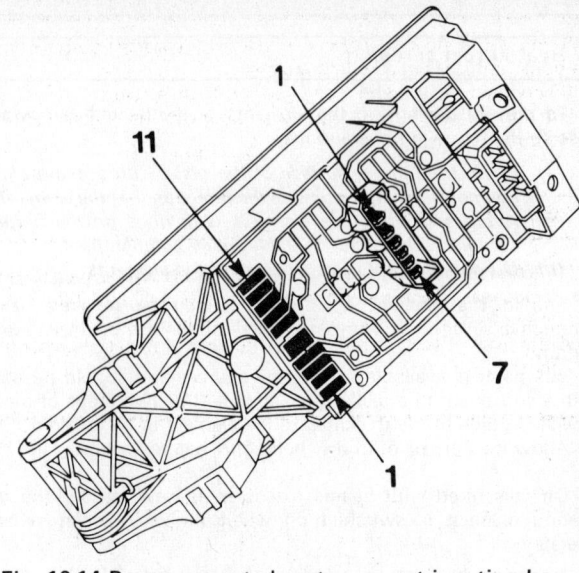

Fig. 12.14 Power operated seats armrest junction box connections (Sec 13)

Eleven-way connector
1 Driver's seat full lowering
2 Rear switch (driver's side)
3 Driver's seat full raising
4 To rear switch (driver's side)
5 Driver's seat lowering – finger pressure
6 Safety notch
7 Switch illumination
8 To rear switch (passenger's side)
9 To rear switch (passenger's side)
10 To passenger switch
11 Driver's seat raising – finger pressure

Seven-way connector
1 Spare
2 Positive (+) after ignition switch (passenger's side)
3 Earth
4 Earth
5 Earth after interdiction switch
6 Positive (+) after ignition switch (driver's side)
7 To switch (passenger's)

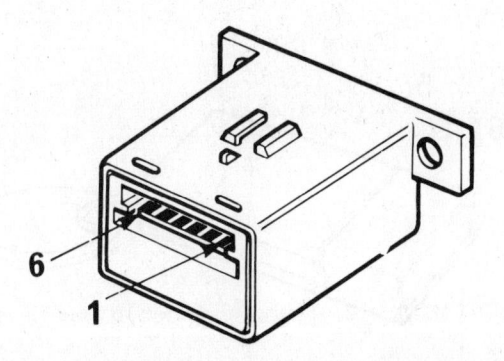

Fig. 12.15 Windscreen wiper timing relay (Sec 14)

1 Earth
2 Washer pump
3 Timed sequence
4 Motor 'park'
5 Position (+) after ignition switch
6 Normal wiping speed

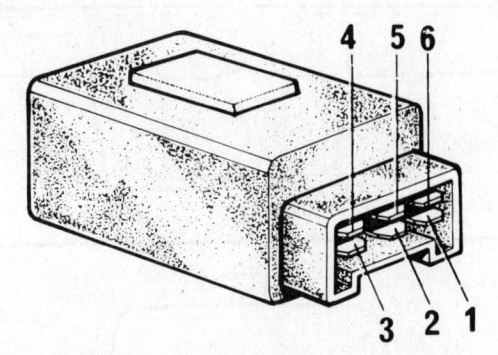

Fig. 12.16 Door lock timer relay (Sec 14)

1 Closing signal
2 Earth
3 Opening signal
4 Motor
5 Positive (+) before ignition switch
6 Motor

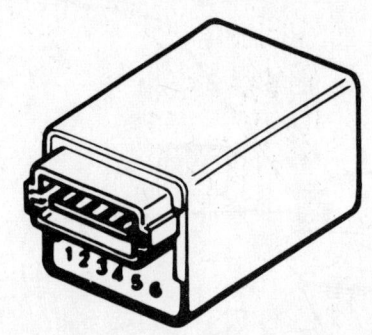

Fig. 12.17 Interior lamps timer relay (Sec 14)

1 Positive (+) before ignition switch
2 Common earth
3 Earth
4 Spare
5 Interior lamps switch
6 Infra red remote control

15 Heated rear screen

1 To prevent damage to the elements of the heated rear window, observe the following precautions:

(a) Clean the interior surface of the glass with a damp cloth or chamois leather, rubbing in the direction that the elements run
(b) Avoid scratching the elements with rings on the fingers or contact with articles in the luggage compartment
(c) Do not stick adhesive labels over the elements

2 Should the element be broken, it can be repaired using a conductive silver paint, without the need to remove the glass from the window.
3 The paint is available from many sources and should be applied with a soft brush to a really clean surface. Use two strips of masking tape as a guide to the thickness of the element to be repaired.
4 Allow the new paint to dry thoroughly before switching the heater on.
5 On cars fitted with heated exterior rear view mirrors, the mirror heating element is switched on when the rear screen switch is operated.

16 Tailgate assisted-closure system

1 This is fitted to six cylinder models and incorporates a control box as shown in fig. 12.18.

17 Vehicle door locking – alternative systems

1 Each system enables the four doors, tailgate and fuel filler cap to be unlocked.

Manual
2 Use the door key in the normal way to unlock one front door then unlock the remaining doors by pulling up the locking knob.
3 The key opens the tailgate and fuel filler flap.

Remote control
4 The door is unlocked by pointing the remote control device at the infra red receiver which is mounted above the rear view interior mirror. The device has its own exclusive code for security reasons.
5 If the tell-tale lamp on the hand held sender unit does not illuminate, renew the batteries.
6 The fuel filler flap and the tailgate are unlocked using the key.

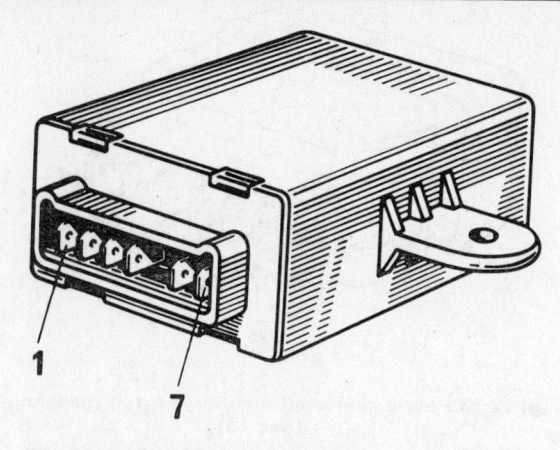

Fig. 12.18 Tailgate closure control box (Sec 16)

1 Motor
2 Closure
3 Opening
4 Earth
5 Spare
6 Positive (+) before ignition switch
7 Motor

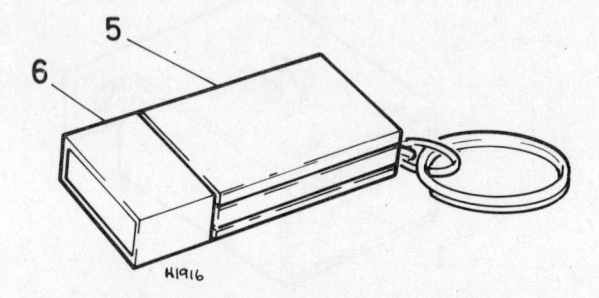

Fig. 12.19 Door lock remote control (Sec 17)

5 Sender unit casing   6 Illuminated tell-tale

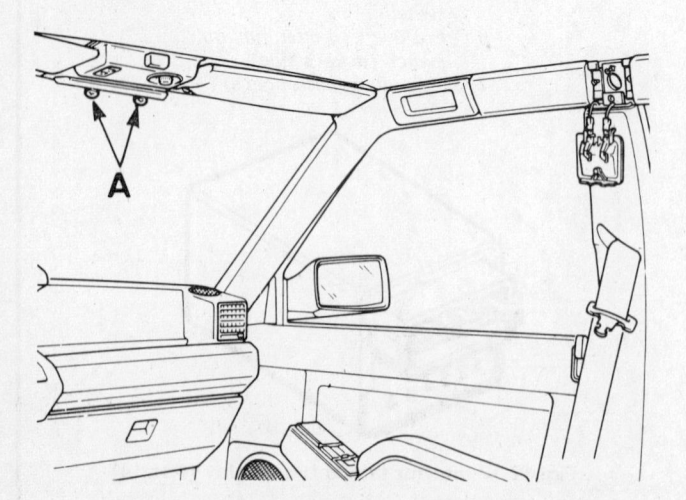

Fig. 12.20 Door locking system infra red receiver (Sec 17)

A Fixing bolts

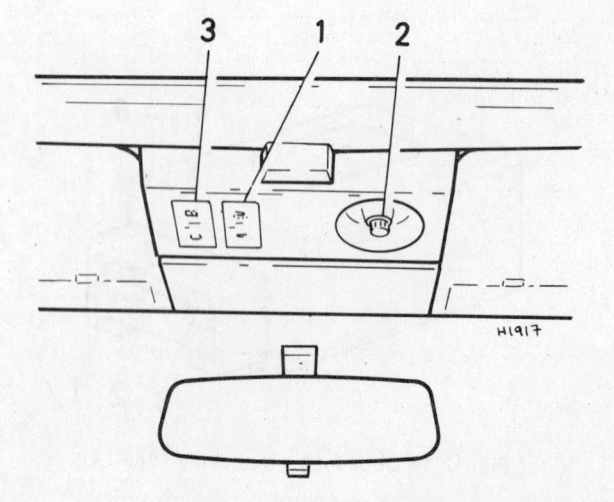

Fig. 12.21 Roof-mounted switches (Sec 17)

1 Map reading lamp switch   3 Sunroof switch
2 Map reading lamp

Chapter 12 Electrical system   301

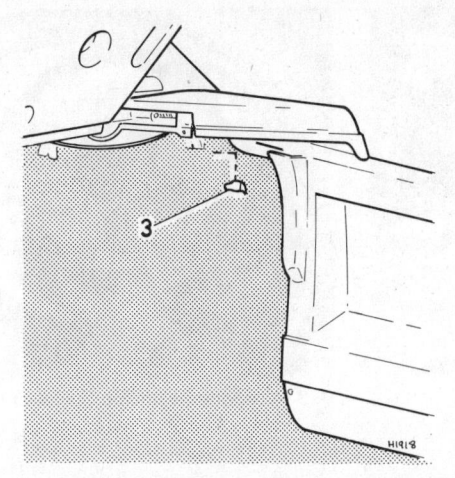

Fig. 12.22 Fuel filler flap emergency release lever (3) (Sec 17)

Electric central door locking
7 This system enables all doors, fuel filler flap and tailgate (certain versions) to be locked or unlocked using the key in a front door lock or the infra red remote control sender unit.
8 Alternatively, the doors can be locked or unlocked from inside the car using the switch on the centre console.
9 Should a fault develop which prevents the fuel filler flap opening, it may be unlocked manually by pulling the lever (3 – Fig. 12.22) which is hidden behind the trim panel at the side of the luggage compartment.
10 Access to the central locking system lock solenoids is described in Chapter 11.

18 Power operated windows

1 Electrically-operated front windows are fitted to all models except the TS.
2 Electrically-operated front and rear windows are fitted to all six-cylinder versions.
3 The control switches are located in the front door armrests and in the hot air outlet grille panel at the end of the rear centre console.
4 Access to the window operating motors is described in Chapter 11.

19 Bulbs (exterior) – renewal

Headlamp bulb
1 Open the bonnet, turn the plastic cover on the rear of the headlamp through a quarter of a turn and remove it (photo).
2 Pull off the wiring plug (photo).
3 Release the bulbholder spring clip and withdraw the bulbholder assembly (photo).
4 When fitting the new halogen type bulb avoid touching it with the fingers. if it is inadvertently touched, clean it with a tissue soaked in methylated spirit.
5 Refitting is a reversal of removal.

Front parking lamp bulb
6 The bulb and holder are mounted in the headlamp reflector. Access is obtained as described for the headlamp bulb (photo).

Front direction indicator bulb (four-cylinder models)
7 The bulb and holder are mounted in the direction indicator lamp reflector at the side of the headlamp. Twist the bulbholder through a quarter of a turn to remove it.

Front direction indicator bulb (six-cylinder models)
8 The lamps are located in the front bumper.
9 Push one side of the lamp lens in and then slip a thin blade

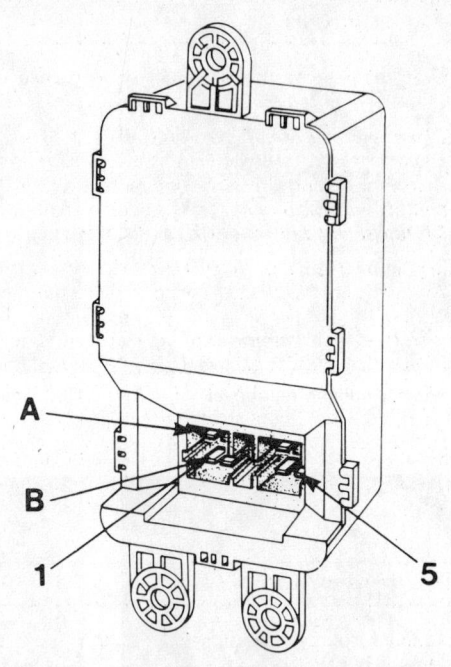

Fig. 12.23 Power operated window finger pressure control box (Sec 18)

Connector plug (A)
1 Motor
2 Earth
4 Motor
5 Positive (+) after ignition

Connector plug (B)
1 Single pressure (raising)
2 Single pressure (lowering)
4 Continuous pressure (lowering)
5 Continuous pressure (raising)

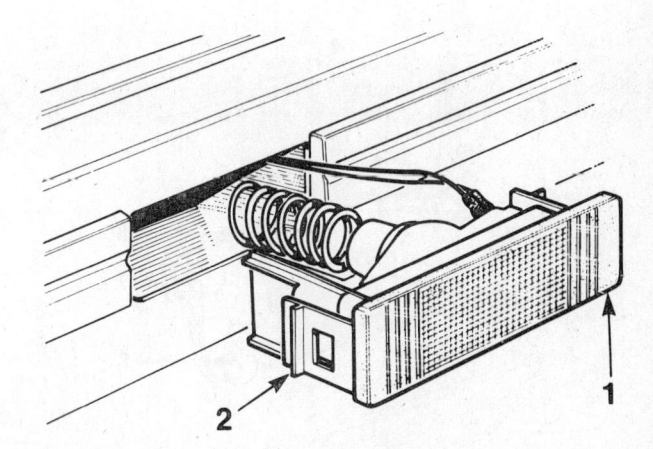

Fig. 12.24 Front bumper-mounted direction indicator lamp (Sec 19)

1 Lens   2 Clip

between the lens and the bumper. Release the lamp and withdraw it, Remove the bulbholder from the rear of the lamp.

Rear lamp bulbs
10 Open the tailgate and unscrew the thumbscrews from the luggage area rear panel (photo).
11 Withdraw the rear lamp assembly.
12 Release the retaining clips, and pull off the lens to expose the bulbs (photos). The lamp unit can be removed completely if the wiring plug is disconnected.

302 Chapter 12 Electrical system

19.1 Removing headlamp rear cover

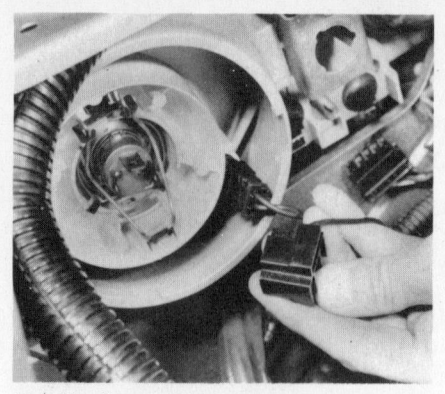

19.2 Headlamp wiring plug

19.3 Removing headlamp bulb and holder

19.6 Front parking lamp bulb and holder

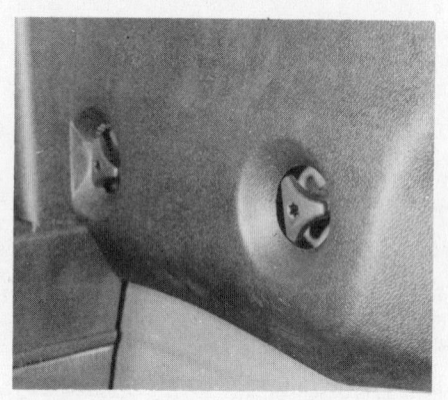

19.10 Luggage area rear panel thumbscrews

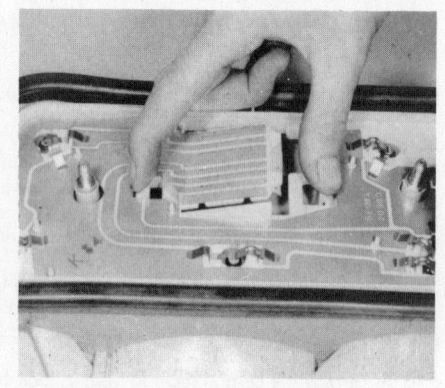

19.12A Releasing rear lamp bulbholder clips

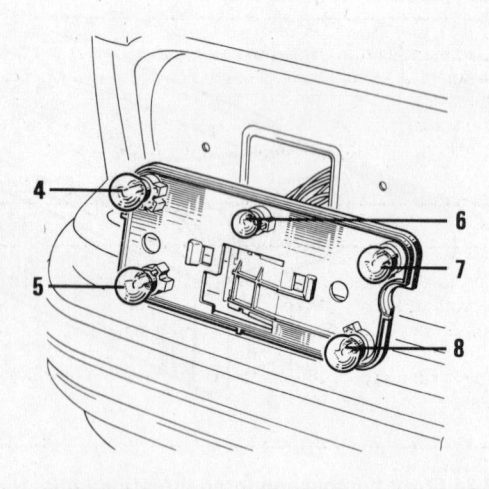

Fig. 12.25 Rear lamp bulb identification (Sec 19)

4 Tail/stop
5 Direction indicator
6 Tail
7 Rear fog
8 Reversing

Rear number plate lamp bulb
13 Extract the screw and remove the bulbholder with the festoon type bulb.

Side repeater lamp bulb
14 Grip the lamp lens and pull the lamp from the wing (photo).
15 Pull the lamp holder from the lens assembly.

20 Bulbs (interior) – renewal

Pillar lamp bulb
1 Pull the lamp lens, lift and pull again to release it from the pillar. Withdraw the festoon type bulb (photo).

Luggage area lamp bulb
2 Prise the lamp from its location and remove the festoon type bulb (photo).

Instrument panel bulbs
3 Remove the instrument panel, as described in Section 31.
4 The bulbholders can then be removed by twisting them and the wedge type bulbs removed simply by pulling them from their holders (photo).

21 Headlamp – removal and refitting

1 Removal of the single (four-cylinder) or dual (six-cylinder) headlamps is similar.
2 Disconnect the wiring plugs from the rear of the headlamps.
3 Disconnect the front parking lamp bulbholder.
4 Unscrew and remove the headlamp mounting nuts from inside the engine compartment and withdraw the headlamp unit forward (photo).
5 Refitting is a reversal of removal but, if a new unit has been fitted, adjust the headlamp beams as described in Section 23.

22 Headlamp beam load adjuster

1 A sealed hydraulic, beam adjusting device is fitted so that the headlamp beams may be adjusted (over and above the normal basic

Chapter 12 Electrical system

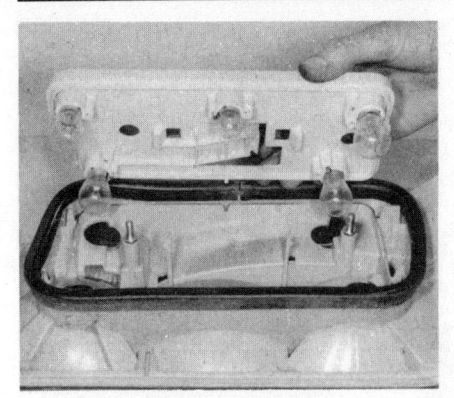

19.12B Removing bulbholder

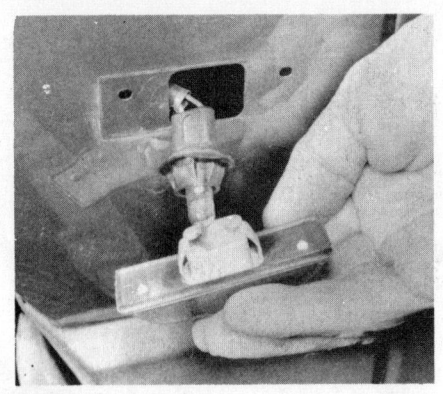

19.14 Side repeater lamp

20.1 Interior pillar lamp, lens removed

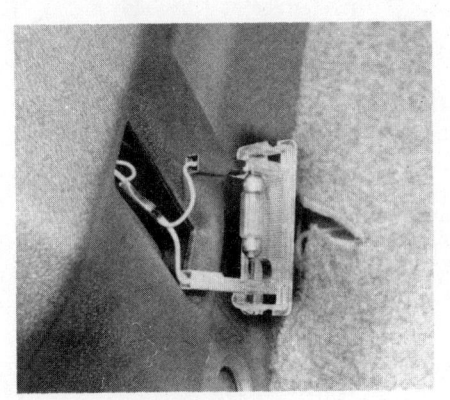

20.2 Luggage area lamp

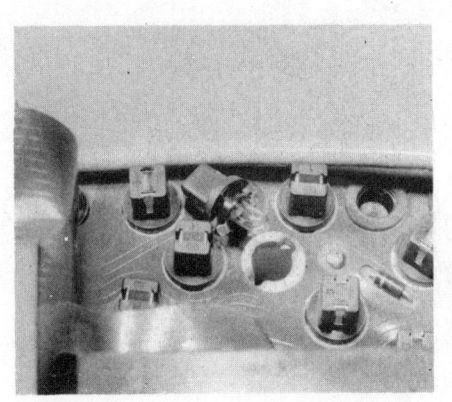

20.4 Instrument panel bulbs and holders

21.4 Withdrawing headlamp

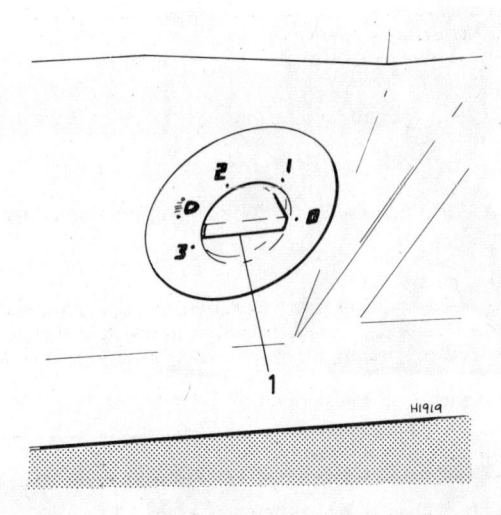

Fig. 12.26 Headlamp beam load adjuster control knob (1) (Sec 22)

setting) to compensate for variations in the vehicle load and so prevent dazzle and improve road illumination.
2 The device consists of a rotary adjustment knob and an actuating unit at the headlamp (photos).
3 The control knob should be set in accordance with the following conditions.

- 0 Driver alone or with front passenger, no luggage
- 2 Driver and four passengers, no luggage
- 3 Driver and four passengers plus luggage
- 3 Driver with luggage compartment full

23 Headlamp beam – alignment

1 It is recommended that the headlamp beam alignment is carried out by your dealer or a service station having the necessary optical beam setting equipment.
2 As a temporary measure the headlamp beam adjusting screws may be turned making sure that the load adjuster knob is set to 0.

24 Horns

1 The horns are of wind tone type and mounted at the forward end of the engine compartment (photo).
2 A compressor is used to generate the necessary air pressure (photo).
3 The filter on the end of the compressor intake pipe should be kept clean (photo).

25 Striplights

1 On Limousine models, two striplights are fitted in the interior of the vehicle.
2 The necessary voltage is produced by a converter which transforms the 12 V direct current into 220 V alternating current.
3 The converter is located within the rear wing and it is accessible from within the luggage compartment after removal of the trim panel.

26 Electrically-controlled seats

Front seats
1 This facility is available on six-cylinder models and provides control of the seat fore and aft movement, and the rake of the seat back. The control switches are mounted in the centre console.

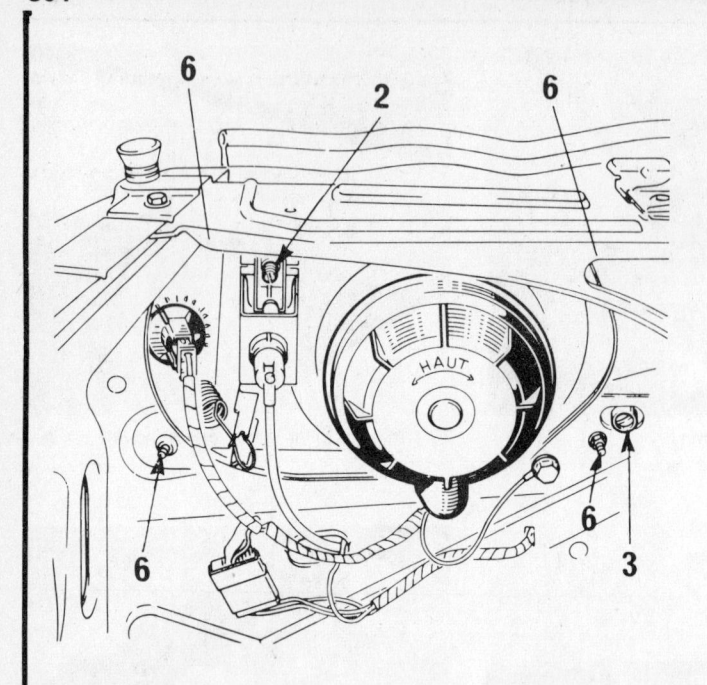

Fig. 12.27 Single headlamp (four-cylinder models) (Sec 23)

2 Beam adjuster screw (vertical)
3 Beam adjuster screw (horizontal)
6 Headlamp mounting screws

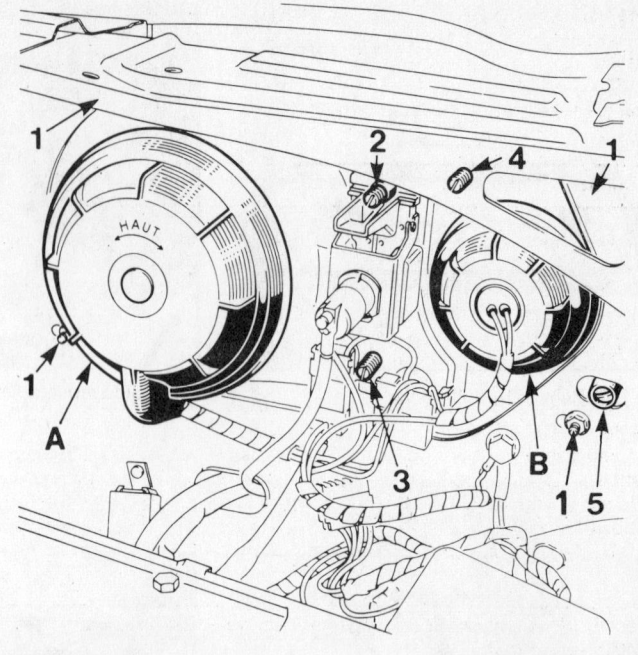

Fig. 12.28 Twin headlamps (six-cylinder models) (Sec 23)

A Main beam/dipped beam unit
B Main beam unit
1 Mounting nuts
2 Beam adjuster screw (vertical)
3 Beam adjuster screw (horizontal)
4 Beam adjuster screw (vertical)
5 Beam adjuster screw (horizontal)

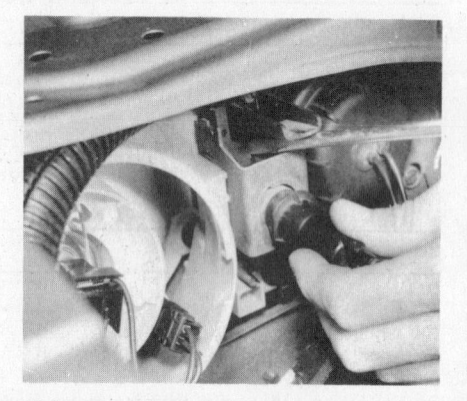

22.2A Releasing load adjuster unit

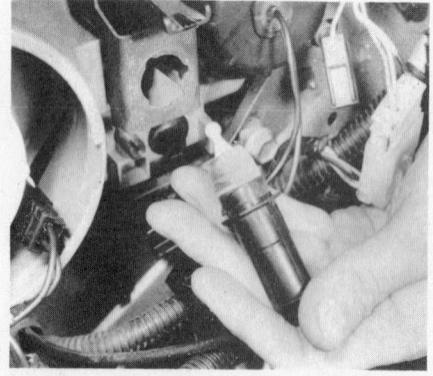

22.2B Removing load adjuster unit

24.1 Horn

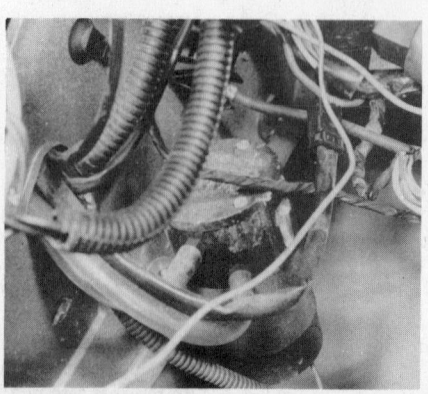

24.2 Horn compressor

24.3 Horn compressor air intake filter

Chapter 12 Electrical system

Rear seats

2 Certain versions are equipped with an electrically-operated rear seat.
3 The system incorporates a rear travel cut-off safety switch, relays and diodes to prevent circuit current interchange.

27 Wiper blades and arms – removal and refitting

1 The wiper blade is a snap fit in the groove at the end of the wiper arm.
2 Pull the wiper arm/blade away from the windscreen or tailgate glass until it locks and pull the blade from the arm by giving the blade a sharp jerk (photo).
3 The wiper arm may be removed from its splined spindle by flipping up the nut cover, unscrewing the nut (photo) and prising the arm from the spindle. If the arm is tight, use two screwdrivers at opposite points under its eye as levers.
4 Refitting is a reversal of removal, but make sure that the alignment of the wiper blades and arms is correct – parallel with the bottom of the screen and the rubber insert approximately 25.4 mm (1.0 in) from the screen frame.
5 Do not overtighten the wiper arm nuts.

28 Windscreen wiper motor and linkage – removal and refitting

1 Remove the wiper arms and blades as described in the preceding Section.
2 Open the bonnet and disconnect the battery.
3 Extract the scuttle grille fixing screws at the base of the windscreen (photos); fold the scuttle forward.
4 Remove the grille assembly.
5 Extract the support screws and then remove the protective cover (photos).
6 Unplug the wiper motor.
7 Unbolt the wiper motor and the wiper linkage mounting plates (photo).
8 Release the wiper arm (spindle) wheel boxes and withdraw the motor and linkage from the scuttle recess.
9 Refer to Fig. 12.29 and unscrew the driving arm nut (A) and then unbolt the wiper motor by unscrewing the bolts (B).
10 When reassembling, make sure that the links (C) and (D) are in alignment and the motor has been switched off in the 'park' position.

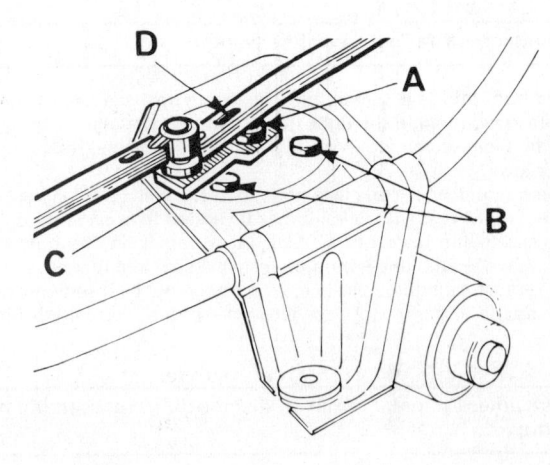

Fig. 12.29 Windscreen wiper motor driving arm nut (A), motor mounting bolts (B) and links (C) and (D) (Sec 28)

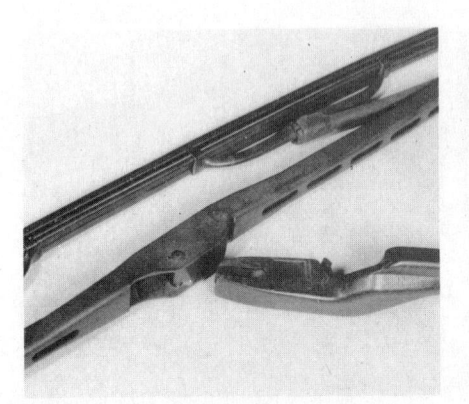

27.2 Wiper blade-to-arm connection

27.3 Unscrewing wiper arm nut

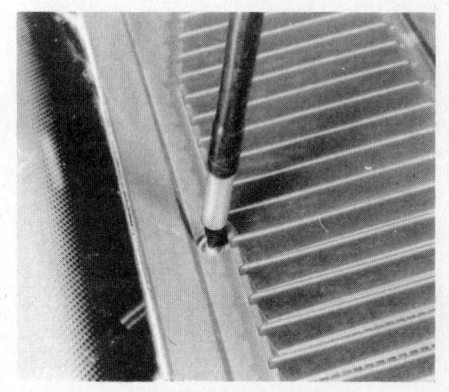

28.3A Extracting scuttle grille screw

28.3B Extracting scuttle cover screw

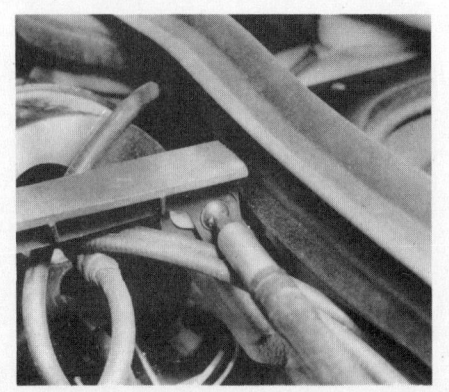

28.5A Scuttle cover support screw

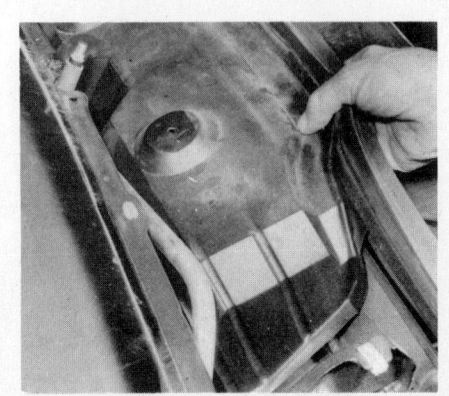

28.5B Removing scuttle flexible cover

Chapter 12 Electrical system

29 Tailgate wiper motor – removal and refitting

1 Remove the wiper arm and blade (Section 27).
2 Open the tailgate and remove the trim panel as described in Chapter 11.
3 Unplug the wiper motor (photo).
4 Unscrew the wiper motor mounting bolts and remove the motor.
5 Refitting is a reversal of removal, but before bolting the motor into place, connect its plug and make sure that it is switched on and then off by means of its control switch so that it will be installed in the 'parked' position.

30 Windscreen/tailgate washer system

1 The front and rear screen and headlamp washer systems operate from a reservoir within the right-hand front wing cavity.
2 To fill the reservoir, open the bonnet and remove the reservoir filler cap (phtoo).
3 Use proprietary screen cleaning fluid in the water and, in very cold weather, a little methylated spirit may be added to prevent freezing.
4 Access to the reservoir is obtained by removing the right-hand front roadwheel, the underwing protective shield and rear cover plate.
5 The adjustment of the washer jets is carried out by inserting a pin in the jet nozzle and moving it to obtain the desired jet pattern on the screen.

31 Instrument panel – removal, dismantling, reassembly and refitting

1 Disconnect the battery.
2 Prise out the trim strip from just below the heater control panel (photo).
3 Remove the steering wheel and upper column shrouds (Chapter 10).
4 Extract the screws from the lower edge of the instrument panel (photos).
5 Withdraw the instrument panel until, if it is the bulbs that are being renewed, their holders can be removed, or if the panel is to be completely removed, the wiring plugs and hoses can be disconnected (photos).
6 If the instrument panel is to be dismantled, observe the following precautions.
7 Cover the work surface with a soft cloth to prevent scratching of the instrument glasses.
8 When extracting the small printed circuits, use a small plastic strip and push it into the circuit fold before pulling the printed circuit from the panel (Fig. 12.30).
9 The instrument panel is in three sections and held together by two clips. Remove the clips before removing the individual instruments which are held to the panel by screws.
10 When reassembling the panel, use the plastic strips to refit the pointed circuits into their connectors. It is very important that the printed circuits are pushed into their connectors the correct way round especially with the fuel consumption module (Fig. 12.31).
11 Make sure that the metal connecting clip between the main printed circuit and the one for the tachometer is in place, as switching on the ignition without it could cause a short circuit.
12 Refitting is a reversal of removal.

32 Bulb filament defect detection unit

1 This is located under the facia panel above the glovebox on the passenger side.
2 When the ignition is switched on, the voice synthesizer will relay a message if the front parking and tail lights are not working or the left or right brake stop-lamps are not working.

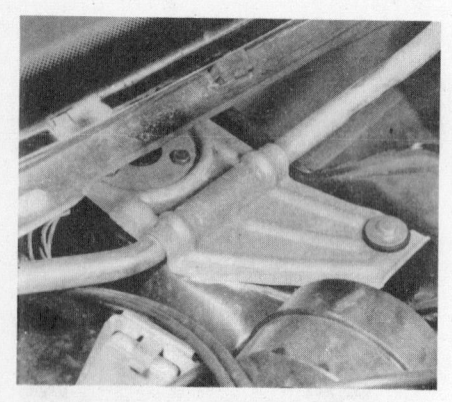

28.7 Wiper motor and linkage mounting plate

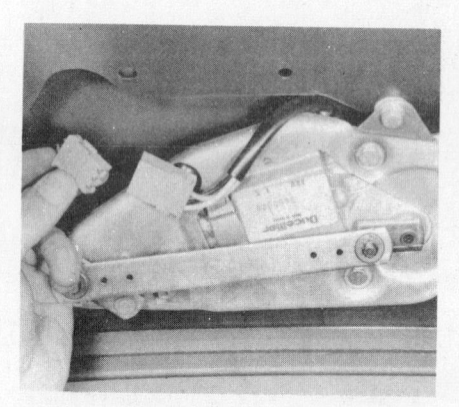

29.3 Tailgate wiper motor

30.2 Filling washer fluid reservoir

31.2 Instrument panel trim strip

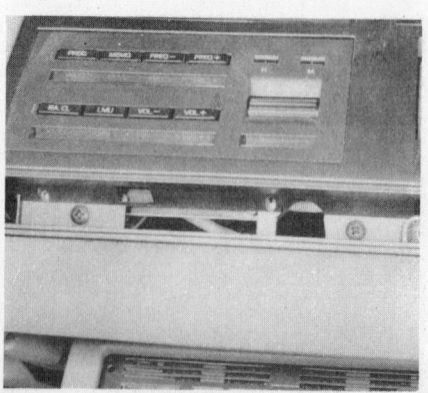

31.4A Instrument panel fixing screws

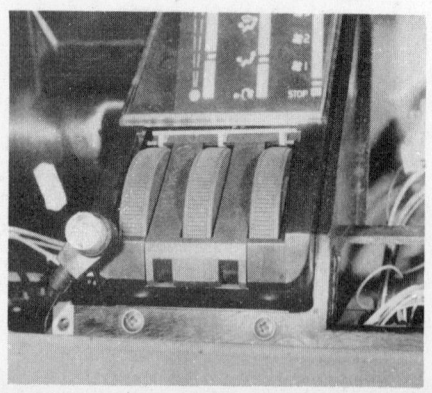

31.4B Instrument panel fixing screws

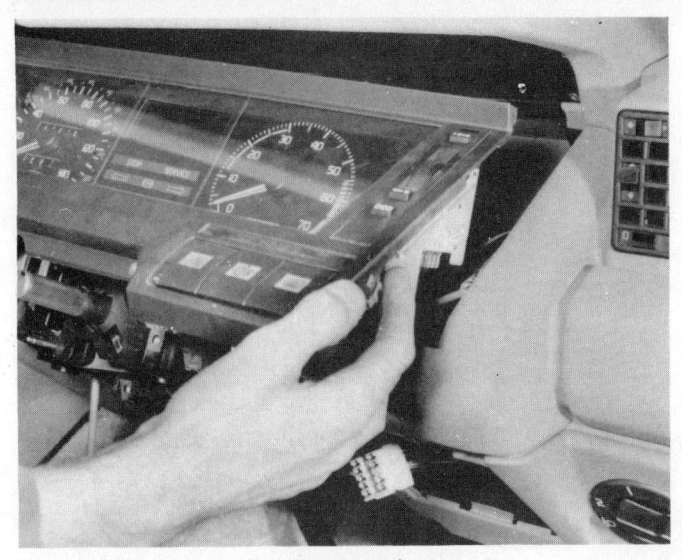

Chapter 12 Electrical system

31.5A Withdrawing the instrument panel

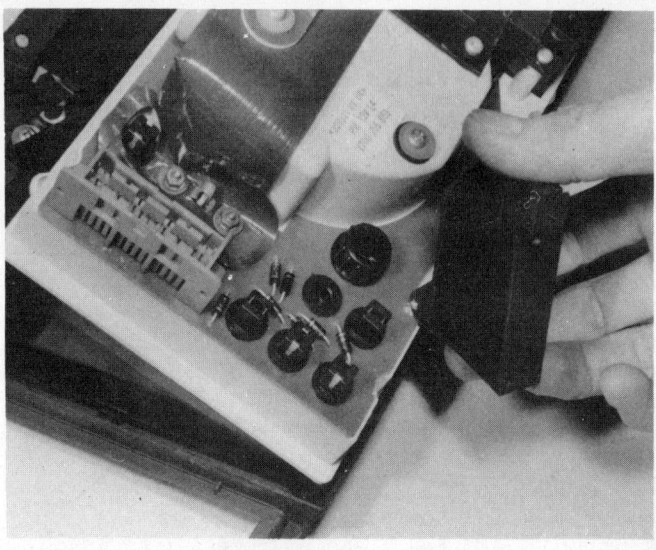

31.5B Instrument panel (bulbholder cover removed)

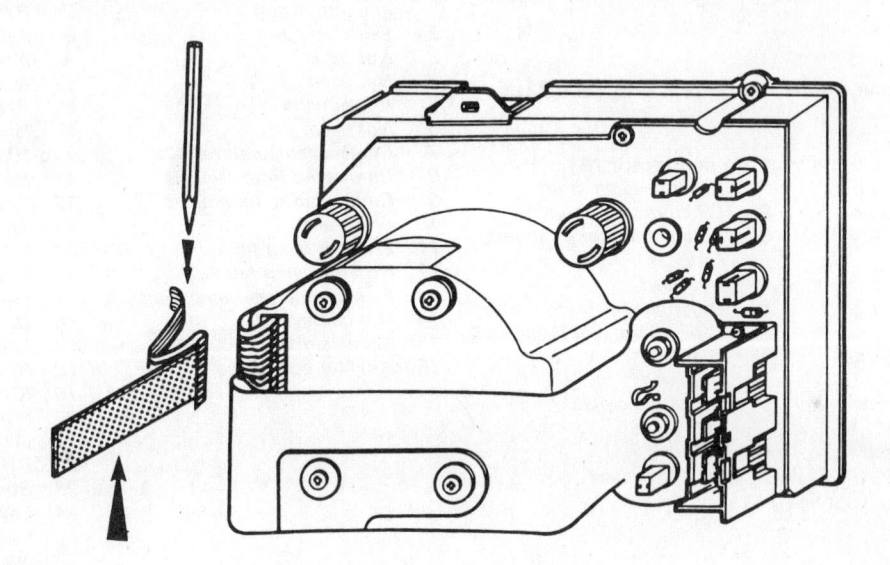

Fig. 12.30 Method of extracting small printed circuit board from instrument panel (Sec 31)

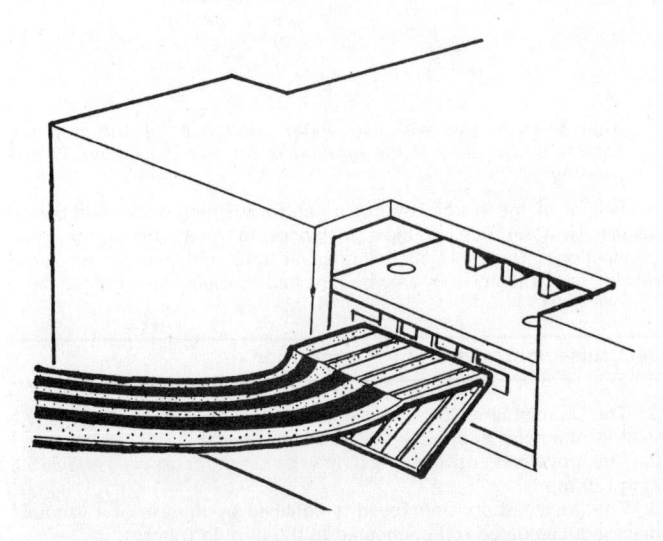

Fig. 12.31 Correct connection of printed circuit board to fuel consumption module (Sec 31)

3 Models operating in certain European territories have a supplementary warning lamp which should flash when the ignition is switched on and extinguish if everything is in order when the brake pedal is applied or the handbrake released.

33 Voice synthesizer

1 This device is fixed behind the facia panel on the passenger side with its speaker located in the lower part of the panel.
2 The purpose of the system is to warn the driver by audible message of twenty faults or omissions covering engine and braking, lighting and door closure.
3 The synthesizer unit obtains its information from sensors and gives a warning, if necessary, following a malfunction – in conjunction with a visual warning lamp for certain components.
4 Three control buttons are provided:

Obliteration button: *this stops the repetition of existing warnings*
Repetition button: *if the 'obliteration' function is on, only new warning messages will be repeated. If the 'obliteration' function is off, all messages will be repeated*

308 Chapter 12 Electrical system

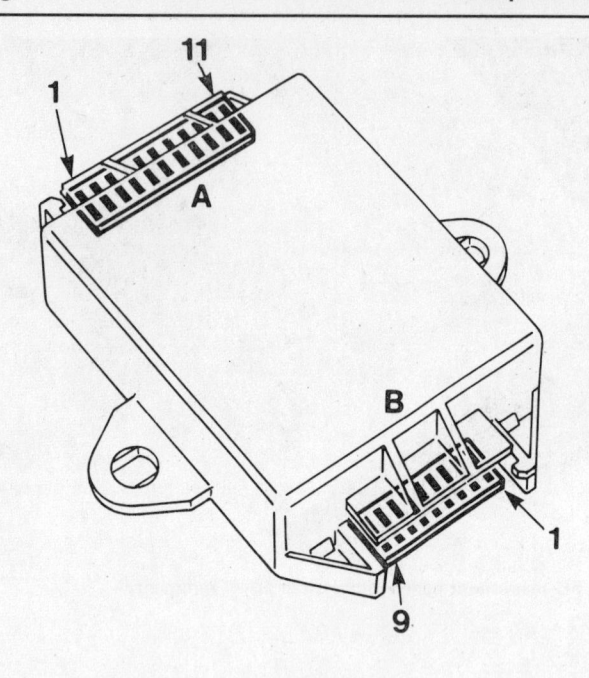

Fig. 13.32 Bulb filament defect detection unit (Sec 32)

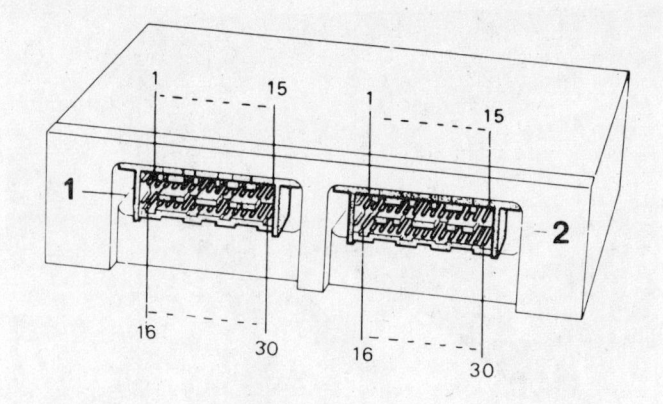

Fig. 12.33 Voice synthesizer (Sec 33)

Plug connector (A)
1. Right-hand front parking
2. Feed after RH parking and rear lamp fuse
3. RH rear lamp
4. LH front parking
5. Connection to terminal 7
6. LH rear lamp
7. Feed after LH parking and rear lamp fuse
8. Rear number plate lamps
9. LH stop-lamp
10. After stop-lamps switch
11. RH stop-lamps

Plug connector (B)
1. RH stop-lamp fault
2. LH stop-lamp fault
3. Before stop-lamps switch
4. Earth
5. Not used
6. Parking lamps before fuse
7. Parking and rear lamps fault
8. Not used
9. Accessories plate (+)

Plug connector (1)
1. Positive (+) before ignition switch
2. Earth
3. Not used
4. Not used
5. Rev counter
6. Not used
7. Coolant temperature
8. Connected to terminal 9
9. Connected to terminal 8
10. Brake pads
11. LH stop-lamp fault
12. RH stop-lamp fault
13. Parking and rear lamp fault
14. Not connected
15. Speed signal
16 to 30 Not connected

Plug connector (2)
1. LH rear door
2. RH rear door
3. Tailgate
4. Not connected
5. Not connected
6. Battery charge warning lamp
7 to 10 Not connected
11. Radio cut-off
12. Speaker
13. Speaker
14. Parking and rear lamps
15. Positive (+) after ignition switch
16. Oil pressure switch
17. Low coolant level
18. Not connected
19. ICP/Nivo code
20. Handbrake
21. LH front door
22. RH front door
23. Bonnet
24. Low fuel level warning lamp
25. Not connected
26. Starter information
27. Not connected
28. Obliteration
29. Repetition
30. Test

Test button: *this will start a run through of all the system's messages, but only if the ignition is on and the engine is not running*

5 Failure of the system by not issuing a warning either audibly or visually will be due in most instances to faulty wiring or poor connections. Generally the synthesizer unit and sensors are very reliable but, if faulty, can only be renewed as sealed assemblies.

34 Cruise control system – description

1 This system enables a constant road speed to be maintained without the need to operate the accelerator pedal.
2 The 'governed cruising speed' may be set at roadspeeds above 50 kmh (30 mph).
3 The governed cruising speed is obtained by the use of a vacuum diaphragm actuated rod connected to the throttle (photo).
4 Partial vacuum in the circuit causes acceleration, while venting to atmosphere causes the car to slow down.

34.3 Cruise control diaphragm unit

Chapter 12 Electrical system

5 Control is carried out by means of a micro-processor, roadspeed sensor and vacuum power module, and pump.
6 The system may be activated by depressing the main on/off switch on the centre console then, with the roadspeed above the prescribed minimum, depress the operating switch which is incorporated in the steering wheel spoke. The current roadspeed will then be maintained without the need to keep the foot on the accelerator.
7 The roadspeed may be altered by depressing the switch (2 – Fig. 12.34) to increase speed or switch (3) to decrease it.
8 An immediate speed increase may be obtained by depressing the accelerator. Once the foot is lifted, the roadspeed will return to the level previously set.
9 The cruise control is made inoperative as soon as the brake or clutch pedals are depressed, or when switch (5) or the main switch on the console are depressed.
10 The previously set roadspeed level can be resumed (provided the main switch has not been turned off) at any time by depressing switch (3).
11 Once the ignition is switched off the cruise control system memory is cancelled.
12 On 1985 and later models, the system has been slightly modified so that the roadspeed adjustment function is temporarily stopped at excessive engine speeds.

35 Cruise control system components – removal and refitting

1 Regularly check the electrical and vacuum hose connections of the system. Testing of the major components is not within the scope of the home mechanic but, where a fault is diagnosed, the unit can be renewed in the following way.

Vacuum power module
2 The unit is located under the right-hand headlamp unit behind the bumper/spoiler.

3 Refer to Chapter 11 and remove the front bumper.
4 Unscrew the module mounting nuts, disconnect the wiring plug and vacuum hose. Remove the module.
5 Extract the screws and remove the cover to expose the vacuum pump and solenoid valves.

Actuating rod/diaphragm unit
6 This can be removed by disconnecting the vacuum hose and the rod and unbolting the diaphragm unit bracket.

Computer
7 This is located behind the facia above the glovebox on the passenger side. The connector wiring details should be noted in Fig. 12.37.

Roadspeed sensor
8 The sensor is attached to the speedometer head which in turn monitors the roadspeed from the target wheel on the differential.
9 Access to the sensor is obtained by removing the instrument panel, as described in Section 31.

Steering wheel switches
10 Refer to Chapter 10 and remove the steering wheel hub cover and steering wheel.
11 Remove the switches from the rear of the spokes after having taken off the spoke rear cover plate.

Refitting
12 Refitting of all components is a reversal of removal, but observe the following special requirements.
13 Clean the steering wheel switch tracks using fine abrasive paper if severely corroded. Smear the tracks with petroleum jelly.
14 Adjust the actuating rod at the throttle so that, with the engine idling, there is a slight clearance, not exceeding 1.5 mm (0.06 in), at the rod-to-diaphragm connection. Adjust by releasing the locknut at the ball socket on the rod and rotate the rod. Retighten the locknut.

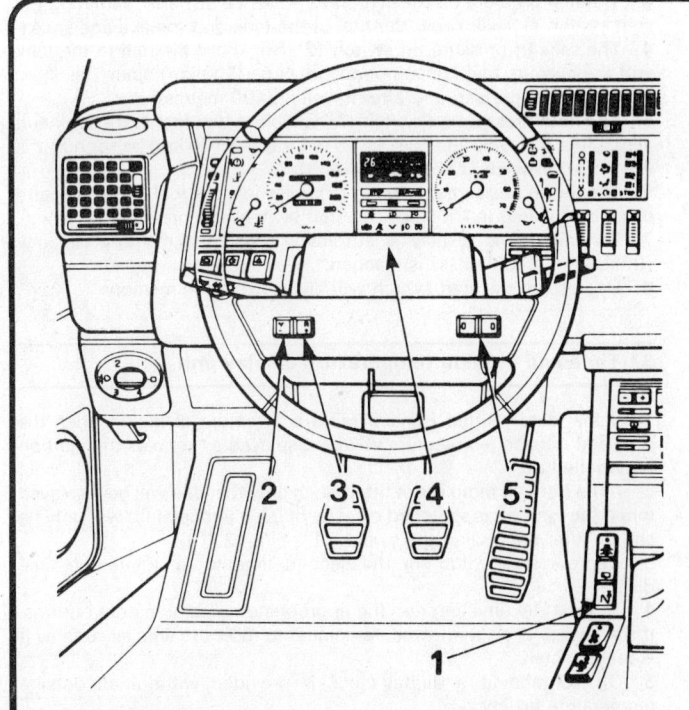

Fig. 12.34 Cruise control switches – LH Drive (Sec 34)

1 Main on/off switch
2 Engagement and speed decrease switch
3 Resume and speed increase switch
4 Warning lamp
5 Off switch

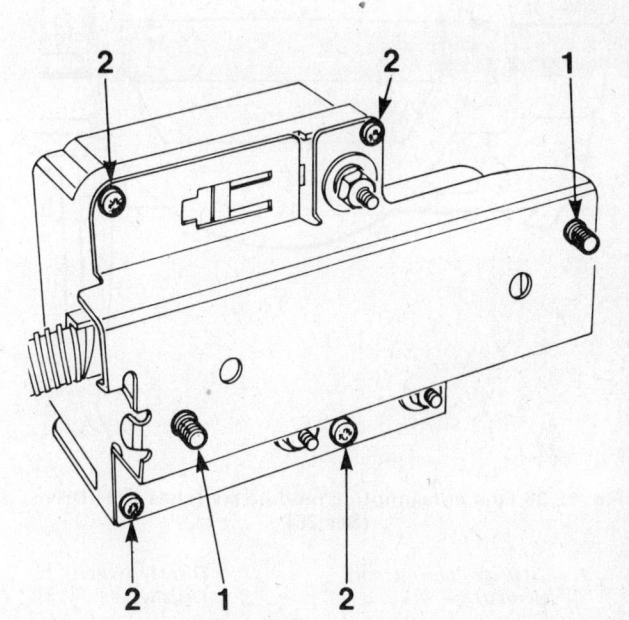

Fig. 12.35 Cruise control vacuum power module (Sec 35)

1 Mounting studs
2 Cover screws

Chapter 12 Electrical system

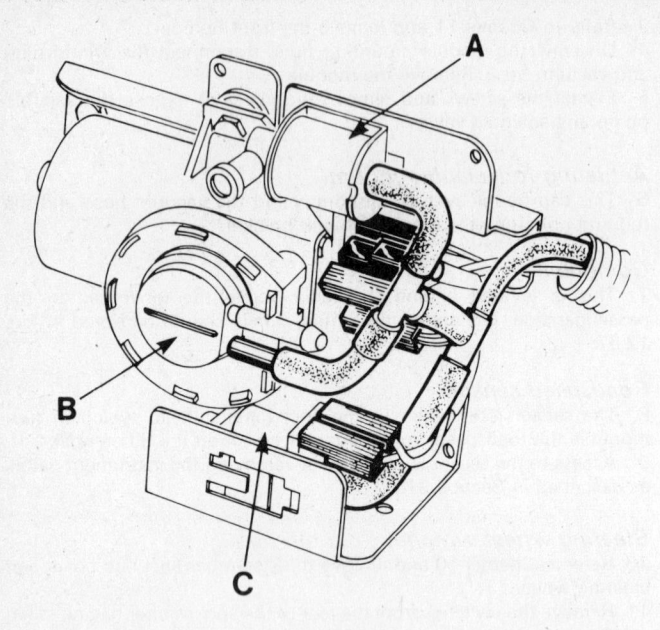

Fig. 12.36 Vacuum power module with cover removed (Sec 36)

A Security solenoid valve
B Vacuum pump
C Governing solenoid valve

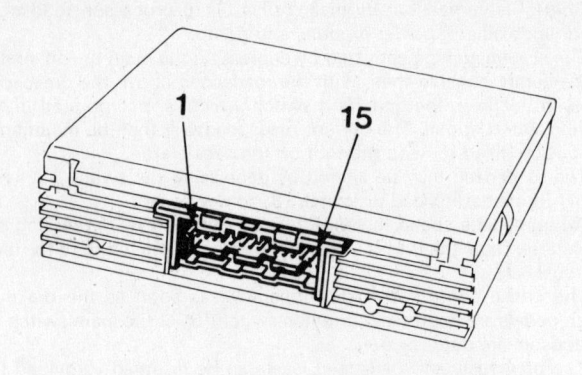

Fig. 12.37 Cruise control computer (Sec 36)

Plug connector
1 Vacuum pump
2 Diagnostic socket
3 Governing solenoid valve
4 Security solenoid valve
5 Warning lamp
6 Roadspeed information input
7 Security information (stop lamp)
8 Steering wheel inner track

Plug connector
9 Steering wheel outer track
10 Auto trans info (1984 models only)
11 Feed to pump and solenoid valves
12 Positive (+) feed
13 Earth
14 Tachometer (earth)
15 To terminal 14

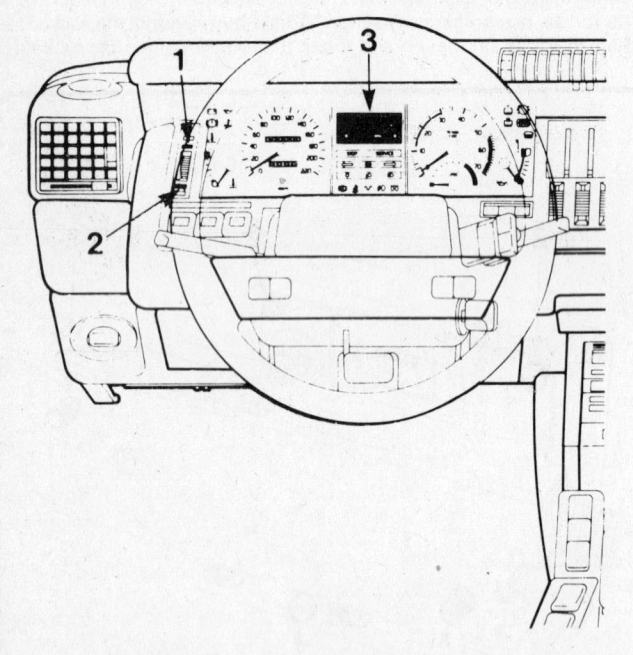

Fig. 12.38 Fuel consumption module switches – LH Drive (Sec 36)

1 Start switch (memory zeroed)
2 Display switch
3 Display

36 Fuel consumption module

1 This device is fitted to certain upper range models and provides the driver with a variety of fuel capacity and consumption information gathered through sensors which include a flowmeter, the fuel gauge and the speedometer/odometer.

2 As soon as the ignition is switched on, the quantity of fuel in the tank is displayed, together with the estimated distance that the car can cover on this amount of fuel. No display is made if the fuel in the tank is less than 5.0 litres (1.1 gallon).
3 The first pressure on the switch (2 – Fig. 12.38) after actuating the start switch (1) will show the fuel in the tank and the average speed.
4 The second pressure on switch (2) will show the fuel in the tank and the average fuel consumption (litres per 100 km) since the 'start' switch was depressed and after travelling 400 metres.
5 The third pressure on switch (2) will show the fuel in the tank and actual fuel consumption (displayed in litres per 100 km) as soon as the roadspeed exceeds 30 km/h (19 mph).
6 The fourth pressure on switch (2) will show the fuel in the tank and distance covered in km since the start switch was pressed.
7 Zeroing of the display is automatic when the memory capacity (9999 km – 6250 miles) is reached.
8 Depressing the start switch will always zero the memory.

37 Time and outside temperature display unit

1 If the factory-fitted Renault radio is a feature of the car then the time and outside temperature will be displayed as soon as the ignition is switched on.
2 If the Renault radio is not fitted, then only the time will be displayed when the ignition is switched on. The outside temperature will only be shown after depressing the switch (T – Fig. 12.40).
3 Depress the switch for the second time to re-instate the time display.
4 To reset the time, depress the appropriate hours or minute buttons. If the battery is disconnected, remember to reset the time as soon as it is reconnected.
5 On TS models, a digital clock is provided without an outside temperature facility.

38 Radio/cassette player – general

1 Only certain upper range models are equipped with in-car entertainment, but all versions are wired and have an aerial fitted as standard. Speaker grilles are also provided in the front doors.

Chapter 12 Electrical system

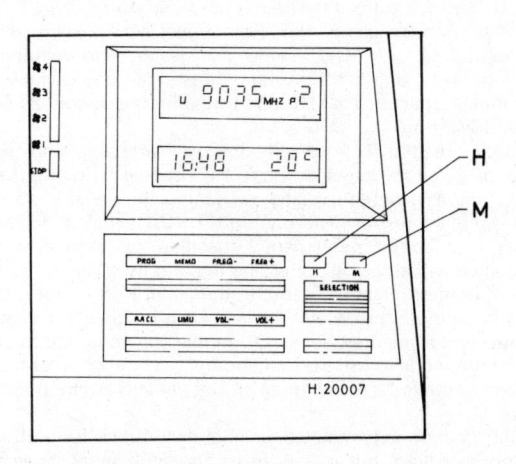

Fig. 12.39 Time and outside temperature display with Renault radio fitted (Sec 37)

H   Hours reset button   M   Minutes reset button

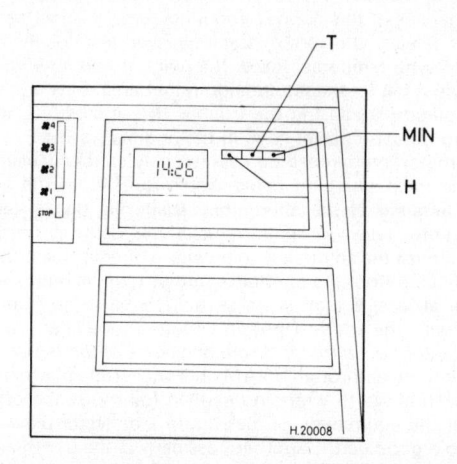

Fig. 12.40 Time and outside temperature display without Renault radio (Sec 37)

H   Hour reset button   T   Display button
MIN   Minute reset button

39 Radio/cassette player (Renault type) – removal and refitting

1   Remove the ashtray from the centre console (photo).
2   Using two thin rods inserted into the holes in the front panels of the radio or cassette player, release the spring clips and at the same time withdraw the unit.
3   Withdraw the unit sufficiently for to be able to disconnect the feed, earth, aerial and speaker leads.
4   Refitting is a reversal of removal.

40 Radio/cassette player – fitting

1   Provision is made for standard size equipment to be installed in the centre console on those vehicles not factory-fitted with in-car entertainment facilities.
2   Extract the centre console fixing screws, as described in Chapter 11, and pull the console slightly away from the facia panel.
3   Pull the tidy box from the console to enable the radio or cassette player to take its place.
4   The aerial, feed and speaker connecting plugs are to be found underneath the tidy box location.
5   A fitting kit will be required to support the radio receiver or cassette player in its console recess. This is obtainable from your Renault dealer or car radio specialist.
6   Fit the radio, connect the leads.
7   A speaker (165.0 mm diameter) must now be fitted into each front door trim panel. To do this, turn the grille anti-clockwise and remove it, and take out the cardboard blanking plate.
8   Connect the leads (which will be found already in position) to the speaker.
9   Insert the speaker fixing screws and refit the grille.
10 Speakers may be fitted at each end of the facia panel and provision is made for locating a treble speaker in the upper part of the facia.
11 Once installed, trim the radio aerial as described in the radio manufacturer's instructions.
12 If more sophisticated radio equipment is required in the vehicle, refer to Section 42.

41 Radio aerial – removal and refitting

1   The radio aerial is roof-mounted above the windscreen and is a standard fitting.
2   If it must be removed, first detach the lock infra red receiver unit by unscrewing the fixing bolts.

39.1 Removing ashtray from centre console

3   Working through the small aperture in the headlining, unscrew the aerial mounting nuts while an assistant lifts off the aerial from outside the car.
4   The aerial lead can be withdrawn from the windscreen pillar once the lead is detached from the radio.
5   It will facilitate refitting if string is taped to the end of the aerial lead before withdrawing it from the pillar.
6   Refitting is a reversal of removal, trim the aerial on completion.

42 Radio interference and CB equipment

Radio/cassette case breakthrough

Magnetic radiation from dashboard wiring may be sufficiently intense to break through the metal case of the radio/cassette player. Often this is due to a particular cable routed too close and shows up as ignition interference on AM and cassette play and/or alternator whine on cassette play.
The first point to check is that the clips and/or screws are fixing all parts of the radio/cassette case together properly. Assuming good

earthing of the case, see if it is possible to re-route the offending cable – the chances of this are not good, however, in most cars.

Next release the radio/cassette player and locate it in different positions with temporary leads. If a point of low interference is found, then if possible fix the equipment in that area. This also confirms that local radiation is causing the trouble. If re-location is not feasible, fit the radio/cassette player back in the original position.

Alternator interference on cassette play is now caused by radiation from the main charging cable which goes from the battery to the output terminal of the alternator, usually via the + terminal of the starter motor relay. In some vehicles this cable is routed under the dashboard, so the solution is to provide a direct cable route. Detach the original cable from the alternator output terminal and make up a new cable of at least 6 mm² cross-sectional area to go from alternator to battery with the shortest possible route. *Remember – do not run the engine with the alternator disconnected from the battery.*

Ignition breakthrough on AM and/or cassette play can be a difficult problem. It is worth wrapping earthed foil round the offending cable run near the equipment, or making up a deflector plate well screwed down to a good earth. Another possibility is the use of a suitable relay to switch on the ignition coil. The relay should be mounted close to the ignition coil; with this arrangement the ignition coil primary current is not taken into the dashboard area and does not flow through the ignition switch. A suitable diode should be used since it is possible that at ignition switch-off the output from the warning lamp alternator terminal could hold the relay on.

Connectors for suppression components

Capacitors are usually supplied with tags on the end of the lead, while the capacitor body has a flange with a slot or hole to fit under a nut or screw with washer.

Connections to feed wires are best achieved by self-stripping connectors. These connectors employ a blade which, when squeezed down by pliers, cuts through cable insulation and makes connection to the copper conductors beneath.

Chokes sometimes come with bullet snap-in connectors fitted to the wires, and also with just bare copper wire. With connectors, suitable female cable connectors may be purchased from an auto-accessory shop together with any extra connectors required for the cable ends after being cut for the choke insertion. For chokes with bare wires, similar connectors may be employed together with insulation sleeving as required.

VHF/FM broadcasts

Reception of VHF/FM in an automobile is more prone to problems than the medium and long wavebands. Medium/long wave transmitters are capable of covering considerable distances, but VHF transmitters are restricted to line of sight, meaning ranges of 10 to 50 miles, depending upon the terrain, the effects of buildings and the transmitter power.

Because of the limited range it is necessary to retune on a long journey, and it may be better for those habitually travelling long distances or living in areas of poor provision of transmitters to use an AM radio working on medium/long wavebands.

When conditions are poor, interference can arise, and some of the suppression devices described previously fall off in performance at very high frequencies unless specifically designed for the VHF band. Available suppression devices include reactive HT cable, resistive distributor caps, screened plug caps, screened leads and resistive spark plugs.

For VHF/FM receiver installation the following points should be particularly noted:

(a) Earthing of the receiver chassis and the aerial mounting is important. Use a separate earthing wire at the radio, and scrape paint away at the aerial mounting.
(b) If possible, use a good quality roof aerial to obtain maximum height and distance from interference generating devices on the vehicle.
(c) Use of a high quality aerial downlead is important, since losses in cheap cable can be significant.
(d) The polarisation of FM transmissions may be horizontal, vertical, circular or slanted. Because of this the optimum mounting angle is at 45° to the vehicle roof.

Citizens' Band radio (CB)

In the UK, CB transmitter/receivers work within the 27 MHz and 934 MHz bands, using the FM mode. At present interest is concentrated on 27 MHz where the design and manufacture of equipment is less difficult. Maximum transmitted power is 4 watts, and 40 channels spaced 10 kHz apart within the range 27.60125 to 27.99125 MHz are available.

Aerials are the key to effective transmission and reception. Regulations limit the aerial length to 1.65 metres including the loading coil and any associated circuitry, so tuning the aerial is necessary to obtain optimum results. The choice of a CB aerial is dependent on whether it is to be permanently installed or removable, and the performance will hinge on correct tuning and the location point on the vehicle. Common practice is to clip the aerial to the roof gutter or to employ wing mounting where the aerial can be rapidly unscrewed. An alternative is to use the boot rim to render the aerial theftproof, but a popular solution is to use the 'magmount' – a type of mounting having a strong magnetic base clamping to the vehicle at any point, usually the roof.

Aerial location determines the signal distribution for both transmission and reception, but it is wise to choose a point away from the engine compartment to minimise interference from vehicle electrical equipment.

The aerial is subject to considerable wind and acceleration forces. Cheaper units will whip backwards and forwards and in so doing will alter the relationship with the metal surface of the vehicle with which it forms a ground plane aerial system. The radiation pattern will change correspondingly, giving rise to break-up of both incoming and outgoing signals.

Interference problems on the vehicle carrying CB equipment fall into two categories:

(a) Interference to nearby TV and radio receivers when transmitting.
(b) Interference to CB set reception due to electrical equipment on the vehicle.

Problems of break-through to TV and radio are not frequent, but can be difficult to solve. Mostly trouble is not detected or reported because the vehicle is moving and the symptoms rapidly disappear at the TV/radio receiver, but when the CB set is used as a base station any trouble with nearby receivers will soon result in a complaint.

It must not be assumed by the CB operator that his equipment is faultless, for much depends upon the design. Harmonics (that is, multiples) of 27 MHz may be transmitted unknowingly and these can fall into other user's bands. Where trouble of this nature occurs, low pass filters in the aerial or supply leads can help, and should be fitted in base station aerials as a matter of course. In stubborn cases it may be necessary to call for assistance from the licensing authority, or, if possible, to have the equipment checked by the manufacturers.

Interference received on the CB set from the vehicle equipment is, fortunately, not usually a severe problem. The precautions outlined previously for radio/cassette units apply, but there are some extra points worth noting.

It is common practice to use a slide-mount on CB equipment enabling the set to be easily removed for use as a base station, for example. Care must be taken that the slide mount fittings are properly earthed and that first class connection occurs between the set and slide-mount.

Vehicle manufacturers in the UK are required to provide suppression of electrical equipment to cover 40 to 250 MHz to protect TV and VHF radio bands. Such suppression appears to be adequately effective at 27 MHz, but suppression of individual items such as alternators/dynamos, clocks, stabilisers, flashers, wiper motors, etc, may still be necessary. The suppression capacitors and chokes available from auto-electrical suppliers for entertainment receivers will usually give the required results with CB equipment.

Other vehicle radio transmitters

Besides CB radio already mentioned, a considerable increase in the use of transceivers (ie combined transmitter and receiver units) has taken place in the last decade. Previously this type of equipment was fitted mainly to military, fire, ambulance and police vehicles, but a large business radio and radio telephone usage has developed.

Generally the suppression techniques described previously will suffice, with only a few difficult cases arising. Suppression is carried out to satisfy the 'receive mode', but care must be taken to use heavy duty chokes in the equipment supply cables since the loading on 'transmit' is relatively high.

43 Cigar lighter – removal and refitting

1 Two cigar lighters are fitted. One under the front ashtray cover and the other in the rear face of the rear centre console.
2 Disconnect the battery and pull out the lighter element.
3 Using a small screwdriver blade, prise out the cigar lighter body and then disconnect the leads and illuminating bulb holder from it.
4 Refitting is a reversal of removal.

44 Warning buzzers

Depending upon the model, one or more of the following warning systems may be fitted.

Headlamps left on warning: *A buzzer sounds if the door is opened without having switched off the headlamps*
Ignition key not removed warning: *If the ignition key is left in place and a door is opened, a warning buzzer sounds as a reminder to remove the key*
Seat belts not buckled warning: *If the ignition key is turned for starting the engine without the front seat belts having been fastened, a reminder buzzer will sound for a limited period.*

45 Fault diagnosis – electrical system

Symptom	Reason(s)
No voltage at starter motor	Battery discharged Battery defective internally Battery terminals loose or earth lead not securely attached to body Loose or broken connections in starter motor circuit Starter motor switch or solenoid faulty
Voltage at starter motor – faulty motor	Starter brushes badly worn, sticking, or brush wires loose Commutator dirty, worn or burnt Starter motor armature faulty Field coils earthed
Starter motor noisy or rough in enagement	Pinion or flywheel gear teeth broken or worn Starter motor retaining bolts loose
Alternator not charging*	Drivebelt loose and slipping, or broken Brushes worn, sticking, broken or dirty Brush springs weak or broken

If all appears to be well but the alternator is still not charging, take the car to an automobile electrician for checking of the alternator

Battery will not hold charge for more than a few days	Battery defective internally Electrolyte level too low or electrolyte too weak due to leakage Plate separators no longer fully effective Battery plates severely sulphated Drivebelt slipping Battery terminal connections loose or corroded Alternator not charging properly Short in lighting circuit causing continual battery drain
Ignition light fails to go out, battery runs flat in a few days	Drivebelt loose and slipping, or broken Alternator faulty

Failure of individual electrical equipment to function correctly is dealt with alphabetically below

Fuel gauge gives no reading	Fuel tank empty Electric cable between tank sender unit and gauge earthed or loose Fuel gauge case not earthed Fuel gauge supply cable interrupted Fuel gauge unit broken
Fuel gauge registers full all the time	Electric cable between tank unit and gauge broken or disconnected
Horn operates all the time	Horn push either earthed or stuck down Horn cable to horn push earthed
Horn fails to operate	Blown fuse Cable or cable connection loose, broken or disconnected Horn has an internal fault
Horn emits intermittent or unsatisfactory noise	Cable connections loose Horn incorrectly adjusted
Lights do not come on	If engine not running, battery discharged Light bulb filament burnt out or bulbs broken Wire connections loose, disconnected or broken Light switch shorting or otherwise faulty

Chapter 12 Electrical system

Symptom	Reason(s)
Lights come on but fade out	If engine not running, battery discharged
Lights give very poor illumination	Lamp glasses dirty Reflector tarnished or dirty Lamps badly out of adjustment Incorrect bulb with too low wattage fitted Existing bulbs old and badly discoloured Electrical wiring too thin not allowing full current to pass
Lights work erratically, flashing on and off, especially over bumps	Battery terminals or earth connections loose Contacts in light switch faulty
Wiper motor fails to work	Blown fuse Wire connections loose, disconnected or broken Brushes badly worn Armature worn or faulty Field coils faulty
Wiper motor works very slowly and takes excessive current	Commutator dirty, greasy or burnt Drive to spindles bent or unlubricated Drive spindle binding or damaged Armature bearings dry or unaligned Armature badly worn or faulty
Wiper motor works slowly and takes little current	Brushes badly worn Commutator dirty, greasy or burnt Armature badly worn or faulty
Wiper motor works but wiper blades remain static	Linkage disengaged or faulty Drive spindle damaged or worn Wiper motor gearbox parts badly worn

---

Wiring diagrams – general information

Each wire is coded. These codes are deciphered as follows:
eg 133 – Or/N2 – 41

133	Wire number	N	End of wire colour	41	The component to which the wire goes
Or	Wire basic colour	2	Wire diameter		

Colour Code

B	Blue	J	Yellow	R	Red
Bc	White	M	Brown	S	Pink
Be	Beige	N	Black	V	Green
C	Clear	Or	Orange	Vi	Violet
G	Grey				

Wire diameter

Code	Diameter (mm)	Code	Diameter (mm)	Code	Diameter (mm)
1	0.7	5	1.6	9	4.5
2	0.9	6	2.0	10	5.0
3	1.0	7	2.5	11	7.0
4	1.2 or 1.4	8	3.0	12	8.0

Key to main circuits

	E1	E2	E2+	E3	2 litre Engine	2.2 litre Engine	V6 Engine
Air conditioning	–	15	15	15	–	–	–
Anti-pollution	–	–	–	–	24	–	–
Anti-stall device	–	–	–	–	–	–	23
Automatic transmission	–	–	–	–	20	20	23
Brake pad wear warning light	03	03	22	22	–	–	–
Boot illumination	05	05	25	26	–	–	–
Caravan feed	02	02	02	02	–	–	–
Car-borne computer	–	–	–	–	–	18	23
Charging circuit	03	03	22	22	–	–	–
Cigar lighter	04	04	25	26	–	–	–
Clock	02	–	–	–	–	–	–
Coolant temperature switch	03	03	22	22	–	–	–
Cruise control	–	–	14	14	–	–	–
Dipped beam headlights	01	12	29	28	–	–	–
Direction indicators	01	12	19	13	–	–	–
Door locks	–	07	07	10	–	–	–
Electric boot release	–	–	–	07	–	–	–
Electric mirrors	–	–	–	11	–	–	–
Electric seats	–	–	–	11	–	–	–
Engine cooling fan motor	–	–	–	–	03	03	06
External temperature sensor	–	08	08	30	–	–	–
Fuel gauge	03	03	22	22	–	–	–
Fuel pump	–	–	–	–	–	18	23
Glovebox light	02	02	25	26	–	–	–
Handbrake	04	06	22	22	–	–	–
"Hazard" warning light	01	12	19	13	–	–	–
Headlight wash/wipe	–	–	–	06	–	–	–
Heating	–	–	–	–	–	–	–
Heating/ventilating	04	15	15	15	–	–	–
Horn	01	12	19	13	–	–	–
Identification plates + switches illuminaton	02	02	25	26	–	–	–
Idle cut-out	–	–	–	–	24	–	–
Ignition	–	–	–	–	24	18	23
Injection	–	–	–	–	–	18	23
Interior lights	02	24	10	10	–	–	–
Main beam headlights	01	12	29	28	–	–	–
Minimum fuel level	03	03	22	22	–	–	–
Minimum coolant level	–	06	06	06	–	–	–
Minimum windscreen washer fluid level	–	06	06	06	–	–	–
Brake hydraulic pressure drop indicator	03	03	22	22	–	–	–
Oil filter pressure switch (clogging)	–	–	–	–	–	–	–
Oil level probe	–	–	–	–	03	22	22
Oil pressure	–	–	–	–	–	–	–
Oil pressure switch	03	03	22	22	–	–	–
Oil temperature	–	–	22	22	–	–	–
Openings (doors, boot, bonnet)	–	–	21	09	–	–	–
Radio	04	04	08	30	–	–	–
Rear foglight	05	05	05	05	–	–	–
Rear foglight switch illumination	01	12	25	26	–	–	–
Rear screen demister	05	16	16	16	–	–	–
Rear screen wiper	–	–	06	06	–	–	–
Rear view mirror defroster	–	–	–	11	–	–	–
Reversing lights	05	05	05	05	–	–	–
Parking lights	01	12	29	28	–	–	–
Speakers	04	–	08	08	–	–	–
Speedometer	20	20	20	20	–	–	–
Starter	–	–	–	–	14	14	23
Stop-lights	05	05	20	20	–	–	–
Sun roof	–	–	17	10	–	–	–
Voice synthesis control	–	–	08	30	–	–	–
Voice synthesizer	–	–	27	27	–	–	–
Window winders	–	21	21	09	–	–	–
Windscreen wash/wipe	04	15	15	15	–	–	–

E1 Vehicle without electric windows
E2 Vehicles with electric windows but without voice synthesizer
E2+ Vehicles with voice synthesizer, but without electric seats
E3 Vehicles with electric seats

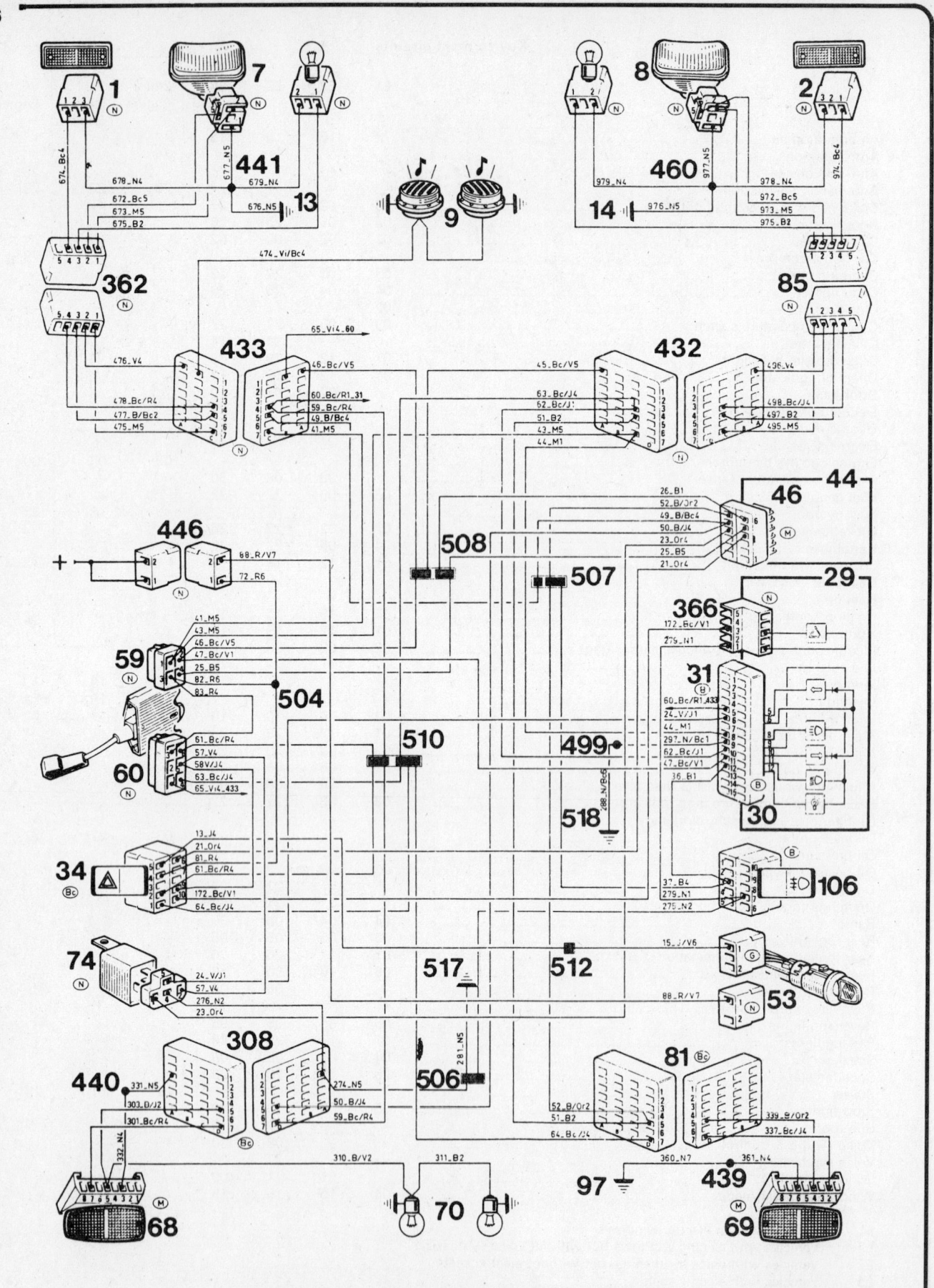

Fig. 12.41 Typical Renault 25 wiring diagram – circuit 1

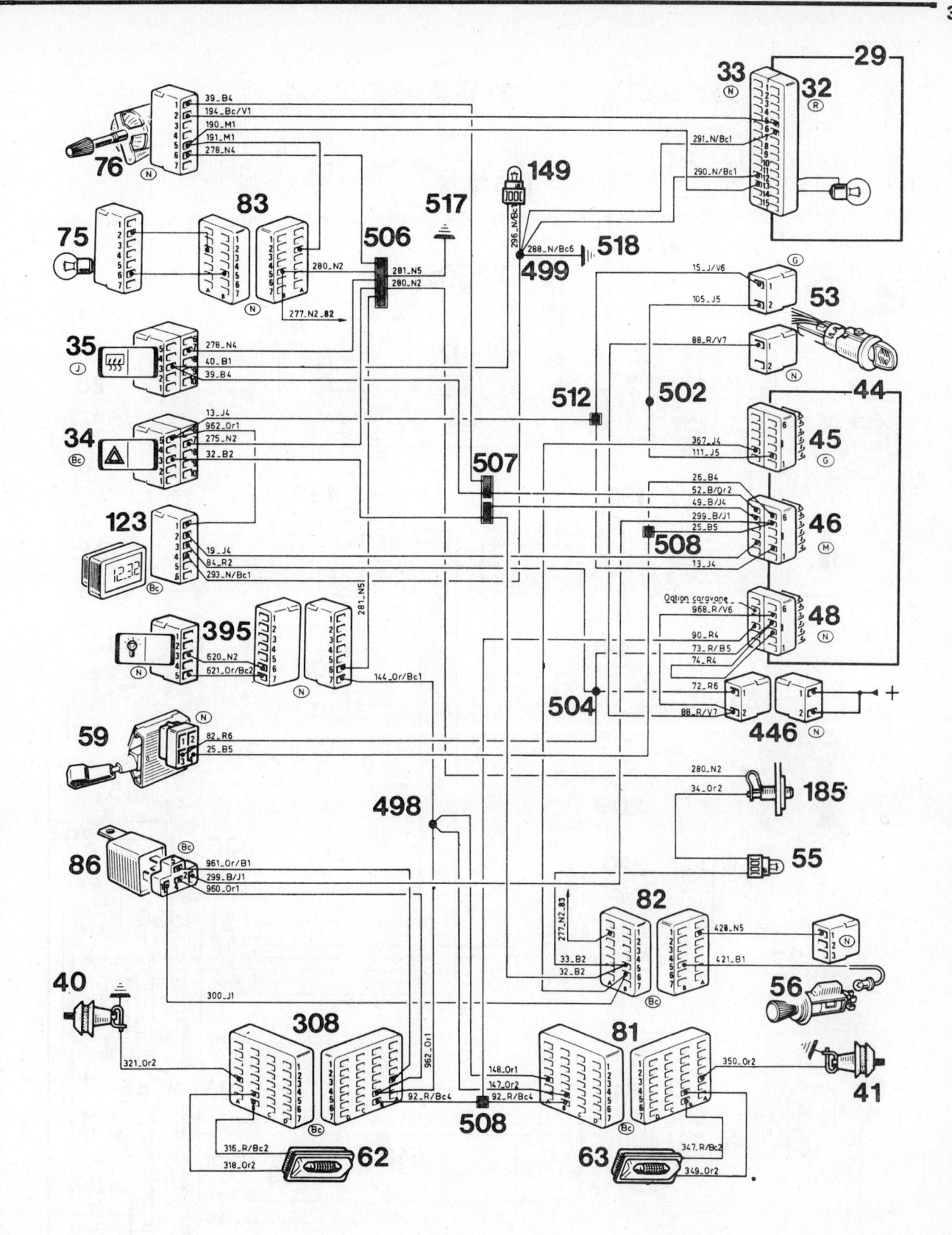

Fig. 12.41 Typical Renault 25 wiring diagram – circuit 2

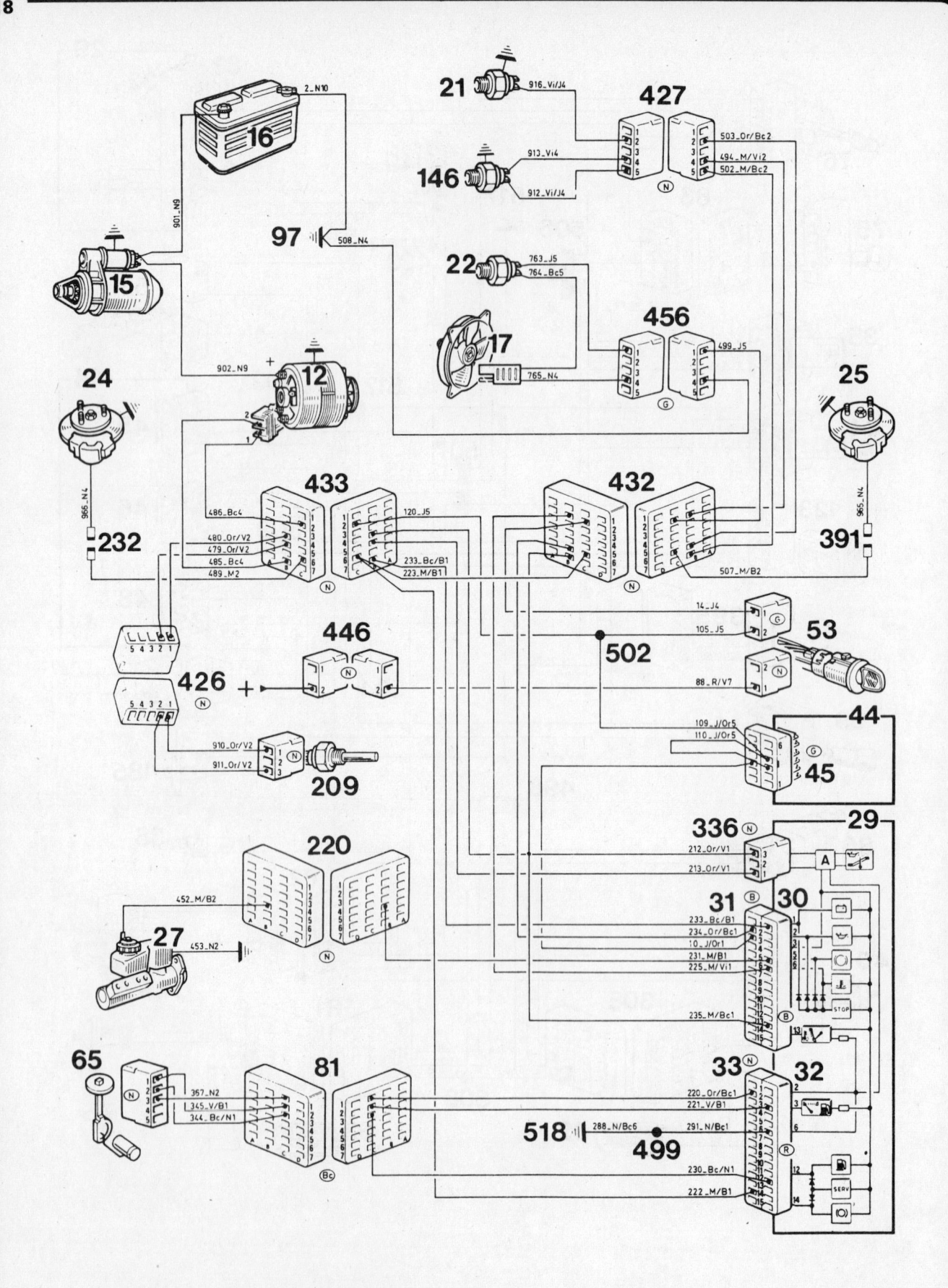

Fig. 12.41 Typical Renault 25 wiring diagram – circuit 3

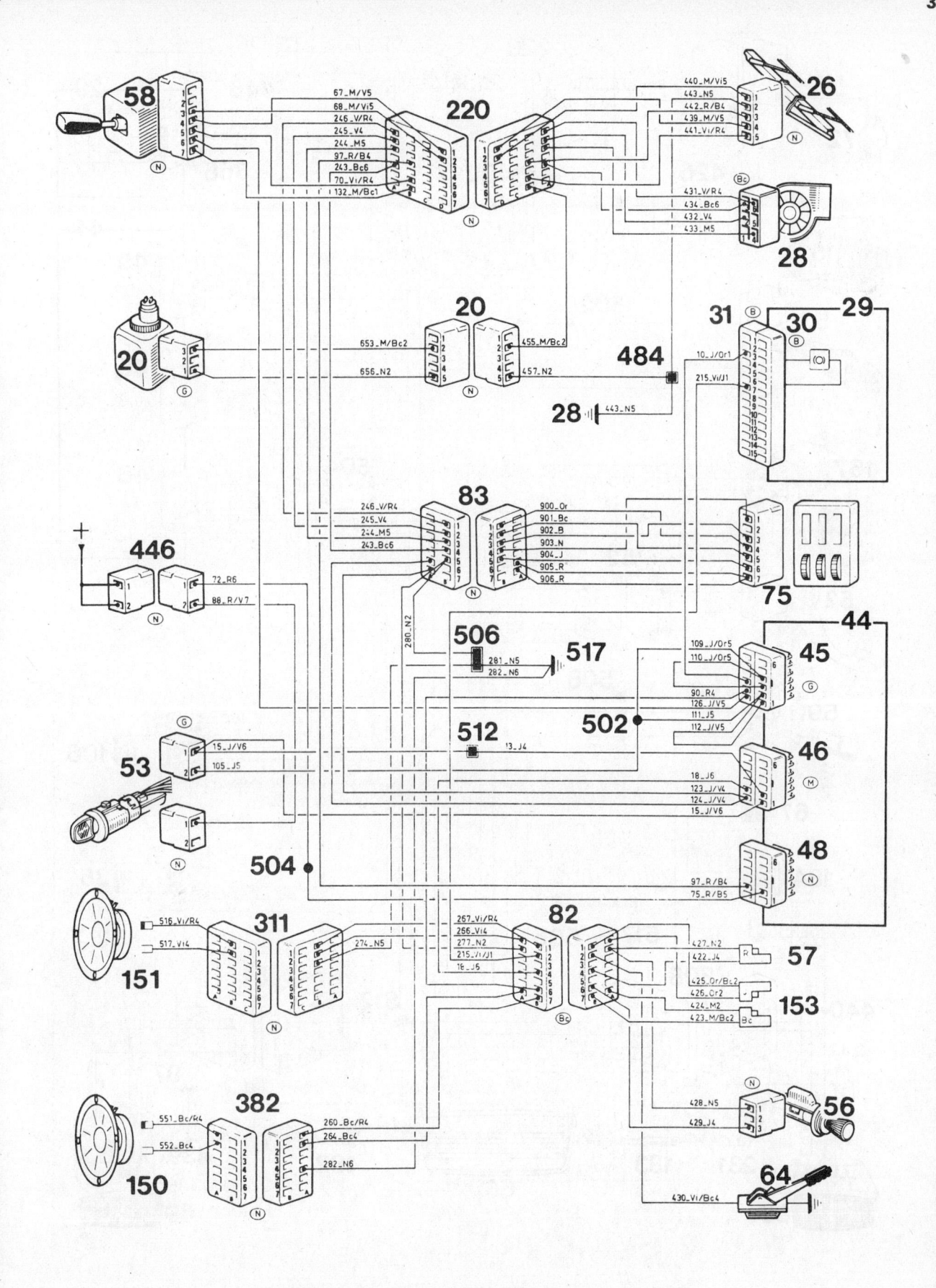

Fig. 12.41 Typical Renault 25 wiring diagram – circuit 4

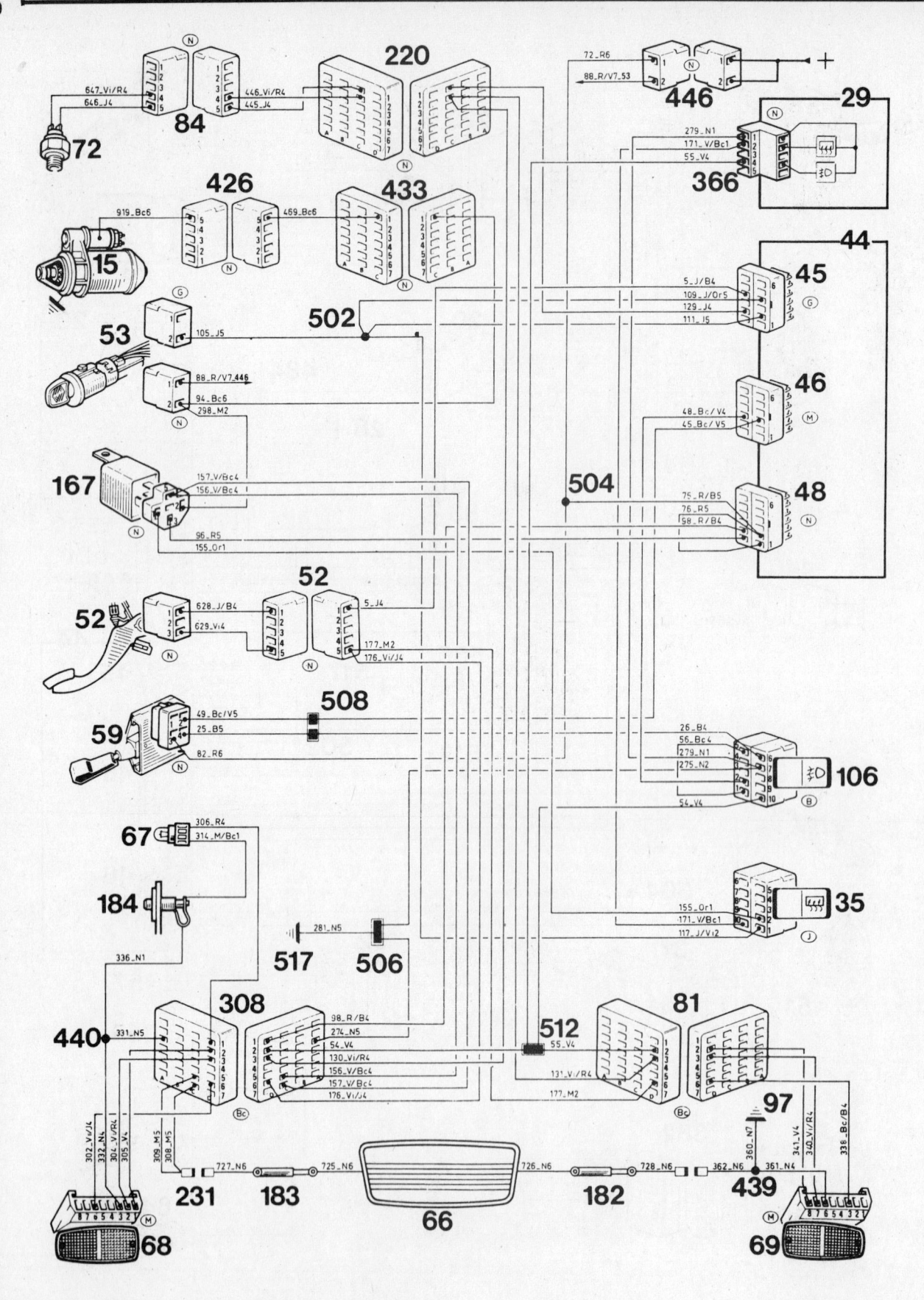

Fig. 12.41 Typical Renault 25 wiring diagram – circuit 5

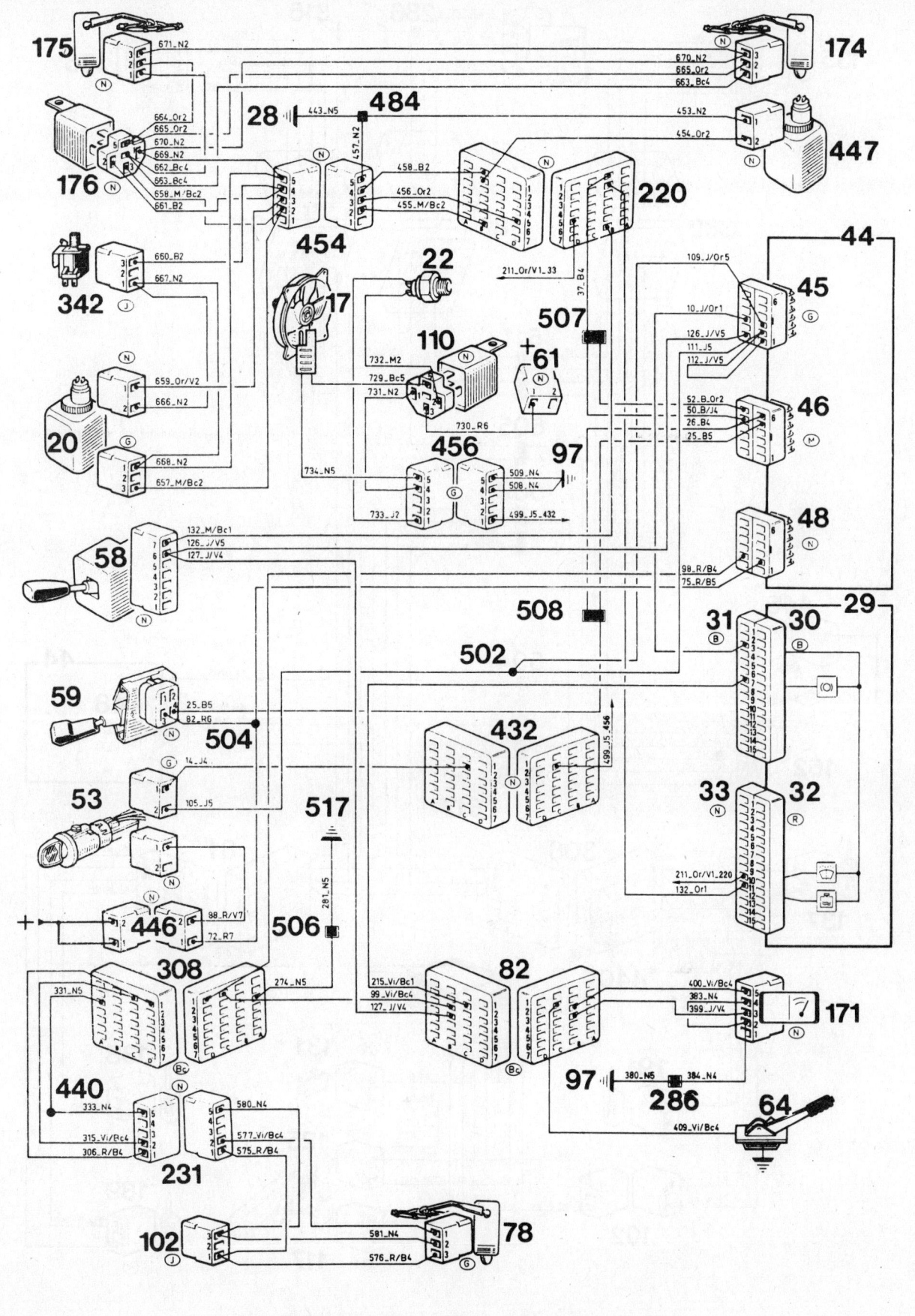

Fig. 12.41 Typical Renault 25 wiring diagram – circuit 6

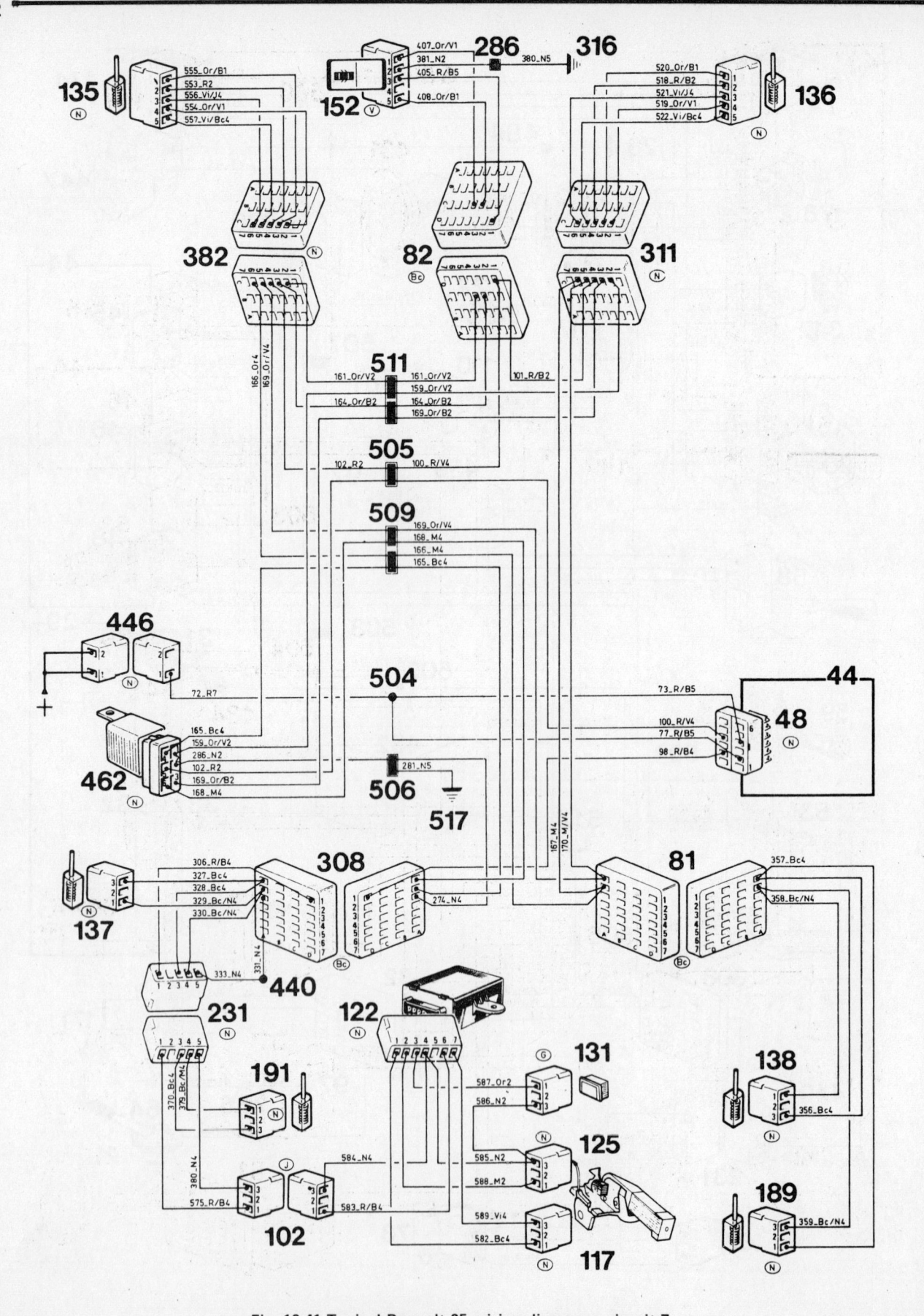

Fig. 12.41 Typical Renault 25 wiring diagram – circuit 7

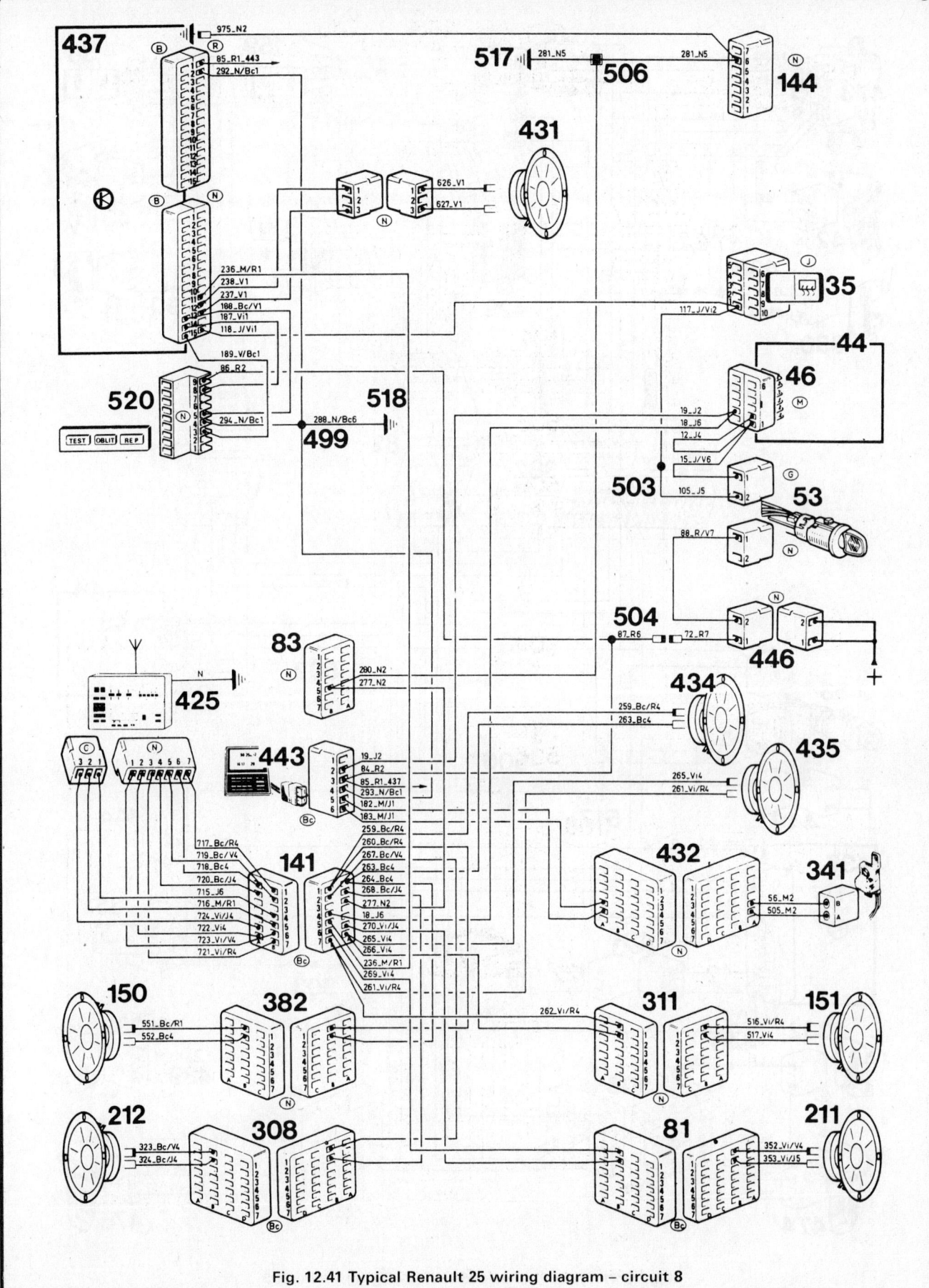

Fig. 12.41 Typical Renault 25 wiring diagram – circuit 8

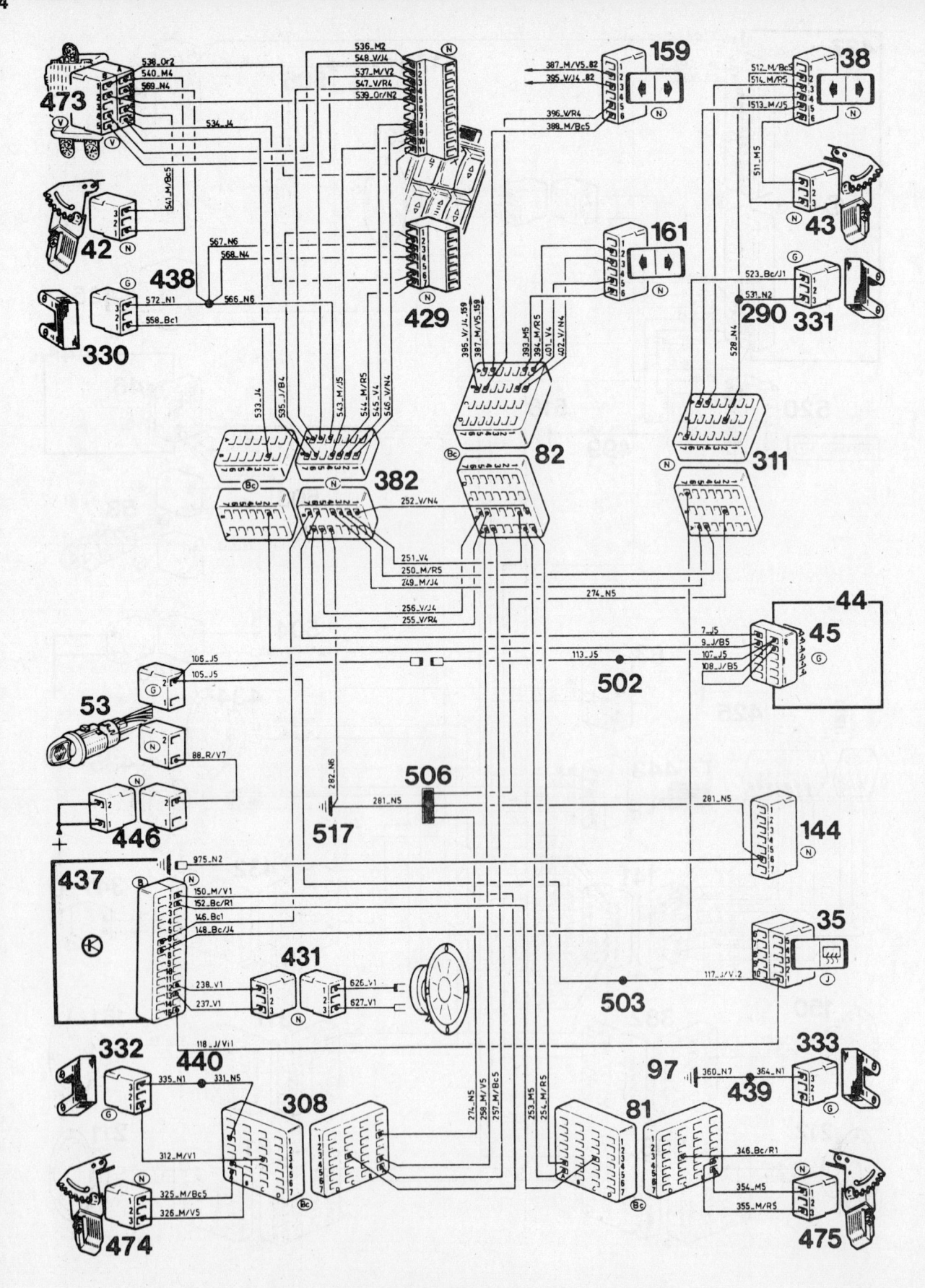

Fig. 12.41 Typical Renault 25 wiring diagram – circuit 9

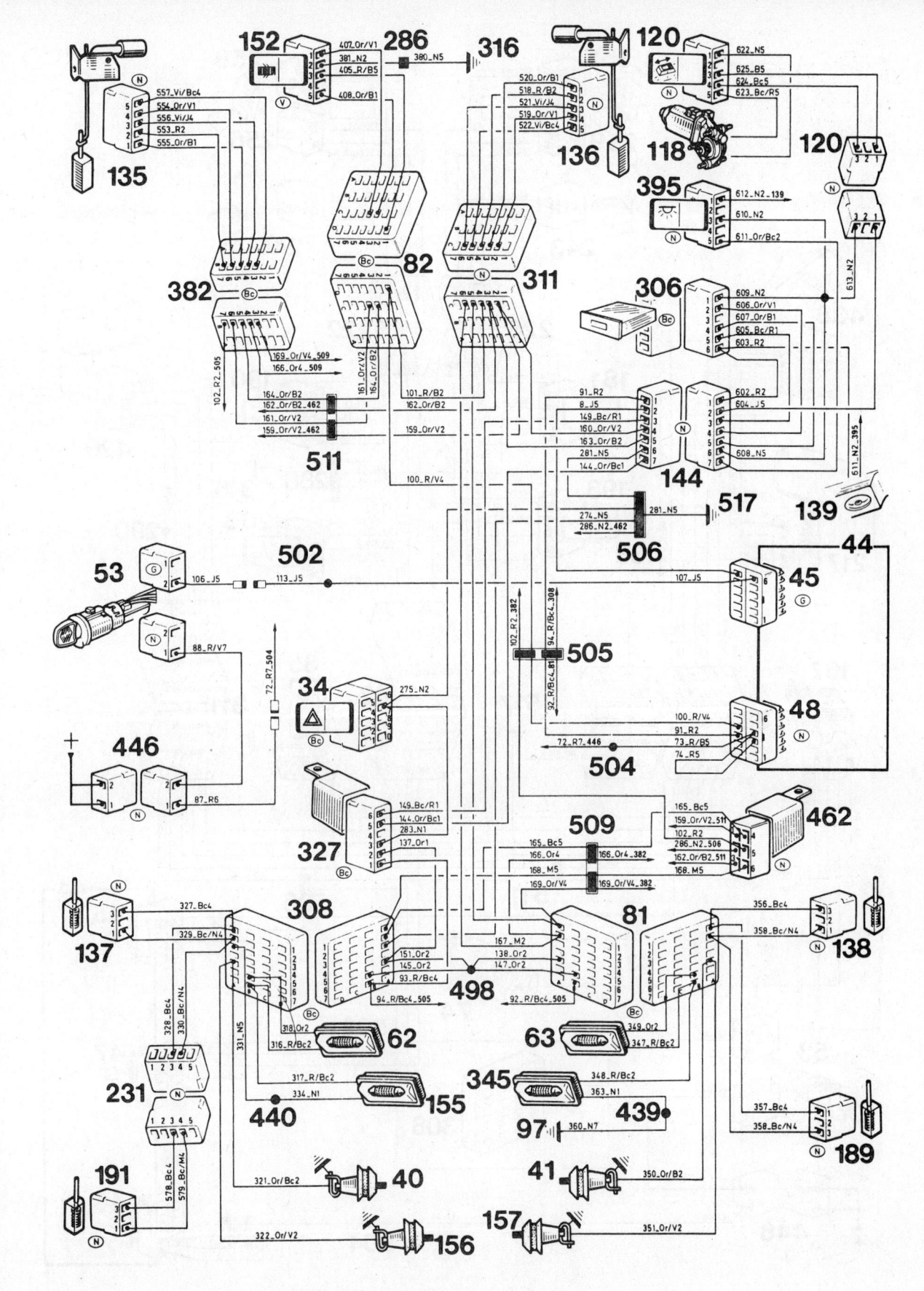

Fig. 12.41 Typical Renault 25 wiring diagram – circuit 10

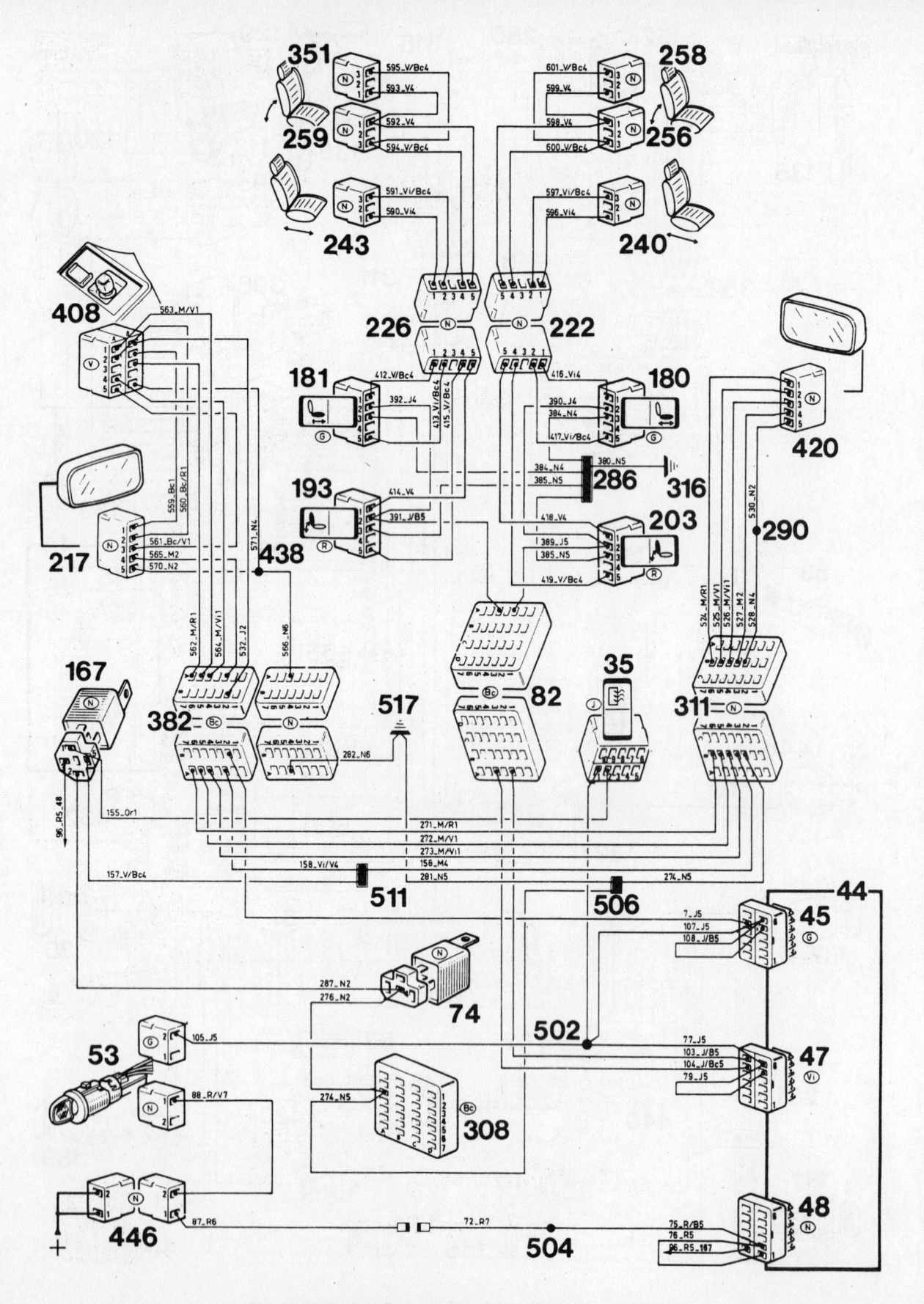

Fig. 12.41 Typical Renault 25 wiring diagram – circuit 11

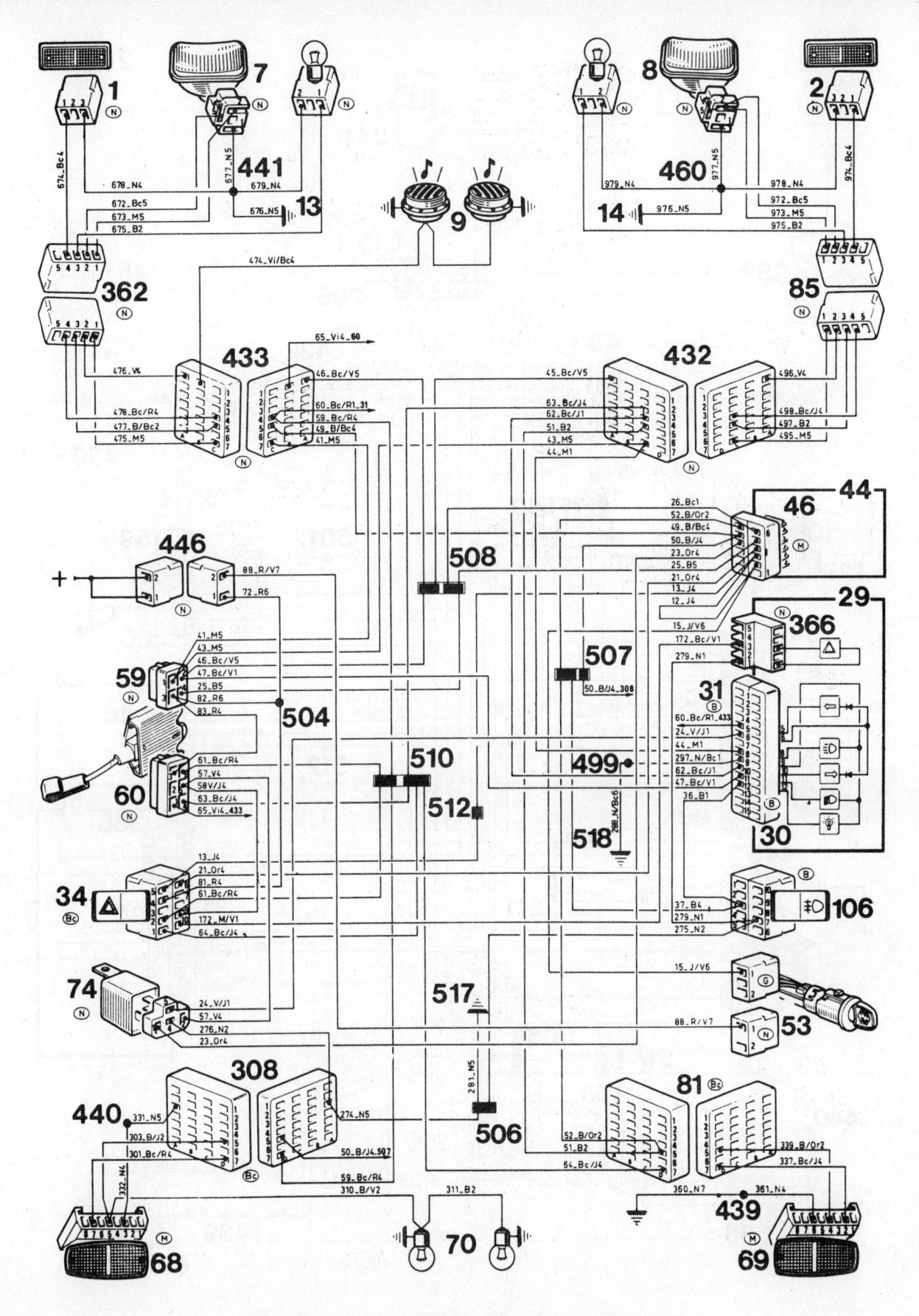

Fig. 12.41 Typical Renault 25 wiring diagram – circuit 12

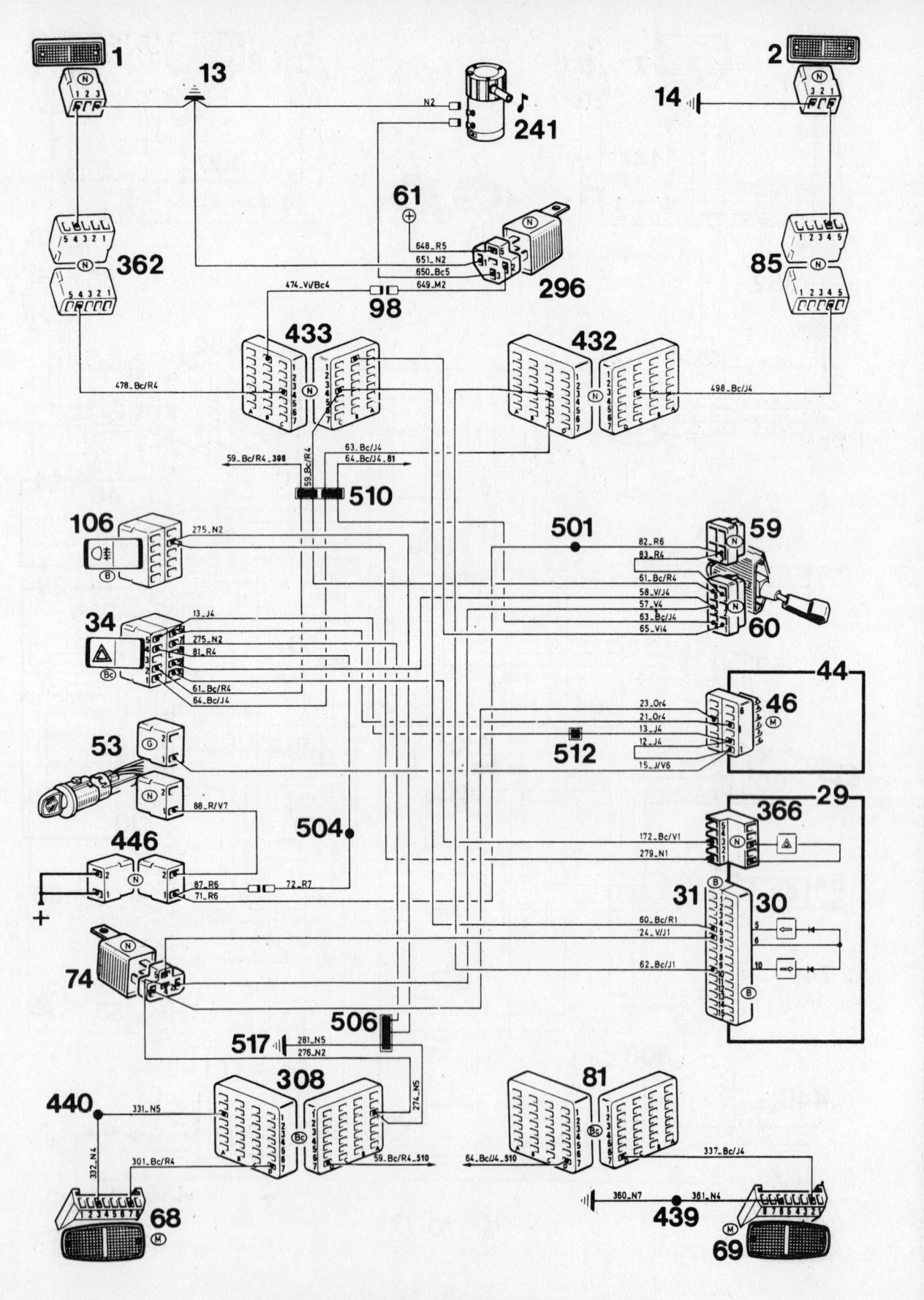

Fig. 12.41 Typical Renault 25 wiring diagram – circuit 13

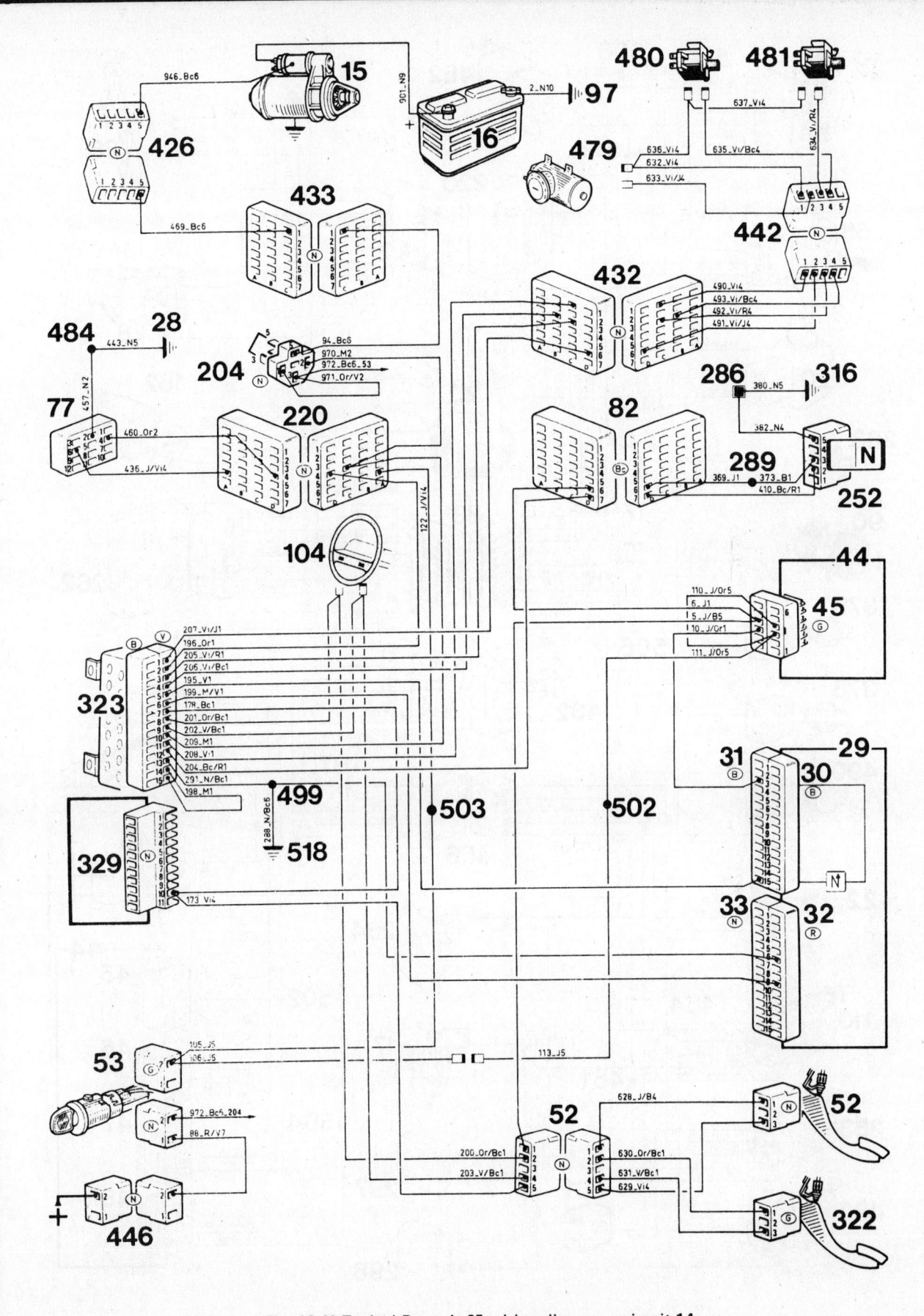

Fig. 12.41 Typical Renault 25 wiring diagram – circuit 14

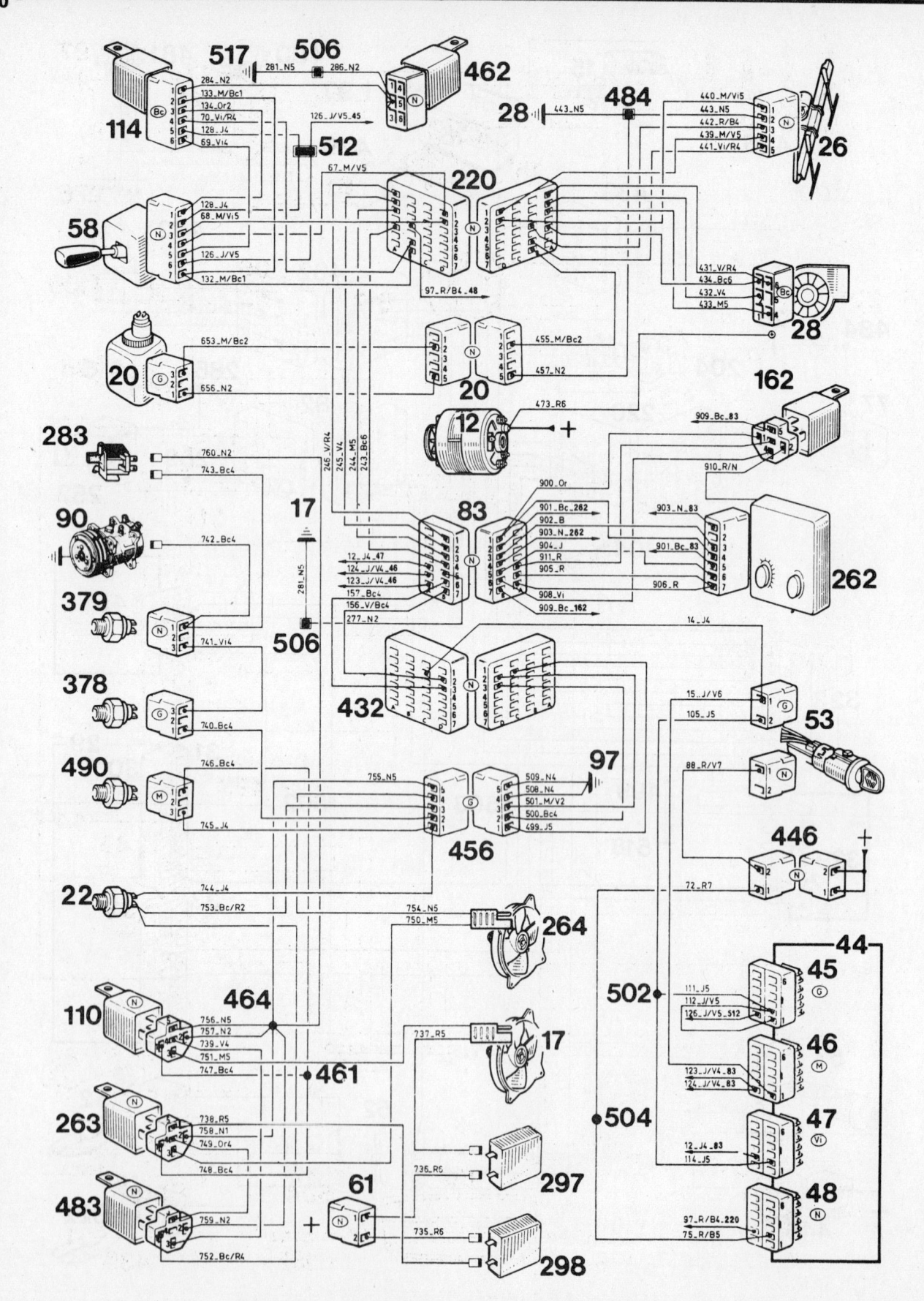

Fig. 12.41 Typical Renault 25 wiring diagram – circuit 15

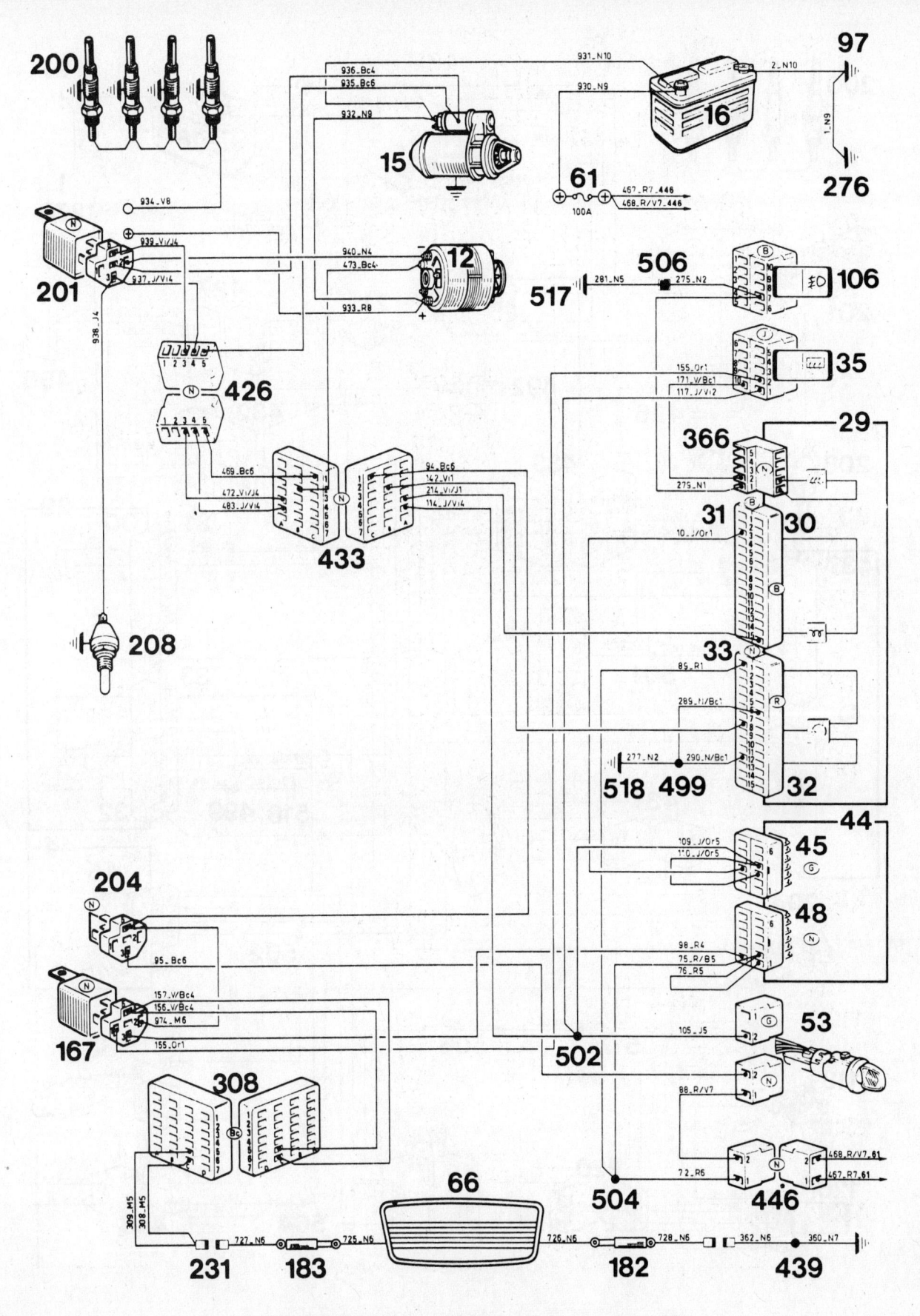

Fig. 12.41 Typical Renault 25 wiring diagram – circuit 16

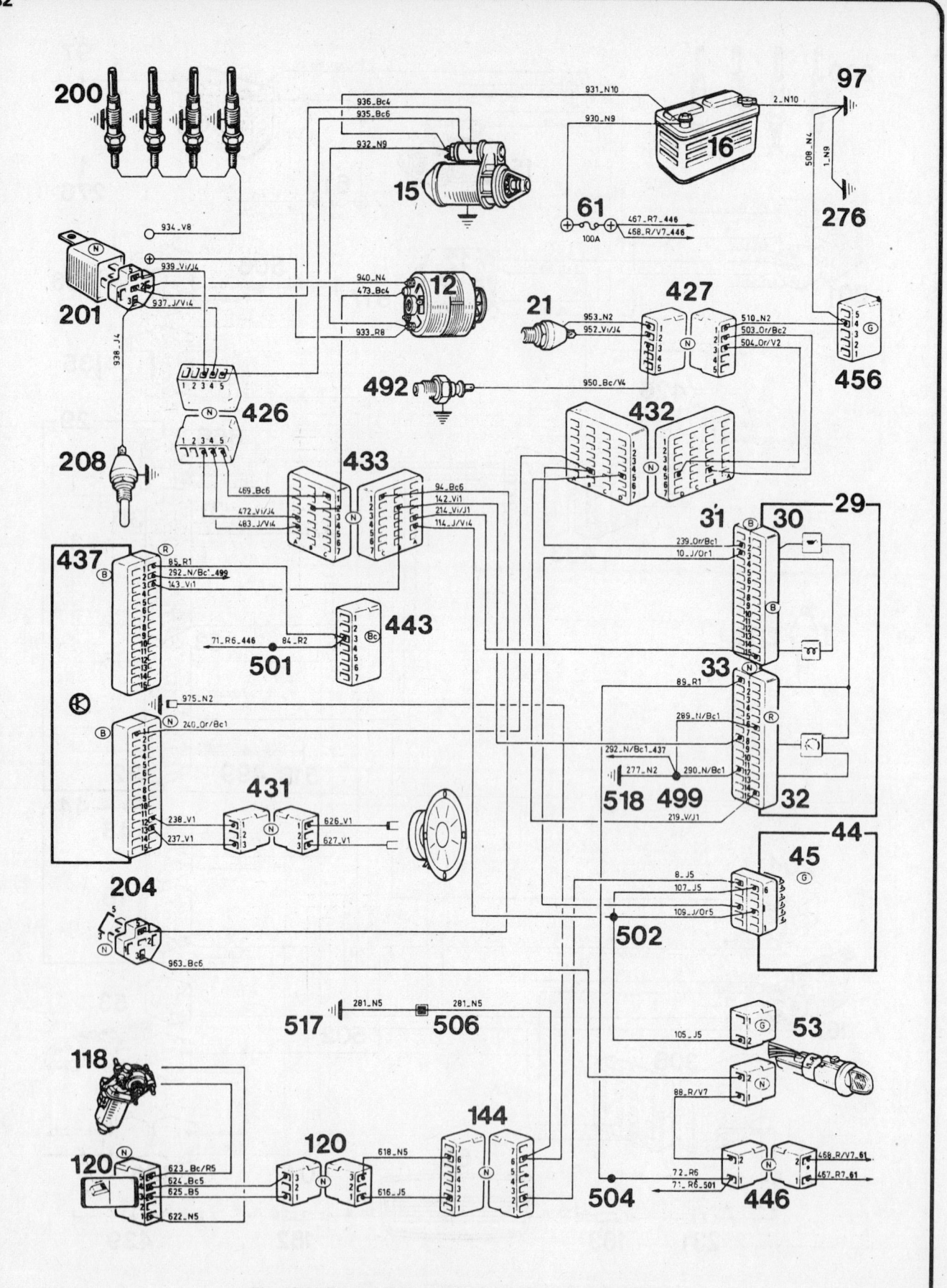

Fig. 12.41 Typical Renault 25 wiring diagram – circuit 17

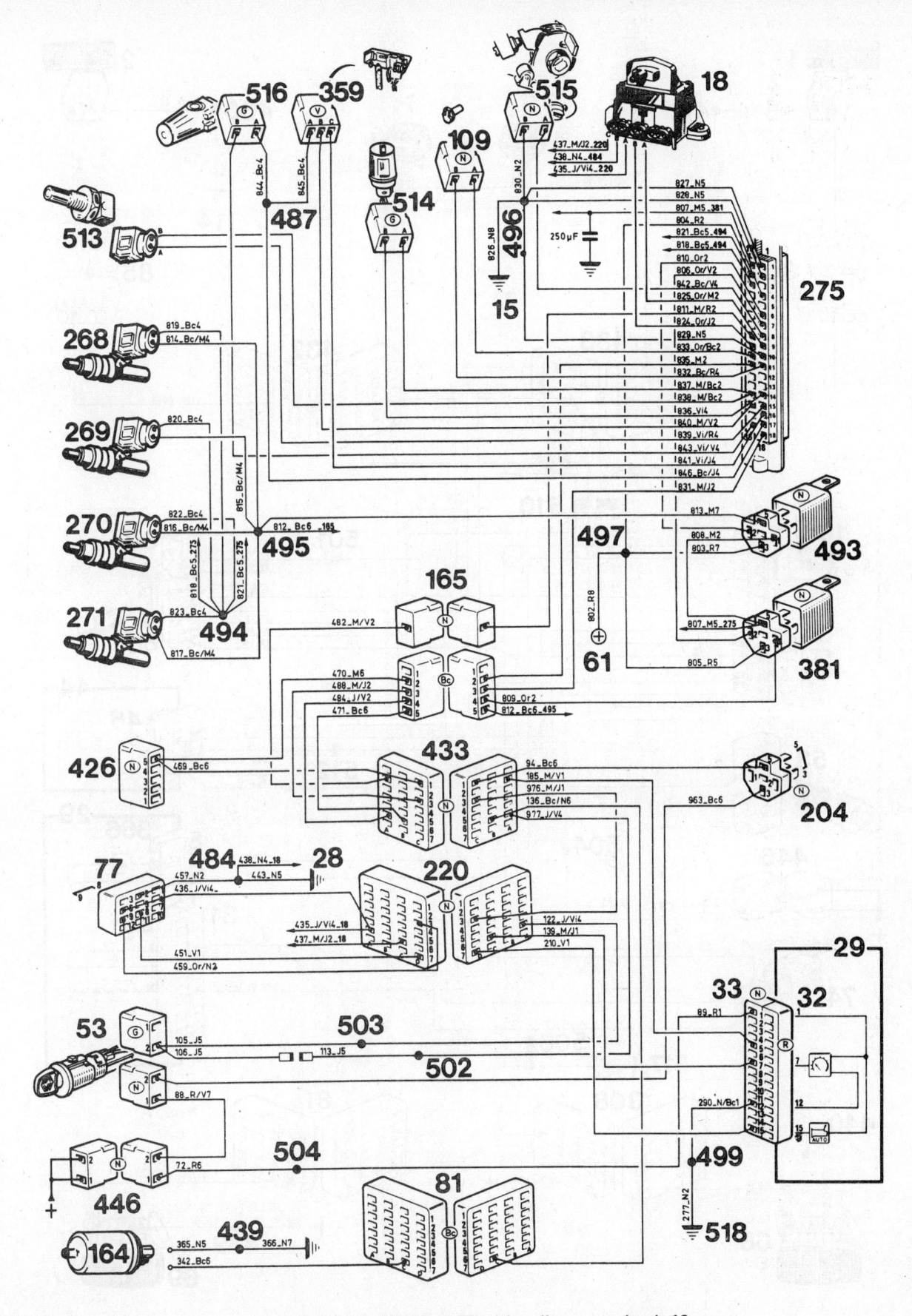

Fig. 12.41 Typical Renault 25 wiring diagram – circuit 18

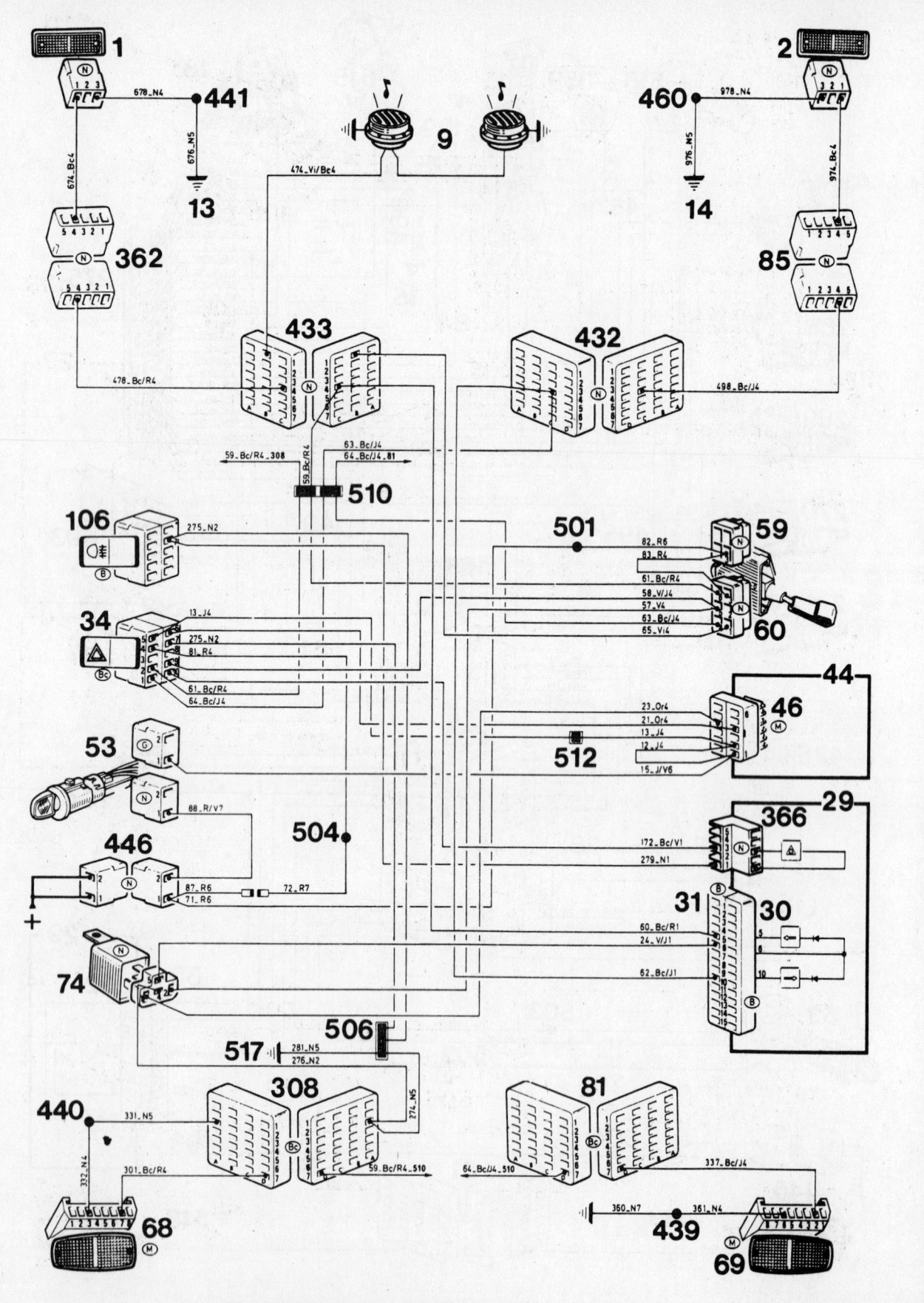

Fig. 12.41 Typical Renault 25 wiring diagram – circuit 19

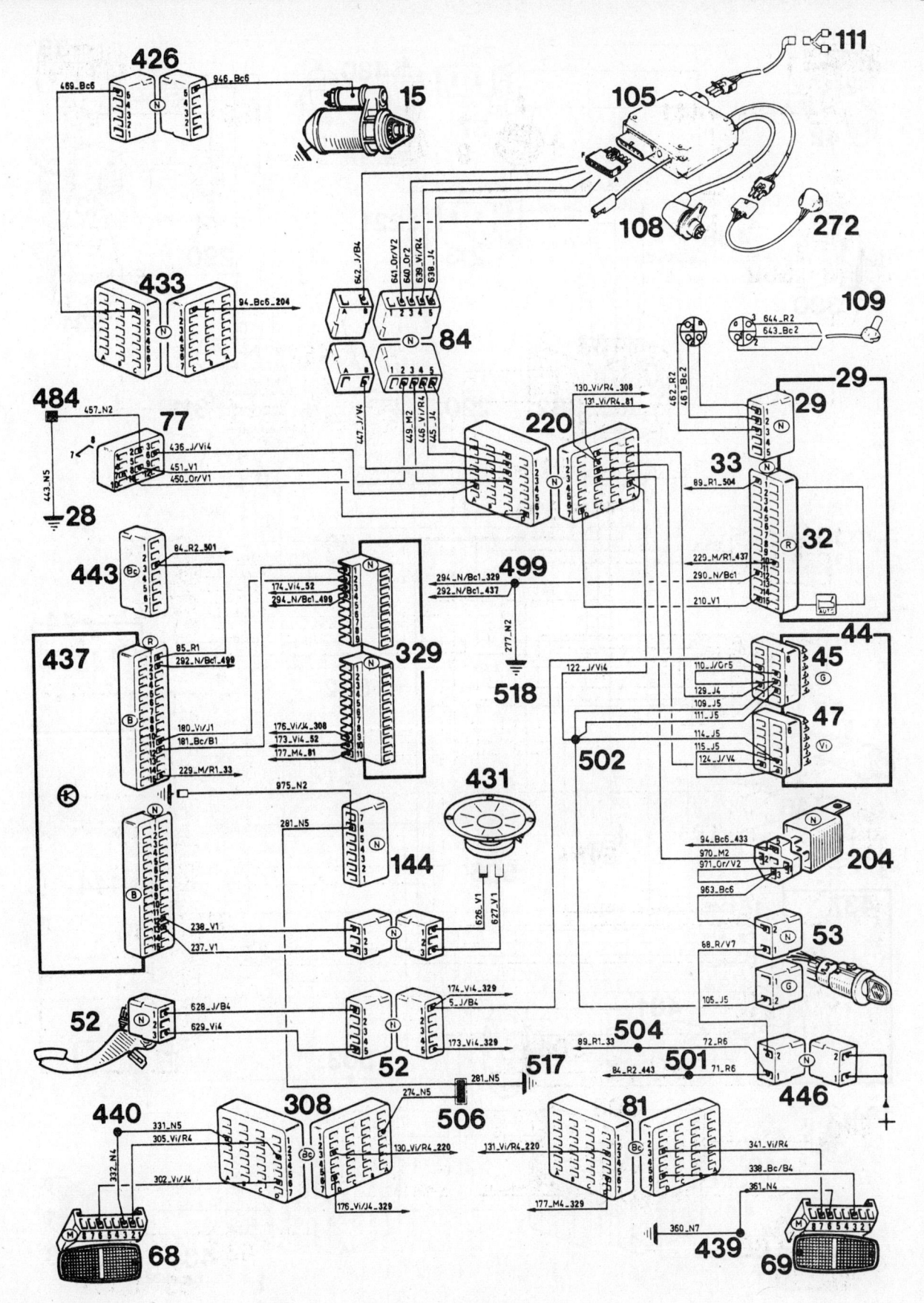

Fig. 12.41 Typical Renault 25 wiring diagram – circuit 20

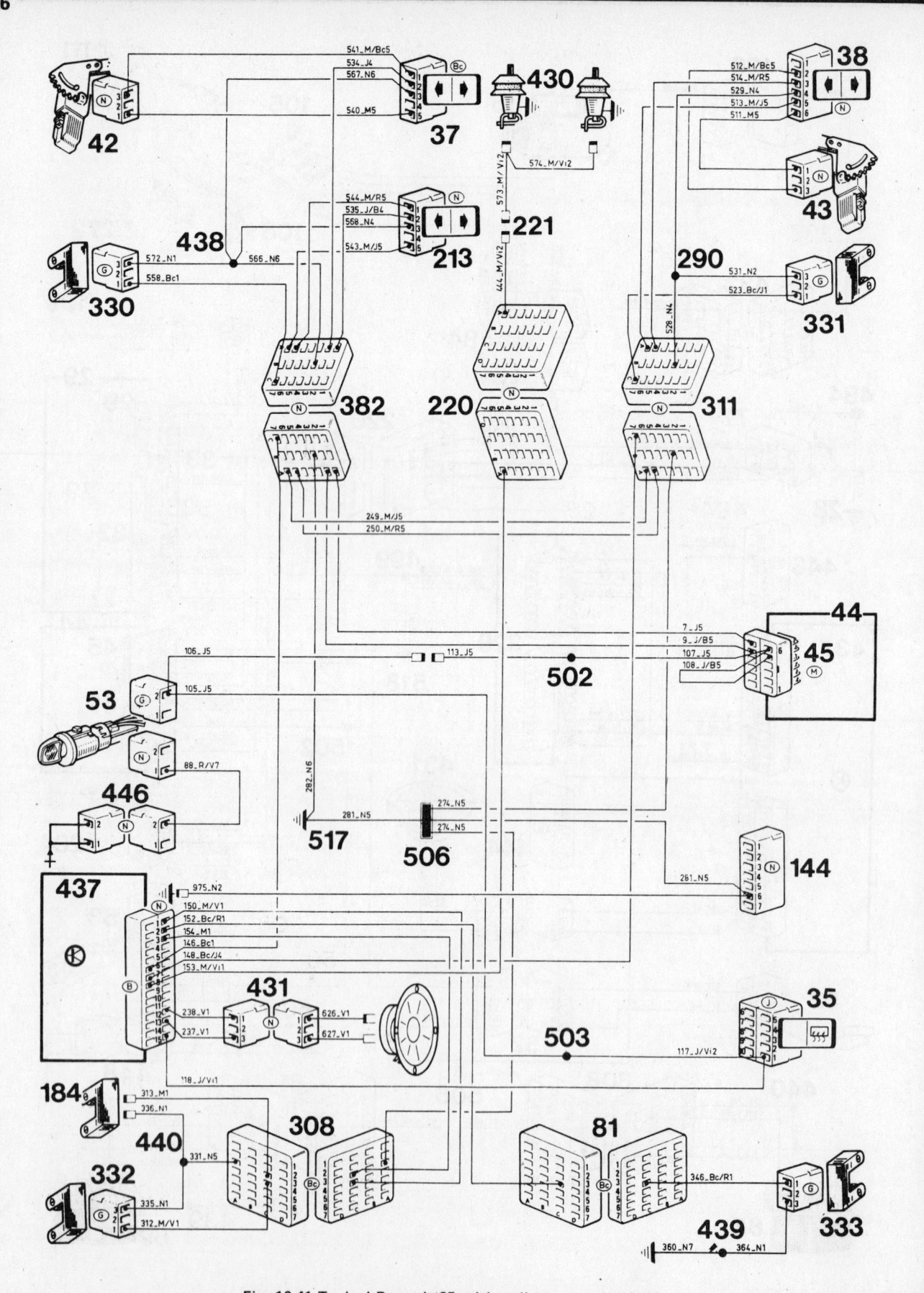

Fig. 12.41 Typical Renault 25 wiring diagram – circuit 21

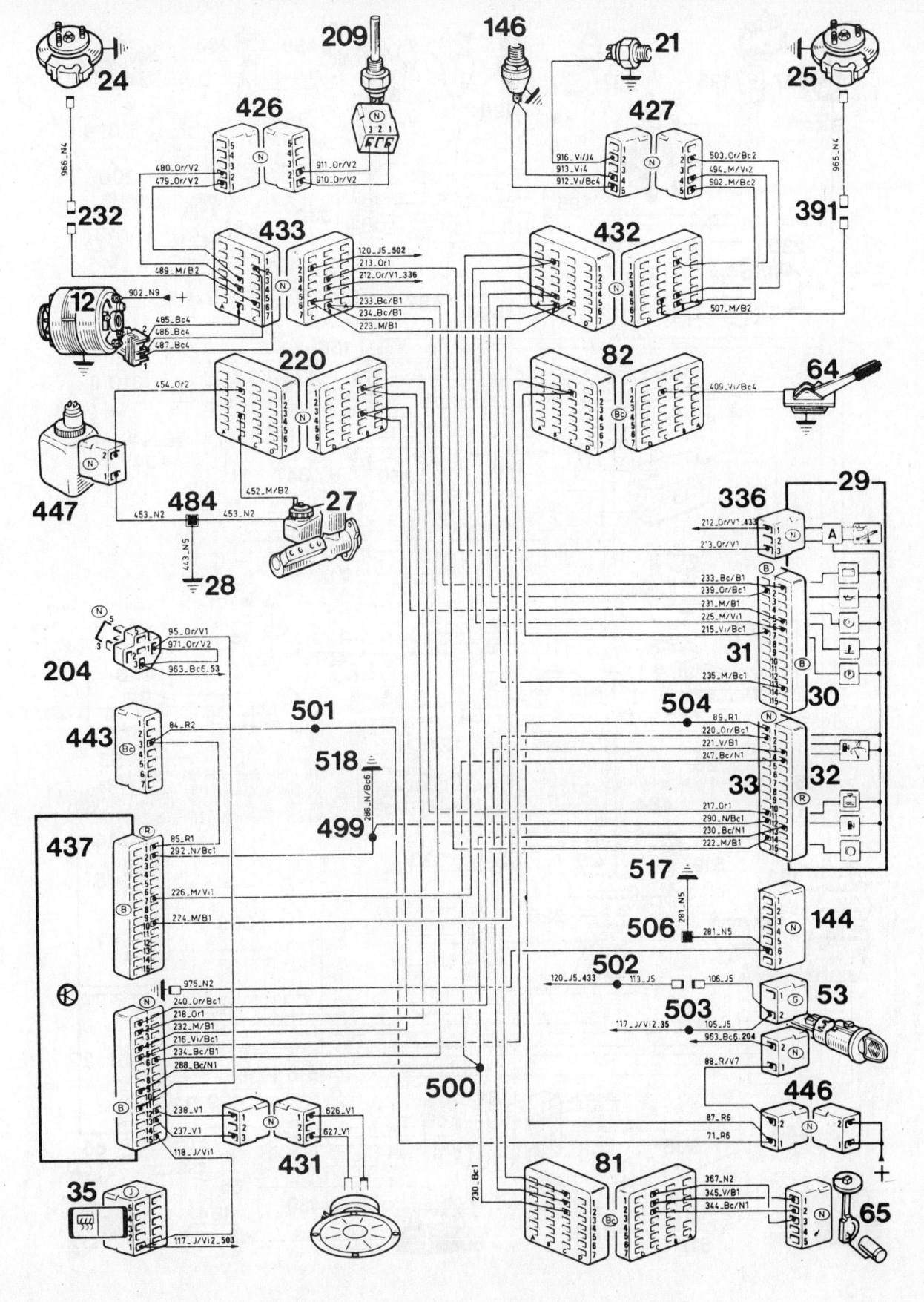

Fig. 12.41 Typical Renault 25 wiring diagram – circuit 22

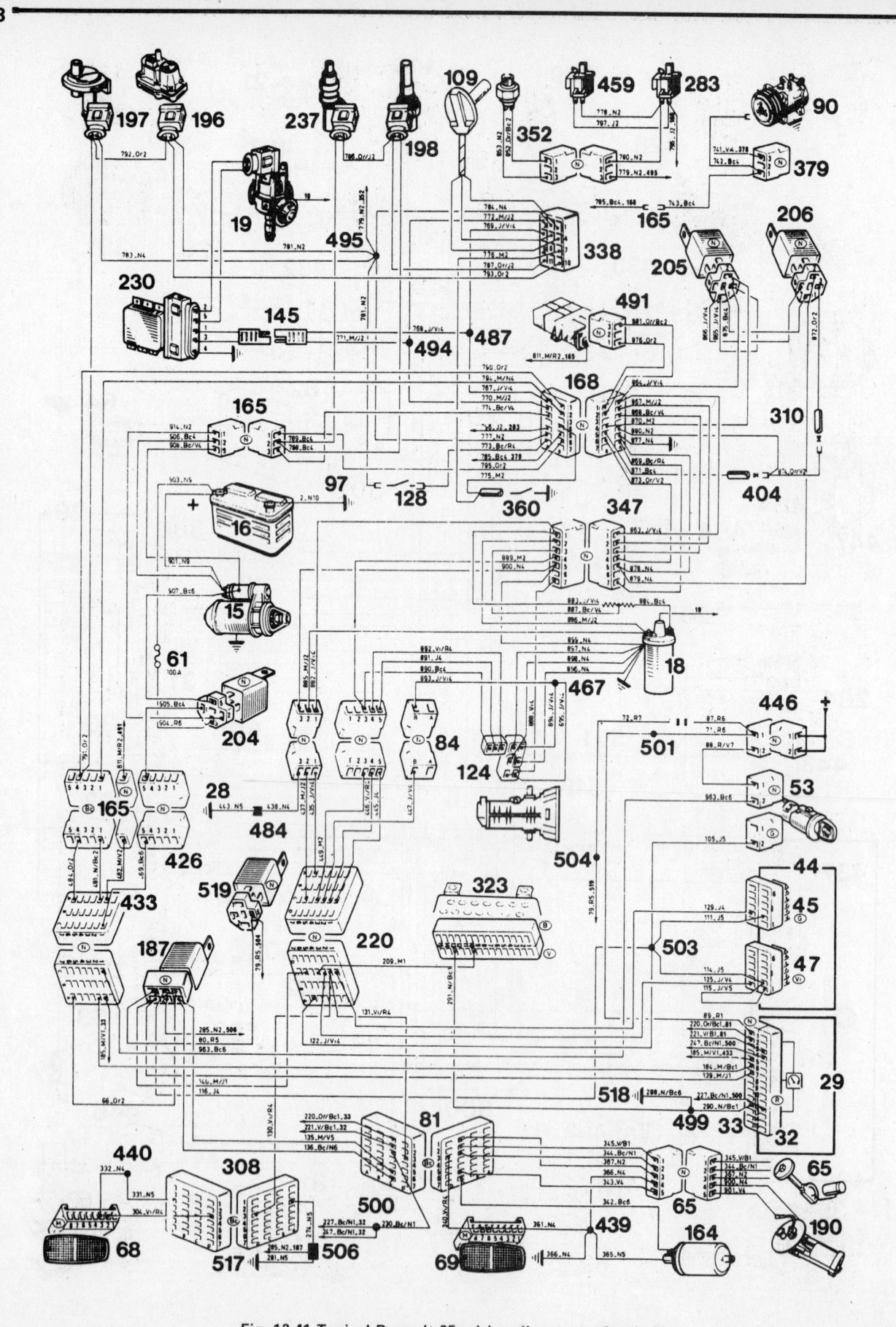

Fig. 12.41 Typical Renault 25 wiring diagram – circuit 23

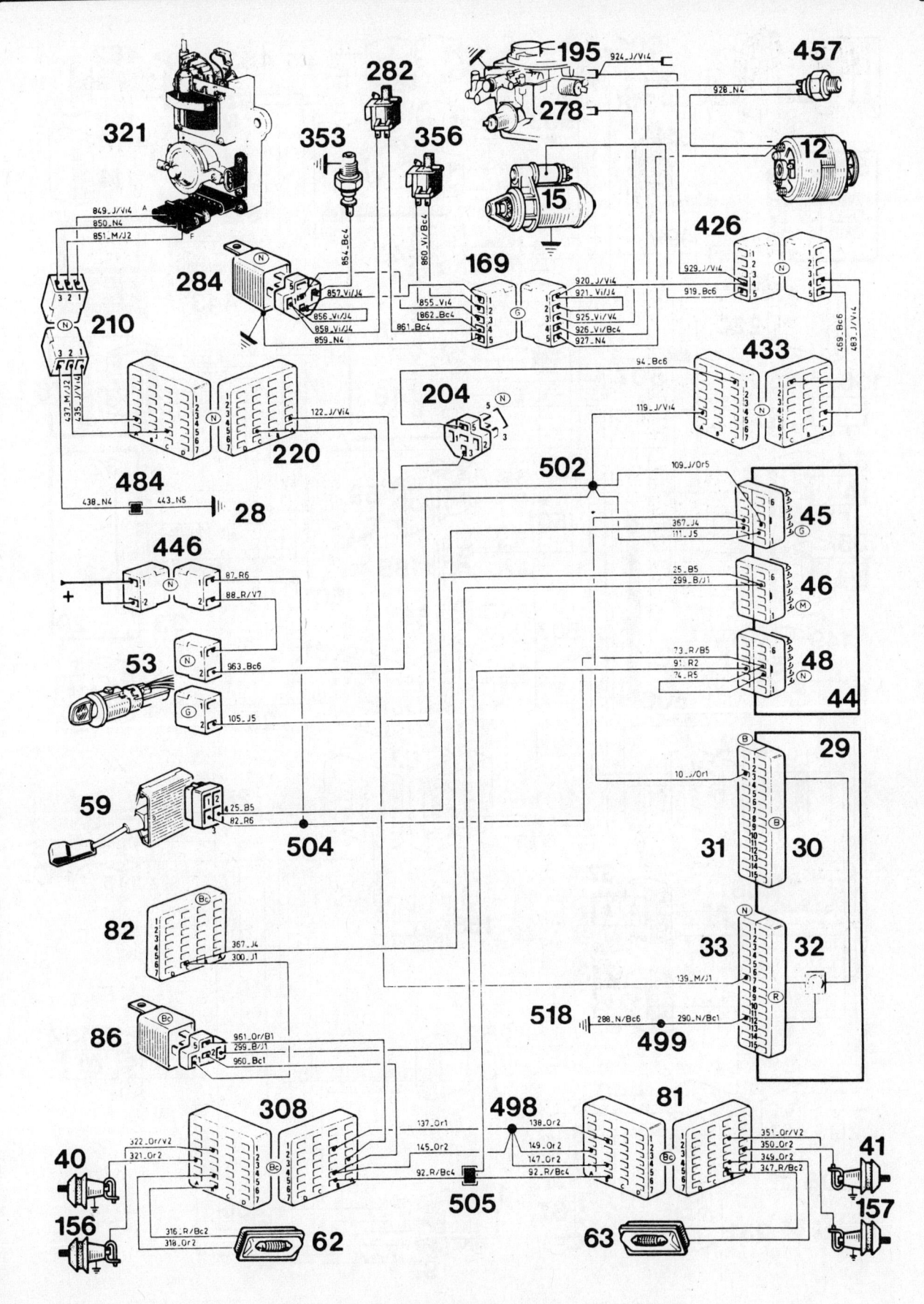

Fig. 12.41 Typical Renault 25 wiring diagram – circuit 24

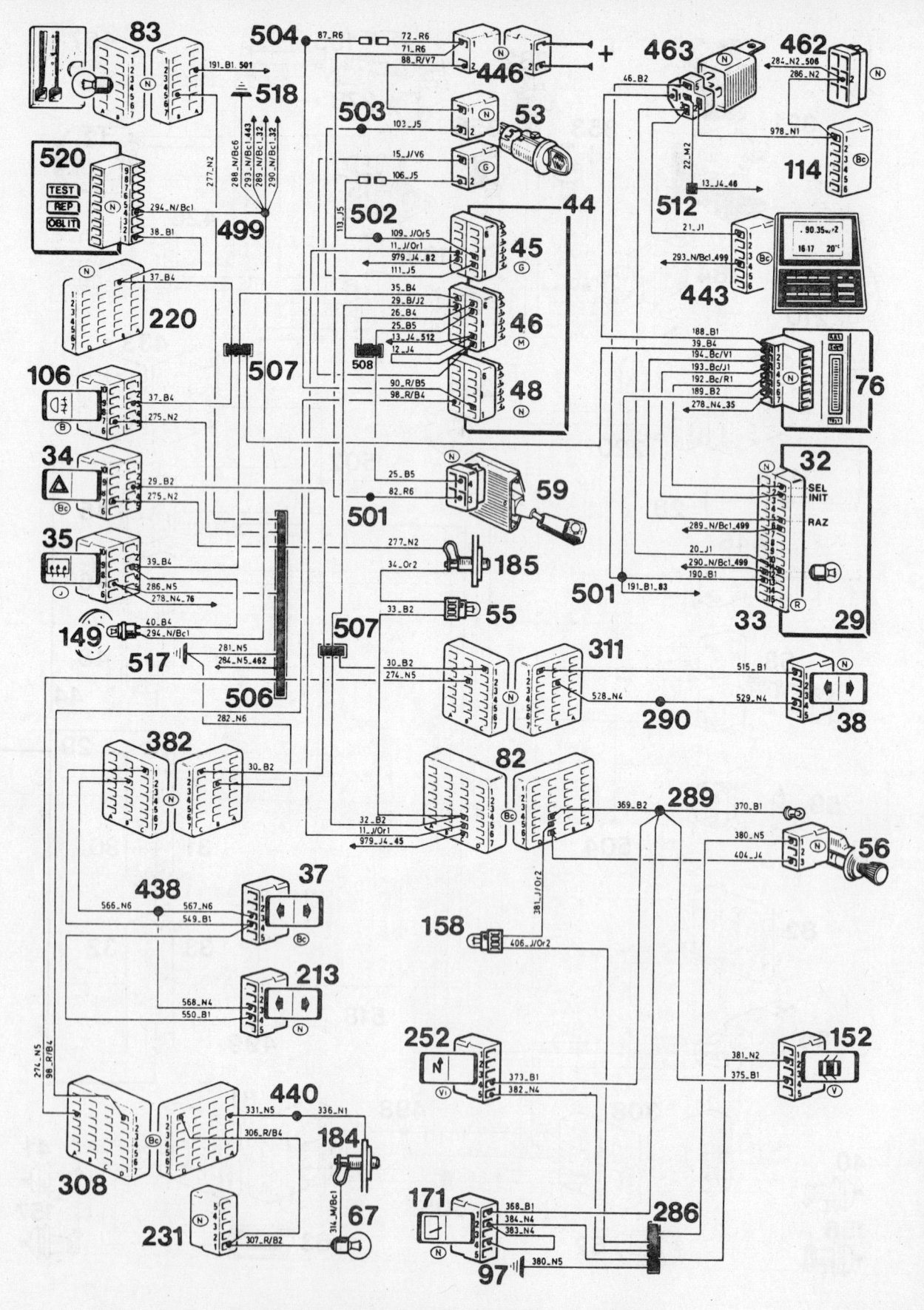

Fig. 12.41 Typical Renault 25 wiring diagram – circuit 25

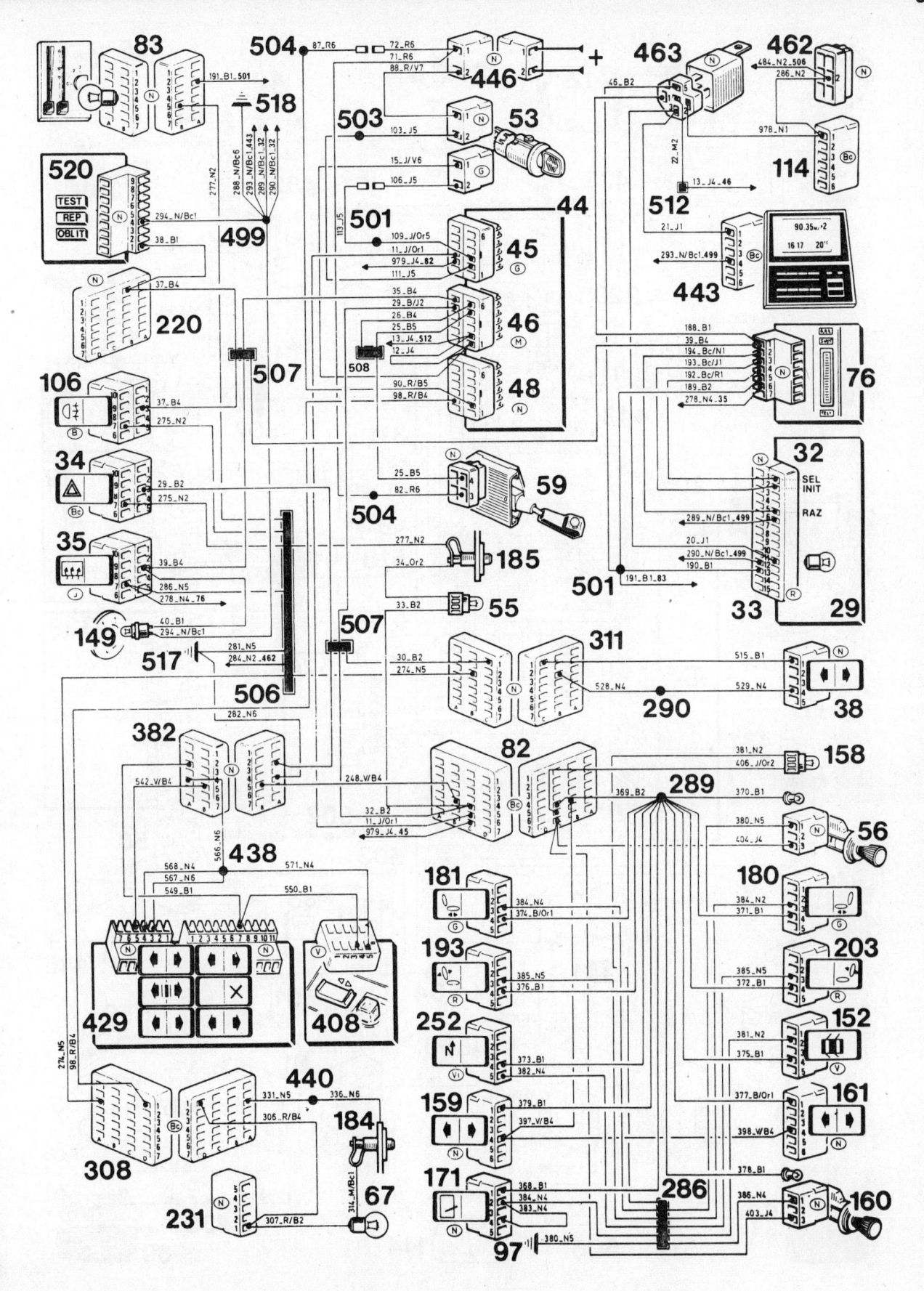

Fig. 12.41 Typical Renault 25 wiring diagram – circuit 26

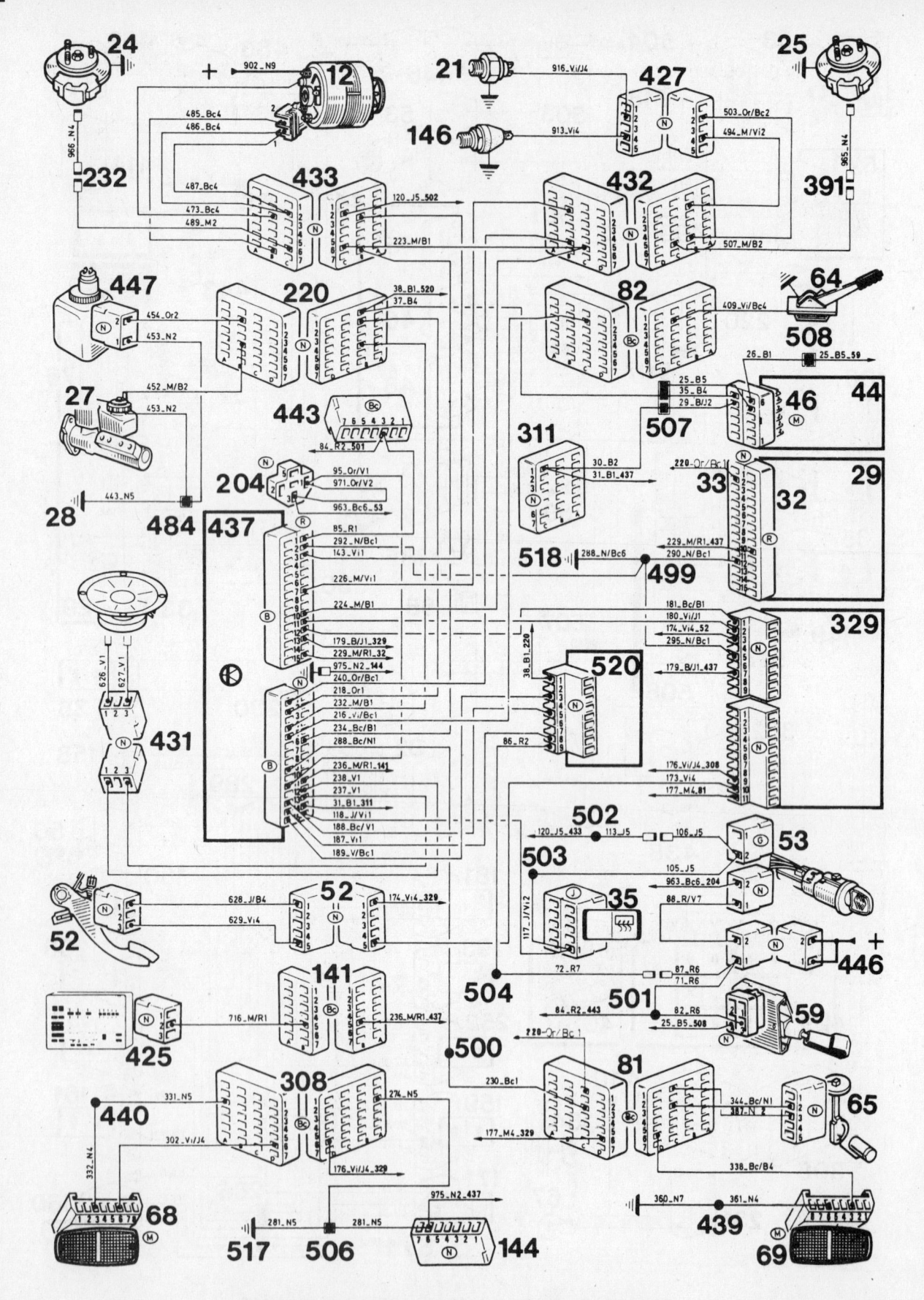

Fig. 12.41 Typical Renault 25 wiring diagram – circuit 27

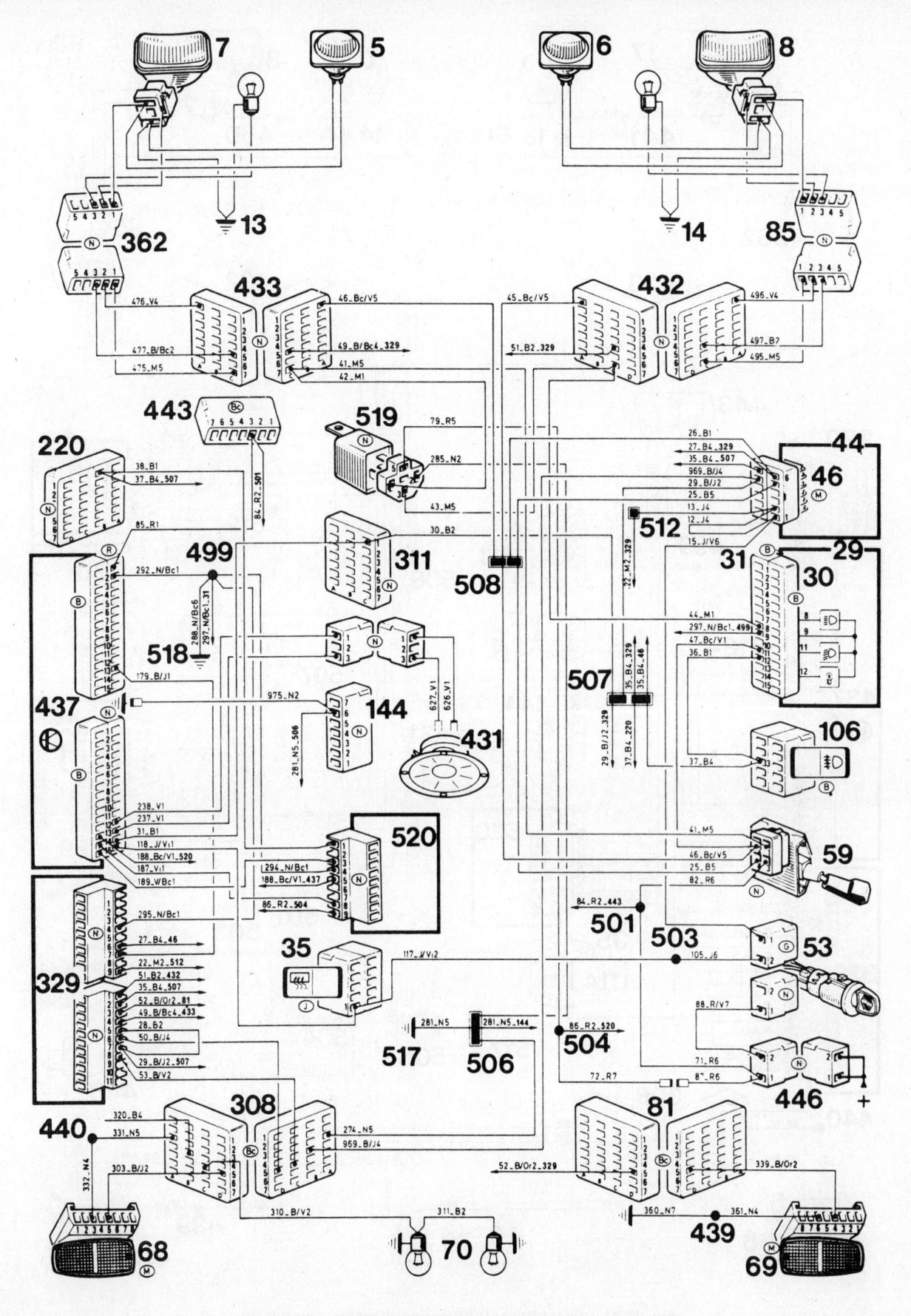

Fig. 12.41 Typical Renault 25 wiring diagram – circuit 28

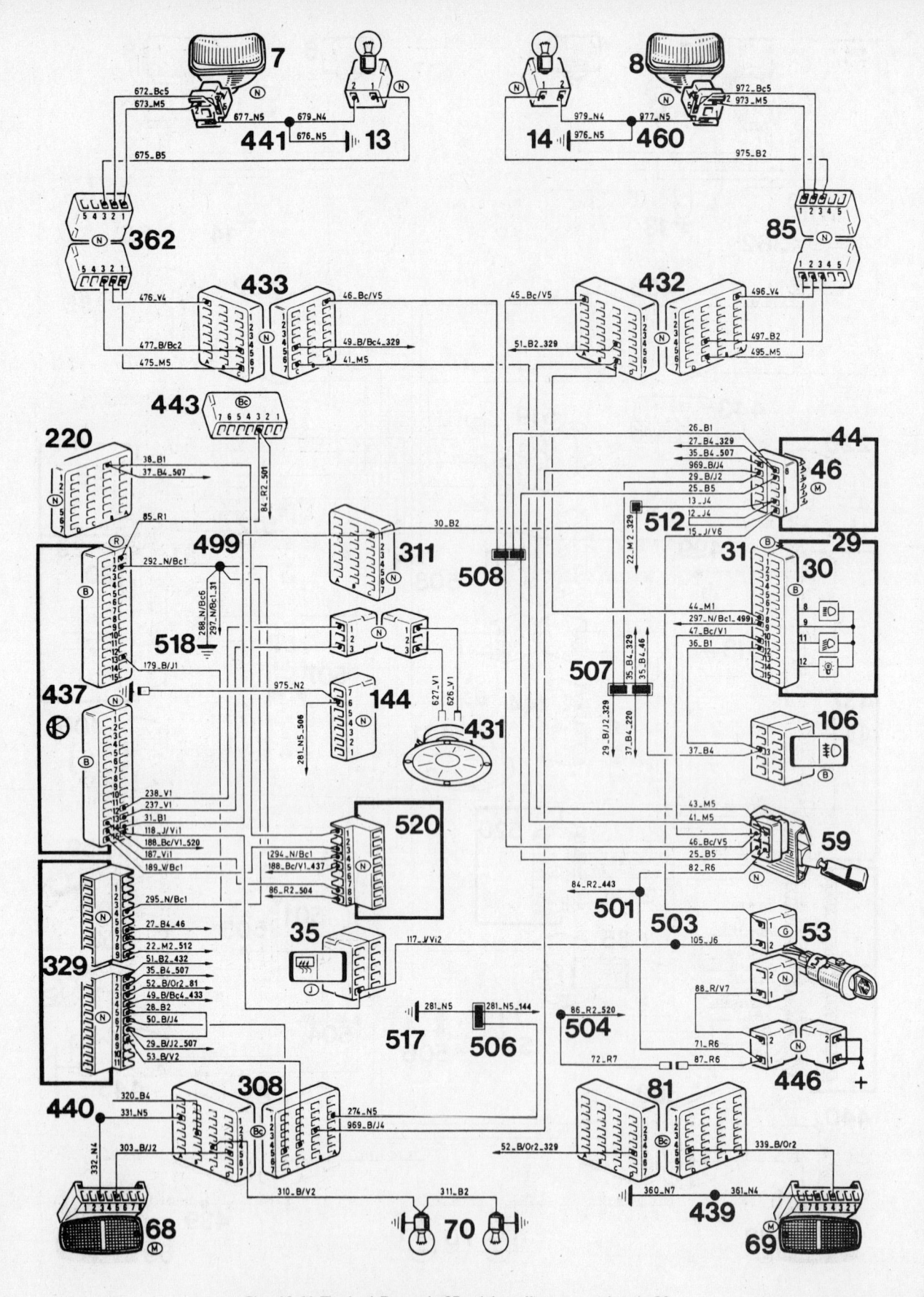

Fig. 12.41 Typical Renault 25 wiring diagram – circuit 29

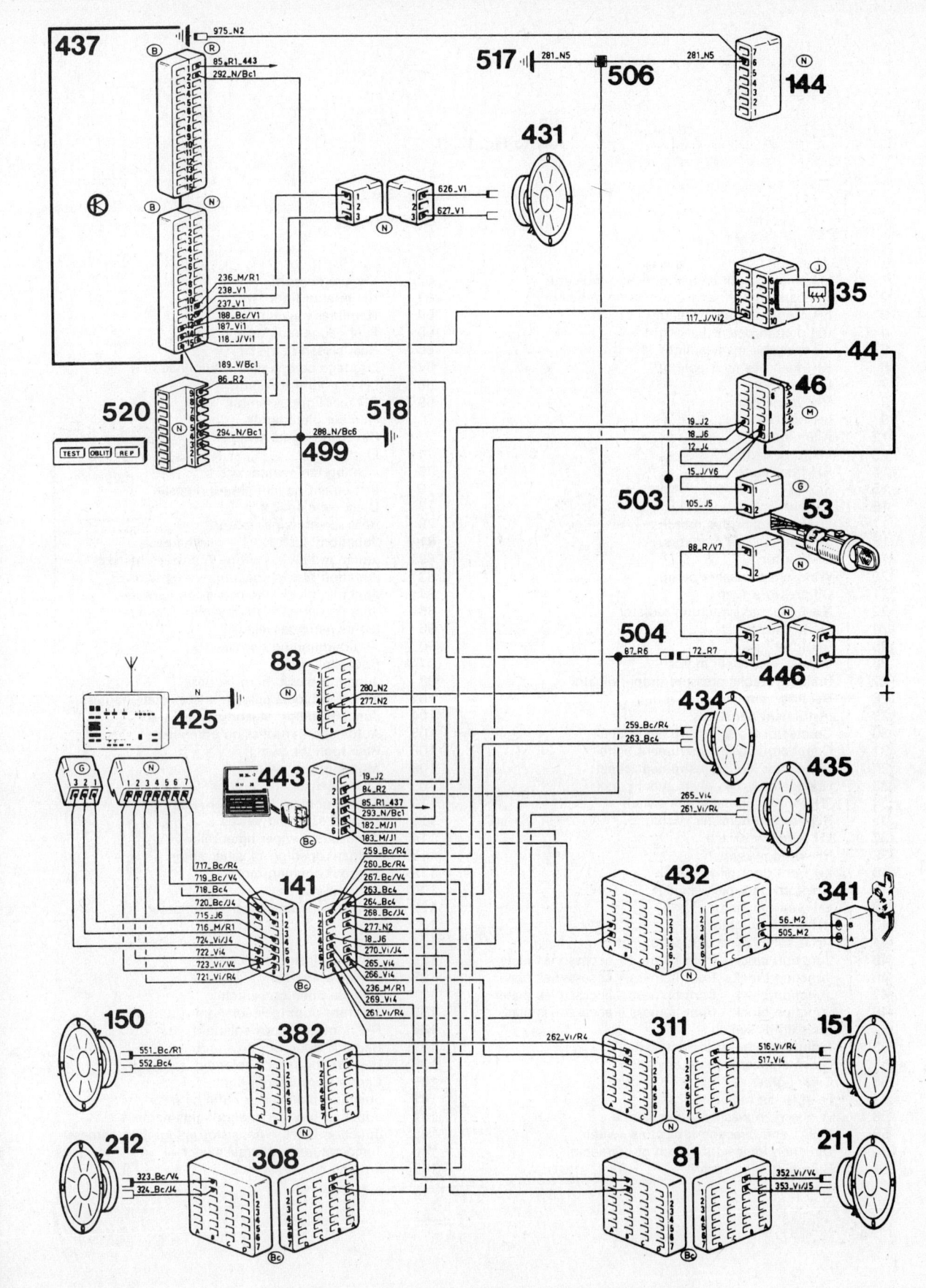

Fig. 12.41 Typical Renault 25 wiring diagram – circuit 30

Key to Fig. 12.41

1	LH parking light and/or direction indicator	62	LH interior light
2	RH parking light and/or direction indicator	63	RH interior light
3	LH dipped beams headlight	64	Handbrake switch
4	RH dipped beams headlight	65	Fuel gauge tank unit
5	LH main beams headlight	66	Rear screen demister
6	RH main beams headlight	67	Luggage compartment illumination
7	LH headlight	68	LH rear light assembly
8	RH headlight	69	RH rear light assembly
9	LH horn	70	Number plate lights
12	Alternator	72	Reversing lights switch
13	LH front earth	74	Flasher unit
14	RH front earth	75	Heating fan switch
15	Starter	76	Instrument panel lighting rheostat
16	Battery	77	Diagnostic socket
17	Engine cooling fan motor	78	Rear screen wiper motor
18	Ignition coil (or mounting)	81	Junction block No 1 – rear harness
19	Distributor	82	Junction block – bridge (console) harness
20	Windscreen washer pump	83	Junction block – heating wiring harness
21	Oil pressure switch	84	Junction block – transmission harness
22	No 1 thermal switch on radiator	85	Junction block – RH headlight harness
24	LH front brake	86	Lights reminder relay
25	RH front brake	90	Air conditioning compressor
26	Windscreen wiper motor	97	Bodyshell earth
27	Brake hydraulic pressure drop indicator	98	Junction block: horn harness
28	Heating-ventilating fan motor	102	Junction block: tailgate intermediate harness
29	Instrument panel	104	Junction block: steering wheel tracks
30	Connector No 1 – instrument panel	105	Automatic transmission computer
31	Connector No 2 – instrument panel	106	Rear foglight switch
32	Connector No 3 – instrument panel	108	Multipurpose switch
33	Connector No 4 – instrument panel	109	Speed sensor
34	"Hazard" warning lights switch	110	Engine cooling fan motor relay
35	Rear screen demister switch	111	1 and 2 solenoid valves
37	LH window switch	114	Windscreen wiper timer relay
38	RH window switch	117	Tailgate opening-closing motor
40	LH front door pillar switch	118	Sunroof opening motor
41	RH front door pillar switch	120	Sunroof switch
42	LH window motor	122	Tailgate opening-closing computer
43	RH window motor	123	Clock
44	Accessories plate or fusebox	124	Gearbox
45	Junction block – front harness – accessories plate	125	Tailgate closing switch
46	Junction block – front harness – accessories plate	128	Kickdown switch
47	Junction block – front harness – accessories plate	131	Tailgate opening switch
48	Junction block – front harness – accessories plate	135	LH front door lock solenoid
52	Stop-lights switch	136	RH front door lock solenoid
53	Ignition-starter-anti-theft switch	137	Rear LH door lock solenoid
55	Glove compartment illumination	138	Rear RH door lock solenoid
56	Cigar lighter	139	Centre front interior light
57	Feed to car radio	141	Junction block – car radio harness
58	Windscreen washer/wiper switch	144	Junction block – interior lights harness
59	Lights and direction indicators switch	145	Junction block – transistorized ignition harness
60	Direction indicators switch or connector	146	Temperature or thermal switch
61	Before ignition switch	147	Ignition coil resistance

Key to Fig. 12.41 (continued)

149	Headlight adjustment control illumination		217	LH electric mirror
150	LH front speaker		220	Junction block – ventilation compartment harness
151	RH front speaker		221	Junction block – bonnet closure switch harness
152	Solenoid-operated locks central switch		222	Junction block – LH seat harness
153	Car radio speaker wires		226	Junction block – RH seat harness
155	Rear or rear LH interior light		230	Ignition starter box
156	Rear LH door pillar switch		231	Junction block – tailgate harness
157	Rear RH door pillar switch		232	Junction block – LH brake pad wear wire
158	Automatic transmission selector illumination		237	Timed temperature switch
159	Rear LH window-winder switch		240	LH seat motor
160	Rear cigar-lighter		241	Horn compressor
161	Rear RH window winder switch		243	RH seat motor
162	No 1 air conditioning relay		252	Cruise control switch
164	Electric petrol pump		256	No 1 LH seatback motor
165	Junction block – injection harness		258	No 2 LH seatback motor
167	Rear screen demister relay		259	No 1 RH seatback motor
168	Junction block – anti-stall harness		261	Junction block – cruise control harness
169	Junction block – solenoid valve harness		262	Air conditioning and heating switch board
171	Rear screen wiper/washer switch		263	No 2 engine cooling fan motor relay
172	Impulse generator		264	No 2 engine cooling fan motor
174	RH headlight wiper motor		268	No 1 cylinder injector
175	LH headlight wiper motor		269	No 2 cylinder injector
176	Headlight wipers timer relay		270	No 3 cylinder injector
180	LH seat movement switch		271	No 4 cylinder injector
181	RH seat movement switch		272	Carburettor throttle spindle switch
182	RH tailgate counterbalance		274	Wire joint No 1
183	LH tailgate counterbalance		275	Injection computer
184	Luggage compartment light switch		276	Engine earth
185	Glove compartment light switch		278	Carburettor
187	Speedo drive relay (fuel pump)		282	Exhaust gas recirculation solenoid valve
189	Fuel filler flap lock solenoid		283	Advance solenoid valve
190	Supercharging pump (diesel)		284	Electric assistance solenoid valve
191	Tailgate lock solenoid		286	Wire joint No 2
193	LH seatback movement switch		289	Wire joint No 3
195	Idle cut-out		290	Wire joint No 4
196	Pressure switch regulator		296	Horn compressor relay
197	Auxiliary air valve		297	Engine cooling fan motor circuit breaker
198	Cold start injector		298	Additional engine cooling fan motor circuit breaker
200	Preheating plugs (diesel)		306	Unlocking remote control
201	Preheating control box (diesel)		308	No 2 junction – rear harness
203	RH seatback movement switch		310	Air conditioning diode
204	Starter relay		311	Junction block – passenger door harness
205	Anti-stall relay		316	Bridge (console) earth
206	Anti-stall timer relay		321	Transistorised ignition module
207	Anti-stall solenoid valve		322	Clutch switch
208	Electrical stop switch		323	Cruise control computer
209	Engine oil level indicator		327	Interior lights timer relay
210	Junction block – transistorised ignition		329	Lamp failure computer
211	RH rear speaker panel		330	Front LH door closure switch
212	LH rear speaker panel		331	Front RH door closure switch
213	LH window winder switch for passenger side		332	Rear LH door closure switch

Key to Fig. 12.41 (continued)

333	Rear RH door closure switch	454	Junction block – headlight wiper harness
336	No 5 dashboard connector	455	Junction block – rear screen wiper harness
338	Injection diagnostic socket	456	Junction block – engine cooling fan motor
341	External temperature sensor	457	60°C temperature switch
342	Headlight washer solenoid valve	460	Wire joint No 9
345	Rear RH interior light	461	Wire joint No 10
347	Junction block – ignition coil harness	462	Door locking timer relay
348	Oil filter pressure switch (clogging)	463	Lighting rheostat relay
351	No 2 seatback motor	464	Wire joint No 11
352	45°C temperature switch	467	Wire joint No 12
353	15°C temperature switch	473	Window winder switch computer
356	Advance solenoid valve (by 15°C temperature switch)	474	Rear LH window winder motor
359	Pressure sensor	475	Rear RH window winder motor
360	Idling switch	479	Electric vacuum pump
362	Junction block – LH headlight harness	480	Security solenoid valve (cruise control)
366	Switch warning lights on dashboard	481	Adjustment solenoid valve (cruise control)
378	Air conditioning high pressure pressure switch	483	No 3 engine cooling fan motor relay
379	Air conditioning low pressure pressure switch	484	Wire joint No 13
381	Injection relay	487	Wire joint No 14
382	Junction block – driver's door harness	490	Air conditioning fan motor pressure switch
391	Junction block – RH brake pad wear wire	491	Converter box
395	Interior light switch	492	Oil pressure temperature switch
396	Rear speaker	493	Fuel pump relay
404	Air conditioning compressor protection diode	494	Wire joint No 15
408	Electric mirror switch	495	Wire joint No 16
420	Electric mirror on passenger side	496	Wire joint No 17
425	Car radio	497	Wire joint No 18
426	Junction block – sidemember harness to No 1 engine harness	498	Wire joint No 19
		499	Wire joint No 20
427	Junction block – sidemember harness to No 2 engine harness	500	Wire joint No 21
		501	Wire joint No 22
428	Junction block – sidemember harness to No 3 engine harness	502	Wire joint No 23
		503	Wire joint No 24
429	Window winder switch support plate	504	Wire joint No 25
430	Bonnect closure switch	505	Wire joint No 26
431	Voice synthesizer speaker	506	Wire joint No 27
432	Junction block – RH front and dashboard harnesses	507	Wire joint No 28
433	Junction block – LH front and dashboard harnesses	508	Wire joint No 29
434	LH speaker (tweeter)	509	Wire joint No 30
435	RH speaker (tweeter)	510	Wire joint No 31
437	Voice synthesizer	511	Wire joint No 32
438	Wire joint No 5	512	Wire joint No 33
439	Wire joint No 6	513	Water temperature sensor
440	Wire joint No 7	514	Intake air temperature sensor
441	Wire joint No 8	515	Full load switch
442	Junction block – sidemember harness to cruise control harness	516	Idling potentiometer
		517	Front LH pillar earth
443	Car radio control module	518	RH front pillar earth
446	No 2 before ignition connection terminal	519	Headlight main beam relay
447	Coolant level detector	520	Voice synthesizer switch

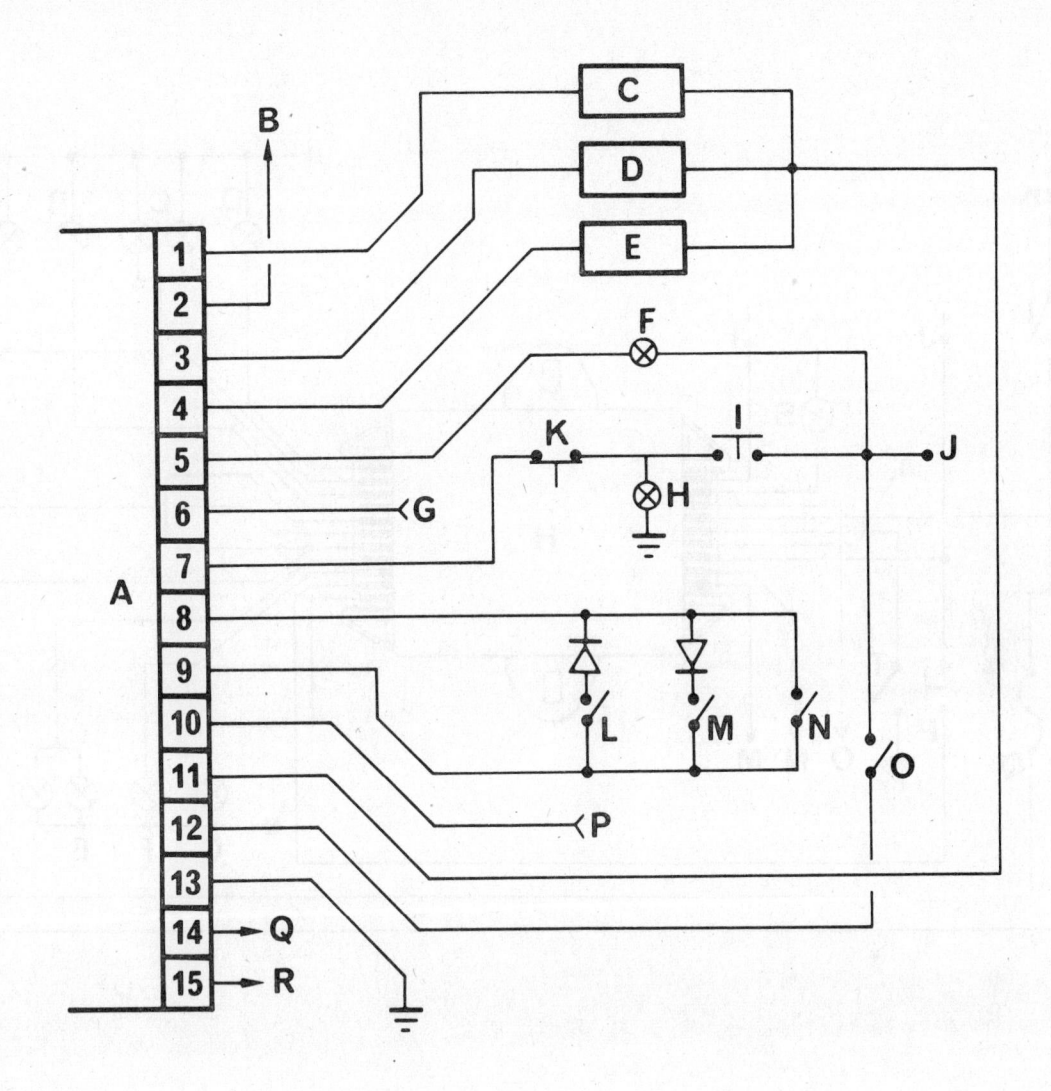

Fig. 12.42 Cruise control system wiring diagram

A	Computer	L	Steering wheel switch
B	To diagnostic socket	M	Steering wheel switch
C	Vacuum pump	N	Steering wheel switch
D	Governing solenoid valve	O	"On/off" switch
E	Security solenoid valve	P	Engine speed information input
F	Warning light on instrument panel	Q	Engine selection :
G	Road speed information input		4-cylinders not connected
H	Stop light bulb		6-cylinders earthed
I	Stop lights switch	R	Speed sensor selection :
J	+ after ignition switch		Halmo not connected
K	Declutching switch		Electric speedometer earthed

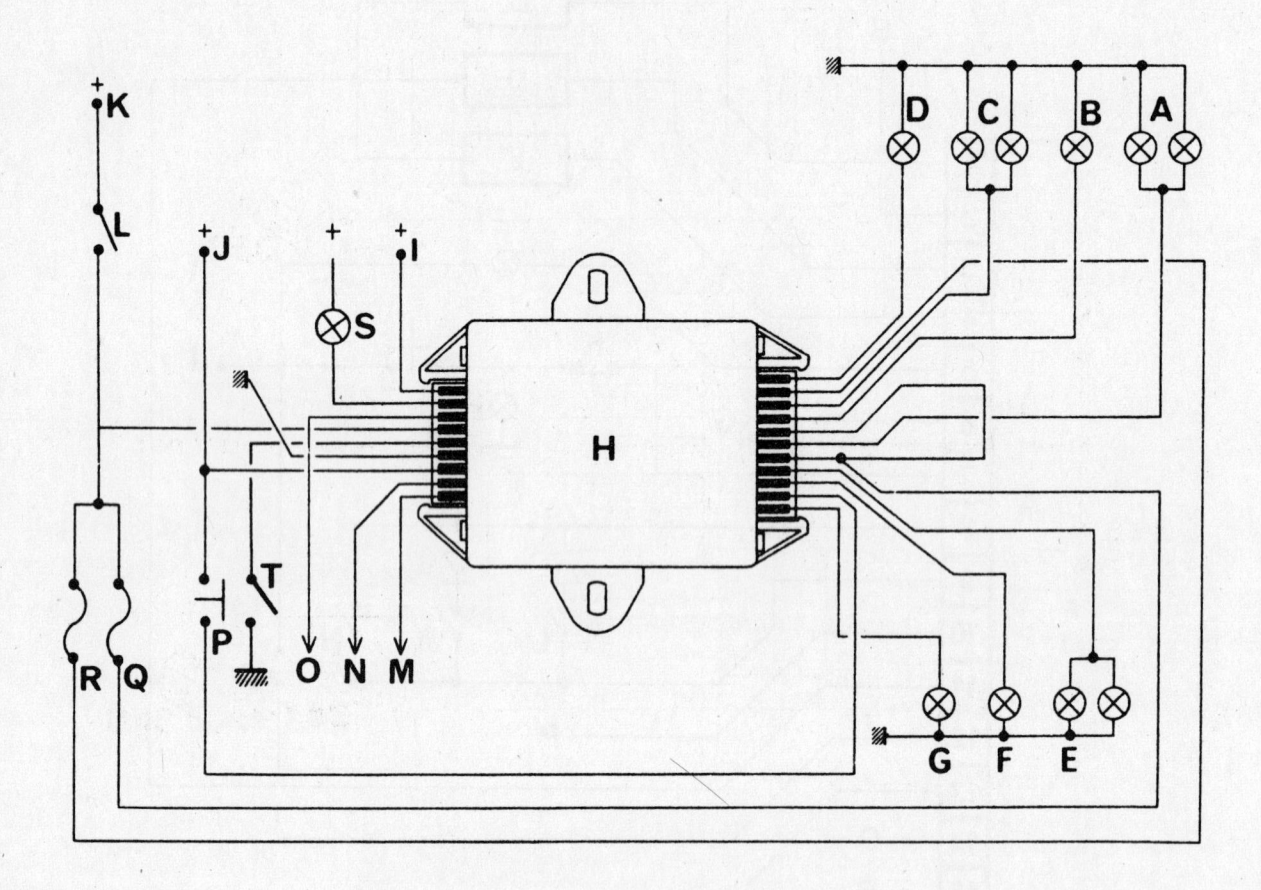

Fig. 12.43 Bulb filament defect detection unit wiring diagram

A	LH rear lights	K	+ input side of ignition switch
B	LH sidelight	L	Side/rear light switch
C	RH rear lights	M	RH stop-light defect
D	RH parking light	N	LH stop-light defect
E	Number plate lights	O	Side/rear light defect
F	LH stop-light bulb	P	Stop-light switch
G	RH stop-light bulb	Q	LH side/rear light fuse
H	Electronic unit	R	RH side/rear light fuse
I	Accessories +	S	Handbrake warning light +
J	+ output side of ignition switch	T	Handbrake switch

Wires S and T are interconnected inside the unit

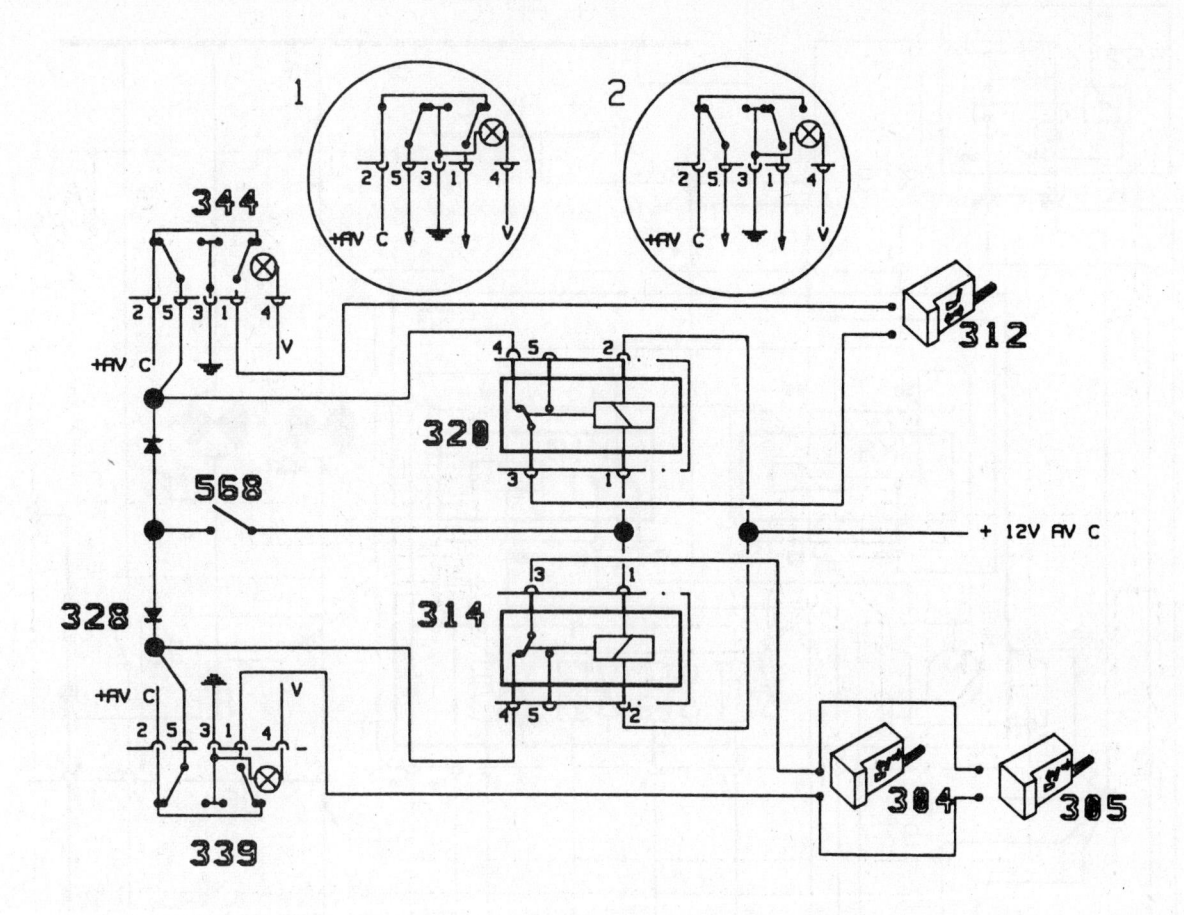

Fig. 12.44 Electrically-operated rear seats wiring diagram

V	Switch lighting +	328	Right-hand seat motor diodes
+ ACV	+ before ignition		(800 volts – 1 amp)
1 – 2	Activated changeover switch	339	Rear right-hand seatback movement changeover switch
304	Rear right-hand seatback motor		
305	Rear right-hand seatback motor	344	Rear right-hand seat movement changeover switch
312	Rear right-hand seat motor	568	Rear right-hand seat motor safety switch (located in seatback)
314	Rear right-hand seatback motor relay		
320	Rear right-hand seat motor relay		

Changeover switches (344) or (328) control motors (312) or (304) and (305).

When changeover switches (344) or (328) are activated : diodes (328) prevent current passing from one circuit to the other.

Motor safety switch (568) stops the seat at the end of its rearward travel via relay (320) or (314) which cut the feed to the motors, only when the "seat movement" switches (344) or (339) are in the "move back" position (see circle no. 1 above).

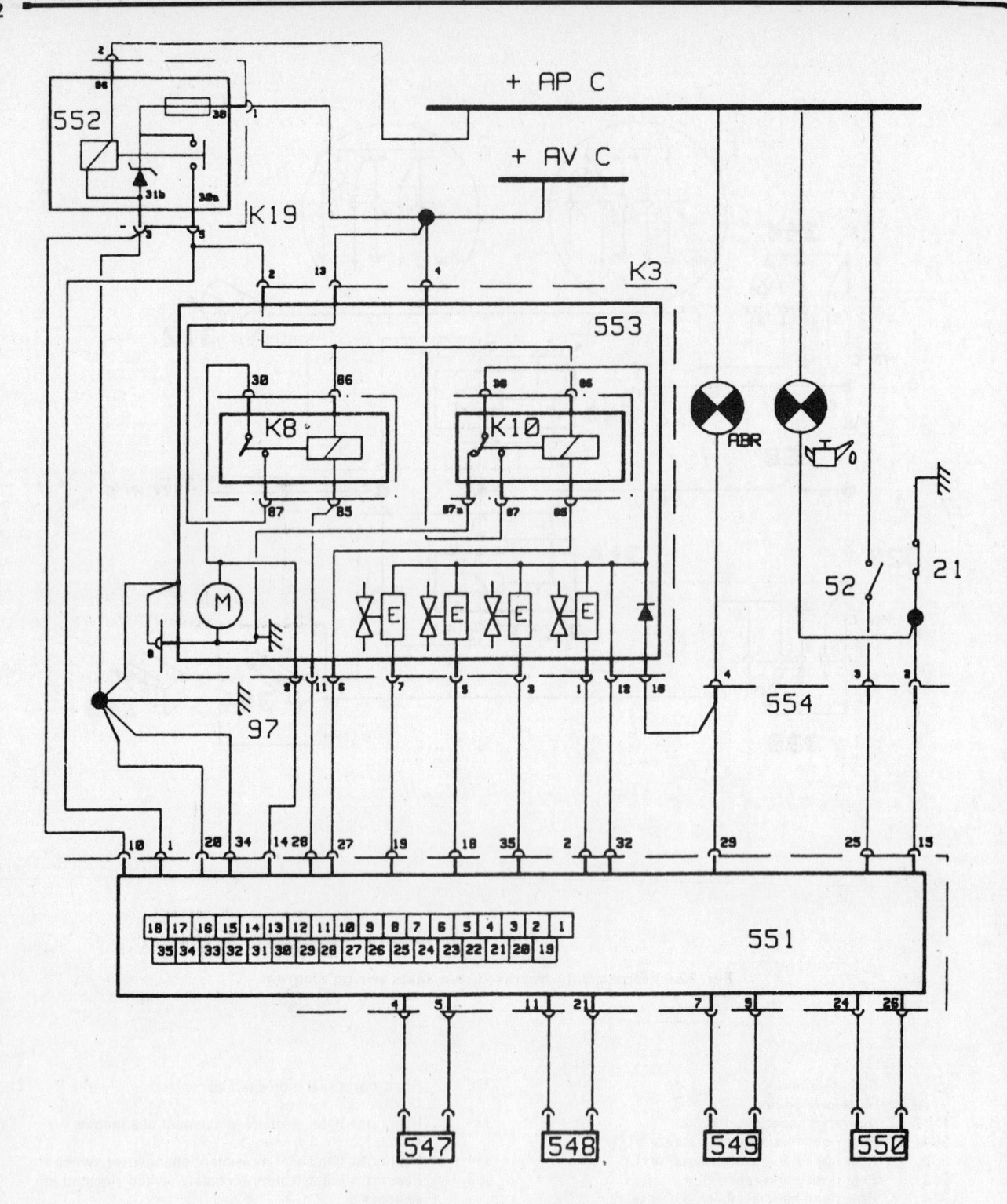

Fig. 12.45 Anti-lock braking system wiring diagram

+ AP C	+ On output side of ignition switch	552	Anti-wheel lock protection relay
+ AV C	+ On input side of ignition switch	553	Anti-wheel lock hydraulic unit
21	Oil pressure switch	554	Connection to anti-wheel lock harness
52	Stop-light switch	K3	Hydraulic unit connector
547	Front LH wheel sensor	K8	Solenoid valve pump relay
548	Front RH wheel sensor	K10	Solenoid valve relay
549	Rear LH wheel sensor	K19	Connector on relay 552
550	Rear RH wheel sensor	M	Pump
551	Anti-wheel lock electronic unit	E	Solenoid valves

Fig. 12.46 Air conditioning system wiring diagram

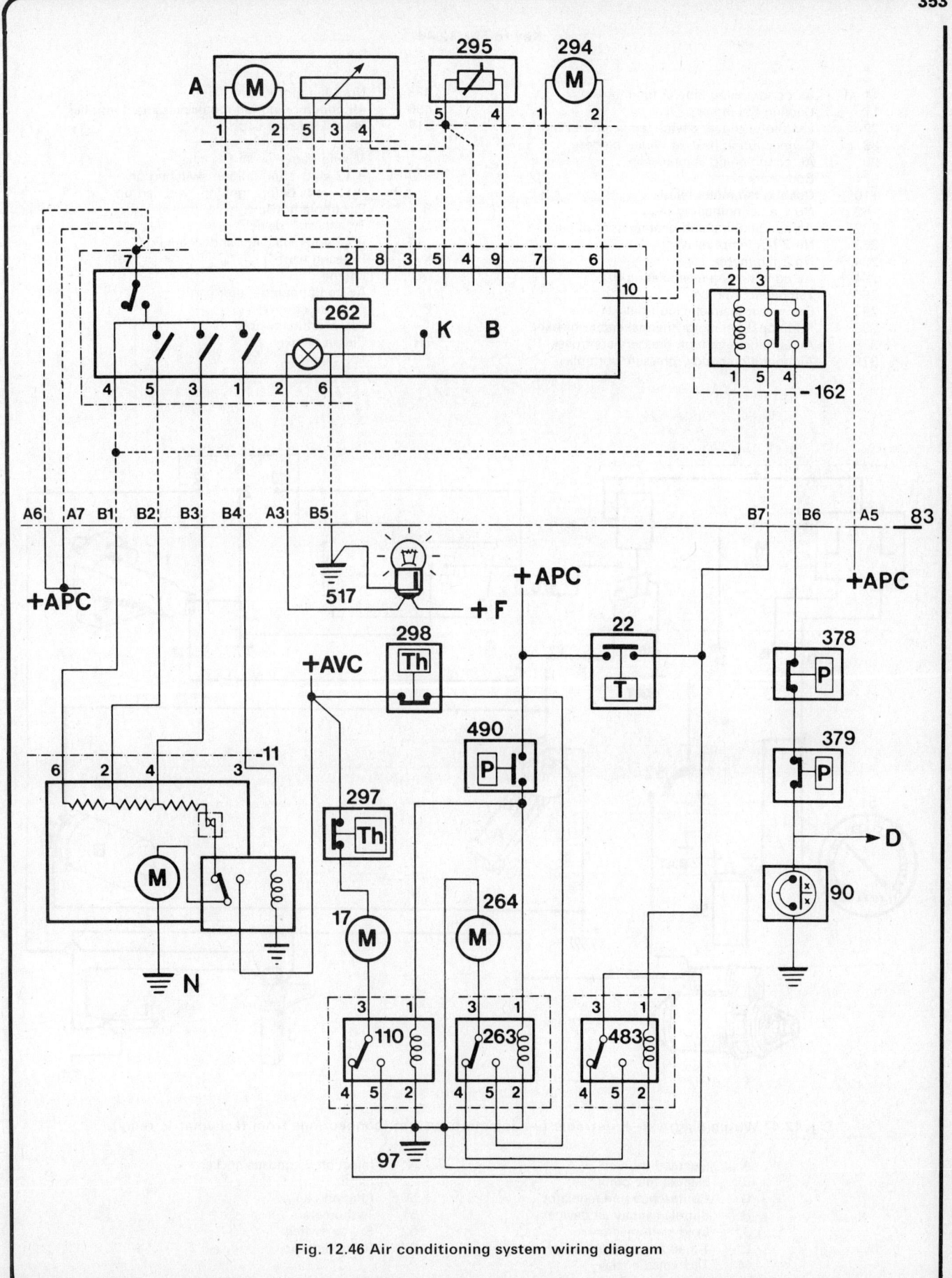

Key to Fig. 12.46

11	Air conditioning blower fan motor	483	No 3 fan motor relay
17	Cooling fan motor	490	Air conditioning fan motor pressure controller
22	Fan motor trigger device temperature switch	517	Front LH lower earth
83	Connection in heating wiring harness	A	Mixing flap
90	Air conditioning compressor	+ APC	+12 volts after ignition switched on
97	Bodywork earth	+ APV	+12 volts before ignition switched on
110	Cooling fan motor relay	B	Electronic housing
162	No 1 air conditioning relay	D	To anti-stall device
262	Air conditioning and heating control panel	+ F	+12 volts indicator and side lights
263	No 2 fan motor relay	K	Housing earth
264	No 2 fan motor	M	Engine
294	Air conditioning recirculating flap	N	Air conditioning motor earth
295	Temperature probe	P	Pressure controller
297	Fan motor thermal circuit breaker	T	Temperature switch
298	Additional fan motor thermal circuit breaker	TH	Circuit breaker
378	Air conditioning high pressure controller		
379	Air conditioning low pressure controller		

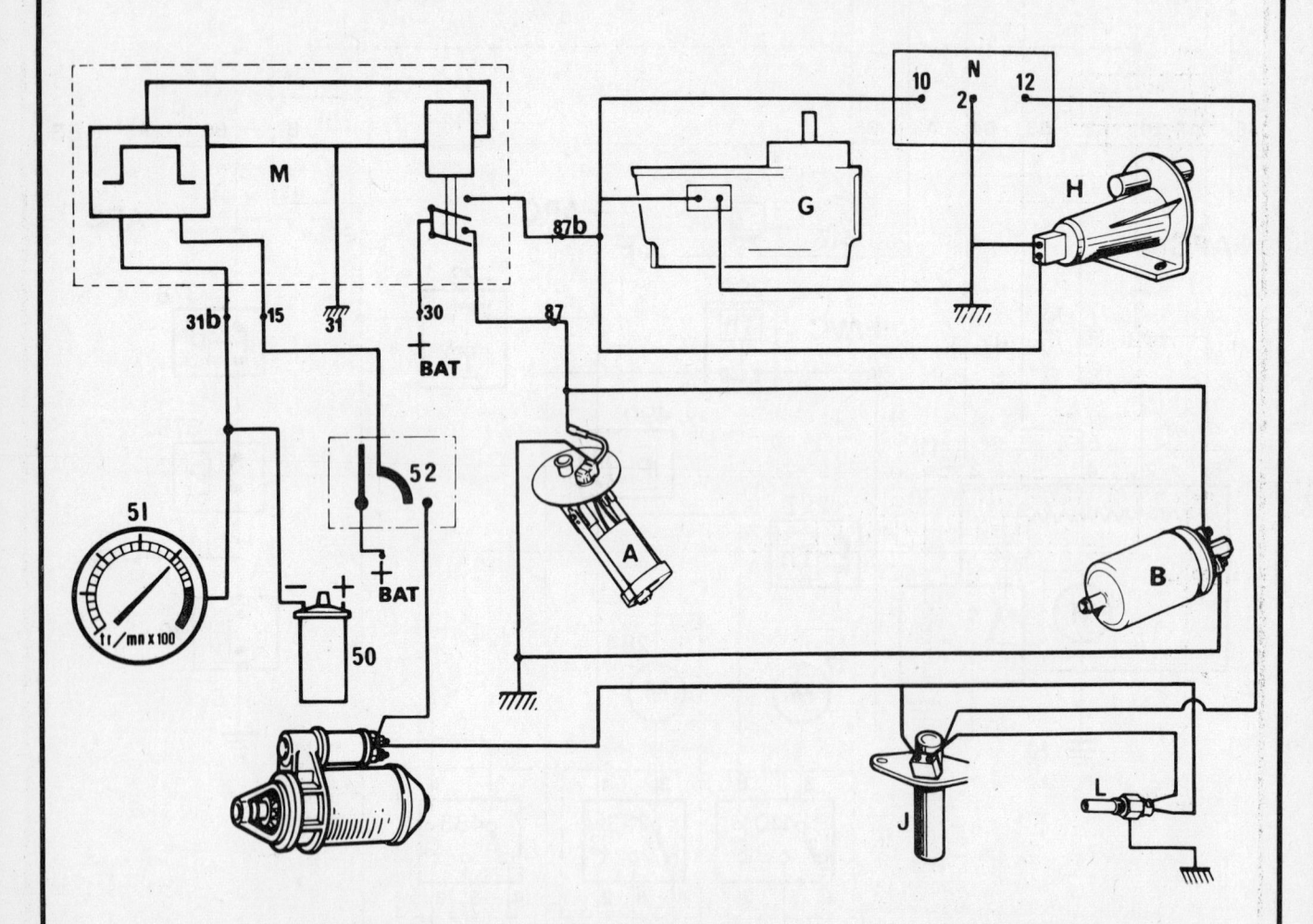

Fig. 12.47 Wiring diagram – K Jetronic fuel injection system (connections from tachymetric relay)

A	Electric lift pump	N	Injection diagnostic socket
B	Electric fuel pump	2	Earth
G	Control pressure regulator	50	Ignition coil
H	Supplementary air device	51	Tachometer
J	Cold start injector	52	Starter switch
L	Timed temperature switch	87	Relay output
M	Tachymetric relay		

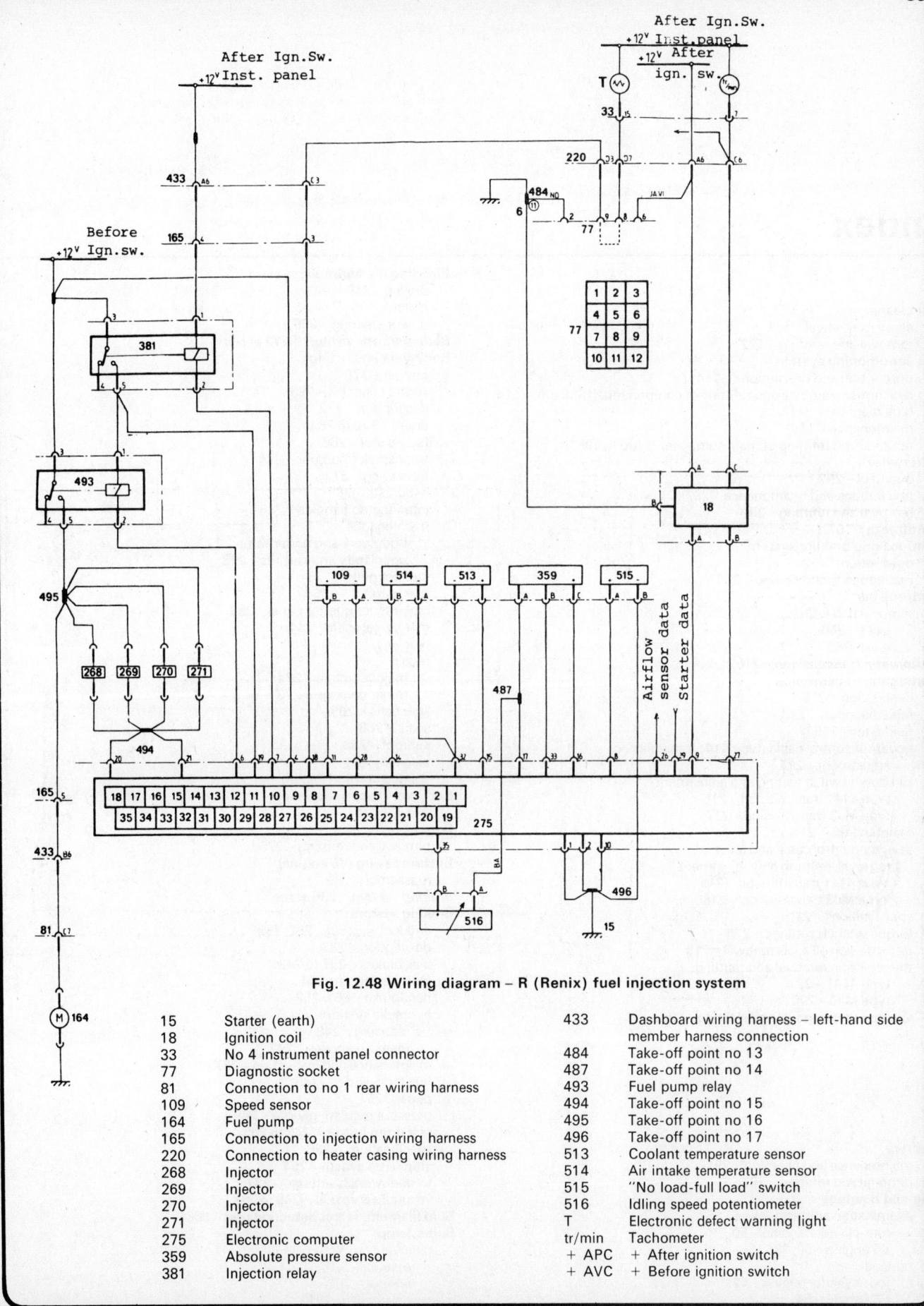

Fig. 12.48 Wiring diagram – R (Renix) fuel injection system

15	Starter (earth)	433	Dashboard wiring harness – left-hand side member harness connection
18	Ignition coil		
33	No 4 instrument panel connector	484	Take-off point no 13
77	Diagnostic socket	487	Take-off point no 14
81	Connection to no 1 rear wiring harness	493	Fuel pump relay
109	Speed sensor	494	Take-off point no 15
164	Fuel pump	495	Take-off point no 16
165	Connection to injection wiring harness	496	Take-off point no 17
220	Connection to heater casing wiring harness	513	Coolant temperature sensor
268	Injector	514	Air intake temperature sensor
269	Injector	515	"No load-full load" switch
270	Injector	516	Idling speed potentiometer
271	Injector	T	Electronic defect warning light
275	Electronic computer	tr/min	Tachometer
359	Absolute pressure sensor	+ APC	+ After ignition switch
381	Injection relay	+ AVC	+ Before ignition switch

Index

A

Air cleaner
    element renewal – 121
    removal and refitting – 122
Air conditioning system
    description and precautions – 114
    description, removal and refitting of components (1986 on) – 114
    fault diagnosis – 117
    maintenance – 114
    removal and refitting of main components (up to 1983) – 114
Alternator
    overhaul – 292
    precautions and maintenance – 292
    removal and refitting – 292
Antifreeze – 107
Anti-locking braking system
    description – 252
    maintenance and testing – 253
Anti-roll bar
    removal and refitting
        front – 256
        rear – 263
Automatic transmission – 210 *et seq*
Automatic transmission
    description – 211
    fault diagnosis – 223
    fluid filter – 215
    governor control cable (type 4141 transmission)
        adjustment – 217
    kickdown switch testing and adjustment
        type 4141 transmission – 215
        type MJ3 transmission – 217
    maintenance – 214
    selector control cable and lever
        removal, refitting and adjustment
        type 4141 transmission – 219
        type MJ3 transmission – 218
    specifications – 210
    torque wrench settings – 210
    transmission oil seals renewal – 219
    transmission removal and refitting
        type 4141 – 221
        type MJ3 – 220
    vacuum unit removal and refitting – 220

B

Battery
    maintenance and charging – 291
    removal and refitting – 292
Big-end bearings
    examination and renovation
        four-cylinder engine – 50
        V6 engine – 50, 98
    renewal
        four-cylinder engine – 42
        V6 engine – 84

Bleeding the hydraulic system
    brakes – 248
    clutch – 178
    power steering – 269
Bodywork and fittings – 272 *et seq*
Bodywork and fittings
    bonnet – 276
    centre consoles – 283
    description – 272
    doors – 278 to 283
    facia panel – 286
    front spoiler/bumper – 274
    front wing – 277
    headrests – 285
    interior grab handles – 286
    maintenance
        bodywork and underframe – 272
        upholstery and carpets – 273
    radiator grille – 275
    rear bumper – 289
    rear interior quarter panel – 283
    rear parcel shelf – 284
    rear view mirrors – 288
    repair
        major damage – 274
        minor damage – 273
    seat belts – 285
    seats – 288
    sunroof – 288
    tailgate – 283
    windscreen and heated rear screen – 283
Bodywork repair sequence – *see colour section between pages 32 and 33*
Bonnet
    removal and refitting – 276
Bottom casing (V6 engine)
    reassembly – 99
Braking system – 236 *et seq*
Braking system
    anti-lock system – 252, 253
    description – 237
    disc brakes – 237 to 242
    fault diagnosis – 254
    handbrake – 251, 252
    hydraulic system
        bleeding – 248
        pipes and hoses – 247
    maintenance and inspection – 237
    master cylinder – 245
    pedal – 252
    pressure regulating valve – 246
    rear drum brakes – 243 to 245
    specifications – 236
    stop-lamp switch – 254
    torque wrench settings – 237
    vacuum servo unit – 249 to 251
Bulb filament defect detection unit – 306
Bulbs, lamp
    renewal
        exterior – 301
        interior – 302
    specifications – 291

Index

C

Camshaft (four-cylinder engine)
    examination and renovation – 50
    removal and refitting – 37
Camshaft front and rear oil seals (V6 engine)
    renewal – 83
Capacities, general – 6
Carburettor
    removal and refitting – 129
    Weber DARA
        description – 129
        idle speed and mixture adjustment – 129
        overhaul and adjustment – 130
Centre consoles
    removal and refitting – 283
Cigar lighter
    removal and refitting – 313
Clutch – 174 *et seq*
Clutch
    cable removal, refitting and adjustment – 174
    description – 174
    fault diagnosis – 182
    hydraulic system bleeding – 178
    inspection – 180
    master cylinder – 175
    pedal – 179
    refitting – 180
    release mechanism – 180
    removal – 180
    slave cylinder – 177
    specifications – 174
Coil spring
    front – 258
    rear – 262
Connecting rods
    examination and renovation
        four-cylinder engine – 50
        V6 engine – 97
Conversion factors – 20
Coolant pump
    removal and refitting – 111
Coolant temperature switch
    testing – 111
Cooling, heating and air conditioning – 103 *et seq*
Cooling system
    coolant mixtures – 107
    coolant pump – 111
    coolant temperature switch – 111
    description – 104
    draining, flushing and refilling – 106
    drivebelts – 109
    maintenance – 106
    radiator – 108
    radiator fan and switch – 109
    specifications – 103
    thermostat – 107
    torque wrench settings – 103
Crankcase (four-cylinder engine)
    examination and renovation – 50
Crankcase ventilation system
    four-cylinder engine – 34
    V6 engine – 69
Crankshaft (four-cylinder engine)
    examination and renovation – 50
    reassembly – 52
Crankshaft (V6 engine)
    front oil seal renewal – 83
    pulley refitting – 74
    reassembly – 98
Cruise control system
    components removal and refitting – 309
    description – 308
Cylinder block (V6 engine)
    examination and renovation – 97

Cylinder head (four-cylinder engine)
    dismantling and decarbonising – 48
    examination and renovation – 50
    removal and refitting – 39
Cylinder heads (V6 engine)
    bolts re-tightening – 102
    dismantling and decarbonising – 94
    reassembly – 100
    removal and refitting – 74
Cylinder liners (V6 engine)
    examination and renovation – 96

D

Decarbonising
    four-cylinder engine – 48
    V6 engine – 94
Dimensions, vehicle – 6
Disc brakes
    caliper removal, overhaul and refitting
        front – 239
        rear – 240
    disc inspection, removal and refitting – 241
    pads inspection and renewal
        front – 237
        rear – 238
Distributor
    integral electronic ignition system
        removal and refitting – 168
    transistorised electronic ignition system
        removal, overhaul and refitting – 169
Doors
    front
        dismantling and reassembly – 278
        trim panel removal and refitting – 278
    locking: alternative systems – 300
    rear
        dismantling and reassembly – 281
        trim panel removal and refitting – 281
    removal and refitting – 283
Drivebelts
    tensioning, removal and refitting – 109
Driveshafts
    bellows and joints – 226
    bellows renewal
        inboard end (type G176) – 226
        inboard end (type G182) – 228
        outboard end (type GE86) – 228
    description and maintenance – 225
    removal and refitting
        with anti-lock braking system – 225
        without anti-lock braking system – 225
    specifications – 224
    torque wrench settings – 225
Driveshafts, hubs, roadwheels and tyres – 224 *et seq*
Drum brakes, rear
    drum inspection and renovation – 245
    shoes inspection and renewal
        Bendix type – 243
        Girling type – 244
    wheel cylinder removal, overhaul and refitting – 245

E

Electrical system – 290 *et seq*
Electrical system
    alternator – 292
    battery – 291, 292
    bulb filament defect detection unit – 306
    bulbs – 291, 301, 302
    cigar lighter – 313
    cruise control system – 308, 309
    description – 291
    door locking – 300
    electrically-controlled seats – 303

fault diagnosis – 21, 313
fuel consumption module – 310
fuses – 291, 296
headlamps – 302, 303
heated rear screen – 300
horns – 303
instrument panel – 306
power-operated windows – 301
radio aerial – 312
radio/cassette player – 310, 311
radio interference and CB equipment – 311
relays – 296
specifications – 290
starter motor – 293 to 295
steering column switches – 296
striplights – 303
switches: general – 298
tailgate
   assisted-closure system – 300
   wiper motor – 306
time and outside temperature display unit – 310
timers – 298
voice synthesizer – 307
warning buzzers – 313
windscreen/tailgate washer system – 306
windscreen wiper – 305
wiper blades and arms – 305
wiring diagrams – 314 to 355

Emission control systems
  AI system description and testing
    four-cylinder carburettor engine – 156
    V6 engine – 158
  EGR system
    four-cylinder carburettor engine – 155
    V6 engine – 156

Engine – 25 *et seq*
Engine (four-cylinder)
  ancillary components
    refitting – 59, 62
    removal – 45
  big-end bearings – 42, 50
  camshaft – 37, 50
  connecting rods – 50
  crankcase – 50
  crankcase ventilation system – 34
  crankshaft – 50, 52
  cylinder block – 50
  cylinder head – 39, 48
  decarbonising – 48
  description – 30
  dismantling – 45
  fault diagnosis – 23, 63
  firing order – 26
  flywheel/driveplate – 50, 54
  intermediate shaft – 40, 50, 57
  main bearings – 50, 52
  mountings – 43
  oil and filter – 30
  oil pump – 42, 50, 57
  operations possible with engine in car – 34
  piston/connecting rod/cylinder liner assembly – 42, 54
  piston rings – 42
  pistons – 50
  reassembly – 52
  reconnection to and refitting with
    automatic transmission – 63
    manual transmission – 62
  refitting
    automatic transmission in car – 63
    manual transmission in car – 62
  removal
    leaving automatic transmission in car – 44
    leaving manual transmission in car – 43
    method – 43
    with and separation from automatic transmission – 45
    with and separation from manual transmission – 44

rocker gear – 50
specifications – 26
starter ring gear – 50
start-up after overhaul – 63
sump pan – 42, 58
timing belt – 35, 50
timing belt tensioner – 50, 57
torque wrench settings – 27
valves – 33

Engine (V6)
  ancillary components
    refitting – 101
    removal – 91
  big-end bearings – 84
  bottom casing – 99
  camshaft front and rear oil seals – 83
  connecting rods – 97
  crankcase ventilation system – 69
  crankshaft – 98
  crankshaft front oil seal – 83
  cylinder block – 97
  cylinder heads – 74, 94, 100, 102
  cylinder liners – 96
  decarbonising – 94
  description – 64
  dismantling – 91
  examination and renovation – 96
  fault diagnosis – 23, 63, 102
  firing order – 28
  flexible mountings – 98
  flywheel/driveplate – 100
  gudgeon pins – 97
  main bearings – 98, 99
  mountings – 86
  oil and filter – 67
  oil pump – 70, 97
  operations possible with engine in car – 70
  piston/connecting rod/cylinder liner assembly – 84
  piston rings – 84, 97
  pistons – 97
  reassembly – 98
  reconnection to and refitting with
    automatic transmission – 101
    manual transmission – 101
  refitting:
    automatic transmission in car – 101
    manual transmission in car – 101
  removal
    leaving automatic transmission in car – 91
    leaving manual transmission in car – 88
    method – 86
    with and separation from automatic transmission – 91
    with and separation from manual transmission – 90
  rocker assemblies – 96
  rocker shafts – 81
  specifications – 28
  start-up after overhaul – 101
  sump pan and anti-emulsion plate – 82
  timing chains and sprockets – 71, 97, 100
  timing cover – 71
  torque wrench settings – 29
  valves – 68

Exhaust system
  general – 160
  torque wrench settings – 120

F

Facia panel
  removal and refitting – 286
Fault diagnosis – 21 *et seq*
Fault diagnosis
  automatic transmission – 223
  braking system – 254
  clutch – 182

Index

cooling, heating and air conditioning – 116
driveshafts, hubs, roadwheels and tyres – 235
electrical system – 21, 313
engine – 23, 63
fuel system
    carburettor – 133
    fuel injection system (Bosch K Jetronic) – 148
    fuel injection system (type R) – 140
ignition system – 173
manual transmission – 209
suspension and steering – 271
Firing order
    four-cylinder engine – 26
    V6 engine – 28
Flywheel/driveplate
    examination and renovation
        four-cylinder engine – 50
        V6 engine – 50, 98
    reassembly
        four-cylinder engine – 54
        V6 engine – 100
Front spoiler/bumper
    removal and refitting – 274
Front wing
    removal and refitting – 277
Fuel consumption module – 310
Fuel filter
    renewal – 124
Fuel injection system (Bosch K Jetronic)
    airflow sensor removal and refitting – 143
    cold start injector removal and refitting – 147
    components
        removal – 143
        testing – 147
    control pressure regulator removal and refitting – 147
    description – 140
    fault diagnosis – 148
    fuel injectors removal and refitting – 146
    idle speed and mixture adjustment – 140
    intake manifold removal and refitting – 146
    lower air casing removal and refitting – 143
    metering/distributor unit removal and refitting – 143
    specifications – 120
    supplementary air valve removal and refitting – 147
Fuel injection system (Type R – Renix)
    components
        removal – 136
        testing – 139
    computer removal and refitting – 137
    description – 133
    fault diagnosis – 140
    fuel injection manifold removal and refitting – 136
    fuel pressure regulator removal and refitting – 137
    idle mixture potentiometer removal and refitting – 138
    idle speed and mixture adjustment – 133
    no load/full load switch
        removal, refitting and adjustment – 138
    sensors removal and refitting – 137
    specifications – 119
Fuel level transmitter
    removal, testing and refitting – 126
Fuel pump
    carburettor models
        cleaning, testing, removal and refitting – 124
    type R fuel injection models: removal and refitting – 125
    K Jetronic fuel injection models: removal and refitting – 125
Fuel system (carburettor)
    carburettor – 129 to 132
    fault diagnosis – 133
    specifications – 118, 119
Fuel systems – 118 *et seq*
Fuel systems (general maintenance)
    air cleaner – 122
    air cleaner element – 121
    description – 120
    fuel filter – 124

fuel level transmitter – 126
fuel pump
    carburettor models – 124
    K Jetronic fuel injection models – 125
    type R fuel injection models – 125
fuel tank – 127
throttle cable – 128
Fuel system (Turbocharger)
    air intercooler removal and refitting – 149
    description – 149
    specifications – 120
    throttle casing removal and refitting – 149
    turbocharger pressure regulator
        adjusting, removal and refitting – 153
    turbocharger removal and refitting – 153
Fuel tank
    removal, repair and refitting – 127
Fuses
    general – 296
    specifications – 291

G

Gudgeon pins (V6 engine)
    examination and renovation – 97

H

Handbrake
    adjustment – 251
    cables renewal – 252
Headlamps
    beam alignment – 303
    beam load adjuster – 302
    bulb renewal – 301
    removal and refitting – 302
Head rests – 285
Heater
    fault diagnosis – 117
    rear passenger compartment heater – 114
    removal and refitting – 112
Heating and ventilation system – 112
High tension (HT) leads – 171
Horns – 303
Hub bearings
    renewal
        front – 232
        rear – 234
Hydraulic system
    brakes
        bleeding – 248
        pipes and hoses inspection, removal and refitting – 247
    clutch: bleeding – 178
    power steering: bleeding – 269

I

Ignition system – 163 *et seq*
Ignition system
    description – 164
    distributor
        integral electronic system – 168
        transistorized system – 169
    fault check
        integral electronic system – 167
        transistorized system – 164
    fault diagnosis – 173
    high tension (HT) leads – 171
    maintenance – 164
    sensors description, removal and refitting – 171
    spark plugs – 171, 171
    specifications – 163
    timing (transistorized system) – 167

Instrument panel
   removal, dismantling, reassembly and refitting – 306
Interior grab handles – 286
Intermediate shaft (four-cylinder engine)
   examination and renovation – 50
   reassembly – 57
   removal and refitting – 40

J

Jacking – 7

L

Lubricants and fluids, recommended – 19
Lubrication chart – 19

M

Main bearings
   four-cylinder engine
      examination and renovation – 50
      reassembly – 52
   V6 engine
      examination and renovation – 50, 98
      reassembly – 98, 99
Maintenance, routine
   automatic transmission
      final drive oil level check/top up (type 4141) – 13, 214
      final drive oil renewal (type 4141) – 13, 215
      fluid level check/top up – 13, 214
      fluid renewal – 13, 215
   bodywork and fittings
      bodywork and underframe – 272
      upholstery and carpets – 273
   braking system
      disc pads wear check – 13, 237, 238
      fluid level check/top up – 13, 237
      fluid renewal – 13, 248
      handbrake cable adjustment check – 13
      pipes and hoses condition check – 13, 247
      rear brake shoes wear check – 13, 243, 244
   clutch cable adjustment check – 13, 174
   cooling system
      coolant level check/top up – 13, 106
      coolant renewal – 13, 107
      drivebelt tension check/adjust – 13, 109
   driveshaft bellows condition check – 13, 226, 228, 230
   electrical system
      equipment operation check – 13
      headlamps beam alignment check – 13, 303
      washer fluid level check/top up – 13
   engine
      crankcase ventilation system hoses and jets cleaning – 13
      oil and filter renewal – 13, 31, 67
      oil level check/top up – 13, 31, 67
      rocker shaft oil filters renewal – 13
      timing belt renewal – 13, 35
      valve clearances check/adjust – 13, 33, 68
   exhaust system condition check – 13
   fuel system and emission control system
      air cleaner element renewal – 13, 121, 122
      EGR (emission control system) nozzles cleaning/valve renewal – 13
      fuel filter renewal – 13, 124
      fuel pump cleaning – 13, 124
      idle speed and exhaust CO content check – 13, 129, 133, 140
   ignition system
      distributor cap cleaning/contacts and rotor check – 13
      spark plugs clean/regap/renew – 13, 171
   manual transmission
      oil level check/top up – 13, 184
      oil renewal – 13, 184

   safety precautions – 12
   schedules – 13
   steering and suspension
      balljoints, bushes and bellows condition check – 13, 256
      front wheel alignment check – 13
      power steering fluid level check/top up – 13, 256
   tyres
      condition check – 234
      pressures check/adjust – 13, 224
   wheels condition check – 234
Manifolds
   removal and refitting
      induction and injection (2165 cc engine) – 42, 136
      inlet and exhaust (1995 cc engine) – 158, 159
Manual transmission (all types)
   fault diagnosis – 209
   specifications – 183
   torque wrench settings – 184
Manual transmission (type NG3)
   description – 184
   differential/final drive overhaul – 191
   differential oil seals renewal (without removing transmission) – 193
   gearchange linkage removal and refitting – 185
   maintenance – 184
   primary shaft overhaul – 186
   rear cover dismantling and reassembly – 191
   reverse idler shaft and gear dismantling and reassembly – 189
   secondary shaft overhaul – 187
   selector forks and shafts dismantling and reassembly – 189
   transmission
      housing inspection – 191
      overhaul – 186
      reassembly – 192
      removal and refitting – 185
      removal of major assemblies – 186
Manual transmission (type UN1)
   description and maintenance – 184, 193
   differential/final drive overhaul – 204
   gearchange linkage removal and refitting – 185, 193
   primary shaft overhaul – 196
   rear cover dismantling and reassembly – 203
   reverse idler shaft and gear dismantling and reassembly – 201
   secondary shaft overhaul – 198
   selector forks and shafts dismantling and reassembly – 201
   transmission
      housing inspection – 191, 203
      reassembly – 204
      removal and refitting – 193
      removal of major assemblies – 194
Master cylinder
   braking system: removal and refitting – 245
   clutch: removal, overhaul and refitting – 175
Mirrors, rear view – 288
Mountings, engine/transmission
   renewal
      four-cylinder engine – 43
      V6 engine – 86
Mountings, flexible (V6 engine)
   examination and renovation – 98

O

Oil and filter
   four-cylinder engine – 31
   V6 engine – 67
Oil pump
   four-cylinder engine
      examination and renovation – 51
      reassembly – 57
      removal and refitting – 42
   V6 engine
      examination and renovation – 97
      removal and refitting – 70

Index

P

Pedal
    removal and refitting
        brake – 252
        clutch – 179
Piston/connecting rod/cylinder liners assembly
    four-cylinder engine
        reassembly – 54
        removal and refitting – 42
    V6 engine: removal, refitting and reassembly – 84
Piston rings
    four-cylinder engine: renewal – 42
    V6 engine
        examination and renovation – 97
        renewal – 42, 84
Pistons
    examination and renovation
        four-cylinder engine – 50
        V6 engine – 97
Pressure regulating valve (braking system) – 246

R

Radiator
    removal, renovation and refitting – 108
Radiator fan and switch
    removal and refitting – 109
Radiator grille
    removal and refitting – 275
Radio
    aerial removal and refitting – 311
    interference and CB equipment – 311
Radio/cassette player
    fitting – 311
    general – 310
    removal and refitting (Renault type) – 311
Rear bumper
    removal and refitting – 289
Rear interior quarter panel
    removal and refitting – 283
Rear parcel shelf – 284
Rear screen, heated
    general – 300
    renewal – 283
Relays – 296
Repair procedures, general – 9
Rocker assemblies (V6 engine)
    examination and renovation – 96
Rocker gear (four-cylinder engine)
    examination and renovation – 50
    removal and refitting – 39
Rocker shafts (V6 engine)
    removal and refitting – 81
Routine maintenance see **Maintenance, routine**

S

Safety precautions – 12
Seat belts
    maintenance, removal and refitting – 285
Seats
    electrically-controlled – 303
    removal and refitting
        front – 288
        rear – 288
Shock absorber
    front – 258
    rear – 262
Slave cylinder, clutch
    removal, overhaul and refitting – 177
Spare parts
    buying – 8
    to carry in car – 21
Spark plugs
    conditions – *see colour section between pages 32 and 33*
    general – 171
Starter motor
    description and testing – 293
    overhaul – 295
    removal and refitting – 294
Starter ring gear (four-cylinder engine)
    examination and renovation – 50
Steering
    column lock/ignition switch removal and refitting – 271
    column removal, overhaul and refitting – 266
    description – 256
    fault diagnosis – 271
    gear removal and refitting
        manual – 268
        power-assisted – 268
    maintenance – 256
    power-assisted steering
        gear and pump overhaul – 269
        hydraulic circuit bleeding – 269
        pump removal and refitting – 269
    rack
        bellows renewal – 265
        slipper adjustment – 267
    rear wheel alignment – 270
    specifications – 255
    steering angles and front wheel alignment – 269
    tie-rod and balljoint renewal – 263
    torque wrench settings – 256
    universally-jointed shaft removal and refitting
        manual steering – 266
        power-assisted steering – 266
    wheel
        height adjuster – 270
        removal and refitting – 265
Striplights – 303
Sump pan (four-cylinder engine)
    reassembly – 58
    removal and refitting – 42
Sump pan and anti-emulsion plate (V6 engine)
    removal and refitting – 820
Sunroof
    dismantling and reassembly – 288
Suspension and steering – 255 *et seq*
Suspension, front
    anti-roll bar – 256
    description – 256
    fault diagnosis – 271
    hub carrier removal and refitting – 261
    lower track control arm removal and refitting – 260
    maintenance – 256
    shock absorber and coil spring removal and refitting – 258
    specifications – 255
    torque wrench settings – 256
    upper track control arm
        balljoint renewal – 260
        removal and refitting – 259
Suspension, rear
    anti-roll bar – 263
    coil spring removal and refitting – 262
    description – 256
    fault diagnosis – 271
    maintenance – 256
    radius rod removal and refitting – 263
    shock absorber removal and refitting – 262
    specifications – 255
    torque wrench settings – 256
    track control arm – 263
Switches
    general – 307
    steering column: removal and refitting – 296

T

Tailgate
   assisted-closure system – 300
   removal and refitting – 283
   wiper motor removal and refitting – 306
Thermostat
   removal, testing and refitting – 107
Throttle cable
   removal, refitting and adjustment – 128
Time and outside temperature display unit – 310
Timers – 298
Timing belt and tensioner (four-cylinder engine)
   examination and renovation – 50
   removal and refitting – 35
Timing belt tensioner (four-cylinder engine)
   reassembly – 57
Timing chains, tensioners and sprockets (V6 engine)
   examination and renovation – 97
   reassembly – 100
Timing cover, chains and sprockets (V6 engine)
   removal and refitting – 71
Timing (transistorized ignition system) – 167
Tools
   general – 10
   to carry in car – 21
Towing – 7
Turbocharger – 149
Transmission see **Automatic transmission and Manual transmission**
Tyres
   care and maintenance – 234
   pressures – 224
   specifications – 224

U

Underbody
   maintenance – 272

Upholstery
   maintenance – 273

V

Vacuum servo unit (braking system)
   air filter renewal – 250
   description and testing – 249
   non-return valve renewal – 250
   removal and refitting – 251
Valve clearances
   adjustment
      four-cylinder engine – 33
      V6 engine – 68
Vehicle identification numbers – 8
Voice synthesizer – 307

W

Warning buzzers – 313
Weights, vehicle – 6
Wheels
   care and maintenance – 234
   changing – 7
   specifications – 224
Windows, power-operated – 301
Windscreen
   renewal – 283
Windscreen/tailgate washer system – 306
Windscreen wiper motor and linkage
   removal and refitting – 305
Wiper blades and arms
   removal and refitting – 305
Wiring diagrams – 314 to 355
Working facilities – 11